MATHEMATICS RESEARCH DEVELOPMENTS

EVOLUTION EQUATIONS

MATHEMATICS RESEARCH DEVELOPMENTS

Additional books in this series can be found on Nova's website
under the Series tab.

Additional E-books in this series can be found on Nova's website
under the E-book tab.

MATHEMATICS RESEARCH DEVELOPMENTS

EVOLUTION EQUATIONS

ARTHUR L. CLAES
EDITOR

Nova Science Publishers, Inc.
New York

LIBRARY OF CONGRESS CATALOGING-IN-PUBLICATION DATA

Evolution equations / editor, Arthur L. Claes.
p. cm.
Includes bibliographical references and index.
ISBN 978-1-61209-652-0 (hardcover)
1. Evolution equations. I. Claes, Arthur L.
QA377.3.E96 2011
515'.39--dc22

2011001537

Published by Nova Science Publishers, Inc. † New York

CONTENTS

PREFACE

This new book presents topical research in the study of evolution equations. Topics discussed include understanding the theory of noisy dynamical systems in Markovian and non-Markovian settings using evolution equations; Einstein's evolution equation for fluid with exponential free paths distribution; a theorem on the optical cubic-quintic Schrodinger equation; methods for solving nonlinear partial differential equations based on Riccati equations; homogeneous and nonlinear generalized master equations and the cosmological linear and nonlinear growth in the accelerating Universe.

Short Communication - There are two fundamental invariant nonlinear functionals associated with a nonlinear evolution equation which is certainly notorious in mathematical physics. The authors refer to the optical cubic-quintic Schrödinger equation. With respect to the lossless equation, the norm of a functional which is a linear combination of the two above functionals is introduced. By establishing a suitable theorem, the authors prove that the above functional is bounded.

Chapter 1 - A real classical topic of research for evolving systems of microstructures with grains like metals and ceramics is grain growth and coarsening (Ostwald Ripening). Both phenomena are driven by the decrease in total surface energy, while the chemical composition of the individual objects remains unchanged. Only the morphology of the system evolves and gives the impression that smaller objects are "eaten" by the larger ones. Evolution equations for these phenomena were published in the last century mostly as phenomenological equations. The authors show, however, that the evolution equations can be derived from the Thermodynamic Extremal Principle (TEP). This principle, based on the works of L. Onsager (1931, 1945) and later of H. Ziegler (1977), allows calculation of the evolution equations of the state parameters, e.g., the effective radii of individual grains, precipitates, etc., by maximizing the dissipation, expressed in a quadratic form of the kinetic parameters, with the constraint that the dissipation is equal to the negative rate of Gibbs energy in the isothermal case. Later studies have allowed working with distribution functions of the effective radii of objects and the evolution of those to stationary distribution functions. Also, the inverse problem has been solved, by which the evolution equations of the grain or precipitate effective radii can be derived from steady state distribution functions. This concept was extended from one-parameter distribution functions to various two-parameter distribution functions. The concept of TEP also allows successful simulations of the microstructure evolution in multi-component multiple particle systems. Practical applications of this concept of evolution equations for multi-component materials are published and used in the metals processing industry—see the program package MatCalc, http://matcalc.wkmp.tuwien.ac.at/.

Chapter 2 – This work is dedicated to the study of existence and uniqueness of classical solution for the following multidimensional mixed problem:

$$\begin{cases} u_{tt}(t,x) + \mathcal{A}(u(t,x)) \equiv u_{tt}(t,x) + \sum\limits_{|\alpha|\le 2m} a_\alpha(x) D^\alpha u(t,x) = \\ = F(t,x,u(t,x), Du(t,x),\dots, D^m u(t,x), u_t(t,x)) \quad (t\in[0,T], x\in\overline{\Omega}), & (0.1) \\ u(0,x) = \varphi(x) \quad (x\in\overline{\Omega}), \quad u_t(0,x) = \phi(x) \quad (x\in\overline{\Omega}), & (0.2) \\ B_j(u(t,x))\big|_\Gamma = 0 \quad (j=1,\dots,m), & (0.3) \end{cases}$$

Chapter 3 - The dynamical system theory in combination with the notion of feedback formalizes the concept of mathematical control theory. The structure of dynamical systems is adjudged on the basis of dimensionality, time-invariance/time varying as well as driving input. The stochastic differential equation resulting from the phase space analysis of the fluctuation equation describing noisy dynamical systems has received remarkable success in different branches of sciences. Consider the stochastic differential system of the form $\dot{x}_t = f(x_t) + g(x_t)\xi_t$, where the terms $f(x_t)$, $g(x_t)\xi_t$ have interpretations as the system non-linearity and the random forcing term respectively. For the input noise process $\xi_t = \dot{B}_t$, where B_t is the Brownian motion, the resulting stochastic differential system would be the Itô stochastic differential system in which the state vector is Markovian. More over, the Ornstein-Uhlenbeck (OU) process-driven stochastic differential system can be recast as the Itô stochastic differential system by writing down the evolution equation of the OU process, i.e. $d\xi_t = -\beta\xi_t dt + \beta\, dB_t$, $\beta > 0$. The contribution to the stochastic evolution of the state vector comes from the system non-linearity as well as the dispersion matrix, which introduces slight modifications into the motion of the dynamical system. This book chapter encompasses *five* stochastic evolution equations as well as their special cases, i.e. classical evolution equations, which are regarded as the cornerstone formalism for analysing and accomplishing filtering for stochastic differential systems. The stochastic evolution equations of the chapter are about the filtering density evolution, the exact stochastic evolution of conditional moment as well as its approximate counterparts, the stochastic evolution of the scalar function of the multi-dimensional state vector. This chapter discusses evolution equations for the non-Markovian stochastic differential system as well, which are not widely available in literature.

The evolution equations of the chapter will be of interest to researchers looking for understanding the theory of noisy dynamical systems in Markovian as well as non-Markovian setting.

Chapter 4 - The evolution of fluid density in time and space is considered in the framework of Einstein's theory involving expansions of the density distribution in powers of displacement and time. By truncating these expansions in the evolution equation, Einstein derived the classical model of diffusion. Here, the authors consider Einstein's model without truncating the density expansions. This demonstrates limits to Einstein's truncations and demonstrates non-classical behavior for diffusion phenomenology in fluids with large mean-free paths, i.e. for fluids between the Knudsen and Fickian limits.

An exact solution to Einstein's evolution equation for a fluid is presented. This solution shows that the classical diffusion model is valid for fluids only in the limit of small mean-free paths, λ.

Here the authors demonstrate that, at large λ, Einstein's evolution equation predicts not only significant quantitative deviations from classical diffusion profiles, but also qualitative metamorphoses in diffusion mechanisms, such as co-existence of "laminar" and "turbulent" fluxes and "cooperative" transitions between molecular chaos and predominantly ballistic motion of molecules.

This has implications for fundamentals of fluids between the Knudsen and Fickian limits, and for a variety of fields where evolution of a system includes random, multi-scale displacement of particles, such as nanotechnology, vacuum techniques, biological systems, turbulence, and astrophysics.

Chapter 5 - Throughout this chapter, the authors investigate the singularity structure analysis of the (2+1)-dimensional coupled nonlinear Schrödinger (CNLS) equations, and they show that these equations are Painlevé (P)-integrable. By means of the Weiss et al.'s methodology, the authors show the arbitrariness of the expansion coefficients and the consistency of the truncation corresponding to a special B˜acklund transformation (BT) of these CNLS equations. In the wake of such transformation, following the Hirota's formalism, the authors derive a one-soliton solution. Besides, by using the Zakharov-Shabat (ZS) scheme which provides a general Lax-representation of an evolution system, the authors show that the (2+1)-dimensional CNLS system under interests is completely integrable. Furthermore, using the arbitrariness of the above coefficients, the authors unearth and investigate a typical spectrum of localized and periodic coherent structures. As a result, they depict the elastic and nonelastic interactions amongst such structures and also a splitting phenomenon occurring during their motion.

Chapter 6 - In this chapter the authors present some ideas for solving nonlinear partial differential equations which are based on the idea of the projective Riccati equations. The authors show how to derive well known methods such as Conte's projective Riccati equations method, tanh-coth method, He's exp function method and a new method the authors called sn-ns method. The authors illustrate the effectiveness of the sn-ns method in the problem of finding new exact solutions to some relevant nonlinear pde's with the aid of Maple 14 and Mathematica 7.

Chapter 7 - Deriving the kinetic (irreversible) equations from the reversible microscopic dynamics of the many-body systems remains one of the principal tasks of statistical physics. One of the key assumptions used to deriving the Boltzmann equation is the Bogoliubov principle of weakening of initial correlations which implies that on a sufficiently large time scale all initial correlations (existing at the initial instant) are damped. This approach leads to the evolution equations that are not valid on all time scales and do not allow considering all the stages of the system evolution, in particular, the short-term and transient regimes, which are of scientific and practical interest for considering the ultrafast relaxation and non-Markovian processes, decoherence phenomena and onset of irreversibility. To take initial correlations into account, a method, based on the conventional time-independent projection operator technique, that allows converting the conventional linear inhomogeneous (containing a source caused by initial correlations) time-convolution generalized master equation (TC-GME) and time-convolutionless GME (TCL-GME) into the homogeneous form exactly, is proposed. This approach results in the exact linear time-convolution and time-convolutionless homogeneous generalized master equations (TC-HGME and TCL-HGME) which take the dynamics of initial correlations into account via modified memory kernels governing the evolution of the relevant part of a distribution function of a many-particle system. These equations describe the evolution of the relevant part of a distribution function on all time scales including the initial stage when the initial correlations matter. The obtained TC-HGME is applied to the spatially homogeneous dilute gas of classical and quantum particles. However, to derive the desired nonlinear equations (the Boltzmann equation in particular) from this actually linear equation, we should make an additional approximation neglecting the time-retardation of a one-particle distribution function which restricts the time scale by times much smaller than the relaxation time for a one-particle distribution function. To obtain the actually nonlinear evolution equations and avoid any restrictions on the time scales, the authors develop a new method, based on using a time-dependent operator (generally not a projection operator) converting a distribution function of a total system into the relevant form, that allows deriving the new exact nonlinear generalized master equations. The derived inhomogeneous nonlinear GME is a generalization of the linear Nakajima-Zwanzig TC-GME and can be viewed as an alternative to the BBGKY chain. To include the initial correlations into consideration, the authors convert the obtained inhomogeneous nonlinear GME into the homogeneous form by the method which they used for conventional linear GMEs. The obtained exact homogeneous nonlinear GME describes all evolution stages of the system of interest and treats initial correlations on an equal footing with collisions by means of the modified memory kernel. As an application, the authors obtain a new homogeneous nonlinear equation retaining initial correlations for a one-particle distribution function of the spatially inhomogeneous nonideal gas of classical particles.

The authors show that on the kinetic time scale, the time-reversible terms resulting from initial correlations vanish (if the particle dynamics have the ergodic mixing property) and this equation can be converted into the Vlasov-Landau and Boltzmann equations without any additional commonly used approximations.

Chapter 8 - The present cosmological observations have confirmed that the present Universe is expanding at an accelerating rate. One of the most popular solutions to this conundrum is introducing the so called "dark energy"" which is the dominant energy contribution to the present energy of the Universe with its equation of state, $\omega_{de} \cong 1$. The effects of a dark energy component can be obtained mainly from both the cosmic microwave background (CMB) observations and the large scale structure formation ones. Even though early time properties of the dark energy dynamics can be extracted from the acoustic feature of the CMB anisotropy, the relevant characters of it have to be proved by investigating physical processes occurring at low redshifts when it starts dominating the cosmic expansion. Thus, the dark energy consequences on the large scale structure formation is the one of the most powerful tools. In this chapter, the authors investigate the cosmological implications of dark energy on both the sub-horizon scale growth of the matter density perturbation and the clustering by using the so called "spherical collapse model". The authors provide both the exact and approximate solutions for linear and non-linear perturbations. The authors provide the differences on the large scale structure formations between different dark energy models. The authors also investigate the cluster number calculations based on this specific model.

Chapter 9 - The authors present the detailed account of our recent results on group classification of nonlinear evolution equations in one spatial variable. In this publication they focus on nonlinear evolution equations that admit semi-simple Lie algebras. The authors give exhaustive description of second-order evolution equations whose symmetry algebras contain semi-simple subalgebras. The same problem is solved for the class of third-order quasi-linear (KdV type) evolution equations. The authors present the exhaustive classification of fourth-order quasi-linear evolution equations invariant under semi-simple Lie algebras.

Chapter 10 - The study of the origin and evolution of microstructures in crystalline solids has gained much interest, virtue of the essential influence of such microscale structures on the macroscopic mechanical properties of the material. The formation of regular, fine-scale patterns on the material's microlevel has been reasoned to stem from non-(quasi)convex energy potentials, which result in a lack of homogeneous solution states to the underlying thermodynamic extremum principles of minimum potential energy and maximum dissipation. The theory of relaxation has provided beneficial tools for understanding and analyzing the origin and subsequent development of microstructures in finite-strain plasticity models, whose prediction and simulation is commonly based on condensed potentials to overcome the non-(quasi)convexity and to render the problem well-posed while considering all admissible microfluctuations, i.e., microstructures. The authors outline an incremental variational alternative to the application of condensed potentials, which allows to account for more physical insight and which greatly expands the regime of applicability of material models. Also, the authors illustrate the incremental approach by its application to an incompressible Neo-Hookean solid with one and multiple active slip systems, but the method can readily be generalized to problems of microstructure evolution.

Chapter 11 - One of the most important tasks in solving problems of control technical systems is the evolutionary behavior formalization of the object. These models are required for the formation control algorithms. The engineers operate only with the basic linear models, leaving the solution of problems of regulation of nonlinear phenomena in industrial controllers. As a result, there are actually functioning control systems that work not only optimal, but, even in manual mode set up control processes.

The authors can distinguish three classical approaches to modeling of nonlinear phenomena in technical systems. The first is to use stochastic models; it is assumed that the nonlinear oscillations are realization of a random process whose characteristics are to be found. Obviously, in this case cannot be rigorously justified assumptions, and assumptions that are not explicitly probabilistic nature, and the observed behavior is possible for a whole class of nonlinear objects. Assessing the adequacy of the construction of stochastic models is to test the hypotheses match the original assumptions. The second approach - the construction of models based on the physical properties of all occurring phenomena in technical systems and the withdrawal of evolution equations. Despite the evidence of such a decision, it is almost impossible for technical systems, since closed and structural complexity of structures defines a knowledge-intensive and cumbersome nature of the physical simulation. Obtaining necessary to control the evolution equations of the models is a separate complex challenge. The third - for the use of an approach to modeling is to choose the kind of mathematical model as an evolution equation and the subsequent identification of parameters or nonparametric model identification. Model is considered adequate if the evaluation of the adequacy of a given criterion, calculated as the residual model dependence of the experimental data is within acceptable limit.

Chapter 12 - Invariant properties of the polynomials are studied. Class of polynomials with congruent set of eigenvalues is introduced. Evolution equations for eigenvalues and coefficients remaining the polynomial within proper class of polynomials are formulated. Evolution equations are given by equations for Weierstrass and Jacobian hyper-elliptic functions. The link between evolution equations for the coefficients of the polynomials and dynamical evolution equations of physical systems is established. Within the framework of generalized mechanics the inner and outer coordinates of momenta are interrelated as roots and coefficients of the characteristic polynomials. In particularly, in the case of $n=2$ degree the authors deal with the relativistic dynamics. The case of $n=3$ degree the authors arrive to new dynamics beyond the relativistic one.

As an application the algorithm of finding of roots of the polynomials based on the evolution process is built.

Chapter 13 - The stochastic processes, which are exploited to analyze the dynamical system in noisy environments, are white noise process, the Brownian motion process, the Ornstein-Uhlenbeck (OU) process. More notably, the OU process is a coloured noise process. In 1991, Karatzas and Shreve published an authoritative book discussing the Brownian motion process as well as the Brownian motion process-driven stochastic differential system in the rigorous mathematical framework, especially in the Itô setting. The estimation algorithm for the Itô stochastic differential equation (SDE) is developed using the Fokker-Planck equation, the master equation for the Itô SDE setting. The power spectral density of the stochastic process associated with the stochastic problems in physics, digital wireless communications is the frequency-dependent that justifies the usefulness of coloured noise processes in dynamical systems. The problem of analyzing the coloured noise-driven stochastic differential system can be accomplished by generalizing the Fokker-Planck equation. Master equations for coloured noise-driven stochastic differential systems involve the more general structures in contrast to the Fokker-Planck equation. For this reason, it seems quite worthwhile to summarize available results in literature, which are widespread, about the master equations. This paper encompasses six different master equations, which can be exploited to accomplish the estimation procedures for stochastic differential systems, especially where observations are not available. The structure of a master equation is dictated on the basis of the noise process associated with the dynamical system. The proofs of master equations of this paper can be accomplished using the functional calculus and perturbation-theoretic approaches to the system non-linearity-process noise coupling terms. Furthermore, this paper discusses briefly the general theory of two mathematical problems related to the estimation theory for coloured noise processes.

Chapter 14 - Based on a general model for the Yosida approximation, the solvability of a general class of first-order nonlinear evolution inclusions is investigated. The notion of the H-maximal monotonicity plays a greater role in developing a general framework for the existing Yosida approximations and its characterizations in the literature.

Chapter 15 - In this chapter the authors study a non-autonomous neutral functional differential equation in a Banach space. Using the theory of semigroups of operators to evolution equations and contraction mapping theorem the authors establish the existence and uniqueness of a weighted pseudo almost automorphic solution of the problem under consideration.

Chapter 16 - The goal of this chapter is to put together some recent results concerning applications of monotone second order differential equations to singularly perturbed problems of elliptic - parabolic and elliptic - hyperbolic type. More exactly, the solution v of the heat equation or of the telegraph system is compared with the solution v_ε of an elliptic regularization. This elliptic regularization is a perturbed problem written with the aid of a small parameter $\varepsilon > 0$. It is a particular case of some second order differential equations governed by a maximal monotone operator in the Hilbert space $L^2(\Omega)$. Under some specific hypotheses, the authors construct a zero order asymptotic approximation for v_ε making use of the boundary layer function method of Vishik and Lyusternik. The higher order regularity of the solutions to both perturbed and unperturbed problems is investigated. The order of accuracy of the difference v_ε - v is also established in some appropriate function spaces. Thus, the solution v of the heat equation (or telegraph system) is approximated by the solution v_ε of its elliptic regularization, which is a more regular function. This is a motivation for the study of the above mentioned second order evolution equations associated with monotone operators. This study can involve different unperturbed problems: semilinear heat equation, linear heat equation with nonlinear boundary conditions, semilinear telegraph system, nonlinear telegraph system with nonlinear boundary conditions, etc.

In: Evolution Equations
Editor: Arthur L. Claes
ISBN: 978-1-61209-652-0

Short Communication

A Theorem on the Optical Cubic-Quintic Schrödinger Equation

M. A. Grado-Caffaro* and M. Grado-Caffaro
Madrid, Spain

Abstract

There are two fundamental invariant nonlinear functionals associated with a nonlinear evolution equation which is certainly notorious in mathematical physics. We refer to the optical cubic-quintic Schrödinger equation. With respect to the lossless equation, the norm of a functional which is a linear combination of the two above functionals is introduced. By establishing a suitable theorem, we prove that the above functional is bounded.

Keywords: Optical cubic-quintic Schrödinger equation; Invariant functionals; Norm; Bounded functional.

1. Introduction

Evolution equations are currently partial differential equations which play a great role in significantly relevant aspects of mathematical physics. Among these equations, nonlinear partial differential equations occupy an important place in quantum mechanics, quantum field theory, nonlinear optics, physics of fluids, etc. In nonlinear optics, we can mention a remarkable example: the optical cubic-quintic Schrödinger equation, whose solutions are solitary waves (see, for instance, refs.[1-3]). As a matter of fact, relevant functionals associated with the aforementioned equation will be the subject of the present communication. More precisely, starting from considering two fundamental invariant quantities associated with the optical lossless cubic-quintic Schrödinger equation, we shall

* www.sapienzastudies.com; e-mail address: ma.grado-caffaro@sapienzastudies.com

treat an invariant quantity which is a real-valued nonlinear functional (see, for example, refs.[1-3]).

In fact, after defining the norm of the above functional, we shall prove a useful theorem by which the functional in question is bounded.

2. THEORY

The optical lossless cubic-quintic Schrödinger equation referred to dimensionless space s and time τ reads [1-3]:

$$\frac{\partial^2 \psi}{\partial \tau^2} + 2i\frac{\partial \psi}{\partial s} \pm 4|\psi|^2 \psi + 2\theta|\psi|^4 \psi = 0 \tag{1}$$

where i is the imaginary unit and θ is a real nonlinearity parameter which can be positive or negative.

The solutions to eq.(1) are solitary waves as follows:

$$\psi(s,\tau) = \varphi(\tau - bs)\exp[i(as + b\tau)] \tag{2}$$

where $\varphi(u)$, with $u \equiv \tau - bs$, is a C^2-class positive real-valued function which represents the amplitude of a given wavefunction $\psi(s,\tau)$ while a and b are real constants.

On the other hand, there exist two fundamental invariant quantities represented by the following nonlinear functionals [1-3]:

$$I_1(\varphi) \equiv \int_{-\infty}^{+\infty} \varphi^2(u)du \tag{3}$$

$$I_2(\varphi,\varphi') \equiv \int_{-\infty}^{+\infty} \left[(\varphi')^2 + b^2\varphi^2 \mp 2\varphi^4 - \frac{2}{3}\theta\varphi^6\right]du \tag{4}$$

where the prime denotes derivative.

Another invariant quantity is the functional $I_3(\varphi,\varphi') \equiv I_2(\varphi,\varphi') + 2aI_1(\varphi)$ [3]. Since there exists a relationship between $\varphi(u)$ and $\varphi'(u)$ that will be found later, we consider $I_2(\varphi)$ so one has $I_3(\varphi)$. Then we have the following:

Definition.- The norm of each above functional is defined as:

$$\|I_k\| \equiv \sup_{\varphi \neq 0} \frac{|I_k(\varphi)|}{\|\varphi\|^2} \tag{5}$$

where $k = 1,2,3$ and $\|\varphi\|$ relative to the standard scalar product for square-integrable functions so $\|\varphi\| \equiv \sqrt{\int_{-\infty}^{+\infty} \varphi^2(u)du}$.

Now we have:

Theorem. Functional $I_3(\varphi)$ is bounded.

Proof: we have to prove that there exists $\|I_k\|$ and that $\|I_k\| < +\infty$ (see relationship (5)) so, first by formulae (3) and (5) and the expression above for $\|\varphi\|$, one has that $\|I_1\| = 1$ so the functional $I_1(\varphi)$ is bounded. Now let us examine the functional $I_2(\varphi, \varphi')$. In this respect, we begin by obtaining the following nonlinear ordinary differential equation after substituting (2) into (1):

$$\varphi'' - (2a + b^2)\varphi \pm 4\varphi^3 + 2\theta\varphi^5 = 0 \tag{6}$$

Multiplying both sides of (6) by $d\varphi$, integrating the resulting expression and taking into account that $\varphi \to 0$ and $\varphi' \to 0$ as $u \to \infty$ [1,2], we get:

$$(\varphi')^2 = (2a + b^2)\varphi^2 \ \mu\ 2\varphi^4 - \frac{2}{3}\theta\varphi^6 \tag{7}$$

Inserting (7) into (4), functional $I_2(\varphi, \varphi')$ becomes:

$$I_2(\varphi) \equiv 2\int_{-\infty}^{+\infty} \left[(a + b^2)\varphi^2 \ \mu\ 2\varphi^4 - \frac{2}{3}\theta\varphi^6 \right] du \tag{8}$$

On the other hand, we have [1,2]:

$$\varphi(u) = \frac{(2a + b^2)^{1/2}}{\left\{ \sqrt{\left[1 + \frac{2}{3}\theta(2a + b^2)\right]} \cosh\left(2\sqrt{2a + b^2}\,u\right) \pm 1 \right\}^{1/2}} \tag{9}$$

Looking at eq.(9), we see that $\varphi \to 0$ as $u \to \infty$ and that $\varphi(u)$ is bounded above so that $\varphi_{\max} = \varphi(0)$. Let us denote $\varphi_{\max}$ by Φ so $\varphi(u) \leq \Phi$. Then, by formula (9), it follows:

$$\Phi^2 = \frac{2a + b^2}{\left[1 + \frac{2}{3}\theta(2a + b^2)\right]^{1/2} \pm 1} \tag{10}$$

By using relationship (8), now we have:

$$\frac{|I_2(\varphi)|}{\|\varphi\|^2} \le \frac{2\int_{-\infty}^{+\infty}\left(|a| + b^2 + 2\varphi^2 + \frac{2}{3}|\theta|\varphi^4\right)\varphi^2 du}{\int_{-\infty}^{+\infty}\varphi^2 du} \le 2\left(|a| + b^2 + 2\Phi^2 + \frac{2}{3}|\theta|\Phi^4\right) < +\infty \quad (11)$$

so $I_2(\varphi)$ is bounded. Consequently, $I_3(\varphi)$ is also bounded •

The theorem has been proved. By relations (5) and (11), one has the following estimation for $\|I_3\|$ namely $\|I_3\| \le 2\left(2|a| + b^2 + 2\Phi^2 + \frac{2}{3}|\theta|\Phi^4\right)$ with Φ from formula (10). Therefore, it is obvious that $\|I_3\| < +\infty$.

Conclusion

We have proved a useful proposition which can be the source for deducing systematically, in future studies, other significant results. The three fundamental nonlinear functionals considered in this communication have been found to be bounded. In particular, one has that $\|I_1\| = 1$ while both $\|I_2\|$ and $\|I_3\|$ have been estimated. The integrand of each nonlinear functional involved in our preceding study is an even polynomial in φ. Each integrand is the density of its respective functional so that the density of I_1 has degree 2 whereas the densities of both I_2 and I_3 have degree 6.

References

[1] S. Cowan, R.H. Enns, S.S. Rangnekar, S.S. Sanghera: Quasi-soliton and other behaviour of the nonlinear cubic-quintic Schrödinger equation, *Can. J. Phys.* 64 (1986) 311-315.

[2] M.A. Grado-Caffaro, M. Grado-Caffaro: Optical quasi-solitons from the cubic-quintic Schrödinger equation, *J. Optics* (Paris) 23 (1992) 229-232.

[3] M.A. Grado-Caffaro, M. Grado-Caffaro: On the two fundamental invariants associated with the lossless cubic-quintic Schrödinger equation, *Optik* 108 (1998) 48-49.

In: Evolution Equations
Editor: Arthur L. Claes
ISBN: 978-1-61209-652-0

Chapter 1

EVOLUTION EQUATIONS FOR GRAIN GROWTH AND COARSENING

***F. D. Fischer*[1] *and J. Svoboda*[2]**
[1]Institute of Mechanics, Montanuniversität Leoben,
Franz-Josef-Straße 18, A-8700 Leoben, Austria
[2]Institute of Physics of Materials, Academy of Sciences of the Czech Republic,
Žižkova 22, CZ-616 62, Brno, Czech Republic

ABSTRACT

A real classical topic of research for evolving systems of microstructures with grains like metals and ceramics is grain growth and coarsening (Ostwald Ripening). Both phenomena are driven by the decrease in total surface energy, while the chemical composition of the individual objects remains unchanged. Only the morphology of the system evolves and gives the impression that smaller objects are "eaten" by the larger ones. Evolution equations for these phenomena were published in the last century mostly as phenomenological equations. The authors show, however, that the evolution equations can be derived from the Thermodynamic Extremal Principle (TEP). This principle, based on the works of L. Onsager (1931, 1945) and later of H. Ziegler (1977), allows calculation of the evolution equations of the state parameters, e.g., the effective radii of individual grains, precipitates, etc., by maximizing the dissipation, expressed in a quadratic form of the kinetic parameters, with the constraint that the dissipation is equal to the negative rate of Gibbs energy in the isothermal case. Later studies have allowed working with distribution functions of the effective radii of objects and the evolution of those to stationary distribution functions. Also, the inverse problem has been solved, by which the evolution equations of the grain or precipitate effective radii can be derived from steady state distribution functions. This concept was extended from one-parameter distribution functions to various two-parameter distribution functions. The concept of TEP also allows successful simulations of the microstructure evolution in multi-component multiple particle systems. Practical applications of this concept of evolution equations for multi-component materials are published and used in the metals processing industry—see the program package MatCalc, http://matcalc.wkmp.tuwien.ac.at/.

The main progress of using the TEP to derive evolution equations can be seen in a rigorous variational approach compared to the numerous heuristic approaches to grain growth and coarsening.

Keywords: grain growth; modelling; thermodynamics; simulation; extremal principle.

1. Introduction

Both grain growth and precipitate coarsening (Oswald ripening) are dissipative processes driven by the decrease of the total Gibbs energy of grain boundaries or of matrix/precipitate interface for systems at constant temperature and pressure. The problem of grain growth (more precisely, grain coarsening) concerns a wide field of materials and has been a favourite topic of research during the last 50 years [1-5]. The term "grain coarsening" reflects much better the fact that during the process only the large grains grow, and the small grains shrink and disappear. As, however, the term "grain growth" has mainly been used for this phenomenon in the literature, we keep using the term grain growth. Also, the problem of bulk diffusion controlled precipitate coarsening concerns a wide field of materials and has been a favourite topic of research during the last 50 years [6-12]. Both the grain growth and precipitate coarsening have usually a rather negative influence on mechanical properties and, thus, modelling of kinetics of these processes is of highly relevant industrial importance.

The grain boundary migration controlled grain growth is often described by classical Hillert (H) theory [1], and the diffusion controlled precipitate coarsening is usually treated by classical Lifshitz-Slyozov-Wagner (LSW) theory [6-8]. In both cases, each object is characterized by its effective radius R_i. The number of grain edges, as in the Neumann-Mullins theory, is not included in this treatment; for discussion, see [13]. The evolution equations for the radii R_i are deduced in the frame of both theories. The evolution equations contain the critical radius R_C as a parameter calculated from the radii distribution function. The objects with radii $R_i < R_C$ shrink, and objects with radii $R_i > R_C$ grow. Both theories provide the steady-state radii distribution functions. If these steady-state distribution functions are normalized to the fixed total volume of objects, and the R coordinate is normalized to R_C or to the mean radius $\overline{R}$, then the normalized "steady-state" distribution functions are time-independent. For the steady-state distribution in both LSW- and H- theories, the critical radius R_C has a well-defined relation to the mean radius $\overline{R}$; see, e.g., the treatment in [14]. In addition, the evolution equations for R_C and $\overline{R}$ are at disposal.

Streitenberger [15] and Streitenberger and Zöllner [16] could unify as well as generalize both the LSW- and H- theory by introducing two parameters, namely β and $\rho_0 = R_0/R_c$ with R_0 being a cut-off radius. If $\beta = 0$, the Gibbs energy of the system is dissipated by grain boundary mobility controlled grain growth. If $\beta = 1$, bulk diffusion controlled particle coarsening is considered to be the dominating dissipative process. Both cases—grain growth and precipitate coarsening—are dealt with in detail in [15] considering a cut-off parameter

ρ_0 being in the range $\rho_H \leq \rho_0 < \infty$ or $\rho_{LSW} \leq \rho_0 < \infty$. As ρ_{LSW} the cut-off parameter of the LSW-theory $(\rho_{LSW} = 1.5)$ and as ρ_H the cut-off parameter of the H-theory $(\rho_H = 2.0)$ are understood. Of specific interest is the case for $\rho_0 \to \infty$ yielding a Weibull-type distribution function, which corresponds for $\beta = 0$ with the classical Rayleigh distribution function proposed by Louat (1974), for reference, see [15]) and for $\beta = 1$ with a particle growth analogue to Louat's distribution function. The general type of the evolution equation according to the generalized LSW- and H-theory can be taken from [15-16] with $\dot{R} = dR/dt$ and t being the time as

$$. R^{\beta+1}\dot{R} = K\left[\frac{R}{R_C}\cdot c\left(R/R_C, R_0/R_C\right) - 1\right] \tag{1.1}$$

The quantity K is a constant reflecting the geometry of objects and the physical and chemical processes behind the evolution. The size correlation function $c\left(R/R_C, R_0/R_C\right)$ can be taken from [15], Eq. (10); both classical LSW- and H-theories have selected $c \equiv 1$. However, Monte Carlo simulations, see Zöllner and Streitenberger [17] and the previous literature cited there, have revealed that $c\left(R/R_C, R_0/R_C\right)$ is not a constant and could be approximated by a polynomial function in R/R_C. Already, a quadratic polynomial yields, however, very complicated mathematical forms of the steady-state distribution functions.

The models on grain growth have been refined, and the treatment has become more and more complicated. At this place, it is necessary to mention the approaches based on irreversible thermodynamics, e.g., Rios et. al [18-20], an ongoing discussion about the grain size distribution with important results from three-dimensional computer simulations and experimental grain size distributions, see, e.g., Rios et al. [21], and questioning about the usual spherical grain approximation by introduction of grain topology, see Glicksman's [22], Rios and Glicksman [23-24].

Non-steady-state distributions of grains are also a topic of current research. These distributions have been constructed by the multi-phase-field (MPF) model or by the Monte Carlo Potts (MCP) model.

A set of continuous field variables $\phi_i\left(\underline{r},t\right)$, depending on the position $\underline{r}$ and time t is defined within the MPF model. These field variables ϕ_i, used as order parameters, overtake the role of the volume fraction of a distinct orientation. A free energy functional must now be proposed, including additional weighted gradient terms of the set of ϕ_i. The MPF model involves a remarkable set of de-facto unknown parameters; for details see, e.g., Suwa et al. [25-26]. A computationally improved version was recently presented in [27] and applied to simulations of normal grain growth reflecting the H-theory. For a comparative study of MPF models for grain growth, the reader is referred to [28].

Zöllner and Streitenberger [17] used the MCP model, which is computationally very efficient and compares the energy state of a lattice site with a possible neighboring state.

Then, a probability is defined for a flip to this neighboring state. This concept needs the definition of the energy states as functions of the orientation of a lattice site and its neighbors; of course, a significant number of parameters appear again. A proper selection allows also reflecting the H-theory for normal grain growth.

It should be noted that a similar regime, as in grain growth, happens in coarsening. In this case, second-phase precipitates are distributed in a matrix and change their morphology by dissolving small precipitates through transfer of mass to larger precipitates ('Ostwald ripening'). Furthermore, in the case of precipitates, their coarsening is accompanied by a diffusive process in the matrix, compensating the different chemical compositions of the precipitates and the matrix. Thus, for a description of the kinetics of coarsening, it is necessary to know the concentrations of all components in the precipitates and in the matrix and the diffusion coefficients of all components in the matrix.

To the knowledge of the present authors, the pioneering work on precipitate coarsening was published by Greenwood [29], Lifshitz and Slyozov [6-7] and Wagner [8], who already incorporated the coupling of reaction at the interface and diffusion in the matrix. Within the framework of mean-field theory, they derived the equation of evolution of a single precipitate controlled by diffusion in the matrix of the component dominating in the precipitate with the same relation as (1.1), however, with $\beta = 1$, $c = 1$ and K replaced by γ / ε with γ being the interface energy and ε a coefficient, which is introduced in subsection 3.1. of this paper and in §2.2.6 and 2.2.7 of [5]. Later, Voorhees and Glicksman [30-31] presented a sophisticated concept of how to investigate diffusion amongst multiple particles in a binary system. This allows us to account for interaction between precipitates of higher volume fractions. Recent results, also including the effect of a stress state on the evolving shape of a precipitate, have been reported by Voorhees [10] and Thornton et al. [32]. The most comprehensive explanation of the problem of diffusion-controlled coarsening has been given in Chapter 21 of the book by Glicksman [33]. An equivalent concept to those of Lifshitz and Slyozov [6-7] and Wagner [8] was applied by Ratke et al. [34] to coarsening of precipitates in the melt controlled by thermal conduction in the melt. The treatment of coarsening of precipitates in multi-component systems has been treated first in 2003, [5], at least to the knowledge of authors, and was extended by Kozeschnik and Svoboda et al. in [11-12] to multi-component, multi-particle, multi-phase systems. Most recently, experimental studies appeared for two different precipitates (e.g., Au and Ag precipitates); see [35], which are in agreement with the current concept. Finally, a study by Niethammer [41] should be mentioned; Niethammer investigated the mathematical background of the LSW theory with respect to higher order corrections and fluctuations. Also, the morphological change of precipitates can be taken into account by adding the elastic strain contribution to the Gibbs energy. Introducing shape factors allows describing precipitates by ellipsoidal bodies with evolving axes; see [37-38]. Already, practical applications of this concept for shape-changing precipitates in a multi-component matrix have shown successful agreement with measurements, see, e.g., [39-40]. With respect to the morphological and topological evolution of the microstructure, nowadays, extremely complicated configurations for binary alloys can be studied; see Fife and Voorhees [36].

The second law of thermodynamics admits all spontaneous processes with positive entropy production (positive dissipation) but does not allow the determination of a distinct unambiguous evolution path of the system [42]. In 1931, L. Onsager formulated an extremal principle [43] for heat conduction and in 1945, for diffusion [44], according to which the unambiguous path of the thermodynamic system corresponds to a constrained maximum of dissipation in any part of the system at any time. From the thermodynamic point of view, Onsager's principle can be understood as a strong form of the second law of thermodynamics, allowing the determination of the unambiguous evolution path of the system.

In 1991, Onsager's principle was first formulated in terms of discrete parameters characterizing the state of the system [45] and named as Thermodynamic Extremal Principle (TEP). One can choose just the parameters, which characterize the system in the best way with a sufficient accuracy. The evolution of these parameters is then aimed to describe the system evolution in a natural way. The application of the TEP provides a systematic concept for the derivation of the evolution equations for these parameters for a wide class of problems in the context of linear non-equilibrium thermodynamics [46].

The theoretical treatment of grain growth and precipitate coarsening is possible only by accepting some model assumptions, which simplify the reality, but still conserve the important features of grain growth and precipitate coarsening. One of these usual model assumptions is the approximation of each object by a volume-equivalent sphere. Then, the system can be described by the effective radii R_k of individual objects, and the evolution of R_k can be treated by application of TEP [5]. This treatment is called "the multi-object concept", and the evolution equations for R_k obtained from TEP are equivalent to those derived first by Hillert [1] for the grain growth or by Lifshitz, Slyozov and Wagner [6-8] for precipitate coarsening by means of heuristic considerations more than 50 years ago.

The structure is usually characterized by an object radii distribution function $f(R) = dN / dR$, where N is the number of objects in the system with radii $\leq R$. Very recently, the TEP was employed to distribution functions and called "the distribution concept" [47], which provides the evolution equations for the time-dependent parameters of the distribution function and, thus, allows also the description of the non-steady grain growth and precipitate coarsening. If the distribution function f depends only on one free parameter a, one can always find a relation between the mean radius $\bar{R}$ and the parameter a directly from the definition of $\bar{R}$ as

$$\bar{R} = \int_0^\infty f(R', a) R' dR' / \int_0^\infty f(R', a) dR' . \tag{1.2}$$

Thus, any one-parametric distribution function may be a candidate for a steady-state distribution function, and the distribution concept allows to derive an evolution equation for the object radius, $\dot{R} = \dot{R}(R)$, from the one-parametric steady state distribution function [48]. Thereby, problems can be solved being inverse to those treated, e.g., by Streitenberger et al. [15-17].

The concepts based on TEP require the proper formulation of the total dissipation Q and of the total Gibbs energy G of the system, both in the case of grain growth and precipitate coarsening; see [13]. The Gibbs energy G contains the total grain boundary (or interface) energy, which, unlike the other contributions to the total Gibbs energy G, may change with time. Svoboda et al. could show in [47], how this concept can be applied to modeling of non-steady grain growth. Two possible approaches were introduced. The evolution equations for the radii of individual grains were obtained by using the multigrain concept and, for the parameters of the distribution functions, by means of the distribution concept. The latter concept has been demonstrated for Rayleigh's distribution [47] with the mean grain size as parameter. The authors have shown in [48] how the distribution concept can be used to derive the evolution equation for the object radius scaling parameter $a(t)$, as well as for the radius $R(t)$ of the individual grains for several prominent one-parametric distribution functions. The value of the critical radius R_C follows then automatically from the condition $\dot{R}(R_C) = 0$.

The distribution concept is also supported by very recent computational analysis methods for three-dimensional polycrystalline microstructures as presented by Groeber et al. [49-50]. This gives a strong motivation to look deeper into the analysis of distribution functions with multiple parameters.

The goals of this chapter are the following:

1. To introduce the thermodynamic concept and general theory of the TEP.
2. To select different sets of parameters describing the system and to derive the evolution equations for grain growth and to precipitate coarsening in the frame of one object concept, multi-object concept and distribution concept by application of TEP.
3. To apply the derived equations for simulation of the kinetics of systems with monomodal and bimodal object radii distributions.
4. To investigate several established one parametric distribution functions and to derive the object radius evolution equations from the distribution functions for grain growth and coarsening.
5. To investigate the evolution of non-steady-state two parametric distribution functions with an object radius scaling parameter $a(t)$, directly related to the critical radius R_C , and a dimension-free parameter $\beta(t)$, determining the shape of the distribution function for both the grain growth and coarsening case.

2. Thermodynamic Concept and Thermodynamic Extremal Principle (TEP)

2.1. General Remarks on the Thermodynamic Concept

The evolution of solid-state systems at elevated temperature occurs predominantly by diffusion of individual components and migration of interfaces (boundaries between individual phases or grains of the same phase with different orientation). Although these processes are based on weakly directed stochastic motion of individual atoms, from the macroscopic point of view, the stochasticity seems to be unimportant and the directed motion prevails. Thus, the processes at elevated temperatures, like diffusion and/or motion of the interfaces, can be described phenomenologically by linear non-equilibrium thermodynamics. The simplest diffusion law (Fick law) has the form

$$\vec{j}_k = -D_k grad c_k , \tag{2.1}$$

which relates the diffusional flux $\vec{j}_k$ of component k with its driving force $grad c_k$ (c_k is the concentration of component k, $k = 1,...,n_C$). The proportionality kinetic coefficient D_k is the diffusion coefficient of component k. Similar (sometimes more complicated) proportionality relations as given by Eq. (2.1) can be written for many other phenomena in non-equilibrium systems near the thermodynamic equilibrium.

If the phenomenological equations are completed by conservation laws like

$$\frac{\partial c_k}{\partial t} = -div \vec{j}_k \tag{2.2}$$

and by initial, boundary and contact conditions, then the evolution of the system is given by a set of partial differential equations. Their solution is, however, rather complicated, even for quite simple systems.

Material science often treats complex systems, and only a limited number of parameters can be used for the characterization of the system. The choice and the number of the state parameters depend on requirements regarding the accuracy of the description and the complexity of the system. Then, if we follow the classical way of simulations, we solve phenomenological equations under some simplified conditions in the first step, and then we extract the time evolution of the parameters from the solution in the second step. If, however, the system is rather complex, the solution of the phenomenological equations cannot be managed, and/or the important features of the problem are lost.

2.2. Thermodynamic Extremal Principle (TEP)

As already mentioned in the Introduction, a variational formulation of a thermodynamic problem was presented first by Onsager in *Physical Review* in 1931 [43]. He derived the equations for heat conduction in an anisotropic material from the assertion of the constrained maximum of entropy production. In an analogous way, he also derived equations for diffusion in multi-component system from his principle [44].

The concept can be shown exemplarily by deriving Ohm's law. Let us assume a single electric circuit with one battery of voltage U and one resistor. The electric current in the circuit is denoted by I. We assume that the resistor is sufficiently cooled and kept at a constant temperature. The rate $\dot{G}$ of the total Gibbs energy G in the electrical system amounts to $\dot{G} = -UI$. The heat generation rate given by Joule's heat corresponds to the dissipation $Q = RI^2$, with R being a proportionality constant. We maximize now Q with respect to I subjected to the constraint $Q = -\dot{G}$ (all released Gibbs energy from the battery is supposed to be dissipated in the resistor). Then, we have with the Lagrange parameter λ

$$\frac{\partial}{\partial I}\left(RI^2 + \lambda\left(RI^2 - UI\right)\right) = 0 \tag{2.3}$$

yielding

$$2RI\left(1+\lambda\right) - \lambda U = 0 . \tag{2.4}$$

Multiplication of this equation with I and keeping in mind $UI = RI^2$ yield immediately λ = -2 and consequently Ohm's law $RI = U$, with R identified as electrical resistance of the resistor.

Due to the identity of $Q = -\dot{G}$, one can also maximize $-\dot{G} = UI$ with the constraint $Q = -\dot{G}$, and with the Lagrange parameter μ this provides

$$\frac{\partial}{\partial I}\left(UI - \mu\left(RI^2 - UI\right)\right) = 0 , \tag{2.5}$$

which yields

$$U\left(1+\mu\right) - 2\mu RI = 0 . \tag{2.6}$$

Keeping in mind again that $UI = RI^2$, it provides immediately $\mu = 1$ and the reproduction of the Ohm's law. Concluding, one has thus found a well-known constitutive law from an extremal principle.

It is interesting to note that Onsager's work in Physical Review [45] was cited more than 2,000 times, but nearly exclusively due to his derivation of the famous Onsager's reciprocity relations from the theorem of microscopical reversibility, for which Onsager received the Nobel prize. It is quite surprising that Onsager did not disseminate his principle intensively. Maybe, Onsager himself did not see any advantage of the variational formulation of a thermodynamic problem: from the principle, one can only derive the already known phenomenological equations, as shown above, which must be solved in a standard way. This really represents no advantage for the solution of problems. The next 60 years after 1931 can be characterized as a period in which Onsager's principle was nearly forgotten, and there is, up to now, no reference of the principle in classical textbooks of thermodynamics, except to Gyarmati's textbook [51]. Only H. Ziegler's work from 1975 [52], sect. 15.1, should be mentioned; Ziegler formulated a constrained extremal problem exactly in the form of the last two equations (2.5) and (2.6). But his goal was to investigate the case of maximum dissipation by looking for orthogonality conditions with respect to a dissipation surface $Q = Q_0$, saying that the thermodynamic forces are parallel to the gradient of Q_0. Kestin and Rice

[53] criticized this assumption, already with respect to an earlier version of it, unless Q happens to be a homogeneous function in the rates of the internal variables, or more general, a function of a homogeneous function of them.

The view on the principle drastically changed in 1991, when the principle was formulated directly in terms of state parameters and the application of the principle provided a systematic way to explain how the evolution equations can be derived without the necessity to solve the partial differential equations [45]—see the next subsection.

2.3. Formulation of the Thermodynamic Extremal Principle in Terms of State Parameters

Let us, for simplicity, assume a closed system at constant temperature and pressure, in which the total Gibbs energy G represents the characteristic potential, which is dissipated during the system evolution. Let the state parameters be denoted by $q_i, (i = 1,....N)$ and their rates (kinetic parameters) by $\dot{q}_i, (i = 1,....N)$. The time evolution of the system is assumed to be caused by diffusive fluxes and motion of the migrating interfaces, which lead to the change of the state parameters $q_i, (i = 1,....N)$ expressed by their rates $\dot{q}_i, (i = 1,....N)$. One can assume that the total Gibbs energy G of the system can be expressed as a function of state parameters, $G = G(q_1, q_2, ..., q_N)$, and, thus, its rate $\dot{G}$ is given by the chain rule

$$\dot{G} = \sum_{i=1}^{N} \frac{\partial G}{\partial q_i} \dot{q}_i . \tag{2.7}$$

To be able to calculate the total dissipation, one must perform a key step, which allows (under reasonable simplifications, assumptions on the system geometry and by application of conservation laws) expressing all diffusive fluxes $\vec{j}_k$ and the interface velocity v as functions of state parameters and their rates. Thus, one looks for the expressions

$$\vec{j}_k = \sum_{i=1}^{N} \vec{f}_{ik}\left(q_1, q_2, ...q_N\right) \dot{q}_i, \quad \left(k = 1, ..., n_C\right), \tag{2.8}$$

$$v = \sum_{i=1}^{N} h_i\left(q_1, q_2, ..., q_N\right) \dot{q}_i , \tag{2.9}$$

where n_C is the number of components in the system and $\vec{j}_k$ and v are linear forms of $\dot{q}_i, (i = 1,....N)$.

The total dissipation in the isotropic system can be expressed by [46]

$$Q = \sum_{k=1}^{n_C} \int_V \frac{R_g T}{c_k D_k} \vec{j}_k^2 dV + \int_{\text{Interfaces}} \frac{1}{M} v^2 \, dA, \tag{2.10}$$

where R_g is the gas constant, T the absolute temperature, c_k the concentration, D_k the tracer diffusion coefficient of component k, and M is the grain boundary or interface mobility. Then, one can insert Eqs. (2.8) and (2.9) into Eq. (2.10) and perform the integrals with the result

$$Q = \sum_{i,j=1}^{N} U_{ij} (q_1, q_2, \ldots q_N) \dot{q}_i \dot{q}_j, \tag{2.11}$$

where U_{ij} is a symmetric and positive definite matrix of global kinetic coefficients of the system and takes into account material properties and geometry of the system.

In many cases, the chosen state parameters q_i, $(i = 1, \ldots . N)$ are not free, and some constraints exist amongst them, which can generally be expressed in the form

$$F_l (q_1, q_2, \ldots, q_N) = 0, \quad (l = 1, \ldots, m). \tag{2.12}$$

If equations (2.12) are differentiated in time, one obtains

$$\sum_{j=1}^{N} a_{jl} (q_1, q_2, \ldots, q_N) \dot{q}_j = 0, \quad (l = 1, \ldots, m) \tag{2.13}$$

with $a_{jl} = \dfrac{\partial F_l}{\partial q_j}$.

Now one can formulate the thermodynamic extremal principle (TEP):

A closed system under constant temperature and pressure evolves in the way corresponding to the maximum dissipation Q *constrained by the balance equation* $Q + \dot{G} = 0$ *and Eqs. (2.13).*

The TEP can be understood as a "strong" form of the second law of thermodynamics. The second law of thermodynamics allows an infinite number of paths of the system with positive dissipation Q. The TEP selects only that one corresponding to the constrained maximum of dissipation Q.

The necessary conditions for the maximum of dissipation Q constrained by $Q + \dot{G} = 0$ and Eqs. (2.13) is given by

$$\frac{\partial}{\partial \dot{q}_i}\left[Q+\alpha\left(Q+\dot{G}\right)+2\sum_{l=1}^{m}\beta_l\sum_{j=1}^{N}a_{jl}\dot{q}_j\right]=0, \quad \left(i=1,\ldots,N\right), \tag{2.14}$$

with α and $\beta_l,\left(l=1,\ldots,m\right)$ being Lagrange multipliers. One can insert $\dot{G}$ and Q from Eqs. (2.7) and (2.11), respectively, and evaluate the derivatives with the result

$$2\sum_{j=1}^{N}U_{ij}\dot{q}_j+\alpha\left(2\sum_{j=1}^{N}U_{ij}\dot{q}_j+\frac{\partial G}{\partial q_i}\right)+2\sum_{l=1}^{m}\beta_l a_{il}=0, \quad \left(i=1,\ldots,N\right). \tag{2.15}$$

If Eqs. (2.15) are multiplied by $\dot{q}_i$, summed over i, $\left(i=1,\ldots,N\right)$, and Eqs. (2.7), (2.11) and (2.9) are used, one obtains

$$2Q+\alpha\left(2Q+\dot{G}\right)=0, \tag{2.16}$$

which, together with the constraint $Q+\dot{G}=0$, allows for the evaluation of the value of α as $\alpha=-2$. Then, Eqs. (2.15) can be rewritten into the final form as

$$\sum_{j=1}^{N}U_{ij}\dot{q}_j+\sum_{l=1}^{m}a_{il}\beta_l=-\frac{\partial G}{\partial q_i}, \quad \left(i=1,\ldots,N\right).$$

(2.17)

It is advantageous to note that the matrix U_{ij} can be calculated from the dissipation function Q as

$$U_{ij}=\frac{1}{2}\frac{\partial^2 Q}{\partial \dot{q}_i \partial \dot{q}_j}, \quad \left(i,j=1,\ldots,N\right) \tag{2.18}$$

and that

$$\frac{\partial \dot{G}}{\partial \dot{q}_i}=\frac{\partial G}{\partial q_i}, \quad \left(i=1,\ldots,N\right) \tag{2.19}$$

follows directly from Eq. (2.7).

If the problem is properly formulated, equations (2.17), together with the constraints (2.13), can always be solved with respect to $\dot{q}_i$, $(i=1,\ldots.N)$, and they can be integrated with respect to time to obtain $q_i=q_i\left(t\right)$, $(i=1,\ldots.N)$, which is the aim of our efforts.

The TEP seems to be a handy tool in modelling of thermodynamic processes, and it has been successfully used in development of models for sintering [54-61], for creep in super-alloy single crystals [62-65] and for diffusional phase transformations [66-73].

3. Thermodynamic Concepts for the Treatment of Grain Growth and Precipitate Coarsening

3.1. General Features

Let us first introduce (or repeat) the common features of all presented thermodynamic concepts for the description of the evolution of the set of objects:

- Each object (grain or precipitate) is approximated by a volume-equivalent sphere and characterized by its effective radius R_i, $(i = 1,....N)$.
- In the case of precipitate coarsening, we assume that the system is chemically in equilibrium, and, thus, the total volume of precipitates remains constant. The only driving force in the system is the change in the total interface energy. Analogous assumptions are made for grain growth, in the case of which a constant volume of grains is assumed, and the only driving force in the system is the change in the total grain boundary energy.

If the system consists of N objects of effective radii R_i, $(i = 1,....N)$, and the specific grain boundary energy (in the case of grain growth) or the interface energy (in the case of precipitate coarsening) is denoted by γ, then, the total Gibbs energy G of the system is given by

$$G = 4\pi\xi\gamma\sum_{i=1}^{N} R_i^2 . \tag{3.1}$$

The factor ξ is $\xi = 1/2$ for grain growth (each grain boundary is shared by two grains) and $\xi = 1$ for precipitate coarsening. Of course, the interface energy γ can be replaced by a quantity, which is a function of the local orientation, e.g., considering specific crystal planes as boundaries; see, e.g., Elsey et al. [74] for a recent study. However, for sake of simplicity, γ is assumed to be a constant, average value. Also, second phase particles may play a role in grain growth. These particles may cluster along the grain boundary and hinder grain growth with a deviation from an evolution law as (1.1); see, e.g., [75]. Also, this problem is not treated here.

To calculate the total dissipation in the system, one may start with Eq. (2.10). In case of grain growth, one can assume that the grain boundary velocity of grain i equals the rate $\dot{R}_i$ of the radius R_i and no diffusive fluxes occur in the system. The total dissipation Q_M due to migration of grain boundaries is then given by

$$Q_M = \frac{1}{2}\frac{4\pi}{M}\sum_{i=1}^{N} R_i^2 \dot{R}_i^2 . \tag{3.2}$$

In case of coarsening of precipitates, we assume that no dissipation occurs due to interface migration, and the total dissipation is given exclusively by diffusion. For application of the TEP, it is necessary to express the diffusive fluxes by means of the rates $\dot{R}_i$.

Let each spherical precipitate with the radius R_i be surrounded by a diffusion zone of the radius Z_i. In each zone, only the radial diffusive fluxes j_{ki} caused by the rate $\dot{R}_i$ and uniform deposition or collection of all components k are supposed in the diffusion zone. The radii Z_i are assumed to be large enough, such that the diffusion zones are overlapping. The matrix does not change its mean chemical composition given by concentrations c_k, and negligible concentration gradients are assumed in the matrix. The concentration of component k in the precipitates is denoted by c_k^*. No radial flux is assumed inside the precipitate $j_{ki}(r) \equiv 0$ for $0 \le r < R_i$. The radial flux $j_{ki}(r)$ of component k outside of the precipitate i in the distance r from its center is given by

$$j_{ki}(r) = \dot{R}_i (c_k - c_k^*) \frac{R_i^2}{r^2} \cdot \frac{Z_i^3 - r^3}{Z_i^3 - R_i^3}, \quad R_i \le r \le Z_i . \tag{3.3}$$

Then the total dissipation Q_D, given by Eq. (2.10), can be expressed as

$$Q_D = \sum_{k=1}^{n_C}\sum_{i=1}^{N} 4\pi \int_{R_i}^{Z_i} \frac{R_g T}{c_k D_k} j_{ki}^2 r^2 dr . \tag{3.4}$$

Assuming $R_i << Z_i$, Eq. (3.4) can be rewritten as

$$Q_D = \sum_{i=1}^{N} 4\pi R_g T R_i^3 \dot{R}_i^2 \sum_{k=1}^{n_C} \frac{\left(c_k^* - c_k\right)^2}{c_k D_k} = 4\pi\varepsilon \sum_{i=1}^{N} R_i^3 \dot{R}_i^2 , \quad \varepsilon = R_g T \sum_{k=1}^{n_C} \frac{\left(c_k^* - c_k\right)^2}{c_k D_k} . \tag{3.5}$$

Thus, Q_D is, in first approximation, independent of the size of the diffusion zones, being large compared to the precipitate sizes (decisive dissipation occurs in the nearest vicinity of the precipitates).

The assumption of constant total volume of all objects yields

$$\frac{4\pi}{3}\sum_{i=1}^{N} R_i^3 = const \quad \Rightarrow \quad \sum_{i=1}^{N} R_i^2 \dot{R}_i = 0 . \tag{3.6}$$

3.2. Evolution Equation for the Mean Radius of Objects — The One-Object Concept

In this case, let us assume that the whole system of objects is described by only one parameter—the mean grain radius $\bar{R}$. If we denote by $\bar{N}$ the number of objects in the system, the total volume of the system is given by $\bar{N}\bar{R}^3$ and the requirement of its time independence leads to

$$\dot{\bar{N}}\bar{R}^3 + 3\bar{N}\bar{R}^2\dot{\bar{R}} = 0 \Rightarrow \dot{\bar{N}} = -3\bar{N}\dot{\bar{R}} / \bar{R} . \qquad (3.7)$$

In agreement with Eq. (3.1), the total Gibbs energy G of the system is given by

$$G = 4\pi\xi\bar{N}\gamma\bar{R}^2 , \qquad (3.8)$$

and its rate $\dot{G}$ reads by using Eq. (3.7) as

$$\dot{G} = 4\pi\xi\gamma\dot{\bar{N}}\bar{R}^2 + 8\pi\xi\gamma\bar{N}\bar{R}\dot{\bar{R}} = -4\pi\xi\gamma\bar{N}\bar{R}\dot{\bar{R}} . \qquad (3.9)$$

The dissipation in the system reads according to Eq. (3.2)

$$Q = \omega\frac{2\pi}{M}\bar{N}\bar{R}^2\dot{\bar{R}}^2 \qquad (3.10)$$

for grain growth, and, according to Eq. (3.5),

$$Q = \omega 4\pi\varepsilon\bar{N}\bar{R}^3\dot{\bar{R}}^2 \qquad (3.11)$$

for coarsening. The factor ω expresses the fact that the dissipation is, in reality, larger than that expressed by Eqs. (3.2) or (3.5), because some objects in the system shrink and some objects grow with much larger rate than $\dot{\bar{R}}$. The factor can be estimated as $\omega \approx 2$.

The system is described by only one state parameter $\bar{R}$, thus, only the constraint $Q + \dot{G} = 0$ is sufficient for the determination of $\dot{\bar{R}}$ as

$$\dot{\bar{R}} = \frac{\gamma M}{\omega\bar{R}} \qquad (3.12)$$

for grain growth (Eqs. (3.9) and (3.10) are used) and

$$\dot{\bar{R}} = \frac{\gamma}{\omega\varepsilon\bar{R}^2} \qquad (3.13)$$

for coarsening (Eqs. (3.9) and (3.11) are used).

One can see that already a very rough approach provides a quite good qualitative and often sufficient quantitative insight into the treated problem.

3.3. Evolution Equations for Radii of Individual Objects — The Multi-object Concept

Let a system be composed of a set of grains approximated by spheres of radii R_i, $i = 1,\ldots,N$ being chosen as state parameters. According to TEP, the evolution path of the system corresponds to the maximum of the total dissipation in the system, Q, constrained by $\dot{G} + Q = 0$ and other constraints; in this case only one constraint exists concerning the time independence of the total volume, see Eq. (3.6). The necessary condition for the constrained maximum of Q is given by

$$\frac{\partial}{\partial \dot{R}_i}\left(Q + \alpha\left(Q + \dot{G}\right) + \lambda 4\pi \sum_{j=1}^{N} R_j^2 \dot{R}_j \right) = 0\,, i = 1,\ldots,N\,, \tag{3.14}$$

with $\dot{G}$ following from Eq. (3.1) as

$$\dot{G} = 8\pi\xi\gamma \sum_{i=1}^{N} R_i \dot{R}_i \,. \tag{3.15}$$

The dissipation Q is given by Eq. (3.2) for grain growth and by Eq. (3.5) for coarsening, and α and λ are Lagrange multipliers. Following the general procedure leading to Eq. (2.17), Eq. (3.14) yields

$$R_i^2 \dot{R}_i = 2M\left(-\gamma R_i + \lambda R_i^2\right), i = 1,\ldots,N\,, \tag{3.16}$$

for grain growth and

$$R_i^2 \dot{R}_i = \frac{2}{\varepsilon}\left(-\gamma + \lambda R_i\right),\ i = 1,\ldots,N\,, \tag{3.17}$$

for coarsening.

Insertion of Eqs. (3.16) into Eq. (3.6) allows for the evaluation of λ, and then the evolution equations for individual grains during grain growth read

$$\dot{R}_i = 2\gamma M\left(\frac{1}{R_{CG}} - \frac{1}{R_i}\right),\ i = 1,\ldots,N\,, \tag{3.18}$$

with $R_{CG} = \sum_{j=1}^{N} R_j^2 \Big/ \sum_{j=1}^{N} R_j$ being a critical radius. Grains with radius R_{CG} neither grow nor shrink.

One can observe that large (super-critical) grains grow, while small (sub-critical) grains shrink and must be eliminated from the system if they reach zero radius. As the total number N of grains decreases and the total volume of grains remains constant during the process, the mean volume as well as the mean radius $\bar{R}$ of the grains increase; a detailed treatment is presented in Fischer et al. [14].

Analogously, insertion of Eqs. (3.17) into Eq. (3.6) allows the evaluation of λ, and then the evolution equations for individual precipitates during coarsening read

$$\dot{R}_i = \frac{2\gamma}{\varepsilon R_i}\left(\frac{1}{R_{CC}} - \frac{1}{R_i}\right),\ i = 1,\ldots,N, \tag{3.19}$$

with $R_{CC} = \sum_{j=1}^{N} R_j / N = \bar{R}$ being a critical radius.

Let $f(R,t)$, for $t = 0$, be a starting grain radii distribution function. Then, one can generate a sufficiently large set of grain radii R_i and integrate equations (3.18) or (3.19) in time. At any time, t, the distribution function $f(R,t)$, can be constructed from the values of $R_i(t)$.

3.4. Evolution Equations for Parameters of Distribution Function — The Distribution Concept

One can also apply the TEP directly to the distribution function itself, see [47]. For this sake, let the distribution function depend on n parameters $a_1(t),\ldots,a_n(t)$, which determine the evolution of the distribution function

$$f(R,t) = f(R, a_1(t),\ldots,a_n(t)). \tag{3.20}$$

To avoid taking into account the condition of a fixed volume of the system as a constraint, we assume that all considered distribution functions f meet a condition analogous to Eq. (3.6) as

$$\int_0^\infty R^3 f dR = 1 \tag{3.21}$$

at any time, i.e., the system has a fixed volume $4\pi/3$.

The well-known continuity relation of the distribution function f in the radii space reads as

$$\frac{\partial f}{\partial t}+\frac{\partial (f\dot{R})}{\partial R}=0. \tag{3.22}$$

Equation (3.22) can be integrated at a fixed time as

$$\int_R^\infty \frac{\partial f}{\partial t}dR' = \left(f\dot{R}\right)\Big|_R - \left(f\dot{R}\right)\Big|_\infty = \left(f\dot{R}\right)\Big|_R, \tag{3.23}$$

since $\left(f\dot{R}\right)\Big|_\infty = 0$, due to a finite value $\dot{R}\,\Big|_\infty$ and $f\Big|_\infty = 0$.

Differentiation of Eq. (3.20) with respect to time yields

$$\frac{\partial f}{\partial t}=\sum_{i=1}^n \frac{\partial f}{\partial a_i}\dot{a}_i, \tag{3.24}$$

and insertion of equation (3.24) into equation (3.23) enables us to calculate $\dot{R}$ as

$$\dot{R}(R)=\frac{1}{f}\sum_{i=1}^n \dot{a}_i\int_R^\infty \frac{\partial f}{\partial a_i}dR'. \tag{3.25}$$

Equation (3.25) represents a key equation of the treatment, as it allows expressing the total dissipation in the system by means of $\dot{a}_1, \ldots, \dot{a}_n$.

Similar to Eq. (3.1), the total Gibbs energy of the system is given by

$$G=4\pi\xi\gamma\int_0^\infty R^2 f dR. \tag{3.26}$$

The total dissipation in the system due to grain boundary migration during grain growth follows as (in analogy to Eq. (3.2))

$$Q=\frac{2\pi}{M}\int_0^\infty R^2 f\dot{R}^2 dR=\frac{2\pi}{M}\int_0^\infty \frac{R^2}{f}\left(\sum_{i=1}^n \dot{a}_i\int_R^\infty \frac{\partial f}{\partial a_i}dR'\right)^2 dR \tag{3.27}$$

and, due to diffusion during coarsening (analogy to Eq. (3.5)), as

$$Q = 4\pi\varepsilon \int_0^\infty R^3 f \dot{R}^2 dR = 4\pi\varepsilon \int_0^\infty \frac{R^3}{f} \left(\sum_{i=1}^{n} \dot{a}_i \int_R^\infty \frac{\partial f}{\partial a_i} dR' \right)^2 dR . \tag{3.28}$$

According to TEP, the evolution of the system corresponds to the maximum of the total dissipation in the system Q constrained by $\dot{G} + Q = 0$ (no other constraints exist for $\dot{a}_1$,..., $\dot{a}_n$). The treatment, analogous to that presented in the previous section, leads to evolution equations of the system given by a set of linear equations with respect to $\dot{a}_i$, $i = 1,...,n$ (for details, see [47])

$$\sum_{j=1}^{n} \frac{1}{2} \frac{\partial^2 Q}{\partial \dot{a}_i \partial \dot{a}_j} \dot{a}_j = - \frac{\partial G}{\partial a_i} , \; i = 1,...,n \tag{3.29}$$

with

$$\frac{\partial G}{\partial a_i} = 4\pi\xi\gamma \int_0^\infty R^2 \frac{\partial f}{\partial a_i} dR , i = 1,...,n, \tag{3.30}$$

$$\frac{1}{2} \frac{\partial^2 Q}{\partial \dot{a}_i \partial \dot{a}_j} = \frac{2\pi}{M} \int_0^\infty \frac{R^2}{f} \int_R^\infty \frac{\partial f}{\partial a_i} dR' \int_R^\infty \frac{\partial f}{\partial a_j} dR' dR , i, j = 1,...,n, \tag{3.31}$$

for grain growth and

$$\frac{1}{2} \frac{\partial^2 Q}{\partial \dot{a}_i \partial \dot{a}_j} = 4\pi\varepsilon \int_0^\infty \frac{R^2}{f} \int_R^\infty \frac{\partial f}{\partial a_i} dR' \int_R^\infty \frac{\partial f}{\partial a_j} dR' dR , \; i, j = 1,...,n . \tag{3.32}$$

for coarsening.

3.5. Examples of Grain Growth Kinetics

Kinetics in System with Initial Rayleigh's Distribution

The measured grain radii distribution functions can often be well approximated by Rayleigh's distribution, see, e.g., [47],

$$f_R\left(R, c(t)\right) = \frac{8}{3\sqrt{\pi}} c^{5/2} R \exp(-cR^2) = K c^{5/2} R \exp(-cR^2) . \tag{3.33}$$

The distribution function is normalized to ensure equation (3.21) for arbitrary positive values of $c(t)$ standing for $a_1(t)$, $n=1$. It can easily be shown that grain growth corresponds to a decreasing value of c. The distribution functions resembling Rayleigh's distribution have been derived also theoretically by taking into account the spatial grain size correlation and topology effects [76], sect. 9.25.

Denoting the dissipation in the system with Rayleigh's distribution by Q_R, equation (3.31) yields

$$\frac{1}{2}\frac{d^2Q_R}{d\dot{c}^2} = \frac{2\pi}{M}\int_0^\infty \frac{R^2}{f_R}\left(\int_R^\infty \frac{df_R}{dc}dR'\right)^2 dR = \frac{5\pi K}{16Mc^{5/2}}. \tag{3.34}$$

Denoting the total Gibbs energy in the system with Rayleigh's distribution by G_R, equation (3.26) provides

$$\frac{dG_R}{dc} = 2\pi\gamma\int_0^\infty R^2\frac{df_R}{dc}dR = \frac{\pi\gamma K}{2c^{1/2}}. \tag{3.35}$$

The kinetics of the system follows from Eq. (3.29) as

$$\dot{c} = -\frac{8}{5}\gamma Mc^2. \tag{3.36}$$

The mean grain radius $\overline{R}_R$ corresponding to the Rayleigh distribution is given by

$$\overline{R}_R = \int_0^\infty Rf_R dR \Big/ \int_0^\infty f_R dR = \frac{\sqrt{\pi}}{2c^{1/2}}. \tag{3.37}$$

Then the kinetics of the mean grain radius $\overline{R}_R$ results from combining equations (3.36) and (3.37) as

$$\overline{R}_R\dot{\overline{R}}_R = -\frac{\sqrt{\pi}}{2c^{1/2}}\frac{\sqrt{\pi}}{4c^{3/2}}\dot{c} = -\frac{\pi}{5}\frac{\dot{c}}{c^2} = \frac{\pi}{5}\gamma M \text{ or } \overline{R}_R^2\Big|_t - \overline{R}_R^2\Big|_0 = \frac{2\pi}{5}\gamma Mt, \tag{3.38}$$

which is consistent with the well-known parabolic law for grain growth.

In this one-parametric case, c can be replaced by $\overline{R}_R$ (use equation (3.37)) in equation (3.33). After a proper normalization of the grain radii distribution function f_R and of the grain radius R by $\bar{R}_R$, the general non-steady distribution concept degenerates to the well-known steady-state concept.

Using equations (3.34)-(3.38), the total dissipation in the system, described by Rayleigh's distribution and characterized by the mean grain radius $\bar{R}_R$, is given by

$$Q_R = \frac{1}{2}\frac{d^2 Q_R}{d\dot{c}^2}\dot{c}^2 = -\frac{dG_R}{dc}\dot{c} = \frac{4\pi^2\gamma^2 M}{15\bar{R}_R^3} . \qquad (3.39)$$

To allow a simple demonstration of the simulation results, values of $\gamma = 1$ and $M = 1$ are used in all following simulations. Two studies of the system evolution are presented with the same starting Rayleigh's distribution given by $c(0) = 1$ in Figure 1. The first simulation covers the distribution concept (solid lines), and the second one, the multigrain concept for 25,000 starting grains (dotted lines). The individual curves in Figure 1a) correspond to grain radii distributions for times $t = 0, 0.2, 0.4, 0.6, 0.8$ and 1, the curves in the detailed Figure 1b) correspond only to the time $t = 1$. In Figure 2, the time dependence of the total Gibbs energy G (Figure 2 a) and b)) and the total dissipation $Q \equiv -\dot{G}$ (Figure 2 c)) are plotted for both simulations. It is evident from the Figure 2c) that the initial dissipation, according to the multigrain concept, is higher than that according to the distribution concept. This is in agreement with expectations, since a higher number of degrees of freedom allows a finer description of the system evolution and, consequently, a higher dissipation. However, in advanced stages, when the grain radii distributions for both concepts begin to deviate more and more, the multigrain concept provides a lower dissipation (see Figure 2a)). As discussed later, this is due to a narrowing of the distribution function treated by the multigrain concept (see Figure 1b)).

Kinetics in Systems with Initial Bimodal Distribution

For the treatment of systems with a bimodal grain radii distribution, see, e.g., [77], one can use a linear combination of two Rayleigh's distributions as

$$f_B = KR\left[Ac_1^{5/2}\exp(-c_1R^2) + (1-A)c_2^{5/2}\exp(-c_2R^2)\right], \qquad (3.40)$$

meeting again the condition (3.21) of a constant volume of the system for arbitrary values of parameters from the region $0 \le A \le 1$, $c_1 > 0$, $c_2 > 0$.

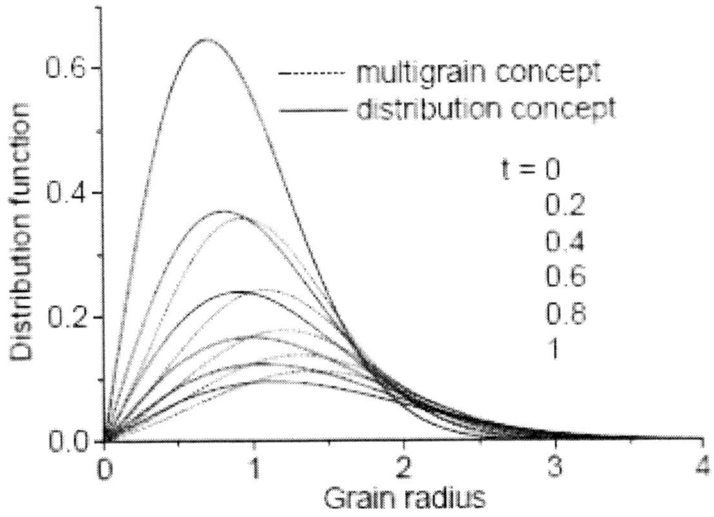

a)

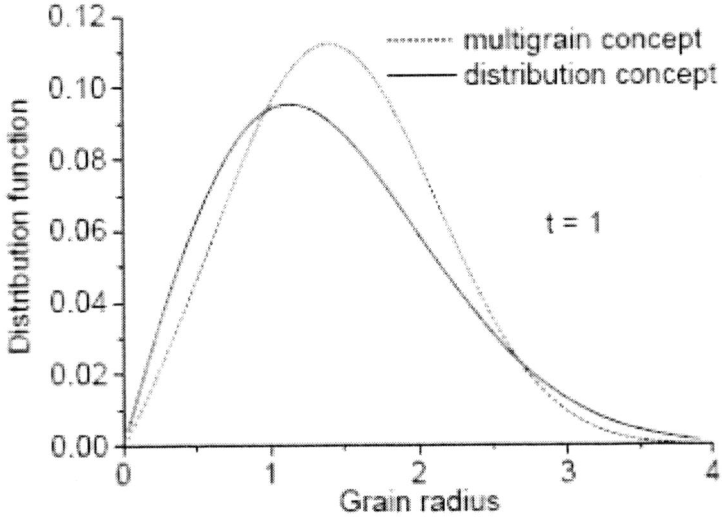

b)

Figure 1. Comparison of the evolution of monomodal grain radii distribution functions by application of multi-object concept and distribution concept. a) Individual curves correspond to times denoted in the figure. b) Detailed plot of both distribution functions for time $t = 1$. Reprinted from Acta Materialia 55, J. Svoboda, F.D. Fischer, A new approach to modelling of non-steady grain growth, 4457-4474, 2007, with permission from Elsevier.

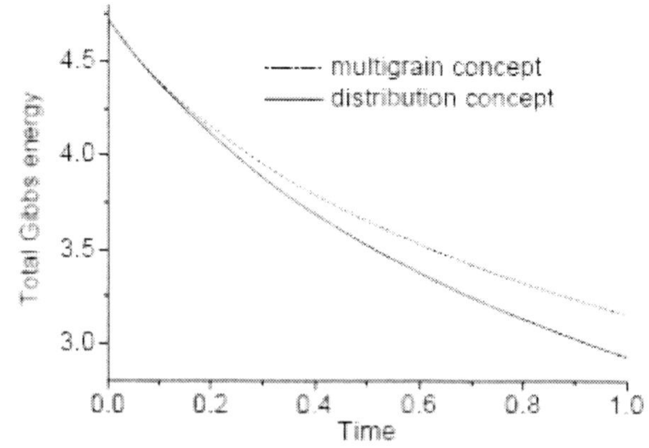

a)

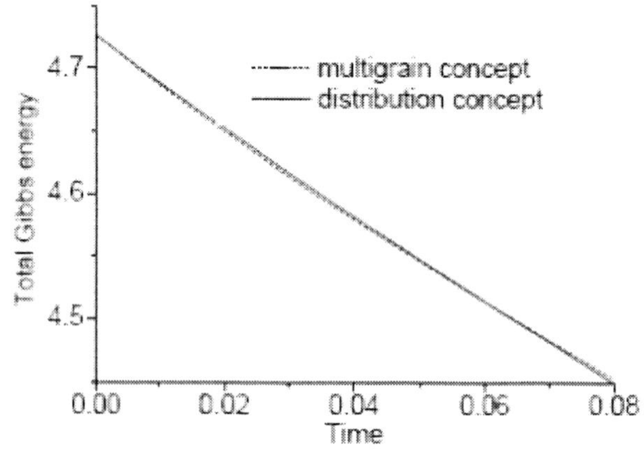

b)

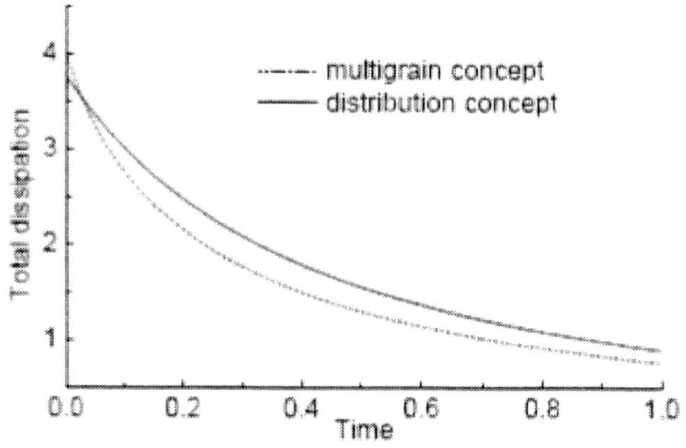

c)

Figure 2. Time dependence of the total Gibbs energy and the total dissipation corresponding to simulations presented in Figure 1. a) Total Gibbs energy. b) Detailed plot of the total Gibbs energy for small times. c) Total dissipation. See Figure 1 for permission.

Using equations (3.30), one can write

$$\frac{\partial G_B}{\partial A} = 2\pi\gamma K \int_0^\infty R^3 \left[c_1^{5/2} \exp(-c_1 R^2) - c_2^{5/2} \exp(-c_2 R^2) \right] dR = \pi\gamma K \left[c_1^{1/2} - c_2^{1/2} \right], \quad (3.41)$$

$$\frac{\partial G_B}{\partial c_1} = 2\pi\gamma A K c_1^{5/2} \int_0^\infty R^3 \left(\frac{5}{2c_1} - R^2 \right) \exp(-c_1 R^2) dR = \frac{\pi\gamma A K}{2c_1^{1/2}}, \quad (3.42)$$

$$\frac{\partial G_B}{\partial c_2} = 2\pi\gamma (1-A) K c_2^{5/2} \int_0^\infty R^3 \left(\frac{5}{2c_2} - R^2 \right) \exp(-c_2 R^2) dR = \frac{\pi\gamma (1-A) K}{2c_2^{1/2}}. \quad (3.43)$$

Equations (3.31) yield

$$\frac{1}{2}\frac{\partial^2 Q_B}{\partial \dot{A}^2} = \frac{\pi K}{2M} \int_0^\infty \frac{R\left[c_1^{3/2} \exp(-c_1 R^2) - c_2^{3/2} \exp(-c_2 R^2) \right]^2}{A c_1^{5/2} \exp(-c_1 R^2) + (1-A) c_2^{5/2} \exp(-c_2 R^2)} dR, \quad (3.44)$$

$$\frac{1}{2}\frac{\partial^2 Q_B}{\partial \dot{A} \partial \dot{c}_1} = \frac{\pi A K c_1^{1/2}}{4M} \int_0^\infty \frac{R\left[c_1^{3/2} \exp(-c_1 R^2) - c_2^{3/2} \exp(-c_2 R^2) \right] \left(3 - 2c_1 R^2\right) \exp(-c_1 R^2)}{A c_1^{5/2} \exp(-c_1 R^2) + (1-A) c_2^{5/2} \exp(-c_2 R^2)} dR, \quad (3.45)$$

$$\frac{1}{2}\frac{\partial^2 Q_B}{\partial \dot{A} \partial \dot{c}_2} = \frac{\pi (1-A) K c_2^{1/2}}{4M} \int_0^\infty \frac{R\left[c_1^{3/2} \exp(-c_1 R^2) - c_2^{3/2} \exp(-c_2 R^2) \right] \left(3 - 2c_2 R^2\right) \exp(-c_2 R^2)}{A c_1^{5/2} \exp(-c_1 R^2) + (1-A) c_2^{5/2} \exp(-c_2 R^2)} dR, \quad (3.46)$$

$$\frac{1}{2}\frac{\partial^2 Q_B}{\partial \dot{c}_1^2} = \frac{\pi A^2 K c_1}{8M} \int_0^\infty \frac{R\left(\left(3 - 2c_1 R^2\right) \exp(-c_1 R^2) \right)^2}{A c_1^{5/2} \exp(-c_1 R^2) + (1-A) c_2^{5/2} \exp(-c_2 R^2)} dR, \quad (3.47)$$

$$\frac{1}{2}\frac{\partial^2 Q_B}{\partial \dot{c}_1 \partial \dot{c}_2} = \frac{\pi A (1-A) K c_1^{1/2} c_2^{1/2}}{8M} \int_0^\infty \frac{R\left(3 - 2c_1 R^2\right)\left(3 - 2c_2 R^2\right) \exp(-(c_1 + c_2) R^2)}{A c_1^{5/2} \exp(-c_1 R^2) + (1-A) c_2^{5/2} \exp(-c_2 R^2)} dR, \quad (3.48)$$

$$\frac{1}{2}\frac{\partial^2 Q_B}{\partial \dot{c}_2^2} = \frac{\pi (1-A)^2 K c_2}{8M} \int_0^\infty \frac{R\left(\left(3 - 2c_2 R^2\right) \exp(-c_2 R^2) \right)^2}{A c_1^{5/2} \exp(-c_1 R^2) + (1-A) c_2^{5/2} \exp(-c_2 R^2)} dR. \quad (3.49)$$

The integrals in equations (3.44) to (3.49) must be calculated numerically. The kinetics of the system is then given by equations (3.29).

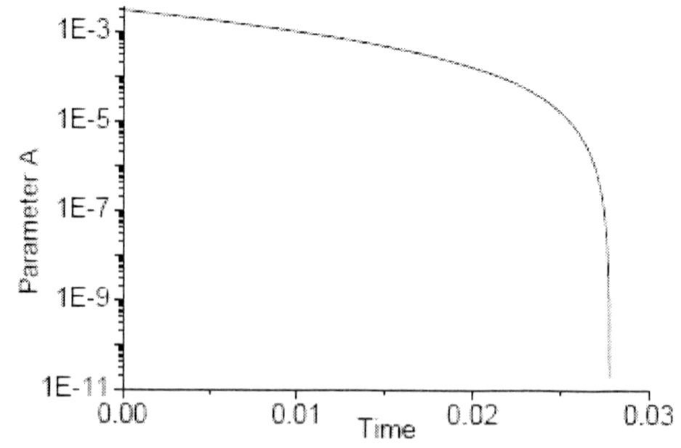

a)

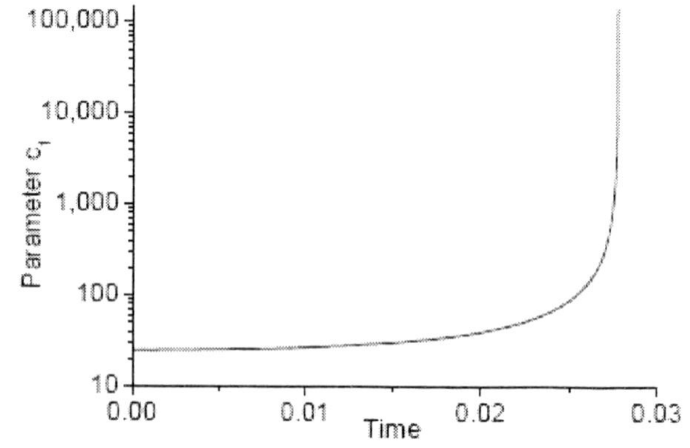

b)

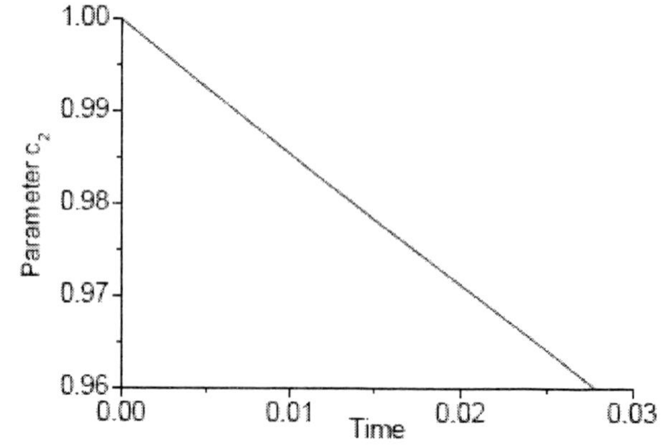

c)

Figure 3. Evolution of parameters of the bimodal distribution function calculated up to the time 0.027741. a) Parameter A . b) Parameter c_1 . c) Parameter c_2 . See Figure 1 for permission.

An initial bimodal grain radii distribution can be obtained by choosing $A(0) = 0.003$, $c_1(0) = 25$ and $c_2(0) = 1$.

The evolution of the parameters is depicted in Figure 3. One can observe that the contribution corresponding to c_1 gradually decreases and disappears at time $t = 0.027741$ ($A \rightarrow 0$ at that time – see Figure 3a)).

The value of the parameter c_1 increases drastically at the same time (see Figure 3b)), which correlates with a significant refinement of grains corresponding to the c_1-contribution. The c_2-contribution exhibits gradual growth (decreasing of the value of c_2 – see Figure 3c)) and survives as the only one. Starting with $t = 0.027741$, the evolution of the system can be described by equation (3.36).

The time evolution of the distribution corresponding to the evolution of the parameters A, c_1 and c_2 (calculated along the distribution concept) is depicted by the solid lines for times $t =$ 0, 0.005, 0.01, 0.015, 0.02, 0.025 and 0.027741 in Figure 4. This evolution is compared with that based on the multigrain concept for the same starting bimodal distribution and the same plot times (dotted lines).

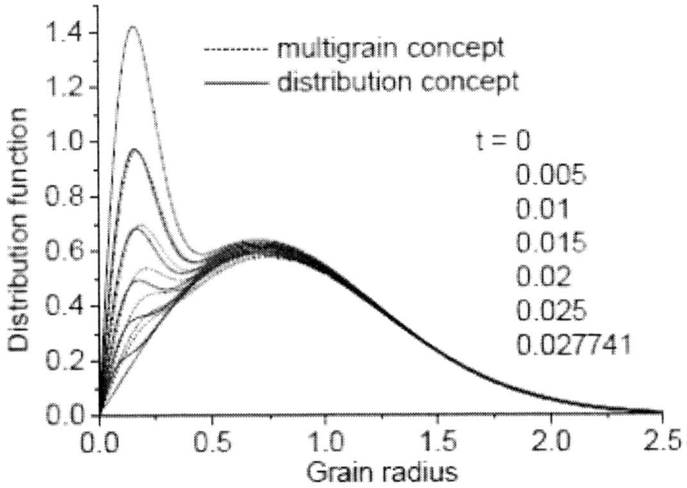

Figure 4. Comparison of the evolution of bimodal grain radii distribution function by application of multi-object concept and distribution concept. Individual curves correspond to times denoted in the figure. See Figure 1 for permission.

Kinetics in System with Hillert's Distribution

The well-known Hillert's theory [1] provides the distribution function f_H being the steady-state solution of equations (3.18) as

$$f_H\left(R, R_{CG}(t)\right) = \frac{L}{R_{CG}^4} \frac{R/R_{CG}}{\left(2 - R/R_{CG}\right)^5} \exp\left(\frac{-6}{2 - R/R_{CG}}\right), 0 \le R < 2R_{CG}, \quad \text{otherwise}$$

$$f_H = 0, \tag{3.50}$$

with $L = 503.88289$ to conform to equation (3.21). The kinetics of the steady-state grain growth is then given by the evolution equations for the critical radius $R_{CG}(t)$ and/or the mean radius $\bar{R}_H = 8R_{CG}/9$, see Hillert [1], as

$$\dot{R}_{CG} = \frac{\gamma M}{2R_{CG}} \text{ and } \dot{\bar{R}}_H = \frac{32\gamma M}{81\bar{R}_H}. \tag{3.51}$$

It is well known that equations (3.50) and (3.51) describe perfectly the evolution of grain radii distributions obtained by the present multi-object concept in the steady state. If any different reasonable grain radii distribution function is used as the starting one, the distribution treated by the multi-object concept approaches the distribution given by equation (3.50), see the discussion by Fischer et al. [14]. In other words, the grain radii distribution function given by equations (3.50) and (3.51) is a long-term solution of the grain growth treated by the multi-object concept.

The distribution concept can also be applied to Hillert's distribution given by equation (3.50). The function involves only one parameter, R_{CG}. Denoting the total Gibbs energy in the system with Hillert's distribution by G_H and its dissipation by Q_H, equations (3.30) and (3.31) deliver

$$\frac{dG_H}{dR_{CG}} = 2\pi\gamma \int_0^{2R_C} R^2 \frac{df_H}{dR_{CG}} dR = -\frac{2\pi L\gamma}{27e^3 R_{CG}^2}, \tag{3.52}$$

$$\frac{1}{2}\frac{d^2 Q_H}{d\dot{R}_{CG}^2} = \frac{2\pi}{M} \int_0^{2R_{CG}} \frac{R^2}{f_H} \left(\int_R^{2R_{CG}} \frac{df_H}{dR_{CG}} dR' \right)^2 dR = \frac{4\pi L}{27e^3 M R_{CG}}. \tag{3.53}$$

Using equation (3.29), Hillert's equation (3.51) is reproduced.

Applying Eqs. (3.51) to (3.53), the dissipation in the system described by Hillert's distribution and characterized by the mean grain radius $\bar{R}_H$, is given by

$$Q_H = \frac{1}{2}\frac{dQ_H}{d\dot{R}_{CG}} \dot{R}_{CG}^2 = -\frac{dG_H}{dR_{CG}} \dot{R}_{CG} = \frac{512\pi L\gamma^2 M}{(27e)^3 \bar{R}_H^3}. \tag{3.54}$$

Comparison of Kinetics in Systems with Rayleigh's and Hillert's Distributions

The total dissipation, expressing the rate of decrease of the total Gibbs energy in the system, provides valuable information on the kinetics of the system. This acts as a motivation to compare the total dissipation in systems with Rayleigh's distribution, Q_R, and Hillert's distribution, Q_H, by using equations (3.39) and (3.54). First, we assume identical mean grain radii in both systems, i.e., $\overline{R}_R = \overline{R}_H$ (see equations (3.37) and (3.51)), and we obtain $Q_R/Q_H = 1.28379$. The ratio differs in a surprising amount from the ideal value of 1. Therefore, we perform a second evaluation for systems with the same volume and the same grain boundary area by using the additional condition

$$\int_0^\infty R^2 f \, dR = 1 \tag{3.55}$$

for $f \equiv f_H$ (equation (3.40)) and $f \equiv f_R$ (equation (3.33)). One finds the values of $R_{CG} = 0.92914266$ and $c = 9\pi/16$, to which the mean radii $\overline{R}_H = 0.82590459$ and $\overline{R}_R = 2/3$ of Hillert's and Rayleigh's distributions, respectively, correspond. Both distributions are compared in Figure 5. After the insertion of the values of $\overline{R}_H$ and $\overline{R}_R$, one obtains the ratio $Q_R/Q_H = 2.44094$, which is drastically different from 1. Thus, one can conclude that the grain radii distribution has a strong impact on the grain growth kinetics.

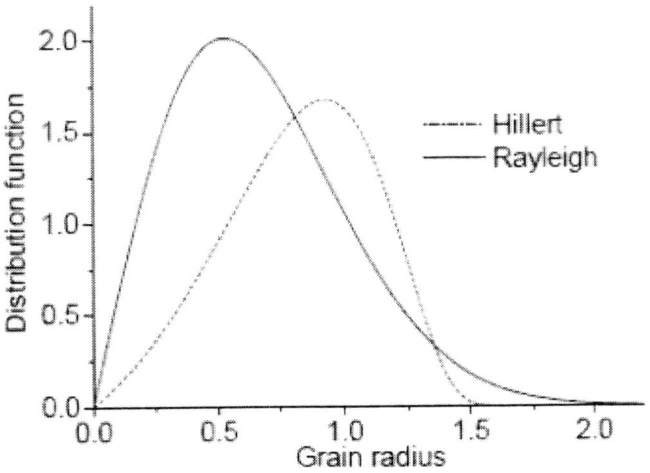

Figure 5. Comparison of Rayleigh's and Hillert's distribution function for systems of the same volume and the same grain boundary area. See Figure 1 for permission.

The reason that the kinetics of the system with Rayleigh's distribution is significantly faster than that with Hillert's is that one evidently originates from the fact that Rayleigh's distribution is markedly wider than Hillert's distribution (see Figure 5). The narrower that a distribution function is, the smaller is the deviation of each grain radius from the critical one and the lower the magnitude of the rate of the radius (see equations (3.19)) and the total dissipation.

The rather complicated interaction amongst grains is reflected in the grain radii distribution function. An advantage of the distribution concept lies in the fact that it reflects the interaction amongst grains purely phenomenologically, only by choosing the proper family of distribution functions. With a higher dissipation in the case of Rayleigh's distribution compared to Hillert's, one could support the experimental observation that this distribution often fits the real grain structures better than Hillert's one. On the other hand, one cannot overestimate the role of the distribution concept by presenting it as a tool for understanding the mathematics and physics behind the complicated interaction amongst objects. The distribution concept is a handy tool for description of grain growth and precipitate coarsening, being much more flexible than recent phenomenological concepts.

4. From Steady-State Distribution Functions to Evolution Equations

Equation (3.25) provides the relation between the rate $\dot{R}$ of the object radius R, the distribution function itself and the rates $\dot{a}_i$ of the parameters a_i of the distribution function. In the steady state, only one parameter a related to the mean object radius $\overline{R}$ evolves in time, and its rate $\dot{a}$ can be determined from the distribution concept. If the value $\dot{a}$ is inserted into Eq. (3.25), then Eq. (3.25) represents the evolution equation for the object radius R depending only on the respective kinetic parameter and the steady-state distribution function. The goal of this section is to investigate several established distribution functions and to derive the object radius evolution equations from the distribution functions for grain growth and precipitate coarsening.

4.1. Theoretical Treatment

For the case of a steady-state distribution function, in which only one parameter develops $f(R,t) = f(R,a(t))$, the rate $\dot{R}$ of the object radius R follows from Eq. (3.25) as

$$\dot{R}(R) = \frac{\dot{a}}{f}\int_R^\infty \frac{\partial f}{\partial a} dR' . \tag{4.1}$$

The rate $\dot{G}$ of the total Gibbs energy of the system then follows from Eq. (3.26) as

$$\dot{G} = 4\pi\xi\gamma\dot{a}\int_0^\infty R^2 \frac{\partial f}{\partial a} dR . \tag{4.2}$$

The total dissipation in the system has, in general, the form, see Eqs. (3.27) and (3.28),

$$Q = \int_0^\infty B(R) f \dot{R}^2 dR = \dot{a}^2 \int_0^\infty \frac{B(R)}{f} \left(\int_R^\infty \frac{\partial f}{\partial a} dR' \right)^2 dR . \tag{4.3}$$

The following functions can be selected for $B(R)$:

- The function $B(R)$ is given as $B(R) = 2\pi R^2 / M$ for grain boundary migration controlled grain growth
- The function $B(R)$ is given as $B(R) = 4\pi\varepsilon R^3$ for diffusion controlled coarsening of precipitates in multi-component systems.
- The function $B(R) = 4\pi R^2 / M + 4\pi\varepsilon R^3$ is valid for coarsening of precipitates controlled by both interface migration and diffusion (this case is not treated here).

If no other dissipative process is assumed to be present in the system, the rate $\dot{a}$ of parameter a can be obtained by equating the negative rate $-\dot{G}$ of the total Gibbs energy G with the dissipation Q, yielding, after using Eqs. (4.2) and (4.3),

$$\dot{a} = -\xi 4\pi\gamma \int_0^\infty R^2 \frac{\partial f}{\partial a} dR \Bigg/ \int_0^\infty \frac{B(R)}{f} \left(\int_R^\infty \frac{\partial f}{\partial a} dR' \right)^2 dR . \tag{4.4}$$

Insertion of Eq. (4.4) into Eq. (4.1) provides the complete evolution equation for the object radius expressed by means of the material and geometry-related parameters. The critical radius R_C (representing R_{CG} for grain growth or R_{CC} for coarsening) follows from the condition $\dot{R}(R_C) = 0$. Insertion of Eq. (4.4) into Eq. (4.2) allows the evaluation of $-\dot{G}$ being equal to the dissipation Q (also Eq. (4.3) can be used for the direct calculation of Q in a more complicated way).

4.2. Calculation of Evolution Equations from Most Common Steady-State Distribution Functions

In the case of Hillert's theory of grain growth as well as in the case of LSW- theory of precipitate coarsening, both the steady-state distributions and the evolution equations for the object radius are known. Thus, the theories can be used for testing the inverse-problem solution. The mathematics processor MAPLE (http://www.scientific.de/maple.html) is used for calculations of complicated integrals.

Calculation of Evolution Equations from Hillert's Distribution

The Hillert's (H) -distribution is given by Eq. (3.50), where R_{CG} is replaced by a. Using Eqs. (4.1) and (4.4), and the values of $\xi = 1/2$ and $B(R) = 2\pi R^2/M$ for the grain boundary migration controlled grain growth, one obtains, after the analytical evaluation of integrals, the evolution equation for the grain radius in the form

$$\dot{a} = \frac{\gamma M}{2a}, \quad R\dot{R} = 2\gamma M\left(\frac{R}{a} - 1\right). \tag{4.5}$$

Equation (4.5) represents Hillert's classical equation for grain growth. Of course, one can also investigate the evolution of a system with the H -distribution for the coarsening case $\left(\xi = 1,\ B = 4\pi\varepsilon R^2\right)$. By exact integration, one obtains

$$\dot{a} = \frac{0.820\gamma}{a^2\varepsilon}, \quad R^2\dot{R} = \frac{3.278\gamma R}{\varepsilon a}\left(\frac{R}{a} - 1\right). \tag{4.6}$$

Calculation of Evolution Equations from LSW-Distribution

The Lifshitz-Slyozov-Wagner (LSW) -distribution, taken from, e.g., [14], sect. 2.2.7, is given by

$$f_{LSW}\left(R, a(t)\right) = \frac{L_{LSW}}{a^4}\frac{(R/a)^2}{(3+R/a)^{\frac{7}{3}}(3/2-R/a)^{\frac{11}{3}}}\exp\left(\frac{-3}{3-2R/a}\right), \quad \begin{array}{l} 0 \le R < 1.5a \\ \text{otherwise } f_{LSW} = 0 \end{array}, \tag{4.7}$$

with the normalization factor $L_{LSW} = 61.396$ obtained by fulfilling Eq. (3.21). The integral $\int_R^\infty \left(\partial f_{LSW}/\partial a\right) dR'$ can be evaluated as

$$\int_R^\infty \frac{\partial f_{LSW}}{\partial a} dR' = -\frac{27}{4}\cdot\frac{L_{LSW}}{a^4}\cdot\frac{(1-R/a)}{(3+R/a)^{\frac{7}{3}}(3/2-R/a)^{\frac{11}{3}}}\exp\left(\frac{-3}{3-2R/a}\right). \tag{4.8}$$

Using Eq. (4.2) with Eqs. (4.1) and (4.8), as well as the set $\xi = 1$, $B(R) = 4\pi R^3\varepsilon$ for the coarsening case, one obtains after the evaluation of the integrals in Eq. (4.4)

$$\dot{a} = 0.296\frac{\gamma}{a^2\varepsilon}, \quad R^2\dot{R} = \frac{2\gamma R}{a\varepsilon}\left(1 - \frac{a}{R}\right) = \frac{2\gamma}{\varepsilon}\left(\frac{R}{a} - 1\right). \tag{4.9}$$

The evolution equation (4.9) agrees with that given by LSW-theory; see, e.g., [14], sect. 2.2.7, or [15].

If the LSW-distribution is applied to grain growth $\left(\xi=\frac{1}{2},\ B=2\pi R^2/M\right)$, its kinetics is given by

$$\dot{a}=0.105\frac{\gamma M}{a},\quad R\dot{R}=0.708\gamma M\left(1-\frac{a}{R}\right). \tag{4.10}$$

Calculation of Evolution Equations from a Weibull-Type Distribution Function

As mentioned in the Introduction, an evolution equation of type (1.1) should also exist for a Weibull (W) -type distribution with an infinite cut-off-parameter, [15], as

$$f_W\left(R,a(t)\right)=\frac{L_W}{a^4}\left(\frac{R}{a}\right)^{\beta+1}\exp\left(-\frac{3}{\beta+2}\left(\frac{R}{a}\right)^{\beta+2}\right), \tag{4.11}$$

with the normalization factor L_W following from the normalization equation (3.21) as

$$L_W=\frac{(\beta+2)^{-\frac{3}{\beta+2}}\cdot 3^{\frac{\beta+5}{\beta+2}}}{\Gamma\left((\beta+5)/(\beta+2)\right)}. \tag{4.12}$$

The parameter β is taken to be in the interval $0\le\beta\le1$ according to [15] and kept as a fixed quantity.

For grain growth $\left(\xi=\frac{1}{2},\ B=2\pi R^2/M\right)$, Eqs. (4.1) and (4.4), applied to the W-type distribution, result in

$$\dot{a}=\frac{U_{WG}\gamma M}{a},\quad R\dot{R}=U_{WG}\gamma M\left(\left(\frac{R}{a}\right)^2-\left(\frac{R}{a}\right)^{-\beta}\right). \tag{4.13}$$

In the coarsening case $\left(\xi=1,\ B=4\pi\varepsilon R^2\right)$, the relations

$$\dot{a}=U_{WC}\frac{\gamma}{a^2\varepsilon},\quad R^2\dot{R}=\frac{U_{WC}\gamma R}{a\varepsilon}\left(\left(\frac{R}{a}\right)^2-\left(\frac{R}{a}\right)^{-\beta}\right) \tag{4.14}$$

are obtained. The functions $U_{WG}(\beta)$ and $U_{WC}(\beta)$ are smooth functions and can be nearly perfectly approximated as

$$U_{WG}(\beta)=6/5\left(1-0.849\beta+0.210\beta^2\right),\quad U_{WC}(\beta)=1.382\left(1-0.396\beta+0.016\beta^2\right). \quad (4.15)$$

The kinetics of systems with the Rayleigh- or Louat-distribution (for details, see [15] and [76]) are obtained by choosing $\beta = 0$. The case for $\beta = 1$ is denominated by Streitenberger [15] as the "particle growth analogue" to the standard Rayleigh- or Louat-distribution function.

Calculation of Evolution Equations from the Maxwell-Distribution

The prominent Maxwell (M) -distribution function, fulfilling Eq. (3.21), stems from statistical physics, see, e.g., [78], sect. 5.2.10, and reads as

$$f_M\left(R,a(t)\right)=\frac{L_M}{a^4}\left(\frac{R}{a}\right)^2\exp\left(-\left(\frac{R}{a}\right)^2\right), \quad (4.16)$$

with $L_M = 1$.

For grain growth $\left(\xi=\frac{1}{2},\ B=2\pi R^2/M\right)$, Eqs. (4.1) and (4.4), applied to the M – distribution, result in

$$\dot{a}=\frac{0.358\gamma M}{a},\quad R\dot{R}=0.358\gamma M g_M\left(\frac{R}{a}\right), \quad (4.17)$$

with $g_M\left(\frac{R}{a}\right)=\left(\frac{R}{a}\right)^2-\frac{3}{2}\left(1+\frac{\sqrt{\pi}a}{2R}\exp\left(\left(\frac{R}{a}\right)^2\right)\left(1-\mathrm{erf}\left(\frac{R}{a}\right)\right)\right)$.

In the coarsening case $\left(\xi=1,\ B=4\pi\varepsilon R^2\right)$, we have

$$\dot{a}=0.438\cdot\frac{\gamma}{\varepsilon a^2},\quad R^2\dot{R}=0.438\frac{\gamma R}{a\varepsilon}g_M\left(\frac{R}{a}\right). \quad (4.18)$$

Calculation of Evolution Equations from the GSEHL-Distribution

Recently, a team of authors (Gendelman, Shapiro, Estrin, Hellmig and Lekhtmakher), [79], (abbreviation GSEHL) proposed a distribution function of the type

$$f_{GSEHL}\left(R,a(t)\right)=\frac{L_{GSEHL}}{a^4}\left(\frac{R}{a}\right)^{\nu}\exp\left(-\frac{R}{a}\right). \quad (4.19)$$

The quantity ν is a positive exponent. The normalization factor $L_{GSEHL}=1/\Gamma(\nu+4)$ follows from Eq. (3.21).

It can be shown that, in the grain growth case, non-zero evolution laws exist only for $\nu \le 2$ and, in the coarsening case, for $\nu \le 3$; otherwise $\dot{a} \equiv 0$. For the grain growth case $\left(\xi = \frac{1}{2},\ B = 2\pi R^2/M\right)$ with $\nu = 2$, it follows that

$$\dot{a} = 0.07407\frac{\gamma M}{a}, \quad R\dot{R} = 0.07407\gamma M g_{GSEHL}\left(\frac{R}{a}\right) \tag{4.20}$$

with $g_{GSEHL}\left(\frac{R}{a}\right) = \frac{a}{R}\left(\left(\frac{R}{a}\right)^3 - 3\left(\frac{R}{a}\right)^2 - 6\left(\frac{R}{a}\right) - 6\right)$.

For $\nu = 2$, the coarsening case $\left(\xi = 1,\ B = 4\pi\varepsilon R^2\right)$ yields

$$\dot{a} = 0.01550\frac{\gamma}{a^2\varepsilon}, \quad R^2\dot{R} = 0.01550\frac{\gamma}{\varepsilon}\left(\left(\frac{R}{a}\right)^3 - 3\left(\frac{R}{a}\right)^2 - 6\left(\frac{R}{a}\right) - 6\right), \tag{4.21}$$

and, for $\nu = 3$, the coarsening case follows as

$$\dot{a} = 0.00840\frac{\gamma}{a^2\varepsilon}, \quad R^2\dot{R} = 0.00840\frac{\gamma}{\varepsilon}\left(\left(\frac{R}{a}\right)^3 - 3\left(\frac{R}{a}\right)^2 - 9\left(\frac{R}{a}\right) - 18 - 18\left(\frac{a}{R}\right)\right). \tag{4.22}$$

Calculation of Evolution Equations from the Kirkaldy-Distribution Function

In 2002, Kirkaldy (K) [80] presented a distribution function, which reads

$$f_K\left(R, a(t)\right) = \frac{L_K}{a^4\eta^2}\left(\frac{R}{a}\right)^2\left(1 - \frac{1}{\eta^4}\left(\frac{R}{a}\right)^4\right), \quad 0 \le R \le \eta/a, \text{ otherwise } f_K = 0, \tag{4.23}$$

with the argument of some deficiencies in the LSW theory. The normalization factor is determined as $L_K = 15/\eta^4$ from Eq. (3.21). Kirkaldy [80] suggested the value for $\eta = 32/21$ for the coarsening case.

Using the distribution function f_K in combination with the growth case $\left(\xi = \frac{1}{2},\ B = 2\pi R^2/M\right)$ yields

$$\dot{a} = \frac{0.550\gamma M}{a\eta^2}, \quad R\dot{R} = 0.1570\gamma M g_K\left(\frac{R}{a}\right),$$

(4.24)

with $g_K\left(\frac{R}{a}\right)=\frac{5\left(\frac{R}{a}\right)^6+5\eta\left(\frac{R}{a}\right)^5+5\eta^2\left(\frac{R}{a}\right)^4+5\eta^3\left(\frac{R}{a}\right)^3-2\eta^4\left(\frac{R}{a}\right)^2-2\eta^5\frac{R}{a}-2\eta^6}{\eta^2\frac{R}{a}\left(\eta+\frac{R}{a}\right)\left(\eta^2+\left(\frac{R}{a}\right)^2\right)}$.

The coarsening case $\left(\xi=1,\ B=4\pi\varepsilon R^2\right)$ leads to

$$\dot{a}=\frac{1.735\gamma}{a^2\eta^3\varepsilon},\quad R^2\dot{R}=0.4958\frac{\gamma}{\varepsilon}\frac{1}{\eta}\frac{R}{a}g_K\left(\frac{R}{a}\right). \tag{4.25}$$

Calculation of Evolution Equations from the Log-Normal-Distribution Function

The Log-normal (Ln) distribution function, see, e.g., [81], can be written as

$$f_{Ln}\left(R,a(t)\right)=\frac{L_{Ln}}{a^4}\frac{\exp\left(-\left(\ln\left(R/a\right)\right)^2/2S^2\right)}{\left(R/a\right)^I.\sqrt{2\pi}S}. \tag{4.26}$$

This distribution function was applied first by Feltham [82] in the context of coarsening based on a diffusion model. It should be mentioned that the exponent I was set by Feltham [82] to zero, see also a later description of the model in [76]. The traditional form of the distribution, as, e.g., applied in [83, 84] has, however, the exponent I=1. In the following context, we use the traditional form and consider the quantity S (the standard deviation of the $\ln\left(R/a\right)$ values) as a fixed parameter; then we find the normalization factor $L_{Ln}=\exp\left(-9S^2/2\right)$.

The integral relation for Q, Eq. (4.3), however, diverges to infinity. The integral $\int_R^\infty\left(\partial f_{Ln}/\partial a\right)dR'$ in Eq. (4.3) approaches asymptotically to a finite value for $R\to 0$, so one has to analyze only the behavior of $B\left(R\right)/f_{Ln}$. This expression is in the order of magnitude of $B\left(R\right)\exp\left(\left(\ln\left(R/a\right)\right)^2/2S^2\right)$ for $R\to 0$, and it approaches infinity stronger than any function $R^{-n}, n>0$. The function $B\left(R\right)/f_{Ln}$, however, reaches a minimum at $R_{min}/a=\exp\left(-3S^2\right)$ for the growth case and at $R_{min}/a=\exp\left(-4S^2\right)$ for the coarsening case. On the other hand, the distribution function has its maximum at $R_{max}/a=\exp\left(3S^2\right)$. For $S\geq 0.5$, a sufficiently large distance exists between R_{min}/a and R_{max}/a. This limitation seems to be reasonable, since $S\geq 0.5$ is a typical value for structural materials, see, e.g., [85] with S=0.55. The divergence of the integral in Eq. (4.3) can be avoided by performing the integral with the lower integration limit being R_{min} instead of zero. This represents an elimination of objects with the radius smaller or equal R_{min} from the system. This approach leads to reasonable results as far as the value of R_{min} is in the order of magnitude of atomic radii. Objects with radii smaller than the radius of an atom are only numerical constructs with no physical meaning.

We find for the growth case $\left(\xi = \frac{1}{2}, B = 2\pi R^2 / M\right)$

$$\dot{a} = U_{Lng} \frac{\gamma M}{a},\ R\dot{R} = U_{Lng} \gamma M g_{Ln}\left(\frac{R}{a}, S\right) \tag{4.27}$$

with the numerically calculated factor $U_{Lng}(S)$ ($U_{Lng} = 0.7891$ for $S = 0.5$ and $U_{Lng} = 0.5339$ for $S = 0.6$) and the function $g_{Ln}\left(\frac{R}{a}, S\right)$ as

$$g_{Ln}\left(\frac{R}{a} S\right) = \left(\frac{R}{a}\right)^2 \cdot \left(1 - 3\sqrt{\frac{\pi}{2}} S\left(1 - \operatorname{erf}\left(\frac{\ln(R/a)}{\sqrt{2}S}\right)\right) \cdot \exp\left(\frac{(\ln(R/a))^2}{2S^2}\right)\right).$$

In the coarsening case $\left(\xi = 1, B = 4\pi\varepsilon R^2\right)$, the relations

$$\dot{a} = U_{Lnc} \frac{\gamma}{a^2 \varepsilon},\ R^2 \dot{R} = U_{Lnc} \frac{\gamma R}{a\varepsilon} g_{Ln}\left(\frac{R}{a} S\right) \tag{4.28}$$

are obtained with the numerically calculated factor $U_{Lnc}(S)$ ($U_{Lnc} = 0.2850$ for $S = 0.5$ and $U_{Lnc} = 0.1059$ for $S = 0.6$).

4.3. Summary of Results

A possible reason for the different evolution laws for the object radii as well as for the distribution parameter may be due to absence of proper topological parameters as worked out by Rios and Glicksman [24, 86]. Moreover, grain rotation may play a certain role; see, e.g., [87]. These phenomena can significantly change the evolution equations derived in classical theories [1, 6, 7, 8].

This paragraph presents a method for deriving the evolution equations for the object radii from the steady-state distribution functions. This method is applied to the most common distribution functions. The considered steady-state distribution functions are presented in Figure 6. The evolution equations for the object radii as derived from the distributions are presented in Figure 7 for the growth case and in Figure 8 for the coarsening case.

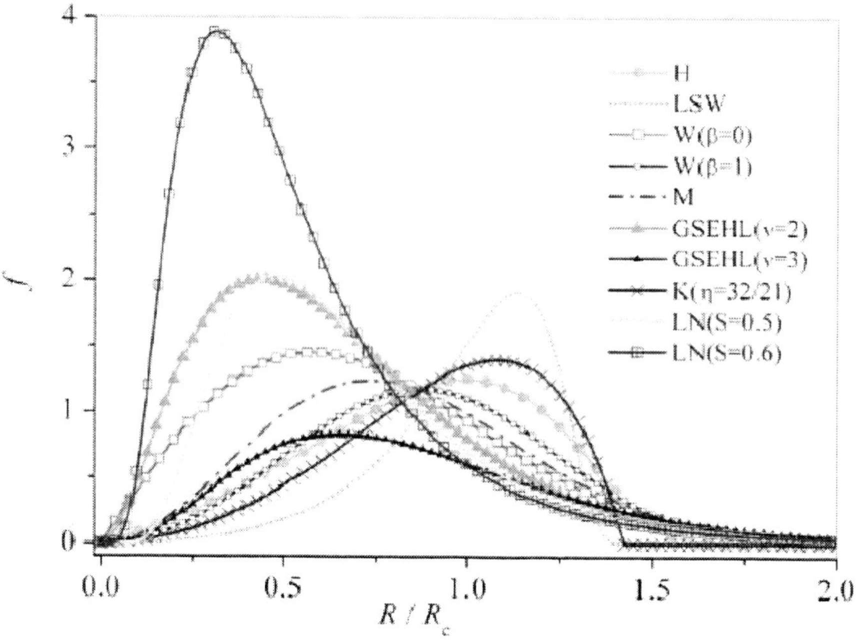

Figure 6. Normalized distribution functions $f = f(R/R_c)$. Reprinted from *Acta Materialia* 56, F.D. Fischer, J. Svoboda, E. Gamsjäger, E.R. Oberaigner, from distribution functions to evolution equations for grain growth and coarsening, 5395-5400, 2008, with permission from Elsevier.

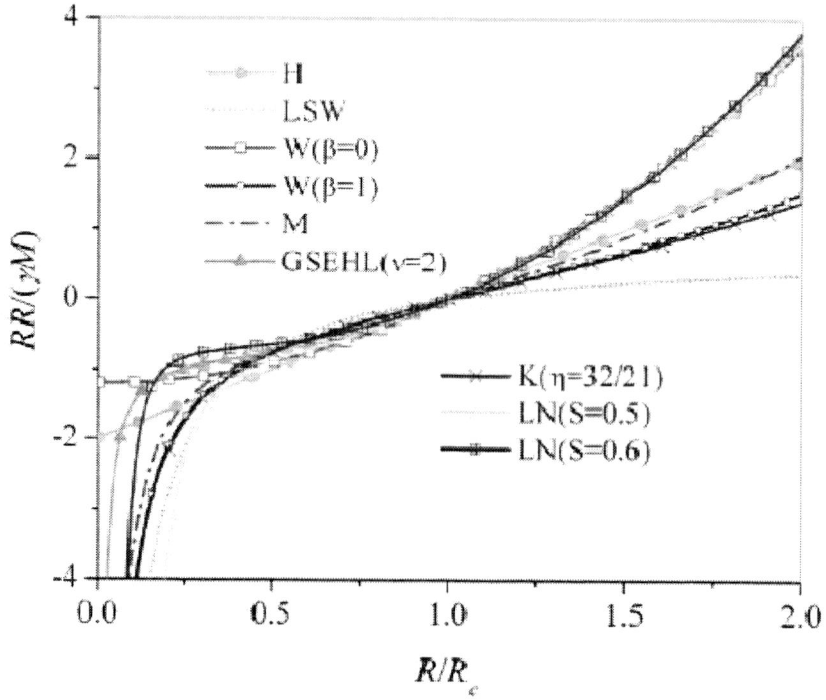

Figure 7. Evolution equations of the object radii for grain growth. See Figure 6 for permission.

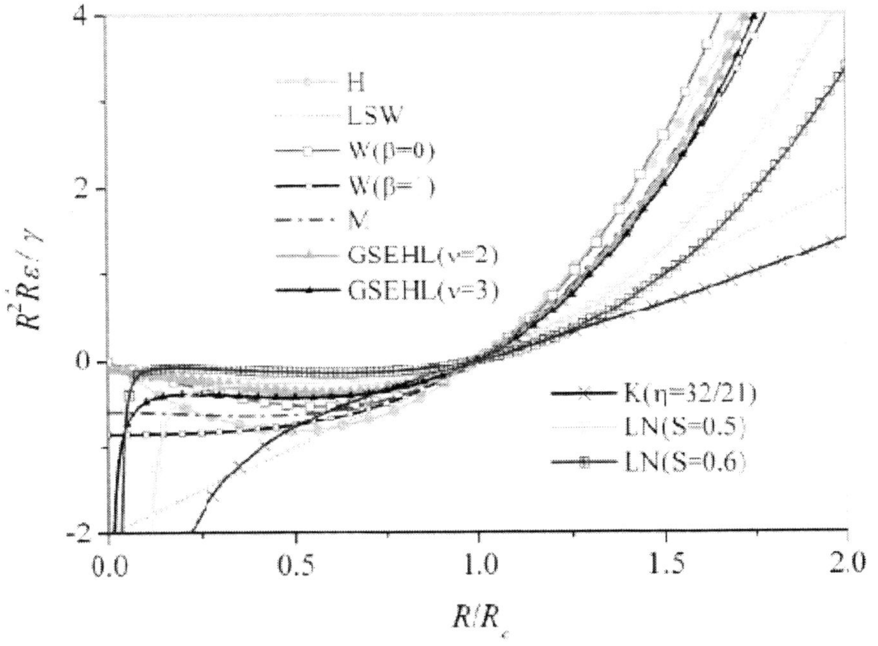

Figure 8. Evolution equations of the object radii for coarsening. See Figure 6 for permission.

In principle, the method can also be applied to experimentally measured steady-state distribution functions, and the evolution equations can be determined by the numerical evaluation of integrals. The obtained evolution equations can be regarded as basis to analyze phenomena active during growth or coarsening.

It is worth pointing to the fact that the ratio R_C/a depends also on the distribution type, but for a certain distribution, it is identical for grain growth as well as particle coarsening. The ratio R_C/a is determined by solving the equation $\dot{R}\left(R_C/a\right)=0$, which is equivalent to using Eq. (4.1) and looking for the solution of the equation

$$\int_{R_C}^{\infty}\frac{\partial f}{\partial a}dR'=0 \tag{4.29}$$

Evidently, no other information than the properties of the distribution function f enters the calculation of R_C/a. Values of R_C/a for different distributions are summarized in Table 1.

For all distributions, one can observe an apparent similarity between the evolution equations for growth and coarsening derived from the same distribution function. This is evidently due to Eq. (4.1), where the only difference stems from the expression for $\dot{a}$, given by Eq. (4.4). It is worth noting that $\dot{a}$ is independent of R.

The kinetics of grain growth or coarsening is usually described by the evolution of the mean radius $\overline{R}$, leading to the parabolic law for grain growth and cubic law for coarsening.

The relation between a and $\overline{R}$, as given by Eq. (1.2) with the coefficients of the parabolic and cubic laws determined by using Eqs. (1.2), (4.1) and (4.4), are summarized in Table 1.

Table 1. Comparison of parameters of different distribution functions

Distribution	R_C / a	$\overline{R} / a$	$\overline{R}\dot{\overline{R}}/(\gamma M)$	$(\overline{R})^2 \dot{\overline{R}} \cdot \varepsilon/\gamma$
H	1	8/9	0.3950	0.5756
LSW	1	1	0.1050	0.2963
W(β=0)	1	0.7236	0.6283	0.5236
W(β=1)	1	0.8930	0.3455	0.6103
M	1.3564	$2/\sqrt{\pi}$	0.4564	0.6297
GSEHL(ν=2)	4.5914	3	2/3	0.4186
GSEHL(ν=3)	5.3989	4	-	0.5374
K(η=32/21)	$0.7017 \cdot \eta$	$21 \cdot \eta/32$	0.2367	0.4904
Ln(S=0.5)	1.6230	1.1866	1.1112	0.4763
Ln(S=0.6)	2.2285	1.2317	0.8100	0.1980

It should be mentioned that the very simplified derivation of the evolution laws of $\overline{R}$ in Sect. 3.2 yields exactly the same structure of evolution equations. Only the coefficients there are estimated, e.g., by $1/\omega = 0.5$ compared to exactly evaluated coefficients in the last two columns of Table 1.

5. Analysis of Non-Steady-State Distribution Functions for Grain Growth and Coarsening

In this Section, a non-steady-state distribution function with an object radius scaling parameter $a(t)$, directly related to the critical radius R_C, and a dimension-free parameter $\beta(t)$, determining the shape of the distribution function, is investigated for both the grain growth and the coarsening case; see [88] for detail. A general derivation of the two equations for $a(t)$ and $\beta(t)$ is given and applied to the Weibull-type distribution function, see, e.g., [48], with the special case of the Rayleigh/Louat distribution function, see, e.g., [76].

5.1. Description of the Distribution Function

Let us deal in detail with two-parametric distribution functions, in which the first parameter a is used for the scaling of the object radius R and the second parameter β influences the shape of the distribution function for the normalized coordinate $r = R/a$.

influences the shape of the distribution function for the normalized coordinate $r = R/a$. Then, one can write the distribution function in the form $f\left(r, a(t), \beta(t)\right)$. As the distribution function must meet the normalization condition for the constant total volume of the objects,

$$\int_0^\infty R^3 f dR = a^4 \int_0^\infty r^3 f dr = 1, \tag{5.1}$$

the distribution function can be written without any loss of generality as

$$f(r, a, \beta) = \frac{1}{a^4} h(r, \beta). \tag{5.2}$$

5.2. Derivation of the Evolution Equations

The growth case and the coarsening case are characterized by a Gibbs energy parameter $A = 4\pi\xi\gamma$ and a dissipation multiplier *B(R)* as outlined in the context with Eq. (4.3). We describe both cases by an indicator 1 and a factor b_1 as following:

$$A = 2^{1+1}\pi\gamma, \quad B = 2^{1+1}\pi b_1 (a\, r)^{1+2}, \tag{5.3}$$

Growth case: $1 = 0$, $b_1 = 1/M$, Coarsening case: $1 = 1$, $b_1 = \varepsilon$.

The Gibbs energy *G* of the system can be expressed by Eq. (3.26) as

$$G = A \int_0^\infty R^2 f dR = A a^3 \int_0^\infty r^2 f dr = \frac{A}{a} \int_0^\infty r^2 h(r, \beta) dr. \tag{5.4}$$

Using the chain rule for calculation of partial derivatives, one obtains

$$\frac{\partial f}{\partial a} = -\frac{1}{a^5}\left(4h + r\frac{\partial h}{\partial r}\right). \tag{5.5}$$

Then, using Eq. (5.3) and applying the integration per parts with $r^3 h = 0$ for $r = 0$ or $r \to \infty$, one obtains the derivatives of *G* as

$$\frac{\partial G}{\partial a} = -\frac{2^{1+1}\pi\gamma}{a^2} \int_0^\infty r^2 h dr, \tag{5.6}$$

$$\frac{\partial G}{\partial \beta} = \frac{2^{\ell+1}\pi\gamma}{a}\int_0^\infty r^2 \frac{\partial h}{\partial \beta} dr . \tag{5.7}$$

The dissipation Q can be found from Eqs. (3.27) and (3.28), with Eqs. (5.3) and (5.5) as

$$\begin{aligned} Q &= \int_0^\infty \frac{B(R)}{f}\left(\dot{a}\int_R^\infty \frac{\partial f}{\partial a} dR' + \dot{\beta}\int_R^\infty \frac{\partial f}{\partial \beta} dR' \right)^2 dR \\ &= 2^{\ell+1}\pi b_\ell a^{\ell+5}\int_0^\infty \frac{r^{\ell+2}}{f}\left(\dot{a}\int_r^\infty \frac{\partial f}{\partial a} dr' + \dot{\beta}\int_r^\infty \frac{\partial f}{\partial \beta} dr' \right)^2 dr \\ &= 2^{\ell+1}\pi b_\ell a^{\ell-1}\int_0^\infty \frac{r^{\ell+2}}{h}\left(-\dot{a}\int_r^\infty \left(4h + r\frac{\partial h}{\partial r} \right) dr' + \dot{\beta}\int_r^\infty a\frac{\partial h}{\partial \beta} dr' \right)^2 dr. \end{aligned} \tag{5.8}$$

The second partial derivatives of Q then read

$$\frac{1}{2}\frac{\partial^2 Q}{\partial \dot{a}^2} = 2^{\ell+1}\pi b_\ell a^{\ell-1}\int_0^\infty \frac{r^{\ell+2}}{h}\left(\int_r^\infty \left(4h + r\frac{\partial h}{\partial r} \right) dr' \right)^2 dr , \tag{5.9}$$

$$\frac{1}{2}\frac{\partial^2 Q}{\partial \dot{\beta}^2} = 2^{\ell+1}\pi b_\ell a^{\ell+1}\int_0^\infty \frac{r^{\ell+2}}{h}\left(\int_r^\infty \frac{\partial h}{\partial \beta} dr' \right)^2 dr , \tag{5.10}$$

$$\frac{1}{2}\frac{\partial^2 Q}{\partial \dot{a}\partial \dot{\beta}} = -2^{\ell+1}\pi b_\ell a^{\ell}\int_0^\infty \frac{r^{\ell+2}}{h}\left(\int_r^\infty \left(4h + r\frac{\partial h}{\partial r} \right) dr' \cdot \int_r^\infty \frac{\partial h}{\partial \beta} dr' \right) dr . \tag{5.11}$$

The system of evolution equations according to Eq. (3.29) reads as

$$\frac{1}{2}\frac{\partial^2 Q}{\partial \dot{a}^2}\dot{a} + \frac{1}{2}\frac{\partial^2 Q}{\partial \dot{a}\partial \dot{\beta}}\dot{\beta} = -\frac{\partial G}{\partial a} , \quad \frac{1}{2}\frac{\partial^2 Q}{\partial \dot{a}\partial \dot{\beta}}\dot{a} + \frac{1}{2}\frac{\partial^2 Q}{\partial \dot{\beta}^2}\dot{\beta} = -\frac{\partial G}{\partial \beta} . \tag{5.12}$$

We introduce the following abbreviations:

$$I_{Ga}(\beta) = \int_0^\infty r^2 h dr , \tag{5.13a}$$

$$I_{G\beta}(\beta) = \int_0^\infty r^2 \frac{\partial h}{\partial \beta} dr \equiv \frac{d}{d\beta} I_{Ga}(\beta) , \tag{5.13b}$$

$$I_{Qaa}(\beta)=\int_0^\infty \frac{r^{\ell+2}}{h}\left(3\int_r^\infty h dr' - rh\right)^2 dr, \tag{5.13c}$$

$$I_{Q\beta\beta}(\beta)=\int_0^\infty \frac{r^{\ell+2}}{h}\left(\int_r^\infty \frac{\partial h}{\partial \beta} dr'\right)^2 dr, \tag{5.13d}$$

$$I_{Qa\beta}(\beta)=\int_0^\infty \frac{r^{\ell+2}}{h}\left(\left(3\int_r^\infty h dr' - rh\right)\cdot \int_r^\infty \frac{\partial h}{\partial \beta} dr'\right) dr. \tag{5.13e}$$

Integration per parts is used in integrals (5.13c) and (5.13e). Then, the evolution of the system according to (5.12) is given by the set of equations

$$2^{\ell+1}\pi b_\ell a^{\ell-1} I_{Qaa}(\beta)\dot{a} - 2^{\ell+1}\pi b_\ell a^\ell I_{Qa\beta}(\beta)\dot{\beta} = \frac{2^{\ell+1}\pi\gamma}{a^2} I_{Ga}(\beta), \tag{5.14}$$

$$-2^{\ell+1}\pi b_\ell a^\ell I_{Qa\beta}(\beta)\dot{a} + 2^{\ell+1}\pi b_\ell a^{\ell+1} I_{Q\beta\beta}(\beta)\dot{\beta} = -\frac{2^{\ell+1}\pi\gamma}{a} I_{G\beta}(\beta), \tag{5.15}$$

and, after simplification, as

$$I_{Qaa}(\beta)\frac{\dot{a}}{a} - I_{Qa\beta}(\beta)\dot{\beta} = \frac{\gamma}{a^{\ell+2} b_\ell} I_{Ga}(\beta), \tag{5.16}$$

$$I_{Qa\beta}(\beta)\frac{\dot{a}}{a} - I_{Q\beta\beta}(\beta)\dot{\beta} = \frac{\gamma}{a^{\ell+2} b_\ell} I_{G\beta}(\beta). \tag{5.17}$$

5.3. Solution of the Evolution Equations

The solution of the set of equations (5.16), (5.17) with respect to $\dot{a}/a$ and $\dot{\beta}$ is given by

$$\frac{\dot{a}}{a} = \frac{\gamma}{a^{\ell+2} b_\ell}\frac{I_{G\beta}(\beta)I_{Qa\beta}(\beta) - I_{Ga}(\beta)I_{Q\beta\beta}(\beta)}{D} = \frac{\gamma}{a^{\ell+2} b_\ell} F_a(\beta), \tag{5.18}$$

$$\dot{\beta} = \frac{\gamma}{a^{\ell+2} b_\ell}\frac{I_{G\beta}(\beta)I_{Qaa}(\beta) - I_{Ga}(\beta)I_{Qa\beta}(\beta)}{D} = \frac{\gamma}{a^{\ell+2} b_\ell} F_\beta(\beta) \tag{5.19}$$

with the determinant D as

$$D = I_{Qaa}(\beta) I_{Q\beta\beta}(\beta) - \left(I_{Qa\beta}(\beta)\right)^2 > 0. \tag{5.20}$$

After dividing Eq. (5.18) by Eq. (5.19), one obtains

$$\frac{\dot{a}}{a} = \frac{F_a(\beta)}{F_\beta(\beta)} \dot{\beta}. \tag{5.21}$$

Equation (5.21) can be integrated in time with the result

$$a = a_0 \exp\left[\int_{\beta_0}^{\beta} \frac{F_a(\beta)}{F_\beta(\beta)} d\beta'\right] = a_0 \exp\left[J(\beta;\beta_0,\ell)\right]. \tag{5.22}$$

The quantities a_0 and β_0 represent the initial values of a and β at time $t = 0$. By inserting of Eq. (5.22) into Eq. (5.19), one finds the differential equation for β as

$$\dot{\beta} = \frac{\gamma}{a_0^{\ell+2} b_\ell} F_\beta(\beta) \exp\left[-(\ell+2) J(\beta;\beta_0,\ell)\right]. \tag{5.23}$$

Equation (5.23) can be integrated with respect to time yielding

$$t = H(\beta; a_0, \beta_0, \ell) \tag{5.24}$$

ensuring that $H(\beta_0; a_0, \beta_0, \ell) = 0$.

The actual solution of β needs an inversion of Eq. (5.24) as

$$\beta = H^{-1}(t; a_0, \beta_0, \ell). \tag{5.25}$$

Insertion of Eq. (5.25) into Eq. (5.22) finally leads to a time evolution $a = a(t; a_0, \beta_0, \ell)$.

5.4. Asymptotic Behaviour of the Solution

The asymptotic value β_{asym} of β can be found from Eq. (5.19) by putting $\dot{\beta} = 0$, which enforces a non-linear equation for β_{asym} as

$$I_{G\beta}\left(\beta_{asym}\right)I_{Qa a}\left(\beta_{asym}\right)=I_{Ga}\left(\beta_{asym}\right)I_{Qa\beta}\left(\beta_{asym}\right). \tag{5.26}$$

If we insert the value β_{asym} into Eq. (5.16) and take $\dot{\beta}=0$ into account, we find an asymptotic evolution equation for a as

$$\dot{a}=\frac{\gamma}{a^{\ell+1}b_\ell}\frac{I_{Ga}\left(\beta_{asym}\right)}{I_{Qaa}\left(\beta_{asym}\right)}. \tag{5.27}$$

This is exactly the evolution equation for a, as it was studied in detail in the previous section for the case that only a is a time-dependent parameter.

5.5. Determination of the Critical Radius R_C

The critical radius R_C is defined in [47], sect. 3, in such a way that $\dot{R}\left(R_C\right)=0$, yielding the equation

$$\dot{a}\int_{R_C}^{\infty}\frac{\partial f}{\partial a}dR' + \dot{\beta}\int_{R_C}^{\infty}\frac{\partial f}{\partial \beta}dR' = f\dot{R},\quad \dot{R}\left(R_C\right)=0\,. \tag{5.28}$$

Equation (5.28) can be reformulated with Eqs. (5.2), (5.5), (5.21) and leads finally to a relation for $r_C = R_C / a$ as

$$\frac{I_{G\beta}(\beta)I_{Qa\beta}(\beta)-I_{Ga}(\beta)I_{Q\beta\beta}(\beta)}{I_{G\beta}(\beta)I_{Qaa}(\beta)-I_{Ga}(\beta)I_{Qa\beta}(\beta)}=\frac{F_a(\beta)}{F_\beta(\beta)}=\left(\int_{r_C}^{\infty}\frac{\partial h}{\partial \beta}dr\right)\Bigg/\left(3\int_{r_C}^{\infty}hdr-r_C h\left(r_C,\beta\right)\right). \tag{5.29}$$

Integration per parts is used in the last integral. The solution of Eq. (5.29) with respect to r_C provides the value of the critical radius $R_C = ar_C\left(\beta\right)$.

5.5. Examples

The Kirkaldy-Distribution Function

The Kirkaldy-distribution function discussed in the previous section is one of the simplest two-parametric distributions reading

$$f_K=\frac{h_K}{a^4},\ h_K=\frac{15}{\beta^4}\left(\frac{r^2}{\beta^2}-\frac{r^6}{\beta^6}\right). \tag{5.30}$$

The quantity β acts in this case as a so-called cut-off dimension-free radius, which was suggested by Kirkaldy [80] to be 32/21 for the coarsening case. One may tend to check whether β_{asym} obtains the value 32/21 by using Eq. (5.26). However, one can show that Eq. (5.26) is trivially fulfilled for any value of β_{asym}. Moreover, also the determinant D becomes 0 for any value of β.

The explanation of both facts can be identified from the distribution function itself, expressed in the variable R as

$$f_K\left(R,a,\beta\right)=\frac{15}{a^4\beta^4}\left(\frac{R^2}{a^2\beta^2}-\frac{R^6}{a^6\beta^6}\right), \tag{5.31}$$

which clearly shows that de facto only one parameter, namely $a\beta$, is involved in f_K, and, thus, Kirkaldy's proposal is a one-parameter distribution function.

The Weibull-Type Distribution Function

As dealt with by Zöllner and Streitenberger [17] and discussed in [48], a widely applicable class of distribution functions is a Weibull-type distribution function. The original version of this distribution includes three parameters, namely a, β and the cut-off radius r_0, which is, e.g., 1.5 in the prominent LSW-distribution and 2.0 in the Hillert-distribution; see also [48], sect. 1. For $r_0 \to \infty$ and $\beta = 0$, we have the classical Rayleigh-distribution function. For $r_0 \to \infty$ and $\beta \neq 0$, the distribution function reads

$$f_W=\frac{h_W}{a^4},\quad h_W=L_W r^{\beta+1}\exp\left(-\frac{3r^{\beta+2}}{\beta+2}\right),\quad L_W=\frac{(\beta+2)^{\frac{-3}{\beta+2}}\cdot 3^{\frac{\beta+5}{\beta+2}}}{\Gamma\left((\beta+5)/(\beta+2)\right)},\quad L_W{}'=\frac{dL_W}{d\beta}. \tag{5.32}$$

In the current case, we do not restrict to a specific value of β and let β develop starting from an initial value β_0. To calculate the second derivatives of Q with respect to $\dot{a}$ and $\dot{\beta}$, we can use exact relations for the integrals

$$\int_r^\infty \frac{\partial f_W}{\partial a}dr'=\frac{L_W}{a^5}\exp\left(-\frac{3r^{\beta+2}}{\beta+2}\right)\left(r^{\beta+2}-1\right), \tag{5.33a}$$

$$\int_r^\infty \frac{\partial f_W}{\partial \beta}dr'=\frac{L_W}{a^4}\exp\left(\frac{-3r^{\beta+2}}{\beta+2}\right)\left[\frac{1}{3}\frac{L_W{}'(\beta)}{L_W(\beta)}+\frac{r^{\beta+2}}{(\beta+2)}\left(\frac{1}{\beta+2}-\ln r\right)\right]. \tag{5.33b}$$

All further integrals with respect to Q in Eqs. (5.13c-e) can also be performed analytically, but the integration procedure leads to extremely lengthy expressions in variable β.

The integrals (5.13a,b) can also be evaluated analytically and read as

$$I_{Ga}(\beta)=3^{\frac{1}{\beta+2}}(\beta+2)^{\frac{\beta+1}{\beta+2}}\Gamma\left(\frac{\beta+4}{\beta+2}\right)\Big/\Gamma\left(\frac{\beta+5}{\beta+2}\right), \tag{5.34a}$$

$$I_{G\beta}(\beta)=\frac{d}{d\beta}I_{Ga}(\beta). \tag{5.34b}$$

We can calculate β_{asym} from the steady-state condition (5.26) yielding for the

Growth case: $\beta_{asym}=0.50703$, Coarsening case: $\beta_{asym}=1.86659$.

The functions $F_a(\beta),F_\beta(\beta)$ in Eqs. (5.18), (5.19) can nearly exactly be fitted by quadratic polynomials in the variable $(\beta_{asym}-\beta)$ as

Growth case:

$$F_a(\beta)=0.37519+0.30307(\beta_{asym}-\beta)+0.25511(\beta_{asym}-\beta)^2, \tag{5.35a}$$

$$F_\beta(\beta)=5.13085(\beta_{asym}-\beta)-0.05895(\beta_{asym}-\beta)^2; \tag{5.35b}$$

Coarsening case:

$$F_a(\beta)=0.16169+0.07278(\beta_{asym}-\beta)+0.11919(\beta_{asym}-\beta)^2, \tag{5.35c}$$

$$F_\beta(\beta)=7.47000(\beta_{asym}-\beta)-0.87085(\beta_{asym}-\beta)^2. \tag{5.35d}$$

Note that the function $F_\beta(\beta)$ is nearly a linear function in β and $F_\beta(\beta_{asym})=0$. This allows expression of $a(\beta)$, Eq. (5.22), in a quite simple, nearly exact analytical form as following:

$$a(\beta)=a_0\left(\frac{\beta_{asym}}{\beta_{asym}-\beta}\right)^b\cdot C(\beta), C(\beta)=1.0+C_1\beta+C_2\beta^2, \tag{5.36}$$

Growth case: $b = 0.07312$, $C_1 = 0.086$, $C_2 = -0.023$;

Coarsening case: $b = 0.02164$, $C_1 = 0.045$, $C_2 = -0.009$.

This rather simple equation allows finding via Eq. (5.23) the equation for β as a function of the dimension-free time parameter τ, $\tau = \gamma t / \left(a_0^{\ell+2} b_\ell \right)$, with $\beta_0 = 0$ as

$$\int_0^\beta \left(\frac{\beta_{asym}}{\beta_{asym} - \tilde{\beta}} \right)^{b(\ell+2)} \cdot \frac{C\left(\tilde{\beta}\right)^{(\ell+2)}}{F_\beta\left(\tilde{\beta}\right)} d\tilde{\beta} = \tau . \tag{5.37}$$

Since the exponent $b(\ell+2)$ is 0.14624 for growth and 0.06492 for coarsening, the integral above has no analytical solution. Of course, the above relation must be inverted to find to the functional relation $\beta = \beta(\tau)$. Insertion of $\beta(\tau)$ into the according Eq. (5.22) yields finally $a(\tau)$.

Let us study two limiting cases, namely $\beta \to 0$ (short-term solution) and $\beta \to \beta_{asym}$ (long-term solution). For $\beta \to 0$, we can use in Eq. (5.37) the values $C = 1$ and $F_\beta = F_\beta(0)$. Then we find after some short analysis the linear relation

$$\beta = F_\beta(0)\tau . \tag{5.38}$$

It is well known that the parameter scaling the radius obeys the parabolic (cubic) law in the steady state of growth (coarsening) as

$$a^{(\ell+2)} - a_0^{(\ell+2)} = \alpha t \tag{5.39}$$

with α being a material parameter. Using the dimension-free time parameter τ, one can rewrite ($\bar{\alpha}$ being a material parameter again) the law as

$$\frac{a^{(\ell+2)}}{a_0^{(\ell+2)}} = 1 + \bar{\alpha}\tau \tag{5.40}$$

Just this form of equation is obtained by linearization of Eq. (5.36), raised to the power $(\ell+2)$. Insertion of Eq. (5.38) yields finally

$$\frac{a^{(\ell+2)}}{a_0^{(\ell+2)}} = 1 + (\ell+2)\left(\frac{bF_\beta(0)}{\beta_{asym}} + C_1 \right)\tau \text{ or } a = a_0 \left[1 + (\ell+2)\left(\frac{bF_\beta(0)}{\beta_{asym}} + C_1 \right)\tau \right]^{\frac{1}{(\ell+2)}} . \tag{5.41}$$

The direct linearization of Eq. (5.36) and the use of Eq. (5.38) provide

Just this form of equation is obtained by linearization of Eq. (5.36), raised to the power $(1+2)$. Insertion of Eq. (5.38) yields finally

$$\frac{a^{(1+2)}}{a_0^{(1+2)}}=1+(1+2)\left(\frac{bF_\beta(0)}{\beta_{asym}}+C_1\right)\tau \text{ or } a=a_0\left[1+(1+2)\left(\frac{bF_\beta(0)}{\beta_{asym}}+C_1\right)\tau\right]^{\frac{1}{(1+2)}}. \quad (5.41)$$

The direct linearization of Eq. (5.36) and the use of Eq. (5.38) provide

$$a=a_0\left(1+\left(\frac{bF_\beta(0)}{\beta_{asym}}+C_1\right)\tau\right). \quad (5.42)$$

The quantity $bF_\beta(0)/\beta_{asym}$ obtains the value 0.37298 for the growth case and 0.12647 for the coarsening case.

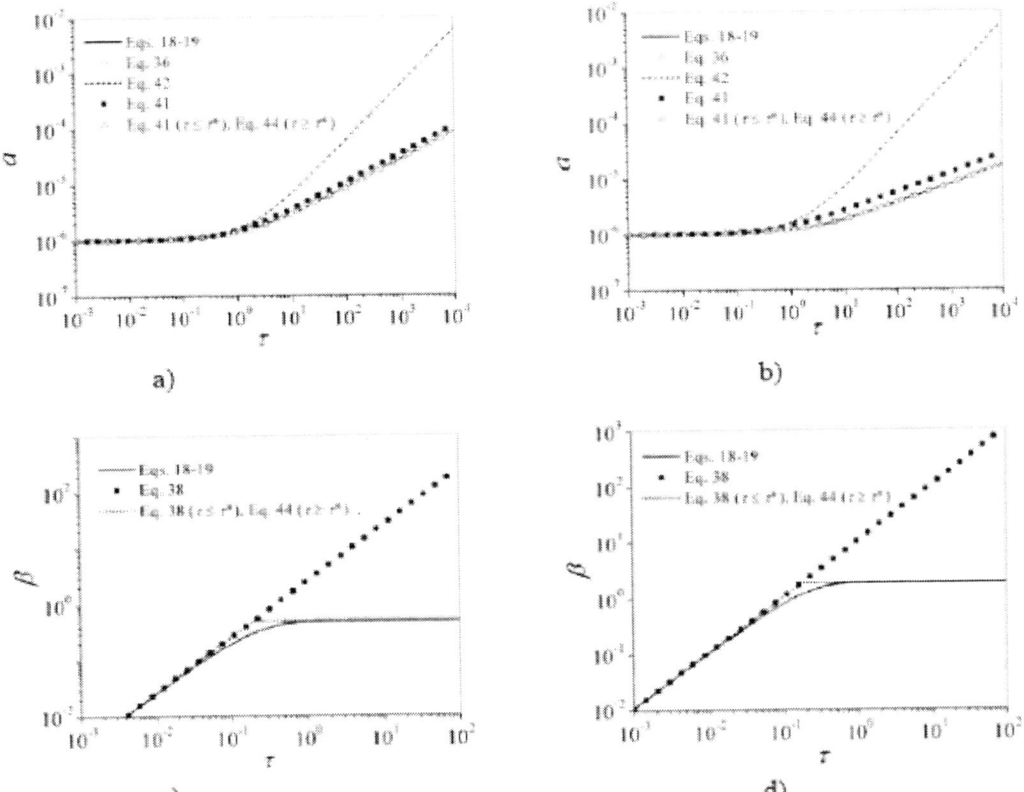

Figure 9. Two-parameter Weibull-type distribution function: Comparison of different solutions $a(\tau)$ and $\beta(\tau)$ for growth case – a) and c) and coarsening case – b) and d). Reprinted from *Philosophical Magazine* 89, F.D. Fischer, J. Svoboda, E. Gamsjäger, Analysis of non-steady-state distribution functions for grain growth and coarsening, 1425-1438, 2009, with permission from Taylor and Francis.

The comparison of the short-term and numerical solutions is presented in Figure 9 for both the growth case (Figure 9 a) and c)) and coarsening case (Figure 9 b) and d)). The solution obtained numerically by direct integration of Eqs. (5.18) and (5.19) can be considered as an exact one and represents a reference solution. The use of the values of β from the numerical solution provides with Eq. (5.36) the solution for a within an accuracy of 1% for both cases. Using Eqs. (5.38) and (5.41), one obtains a very good agreement for $\tau < 0.1$. For larger times, however, the values of a differ by a factor of up to 1.26 for the growth case and by a factor of up to 1.58 for the coarsening case. Moreover, no asymptotic behavior for β can be found. Using Eqs. (5.38) and (5.42), one obtains a very good agreement for $\tau < 0.1$ and a disagreement for larger times.

From Figure 9b) and d), it is evident that the dependence of β on (τ) can be described quite well by relation (5.38) up to the time

$$\tau^{*} = \beta_{asym} / F_{\beta}(0) \tag{5.43}$$

and then by keeping

$$\beta = \beta_{asym} \text{ for } \tau > \tau^{*}. \tag{5.44}$$

The short-term solution for a is given by Eq. (5.41) for $\tau \leq \tau^{*}$, and a reaches the value

$$a^{*} = a_0 \left[1 + (1+2)\left(b + \frac{C_1 \beta_{asym}}{F_{\beta}(0)} \right) \right]^{\frac{1}{(1+2)}} \tag{5.45}$$

for $\tau = \tau^{*}$. For $\tau > \tau^{*}$ one can insert $\beta = \beta_{asym}$ into Eq. (5.18), and its direct integration provides the long-term solution:

$$a^{(1+2)} - a^{*(1+2)} = (1+2)\frac{\gamma}{b_1} tF_a(\beta_{asym}) = a_0^{1+2}(1+2)F_a(\beta_{asym})(\tau - \tau^{*}),\ \tau > \tau^{*}. \tag{5.46}$$

Again, the parabolic (cubic) law can be recognized for $\tau > \tau^{*}$. The solutions $\beta(\tau)$ and $a(\tau)$, given by the combination of the short-term solution (Eqs. (5.38) and (5.41)) for $\tau \leq \tau^{*}$ and the long-term solution (Eqs. (5.44) and (5.46)) for $\tau > \tau^{*}$, are presented by triangles in Figure 9 a) and b). The deviation of $a(\tau)$ from the reference solution is less than 1% in the whole interval $0 \leq \tau < \infty$.

The evolution of $f_W(\tau)$ given by Eq. (5.32) for different values of τ is presented for both the growth and coarsening cases in Figure 10. The function $h_W(\tau)$ stops to evolve for the normalized time $\tau \geq 1$, which corresponds to the values of τ, for which $\beta \approx \beta_{asym}$. The evolution of distribution function $f_W(\tau)$ given by Eqs. (5.2) and (5.32) for different values of τ for both the growth and coarsening case is presented in Figure 11.

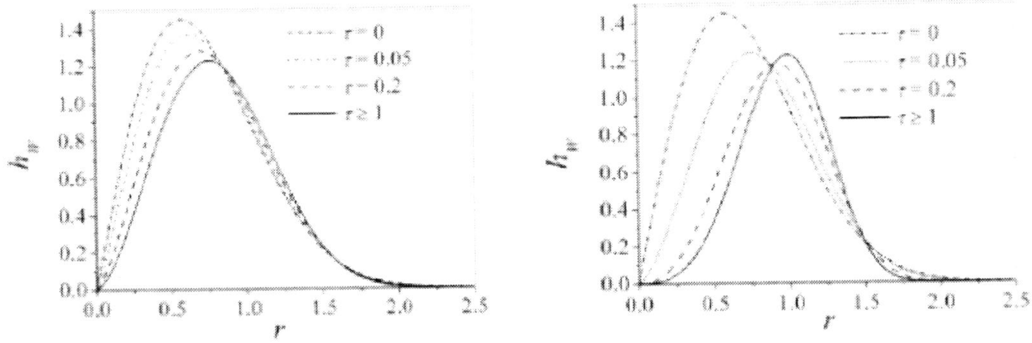

a) b)

Figure 10. Two parameter Weibull-type distribution function: Evolution of $h_W(\tau)$ given by Eq. (5.32) for different values of τ. a) growth case, b) coarsening case. See Figure 9 for permission.

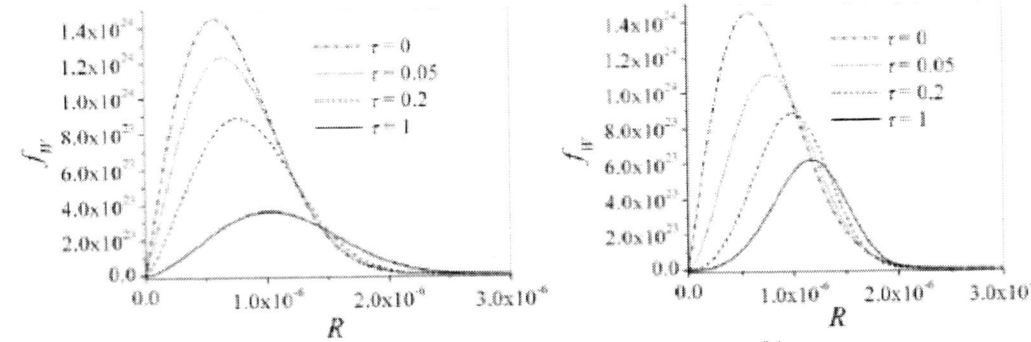

a) b)

Figure 11. Two-parameter Weibull-type distribution function: Evolution of distribution function $f_W(\tau)$ given by Eq. (5.32) for different values of τ. a) growth case, b) coarsening case. See Figure 9 for permission.

The critical radius R_C is investigated according Eq. (5.29). For the left-hand side of Eq. (5.29), we use the fits from Eqs. (5.35a-d). The integrals on the right-hand side of Eq. (5.29) can be found analytically. It can easily be shown that for $\beta = \beta_{asym}$ the critical radius is a. The numerical solution of Eq. (5.29) both for the growth and coarsening cases are shown in

$$R^{\ell+1}\dot{R} = \frac{\gamma}{b_\ell} r^\ell \left\{ F_a(\beta)\left(r^2 - r^{-\beta}\right) + \frac{F_\beta(\beta)}{(\beta+2)} \left[\left(\frac{1}{\beta+2} - \ln r \right) r^2 + \frac{1}{3} \frac{L_W'}{L_W} r^{-\beta} \right] \right\}. \quad (5.47)$$

The first part of the above kinetic expression agrees with the mathematical structure of the kinetics of the one-parameter Weibull-type distribution in [48], Eqs. (5.20) and (5.21), however, with a fixed value of β there. The second contribution to the kinetics refers to the term $\dot{\beta}$ and approaches zero, when $\beta \to \beta_{asym}$. The most important feature of this kinetics is that β is now a function evolving with time.

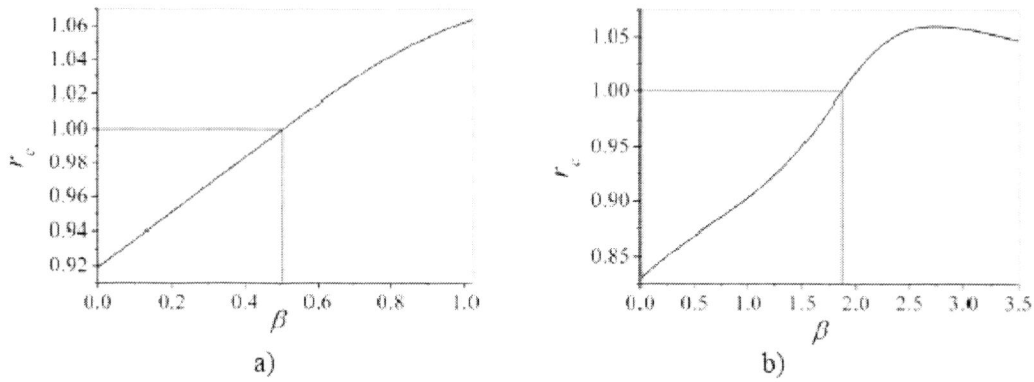

Figure 12. Two-parameter Weibull-type distribution function: Dependence of normalized critical radius r_C on parameter β being a solution of Eq. (5.29) with $h = h_W$. a) growth case, b) coarsening case. See Figure 9 for permission.

5.5. Summary of Results and Outlook

A non-steady object-radius distribution function is described by two parameters. The first parameter, a, scales the object radius R and can be considered as the "representative" object radius. The second parameter, β, is dimension-free and characterizes the shape of the radius distribution function.

One can show that

- β is not a somewhat arbitrary parameter but tends to a distinct value β_{asym}, which differs for the growth and coarsening cases and can be determined from the TEP,
- an explicit evolution equation can be given for $\dot{a}(a,\beta)$ and $\dot{\beta}(a,\beta)$, so finally the evolution of a as well as β can be calculated by numerical integration in time,
- an explicit functional relation can be determined for $a(t)$ and $\beta(t)$.

This offers the possibility of following actual distributions during their evolution with a dimension-free time parameter τ being either $\left(\gamma M / a_0^2\right)t$ in the growth case or $\left(\gamma / \left(\varepsilon a_0^3\right)\right)t$ in the coarsening case.

The study of a Weibull-type distribution function gives clear evidence that the evolution of the parameter scaling the object radius R represented by $a(\tau)$ follows a "parabolic" or "cubic" regime with a changing parameter $\beta(\tau)$.

As future work, the following two tasks seem to be worthwhile to solve:

- Treatment of precipitation controlled by bulk diffusion and additionally by interface reaction. In this case, the dissipation consists of two terms due to bulk diffusion being proportional to $R^3 \dot{R}^2$ and due to interface reaction (analogy to the grain boundary migration) being proportional to $R^2 \dot{R}^2$. Now, the features of grain growth and precipitate coarsening are coupled. The interface reaction controlled mechanism may dominate in the initial stage with fine precipitate structure, and the bulk diffusion controlled mechanism may dominate in later stages for coarse precipitate structure. As the relations of the controlling mechanisms change during the system evolution, one can expect that the parameter β approaches its asymptotic value very slowly.
- A third parameter may be brought into play, e.g., the so-called cut-off parameter r_0, restricting the distribution $f(R)$ to the interval $0 \le r \le r_0 = R_0/a$. The cut-off parameter r_0 is set to 2 in the Hillert-distribution or 1.5 in the LSW-distribution, and it is discussed by Zöllner and Streitenberger [17] for the Weibull-type distribution function. In this case, the matrix of the set of linear equations for $\dot{a}$, $\dot{\beta}$ and $\dot{r}_0$ is a 3x3 matrix with the unchanged original 2x2 sub-matrix corresponding to $\dot{a}$ and $\dot{\beta}$ for $r_0 \to \infty$. The solution of this problem may be tackled in an iterative way, e.g., by considering a finite r_0 as a certain disturbance in relation to $r_0 \to \infty$ in the cases considered in this paper.

Specifically, in the case of two and more parameters, the TEP has proven to be a rational and efficient tool to derive evolution equations in an analytical form. However, the authors assume that in both new tasks, only numerical strategies will lead to results. It should be mentioned that very recently, the evolution equations have been derived without the need to apply the TEP for one-parameter systems; see [89], since only the usage of the total Gibbs energy balance equation is sufficient.

assume that in both new tasks, only numerical strategies will lead to results. It should be mentioned that very recently, the evolution equations have been derived without the need to apply the TEP for one-parameter systems; see [89], since only the usage of the total Gibbs energy balance equation is sufficient.

CONCLUSION

Based on Onsager's principle, the one-object, multi-object and distribution concepts are developed. The equations for the evolution of individual objects derived within the framework of the multi-object concept coincide with Hillert's and Lifshitz-Slyozov-Wagner's classical equations. The new distribution concept allows for the treatment of the evolution of the object radii distribution without the necessity of knowing the evolution equations for individual objects.

The distribution concept is applied to the monomodal and bimodal Rayleigh-type object radii distribution functions for grain growth. The results of simulations along the distribution concept are compared with those calculated by means of the multi-object concept. The distribution concept is also applied to the treatment of steady-state grain growth with Rayleigh's and Hillert's distribution functions. Their comparison clearly indicates that the kinetics of the system depends decisively on the width of the distribution function.

A novel method to derive the evolution equations for the object radii for grain growth and coarsening from steady state distribution functions is presented. It is successfully tested by comparing the results with those of the classical Hillert and Lifshitz-Slyozov-Wagner theory, respectively. The evolution equations of the object radii for grain growth and coarsening are then derived for Weibull-type, Maxwell, Gendelman et al., Kirkaldy and Log-normal distribution functions. In addition, the evolution equations for the mean and critical radii are determined for these distribution functions.

In the simplest case, the non-steady-state evolution of the object radii distribution can be described by a two-parameter function. The evolution equations are derived, and their semi-analytical solution is developed.

The general evolution equations are applied to Kirkaldy and Weibull-type two-parameter distribution functions. The Kirkaldy distribution function is characterized as having two dependent parameters, and the general analysis fails. On the other hand, the Weibull-type distribution function is analyzed in detail. The evolution equations for parameters of the Weibull-type distribution function are integrated numerically with respect to time. This solution is considered as the exact reference solution. Several simple approximate analytical solutions are presented. A combination of a short-term and a long-term solution seems to be very successful.

REFERENCES

[1] Hillert, M. Acta metall. 1965,13, 227-238; *Mater. Sci. Forum* 1996, 204-206.

[2] Atkinson, H.V. *Acta metall.* 1988, 36, 469-491.

[3] Saetre, T.O. *Modelling Simul. Mater. Sci. Eng.* 2004, 12, 1267-1277.

[4] Saetre, T.O. *Modelling Simul. Mater. Sci. Eng.* 2004, 12, 1279-1292.
[5] Fischer, F.D.; Svoboda, J.; Fratzl, P. *Philos. Mag. A* 2003, 83, 1075-1093.
[6] Lifshitz, L.M.; Slyozov, V.V. Soviet Phys. *J. Exp. Theor. Phy.* 1958, 35, 479-491.
[7] Lifshitz, L.M.; Slyozov, V.V. *J. Phys. Chem. Solids* 1961, 19, 35-50.
[8] Wagner, C. Z. *Elektrochem.* 1961, 65, 581-591.
[9] Vorhees, P.W. *J. Stat. Phys.* 1985, 38, 231-252.
[10] Vorhees, P.W. *Annu. Rev. Mater. Sci.* 1992, 22, 197-215.
[11] Svoboda, J.; Fischer, F.D.; Fratzl, P.; Kozeschnik, E. *Mat. Sci. Engng. A* 2004, 385, 166-174.
[12] Kozeschnik, E.; Svoboda, J.; Fratzl, P.; Fischer, F.D. *Mat. Sci. Engng. A* 2004, 385, 157-165.
[13] Kim, B.-N.; Hiragoa, K.; Morita, K. *Mater. Trans.* 2003, 44, 2239-2244.
[14] Fischer, F.D.; Svoboda, J.; Fratzl, P. *Philos. Mag.* 2003, 83, 1075-1093.
[15] Streitenberger, P. *Scripta mater.* 1998, 39, 1719-1724.
[16] Streitenberger, P.; Zöllner, D. *Scripta mater.* 2006, 55, 461-464.
[17] Zöllner D, Streitenberger P, *Scripta mater*, 2006, 54, 1697-1702.
[18] Rios, P.R.; Gottstein, G.; Shvindlerman, L.S. *Scripta Mater.* 2001, 44, 893-897.
[19] Rios, P.R.; Gottstein, G.; Shvindlerman, L.S. *Mat. Sci. Eng.* 2002, A332, 231-235.
[20] Rios, P.R. *Acta Mater.* 2004, 52, 249-256.
[21] Rios, P.R.; Dalpian, T.G.; Brandão Castro, J.A.; Oliveira, A.C.L. *Scripta Mater.* 2006, 54, 1633-1637.
[22] Glicksman, M.E. *Philos. Mag.* 2005, 85, 3-31.
[23] Rios, P.R.; Glicksman, M.E. *Acta Mater.* 2006, 54, 1041-1051.
[24] Rios, P.R.; Glicksman, M.E. *Acta Mater.* 2007, 55, 1565-1571.
[25] Suwai, Y.; Saito, Y. *Mater. Trans.* 2003, 44, 2245-2251.
[26] Suwa, Y.; Saito, Y.; Onodera, H. *Acta mater.* 2007, 55, 6881-6894.
[27] Suwa, Y.; Saito, Y.; Onodera, H. *Mater. Trans.* 2008, 49, 704-709.
[28] Moelans, N.; Wendler, F.; Nestler, B. Comput. *Mater. Sci.* 2009, 46, 479-790.
[29] Greenwood, G.W. *Acta Metall.* 1956, 4, 243-248.
[30] Vorhees, P.W.; Glicksman, M.E. *Acta Metal.* 1984, 32, 2001-2011.
[31] Voorhees, P.W.; Glicksman, M.E. *Acta Metal.* 1984, 32, 2013-2030.
[32] Thornton, K.; Akaiwa, N.; Voorhees, P.W. 2000, *Phase Transformations and Evolution in Materials* (Eds.: Turchi, P.E.A.; Gonis, A. (Warrendale, Pennsylvania: Metallurgical Society of AIME), 73-88.
[33] Glicksman, M.E. *Diffusion in Solids*; Wiley: New York, NY, 2000; chpt. 21.
[34] Redanz, P.; McMeeking, R.M. 2001, *Proceedings of the Fourth International ESAFORM Conference on Material Forming* Vol. 2 (Eds.: Habraken, A.M. (University of Liege)), 749-752.
[35] Lee, D.-K.; Hwang, N.-M. *Scripta Mater.* 2009, 61, 304-307.
[36] Fife, J.L.; Voorhees, P.W. *Scripta Mater.* 2009, 60, 839-842.
[37] Kozeschnik, E.; Svoboda, J.; Fischer, F.D. *Mater. Sci. Engng. A* 2006, 441, 68-72.
[38] Svoboda, J.; Fischer, F.D.; Mayrhofer, P.H. *Acta Mater.* 2008, 56, 4896-4904.
[39] Sonderegger, B.; Kozeschnik, E.; Leitner, H.; Clemens, H.; Svoboda, J.; Fischer, F.D. *Int. J. Mat. Res.* 2008, 99, 410-415.
[40] Sonderegger, B.; Kozeschnik, E.; Leitner, H.; Clemens, H.; Svoboda, J.; Fischer, F.D.; Staron, P. *Steel Research* 2010, 81, 64-73.

[41] Niethammer, B.Z. *Angew. Math. Mech.* 2010, 90, 259-270.

[42] Groot, S.R. de; Mazur, P. *Non-equilibrium thermodynamics*; New York: Dover Publication, 1984.

[43] Onsager, L. *Phys. Rev.* 1931, 37, 405-426.

[44] Onsager, J. *Ann. New. York Acad. Sci.* 1945, 46, 241-265.

[45] Svoboda, J.; Turek, I. *Philos. Mag.* 1991, 64, 749-759.

[46] Svoboda, J.; Turek, I.; Fischer, F.D. *Philos. Mag.* 2005, 85, 3699-3707.

[47] Svoboda, J.; Fischer, F.D. *Acta Mater.* 2007, 55, 4467-4474.

[48] Fischer, F.D.; Svoboda, J.; Gamsjäger, E.; Oberaigner, E.R. *Acta Mater.* 2008, 56, 5395-5400.

[49] Groeber, M.; Ghosh, S.; Uchic, M.D.; Dimiduk, D.M. *Acta Mater.* 2008, 56, 1257-1273.

[50] Groeber, M.; Ghosh, S.; Uchic, M.D.; Dimiduk, D.M. *Acta Mater.* 2008, 56, 1274-1287.

[51] Gyarmati, I. *Non-equilibrium thermodynamics*; Springer-Verlag: Berlin/Heidelberg/ New York, DE, 1970.

[52] Ziegler, H. *An Introduction to Thermodyanmics*; North-Holland Publishing Company: Amsterdam-New York-Oxford, 1977.

[53] Kestin, J.; Rice, J.R. In: *A critical Review of Thermodyanmics*; Stuart, B. et al.; Eds.; Mono Book: Baltimore, MD, 1970, pp 275-298.

[54] Svoboda, J.; Riedel, H. *Acta Metall. Mater.* 1992, 40, 2829-2840.

[55] Riedel, H.; Svoboda, J. *Acta Metall. Mater.* 1993, 41, 1929-1936.

[56] Svoboda, J.; Riedel, H.; Zipse, H. *Acta Metall. Mater.* 1994, 42, 435-443.

[57] Riedel, H.; Zipse, H.; Svoboda, J. *Acta Metall. Mater.* 1994, 42, 445-452.

[58] Riedel, H.; Kozák, V.; Svoboda, J. *Acta Metall. Mater.* 1994, 42, 3093-3103.

[59] Svoboda, J.; Riedel, H. *Acta Metall. Mater.* 1995, 43, 499-506.

[60] Svoboda, J.; Riedel, H. *Acta Metall. Mater.* 1995, 43, 1-10.

[61] Svoboda, J.; Riedel, H.; Gaebel. R. *Acta Mater.* 1996, 44, 3215-3226.

[62] Svoboda, J.; Lukáš, P. *Acta Mater.* 1996, 44, 2557-2565.

[63] Svoboda, J.; Lukáš, P. *Acta Mater.* 1997, 45, 125-135.

[64] Svoboda, J.; Lukáš, P. *Acta Mater.* 1998, 46, 3421-3431.

[65] Svoboda, J.; Lukáš, P. *Acta Mater.* 2000, 48, 2519-2528.

[66] Gamsjäger, E.; Svoboda, J.; Fischer, F.D. *Comput. Mater. Sci.* 2005, 32, 360-369.

[67] Svoboda, J.; Gamsjäger, E.; Fischer, F.D.; Kozeschnik, E. *J. Phase Equilibria Diff.* 2006, 27, 622-628.

[68] Gamsjäger, E.; Svoboda, J.; Fischer, F.D.; Rettenmayr, M. *Acta Mater.* 2007, 55, 2599-2607.

[69] Svoboda, J.; Fischer, F.D.; Vollath, D. *Acta Mater.* 2009, 57, 1912-1919.

[70] Abart, R.; Petrishcheva, E.; Fischer, F.D.; Svoboda, *J. Am. J. Sci.* 2009, 309, 114-131.

[71] Kozeschnik, E.; Svoboda, J.; Radis, R.; Fischer, F.D. Model. *Simul. Mater. Sci. Eng.* 2010, 18, 015011.

[72] Svoboda, J.; Fischer, F.D.; Abart, R. *Acta Mater.* 2010, 58, 2905-2911.

[73] Svoboda, J.; Fischer, F.D. to be published *Acta Mater.*, 2010

[74] Elsey, M.; Esedoglu, S.; Smereka, P. *Proc. R. Soc. A* 2010, doi: 10.1098/rspa. 2010 . 0194, published online

[75] Flores, E.; Cabrera, J.M.; Prado, J.M. *Met. Mater. Trans.* A 2004, 35A, 1097-1103.

[76] Humphreys F.J.; Hatherly M. *Recrystallization and Related Annealing Phenomena*; Pergamon/Elsevier: Oxford, UK, 1995.
[77] Marthinsen K.; Hunderi O.; Ryum N. *Acta Mater*. 1996, 44, 1681-1689.
[78] Meschede D. *Gerthsen Physik*, Springer: Berlin-Heidelberg-New York, 2006.
[79] Gendelman O.V.; Shapiro M.; Estrin Y.; Hellmig R.J.; Lekhtmakher S. *Mat. Sci. Eng. A*, 2006, 434, 88-94.
[80] Kirkaldy J.S. *J. Mater. Sci*. 2002, 37, 65-70.
[81] Crow E.L.; Shimizu K. *Lognormal Distributions Theory and Applications*; Marcel Dekker, Inc.: New York and Basel, 1988.
[82] Feltham P. *Acta Mater*. 1957, 5, 97-105.
[83] Ferry M.; Burhan N. *Acta Mater*. 2007, 55, 3479-3491.
[84] Berbenni S.; Favier V.; Berveiller M. *Int. J. Plasticity* 2007, 23, 114-142.
[85] Azizi-Alizamini H.; Militzer M.; Poole W.J. *Scripta Mater*. 2007, 57, 1065-1068.
[86] Rios P.R.; Glicksman M.E. *Acta Mater*. 2006, 54, 5313-5321.
[87] Moldovan D.; Wolf D.; Phillipot S.R. *Acta Mater*. 2001, 49, 3521-3532.
[88] Fischer F.D.; Svoboda J.; Gamsjäger E. *Philos. Mag*. 2009, 89, 1425-1438.
[89] Jeppsson J.; Agren J.; Hillert M. *Acta Mater*. 2008, 56, 5188-5201.

Reviewed by E. Kozeschnik, Vienna University of Technology.
Email: ernst.kozeschnik@tuwien.ac.at

In: Evolution Equations
Editor: Arthur L. Claes

ISBN: 978-1-61209-652-0

Chapter 2

THE STUDY OF MULTIDIMENSIONAL MIXED PROBLEM FOR NONLINEAR HYPERBOLIC EQUATIONS

K. I. Khudaverdiyev[1] and F. K. Khudaverdiyev[2]
[1]Department of Mechanics and Mathematics,
Baku State University, Baku, Azerbaijan
[2]Institute of Mathematics and Mechanics,
National Academy of Science of Azerbaijan, Baku, Azerbaijan

INTRODUCTION

This work is dedicated to the study of existence and uniqueness of classical solution for the following multidimensional mixed problem:

$$\begin{cases} u_{tt}(t,x) + \text{✌}(u(t,x)) \equiv u_{tt}(t,x) + \sum\limits_{|\alpha| \le 2m} a_\alpha(x) \text{☟}^\alpha u(t,x) = \\ = F(t,x,u(t,x), \text{☟} u(t,x), \dots, \text{☟}^m u(t,x), u_t(t,x)) \quad (t \in [0,T], x \in \overline{\Omega}), & (0.1) \\ u(0,x) = \varphi(x) \quad (x \in \overline{\Omega}), \quad u_t(0,x) = \phi(x) \quad (x \in \overline{\Omega}), & (0.2) \\ B_j(u(t,x))\big|_\Gamma = 0 \quad (j = 1, \dots, m), & (0.3) \end{cases}$$

where $0 < T < +\infty$, Ω is a bounded n-dimensional domain with a boundary $S \in C^{2m}$, $x = (x_1, \dots, x_n)$, $\Gamma = [0,T] \times S$; $\text{☟}^\alpha = \partial^{|\alpha|} / \partial x_1^{\alpha_1} \Lambda \, \partial x_n^{\alpha_n}$, $\alpha = (\alpha_1, \dots, \alpha_n)$ is a multi-index, $|\alpha| = \alpha_1 + \Lambda + \alpha_n$; ☟$u = (\frac{\partial u}{\partial x_1}, \ldots, \frac{\partial u}{\partial x_n}), \ldots,$ ☟$^m u = (\frac{\partial^m u}{\partial x_1^{\ m}}, \ldots, \frac{\partial^m u}{\partial x_n^{\ m}})$; ✌ is a formally self-adjoint positively elliptic operator ([1], p.207) in $\overline{\Omega}$ with real coefficients $a_\alpha(x) \in C^{|\alpha|}(\overline{\Omega})$ $(|\alpha| \le 2m)$;

$$B_j = \sum_{|\beta| \le k_j} b_{\beta j}(x)☟^{\beta} \quad (j = 1, \ldots, m) \tag{0.4}$$

is a normal system ([2], p.98) of boundary operators of orders $k_j \le 2m - 1$ $(0 = k_1 < \ldots < k_m \le 2m - 1)$ with real coefficients $b_{\beta j}(x) \in C^{2m - k_j}(S)$, which is formally self-adjoint with regard to ✌ ([2], p.98) and satisfies covering condition ([2], p.97); F, φ, ϕ are the given functions, and $u(t, x)$ is a sought function. We define the classical solution of problem (0.1)-(0.3) as a function $u(t,x) \in C^{2,2m}_{t,x}(\overline{Q}_T)$ satisfying all the conditions (0.1)-(0.3) in the usual sense, where $Q_T \equiv (0, T) \times \Omega$.

Note that this work is a statement of results obtained in [3], [4] which have a relationship to the classical solution of problem (0.1)-(0.3).

In 1973, K.I.Khudaverdiyev [3] in his doctoral thesis considered, in particular, the following multidimensional mixed problem:

$$\begin{cases} u_{tt}(t,x) - L(u(t,x)) = ☞(u(t,x)) \ \ (t \in [0,T],\ x \in \Omega), & (0.5) \\ u(0,x) = \varphi(x) \ \ (x \in \Omega),\ \ u_t(0,x) = \psi(x) \ \ (x \in \Omega), & (0.6) \\ u(t,x)\big|_{x \in S} = 0 \ \ (t \in [0,T]), & (0.7) \end{cases}$$

where $0 < T < +\infty$, Ω is a bounded n-dimensional domain with a boundary S, $x = (x_1, \ldots, x_n)$;

$$L(u) = \sum_{i,j=1}^{n} \frac{\partial}{\partial x_i}\left(a_{ij}(x) \frac{\partial}{\partial x_i} \right) - a(x)u\,, \tag{0.8}$$

functions $a_{ij}(x)$ $(i, j = \overline{1,n})$ and $a(x)$ are measurable and bounded in Ω and satisfy in Ω the following conditions:

$$a_{ij}(x) = a_{ji}(x),\ a(x) \geq 0,\ \sum_{i,j=1}^{n} a_{ij}(x)\xi_i\xi_j \geq \alpha \cdot \sum_{i=1}^{n} \xi_i^2 \qquad (\alpha = const > 0), \tag{0.9}$$

where ξ_i are arbitrary real numbers; $\varphi(x)$ and $\phi(x)$ are the given functions in Ω; F is some, generally speaking, nonlinear operator, and $u(t,x)$ is a sought function.

Concepts of weakly generalized, strongly generalized, almost everywhere and classical solutions of problem (0.5)-(0.7) are introduced in that work. Uniqueness theorems, existence theorems, existence and uniqueness theorems, theorems on continuous dependence (in a certain sense) on initial functions $\varphi(x)$, $\phi(x)$ and nonlinear operator F for all four types of solutions of problem (0.5)-(0.7) are proved. Fourier method, contracted mappings principle, Schauder's principle, combined principle of Krasnoselski ([5], p.148) which generalizes two previous principles, Schauder's stronger principle ([6], p.206), non-zero rotation principle ([6], p.207), some properties of vector fields which are smooth (not necessarily completely continuous) perturbations of completely continuous vector fields with non-zero rotations, method of successive approximations, imbedding theorems, main properties of superposition (Nemitski) operator, a priori estimates methods are used to explore the problem (0.5)-(0.7). When proving existence theorems by the method of successive approximations, factorial speed of convergence (in corresponding metric for each type of solution) of successive approximations to the precise (weakly generalized, strongly generalized, almost everywhere and classical) solution of problem (0.5)-(0.7) is established.

Note that in all abstract theorems for problem (0.5)-(0.7) with nonlinear operator F dimension *n* is arbitrary. The presence of nonlinear operator F on the right side of equation (0.5) provides a possibility to cover many nonlinear equations both with retarded and advanced time arguments. However, for some specific nonlinear equations, the right side of which is a usual function, these abstract conditions on the right side of equation (0.5) may not be fulfilled. Therefore, there was a necessity in [3] to carry out more detailed research of problem (0.5)-(0.7) in case when F is a function-type operator, i.e. in case when the equation (0.5) has the following form:

$$u_{tt}(t,x) - L(u(t,x)) = F(t,x,u(t,x),u_{x_1}(t,x),\ldots,u_{x_n}(t,x),u_t(t,x)),$$

$$(t \in [0,T], x \in \Omega), \tag{0.10}$$

where $F(t,x,u_0,u_1,\ldots,u_n,u_{n+1})$ is a usual function.

It should be noted that when studying mixed problems like (0.10), (0.6), (0.7), along with the difficulties related to the form, order and sign of nonlinear function

$$F(t,x,u,u_{x_1},\ldots,u_{x_n},u_t)$$

in $u,u_{x_1},\ldots,u_{x_n},u_t$, a particular (and sometimes insuperable) difficulty is presented by dimension *n*. In [3], this difficulty is overcome:

1) for all four types of solutions of problem (0.10), (0.6), (0.7) concerning uniqueness and existence in small,
2) for weakly generalized, strongly generalized and almost everywhere solutions concerning existence in large, and
3) for classical solution concerning existence in large only with $n = 1, \ldots, 6$.

The restriction in latter case is caused by the necessity to obtain corresponding a priori estimates for classical solutions.

Remark 1. Throughout this work the expressions "existence in small" (i.e. for sufficiently small values of T) and "local existence" will be identical, and the expressions "existence in large" (i.e. for any finite value of T) and "global existence" will also be identical.

Remark 2. Note that when studying an existence of a solution (in any sense) of mixed problems for nonlinear equations, the proof of local existence theorems, in addition to being of interest itself, shows what kind of a priori estimate must be obtained (more precisely, is sufficient to obtain) for solution to prove global existence under the conditions of local existence theorem. Specifically, in our case, to prove existence theorems in large it suffices to obtain a priori estimate in a metric of space where the existence in small is already proved, and, it is not advisable to try to obtain a needed a priori estimate at once. It is better to go to it gradually, i.e. at first you should get the first a priori estimate and draw maximum information from it (using imbedding theorems). Then, strengthening conditions on the nonlinear right side of the considered equation, you should obtain the second, stronger a priori estimate, draw maximum information from it, and so on. You continue until you get the needed final a priori estimate.

In [3], the first a priori estimate for the classical solution $u(t,x)$ of problem (0.10), (0.6), (0.7) of the form

$$\|u(t,x)\|_{W_2^1(\Omega)} \leq C_0, \ \|u_t(t,x)\|_{L_2(\Omega)} \leq C_0 \quad \forall t \in [0,T],$$

is obtained with the help of equation (0.10) under rather general one-sided estimate-type conditions on nonlinear right side of equation (0.10) which allows rather strong nonlinearities of special form, order and sign in function $F(t,x,u,u_{x_1},\ldots,u_{x_n},u_t)$ with regard to some of (even all of) variables $u, u_{x_1}, \ldots, u_{x_n}$ and u_t. Next, the second obtained a priori estimate has the following form:

$$\|u(t,x)\|_{W_2^2(\Omega)} \leq C_0, \ \|u_t(t,x)\|_{W_2^1(\Omega)} \leq C_0 \quad \forall t \in [0,T].$$

Continuing in this way, the final a priori estimate is obtained in the following form:

$$\|u(t,x)\|_{W_2^{\left[\frac{n}{2}\right]+3}(\Omega)} \leq C_0, \ \|u_t(t,x)\|_{W_2^{\left[\frac{n}{2}\right]+2}(\Omega)} \leq C_0 \quad \forall t \in [0,T],$$

where the constant C_0 does not depend neither on u nor on $t \in [0,T]$. Note that the second and all the next a priori estimates are obtained by means of countable system of nonlinear integro-differential equations, to solution of which the finding of unknown Fourier coefficients of the sought classical solution $u(t,x)$ of problem (0.10), (0.6), (0.7) is reduced.

Thus, the following results are obtained for problem (0.10), (0.6), (0.7) in [3]:

1) for any n, uniqueness in large theorems for all four types of solutions of problem (0.10), (0.6), (0.7) are proved;
2) for any n, existence in small theorems for all four types of solutions of problem (0.10), (0.6), (0.7) are proved;
3) for any n, existence in large theorems for weakly generalized, strongly generalized and almost everywhere solutions of problem (0.10), (0.6), (0.7) are proved;
4) for $n = 1, \ldots, 6$, existence in large theorems for classical solution of problem (0.10), (0.6), (0.7) are proved.

Note that the results of [3] related to problems (0.5)-(0.7) and (0.10), (0.6), (0.7), are announced in the works of K.I.Khudaverdiyev [7]-[15].

In 1987, F.K.Khudaverdiyev [4] in his Ph.D. thesis, studied existence (both in small and in large) and uniqueness (in large) of almost everywhere and classical solutions of problem (0.1)-(0.3) by means of methods and ideas used in [3] to explore problem (0.10), (0.6), (0.7). The following results are obtained:

1) for any n, uniqueness theorems for almost everywhere and classical solutions of problem (0.1)-(0.3) are proved;
2) for any n, existence in small theorems for almost everywhere and classical solutions of problem (0.1)-(0.3) are proved;
3) for any n, existence in large theorem for almost everywhere solution of problem (0.1)-(0.3) is proved;
4) for $n \leq 4m + 2$, existence in large theorems for classical solution of problem (0.1)-(0.3) are proved, where $2m$ is an order of elliptic operator A appearing in the principal part of equation (0.1).

Results are announced in [16]-[19] by F.K.Khudaverdiyev.

Usually, when you generalize some problem to more general case, you have to sacrifice something. But in this work, the passage from problem (0.10), (0.6), (0.7) to more general problem (0.1)-(0.3) is effectuated without any loss.

In view of unhandiness of proofs and space limit, we decided, having combined results of [3], [4] related to problems (0.10), (0.6), (0.7) and (0.1)-(0.3), to state in this work (for the first time) results with full proofs only for classical solution of problem (0.1)-(0.3).

Thus, this work is dedicated to the study of existence (both in small and in large) and uniqueness for classical solution of problem (0.1)-(0.3).

Note that the theory of mixed problems for hyperbolic equations is one of the most intensively developing fields of modern theory of partial differential equations. For example, the works by Z.I.Halilov [21]-[24], O.A.Ladizhenskaya [25], Y.B.Lopatinski [26], T.Kato [27], M.I.Vishik and O.A.Ladizhenskaya [28], Y.F.Korobeynik [29], B.M.Levitan [30], V.A.Ilyin [31], N.I.Brish [32]-[34], N.I.Brish and I.N.Valeshkevich [35]-[36], I.N.Valeshkevich [37], H.O.Kreiss [38], T.Shirota and K.Asano [39], Ikawa Mitsuru [40]-

[41], R.Sakamato [42]-[43], Wolf von Wahl [44] and many others are dedicated to the mixed problem for linear hyperbolic equations.

As to the works dedicated to the mixed problems for nonlinear hyperbolic equations, they are too much in number, and it is almost impossible to mention all of them here. We will confine ourselves to mentioning just the following works known to us, which treat existence and uniqueness for solutions of mixed problems in finite (with regard to spatial variables) domain for nonlinear non-degenerate hyperbolic equations:

1) In [45]-[55], one-dimensional mixed problems for nonlinear second order hyperbolic equations have been studied;
2) In [56]-[85], multidimensional mixed problems have been studied, with

   a) [56]-[78] being dedicated to the multidimensional mixed problems for nonlinear ordinary second order hyperbolic equations (i.e. for equations of second order both in time and spatial variables), and
   b) [79]-[85] being dedicated to the multidimensional mixed problems for nonlinear abstract second order hyperbolic equations (i.e. for nonlinear second order hyperbolic equations in abstract formulation with self-adjoint positively definite principal elliptic part of arbitrary even order $2m$).

Below we will analyze only [79]-[85].

In 1970, J.C.Clements [79] considered the problem

$$\begin{cases} u_{tt}(t,x) + \sum\limits_{|\alpha|\le m} D^{\alpha} A_{\alpha}(t,x,u(t,x),\ldots,D^{m}u(t,x)) + d\cdot u_t(t,x) = f(t,x) \\ \qquad\qquad\qquad\qquad (t\in[0,T],\ x\in\Omega), \\ u(0,x)=\varphi(x),\ u_t(0,x)=\psi(x)\ \ (x\in\Omega), \\ D^{\alpha}u(t,x)\big|_{[0,T]\times\partial\Omega} = 0\ \ (|\alpha|\le m-1), \end{cases} \tag{0.11}$$

and, under some conditions on initial data, using Galerkin method, compactness method and monotony method, proved existence theorem for weak solution of this problem with the following properties:

$$u\in L^{\infty}\left([0,T];\ \overset{0}{W}{}_{p}^{m}(\Omega)\right),\quad u_t\in L^{\infty}\left([0,T];\ L_2(\Omega)\right),$$

where d is a real constant (damping coefficient), and $p\ge 2$ is some positive integer.

In 1970, S.Y.Yakubov [80] proved existence (both local and global) and uniqueness theorems for strong solution of Cauchy problem for quasi-linear abstract second order hyperbolic equations in Hilbert and Banach spaces using methods of the theory of semigroups. Abstract results obtained in this work imply, in particular, the following results for the mixed problem (0.1)-(0.3):

1) Local existence and uniqueness theorems for almost everywhere solution of problem (0.1)-(0.3) are proved for $1 \le n < 2m$ when $F = f(t,x,u(t,x),..., D^m u(t,x), u_t(t,x))$ and for $2m \le n < 4m$ when $F = f(t,x,u(t,x),..., D^\beta u(t,x) \cdot u_t(t,x) + \Phi(t,x,u(t,x),..., D^\beta u(t,x))$ with $|\beta| < \frac{4m-n}{2}$;
2) Global existence and uniqueness theorems for almost everywhere solution of problem (0.1)-(0.3) are proved for $1 \le n < 2m$ when $F = f(t,x,|u|^2,...,|D^\beta u|^2) \cdot u_t + \Phi'(|u|^2) \cdot u + f_0(t,x)$ with $|\beta| < \frac{2m-n}{2}$, and for $2m \le n < 4m$ when $F = \Phi'(|u|^2) \cdot u + f(t,x)$.

Remark 3. It should be especially noted that in general case of equation (0.1) and for any value of n existence theorems (both in small and in large) for almost everywhere solution of problem (0.1)-(0.3) have been announced in [16], [17] and proved in detail in [18] by F.K.Khudaverdiyev.

Remark 4. Also note that in all the works dedicated to the study of Cauchy problem for nonlinear abstract second order hyperbolic equations (including that of S.Y.Yakubov [80]) the found strong abstract solutions are not, generally speaking, usual classical solutions (as defined for problem (0.1)-(0.3)) even in one-dimensional case.

In 1974, Hartmut Pecher [81] considered a special case of problem (0.1), (0.2), (0.11) with $F = -u^2$ and $n \le 4m$. He proved a theorem of existence of almost everywhere solution for initial functions $\varphi(x)$, $\phi(x)$ with sufficiently small (in a certain metric) norms.

In 1975, Wolf von Wahl [82] considered problem (0.1), (0.2), (0.11) with $F = -f'(u^2) \cdot u$ and $n < 4m$. He proved existence in large theorem for classical solution.

In 1975, S.I.Pokhozhayev [83] considered the problem

$$\begin{cases} u_{tt}(t,x) + (-1)^m \cdot a\left(\int\limits_\Omega |\nabla^m u(t,x)|^2 dx \right) \Delta^m u(t,x) = f(t,x) \quad (t \in [0,T],\ x \in \Omega), & (0.12) \\ u(0,x) = \varphi(x),\ u_t(0,x) = \psi(x) \quad (x \in \Omega), & (0.13) \\ \left.\dfrac{\partial^k u(t,x)}{\partial \nu^k}\right|_\Gamma = 0 \quad (k = 0,1,..., m-1), & (0.14) \end{cases}$$

where γ is an external normal to the lateral surface $\Gamma \equiv [0,T] \times \partial\Omega$;

$$\left|\nabla^m u\right| = (\Delta^{\frac{m}{2}} u)^2 \text{ for even } m,$$

$$\left|\nabla^m u\right| = \left|\nabla(\Delta^{\frac{m-1}{2}} u)\right|^2 \text{ for odd } m,$$

$|\nabla v|^2 = \sum_{i=1}^{n} (\frac{\partial v}{\partial x_i})^2$, Δ is a Laplace operator; $a(s) \in C^1([0,+\infty))$ and $a(s) \geq a_0 > 0 \ \forall s \geq 0$. By means of Galerkin method, the author proved existence and uniqueness theorem for the solution of problem (0.12)-(0.14) in a certain class of functions.

In 1979, A.B.Aliyev [84] proved global existence and uniqueness theorem for almost everywhere solution of problem (0.1)-(0.3) when $F = \Phi'(|u|^2) \cdot u + G(t,x,u)$ and $2m \leq n < 6m$.

In 1981, Philip Brenner and Wolf von Wahl [85] proved existence in large theorem for almost everywhere solution of (0.1), (0.2), (0.11) when $F = f(u^2) \cdot u$ and $n \leq 6m + 1$.

In 1986, F.K.Khudaverdiyev [19] obtained the following results for the classical solution of problem (0.1)-(0.3) when $F = F(t,x,u)$:

- uniqueness in large theorem is proved for any n;
- existence in small theorem is proved for any n;
- existence in large theorem is proved for $n \leq 4m + 2$.

Finally, in 1986, F.K.Khudaverdiyev [20] obtained the following results for the classical solution of problem (0.1)-(0.3):

1) uniqueness in large theorem is proved for any n;
2) existence in small theorem is proved for any n;
3) existence in large theorem is proved for $n < 2m$;
4) in case $F = F(t,x,u)$, existence in large theorem is proved for $n \leq 4m + 2$.

And now we proceed to detailed proofs of latter four results, to which this work is actually dedicated.

1. AUXILIARIES

In this paragraph, we introduce a number of concepts, notations and facts to be used later.

1) For $l > k_j$ $(j = 1,..., p)$, we denote by $W_2^p(\Omega, \{B_j\}_{j=1}^p)$ the subspace of all functions $u(x)$ from $W_2^l(\Omega)$ satisfying boundary conditions $B_j(u(x))\big|_S = 0$ $(j = 1,..., p)$, where $S = \partial\Omega$. Note that the set $\{B_j\}_{j=1}^p$ may be empty. In that case we will assume that $W_2^p(\Omega, \{B_j\}_{j=1}^p) = W_2^l(\Omega)$.

2) Suppose that there exists an Hermitian integro-differential bilinear form

$$a(u,v) = \int\limits_{\Omega} \sum_{|\alpha|,|\beta| \le m} a_{\alpha\beta}(x) \partial^\alpha u(x) \partial^\beta v(x) dx \tag{1.1}$$

such that for $u, v \in W_2^{2m}(\Omega, \{B_j\}_{j=1}^m)$

$$a(u,v) = (A u, v)_{L_2(\Omega)} \tag{1.2}$$

and the corresponding quadratic form $a(u,u)$ is coercive ([1], p.184) in $W_2^m(\Omega, \{B_j\}_{j=1}^r)$, where $k_r \le m-1$, $k_{r+1} \ge m$ and $a_{\alpha\beta}(x)$ are measurable bounded functions in Ω. The latter implies the existence of constants $C_0 > 0$ and $C_1 \ge 0$ such that $\forall u \in W_2^m(\Omega, \{B_j\}_{j=1}^r)$

$$a(u,v) \equiv a(u) \ge C_0 \cdot \|u\|^2_{W_2^m(\Omega)} - C_1 \cdot \|u\|^2_{L_2(\Omega)}. \tag{1.3}$$

Without loss of generality we assume here that $C_1 = 0$.

Let's agree to assume throughout this work that all the conditions imposed in the statement of problem (0.1)-(0.3) and assumption in section 2 above are satisfied. In every single case we will only state additional conditions.

3) We call function $v \in W_2^{2m}(\Omega, \{B_j\}_{j=1}^m)$ an eigenfunction of problem

$$\begin{cases} A(v) = \lambda^2 v, & (1.4) \\ B_j(v)\big|_S = 0 \ (j = 1,..., m), & (1.5) \end{cases}$$

if it is not identically zero and

$$a(v,\eta) = \lambda^2 \cdot (v,\eta)$$

for any function $\eta \in W_2^{2m}(\Omega, \{B_j\}_{j=1}^m)$. And we call number λ^2 appearing in last equality an eigenvalue corresponding to an eigenfunction *v(x)* , where *A* is the operator appearing in equation (0.1).

As $S \in C^{2m}$, $a_\alpha(x) \in C^{|\alpha|}(\overline{\Omega})$ $(|\alpha| \le 2m)$ and $b_{\beta j}(x) \in C^{2m-k_j}(S)$ $(|\beta| \le k_j, k_j \le 2m-1)$, it is known (see e.g. [36], p.395) that the problem (1.4), (1.5), i.e. differential operator generated by differential expression *A(u)* and boundary conditions (0.3), has a complete in $L_2(\Omega)$, $W_2^m(\Omega, \{B_j\}_{j=1}^r)$ and $W_2^{2m}(\Omega, \{B_j\}_{j=1}^m)$ orthonormal system $\{v_s(x)\}_{s=1}^\infty$ of eigenfunctions $v_s(x)$ corresponding to eigenvalues λ_s^2 numbered in order of increase, i.e.

$$0 < \lambda_1^2 \le \lambda_2^2 \le \ldots \le \lambda_s^2 \le \ldots \quad (0 < \lambda_s \to +\infty \text{ as } s \to \infty).$$

4) We denote by $B_{\beta_0,\ldots,\beta_l,T}^{\alpha_0,\ldots,\alpha_l}$ a totality of all the functions of the form $u(t,x) = \sum_{s=1}^{\infty} u_s(t) v_s(x)$ considered in $Q_T = (0,T)\times\Omega$, where $u_s(t) \in C^{(l)}([0,T])$ for all *s* and

$$J_T(u) \equiv \sum_{i=0}^{l} \left\{ \sum_{s=1}^{\infty} \left(\lambda_s^{\alpha_i} \cdot \max_{0\le t\le T} \left| u_s^{(i)}(t) \right| \right)^{\beta_i} \right\}^{\frac{1}{\beta_i}} < +\infty,$$

with $\alpha_i \ge 0, 1 \le \beta_i \le 2$ $(i = \overline{0,l})$. We define the norm in this set as $\|u\| = J_T(u)$. It is evident that all these spaces are Banach spaces.

For every function $u(t,x) = \sum_{s=1}^{\infty} u_s(t) v_s(x)$, we call function $u_s(t)$ its *s*-th component. Let *M* be any nonempty set in $B_{\beta_0,\ldots,\beta_l,T}^{\alpha_0,\ldots,\alpha_l}$. We denote by M_s the totality of *s*-th components of all the elements of set *M*. Then the following theorem is true:

Theorem 1. The following two conditions are necessary and sufficient for compactness of set $M \subset B_{\beta_0,\ldots,\beta_l,T}^{\alpha_0,\ldots,\alpha_l}$ in $B_{\beta_0,\ldots,\beta_l,T}^{\alpha_0,\ldots,\alpha_l}$:

a. for every s $(s = 1,2,...)$ set M_s is compact in $C^{(l)}([0,T])$;

b. for any $\varepsilon > 0$ there exists a number n_ε, the same for all

$$u(t,x) = \sum_{s=1}^{\infty} u_s(t) v_s(x) \in M \text{, such that}$$

$$\sum_{i=0}^{l}\left\{\sum_{s=n_\varepsilon}^{\infty}\left(\lambda_s^{\alpha_i}\cdot\max_{0\le t\le T}\left|u_s^{(i)}(t)\right|\right)^{\beta_i}\right\}^{\frac{1}{\beta_i}} < \varepsilon \quad \forall u \in M\cdot$$

Proof of this theorem is absolutely similar to that of theorem on compactness criterion for sets in space l_p $(1 \le p < +\infty)$ (see, e.g. [86], pp.295-296).

In what follows we will use the following notations for functions $u(t,x) = \sum_{s=1}^{\infty} u_s(t) v_s(x) \in B^{\alpha_0,\dots,\alpha_l}_{\beta_0,\dots,\beta_l,T}$:

$$\|u\|_{B^{\alpha_0,\dots,\alpha_l}_{\beta_0,\dots,\beta_l,t}} \equiv \sum_{i=0}^{l}\left\{\sum_{s=1}^{\infty}\left(\lambda_s^{\alpha_i}\cdot\max_{0\le\tau\le t}\left|u_s^{(i)}(\tau)\right|\right)^{\beta_i}\right\}^{\frac{1}{\beta_i}} \quad (0 \le t \le T). \tag{1.6}$$

5) Let $f(x) \in W_2^m(\Omega, \{B_j\}_{j=1}^r)$. Then it is obvious that the series $f(x) = \sum_{s=1}^{\infty} f_s \cdot v_s(x)$, where $f_s = \int_\Omega f(x) v_s(x)dx$, converges in $W_2^m(\Omega)$. Due to the coerciveness inequality (1.3) (where, as agreed above, we assume that $C_1 = 0$), we have:

$$\|f(x)\|^2_{W_2^m(\Omega)} \le \frac{1}{C_0}\cdot a(f(x)) = C_0^{-1}\cdot a(f(x), f(x))\cdot$$

Next, it is evident that

$$a(f,f) = \lim_{N\to\infty} a\left(\sum_{s=1}^{N} f_s v_s, \sum_{s=1}^{N} f_s v_s,\right) = \lim_{N\to\infty}\left(\mathrm{A}\sum_{s=1}^{N} f_s v_s, \sum_{s=1}^{N} f_s v_s\right)_{L_2(\Omega)} =$$

$$= \lim_{N\to\infty}\left(\sum_{s=1}^{N}\lambda_s^2 f_s v_s, \sum_{s=1}^{N} f_s v_s\right)_{L_2(\Omega)} = \lim_{N\to\infty}\sum_{s=1}^{\infty}\lambda_s^2 f_s^2 = \sum_{s=1}^{\infty}\lambda_s^2 f_s^2\,. \tag{1.7}$$

Thus, $\forall f(x) \in W_2^m(\Omega, \{B_j\}_{j=1}^r)$ we have

$$\|f(x)\|^2_{W_2^m(\Omega)} \le \frac{1}{C_0} \cdot \sum_{s=1}^{\infty} \lambda_s^2 f_s^2 . \qquad (1.8)$$

Note that the estimate (1.8) will be considerably used throughout this work.

Besides, we will also often use the following two lemmas from [36] by N.I.Brish and I.N.Valeshkevich:

Lemma 1 (see [36], p.396, lemma 3 and remark after its statement). Let the following conditions be satisfied for a natural number $k \ge 2$:

a. Ω is a finite domain bounded by a surface S of C^{km};

b. $a_\alpha(x) \in C^{|\alpha|+(k-2)m}(\overline{\Omega})$ $(|\alpha| \le 2m)$, $b_{\beta j}(x) \in C^{km-k_j}(S)$

$(|\beta| \le k_j \le 2m-1;\ j=1,\dots,m)$;

c. $f(x) \in W_2^{km}(\Omega)$, $f(x), \mathrm{A} f(x), \dots, \mathrm{A}^{[\frac{k-2}{2}]}(f(x)) \in W_2^{2m}(\Omega, \{B_j\}_{j=1}^m)$, and,

for odd values of k, $\mathrm{A}^{[\frac{k-1}{2}]}(f(x)) \in W_2^m(\Omega, \{B_j\}_{j=1}^r)$.

Then the following inequalities hold:

$$\|f(x)\|^2_{W_2^{km}(\Omega)} \le \begin{cases} C_1 \cdot \left\| \mathrm{A}^{\frac{k}{2}}(f) \right\|^2_{L_2(\Omega)}, & \text{if } k \text{ is even,} \\ C_1 \cdot a\left(\mathrm{A}^{\frac{k-1}{2}}(f) \right), & \text{if } k \text{ is odd,} \end{cases}$$

where $C_1 = 0$ is some constant not depending on f.

Lemma 2 (see [36], p.395). If the conditions of Lemma 1 are satisfied, then the following equalities hold:

$$\sum_{s=1}^{\infty}\lambda_s^{2k}\cdot f_s^2=\begin{cases}\left\|\mathrm{A}^{\frac{k}{2}}(f)\right\|^2_{L_2(\Omega)}, & \text{if } k \text{ is even},\\ a\left(\mathrm{A}^{\frac{k-1}{2}}(f)\right), & \text{if } k \text{ is odd},\end{cases}$$

where $f_s = \int\limits_{\Omega} f(x)v_s(x)dx \ (s = 1,2,...)$.

From Lemmas 1 and 2 it follows, in particular, that for every function $f(x)\in W_2^{2m}(\Omega,\{B_j\}_{j=1}^m)$ we have

$$\|f(x)\|^2_{W_2^{2m}(\Omega)} \le C_1\cdot\|\mathrm{A}(f(x))\|^2_{L_2(\Omega)} = C_1\cdot\sum_{s=1}^{\infty}\lambda_s^4 f_s^2 \cdot \tag{1.9}$$

Now we establish another inequality which will be used in what follows. Let $u(x)\in W_2^m(\Omega)$ and let N be any natural number. Then it is evident that

$$0\le a\left(u(x)-\sum_{s=1}^{N}a\left(u(x),\frac{1}{\lambda_s}v_s(x)\right)\cdot\frac{1}{\lambda_s}v_s(x)\right)=a\left(u(x)-\sum_{s=1}^{N}a\left(u(x),\frac{1}{\lambda_s}v_s(x)\right)\cdot\right.$$

$$\left.\cdot\frac{1}{\lambda_s}v_s(x),u(x)-\sum_{s=1}^{N}a\left(u(x),\frac{1}{\lambda_s}v_s(x)\right)\cdot\frac{1}{\lambda_s}v_s(x)\right)=$$

$$=a(u(x),u(x))-2\sum_{s=1}^{N}\left\{a\left(u(x),\frac{1}{\lambda_s}v_s(x)\right)\right\}^2+$$

$$+\sum_{s,r=1}^{N}a\left(u(x),\frac{1}{\lambda_s}v_s(x)\right)\cdot a\left(u(x),\frac{1}{\lambda_r}v_r(x)\right)\cdot a\left(\frac{1}{\lambda_s}v_s(x),\frac{1}{\lambda_r}v_r(x)\right).$$

From here, using relations

$$a\left(\frac{1}{\lambda_s}v_s,\frac{1}{\lambda_r}v_r\right)=\frac{1}{\lambda_s}\cdot\frac{1}{\lambda_r}\cdot a(v_s,v_r)=\frac{1}{\lambda_s}\cdot\frac{1}{\lambda_r}\cdot(\mathrm{A}(v_s,v_r))_{L_2(\Omega)}=$$

$$=\frac{1}{\lambda_s}\cdot\frac{1}{\lambda_r}\cdot\lambda_s^2\cdot(v_s,v_r)_{L_2(\Omega)}=\begin{cases}0, & \text{if } s\neq r,\\ 1, & \text{if } s=r,\end{cases}$$

we obtain:

$$0\leq a(u,u)-\sum_{s=1}^{N}\left\{a\left(u,\frac{1}{\lambda_s}v_s\right)\right\}^2.$$

Consequently, $\forall u(x)\in W_2^m(\Omega)$

$$\sum_{s=1}^{\infty}\left\{a\left(u(x),\frac{1}{\lambda_s}v_s(x)\right)\right\}^2\leq a(u(x),u(x)).\leq C\cdot\|u(x)\|_{W_2^m(\Omega)}^2, \tag{1.10}$$

where $C>0$ is some constant not depending on $u(x)\in W_2^m(\Omega)$.

6) As the system $\{v_s(x)\}_{s=1}^{\infty}$ is complete orthonormal in $L_2(\Omega)$, then it is evident that every classical solution of problem (0.1)-(0.3) has the following form:

$$u(t,x)=\sum_{s=1}^{\infty}u_s(t)v_s(x), \tag{1.11}$$

where

$$u_s(t)=\int_{\Omega}u(t,x)v_s(x)dx \quad (s=1,2,...). \tag{1.12}$$

Then, after applying Fourier method, the finding of unknown Fourier coefficients $u_s(t)$ $(s=1,2,...)$ of the sought classical solution $u(t,x)$ of problem (0.1)-(0.3) is reduced to the solution of the following countable system of non-linear integro-differential equations:

$$u_s(t)=\varphi_s\cdot\cos\lambda_s t+\frac{\psi_s}{\lambda_s}\cdot\sin\lambda_s t+\frac{1}{\lambda_s}\cdot\int_0^t\int_{\Omega}\mathsf{F}(u(\tau,x))v_s(x)\cdot\sin\lambda_s(t-\tau)dxd\tau$$

$$(s=1,2,...;\ t\in[0,T]), \tag{1.13}$$

where

$$\varphi_s = \int_\Omega \varphi(x) v_s(x)dx \,, \quad \psi_s = \int_\Omega \psi(x) v_s(x)dx \quad (s = 1,2,\ldots)\,, \tag{1.14}$$

$$☞(u(t,x)) \equiv F(t,x,u(t,x),☟u(t,x),\ldots,☟^{m} u(t,x),u_t(t,x))\,, \tag{1.15}$$

with (1.11) and (1.12) taken into account.

Proceeding from the definition of classical solution of problem (0.1)-(0.3), it is easy to prove the following

Lemma 3. If $u(t,x) = \sum_{s=1}^{\infty} u_s(t)v_s(x)$ is any classical solution of problem (0.1)-(0.3), then functions $u_s(t)$ $(s = 1,2,\ldots)$ satisfy system (1.13).

Proof. Let $u(t,x) = \sum_{s=1}^{\infty} u_s(t)v_s(x)$ be any classical solution of problem (0.1)-(0.3). According to the definition of classical solution of problem (0.1)-(0.3), $u(t,x) \in C^{2,2m}_{t,x}(\overline{Q}_T)$ and all the conditions (0.1)-(0.3) are satisfied in the usual sense. Then, multiplying both sides of equation (0.1) by function $v_s(x)$, integrating the obtained equality over Ω and using conditions (0.2), relations $\mathsf{A}(v_s(x)) = \lambda_s^2 v_s(x)$ and notations

$$\mathsf{F}_s(u;t) = \int_\Omega \mathsf{F}(u(t,x))v_s(x)dx \quad (s = 1,2,\ldots;\ t \in [0,T])\,, \tag{1.16}$$

we get the following Cauchy problem:

$$\begin{cases} u_s''(t) + \lambda_s^2 u_s(t) = \mathsf{F}_s(u;t) & (s = 1,2,\ldots;\ t \in [0,T]), \\ u_s(0) = \varphi_s, u_s'(0) = \psi_s & (s = 1,2,\ldots), \end{cases}$$

which is obviously equivalent to system (1.13). Lemma is proved.

7) To study existence of classical solution of problem (0.1)-(0.3), we consider along with system (1.13) the following system obtained from (1.13) by the formal (for the present) integration by parts in τ once on its right side:

$$u_s(t) = \varphi_s \cdot \cos\lambda_s t + \frac{\psi_s}{\lambda_s} \cdot \sin\lambda_s t + \frac{1}{\lambda_s^2} \cdot (1 - \cos\lambda_s t) \cdot \int_\Omega f(x)v_s(x)dx + \tag{1,17}$$

$$+ \frac{1}{\lambda_s^2} \cdot \int_0^t \int_\Omega \frac{\partial}{\partial\tau} \{\mathsf{F}(u(\tau,x))\} v_s(x) \cdot [1 - \cos\lambda_s(t-\tau)] dx d\tau \quad (s = 1,2,\ldots;\ t \in [0,T]),$$

where

$$f(x) \equiv F(0, x, \varphi(x), \text{☟}\,\varphi(x), \ldots, \text{☟}^{m}\varphi(x), \psi(x)). \qquad (1.18)$$

Next, we formally (for the present) transform system (1.17) to the following forms:

$$u_s(t) = \varphi_s \cdot \cos \lambda_s t + \frac{\psi_s}{\lambda_s} \cdot \sin \lambda_s t + \frac{1}{\lambda_s^k} \cdot (1 - \cos \lambda_s t) \cdot \int_\Omega \mathsf{A}^{l}(f(x)) v_s(x) dx +$$

$$+ \frac{1}{\lambda_s^k} \cdot \int_0^t \int_\Omega \mathsf{A}^{l}\left(\frac{\partial}{\partial \tau} \mathsf{F}\,(u(\tau, x)) \right) v_s(x) \cdot \left[1 - \cos \lambda_s (t - \tau)\right] dx d\tau \qquad (1.19)$$

$$(s = 1,2,\ldots; k = 2 + 2l,\ l = 0,1,\ldots;\quad t \in [0,T]),$$

$$u_s(t) = \varphi_s \cdot \cos \lambda_s t + \frac{\psi_s}{\lambda_s} \cdot \sin \lambda_s t + \frac{1}{\lambda_s^k} \cdot (1 - \cos \lambda_s t) \cdot a\left(\mathsf{A}^{l}(f(x)), \frac{1}{\lambda_s} \cdot v_s(x) \right) +$$

$$+ \frac{1}{\lambda_s^k} \cdot \int_0^t a\left(\mathsf{A}^{l}\left(\frac{\partial}{\partial \tau} \mathsf{F}\,(u(\tau, x)) \right), \frac{1}{\lambda_s} \cdot v_s(x) \right) \cdot \left[1 - \cos \lambda_s (t - \tau)\right] d\tau$$

$$(s = 1,2,\ldots; k = 2l + 3,\ l = 0,1,\ldots;\quad t \in [0,T]), \qquad (1.20)$$

where bilinear form $a(u, v)$ is defined by (1.1).

Besides, we will use the following notations in this work:

$$z_k(t,x) = \sum_{s=1}^{\infty} \left\{ \varphi_s \cdot \cos \lambda_s t + \frac{\psi_s}{\lambda_s} \cdot \sin \lambda_s t + \frac{1}{\lambda_s^k} \cdot (1 - \cos \lambda_s t) \cdot \int_\Omega \mathsf{A}^{l}(f(x)) v_s(x) dx \right\} v_s(x)$$

$$\textit{for} \quad k = 2 + 2l \ (l = 0,1,\ldots), \qquad (1.21)$$

$$z_k(t,x) = \sum_{s=1}^{\infty} \left\{ \varphi_s \cdot \cos \lambda_s t + \frac{\psi_s}{\lambda_s} \cdot \sin \lambda_s t + \frac{1}{\lambda_s^k} \cdot (1 - \cos \lambda_s t) \cdot a\left(\mathsf{A}^{l}(f(x)), \frac{1}{\lambda_s} \cdot v_s(x) \right) \right\} v_s(x)$$

$$\textit{for} \quad k = 2l + 3 \ (l = 0,1,\ldots); \qquad (1.22)$$

$$\mathsf{P}_k(u(t,x)) = \sum_{s=1}^{\infty} \frac{1}{\lambda_s^k} \cdot \int_0^t \int_\Omega u(\tau, x)) v_s(x) \cdot [1 - \cos \lambda_s (t - \tau)] dx d\tau \cdot v_s(x)$$

$$\textit{for even values of } k; \qquad (1.23)$$

$$\mathsf{P}_k(u(t,x)) = \sum_{s=1}^{\infty} \frac{1}{\lambda_s^k} \cdot \int_0^t a\left(u(\tau,x)), \frac{1}{\lambda_s} \cdot v_s(x) \right) \cdot [1 - \cos \lambda_s (t-\tau)] d\tau \cdot v_s(x)$$

for odd values of k; (1.24)

8) We agree to assume that all the quantities throughout this work are real, all the functions are real-valued, and all the integrals are understood in the sense of Lebesgue.

Concerning numbering, the first number in brackets will indicate the number of current paragraph, while the second one will indicate the sequence number inside that paragraph.

2. Study of Uniqueness of the Classical Solution of Problem (0.1)-(0.3)

In this paragraph, the following uniqueness theorem for classical solution of problem (0.1)-(0.3) is proved using Bellman's inequality (see, e.g. [87], pp.188-189).

Theorem 2. Let

1. $F(t,x,u_1,...,u_N) \in C\left([0,T] \times \overline{\Omega} \times (-\infty,\infty)^N\right)$.
2. $\forall R > 0$ in $[0,T] \times \overline{\Omega} \times [-R,R]^N$

$$\left|F(t,x,u_1,...,u_N) - F(t,x,\widetilde{u}_1,...,\widetilde{u}_N)\right| \le C_R \cdot \sum_{i=1}^{N} \left|u_i - \widetilde{u}_i\right|, \tag{2.1}$$

where $C_R > 0$ is a constant.

Then problem (0.1)-(0.3) can not have more than one classical solution.

Proof. Let $u(t,x) = \sum_{s=1}^{\infty} u_s(t) v_s(x)$ and $\widetilde{u}(t,x) = \sum_{s=1}^{\infty} \widetilde{u}_s(t) v_s(x)$ be two arbitrary classical solutions of problem (0.1)-(0.3). Then, due to lemma 3 proved in section 6 of §1, functions $u_s(t)$ $(s = 1,2,...)$ and $\widetilde{u}_s(t)$ $(s = 1,2,...)$ satisfy system (1.13) (in case of $\widetilde{u}_s(t)$ $(s = 1,2,...)$, $u(\tau,x)$ must be replaced by $\widetilde{u}(\tau,x)$ on the right side of (1.13)). Using Parseval's equality, it is easy to obtain from system (1.13) that $\forall t \in [0,T]$:

$$\left\|u - \widetilde{u}\right\|_{B_{2,t}^1}^2 \equiv \sum_{s=1}^{\infty} \left(\lambda_s \cdot \max_{0 \le \tau \le t} \left|u_s(\tau) - \widetilde{u}_s(\tau)\right| \right)^2 \le t \cdot \int_0^t \left\|\mathsf{F}(u(\tau,x)) - \mathsf{F}(\widetilde{u}(\tau,x))\right\|_{L_2(\Omega)}^2 d\tau,$$

$$\|u_t-\tilde{u}_t\|^2_{B^0_{2,t}} \equiv \sum_{s=1}^{\infty}\left(\max_{0\le\tau\le t}|u'_s(\tau)-\tilde{u}'_s(\tau)|\right)^2 \le t\cdot\int_0^t\|\mathsf{F}(u(\tau,x))-\mathsf{F}(\tilde{u}(\tau,x))\|^2_{L_2(\Omega)}d\tau;$$

$$\|u-\tilde{u}\|^2_{B^{1,0}_{2,2,t}} \equiv \left(\|u-\tilde{u}\|_{B^1_{2,t}}+\|u_t-\tilde{u}_t\|_{B^0_{2,t}}\right)^2 \le 2\|u-\tilde{u}\|^2_{B^1_{2,t}}+2\|u_t-\tilde{u}_t\|^2_{B^0_{2,t}} \le$$

$$\le 4T\cdot\int_0^t\|\mathsf{F}(u(\tau,x))-\mathsf{F}(\tilde{u}(\tau,x))\|^2_{L_2(\Omega)}d\tau, \tag{2.2}$$

where operator F is defined by (1.15).

On the other hand, it is evident that there is a number $R_0>0$ such that in $\overline{Q}_T=[0,T]\times\overline{\Omega}$:

$$|\mathsf{D}^\alpha u(t,x)|\ (0\le|\alpha|\le m),\ |u_t(t,x)|,\ |\mathsf{D}^\alpha\tilde{u}(t,x)|\ (0\le|\alpha|\le m),\ |\tilde{u}_t(t,x)|\le R_0.$$

Then, using inequality (2.1) for $R=R_0$ and elementary inequality

$$(a_1+\ldots+a_N)^2\le N\cdot(a_1^2+\ldots+a_N^2), \tag{2.3}$$

we obtain from (1.26) that $\forall t\in[0,T]$:

$$\|u-\tilde{u}\|^2_{B^{1,0}_{2,2,t}} \le 4T\cdot C^2_{R_0}\cdot N\cdot\int_0^t\Big\{\|u(\tau,x))-\tilde{u}(\tau,x)\|^2_{W_2^m(\Omega)}+$$

$$+\|u_\tau(\tau,x))-\tilde{u}_\tau(\tau,x)\|^2_{L_2(\Omega)}\Big\}d\tau. \tag{2.4}$$

Next, due to the estimate (1.8), $\forall\tau\in[0,T]$ we have:

$$\|u(\tau,x))-\tilde{u}(\tau,x)\|^2_{W_2^m(\Omega)} \le C_0^{-1}\cdot\sum_{s=1}^{\infty}\lambda_s^2\cdot(u_s(\tau)-\tilde{u}_s(\tau))^2 \le$$

$$\le C_0^{-1}\cdot\sum_{s=1}^{\infty}\left(\lambda_s^2\cdot\max_{0\le\eta\le\tau}|u_s(\eta)-\tilde{u}_s(\eta)|\right)^2 =$$

$$= C_0^{-1} \cdot \|u - \tilde{u}\|^2_{B^1_{2,\tau}} \le C_0^{-1} \cdot \|u - \tilde{u}\|^2_{B^{1,0}_{2,2,\tau}} . \quad (2.5)$$

Besides, it is evident that $\forall \tau \in [0, T]$

$$\|u_\tau(\tau, x)) - \tilde{u}_\tau(\tau, x)\|^2_{L_2(\Omega)} = \sum_{s=1}^{\infty} (u_s'(\tau) - \tilde{u}_s'(\tau))^2 \le$$

$$\le \sum_{s=1}^{\infty} \left(\max_{0 \le \eta \le \tau} |u_s'(\eta) - \tilde{u}_s'(\eta)| \right)^2 = \|u_\tau - \tilde{u}_\tau\|^2_{B^0_{2,\tau}} \le \|u - \tilde{u}\|^2_{B^{1,0}_{2,2,\tau}} . \quad (2.6)$$

Now, using estimates (2.5) and (2.6), we obtain from (2.4) that $\forall t \in [0, T]$:

$$\|u - \tilde{u}\|^2_{B^{1,0}_{2,2,t}} \le 4T \cdot C^2_{R_0} \cdot N \cdot (C_0^{-1} + 1) \cdot \int_0^t \|u - \tilde{u}\|^2_{B^{1,0}_{2,2,\tau}} d\tau \cdot$$

From here, on applying Bellman's inequality, we obtain that $\forall t \in [0, T]$ $\|u - \tilde{u}\|^2_{B^{1,0}_{2,2,t}} = 0$. Hence, $u = \tilde{u}$. Theorem is proved.

3. Study of Existence In Small for the Classical Solution of Problem (0.1)-(0.3)

In this paragraph, using contracted mappings principle and Schauder's principle, the following two existence in small theorems for the classical solution of problem (0.1)-(0.3) are proved for any n :

Theorem 3. Let the following conditions hold for $k = \left[\frac{n}{2m} \right] + 3$:

1. a) $\partial\Omega = S \in C^{km}$;

   b) $a_\alpha(x) \in C^{|\alpha| + (k-2)m}(\overline{\Omega})$ $(|\alpha| \le 2m)$,

   $b_{\beta j}(x) \in C^{km - k_j}(S)$ $(|\beta| \le k_j,\ j = 1, \ldots, m)$;

c) $\varphi(x) \in W_2^{km}(\Omega)$, $\varphi(x)$, $\mathrm{A}\varphi(x)$, ... , $\mathrm{A}^{\left[\frac{k-2}{2}\right]}\varphi(x) \in W_2^{2m}\left(\Omega, \{B_j\}_{j=1}^{m}\right)$

and besides, for odd values of k $\mathrm{A}^{\left[\frac{k-1}{2}\right]}\varphi(x) \in W_2^{m}\left(\Omega, \{B_j\}_{j=1}^{r}\right)$;

d) $\psi(x) \in W_2^{(k-1)m}(\Omega)$, $\psi(x)$, $\mathrm{A}\psi(x)$, ... , $\mathrm{A}^{\left[\frac{k-3}{2}\right]}\psi(x) \in W_2^{2m}\left(\Omega, \{B_j\}_{j=1}^{m}\right)$

and besides, for even values of k $\mathrm{A}^{\frac{k-2}{2}}\psi(x) \in W_2^{m}\left(\Omega, \{B_j\}_{j=1}^{r}\right)$.

2. a) $\partial^s F(t,x,u_1,...,u_N) / \partial t^\alpha \partial x^\beta \partial u_1^{\gamma_1} ... \partial u_N^{\gamma_N} \in C(\overline{Q}_T \times (-\infty,\infty)^N)$

$$\left(0 \le \alpha + |\beta| + \sum_{i=1}^{N} \gamma_i = s \le (k-2)m + 1 = \left(\left[\frac{n}{2m}\right] + 1\right)m + 1;\ \alpha \le 1;\ |\beta| \le (k-2)m\right);$$

b) function $F(t,x,u_1,..., u_N)$ is equal to zero in all points $(t,x,0,u_2,..., u_{N-1},0)$ where $t \in [0,T]$, $x \in S$ and $u_2,..., u_{N-1} \in (-\infty,\infty)$, together with all its derivatives of the form mentioned in item a) up to the order $(k-4)m + 1 + k_m = \left(\left[\frac{n}{2m}\right] - 1\right)m + 1 + k_m$ for odd values of $\left[\frac{n}{2m}\right]$ and up to the order $(k-3)m + 1 + k_r = \left[\frac{n}{2m}\right] \cdot m + 1 + k_r$ for even values of $\left[\frac{n}{2m}\right]$;

c) $\forall R > 0$ in $Q_T \times [-R,R]^N$

$$|\partial^s F(t,x,u_1,...,u_N) / \partial t^\alpha \partial x^\beta \partial u_1^{\gamma_1} ... \partial u_N^{\gamma_N} - \\ - \partial^s F(t,x,\tilde{u}_1,...,\tilde{u}_N) / \partial t^\alpha \partial x^\beta \partial u_1^{\gamma_1} ... \partial u_N^{\gamma_N} | \le C_R \cdot \sum_{i=1}^{N} |u_i - \tilde{u}_i|, \tag{3.1}$$

where $\alpha + |\beta| + \sum_{i=1}^{N} \gamma_i = s = (k-2)m + 1 = \left(\left[\frac{n}{2m}\right] + 1\right)m + 1;\ \alpha \le 1;\ |\beta| \le (k-2)m$, and $C_R > 0$ is a constant.

Then there exists in small (i.e. for sufficiently small values of T) a unique in large (i.e. for any finite value of T) classical solution of problem (0.1)-(0.3).

Proof. Let k be a number appearing in the statement of this theorem, i.e. $k=\left[\frac{n}{2m}\right]+3$.

We consider the following operator Q in space $B_{2,2,2,T}^{k,k-1,k-2}$:

$$Q(u(t,x))=Z_2(t,x)+\mathsf{P}_2\left(\frac{\partial}{\partial t}\{\mathsf{F}\,(u(t,x))\}\right), \tag{3.2}$$

where function $Z_2(t,x)$ is defined by (1.21), operator P_2 is defined by (1.23), and operator F is defined by (1.15). It is easy to see that under the conditions of this theorem $\forall u(t,x)\in B_{2,2,2,T}^{k,k-1,k-2}$:

$$Q(u(t,x))=Z_k(t,x)+\mathsf{P}_k\left(\mathsf{A}^l\left(\frac{\partial}{\partial t}\{\mathsf{F}\,(u(t,x))\}\right)\right), \tag{3.3}$$

where $l=\dfrac{k-2}{2}$ for even k, $l=\dfrac{k-3}{2}$ for odd k, function $Z_k(t,x)$ is defined by (1.21) and (1.22), and operator P_k is defined by (1.23) and (1.24), respectively. To ease our writings, we will use the following notations:

$$\|u\|_{B_{2,2,2,t}^{k,k-1,k-2}}=\|u\|_{E_t}\qquad (t\in[0,T]). \tag{3.4}$$

Now we consider operator Q defined by (3.3) in a closed sphere K_r of space $E_T\equiv B_{2,2,2,T}^{k,k-1,k-2}$ with a center in zero and a radius $r>\|Z_k(t,x)\|_{E_T}$. Note that the belonging of function $Z_k(t,x)$ to the space E_T follows from conditions 1_c, 1_d and 2_a of this theorem and is established by means of S.L.Sobolev's imbedding theorems, Bessel's inequality and inequality (1.10), where function $u(x)\in W_2^m(\Omega)$ must be replaced by function $\mathsf{A}^l(f(x))$. Then, using Bessel's inequality (in case of even k) and inequality (1.10) (in case of odd k), where function $u(x)$ must be replaced by $\mathsf{A}^l\left(\frac{\partial}{\partial\tau}\{\mathsf{F}\,(u(\tau,x))\}\right)$, we obtain from (3.3), (1.23) and (1.24) that $\forall u\in B_{2,2,2,T}^{k,k-1,k-2}=E_T$:

$$\|Q(u)\|_{E_T} \le \|Z_k(t,x)\|_{E_T} + 4\sqrt{T}\cdot\left\|\mathsf{A}^l\left(\frac{\partial}{\partial t}\{\mathsf{F}(u(t,x))\}\right)\right\|_{L_2(Q_T)} \le$$

$$\le \|Z_k\|_{E_T} + C\cdot\sqrt{T}\cdot\left\|\frac{\partial}{\partial t}\{\mathsf{F}(u(t,x))\}\right\|_{W^{0,2ml}_{t,x,2}(Q_T)} =$$

$$= \|Z_k\|_{E_T} + C\cdot\sqrt{T}\cdot\left\|\frac{\partial}{\partial t}\{\mathsf{F}(u(t,x))\}\right\|_{W^{0,(k-2)m}_{t,x,2}(Q_T)} \qquad for\ k = 2+2l, \tag{3.5}$$

$$\|Q(u)\|_{E_T} \le \|Z_k\|_{E_T} + C\cdot\sqrt{T}\cdot\left\|\mathsf{A}^l\left(\frac{\partial}{\partial t}\{\mathsf{F}(u(t,x))\}\right)\right\|_{W^{0,m}_{t,x,2}(Q_T)} \le$$

$$\le \|Z_k\|_{E_T} + C\cdot\sqrt{T}\cdot\left\|\frac{\partial}{\partial t}\{\mathsf{F}(u(t,x))\}\right\|_{W^{0,2ml+m}_{t,x,2}(Q_T)} =$$

$$= \|Z_k\|_{E_T} + C\cdot\sqrt{T}\cdot\left\|\frac{\partial}{\partial t}\{\mathsf{F}(u(t,x))\}\right\|_{W^{0,(k-2)m}_{t,x,2}(Q_T)} \qquad for\ k = 3+2l, \tag{3.6}$$

where $C > 0$ is some constant.

Thus, $\forall u \in B^{k,k-1,k-2}_{2,2,2,T} = E_T$:

$$\|Q(u)\|_{E_T} \le \|Z_k\|_{E_T} + C\cdot\sqrt{T}\cdot\left\|\frac{\partial}{\partial t}\{\mathsf{F}(u(t,x))\}\right\|_{W^{0,(k-2)m}_{t,x,2}(Q_T)}. \tag{3.7}$$

Using Sobolev's imbedding theorems, coerciveness condition (1.3) with $C_1 = 0$, Lemmas 1, 2 from section 5, §1, and the structure of space E_T, we have for every $u \in E_T$ and $t \in [0,T]$:

$$\left\|\mathsf{D}_t^{\alpha}\mathsf{D}^{\beta}u(t,x)\right\|_{L_2(\Omega)} \le C\cdot\|u\|_{E_t} \le C\cdot\|u\|_{E_T} \quad \left(\alpha \le 2;\ 0 \le \alpha\cdot m + |\beta| \le m\left(\left[\frac{n}{2m}\right]+3\right) = km\right) \tag{3.8}$$

$$\left\| D_t^{\alpha} D^{\beta} u(t,x) \right\|_{C(\overline{Q}_T)} \le C \cdot \|u\|_{E_T} \quad \left(0 \le \alpha \cdot m + |\beta| \le 2m \right), \tag{3.9}$$

$$\left\| D_t^{\alpha} D^{\beta} u(t,x) \right\|_{L_q(\Omega)} \le C_q \cdot \|u\|_{E_t} \le C_q \cdot \|u\|_{E_T} \left(\alpha \le 2;\, 0 \le \alpha \cdot m + |\beta| \le m\left(\left[\frac{n}{2m} \right] + 3 \right) \right), \tag{3.10}$$

where $C > 0,\ C_q > 0$ are some constants not depending on u and t, and

$$1 \le q \le \frac{2n}{n - 2\left\{ \left(\left[\frac{n}{2m} \right] + 3 \right) m - \alpha \cdot m - |\beta| \right\}},\ q < +\infty \cdot \tag{3.11}$$

Due to estimates (3.9) and condition 2_a of this theorem, $\forall u \in K_r$:

$$\| \partial^s F(t,x,u(t,x), Du(t,x), \dots, D^m u(t,x), u_t(t,x)) / \partial t^{\alpha} \partial x^{\beta} \partial u_1^{\gamma_1} \dots \partial u_N^{\gamma_N} \|_{C(\overline{Q}_T)} \le A_r$$

$$\left(0 \le \alpha + |\beta| + \sum_{i=1}^{N} \gamma_i = s \le (k-2)m + 1 = \left(\left[\frac{n}{2m} \right] + 1 \right) m + 1;\ \alpha \le 1;\ |\beta| \le (k-2)m \right), \tag{3.12}$$

where $A_r > 0$ is some constant depending on the radius of closed sphere K_r in space $B_{2,2,2,T}^{k,k-1,k-2} = E_T$ with a center in zero and a radius r. Next, due to estimates (3.12) and (3.9), to have an estimate for

$$\frac{\partial^{(k-2)m}}{\partial x_1^{\alpha_1} \dots \partial x_n^{\alpha_n}} \frac{\partial}{\partial t} \{ F(u(t,x)) \} = \frac{\partial^{\left(\left[\frac{n}{2m} \right] + 1 \right) m + 1}}{\partial t \partial x_1^{\alpha_1} \dots \partial x_n^{\alpha_n}} \{ F(u(t,x)) \} \tag{3.13}$$

with $u \in K_r$, it suffices to estimate in $L_2(Q_T)$ the products of the following form:

$$\prod_{i=1}^{l} D_t^{\alpha_{il}} D^{\beta_{il}} u(t,x) \ (l = \overline{1, k-2};\ 2m+1 \le \alpha_{il} \cdot m + |\beta_{il}|;$$

$$\sum_{i=1}^{l} \left(\alpha_{il} \cdot m + |\beta_{il}| \right) \le (k+l-1)m;\ \alpha_{il} \le 2$$

with $\alpha_{il} = 2$ only for one value of i). (3.14)

Using estimate (3.10) and taking into account (3.11), for products of the form (3.14) with any $u \in \mathrm{K}_r$ and $t \in [0,T]$ we have:

$$\left\| \prod_{i=1}^{l} \mathrm{D}_t^{\alpha_{il}} \mathrm{D}^{\beta_{il}} u(t,x) \right\|_{L_2(\Omega)} \leq \prod_{i=1}^{l} \left\| \mathrm{D}_t^{\alpha_{il}} \mathrm{D}^{\beta_{il}} u(t,x) \right\|_{L_{2p_i}(\Omega)} \leq C \cdot \|u\|_{E_t}^{l} \leq C \cdot \|u\|_{E_T}^{l} \leq C \cdot r^{l}, \quad (3.15)$$

where $C > 0$ is some constant and

$$1 \leq p_i \leq \frac{n}{n - 2\left\{\left(\left[\frac{n}{2m}\right] + 3\right)m - \alpha_{il} \cdot m - |\beta_{il}|\right\}}, \quad p_i < +\infty, \sum_{i=1}^{l} \frac{1}{p_i} = 1,$$

(3.16)

the possibility of last equality following from the relation below:

$$\sum_{i=1}^{l} \frac{n - 2\left\{\left(\left[\frac{n}{2m}\right] + 3\right)m - \alpha_{il} \cdot m - |\beta_{il}|\right\}}{n} = \frac{1}{n} \cdot \left\{\left\{n - 2\left(\left[\frac{n}{2m}\right] + 3\right)m\right\} \cdot l + 2\sum_{i=1}^{l} (\alpha_{il} \cdot m + |\beta_{il}|)\right\} \leq$$

$$\leq \frac{1}{n} \cdot \left\{\left\{n - 2\left(\left[\frac{n}{2m}\right] + 3\right)m\right\} \cdot l + 2(k + l - 1)m\right\} = \frac{1}{n} \cdot \left\{\left\{n - 2\left(\left[\frac{n}{2m}\right] + 3\right)m\right\} \cdot l + 2\left(\left[\frac{n}{2m}\right] + 3 + l - 1\right)m\right\} =$$

$$= \frac{1}{n} \cdot \left\{n + (l-1) \cdot \left(n - 2m \cdot \left[\frac{n}{2m}\right] - 4m\right)\right\} \leq 1.$$

(3.17)

When $l \geq 2$, the latter part of (3.17) becomes a strict inequality because $n - 2m \cdot \left[\frac{n}{2m}\right] - 4m < 0$.

Now, using estimates (3.12), (3.9) and (3.15), we obtain from (3.7) that $\forall u \in \mathrm{K}_r$:

$$\|Q(u)\|_{E_T} \leq \|Z_k\|_{E_T} + \sqrt{T} \cdot C_r, \quad (3.18)$$

where $C_r > 0$ is some number depending on r.

Next, similar to (3.7), $\forall u, \tilde{u} \in \mathrm{K}_r$ we have:

$$\|Q(u) - Q(\tilde{u})\|_{E_T} \leq C \cdot \sqrt{T} \cdot \left\| \frac{\partial}{\partial t}\{\mathrm{F}(u(t,x))\} - \frac{\partial}{\partial t}\{\mathrm{F}(\tilde{u}(t,x))\} \right\|_{W_{t,x,2}^{0,(k-2)m}(Q_T)}. \quad (3.19)$$

Due to estimates (3.9) and condition 2_c of this theorem, $\forall u, \tilde{u} \in \mathsf{K}_r$:

$$\Big\| \partial^s F(t,x,u(t,x),\mathsf{D}u(t,x),...,\mathsf{D}^m u(t,x),u_t(t,x))/\partial t^\alpha \partial x^\beta \partial u_1^{\gamma_1}...\partial u_N^{\gamma_N} - \partial^s F(t,x,\tilde{u}(t,x),$$

$$\mathsf{D}\tilde{u}(t,x),...,\mathsf{D}^m\tilde{u}(t,x),\tilde{u}_t(t,x))/\partial t^\alpha \partial x^\beta \partial u_1^{\gamma_1}...\partial u_N^{\gamma_N} \Big\|_{C(\overline{Q}_T)} \le \tilde{\mathsf{A}}_r \cdot \|u - \tilde{u}\|_{E_T}$$

$$\left(0 \le \alpha + |\beta| + \sum_{i=1}^{N} \gamma_i = s \le (k-2)m + 1 = \left(\left[\frac{n}{2m} \right] + 1 \right) m + 1;\ \alpha \le 1;\ |\beta| \le (k-2)m \right), \tag{3.20}$$

where $\tilde{\mathsf{A}}_r > 0$ is some constant depending on r.

Next, using estimate (3.10) and taking into account (3.11), similar to (3.15) we have for every $t \in [0,T]$:

$$\left\| \prod_{i=1}^{l} \mathsf{D}_t^{\alpha_{il}} \mathsf{D}^{\beta_{il}} u_i(t,x) \right\|_{L_2(\Omega)} \le \prod_{i=1}^{l} \left\| \mathsf{D}_t^{\alpha_{il}} \mathsf{D}^{\beta_{il}} u_i(t,x) \right\|_{L_{2p_i}(\Omega)} \le$$

$$\le C \cdot \prod_{i=1}^{l} \|u_i\|_{E_t} \le C \cdot \prod_{i=1}^{l} \|u_i\|_{E_T} \le C \cdot r^{l-1} \cdot \|u - \tilde{u}\|_{E_T}, \tag{3.21}$$

where the conditions (3.16) are satisfied, one of functions u_i $(i = \overline{1,l})$ is equal to $u - \tilde{u}$ while the other u_i's are equal to u and $\tilde{u}$, with $u, \tilde{u} \in \mathsf{K}_r$.

Now, using estimates (3.20), (3.9) and (3.21) (in (3.9) u must be replaced by $u - \tilde{u}$), we obtain from (3.19) that $\forall u \in \mathsf{K}_r$:

$$\|Q(u) - Q(\tilde{u})\|_{E_T} \le \sqrt{T} \cdot \tilde{C}_r \cdot \|u - \tilde{u}\|_{E_T}, \tag{3.22}$$

where $\tilde{C}_r > 0$ is some number depending on r.

It can be seen from the inequalities (3.18) and (3.22) that for sufficiently small values of T the operator Q is a contraction in sphere K_r, and, consequently, has a unique fixed point $u(t,x)$ in K_r. Then it is evident that

$$u(t,x) = Q(u(t,x)) = Z_k(t,x) + \mathsf{P}_k\left(\mathsf{A}^l \left(\frac{\partial}{\partial t} \{\mathsf{F}(u(t,x))\} \right) \right) =$$

$$= Z_2(t,x) + \mathrm{P}_2\left(\frac{\partial}{\partial t}\{\mathrm{F}\,(u(t,x))\}\right). \qquad (3.23)$$

Thus, Fourier coefficients $u_s(t)$ of the found function

$$u(t,x) \in \mathrm{K}_r \subset E_T = B^{k,k-1,k-2}_{2,2,2,T} = B^{\left[\frac{n}{2m}\right]+3,\left[\frac{n}{2m}\right]+2,\left[\frac{n}{2m}\right]+1}_{2,\quad 2,\quad 2,\quad T} \qquad (3.24)$$

satisfy not only systems (1.19) and (1.20) for even and odd values of k, respectively, but also system (1.17), and hence, system (1.13).

And now let's show that the function $u(t,x)$ is a classical solution of problem (0.1)-(0.3). As

$$\sum_{s=1}^{\infty} u_s(t)v_s(x) = u(t,x) \in B^{k,k-1,k-2}_{2,2,2,T} = B^{\left[\frac{n}{2m}\right]+3,\left[\frac{n}{2m}\right]+2,\left[\frac{n}{2m}\right]+1}_{2,\quad 2,\quad 2,\quad T},$$

then it is evident that the functions $u_{pq}(t,x) = \sum_{s=p}^{q} u_s(t)v_s(x)$ $(1 \le p \le q)$ satisfy conditions of Lemma 1. Then $\forall t \in [0,T]$ we have:

$$\|u_{pq}(t,x)\|^2_{W_2^{km}(\Omega)} \le \begin{cases} C \cdot \left\| \mathrm{A}^{\frac{k}{2}}(u_{pq}(t,x)) \right\|^2_{L_2(\Omega)}, & \text{if } k \text{ is even,} \\ C \cdot a\left(\mathrm{A}^{\frac{k-1}{2}}(u_{pq}(t,x))\right), & \text{if } k \text{ is odd,} \end{cases}$$

where $C > 0$ is some constant.

From here, using Lemma 2 we obtain $\forall t \in [0,T]$:

$$\|u_{pq}(t,x)\|^2_{W_2^{km}(\Omega)} \le C \cdot \sum_{s=p}^{q} \lambda_s^{2k} \cdot u_s^2(t) \le C \cdot \sum_{s=p}^{q} \left(\lambda_s^k \cdot \max_{0 \le t \le T} |u_s(t)|\right)^2.$$

From convergence of series

$$\sum_{s=1}^{\infty}\left(\lambda_s^k \cdot \max_{0\le t\le T}|u_s(t)|\right)^2 = \sum_{s=1}^{\infty}\left(\lambda_s^{\left[\frac{n}{2m}\right]+3} \cdot \max_{0\le t\le T}|u_s(t)|\right)^2$$

it follows that $\|u_{pq}(t,x)\|_{W_2^{\left(\left[\frac{n}{2m}\right]+3\right)m}(\Omega)} \to 0$ uniformly with regard to $t\in[0,T]$ as $p,q\to\infty$.

Similarly to the foregoing, it is easy to show that

$$\left\|\frac{\partial u_{pq}(t,x)}{\partial t}\right\|_{W_2^{\left(\left[\frac{n}{2m}\right]+2\right)m}(\Omega)} \to 0, \quad \left\|\frac{\partial^2 u_{pq}(t,x)}{\partial t^2}\right\|_{W_2^{\left(\left[\frac{n}{2m}\right]+1\right)m}(\Omega)} \to 0$$

uniformly with regard to $t\in[0,T]$ as $p,q\to\infty$, because numerical series

$$\sum_{s=1}^{\infty}\left(\lambda_s^{\left[\frac{n}{2m}\right]+2} \cdot \max_{0\le t\le T}|\dot{u}_s(t)|\right)^2 \text{ и } \sum_{s=1}^{\infty}\left(\lambda_s^{\left[\frac{n}{2m}\right]+1} \cdot \max_{0\le t\le T}|u_s''(t)|\right)^2$$

are convergent.

So we got that the series

$$\sum_{s=1}^{\infty} u_s(t)v_s(x),\ \sum_{s=1}^{\infty} u_s'(t)v_s(x),\ \sum_{s=1}^{\infty} u_s''(t)v_s(x)$$

and the ones obtained from them by differentiating them with regard to $x_1,\dots,x_n$ up to $\left(\left[\frac{n}{2m}\right]+3\right)m$, $\left(\left[\frac{n}{2m}\right]+2\right)m$ and $\left(\left[\frac{n}{2m}\right]+1\right)m$ times, respectively, converge in $L_2(\Omega)$ uniformly with regard to $t\in[0,T]$. Consequently, function $u(t,x)$ has quadratically summable in Ω (with $t\in[0,T]$) generalized derivatives of the following form:

$$\frac{\partial^{|\alpha|}u(t,x)}{\partial x_1^{\alpha_1}\dots\partial x_n^{\alpha_n}}\ \left(0\le|\alpha|\le\left(\left[\frac{n}{2m}\right]+3\right)m\right),\quad \frac{\partial^{1+|\alpha|}u(t,x)}{\partial t\partial x_1^{\alpha_1}\dots\partial x_n^{\alpha_n}}\ \left(0\le|\alpha|\le\left(\left[\frac{n}{2m}\right]+2\right)m\right),$$

$$\frac{\partial^{2+|\alpha|}u(t,x)}{\partial t^2\partial x_1^{\alpha_1}\dots\partial x_n^{\alpha_n}}\ \left(0\le|\alpha|\le\left(\left[\frac{n}{2m}\right]+1\right)m\right).$$

Then, due to S.L.Sobolev's imbedding theorems, we obtain that each of functions

$$\frac{\partial^{|\alpha|}u(t,x)}{\partial x_1^{\alpha_1}...\partial x_n^{\alpha_n}}\ (0\le|\alpha|\le 2m),\ \frac{\partial^{1+|\alpha|}u(t,x)}{\partial t\partial x_1^{\alpha_1}...\partial x_n^{\alpha_n}}\ (0\le|\alpha|\le m),\ u_{tt}(t,x)$$

is continuous in closed domain $\overline{Q}_T$,

$$\mathsf{F}(u(t,x))=\sum_{s=1}^{\infty}\left(\int_{\Omega}\mathsf{F}(u(t,x))v_s(x)dx\right)v_s(x),$$

$$u_{tt}(t,x)=\mathsf{F}(u(t,x))-\mathsf{A}(u(t,x)).$$

Obviously, function $u(t,x)$ satisfies boundary conditions (0.3) in the usual sense. It is easy to verify that the function $u(t,x)$ satisfies both of initial conditions (0.2) in the usual sense, too. Moreover,

$$\|u(t,x)-\varphi(x)\|_{W_2^{\left(\left[\frac{n}{2m}\right]+3\right)m}(\Omega)}\to 0,$$

$$\|u_t(t,x)-\psi(x)\|_{W_2^{\left(\left[\frac{n}{2m}\right]+2\right)m}(\Omega)}\to 0 \text{ as } t\to +0.$$

Thus, function $u(t,x)$ is a classical solution of problem (0.1)-(0.3). As to the uniqueness (in large) of the classical solution of problem (0.1)-(0.3), it follows from Theorem 2. Theorem is proved.

Theorem 4. Let

1. condition 1 of Theorem 3 be satisfied for $k=\left[\frac{n}{2m}\right]+3$.

2. $F(t,x,u,\mathsf{D}u,...,\mathsf{D}^m u,u_t)=F(t,x,u)$, and:

a) $\partial^s F(t,x,u)/\partial t^\alpha\partial x^\beta\partial u^\gamma\in C(\overline{Q}_T\times(-\infty,\infty))$

$$\left(0\le\alpha+|\beta|+\gamma=s\le(k-2)m+1=\left(\left[\frac{n}{2m}\right]+1\right)m+1;\ \alpha\le 1;\ |\beta|\le(k-2)m\right);$$

b) function $F(t,x,u)$ be equal to zero in all points $(t,x,0)$ where $t \in [0,T]$ и $x \in S$, together with all its derivatives of the form mentioned in item a) up to the order $(k-4)m+1+k_m = \left(\left[\frac{n}{2m}\right]-1\right)m+1+k_m$ for odd values of $\left[\frac{n}{2m}\right]$ and up to the order $(k-3)m+1+k_r = \left[\frac{n}{2m}\right]\cdot m+1+k_r$ for even values of $\left[\frac{n}{2m}\right]$.

Then there exists in small a unique in large classical solution of problem (0.1)-(0.3).

Proof. Let k be a number appearing in the statement of this theorem, i.e. $k = \left[\frac{n}{2m}\right]+3$. We consider in space $B_{2,2,2,T}^{k,k-1,k-2}$ the operator Q, defined by relation (3.2) where $\mathsf{F}\,(u(t,x)) = F(t,x,u(t,x))$. It is evident that $\forall u \in B_{2,2,2,T}^{k,k-1,k-2}$ the equality (3.3) holds.

Let $\{u_j(t,x)\}_{j=1}^{\infty}$ be the arbitrary succession which is bounded in $B_{2,2,2,T}^{k,k-1,k-2}$ and converges in $B_{2,2,T}^{k-1,k-2}$ to the function $u_0(t,x)$. Then, using Sobolev's imbedding theorems and one well-known fact from the theory of real-valued functions (see, e.g., [88], p.80, lemma 3.2), it is easy to obtain that

$$\left\|\mathsf{D}_t^{\alpha}\mathsf{D}^{\beta}u_j(t,x)-\mathsf{D}_t^{\alpha}\mathsf{D}^{\beta}u_0(t,x)\right\|_{C(\overline{Q}_T)} \to 0 \text{ as } j \to \infty$$

$$\left(\alpha=0,1;\ 0 \le \alpha\cdot m+|\beta| \le i_0{}^{*)} \equiv (k-1)m-\left[\frac{n}{2}\right]-1=\left(\left[\frac{n}{2m}\right]+2\right)m-\left[\frac{n}{2}\right]-1\right), \tag{3.25}$$

$$\left\|\mathsf{D}_t^{\alpha}\mathsf{D}^{\beta}u_j(t,x)-\mathsf{D}_t^{\alpha}\mathsf{D}^{\beta}u_0(t,x)\right\|_{L_q(Q_T)} \to 0 \text{ as } j \to \infty$$

$$\left(\alpha=0,1;\ i_0+1 \le \alpha\cdot m+|\beta| \le i_{00} \equiv \min\{i_0+m,(k-1)m\};\ 1 \le q < +\infty\right), \tag{3.26}$$

$$\left\|\mathsf{D}_t^{\alpha}\mathsf{D}^{\beta}u_j(t,x)-\mathsf{D}_t^{\alpha}\mathsf{D}^{\beta}u_0(t,x)\right\|_{L_p(Q_T)} \to 0 \text{ as } j \to \infty$$

---

*) It is evident that $i_0 \ge m$.

$$\left(\alpha = 0,1;\ i_{00} + 1 \le \alpha \cdot m + |\beta| \le (k-1)m;\ 1 \le p < \frac{2n}{n - 2(km - \alpha \cdot m - |\beta|)}\right)^{**)},$$
(3.27)

$$\left\| D_t^{\alpha} D^{\beta} u_j(t,x) \right\|_{C(\overline{Q}_T)} \le C_0 \left(j = 0,1,2,...;\ \alpha = 0,1;\ 0 \le \alpha \cdot m + |\beta| \le i_0 + m \ge 2m\right),$$
(3.28)

$$\left\| D_t^{\alpha} D^{\beta} u_j(t,x) \right\|_{L_p(Q_T)} \le C_p \left(j = 0,1,2,...;\ \alpha = 0,1;\ i_0 + m + 1 \le \alpha \cdot m + |\beta| \le km;\right.$$

$$\left. 1 \le p \le \frac{2n}{n - 2(km - \alpha \cdot m - |\beta|)},\ p < +\infty\right). \tag{3.29}$$

Now let's establish one inequality needful to us. Let $V_1(t,x),..., V_l(t,x)$, $V(t,x) \in B^{k,k-1,k-2}_{2,2,2,T}$. We consider the following integrals:

$$\int_0^T \int_\Omega \prod_{i=1}^{l} \left\{D^{\alpha_i} V_i(t,x)\right\}^2 \cdot \left\{D^{\alpha_0} V_t(t,x)\right\}^2 dxdt \left(\sum_{i=1}^{l} |\alpha_i| + |\alpha_0| = (k-2)m\right), \tag{3.30}$$

where $l \ge 0$. For $l = 0$ we assume that $\prod_{i=1}^{l} ...$ is 1.

Then it is evident that

$$\left\| \prod_{i=1}^{l} D^{\alpha_i} V_i(t,x) \cdot D^{\alpha_0} V_t(t,x) \right\|^2_{L_2(Q_T)} \le \prod_{i=1}^{l} \left\| D^{\alpha_i} V_i(t,x) \right\|^2_{L_{2p_i}(Q_T)} \cdot \left\| D^{\alpha_0} V_t(t,x) \right\|^2_{L_{2p_0}(Q_T)}, \tag{3.31}$$

where

$$1 \le p_i < \frac{2n}{n - 2(km - |\alpha_i|)}\ (i = \overline{1,l}),\ 1 \le p_0 < \frac{n}{n - 2((k-1)m - |\alpha_0|)},\ \sum_{i=0}^{l} \frac{1}{p_i} = 1,$$
(3.32)

**) For $i_0 + m \ge (k-1)m$ the relations (3.27) are not available.

the possibility of equality $\sum_{i=0}^{l}\frac{1}{p_i}=1$ following from the relation below:

$$\frac{n-2((k-1)m-|\alpha_0|)}{n}+\sum_{i=1}^{l}\frac{n-2(km-|\alpha_i|)}{n}=\frac{n+(n-2km)l-2m}{n}<1,$$

because $(n-2km)\cdot l=\left\{n-2m\cdot\left(\left[\frac{n}{2m}\right]+3\right)\right\}\cdot l\le 0$, and, consequently, $n+(n-2km)\cdot l-2m<n$.

Using the definition of operator Q, i.e. relations (1.21)-(1.24) and (3.3), it is easy to obtain similarly to (3.19) that

$$\left\|Q(u_j)-Q(u_0)\right\|_{B_{2,2,2,T}^{k,k-1,k-2}}\le C\cdot\left\|\frac{\partial}{\partial t}\{F(t,x,u_j(t,x))\}-\frac{\partial}{\partial t}\{F(t,x,u_0(t,x))\}\right\|_{W_{t,x,2}^{0,(k-2)m}(Q_T)}. \tag{3.33}$$

Using relations (3.25)-(3.29) and estimate (3.31), where one of functions $V_1,\ldots,V_l,V$ is equal to u_j-u_0 while the others are equal either to u_j or to u_0, we obtain from (3.33) that

$$\left\|Q(u_j)-Q(u_0)\right\|_{B_{2,2,2,T}^{k,k-1,k-2}}\to 0 \text{ as } j\to\infty. \tag{3.34}$$

Thus, operator Q is mapping every succession, bounded in $B_{2,2,2,T}^{k,k-1,k-2}$ and convergent in $B_{2,2,T}^{k-1,k-2}$, into the succession convergent in $B_{2,2,2,T}^{k,k-1,k-2}$. And as, due to Theorem 1, the inclusion $B_{2,2,2,T}^{k,k-1,k-2}\subset B_{2,2,T}^{k-1,k-2}$ is completely continuous, operator Q is acting in $B_{2,2,2,T}^{k,k-1,k-2}$ completely continuously. On the other hand, similarly to (3.7) and (3.18), $\forall u\in B_{2,2,2,T}^{k,k-1,k-2}$ we have:

$$\left\|Q(u)\right\|_{B_{2,2,2,T}^{k,k-1,k-2}}\le\left\|Z_k\right\|_{B_{2,2,2,T}^{k,k-1,k-2}}+\sqrt{T}\cdot f\left(\left\|u\right\|_{B_{2,2,2,T}^{k,k-1,k-2}}\right), \tag{3.35}$$

where function $f(y)$ is bounded in every bounded part of $[0,+\infty)$. Thus, if K_r is any fixed closed sphere in space $B_{2,2,2,T}^{k,k-1,k-2}$ with a center in zero and a radius

$r > \|Z_k(t,x)\|_{B^{k,k-1,k-2}_{2,2,2,T}}$, then for sufficiently small values of T operator Q is mapping sphere K_r into itself, doing it completely continuously. Consequently, due to Schauder's fixed point principle, for sufficiently small values of T operator Q has a fixed point $u(t,x)$ in $K_r \subset B^{k,k-1,k-2}_{2,2,2,T}$; it is easy to verify (in absolutely the same way as in the Proof of Theorem 3) that this point is a classical solution of problem (0.1)-(0.3). As to the uniqueness (in large) of the classical solution of problem (0.1)-(0.3), it follows from Theorem 2. Theorem is proved.

Remark 5. Note that the Theorems 3 and 4 are independent of each other. Because, despite Theorem 3 covering general case and Theorem 4 covering simplest special case of equation (0.1), the statement of Theorem 3 includes condition 2_c which is absent in the statement of Theorem 4. Besides, Theorems 3 and 4 are proved by different methods. Namely, Theorem 3 is proved using contracted mappings principle, while Theorem 4 is proved with the help of Schauder's fixed point principle.

4. First A Priori Estimate for Classical Solutions of Problem (0.1)-(0.3)

In this paragraph, we establish a priori estimate in

$$C([0,T];W_2^m(\Omega)) \cap C^1([0,T];L_2(\Omega))$$

for all the possible classical solutions of problem (0.1)-(0.3) belonging to $B^{2,1}_{2,2,T}$.

Theorem 5. Let

1. $\varphi(x) \in W_2^{2m}(\Omega,\{B_j\}_{j=1}^m)$, $\psi(x) \in W_2^m(\Omega,\{B_j\}_{j=1}^r)$.

2. $F(t,x,u,Du,...,D^m u,u_t) = f(t,x,u) + \Phi(t,x,u) + H(t,x,u,Du,...,D^m u,u_t)$ (4.1)

with:

a) $f(t,x,u) \in C(\overline{Q}_T \times (-\infty,\infty))$, $\int_0^u f(t,x,\xi)d\xi \equiv g(t,x,u) \le a_0(x) + b_0 \cdot u^2 - g_0(u)$,

where $0 \le a_0(x) \in L(\Omega)$, $b_0 \ge 0$ is a constant, $0 \le g_0(u) \in C(-\infty,\infty)$;

$g_0(u) \le C \cdot e^{u^2}$ for $n = 4m$ and $g_0(u) \le C \cdot \left(1 + |u|^{\frac{2n}{n-4m}}\right)$ for $n > 4m$;

$f_t(t,x,u) \in C(\overline{Q}_T \times (-\infty,\infty))$, $|f_t(t,x,u)| \le a(t,x) + b(t)\cdot|u|$, where $C > 0$ is a constant, $a(t,x) \in L_2(Q_T)$, $b(t) \in L_2(0,T)$;

b) $\Phi(t,x,u) \in C(\overline{Q}_T \times (-\infty,\infty))$, $|\Phi(t,x,u)| \le a_1(t,x) + b_1(t)\cdot\left(|u| + \sqrt{g_0(u)}\right)$,

where $a_1(t,x) \in L_2(Q_T)$, $b_1(t) \in L_2(0,T)$;

c) $H(t,x,u_1,...,u_N) \in C(\overline{Q}_T \times (-\infty,\infty)^N)$ $(N = 2 + n + n^2 + ... + n^m)$,

$$H(t,x,u,\mathrm{D}u,...,\mathrm{D}^m u,u_t)\cdot u_t \le a_2(t,x) + b_2(t)\cdot\left\{u^2 + |\mathrm{D}u|^2 + ... + |\mathrm{D}^m u|^2 + u_t^2 + g_0(u)\right\} \tag{4.2}$$

where $a_2(t,x) \in L(Q_T)$, $b_2(t) \in L(0,T)$, $|\mathrm{D}u| = \sum_{i=1}^{n} |u_{x_i}|, ..., |\mathrm{D}^m u| = \sum_{|\alpha|=m} |\mathrm{D}^\alpha u|$.

Then for every classical solution $u(t,x) \in B_{2,2,T}^{2,1}$ of problem (0.1)-(0.3) the following a priori estimates hold:

$$\|u_t(t,x)\|_{L_2(\Omega)} \le C_0 \quad \forall t \in [0,T], \tag{4.3}$$

$$\|u(t,x)\|_{W_2^m(\Omega)} \le C_0 \quad \forall t \in [0,T], \tag{4.4}$$

Proof. Let $u(t,x)$ be any classical solution of problem (0.1)-(0.3) belonging to the space $B_{2,2,T}^{2,1}$. Then it is evident that

$$\left.\begin{aligned} &u(t,x) \in C([0,T]);\ W_2^{2m}(\Omega,\{B_j\}_{j=1}^m)),\\ &u_t(t,x) \in C([0,T]);\ W_2^m(\Omega,\{B_j\}_{j=1}^r)). \end{aligned}\right\} \tag{4.5}$$

As agreed above in section 2, §1, throughout this work we assume that not only conditions mentioned in the statement of problem (0.1)-(0.3), but also the assumptions of section 2, §1 are fulfilled. Then, using (4.5) we obtain that $\forall t \in [0,T]$:

$$(\mathrm{A}\,(u(t,x)),\, u_t(t,x))_{L_2(\Omega)} = a(u(t,x),u_t(t,x)) = \frac{1}{2}\cdot\frac{d}{dt}a(u(t,x)), \tag{4.6}$$

where $a(u,V)$ is an Hermitian integro-differential bilinear form defined by (1.1), and $a(u,u) \equiv a(u)$ is the corresponding quadratic form.

Now, we multiply both sides of equation (0.1) by function $2u_t(t,x)$ and integrate the obtained equality over the cylinder $Q_t = (0,t) \times \Omega$, where $t \in [0,T]$:

$$\int_0^t \int_\Omega 2u_{\tau\tau}(\tau,x) \cdot u_\tau(\tau,x)dxd\tau + 2\int_0^t \int_\Omega \mathsf{A}(u(\tau,x)) \cdot u_\tau(\tau,x)dxd\tau =$$

$$= 2\int_0^t \int_\Omega F(\tau,x,u(\tau,x),\mathsf{D}u(\tau,x),...,\mathsf{D}^m u(\tau,x),u_\tau(\tau,x)) \cdot u_\tau(\tau,x)dxd\tau \cdot \tag{4.7}$$

It is evident that $\forall t \in [0,T]$:

$$2\int_0^t \int_\Omega u_{\tau\tau}(\tau,x) \cdot u_\tau(\tau,x)dxd\tau = \int_0^t \left\{ \frac{d}{d\tau} \int_\Omega u_\tau^2(\tau,x)dx \right\} d\tau =$$

$$= \int_\Omega u_t^2(t,x)dx - \int_\Omega \psi^2(x)dx \text{,} \tag{4.8}$$

$$2\int_0^t \int_\Omega \mathsf{A}(u(\tau,x)) \cdot u_\tau(\tau,x)dxd\tau = \int_0^t \left\{ \frac{d}{d\tau} a(u(\tau,x)) \right\} d\tau =$$

$$= a(u(t,x)) - a(\varphi(x)) \ge C_0 \cdot \|u(t,x)\|^2_{W_2^m(\Omega)} - a(\varphi(x)),^{*)} \tag{4.9}$$

$$\int_0^t \int_\Omega f(\tau,x,u(\tau,x)) \cdot u_\tau(\tau,x)dxd\tau = \int_0^t \left\{ \frac{d}{d\tau} \int_\Omega g(\tau,x,u(\tau,x))dx \right\} d\tau -$$

$$- \int_0^t \int_\Omega \int_o^{u(\tau,x)} f_\tau(\tau,x,\xi)d\xi dxd\tau = \int_\Omega g(t,x,u(t,x)dx - \int_\Omega g(0,x,\varphi(x))dx -$$

$$- \int_0^t \int_\Omega \int_o^{u(\tau,x)} f_\tau(\tau,x,\xi)d\xi dxd\tau \le \int_\Omega a_0(x)dx + b_0 \int_\Omega u^2(\tau,x)dx - \int_\Omega g_0(u(t,x))dx +$$

---

*) Here we made use of coerciveness condition (1.3), where we agreed to assume $C_1 = 0$.

$$+\int_{\Omega}|g(0,x,\varphi(x))|dx+\int_{0}^{t}\int_{\Omega}a_{\tau}(\tau,x)\cdot|u(\tau,x)|dxd\tau+\frac{1}{2}\int_{0}^{t}b(\tau)\int_{\Omega}u^{2}(\tau,x)dxd\tau\leq$$

$$\leq\int_{\Omega}a_{0}(x)dx+\int_{\Omega}|g(0,x,\varphi(x))|dx+\frac{1}{2}\int_{0}^{T}\int_{\Omega}a^{2}(\tau,x)dxd\tau+\frac{1}{2}\int_{0}^{t}\int_{\Omega}u^{2}(\tau,x)dxd\tau+$$

$$+\frac{1}{2}\int_{0}^{t}b(\tau)\int_{\Omega}u^{2}(\tau,x)dxd\tau+b_{0}\int_{\Omega}u^{2}(\tau,x)dx-\int_{\Omega}g_{0}(u(t,x))dx. \qquad (4.10)$$

On the other hand, it is evident that $\forall t\in[0,T]$ and $x\in\Omega$:

$$u(t,x)=u(0,x)+\int_{0}^{t}u_{\tau}(\tau,x)d\tau=\varphi(x)+\int_{0}^{t}u_{\tau}(\tau,x)d\tau,$$

and consequently,

$$\int_{\Omega}u^{2}(t,x)dx\leq 2\int_{\Omega}\varphi^{2}(x)dx+2T\cdot\int_{0}^{t}\int_{\Omega}u_{\tau}^{2}(\tau,x)dxd\tau. \qquad (4.11)$$

Using estimate (4.11), we obtain from (4.10) that $\forall t\in[0,T]$:

$$\int_{0}^{t}\int_{\Omega}f(\tau,x,u(\tau,x))\cdot u_{\tau}(\tau,x)dxd\tau\leq C_{1}+\frac{1}{2}\int_{0}^{t}[1+b(\tau)]\int_{\Omega}u^{2}(\tau,x)dxd\tau+$$

$$+2T\cdot b_{0}\cdot\int_{0}^{t}\int_{\Omega}u_{\tau}^{2}(\tau,x))dxd\tau-\int_{\Omega}g_{0}(u(t,x))dx\leq C_{1}+\int_{0}^{t}\left\{2T\cdot b_{0}+\frac{1}{2}[1+b(\tau)]\right\}\cdot$$

$$\left\{\int_{\Omega}u_{\tau}^{2}(\tau,x))dx+\int_{\Omega}u^{2}(\tau,x))dx\right\}d\tau-\int_{\Omega}g_{0}(u(t,x))dx, \qquad (4.12)$$

where

$$C_{1}\equiv\int_{\Omega}a_{0}(x)dx+\int_{\Omega}|g(0,x,\varphi(x))|dx+\frac{1}{2}\int_{0}^{T}\int_{\Omega}a^{2}(\tau,x)dxd\tau+2b_{0}\cdot\int_{\Omega}\varphi^{2}(x)dx\cdot$$

Next, $\forall t \in [0,T]$ we have:

$$\int_0^t \int_\Omega \Phi(\tau,x,u(\tau,x)) \cdot u_\tau(\tau,x)dxd\tau \le \frac{1}{2}\int_0^t \int_\Omega u_\tau^2(\tau,x)dxd\tau +$$

$$+\frac{1}{2}\int_0^t \int_\Omega \Phi^2(\tau,x,u(\tau,x))dxd\tau \le \frac{1}{2}\int_0^t \int_\Omega u_\tau^2(\tau,x)dxd\tau +$$

$$+\frac{3}{2}\left\{\int_0^t \int_\Omega a_1^2(\tau,x)dxd\tau + \int_0^t b_1^2(\tau)\int_\Omega u^2(\tau,x))dxd\tau + \int_0^t b_1^2(\tau)\cdot\right.$$

$$\left.\cdot\int_\Omega g_0(u(\tau,x))dxd\tau\right\} \le \frac{3}{2}\int_0^T \int_\Omega a_1^2(\tau,x)dxd\tau + \int_0^t \left[\frac{1}{2}+\frac{3}{2}b_1^2(\tau)\right]\cdot$$

$$\cdot\left\{\int_\Omega u_\tau^2(\tau,x)dx + \int_\Omega u^2(\tau,x))dx + \int_\Omega g_0(u(\tau,x))dx\right\}d\tau, \tag{4.13}$$

$$\int_0^t \int_\Omega H(\tau,x,u(\tau,x),\mathrm{D}u(\tau,x),...,\mathrm{D}^m u(\tau,x),u_\tau(\tau,x)) \cdot u_\tau(\tau,x)dxd\tau \le$$

$$\le \int_0^T \int_\Omega a_2(\tau,x)dxd\tau + n^m \cdot \int_0^t b_2(\tau)\left\{\int_\Omega u_\tau^2(\tau,x)dx +\right.$$

$$\left.+\|u(\tau,x)\|^2_{W_2^m(\Omega)} + \int_\Omega g_0(u(\tau,x))dx\right\}d\tau\cdot \tag{4.14}$$

Now, using relations (4.8), (4.9), estimates (4.12)-(4.14) and representation (4.1), we obtain from (4.7) that $\forall t \in [0,T]$:

$$\int_\Omega u_t^2(t,x)dx + c_0 \cdot \|u(t,x)\|^2_{W_2^m(\Omega)} + 2\int_\Omega g_0(u(t,x))dx \le$$

$$\le \int_\Omega \psi^2(x)dx + a(\varphi(x)) + 2c_1 + 3\int_0^T\int_\Omega a_1^2(\tau,x)dxd\tau + 2\int_0^T\int_\Omega a_2(\tau,x)dxd\tau +$$

$$+\int_0^t \{4T\cdot b_0 + [1+b(\tau)] + [1+3b_1^2(\tau)] + 2n^m \cdot b_2(\tau)\}\cdot$$

$$\cdot\left\{\int_\Omega u_\tau^2(\tau,x))dx + \|u(\tau,x)\|^2_{W_2^m(\Omega)} + \int_\Omega g_0(u(\tau,x))dx\right\}d\tau =$$

$$= \tilde{c} + \int_0^t \tilde{B}(\tau)\left\{\int_\Omega u_\tau^2(\tau,x)dx + \|u(\tau,x)\|^2_{W_2^m(\Omega)} + \int_\Omega g_0(u(\tau,x))dx\right\}d\tau,$$

where $\tilde{c} > 0$ is a constant and $0 \le \tilde{B}(\tau) \in L(0,T)$.

Thus, $\forall t \in [0,T]$:

$$\int_\Omega u_t^2(t,x))dx + \|u(t,x)\|^2_{W_2^m(\Omega)} + \int_\Omega g_0(u(t,x))dx \le$$

$$\le \tilde{\tilde{c}} + \int_0^t \tilde{\tilde{B}}(\tau)\left\{\int_\Omega u_\tau^2(\tau,x)dx + \|u(\tau,x)\|^2_{W_2^m(\Omega)} + \int_\Omega g_0(u(\tau,x))dx\right\}d\tau, \tag{4.15}$$

where $\tilde{\tilde{c}} = \tilde{c}\cdot(\min\{1,c_0\})^{-1}, \tilde{\tilde{B}}(\tau) = (\min\{1,c_0\})^{-1}\cdot\tilde{B}(\tau)$.

On applying Bellman's inequality, we obtain from (4.15) that $\forall t \in [0,T]$:

$$\int_\Omega u_t^2(t,x))dx + \|u(t,x)\|^2_{W_2^m(\Omega)} + \int_\Omega g_0(u(t,x))dx \le C^*, \tag{4.16}$$

where $C^* > 0$ is a constant not depending on u and t.

From (4.16), due to the non-negativity of function $g_0(u)$, it follows that the a priori estimates (4.3) and (4.4) hold. Theorem is proved.

Remark 6. As seen from the conditions 2_a, 2_b and 2_c of the last theorem, to join functions $f(t,x,u)$ and $\Phi(t,x,u)$ with function $H(t,x,u,Du,...,D^m u,u_t)$ is, generally speaking, inadvisable. It is also inadvisable to join function $\Phi(t,x,u)$ with function $f(t,x,u)$; that is, the condition 2_b of this theorem, generally speaking, does not imply the condition 2_a of the same theorem. To make sure about it, it suffices to consider the following example:

$$n \le 4m,\ \ f(t,x,u) = -g_0'(u), \text{ where } g_0(u) = u^8 \cdot \sin^2 u,$$

i.e.

$$f(t,x,u) = -u^8 \sin 2u - 8u^7 \sin^2 u; \quad \Phi(t,x,u) = u^4 \sin u.$$

Then it is evident that the conditions 2_a (with $a_0(x) \equiv 0,\ b_0 = 0$) and 2_b (with $a_1(t,x) \equiv 0,\ b_1(t) \equiv 1$) of Theorem 5 are fulfilled. But, as

$$\int_0^u \{f(t,x,\xi) + \Phi(t,x,\xi)\} d\xi = -u^8 \sin^2 u - u^4 \cos u +$$

$$+ 4u^3 \sin u + 12u^2 \cos u - 24u \sin u - 24 \cos u + 24,$$

the condition of the form

$$\forall u \in (-\infty,\infty)\ \int_0^u \{f(t,x,\xi) + \Phi(t,x,\xi)\} d\xi \le a_0(x) + b_0 \cdot u^2 - g_0(u)$$

does not hold. Because, for example, for $u = (2k-1)\pi \ \ (k = 1,2,3,\ldots)$ it becomes

$$(2k-1)^4 \pi^4 - 12(2k-1)^2 \pi^2 + 48 \le a_0(x) + b_0 \cdot (2k-1)^2 \pi^2.$$

Next, according to imbedding theorems of S.L.Sobolev and S.I.Pokhozhayev, the proved a priori estimates (4.3) and (4.4) imply the following a priori estimates:

$$\left\| \mathrm{D}^\alpha u(t,x) \right\|_{C(Q_T)} \le C_0 \text{ for } n < 2(m - |\alpha|); \tag{4.17}$$

$$\int_\Omega e^{\left| \mathrm{D}^\alpha u(t,x) \right|^2} dx \le C_0 \text{ for } n = 2(m - |\alpha|)\ \forall t \in [0,T]; \tag{4.18}$$

$$\left\|D^{\alpha}u(t,x)\right\|_{L_{\frac{2n}{n-2m+2|\alpha|}}(\Omega)} \le C_0 \text{ for } n > 2(m-|\alpha|)\ (0 \le |\alpha| \le m)\ \forall t \in [0,T], \tag{4.19}$$

$$\left\|u_t(t,x)\right\|_{L_2(\Omega)} \le C_0 \ \forall n \text{ and } t \in [0,T]. \tag{4.20}$$

5. Second a Priori Estimate for Classical Solutions of Problem (0.1)-(0.3)

In this paragraph, we establish a priori estimate in

$$C([0,T];\ W_2^{2m}(\Omega)) \cap C^1([0,T]; W_2^m(\Omega)) \cap C^2([0,T];\ L_2(\Omega))$$

for all the possible classical solutions of problem (0.1)-(0.3) belonging to $B^{2,1,0}_{2,2,2,T}$.

Theorem 6. Let

1. all the conditions of theorem 5 be satisfied;

2. $F(t,u_0,u_1,...,u_N),\ F_{u_i}(t,u_0,u_1,...,u_N)\ (i=\overline{0,N}) \in C([0,T]\times\overline{\Omega}\times(-\infty,\infty)^N)$;

3. **when** $n<2m$ $\forall R > 0$ for $(t,x) \in Q_T$, $\left|D^i u\right| \le R\left(0 \le i \le i_0 \equiv m - \left[\frac{n}{2}\right] - 1\right)$,

$$\left|D^i u\right| < +\infty\ (i = \overline{i_0 + 1, m}),\ |u_t| < +\infty:$$

$$\left|F_t(t,x,u,Du,...,D^m u,u_t)\right| \le C_R \cdot \left\{1 + \sum_{i=i_0+1}^{m} \left|D^i u\right|^{\gamma_i} + |u_t|^{\gamma_m}\right\} \tag{5.1}$$

if n is odd;

$$\left|F_t(t,x,u,Du,...,D^m u,u_t)\right| \le C_R \cdot \left\{e^{\left|D^{i_0+1}u\right|^2} + \sum_{i=i_0+2}^{m} \left|D^i u\right|^{\gamma_i} + |u_t|^{\gamma_m}\right\} \tag{5.2}$$

if n is even (in this case $n - 2m + 2(i_0+1) = 0$),

$$\left|F_{D^j u}(t,x,u,Du,...,D^m u,u_t)\right| \le C_R \cdot \left\{1 + \sum_{i=i_0+1}^{m} \left|D^i u\right|^{\gamma_{i,j}} + \left|u_t\right|^{\gamma_{m,j}}\right\} \qquad (5.3)$$

$$(j = \overline{0, m-1})$$

if n is odd;

$$\left|F_{D^j u}(t,x,u,Du,...,D^m u,u_t)\right| \le$$

$$\le C_R \cdot \left\{e^{\left|D^{i_0+1}u\right|^2} + \sum_{i=i_0+2}^{m} \left|D^i u\right|^{\gamma_{i,j}} + \left|u_t\right|^{\gamma_{m,j}}\right\} (j = \overline{0, m-1}), \qquad (5.4)$$

if n is even (in this case $n - 2m + 2(i_0 + 1) = 0$);

$$\left|F_{D^m u}(t,x,u,Du,...,D^m u,u_t)\right| \le C_R, \left|F_{u_t}(t,x,u,Du,...,D^m u,u_t)\right| \le C_R; \qquad (5.5)$$

and **when $n \ge 2m$** in $Q_T \times (-\infty,\infty)^N$:

$$\left|F_t(t,x,u,Du,...,D^m u,u_t)\right| \le C\left\{e^{u^2} + \sum_{i=1}^{m} \left|D^i u\right|^{\gamma_i} + \left|u_t\right|^{\gamma_m}\right\}, \qquad (5.6)$$

if **$n=2m$**;

$$\left|F_t(t,x,u,Du,...,D^m u,u_t)\right| \le C \cdot \left\{1 + \sum_{i=0}^{m} \left|D^i u\right|^{\gamma_i} + \left|u_t\right|^{\gamma_m}\right\}, \qquad (5.7)$$

if **$n>2m$**;

$$\left|F_{D^j u}(t,x,u,Du,...,D^m u,u_t)\right| \le \le C \cdot \left\{e^{u^2} + \sum_{i=1}^{m} \left|D^i u\right|^{\gamma_{i,j}} + \left|u_t\right|^{\gamma_{m,j}}\right\}, \qquad (5.8)$$

$$(j = \overline{0, m-1})$$

if **$n=2m$**;

$$\left|F_{\mathrm{D}^j u}(t,x,u,\mathrm{D}u,...,\mathrm{D}^m u,u_t)\right| \le C\cdot\left\{1+\sum_{i=0}^{m}\left|\mathrm{D}^i u\right|^{\gamma_{i,j}}+\left|u_t\right|^{\gamma_{m,j}}\right\}\ (j=\overline{0,m-1}) \quad (5.9)$$

if ***n>2m***;

$$\left|F_{\mathrm{D}^m u}(t,x,u,\mathrm{D}u,...,\mathrm{D}^m u,u_t)\right|\le C,\left|F_{u_t}(t,x,u,\mathrm{D}u,...,\mathrm{D}^m u,u_t)\right|\le C, \quad (5.10)$$

where:

for odd $n<2m,\ m-\left[\frac{n}{2}\right]\equiv i_0+1\le i\le m,$ (5.11)

for even $n<2m,\ i_0+2\le i\le m,$ (5.12)

for $n=2m,\ 1\le i\le i_{00}\equiv 2m-\left[\frac{n}{2}\right]-1,$ (5.13)

for $2m<n<4m,\ 0\le i\le i_{00}$ (5.14)

$$0<\gamma_i\le\frac{2n-2m+2i}{n-2m+2i}; \quad (5.15)$$

for $2m\le n<4m,\ i_{00}+1\le i\le m$ and $n\ge 4m,\ 0\le i\le m$ (5.16)

$$0<\gamma_i\le\frac{n+2m}{n-2m+2i}, \quad (5.17)$$

with $0<\gamma_i\le\frac{n+2m}{n-2m+2i}=\frac{n+2m}{2m}$ for $n-4m+2i=0$;

for odd $n<2m,\ i_0+1\le i\le m,\ 0\le j\le i_0,$ (5.18)

for even $n<2m,\ i_0+2\le i\le m,\ 0\le j\le i_0$ (5.19)

$$0<\gamma_{i,j}\le\frac{n}{n-2m+2i}; \quad (5.20)$$

for odd $n<2m,\ i_0+1\le i\le m,\ i_0+1\le j\le m-1,$ (5.21) for even $n < 2m,\ i_0 + 2 \le i \le m,\ i_0 + 1 \le j \le m - 1$, (5.22)

for $n = 2m,\ 1 \le i \le m,\ 0 \le j \le m - 1$, (5.23)

for $n > 2m,\ 0 \le i \le m,\ 0 \le j \le m - 1$ (5.24)

$$0 < \gamma_{i,j} \le \frac{2m - 2j}{n - 2m + 2i}, \tag{5.25}$$

with

$$0 < \gamma_{i,j} < \frac{2m - 2j}{n - 2m + 2i}$$

for $n - 2m + 2j = 0$; and $C > 0,\ C_R > 0$ are constants.

Then, the following a priori estimate is true for all the possible classical solutions $u(t,x) \in B^{2,1,0}_{2,2,2,T}$ of problem (0.1)-(0.3):

$$\|u(t,x)\|_{B^{2,1,0}_{2,2,2,T}} \le C_0. \tag{5.26}$$

Proof. Let $u(t,x) = \sum_{s=1}^{\infty} u_s(t) v_s(x)$ be any classical solution of problem (0.1)-(0.3) belonging to the space $B^{2,1,0}_{2,2,2,T}$. Then, due to the Lemma 3 proved in §1, functions $u_s(t)\ \ (s = 1,2,...)$ satisfy system (1.13), which, under the conditions of this theorem, is equivalent to system (1.17). As the conditions of this theorem include all the conditions of Theorem 5, then, according to Theorem 5, a priori estimates (4.3) and (4.4) hold, and this, in turn, implies the trueness of a priori estimates (4.17)-(4.20).

Now, using system (1.17) we obtain that $\forall t \in [0,T]$:

$$\|u\|^2_{B^{2,1,0}_{2,2,2,t}} \le 2\left\{\|Z(t,x)\|^2_{B^{2,1,0}_{2,2,2,T}} + 16T \cdot \int_0^t \int_\Omega \left[\frac{\partial}{\partial \tau} \mathsf{F}\,(u(\tau,x))\right]^2 dx d\tau\right\} =$$

$$= c_1 + c_2 \cdot \int_0^t \int_\Omega \left\{\frac{\partial}{\partial \tau} \mathsf{F}\,(u(\tau,x))\right\}^2 dx d\tau, \tag{5.27}$$

where

$$Z(t,x)=\sum_{s=1}^{\infty}\left\{\varphi_s\cdot\cos\lambda_s t+\frac{\psi_s}{\lambda_s}\cdot\sin\lambda_s t+\frac{1}{\lambda_s^2}\cdot(1-\cos\lambda_s t)\cdot\int_{\Omega}f(x)v_s(x)dx\right\}\cdot v_s(x)\cdot \tag{5.28}$$

$c_1\equiv 2\|Z\|^2_{B^{2,1,0}_{2,2,2,T}}$, $c_2\equiv 32T$ and

$$\frac{\partial}{\partial\tau}\mathsf{F}(u(\tau,x))\equiv F_\tau(\tau,x,u(\tau,x),\mathsf{D}u(\tau,x),...,\mathsf{D}^m u(\tau,x),u_\tau(\tau,x))+$$

$$+\sum_{0\le|\beta|\le m}F_{\mathsf{D}^\beta u}(\tau,x,u(\tau,x),\mathsf{D}u(\tau,x),...,\mathsf{D}^m u(\tau,x),u_\tau(\tau,x))\cdot\mathsf{D}^\beta u_\tau(\tau,x)+$$

$$+F_{u_\tau}(\tau,x,u(\tau,x),\mathsf{D}u(\tau,x),...,\mathsf{D}^m u(\tau,x),u_\tau(\tau,x))\cdot u_{\tau\tau}(\tau,x). \tag{5.29}$$

Now we introduce the following notation for the case of $n<2m$:

$$i_0\equiv\max\{i:n-2m+2i<0\}=m-\left[\frac{n}{2}\right]-1. \tag{5.30}$$

It is evident that $0\le i_0<m$, and, due to a priori estimates (4.17), the following a priori estimates hold:

$$\left\|\mathsf{D}^\alpha u(t,x)\right\|_{C(\overline{Q}_T)}\le C_0\left(0\le|\alpha|\le i_0=m-\left[\frac{n}{2}\right]-1\right). \tag{5.31}$$

Next, it is evident that for every $\tau\in[0,T]$ and $|\alpha|\equiv i$ in all the cases (5.11)-(5.14) we have:

$$\left\|\mathsf{D}^\alpha u(\tau,x)\right\|_{C(\overline{\Omega})}\le C\cdot\|u\|_{B^{2,1,0}_{2,2,2,\tau}}; \tag{5.32}$$

for $|\alpha|\equiv i$ in cases (5.16) we have:

$$\left\|\mathsf{D}^\alpha u(\tau,x)\right\|_{L_p(\Omega)}\le C_p\cdot\|u\|_{B^{2,1,0}_{2,2,2,\tau}}\ (1\le p<+\infty),\text{ if } n-4m+2|\alpha|=0, \tag{5.33}$$

$$\left\| D^{\alpha} u(\tau,x) \right\|_{L_{\frac{2n}{n-4m+2|\alpha|}}(\Omega)} \le C \cdot \|u\|_{B^{2,1,0}_{2,2,2,\tau}}, \text{ if } n-4m+2|\alpha|>0; \tag{5.34}$$

in case $n<2m,\ 0\le|\beta|\le i_0 = m - \left[\frac{n}{2}\right] - 1$ we have:

$$\left\| D^{\beta} u(\tau,x) \right\|_{C(\overline{\Omega})} \le C \cdot \|u\|_{B^{2,1,0}_{2,2,2,\tau}}; \tag{5.35}$$

in cases $n<2m,\ i_0+1\le|\beta|\le m$ and $n\ge 2m,\ 0\le|\beta|\le m$ we have:

$$\left\| D^{\beta} u_{\tau}(\tau,x) \right\|_{L_p(\Omega)} \le C_p \cdot \|u\|_{B^{2,1,0}_{2,2,2,\tau}} \ (1\le p<+\infty), \text{ if } n-2m+2|\beta|=0, \tag{5.36}$$

$$\left\| D^{\beta} u_{\tau}(\tau,x) \right\|_{L_{\frac{2n}{n-4m+2|\beta|}}(\Omega)} \le C \cdot \|u\|_{B^{2,1,0}_{2,2,2,\tau}}, \text{ if } n-2m+2|\beta|>0; \tag{5.37}$$

and, finally, $\forall n\ge 1$ we have:

$$\left\| u_{\tau\tau}(\tau,x) \right\|_{L_p(\Omega)} \le C \cdot \|u\|_{B^{2,1,0}_{2,2,2,\tau}}, \tag{5.38}$$

where $C>0$ and $C_p>0$ are some constants not depending on u and τ.

Now, using a priori estimate (4.18), $\forall \tau\in[0,T]$ we have:

$$\int_{\Omega} e^{2\left|D^{\alpha} u(\tau,x)\right|^2} dx \le C_0 \text{ for } n-2m+2|\alpha|=0, \tag{5.39}$$

where $C_0>0$ is some number not depending on u and τ.

Next, in cases (5.11)-(5.14), without loss of generality, assuming $\gamma_i = \frac{2n-2m+2i}{n-2m+2i}$ and using estimate (5.32) and a priori estimate (4.19), we obtain that for $|\alpha|\equiv i$ and $\tau\in[0,T]$:

$$\int_{\Omega} \left| D^{\alpha} u(\tau,x) \right|^{2\gamma_i} dx \le \left\| D^{\alpha} u(\tau,x) \right\|^2_{C(\overline{\Omega})} \cdot \int_{\Omega} \left| D^{\alpha} u(\tau,x) \right|^{2(\gamma_i-1)} dx =$$

$$= \left\| \mathrm{D}^{\alpha} u(\tau,x) \right\|^{2(\gamma_i - 1)}_{L_{\frac{2n}{n-4m+2i}}(\Omega)} \cdot \left\| \mathrm{D}^{\alpha} u(\tau,x) \right\|^{2}_{C(\overline{\Omega})} \le C_0 \cdot \|u\|^2_{B^{2,1,0}_{2,2,2,\tau}}; \qquad (5.40)$$

in cases (5.16) for $n - 4m + 2i = 0$, without loss of generality, assuming $1 < \gamma_i < \dfrac{n+2m}{n-2m+2i} = \dfrac{n+2m}{2m} \ge 2$ and using estimate (5.33) and a priori estimate (4.19), we obtain that for $|\alpha| \equiv i$ and $\tau \in [0,T]$:

$$\int_{\Omega} \left| \mathrm{D}^{\alpha} u(\tau,x) \right|^{2\gamma_i} dx = \int_{\Omega} \left| \mathrm{D}^{\alpha} u(\tau,x) \right|^{2} \cdot \left| \mathrm{D}^{\alpha} u(\tau,x) \right|^{2(\gamma_i - 1)} dx \le$$

$$\le \left\| \mathrm{D}^{\alpha} u(\tau,x) \right\|^{2}_{L_{2p}(\Omega)} \cdot \left\| \mathrm{D}^{\alpha} u(\tau,x) \right\|^{2(\gamma_i - 1)}_{L_{2(\gamma_i - 1)q}(\Omega)} \le C_0 \cdot \|u\|^2_{B^{2,1,0}_{2,2,2,\tau}}, \qquad (5.41)$$

where $1 < p < +\infty$, $\dfrac{1}{p} + \dfrac{1}{q} = 1$ and $2(\gamma_i - 1)q = \dfrac{2n}{n-2m+2i} = \dfrac{n}{m}$, the possibility of choosing such $q > 1$ following from the relation below:

$$2(\gamma_i - 1) < 2\left(\frac{n+2m}{n-2m+2i} - 1 \right) = \frac{2(4m-2i)}{n-2m+2i} = \frac{2n}{n-2m+2i};$$

and in cases (5.16) for $n - 2m + 2i > 0$, without loss of generality, assuming $\gamma_i = \dfrac{n+2m}{n-2m+2i}$, and using estimate (5.34) and a priori estimate (4.19), we obtain that for $|\alpha| \equiv i$ and $\tau \in [0,T]$:

$$\int_{\Omega} \left| \mathrm{D}^{\alpha} u(\tau,x) \right|^{2\gamma_i} dx = \int_{\Omega} \left| \mathrm{D}^{\alpha} u(\tau,x) \right|^{2} \cdot \left| \mathrm{D}^{\alpha} u(\tau,x) \right|^{2(\gamma_i - 1)} dx \le$$

$$= \left\| \mathrm{D}^{\alpha} u(\tau,x) \right\|^{2}_{L_{\frac{2n}{n-4m+2i}}(\Omega)} \cdot \left\| \mathrm{D}^{\alpha} u(\tau,x) \right\|^{2(\gamma_i - 1)}_{L_{\frac{2(\gamma_i-1)n}{4m-2i}}(\Omega)} =$$

$$= \left\| \mathrm{D}^{\alpha} u(\tau,x) \right\|^{2(\gamma_i - 1)}_{L_{\frac{2n}{n-4m+2i}}(\Omega)} \cdot \left\| \mathrm{D}^{\alpha} u(\tau,x) \right\|^{2}_{L_{\frac{2n}{n-4m+2i}}(\Omega)} \le C_0 \cdot \|u\|^2_{B^{2,1,0}_{2,2,2,\tau}}; \qquad (5.42)$$

as the functions $D^{\alpha}u(\tau,x)$ ($|\alpha|=m$) and $u_{\tau}(\tau,x)$ for every fixed $\tau\in[0,T]$ are equal in rights in the sense that each of them belongs to the space $W_2^m(\Omega)$, then, absolutely similarly to (5.40)-(5.42), for every $n\geq 1$ and $\tau\in[0,T]$ we have:

$$\int_{\Omega}|u_{\tau}(\tau,x)|^{2\gamma_m}dx\leq C_0\cdot\|u\|^2_{B^{2,1,0}_{2,2,2,\tau}}, \tag{5.43}$$

where $C_0>0$ is some constant not depending on u and τ.

Now, using estimates (5.35)-(5.37) and a priori estimate (4.18), in case of $n<2m$, $n-2m+2|\alpha|=0$ $\forall\tau\in[0,T]$ we have:

$$\int_{\Omega}e^{2|D^{\alpha}u(\tau,x)|^2}\cdot|D^{\beta}u_{\tau}(\tau,x)|^2dx\leq\|D^{\beta}u_{\tau}(\tau,x)\|^2_{C(\overline{\Omega})}\cdot\int_{\Omega}e^{2|D^{\alpha}u(\tau,x)|^2}dx\leq$$

$$\leq C_0\cdot\|u\|^2_{B^{2,1,0}_{2,2,2,\tau}}\quad\text{for } 0\leq|\beta|\leq i_0\equiv m-\left[\frac{n}{2}\right]-1; \tag{5.44}$$

$$\int_{\Omega}e^{2|D^{\alpha}u(\tau,x)|^2}\cdot|D^{\beta}u_{\tau}(\tau,x)|^2dx\leq\|D^{\beta}u_{\tau}(\tau,x)\|^2_{L_{2p_0}(\Omega)}\cdot$$

$$\cdot\left\{\int_{\Omega}e^{2q_0|D^{\alpha}u(\tau,x)|^2}dx\right\}^{1/q_0}\leq C_0\cdot\|u\|^2_{B^{2,1,0}_{2,2,2,\tau}}\quad\text{for } |\beta|=i_0+1, \tag{5.45}$$

where $1<p_0<+\infty$, $\dfrac{1}{p_0}+\dfrac{1}{q_0}=1$ (note that in this case $n-2m+2|\beta|=0$);

$$\int_{\Omega}e^{2|D^{\alpha}u(\tau,x)|^2}\cdot|D^{\beta}u(\tau,x)|^2dx\leq\left\{\int_{\Omega}e^{\frac{2n}{2m-2|\beta|}\cdot|D^{\alpha}u(\tau,x)|^2}dx\right\}^{\frac{2m-2|\beta|}{n}}\cdot$$

$$\cdot\|D^{\beta}u_{\tau}(\tau,x)\|^2_{L_{\frac{2n}{n-2m+2|\beta|}}(\Omega)}\leq C_0\cdot\|u\|^2_{B^{2,1,0}_{2,2,2,\tau}}\quad\text{for } i_0+2\leq|\beta|\leq m, \tag{5.46}$$

note that in these cases $n - 2m + 2|\beta| > 0$.

Similarly, using estimates (5.36), (5.37) and a priori estimate (4.18), in case of $n = 2m$ $\forall \tau \in [0,T]$ we have:

$$\int_\Omega e^{2|u(\tau,x)|^2} \cdot \left|D^\beta u_\tau(\tau,x)\right|^2 dx \le \left\|D^\beta u_\tau(\tau,x)\right\|^2_{L_{2p_0}(\Omega)} \cdot$$

$$\cdot \left\{ \int_\Omega e^{2q_0|u(\tau,x)|^2} dx \right\}^{1/q_0} \le C_0 \cdot \|u\|^2_{B^{2,1,0}_{2,2,2,\tau}} \quad \text{for } |\beta| = 0, \tag{5.47}$$

where $1 < p_0 < +\infty$, $\frac{1}{p_0} + \frac{1}{q_0} = 1$ (note that in this case $n - 2m + 2|\beta| = 0$);

$$\int_\Omega e^{2|u(\tau,x)|^2} \cdot \left|D^\beta u_\tau(\tau,x)\right|^2 dx \le \left\{ \int_\Omega e^{\frac{2n}{2m-2|\beta|}\cdot|u(\tau,x)|^2} dx \right\}^{\frac{2m-2|\beta|}{2n}} \cdot$$

$$\cdot \left\|D^\beta u_\tau(\tau,x)\right\|^2_{L_{\frac{2n}{n-2m+2|\beta|}}(\Omega)} \le C_0 \cdot \|u\|^2_{B^{2,1,0}_{2,2,2,\tau}} \quad \text{for } 1 \le |\beta| \le m, \tag{5.48}$$

note that in these cases $n - 2m + 2|\beta| > 0$; in all these estimates $C_0 > 0$ is some constant not depending on u and τ.

Next, in cases (5.18) and (5.19), without loss of generality, assuming $\gamma_{i,j} = \dfrac{n}{n - 2m + 2i}$ and using estimate (5.35) and a priori estimates (4.19), (4.20), we obtain that for $|\alpha| \equiv i$, $|\beta| \equiv j$ and $\tau \in [0,T]$:

$$\int_\Omega \left|D^\alpha u(\tau,x)\right|^{2\gamma_{i,j}} \cdot \left|D^\beta u_\tau(\tau,x)\right|^2 dx \le \left\|D^\beta u_\tau(\tau,x)\right\|^2_{C(\overline{\Omega})} \cdot \int_\Omega \left|D^\alpha u(\tau,x)\right|^{2\gamma_{i,j}} dx =$$

$$= \left\|D^\alpha u(\tau,x)\right\|^{2\gamma_{i,j}}_{L_{\frac{2n}{n-2m+2|\alpha|}}(\Omega)} \cdot \left\|D^\beta u_\tau(\tau,x)\right\|^2_{C(\overline{\Omega})} \le C_0 \cdot \|u\|^2_{B^{2,1,0}_{2,2,2,\tau}}; \tag{5.49}$$

$$\int_\Omega |u_\tau(\tau,x)|^{2\gamma_{m,j}} \cdot |\mathrm{D}^\beta u_\tau(\tau,x)|^2 dx = \int_\Omega |u_\tau(\tau,x)|^2 \cdot |\mathrm{D}^\beta u_\tau(\tau,x)|^2 dx \le$$

$$\le \|\mathrm{D}^\beta u_\tau(\tau,x)\|^2_{C(\overline{\Omega})} \cdot \int_\Omega u_\tau^2(\tau,x)dx \le C_0 \cdot \|u\|^2_{B^{2,1,0}_{2,2,2,\tau}}, \tag{5.50}$$

where the constant $C_0 > 0$ does not depend on u and τ.

Besides, for even values of $n < 2m,\ i_0 + 2 \le |\alpha| \le m$, $|\beta| = i_0 + 1$ (evidently, in this case $n - 2m + 2|\beta| = 0$), and also for $n = 2m,\ 1 \le |\alpha| \le m$, $|\beta| = 0$ (in this case also $n - 2m + 2|\beta| = 0$), using estimate (5.36) and a priori estimate (4.19), we obtain that $\forall \tau \in [0,T]$:

$$\int_\Omega |\mathrm{D}^\alpha u(\tau,x)|^{2\gamma_{|\alpha|,|\beta|}} \cdot |\mathrm{D}^\beta u_\tau(\tau,x)|^2 dx \le \|\mathrm{D}^\beta u_\tau(\tau,x)\|^2_{L_{2p_0}(\Omega)} \cdot$$

$$\cdot \left\{ \int_\Omega |\mathrm{D}^\alpha u(\tau,x)|^{2q_0\gamma_{|\alpha|,|\beta|}} dx \right\}^{1/q_0} \le C_0 \cdot \|u\|^2_{B^{2,1,0}_{2,2,2,\tau}}, \tag{5.51}$$

where $1 < p_0 < +\infty$, $\dfrac{1}{p_0} + \dfrac{1}{q_0} = 1$ and $2q_0\gamma_{|\alpha|,|\beta|} = \dfrac{2n}{n - 2m + 2|\alpha|}$, the possibility of choosing such $q_0 > 1$ following from the fact that for considered values of β we have $n - 2m + 2|\beta| = 0$ and, consequently, $2\gamma_{|\alpha|,|\beta|} < 2 \cdot \dfrac{2m - 2|\beta|}{n - 2m + 2|\alpha|} = \dfrac{2n}{n - 2m + 2|\alpha|}$. Similarly, for even values of $n < 2m$, $|\beta| = i_0 + 1$ (in this case $n - 2m + 2|\beta| = 0$) and for $n = 2m,\ \ |\beta| = 0$ (in this case also $n - 2m + 2|\beta| = 0$), using estimate (5.36) and a priori estimate (4.20), we obtain that $\forall \tau \in [0,T]$:

$$\int_\Omega |u_\tau(\tau,x)|^{2\gamma_{m,|\beta|}} \cdot |\mathrm{D}^\beta u_\tau(\tau,x)|^2 dx \le \|\mathrm{D}^\beta u_\tau(\tau,x)\|^2_{L_{2p_0}(\Omega)} \cdot$$

$$\cdot\left\{\int_{\Omega}\left|u_{\tau}(\tau,x)\right|^{2q_0\gamma_{m,|\beta|}}dx\right\}^{1/q_0}\leq C_0\cdot\|u\|^2_{B^{2,1,0}_{2,2,2,\tau}}, \tag{5.52}$$

where $1<p_0<+\infty$, $\frac{1}{p_0}+\frac{1}{q_0}=1$ and $2q_0\gamma_{m,|\beta|}=2$, the possibility of choosing such $q_0>1$ following from the fact that for considered values of β we have $n-2m+2|\beta|=0$ and, consequently, $2\gamma_{m,|\beta|}<2\cdot\frac{2m-2|\beta|}{n}=2$.

In other cases of (5.21)-(5.24), i.e.

$$\left.\begin{array}{lll} \textit{for odd} & n<2m, & i_0+1\leq i\leq m,\ i_0+1\leq j\leq m-1, \\ \textit{for even} & n<2m, & i_0+2\leq i\leq m,\ i_0+2\leq j\leq m-1, \\ \textit{for} & n=2m, & 1\leq i\leq m,\ 1\leq j\leq m-1, \\ \textit{for} & n>2m, & 0\leq i\leq m,\ 0\leq j\leq m-1, \end{array}\right\}, \tag{5.53}$$

without loss of generality, assuming $\gamma_{i,j}=\frac{2m-j}{n-2m+2i}$ and using estimate (5.37) and a priori estimates (4.19), (4.20), for $|\alpha|\equiv i$, $|\beta|\equiv j$ and $\tau\in[0,T]$ we have:

$$\int_{\Omega}\left|D^{\alpha}u(\tau,x)\right|^{2\gamma_{i,j}}\cdot\left|D^{\beta}u_{\tau}(\tau,x)\right|^2dx\leq\left\|D^{\beta}u_{\tau}(\tau,x)\right\|^2_{L_{\frac{2n}{n-2m+2|\beta|}}(\Omega)}\cdot$$

$$\cdot\left\|D^{\alpha}u(\tau,x)\right\|^{2\gamma_{i,j}}_{L_{2\gamma_{i,j}\cdot\frac{n}{2m-2|\beta|}}(\Omega)}\leq C_0\cdot\|u\|^2_{B^{2,1,0}_{2,2,2,\tau}}; \tag{5.54}$$

$$\int_{\Omega}\left|u_{\tau}(\tau,x)\right|^{2\gamma_{m,j}}\cdot\left|D^{\beta}u_{\tau}(\tau,x)\right|^2dx=\left|D^{\beta}u_{\tau}(\tau,x)\right|^2_{L_{\frac{2n}{n-2m+2|\beta|}}(\Omega)}\cdot$$

$$=\left\|u_{\tau}(\tau,x)\right\|^{2\gamma_{m,j}}_{L_{2\gamma_{m,j}\cdot\frac{n}{2m+2|\beta|}}(\Omega)}\leq C_0\cdot\|u\|^2_{B^{2,1,0}_{2,2,2,\tau}}, \tag{5.55}$$

where $C_0>0$ is a constant not depending on u and τ.

Now, using representation (5.29), a priori estimate (5.31) and inequalities (5.1)-(5.5) with $R = C_0$ in case $n > 2m$, inequalities (5.6)-(5.10) in case $n \geq 2m$, and then conditions (5.15), (5.17), (5.20) and (5.25), a priori estimate (5.39), estimates (5.40)-(5.52), (5.54), (5.55), and $\forall \tau \in [0,T]$ estimates

$$\int_{\Omega} \{D_t^s D^{\alpha} u(\tau,x)\}^2 dx \leq C_0 \cdot \|u\|^2_{B^{2,1,0}_{2,2,2,\tau}} \quad (0 \leq s \cdot m + |\alpha| \leq 2m), \tag{5.56}$$

we obtain from inequality (5.27) that $\forall t \in [0,T]$:

$$\|u\|^2_{B^{2,1,0}_{2,2,2,t}} \leq C_0 + C_2 \cdot \int_0^t \|u\|^2_{B^{2,1,0}_{2,2,2,\tau}} d\tau, \tag{5.57}$$

where $C_1 > 0$ and $C_2 > 0$ are some constants not depending on u and τ.

On applying Bellman's inequality, we obtain from (5.57) that

$$\|u\|^2_{B^{2,1,0}_{2,2,2,T}} \leq C_1 \cdot e^{C_2 T} \equiv C_0,$$

that is, all the possible classical solutions $u(t,x)$ of problem (0.1)-(0.3) belonging to $B^{2,1,0}_{2,2,T}$ are a priori bounded in $B^{2,1,0}_{2,2,T}$. Theorem is proved.

6. Study Of Existence in Large for the Classical Solution of Problem (0.1)-(0.3) in General Case for n<2m

In this paragraph, using Theorem 3 on existence in small for the classical solution of problem (0.1)-(0.3) proved in §3 and Theorem 6 on a priori boundedness in $B^{2,1,0}_{2,2,2,T}$ of classical solutions of problem (01)-(0.3) proved in §5, we prove the following existence in large theorem for the classical solution of problem (01)-(0.3) for dimensions $n < 2m$.

Theorem 7. Let

1. $n < 2m$.

2. condition 1 of Theorem 3 be fulfilled for $k = 3$, i.e.:

$$\partial\Omega = S \in C^{3m}; \; a_{\alpha}(x) \in C^{|\alpha|+m}(\overline{\Omega}) \; (|\alpha| \leq 2m), \; b_{\beta j}(x) \in C^{3m-k_j}(S)$$

$$\left(|\beta| \le k_j,\ j=1,\dots,m\right);\ \varphi(x) \in W_2^{3m}(\Omega) \cap W_2^{2m}\left(\Omega, \{B_j\}_{j=1}^m\right),$$

$$\mathrm{A}\,(\varphi(x)) \in W_2^{m}\left(\Omega, \{B_j\}_{j=1}^r\right);\ \psi(x) \in W_2^{2m}\left(\Omega, \{B_j\}_{j=1}^m\right).$$

3. condition 2 of Theorem 3 be fulfilled for $k = 3$, i.e.:

a) $\partial^s F(t,x,u_1,\dots,u_N)/\partial t^\alpha \partial x^\beta \partial u_1^{\gamma_1} \dots \partial u_N^{\gamma_N} \in C(\overline{Q}_T \times (-\infty,\infty)^N)$

$$\left(0 \le \alpha + |\beta| + \sum_{i=1}^N \gamma_i = s \le m+1;\ \alpha \le 1;\ |\beta| \le m\right);$$

b) function $F(t,x,u_1,\dots,u_N)$ be equal to zero in all points $(t,x,0,u_2,\dots,u_{N-1},0)$, where $t \in [0,T]$, $x \in S$ and $u_2,\dots,u_{N-1} \in (-\infty,\infty)$, together with all its derivatives of the form mentioned in item a) up to the order $1 + k_r$;

c) $\forall R > 0$ in $Q_T \times [-R,R]^N$

$$|\,\partial^{m+1} F(t,x,u_1,\dots,u_N)/\partial t^\alpha \partial x^\beta \partial u_1^{\gamma_1} \dots \partial u_N^{\gamma_N} -$$

$$- \partial^{m+1} F(t,x,\tilde{u}_1,\dots,\tilde{u}_N)/\partial t^\alpha \partial x^\beta \partial u_1^{\gamma_1} \dots \partial u_N^{\gamma_N} \,| \le C_R \cdot \sum_{i=1}^N |u_i - \tilde{u}_i|,$$

where $\alpha + |\beta| + \sum_{i=1}^N \gamma_i = m+1,\ \alpha \le 1,\ |\beta| \le m$, and $C_R > 0$ is a constant.

4. $F(t,x,u,\mathrm{D}u,\dots,\mathrm{D}^m u,u_t) = f(t,x,u) + \Phi(t,x,u) + H(t,x,u,\mathrm{D}u,\dots,\mathrm{D}^m u,u_t)$

with:

a) $f(t,x,u) \in C(\overline{Q}_T \times (-\infty,\infty))$, $\int_0^u f(t,x,\xi)d\xi \equiv g(t,x,u) \le a_0(x) + b_0 u^2 - g_0(u)$,

where $0 \le a_0(x) \in L(\Omega)$, $b_0 > 0$ is a constant, and $0 \le g_0(u) \in C(-\infty,\infty)$;

$f_t(t,x,u) \in C(\overline{Q}_T \times (-\infty,\infty))$, $|f_t(t,x,u)| \le a(t,x) + b(t) \cdot |u|$, where

$a(t,x) \in L_2(Q_T), \quad b(t) \in L_2(0,T)$;

b) $\Phi(t,x,u) \in C(\overline{Q}_T \times (-\infty,\infty)), \quad |\Phi(t,x,u)| \le a_1(t,x) + b_1(t) \cdot \left(|u| + \sqrt{g_0(u)}\right),$

where $a_1(t,x) \in L_2(Q_T), \quad b_1(t) \in L_2(0,T)$;

c) $H(t,x,u_1,...,u_N) \in C(\overline{Q}_T \times (-\infty,\infty)^N), \quad H(t,x,u,\mathrm{D}u,...,\mathrm{D}^m u,u_t) \cdot u_t \le$

$$\le a_2(t,x) + b_2(t) \cdot \left\{u^2 + |\mathrm{D}u|^2 + ... + |\mathrm{D}^m u|^2 + u_t^2 + g_0(u)\right\},$$

where $a_2(t,x) \in L(Q_T), \quad b_2(t) \in L_2(0,T)$,

$$|\mathrm{D}u| = \sum_{i=1}^{n} |u_{x_i}|, ..., |\mathrm{D}^m u| = \sum_{|\alpha|=m} |\mathrm{D}^\alpha u| .$$

5. for every $R > 0$ and $(t,x) \in Q_T$, $|\mathrm{D}^i u| \le R \left(0 \le i \le i_0 \equiv m - \left[\frac{n}{2}\right] - 1\right)$,

$|\mathrm{D}^i u| < +\infty \; (i_0 + 1 \le i \le m), \; |u_t| < +\infty$:

$$|F_t(t,x,u,\mathrm{D}u,...,\mathrm{D}^m u,u_t)| \le C_R \cdot \left\{1 + \sum_{i=i_0+1}^{m} |\mathrm{D}^i u|^{\gamma_i} + |u_t|^{\gamma_m}\right\}, \text{ if } n \text{ is odd;}$$

$$|F_t(t,x,u,\mathrm{D}u,...,\mathrm{D}^m u,u_t)| \le C_R \cdot \left\{e^{|\mathrm{D}^{i_0+1}u|^2} + \sum_{i=i_0+2}^{m} |\mathrm{D}^i u|^{\gamma_i} + |u_t|^{\gamma_m}\right\},$$

if n is even (in this case $n - 2m + 2(i_0 + 1) = 0$);

$$|F_{\mathrm{D}^j u}(t,x,u,\mathrm{D}u,...,\mathrm{D}^m u,u_t)| \le C_R \cdot \left\{1 + \sum_{i=i_0+1}^{m} |\mathrm{D}^i u|^{\gamma_{i,j}} + |u_t|^{\gamma_{m,j}}\right\}$$

$(j = \overline{0,m-1})$,

if n is odd;

$$\left|F_{\mathrm{D}^i u}(t,x,u,\mathrm{D}u,...,\mathrm{D}^m u,u_t)\right| \le C_R \cdot \left\{ e^{\left|\mathrm{D}^{i_0+1}u\right|^2} + \sum_{i=i_0+2}^{m} \left|\mathrm{D}^i u\right|^{\gamma_{i,j}} + \left|u_t\right|^{\gamma_{m,j}} \right\}$$

$(j=\overline{0,m-1})$,

if n is even (in this case $n-2m+2(i_0+1)=0$);

$$\left|F_{\mathrm{D}^m u}(t,x,u,\mathrm{D}u,...,\mathrm{D}^m u,u_t)\right| \le C_R, \left|F_{u_t}(t,x,u,\mathrm{D}u,...,\mathrm{D}^m u,u_t)\right| \le C_R,$$

where:

for odd values of $n < 2m$, $m - \left[\frac{n}{2}\right] \equiv i_0 + 1 \le i \le m$ and

for even values of $n < 2m$, $i_0 + 2 \le i \le m$

$$0 < \gamma_i \le \frac{2n-2m+2i}{n-2m+2i};$$

for odd values of $n < 2m$, $i_0 + 1 \le i \le m$, $0 \le j \le i_0$ and

for even values of $n < 2m$, $i_0 + 2 \le i \le m$, $0 \le j \le i_0$

$$0 < \gamma_{i,j} \le \frac{n}{n-2m+2i};$$

for odd values of $n < 2m$, $i_0 + 1 \le i \le m$, $i_0 + 1 \le j \le m-1$ and

for even values of $n < 2m$, $i_0 + 2 \le i \le m$, $i_0 + 1 \le j \le m-1$

$$0 < \gamma_{i,j} \le \frac{2m-2j}{n-2m+2i};$$

note that for $n-2m+2j=0$

$$0 < \gamma_{i,j} \le \frac{2m - 2j}{n - 2m + 2i};$$

and $C_R > 0$ is a constant.

Then problem (0.1)-(0.3) has a unique classical solution.

Proof. According to Theorem 3, there exists in small a unique in large classical solution of problem (0.1)-(0.3). Besides, as seen from the proof of Theorem 3, in order to prove the existence in large for the classical solution of problem (0.1)-(0.3), it suffices to show that all the possible classical solutions of problem (0.1)-(0.3) belonging to the space $B^{3,2,1}_{2,2,2,T}$ are a priori bounded in $B^{3,2,1}_{2,2,2,T}$. So, let $u(t.x)$ be any classical solution of problem (0.1)-(0.3) belonging to the space $B^{3,2,1}_{2,2,2,T}$. Due to conditions 3 and 4 of this theorem, according to Theorem 6 of §5, all the possible classical solutions $u(t,x)$ of problem (0.1)-(0.3) belonging to the space $B^{2,1.0}_{2,2,2,T}$, and, moreover, all the possible classical solutions $u(t,x)$ of problem (0.1)-(0.3) belonging to the space $B^{3,2,1}_{2,2,2,T}$, are a priori bounded in $B^{2,1.0}_{2,2,2,T}$:

$$\|u(t,x)\|_{B^{2,1.0}_{2,2,2,T}} \le C_0. \tag{6.1}$$

The a priori estimate (6.1), in turn, implies the following a priori estimates:

$$\|u(t,x)\|_{W_2^{2m}(\Omega)} \le C_1,\ \|u_t(t,x)\|_{W_2^{m}(\Omega)} \le C_1,\ \|u_{tt}(t,x)\|_{L_2(\Omega)} \le C_1$$
$$\forall t \in [0,T], \tag{6.2}$$

where the constant $C_1 > 0$ does not depend on u and t. The a priori estimates (6.2), due to Sobolev's imbedding theorems, imply the following a priori estimates:

$$\left\| \mathsf{D}_t^{\alpha} \mathsf{D}^{\beta} u(t,x) \right\|_{C(\overline{Q}_T)} \le C \left(0 \le \alpha \cdot m + |\beta| \le i_{00} \equiv 2m - \left[\frac{n}{2}\right] - 1 \right), \tag{6.3}$$

$$\left\| \mathsf{D}_t^{\alpha} \mathsf{D}^{\beta} u(t,x) \right\|_{L_q(Q)} \le \mathsf{A}_q \ (i_{00} + 1 \le \alpha \cdot m + |\beta| \le 2m\ ;\ t \in [0,T];$$

$$1 \le q \le \frac{2n}{n - 2(2m - \alpha \cdot m - |\beta|)}, q < +\infty), \tag{6.4}$$

where $C > 0, \mathrm{A}_q > 0$ are the constants which don't depend on u and t, $i_{00} \equiv \max\{i : n < 2(2m - i)\}$, with

$$m \le i_{00} < 2m \tag{6.5}$$

as $n < 2m$.

Besides, using the structure of space $B^{3,2,1}_{2,2,2,T}$ and Sobolev's imbedding theorems, for every $t \in [0,T]$ we have:

$$\left\| \mathrm{D}_t^{\alpha} \mathrm{D}^{\beta} u(t,x) \right\|_{C(\overline{Q}_T)} \le C \cdot \|u\|_{B^{3,2,1}_{2,2,2,t}} \left(0 \le \alpha \cdot m + |\beta| \le i_{000} \equiv 3m - \left[\frac{n}{2}\right] - 1 \right), \tag{6.6}$$

$$\left\| \mathrm{D}_t^{\alpha} \mathrm{D}^{\beta} u(t,x) \right\|_{L_q(Q)} \le \mathrm{A}_q \cdot \|u\|_{B^{3,2,1}_{2,2,2,t}} \ (i_{000} + 1 \le \alpha \cdot m + |\beta| \le 3m;\ \alpha \le 2;$$

$$1 \le q \le \frac{2n}{n - 2(3m - \alpha \cdot m - |\beta|)}, q < +\infty), \tag{6.7}$$

where $C > 0, \mathrm{A}_q > 0$ are the constants which don't depend on u and t, $i_{000} \equiv \max\{i : n < 2(3m - i)\}$, with

$$2m \le i_{000} < 3m \tag{6.8}$$

as $n < 2m$.

Next, under the conditions of this theorem it is evident that the Fourier coefficients $u_s(t) = \int_{\Omega} u(t,x) v_s(x) dx$ of the classical solution $u(t,x) \in B^{3,2,1}_{2,2,2,T}$ of problem (0.1)-(0.3) satisfy not only systems (1.13) and (1.17), but also the system (1.20) with $l = 0$. Then, using inequality (1.10), where $\frac{\partial}{\partial \tau}\{\mathrm{F}(u(\tau,x))\}$ must be taken instead of function $u(x)$, we obtain from system (1.20) (with $l = 0$) that $\forall t \in [0,T]$:

$$\|u\|^2_{B^{3,2,1}_{2,2,2,t}} \le 2\|Z_3\|^2_{B^{3,2,1}_{2,2,2,T}} + C \cdot \int_0^t \left\| \frac{\partial}{\partial \tau}\{\mathrm{F}(u(\tau,x))\} \right\|^2_{W_2^m(\Omega)} d\tau, \tag{6.9}$$

where function $Z_3(t,x)$ is defined by (1.22) with $l=0$; $C>0$ is some constant not depending on u and t, and operator F is defined by (1.15).

Using condition 3_a of this theorem and taking into account (6.5), we obtain from a priori estimates (6.3) the following a priori estimates:

$$\left\| \partial^s F(t,x,u(t,x),\mathsf{D}u(t,x),...,\mathsf{D}^m u(t,x),u_t(t,x)) / \partial t^\alpha \partial x^\beta \partial u_1^{\gamma_1} ... \partial u_N^{\gamma_N} \right\|_{C(\overline{Q}_T)} \le C^*$$

$$\left(0 \le \alpha + |\beta| + \sum_{i=1}^{N} \gamma_i = s \le m+1;\ \alpha \le 1;\ |\beta| \le m \right). \tag{6.10}$$

It is evident that in an extended version of expressions

$$\frac{\partial^s}{\partial x_1^{\alpha_1}...\partial x_n^{\alpha_n}} \frac{\partial}{\partial \tau} \{\mathsf{F}\,(u(\tau,x))\} \equiv \frac{\partial^s}{\partial x_1^{\alpha_1}...\partial x_n^{\alpha_n}} \cdot$$

$$\cdot \frac{\partial}{\partial \tau} \left\{ F(\tau,x,u(\tau,x),\mathsf{D}u(\tau,x),...,\mathsf{D}^m u(\tau,x),u_\tau(\tau,x)) \right\} (s=\overline{0,m}) \tag{6.11}$$

the "most bad" multipliers of the derivatives of function F can only be the products of the following form:

$$\prod_{i=1}^{l} \mathsf{D}_\tau^{\alpha_i} \mathsf{D}^{\beta_i} u(\tau,x)$$

$$\left(l=\overline{1,m+1};\ 1 \le \alpha_1 \cdot m + |\beta_1| \le ... \le \alpha_l \cdot m + |\beta_l| \le 3m; \right.$$

$$\alpha_l \cdot m + |\beta_l| \ge 2m;\ \sum_{i=1}^{l} (\alpha_i \cdot m + |\beta_i|) = (l+2)m;\ \alpha_i \le 2,$$

with $\alpha_i = 2$ only for one value of i). (6.12)

Due to a priori estimates (6.10) and (6.3) (with (6.5) taken into account), in order to estimate

$$\left\| \frac{\partial}{\partial \tau} \{\mathsf{F}\,(u(\tau,x))\} \right\|^2_{W_2^m(\Omega)} \tag{6.13}$$

it suffices to estimate ($\forall \tau \in [0,T]$) the following integrals:

$$\int_{\Omega}\left\{\prod_{i=1}^{l} \mathrm{D}_{\tau}^{\alpha_i}\mathrm{D}^{\beta_i} u(\tau,x)\right\}dx \tag{6.14}$$

$$(l=\overline{1,m+1}; i_{00}+1=2m-\left[\frac{n}{2}\right]\leq \alpha_1\cdot m+|\beta_1|\leq \ldots\leq \alpha_l\cdot m+|\beta_l|\leq 3m;$$

$$\alpha_l\cdot m+|\beta_l|\geq 2m; \sum_{i=1}^{l}(\alpha_i\cdot m+|\beta_i|)=(l+2)m; \quad \alpha\leq 2$$

with $\alpha_i = 2$ only for one value of i). (6.15)

For this purpose, let's consider the following three possible cases:

I. Let $l = 1$. Then, due to the estimates (6.6) and (6.7), it is evident that $\forall \tau \in [0,T]$:

$$\int_{\Omega}\left\{\prod_{i=1}^{l} \mathrm{D}_{\tau}^{\alpha_i}\mathrm{D}^{\beta_i} u(\tau,x)\right\}^2 dx=\int_{\Omega}\left\{\mathrm{D}_{\tau}^{\alpha_1}\mathrm{D}^{\beta_1} u(\tau,x)\right\}^2 dx\leq c\cdot\|u\|^2_{B^{3,2,1}_{2,2,2,\tau}}, \tag{6.16}$$

where $C > 0$ is some constant not depending on u and τ.

II. Let $l \geq 2$ and $2m\leq \alpha_l\cdot m+|\beta_l|\leq i_{000}\equiv 3m-\left[\frac{n}{2}\right]-1$. Then, using estimate (6.6) (for $\alpha=\alpha_l$, $\beta=\beta_l$) and a priori estimate (6.4) (for $\alpha=\alpha_i$, $\beta=\beta_i$, $i=\overline{1,l-1}$), for integrals (6.14) (under conditions (6.15)) with any $\tau\in[0,T]$ we have:

$$\int_{\Omega}\left\{\prod_{i=1}^{l} \mathrm{D}_{\tau}^{\alpha_i}\mathrm{D}^{\beta_i} u(\tau,x)\right\}^2 dx=\left\|\mathrm{D}_{\tau}^{\alpha_l}\mathrm{D}^{\beta_l} u(\tau,x)\right\|^2_{C(\overline{\Omega})}\cdot\int_{\Omega}\left\{\prod_{i=1}^{l-1} \mathrm{D}_{\tau}^{\alpha_i}\mathrm{D}^{\beta_i} u(\tau,x)\right\}^2 dx\leq$$

$$\leq C\cdot\|u\|^2_{B^{3,2,1}_{2,2,2,\tau}}\cdot\prod_{i=1}^{l-1}\left\|\mathrm{D}_{\tau}^{\alpha_i}\mathrm{D}^{\beta_i} u(\tau,x)\right\|^2_{L_{2p_i}(\Omega)}\leq \widetilde{C}\cdot\|u\|^2_{B^{3,2,1}_{2,2,2,\tau}}, \tag{6.17}$$

where $C > 0$ and $\widetilde{C} > 0$ are some constants not depending on u and τ,

$$1\leq p_i\leq\frac{n}{n-2(2m-\alpha_i\cdot m-|\beta_i|)}, p_i<+\infty \ (i=\overline{1,l-1}); \sum_{i=1}^{l=1}\frac{1}{p_i}=1,$$

the possibility of last equality following from the relation below:

$$\sum_{i=1}^{l-1}\frac{n-2(2m-\alpha_i\cdot m-|\beta_i|)}{n}=\frac{(n-4m)(l-1)+2\sum_{i=1}^{l-1}(\alpha_i\cdot m+|\beta_i|)}{n}=$$

$$=\frac{(n-4m)(l-1)+2\{(l+2)m-(m\cdot\alpha_l+|\beta_l|\}}{n}\le$$

$$\le\frac{(n-4m)(l-1)+2(l+2)m-4m}{n}=\frac{(n-2m)(l-1)+2m}{n}\le 1.$$

When $l\ge 2$, the latter part of this inequality becomes a strict inequality because $n<2m$.

III. Let $l>2$ and $i_{000}+1=3m-\left[\frac{n}{2}\right]\le\alpha_l\cdot m+|\beta_l|\le 3m$. Then, using estimate (6.7) (for $\alpha=\alpha_l,\ \beta=\beta_l$) and a priori estimate (6.4) (for $\alpha=\alpha_i,\ \beta=\beta_i,\ i=\overline{1,l-1}$), for integrals (6.14) (under conditions (6.15)) we have:

$$\int_\Omega\left\{\prod_{i=1}^{l}\mathrm{D}_\tau^{\alpha_i}\mathrm{D}^{\beta_i}u(\tau,x)\right\}^2dx\le\prod_{i=1}^{l-1}\left\|\mathrm{D}_\tau^{\alpha_i}\mathrm{D}^{\beta_i}u(\tau,x)\right\|^2_{L_{2p_i}(\Omega)}\cdot$$

$$\cdot\left\|\mathrm{D}_\tau^{\alpha_l}\mathrm{D}^{\beta_l}u(\tau,x)\right\|^2_{L_{2p_l}(\Omega)}\le C\cdot\|u\|^2_{B^{3,2,1}_{2,2,2,\tau}},\tag{6.18}$$

where $C>0$ is some constant not depending on u and τ,

$$1\le p_i\le\frac{n}{n-2(2m-\alpha_i\cdot m+|\beta_i|)},\ p_i<+\infty\ \ (i=\overline{1,l-1}),$$

$$1\le p_l\le\frac{n}{n-2(3m-\alpha_l\cdot m+|\beta_l|)},\ p_l<+\infty,\ \sum_{i=1}^{l}\frac{1}{p_i}=1,$$

the possibility of last equality following from the relation below:

$$\sum_{i=1}^{l-1}\frac{n-2(2m-\alpha_i\cdot m-|\beta_i|)}{n}+\frac{n-2(3m-\alpha_l\cdot m-|\beta_l|)}{n}=$$

$$=\frac{(n-4m)(l-1)+n-6m+2\sum_{i=1}^{l}(\alpha_i\cdot m+|\beta_i|)}{n}=$$

$$=\frac{(n-4m)(l-1)+n-6m+2(l+2)m}{n}=\frac{(n-2m)(l-1)+n}{n}<1,$$

because $n<2m$, $l\geq 2$, and, consequently, $(n-2m)(l-1)<0$.

Now, using a priori estimates (6.10), (6.3) and estimates (6.16)-(6.18), we obtain from (6.9) that $\forall t\in[0,T]$:

$$\|u\|^2_{B^{3,2,1}_{2,2,2,t}}\leq C_1+C_2\cdot\int_0^t\|u\|^2_{B^{3,2,1}_{2,2,2,\tau}}\,d\tau\,, \qquad (6.19)$$

where $C_1>0$ and $C_2>0$ are some constants not depending on u and τ. On applying Bellman's inequality, we obtain from (6.19):

$$\|u(t,x)\|^2_{B^{3,2,1}_{2,2,2,T}}\leq C_1\cdot\exp\{C_2\cdot T\}\equiv C_3,$$

i.e. all the possible classical solutions $u(t,x)\in B^{3,2,1}_{2,2,2,T}$ of problem (0.1)-(0.3) are a priori bounded in $B^{3,2,1}_{2,2,2,T}$. Theorem is proved.

7. Study Of Existence In Large for the Classical Solution of Problem (0.1)-(0.3) in Case $F=F(t,x,u)$, $n\leq 4m+2$

In this paragraph, using Theorem 4 on existence in small for the classical solution of problem (0.1)-(0.3) in case $F=F(t,x,u)$ proved in §3 and Theorem 6 on a priori boundedness in $B^{2,1,0}_{2,2,2,T}$ of classical solutions of problem (01)-(0.3) proved in §5, we prove the following existence in large theorem for the classical solution of problem (01)-(0.3) in case $F=F(t,x,u)$ for dimensions $n\leq 4m+2$.

Theorem 8. Let

1. $1 \le n \le 4m + 2$.

2. all the conditions of Theorem 4 be fulfilled for $k = \left[\dfrac{n}{2m}\right] + 3$.

3. $F(t,x,u) = f(t,x,u) + \Phi(t,x,u)$ with:

a) $f(t,x,u),\ f_t(t,x,u),\ \Phi(t,x,u) \in C(\overline{Q}_T \times (-\infty,\infty))$;

b) in $Q_T \times (-\infty,\infty)$

$$\int_0^u f(t,x,\xi)d\xi \equiv g(t,x,u) \le a_0(x) + C_0 \cdot u^2 - g_0(u),$$

$$|f_t(t,x,u)| \le a(t,x) + b(t) \cdot |u|,$$

$$|\Phi(t,x,u)| \le a(t,x) + b(t) \cdot \left(|u| + \sqrt{g_0(u)}\right),$$

where $0 \le a_0(x) \in L(\Omega)$, $C_0 > 0$ is a constant, $0 \le g_0(x) \in C(-\infty,\infty)$, $a(t,x) \in L_2(Q_T)$,

$b(t) \in L_2(0,T)$.

4. when $2m \le n \le 4m + 2$ in $Q_T \times (-\infty,\infty)$

$$|F_t(t,x,u)| \le C \cdot \left(1 + |u|^\alpha\right),$$

$$|F_u(t,x,u)| \le C \cdot \left(1 + |u|^\beta\right),$$

where

for $2m \le n < 4m$ $0 < \alpha \le \dfrac{2n - 2m}{n - 2m}$;

for $4m \le n \le 4m + 2$ $0 < \alpha \le \dfrac{n + 2m}{n - 2m}$, besides, for $n = 4m$ $0 < \alpha < \dfrac{n+2m}{n-2m}$;

for $2m \le n \le 4m+2$ $0 < \beta \le \dfrac{2m}{n-2m}$.

5. when $2m \le n \le 4m+2$ в $Q_T \times (-\infty, \infty)$

$$\left|\partial^{1+|\alpha|+s} F(t,x,u) / \partial t \partial x^{\alpha} \partial u^{s}\right| \le C \cdot (1 + |u|^{\gamma_{\alpha,s}}) \quad (0 \le |\alpha| + s \le m), \tag{7.1}$$

$$\left|\partial^{1+|\beta|+l} F(t,x,u) / \partial x^{\beta} \partial u^{1+l}\right| \le C \cdot (1 + |u|^{\delta_{\beta,s}}) \quad (0 \le |\beta| + l \le m), \tag{7.2}$$

where $C > 0$ is a constant, and

$$0 < \gamma_{\alpha,0} \le \frac{2n-4m}{n-4m} \text{ for } 4m \le n \le 4m+2,\ 0 \le |\alpha| \le m, \tag{7.3}$$

besides, for $n = 4m+2$, $m = 1$ $0 < \gamma_{\alpha,0} \le \dfrac{2n-4m}{n-4m}$;

$$\text{for } n = 4m+1,\ s = 1,\ \alpha = 0 \tag{7.4}$$

$$0 < \gamma_{\alpha,0} = \gamma_{0,1} \le \frac{n-(n-4m)s+2|\alpha|}{n-4m} = \frac{4m}{n-4m}; \tag{7.5}$$

$$\text{for } n = 4m,\quad s = 1,\ \alpha = 0, \tag{7.6}$$

$$\text{for } n = 4m+2,\ s = 1,\ \alpha = 0, \tag{7.7}$$

$$\text{for } n = 4m+2,\ s = 1,\ \alpha = 1, \tag{7.8}$$

$$\text{for } n = 4m+2,\ s = 2,\ \alpha = 0 \tag{7.9}$$

$$0 < \gamma_{\alpha,s} < \frac{n-(n-4m)s+2|\alpha|}{n-4m}; \tag{7.10}$$

and in all other cases

$$0 < \gamma_{\alpha,s} \le \frac{2n - 6m - (n-4m)s + 2|\alpha| + 2}{n-4m}; \tag{7.11}$$

$$0 \le \delta_{\beta,l} \le \frac{2m - (n-4m)\cdot l + 2|\beta|}{n-4m} \text{ for } 4m \le n \le 4m+2,\ 0 \le |\beta| + l \le m. \tag{7.12}$$

6. when $m = 1,\quad n = 4m + 2$ in $Q_T \times (-\infty, \infty)$

$$\left|\partial^{1+|\alpha|+s} F(t,x,u) / \partial t \partial x^{\alpha} \partial u^{s}\right| \le C \cdot e^{u^2} \quad (0 \le |\alpha| + s \le 2m), \tag{7.13}$$

$$\left|\partial^{1+|\beta|+l} F(t,x,u) / \partial x^{\beta} \partial u^{1+l}\right| \le C \cdot e^{u^2} \quad (0 \le |\beta| + l \le 2m). \tag{7.14}$$

Then problem (0.1)-(0.3) has a unique classical solution.

Proof. According to Theorem 4, there exists in small a unique in large classical solution of problem (0.1)-(0.3). Besides, as seen from the proof of Theorem 4, in order to prove the existence in large for the classical solution of problem (0.1)-(0.3), it suffices to show that all the possible classical solutions of problem (0.1)-(0.3) belonging to the space $B^{k,k-1,k-2}_{2,2,2,T}$, where $k \equiv \left[\frac{n}{2m}\right] + 3$, are a priori bounded in $B^{k,k-1,k-2}_{2,2,2,T}$. So, let $u(t.x)$ be any classical solution of problem (0.1)-(0.3) belonging to the space $B^{k,k-1,k-2}_{2,2,2,T}$. Due to conditions 3 and 4 of this theorem and according to Theorem 6 of §5, all the possible classical solutions $u(t,x)$ of problem (0.1)-(0.3) belonging to the space $B^{2,1.0}_{2,2,2,T}$, and, moreover, all the possible classical solutions $u(t,x)$ of problem (0.1)-(0.3) belonging to the space $B^{k,k-1,k-2}_{2,2,2,T}$, are a priori bounded in $B^{2,1.0}_{2,2,2,T}$:

$$\|u(t,x)\|_{B^{2,1,0}_{2,2,2,T}} \le C_0. \tag{7.15}$$

The a priori estimate (7.15), in particular, implies the following a priori estimates:

$$\|u(t,x)\|_{W_2^{2m}(\Omega)} \le C_1,\ \|u_t(t,x)\|_{W_2^{m}(\Omega)} \le C_1,\ \forall t \in [0,T]. \tag{7.16}$$

Due to Sobolev's imbedding theorems, a priori estimates (7.16), in turn, imply the following a priori estimates:

$$\left\| D^{\alpha} u(t,x) \right\|_{C(\overline{Q}_T)} \leq C \qquad \left(n < 4m;\ 0 \leq |\alpha| \leq i_{00} \equiv 2m - \left[\frac{n}{2} \right] - 1 \right), \tag{7.17}$$

$$\left\| D^{\alpha} u(t,x) \right\|_{L_q(\Omega)} \leq A_q \qquad \left(n < 4m,\ i_{00} + 1 = 2m - \left[\frac{n}{2} \right] \leq |\alpha| \leq 2m \right.$$

$$\text{or } \left. 4m \leq n \leq 4m + 2,\ 0 \leq |\alpha| \leq 2m;\ 1 \leq q \leq \frac{2n}{n - 2(2m - |\alpha|)},\ q < +\infty \right), \tag{7.18}$$

where $C > 0$ and $A_q > 0$ are the constants which don't depend on u and t; $i_{00} \equiv \max\{i : n < 2(2m - i)\}$, with $0 \leq i_{00} < 2m$ in case $n < 2m$.

Besides, using the structure of space $B^{3,2,1}_{2,2,2,T}$ and Sobolev's imbedding theorems, for every $t \in [0,T]$ we have:

$$\left\| D_t^s D^{\alpha} u(t,x) \right\|_{C(\overline{\Omega})} \leq C \cdot \|u\|_{B^{3,2,1}_{2,2,2,t}} \left(n < 6m;\ 0 \leq s \cdot m + |\alpha| \leq i_{000} \equiv 3m - \left[\frac{n}{2} \right] - 1; s \leq 1 \right) \tag{7.19}$$

$$\left\| D_t^s D^{\alpha} u(t,x) \right\|_{L_q(\Omega)} \leq A_q \cdot \|u\|_{B^{3,2,1}_{2,2,2,t}} \quad \left(n < 6m,\ i_{000} + 1 \leq s \cdot m + |\alpha| \leq 3m \right.$$

$$\text{or } \left. n \leq 6m,\ 0 \leq s \cdot m + |\alpha| \leq 3m;\ s \leq 1;\ 1 \leq q \leq \frac{2n}{n - 2(3m - s \cdot m - |\alpha|)},\ q < +\infty \right), \tag{7.20}$$

where $C > 0$ and $A_q > 0$ are the constants which don't depend on u and t; $i_{000} \equiv \max\{i : n < 2(3m - i)\}$, with $0 \leq i_{000} < 3m$ in case $n < 6m$.

Next, it is evident that the estimate (6.9) with $F(u(\tau,x)) = F(\tau,x,u(\tau,x))$ holds also under the conditions of this theorem.

Due to a priori estimate (7.17), the following a priori estimates hold:

$$\left\| \partial^s F(t,x,u(t,x)) / \partial t^{\alpha} \partial x^{\beta} \partial u^{\gamma} \right\|_{C(\overline{Q}_T)} \leq C^*$$

$$\left(n < 4m;\ 0 \leq \alpha + |\beta| + \gamma = s \leq m + 1; \alpha \leq 1; |\beta| \leq m \right). \tag{7.21}$$

Next, it is evident that the expressions

$$\frac{\partial^i}{\partial x_1^{\alpha_1} \ldots \partial x_n^{\alpha_n}} \frac{\partial}{\partial \tau}\{F(\tau, x, u(\tau, x))\} \quad (i = \overline{0, m}) \tag{7.22}$$

consist of finite number of summands of the following form:

$$\frac{\partial^{1+|\alpha|+s} F(\tau, x, u(\tau, x))}{\partial \tau \partial x^\alpha \partial u^s} \cdot \prod_{i=1}^{s} \mathrm{D}^{\alpha_i} u(\tau, x) \quad (0 \le s \le m;\ \sum_{i=1}^{s} |\alpha_i| + |\alpha| \le m), \tag{7.23}$$

$$\frac{\partial^{1+|\beta|+l} F(\tau, x, u(\tau, x))}{\partial x^\beta \partial u^{1+l}} \cdot \prod_{j=1}^{s} \mathrm{D}^{\beta_j} u(\tau, x) \cdot \mathrm{D}^{\beta_0} u_\tau(\tau, x) \quad (0 \le l \le m;\ \sum_{j=0}^{l} |\beta_j| + |\beta| \le m) \tag{7.24}$$

where $1 \le |\alpha_1| \le \ldots \le |\alpha_s| \le m;\ 1 \le |\beta_1| \le \ldots \le |\beta_l| \le m$; besides, for $s = 0$ we assume that $\prod_{i=1}^{s} \ldots$ is 1, and for $l = 0$ we assume that $\prod_{j=1}^{l} \ldots$ is 1.

Therefore, due to a priori estimate (7.21) and estimates (7.1), (7.2), in order to have an estimate ($\forall \tau \in [0, T]$) for the expression

$$\left\| \frac{\partial}{\partial \tau}\{F(\tau, x, u(\tau, x))\} \right\|_{W_2^m(\Omega)}^2,$$

it suffices to have an estimate ($\forall \tau \in [0, T]$) for the following integrals:

$$\int_\Omega \prod_{i=1}^{s} \left|\mathrm{D}^{\alpha_i} u(\tau, x)\right|^2 dx \quad (n < 4m;\ 1 \le s \le m;\ \sum_{i=1}^{s} |\alpha_i| \le m;\ 1 \le |\alpha_1| \le \ldots \le |\alpha_s| \le m), \tag{7.25}$$

$$\int_\Omega \prod_{j=1}^{l} \left|\mathrm{D}^{\beta_j} u(\tau, x)\right|^2 \cdot \left|\mathrm{D}^{\beta_0} u_\tau(\tau, x)\right|^2 dx \quad (n < 4m;\ 0 \le l \le m;\ \sum_{j=0}^{l} |\beta_j| + |\beta| \le m$$

$$1 \le |\beta_1| \le \ldots \le |\beta_l| \le m), \tag{7.26}$$

$$\int_\Omega |u(\tau, x)|^{2\gamma_{\alpha,s}} \cdot \prod_{i=1}^{s} \left|\mathrm{D}^{\alpha_i} u(\tau, x)\right|^2 dx \quad (4m \le n \le 4m + 2;\ 0 \le s \le m;$$

$$\sum_{i=1}^{s}|\alpha_i|+|\alpha|\le m;\ 1\le|\alpha_1|\le\ldots\le|\alpha_s|\le m), \tag{7.27}$$

$$\int_\Omega |u(\tau,x)|^{2\delta_{\beta,l}}\cdot\prod_{j=1}^{l}\left|D^{\beta_j}u(\tau,x)\right|^2\cdot\left|D^{\beta_0}u_\tau(\tau,x)\right|^2dx\quad (4m\le n\le 4m+2;$$

$$0\le l\le m;\ \sum_{j=0}^{l}|\beta_j|+|\beta|\le m;\ 1\le|\beta_1|\le\ldots\le|\beta_l|\le m). \tag{7.28}$$

Due to a priori estimates (7.17) and (7.18), if $n<2m$ (in this case it is evident that $i_{00}=2m-\left[\frac{n}{2}\right]-1\ge m$), then integrals (7.25) and (7.26) are a priori bounded:

$$\int_\Omega\prod_{i=1}^{s}\left|D^{\alpha_i}u(\tau,x)\right|^2dx\le C\quad (n<2m), \tag{7.29}$$

$$\int_\Omega\prod_{j=1}^{l}\left|D^{\beta_j}u(\tau,x)\right|^2\cdot\left|D^{\beta_0}u_\tau(\tau,x)\right|^2dx\le C\int_\Omega\left|D^{\beta_0}u_\tau(\tau,x)\right|^2dx\cdot\le\widetilde{C}\quad(n<2m)\cdot \tag{7.30}$$

Therefore, to have an estimate for all integrals (7.25) and (7.26), it suffices to have an estimate for the following integrals:

$$\int_\Omega\prod_{i=1}^{s}\left|D^{\alpha_i}u(\tau,x)\right|^2dx\qquad (2m\le n<4m;\ 1\le s\le m;$$

$$\sum_{i=1}^{s}|\alpha_i|\le m;\ i_{00}+1\le|\alpha_1|\le\ldots\le|\alpha_s|\le m), \tag{7.31}$$

$$\int_\Omega\prod_{j=1}^{l}\left|D^{\beta_j}u(\tau,x)\right|^2\cdot\left|D^{\beta_0}u_\tau(\tau,x)\right|^2dx\quad (2m\le n<4m;\ 0\le l\le m;$$

$$\sum_{j=0}^{l}|\beta_j|\le m;\ i_{00}+1\le|\beta_1|\le\ldots\le|\beta_l|\le m). \tag{7.32}$$

Now, due to a priori estimates (7.18), we obtain that the integrals (7.31) and (7.32) are also a priori bounded:

$$\int_\Omega \prod_{i=1}^{s} \left| D^{\alpha_i} u(\tau,x) \right|^2 dx \le \prod_{i=1}^{s} \left\| D^{\alpha_i} u(\tau,x) \right\|^2_{L_{2p_i}(\Omega)} \le C, \tag{7.33}$$

$$\int_\Omega \prod_{j=1}^{l} \left| D^{\beta_j} u(\tau,x) \right|^2 \cdot \left| D^{\beta_0} u_\tau(\tau,x) \right|^2 dx \le$$

$$\le \prod_{j=1}^{l} \left\| D^{\beta_j} u(\tau,x) \right\|^2_{L_{2q_j}(\Omega)} \cdot \left\| D^{\beta} u_\tau(\tau,x) \right\|^2_{L_{2q_0}(\Omega)} \le C, \tag{7.34}$$

where $C > 0$ is some constant not depending on u and τ, and

$$1 \le p_i \le \frac{n}{n-2(2m-|\alpha_i|)}, p_i < +\infty \ (i=\overline{1,s}), \ \sum_{i=1}^{s} \frac{1}{p_i} = 1,$$

$$1 \le q_j \le \frac{n}{n-2(2m-|\beta_j|)}, q_j < +\infty \ (j=\overline{1,l}), \ 1 \le q_0 \le \frac{n}{n-2(2m-|\beta_0|)},$$

$$q_0 < +\infty, \ \sum_{j=0}^{l} \frac{1}{q_j} = 1,$$

the possibility of equalities $\sum_{i=1}^{s} \frac{1}{p_i} = 1$ and $\sum_{j=0}^{l} \frac{1}{q_j} = 1$ following from the relations below:

$$\sum_{i=1}^{s} \frac{n-2(2m-|\alpha_i|)}{n} = \frac{(n-4m)s + 2\sum_{i=1}^{s}|\alpha_i|}{n} \le \frac{(n-4m)s+2m}{n} \le \frac{n-4m+2m}{n} < 1,$$

$$\sum_{j=1}^{l} \frac{n-2(2m-|\beta_j|)}{n} + \frac{n-2(2m-|\beta_0|)}{n} = \frac{(n-4m)\cdot l + n - 2m + 2\sum_{j=0}^{l}|\beta_j|}{n} \le$$

$$\leq \frac{(n-4m)l+n-2m+2m}{n} = \frac{n+(n-4m)\cdot l}{n} \leq 1 .$$

When $l > 0$, the latter part of this inequality becomes a strict inequality because $n < 4m$.

Thus, all integrals (7.25) and (7.26) are a priori bounded.

Now let's estimate integrals (7.27) in case $s = 0$. First, let $m \geq 1$, $n = 4m$, $4m+1$ or $m > 1$, $n = 4m+2$. Then it is evident that $n < 6m$. Using a priori estimate (7.18) (for $|\alpha| = 0$), estimate (7.19) (for $s = 0$, $|\alpha| = 0$) and condition (7.3), we have:

$$\int_\Omega |u(\tau,x)|^{2\gamma_{\alpha,0}}\, dx = \int_\Omega |u(\tau,x)|^{2(\gamma_{\alpha,0}-1)} \cdot |u(\tau,x)|^2\, dx \leq$$

$$\leq \|u(\tau,x)\|_{C(\overline{\Omega})}^2 \cdot \int_\Omega |u(\tau,x)|^{2(\gamma_{\alpha,0}-1)}\, dx \leq C \cdot \|u(\tau,x)\|_{C(\overline{\Omega})}^2 \leq \widetilde{C} \cdot \|u(\tau,x)\|_{B^{3,2,1}_{2,2,2,\tau}}^2 , \tag{7.35}$$

because $2(\gamma_{\alpha,0}-1) \leq 2\left(\frac{2n-4m}{n-4m} - 1\right) = \frac{2n}{n-4m}$; here, without loss of generality, we assume that $\gamma_{\alpha,0} > 1$. And if $m = 1$, $n = 4m+2$, i.e. $n = 6m$, then, using a priori estimate (7.18) (for $|\alpha| = 0$), estimate (7.20) (for $s = 0$, $|\alpha| = 0$) and condition (7.3) (i.e. condition $0 < \gamma_{\alpha,0} < \frac{2n-4m}{n-4m} = 4$), we obtain:

$$\int_\Omega |u(\tau,x)|^{2\gamma_{\alpha,0}}\, dx = \int_\Omega |u(\tau,x)|^{2(\gamma_{\alpha,0}-1)} \cdot |u(\tau,x)|^2\, dx \leq$$

$$\leq \|u(\tau,x)\|_{L_{2(\gamma_{\alpha,0}-1)\frac{n}{(\gamma_{\alpha,0}-1)(n-4m)}}(\Omega)}^{2(\gamma_{\alpha,0}-1)} \cdot \|u(\tau,x)\|_{L_{2\frac{n}{n-(\gamma_{\alpha,0}-1)(n-4m)}}(\Omega)}^2 = \|u(\tau,x)\|_{L_{\frac{2n}{n-4m}}(\Omega)}^{2(\gamma_{\alpha,0}-1)} \cdot$$

$$\cdot \|u(\tau,x)\|_{L_{\frac{2n}{n-(\gamma_{\alpha,0}-1)(n-4m)}}(\Omega)}^2 \leq C \cdot \|u(\tau,x)\|_{L_{\frac{2n}{n-(\gamma_{\alpha,0}-1)(n-4m)}}(\Omega)}^2 \leq \widetilde{C} \cdot \|u\|_{B^{3,2,1}_{2,2,2,\tau}}^2 , \tag{7.36}$$

where $C > 0$ and $\widetilde{C} > 0$ are some constants not depending on u and τ; besides, without loss of generality, we assume that $1 < \gamma_{\alpha,0} < \frac{2n-4m}{n-4m} = 4$; and, consequently,

$$n - (\gamma_{\alpha,0}-1)(n-4m) > n - \left(\frac{2n-4m}{n-4m} - 1\right)(n-4m) = n - n = 0 \cdot$$

Thus, for integrals (7.27) in case $s=0$, $0 \le |\alpha| \le m$ the estimate (7.35) or (7.36) is true.

Now let's estimate integrals (7.27) in case $s \ge 1$. We consider the following three possible cases:

I. Let $n-2(3m-|\alpha_s|)=n-6m+2|\alpha_s|>0$. Then: either a) $n=4m+1$, $|\alpha_s|=m$, $s=1$, $|\alpha|=0$, or b) $n=4m+2$, $|\alpha_s|=m$, $s=1$, $|\alpha|=0$. In both cases, using estimate (7.20), a priori estimate (7.18) and conditions (7.5), (7.10), $\forall \tau \in [0,T]$ we have:

$$\int_\Omega |u(\tau,x)|^{2\gamma_{\alpha,s}} \cdot \prod_{i=1}^{s} \left|D^{\alpha_i} u(\tau,x)\right|^2 dx = \int_\Omega |u(\tau,x)|^{2\gamma_{\alpha,s}} \cdot \left|D^{\alpha_s} u(\tau,x)\right|^2 dx \le$$

$$\le \left\|D^{\alpha_s} u(\tau,x)\right\|^2_{L_{2p_s}(\Omega)} \cdot |u(\tau,x)|^{2\gamma_{\alpha,s}}_{L_{2\gamma_{\alpha,s}\cdot q}(\Omega)} \le C \cdot \left\|D^{\alpha_s} u(\tau,x)\right\|^2_{L_{2p_s}(\Omega)} \le \widetilde{C} \cdot \|u\|^2_{B^{3,2,1}_{2,2,2,\tau}} \tag{7.37}$$

where $C>0$, $\widetilde{C}>0$ are some constants not depending on u and τ,

$$p_s = \frac{n}{n-6m+2|\alpha_s|)} = \frac{n}{n-4m},\ q = \frac{n}{6m-2|\alpha_s|)} = \frac{n}{4m},\ \frac{1}{p_s}+\frac{1}{q_s}=1,$$

$$2\gamma_{\alpha,s}\cdot q = \frac{2n}{4m}\cdot\gamma_{\alpha,s} \le \frac{2n}{4m}\cdot\frac{n-(n-4m)\cdot s+2|\alpha|}{n-4m} = \frac{2n}{4m}\cdot\frac{4m}{n-4m} = \frac{2n}{n-4m};$$

here, without loss of generality, we assume that $2\gamma_{\alpha,s}\cdot q \ge 1$.

II. Let $n-6m+2|\alpha_s|=0$. Then: either a) $n=4m$, $|\alpha_s|=m$, $s=1$, $|\alpha|=0$, or b) $n=4m+2$, $|\alpha_s|=m-1$, $s=1$, $|\alpha|=0{,}1$, or c) $n=4m+2$, $|\alpha_s|=m-1$, $s=2$, $|\alpha|=0$, $|\alpha_1|=1$. In cases a) and b), using estimate (7.20), a priori estimate (7.18) (for $|\alpha|=0$) and conditions (7.10), we obtain that $\forall \tau \in [0,T]$ the estimate (7.37) is true with

$$1<p_s<+\infty,\ \frac{1}{p_s}+\frac{1}{q}=1,\ 2\gamma_{\alpha,s}\cdot q \le \frac{2n}{n-4m},$$

the possibility of choosing such a number q $(1<q<+\infty)$ and, consequently, such a number p_s $(1<p_s<+\infty)$ following from the fact that, due to (7.10), in all the cases (7.6), (7.7) and (7.8) we have $\gamma_{\alpha,s} < \dfrac{n}{n-4m}$, i.e. $2\gamma_{\alpha,s} < \dfrac{2n}{n-4m}$.

As to the case c), using estimate (7.20), a priori estimate (7.18) (for $|\alpha| = 0$ and $|\alpha| = |\alpha_1|$) and condition (7.10), $\forall \tau \in [0,T]$ we have:

$$\int_\Omega |u(\tau,x)|^{2\gamma_{\alpha,s}} \cdot \prod_{i=1}^{s} \left|D^{\alpha_i} u(\tau,x)\right|^2 dx \le$$

$$\le \left\|D^{\alpha_s} u(\tau,x)\right\|^2_{L_{2p_s}(\Omega)} \cdot \left\|D^{\alpha_1} u(\tau,x)\right\|^2_{L_{2p_1}(\Omega)} \cdot \|u(\tau,x)\|^{2\gamma_{\alpha,s}}_{L_{2\gamma_{\alpha,s}\cdot q}(\Omega)} \le$$

$$\le C \cdot \left\|D^{\alpha_s} u(\tau,x)\right\|^2_{L_{2p_s}(\Omega)} \le \widetilde{C} \cdot \|u\|^2_{B^{3,2,1}_{2,2,2,\tau}}, \tag{7.38}$$

where $C > 0$ and $\widetilde{C} > 0$ are some constants not depending on u and τ; $s = 2$,

$$1 < p_s < +\infty,\ p_1 = \frac{n}{n-4m+2|\alpha_1|} = \frac{n}{4},\ \frac{n}{n-4} < q < +\infty,$$

$$\frac{1}{p_1} + \frac{1}{p_s} + \frac{1}{q} = 1,\ 2\gamma_{\alpha,s} \cdot q = \frac{2n}{n-4m},$$

the possibility of choosing such q $\left(\frac{n}{n-4} < q < +\infty\right)$ and, consequently, such p_s following, due to (7.10), from the relation below:

$$2\gamma_{\alpha,s} \cdot \frac{n}{n-4} < \frac{2n}{n-4} \cdot \frac{n-(n-4m)\cdot s + 2|\alpha|}{n-4m} = \frac{2n}{n-4} \cdot \frac{8m-n}{n-4m} =$$

$$= \frac{2n}{4m-2} \cdot \frac{4m-2}{n-4m} = \frac{2n}{n-4m},$$

i.e. from the relation $2\gamma_{\alpha,s} \cdot \frac{n}{n-4} < \frac{2n}{n-4m}$.

III. Let $n - 6m + 2|\alpha_s| = 0$. Due to condition 5 of this theorem, in cases $n - 6m + 2|\alpha_s| < 0$ relation (7.11) holds. Then, using estimate (7.19) (for $s = 0$,

$\alpha = \alpha_s$), a priori estimate (7.18) (for $\alpha = \alpha_i$ $(i = \overline{1, s-1})$ and $\alpha = 0$) and condition (7.11), we obtain that $\forall \tau \in [0, T]$:

$$\int_\Omega |u(\tau,x)|^{2\gamma_{\alpha,s}} \cdot \prod_{i=1}^{s} \left| D^{\alpha_i} u(\tau,x) \right|^2 dx \le \left\| D^{\alpha_s} u(\tau,x) \right\|^2_{C(\overline{\Omega})} \cdot \int_\Omega |u(\tau,x)|^{2\gamma_{\alpha,s}} \cdot$$

$$\cdot \prod_{i=1}^{s-1} \left| D^{\alpha_i} u(\tau,x) \right|^2 dx \le \prod_{i=1}^{s-1} \left\| D^{\alpha_i} u(\tau,x) \right\|^2_{L_{2p_i}(\Omega)} \cdot \left\| u(\tau,x) \right\|^{2\gamma_{\alpha,s}}_{L_{2\gamma_{\alpha,s} \cdot q}(\Omega)} \cdot$$

$$\cdot \left\| D^{\alpha_s} u(\tau,x) \right\|^2_{C(\overline{\Omega})} \le C \cdot \left\| D^{\alpha_s} u(\tau,x) \right\|^2_{C(\overline{\Omega})} \le \widetilde{C} \cdot \|u\|^2_{B^{3,2,1}_{2,2,2,\tau}}, \tag{7.39}$$

where $C > 0$ and $\widetilde{C} > 0$ are some constants not depending on u and τ;

$$p_i{}^{*)} = \frac{n}{n - 4m + 2|\alpha_i|} \quad (i = \overline{1, s-1}),$$

$$q = \frac{n}{n - \sum_{i=1}^{s-1} (n - 4m + 2|\alpha_i|)} = \frac{n}{n - (n-4m)(s-1) - 2\sum_{i=1}^{s-1} |\alpha_i|},$$

and it is evident that

$$n - 4m + 2|\alpha_i| \ge 2|\alpha_i| \ge 2 \quad (i = \overline{1, s-1}),$$

$$n - \sum_{i=1}^{s-1} (n - 4m + 2|\alpha_i|) = n - (n-4m)(s-1) - 2\sum_{i=1}^{s-1} |\alpha_i| \ge$$

$$\ge n - (n-4m)(s-1) - 2(m - |\alpha_s| - |\alpha|) = 2n - 6m - (n-4m)s + 2|\alpha| + 2|\alpha_s| \ge$$

$$\ge 2n - 6m - 2m + 2|\alpha| + 2|\alpha_s| = 2n - 8m + 2|\alpha| + 2|\alpha_s| \ge 2|\alpha| + 2|\alpha_s| \ge 2|\alpha_s| \ge 2,$$

$$\sum_{i=1}^{s-1} \frac{1}{p_i} + \frac{1}{q} = 1,$$

---

*) For $s = 1$ these numbers are not available.

$$2\gamma_{\alpha,s}\cdot q\leq\gamma_{\alpha,s}\cdot\frac{2n}{2n-6m-(n-4m)s+2|\alpha|+2|\alpha_s|}\leq$$

$$\leq\frac{2n-6m-(n-4m)s+2|\alpha|+2}{n-4m}\cdot\frac{2n}{2n-6m-(n-4m)s+2|\alpha|+2|\alpha_s|}\leq\frac{2n}{n-4m}.$$

Thus, integrals (7.27) are estimated in all the cases, i.e. the estimates (7.35)-(7.39) are true for them.

Now let's estimate integrals (7.28). We consider the following three possible cases:

1. Let $n=4m+2,\ |\beta|=0,\ l=m$. Consequently, due to (7.12), $\delta_{\beta,l}=0$. Then, using estimate (7.20) (for $s=1$ and $\alpha=\beta_0$) and a priori estimate (7.19) (for $\alpha=\beta_j,\ j=\overline{1,l}$), we obtain that $\forall\tau\in[0,T]$:

$$\int_\Omega|u(\tau,x)|^{2\delta_{\beta,l}}\cdot\prod_{j=1}^{l}\left|D^{\beta_j}u(\tau,x)\right|^2\cdot\left|D^{\beta_0}u_\tau(\tau,x)\right|^2dx=$$

$$=\int_\Omega\prod_{j=1}^{l}\left|D^{\beta_j}u(\tau,x)\right|^2\cdot\left|D^{\beta_0}u_\tau(\tau,x)\right|^2dx\leq\left\|D^{\beta_0}u_\tau(\tau,x)\right\|^2_{L_{2p_0}(\Omega)}\cdot$$

$$\cdot\prod_{j=1}^{l}\left\|D^{\beta_j}u(\tau,x)\right\|^2_{L_{2p_j}(\Omega)}\leq C\cdot\left\|D^{\beta_0}u_\tau(\tau,x)\right\|^2_{L_{2p_0}(\Omega)}\leq\widetilde{C}\cdot\|u\|^2_{B^{3,2,1}_{2,2,2,\tau}},\tag{7.40}$$

where $p_j=\dfrac{n}{n-4m+2|\beta_j|}$ $(j=\overline{0,l})$, and it is evident that

$$\sum_{j=0}^{l}\frac{1}{p_j}=\sum_{j=0}^{l}\frac{n-4m+2|\beta_j|}{n}=\frac{(n-4m)(l+1)+2\sum_{j=0}^{l}|\beta_j|}{n}=$$

$$=\frac{(n-4m)(l+1)+2m}{n}=\frac{2(l+1)+2m}{n}=\frac{2(m+1)+2m}{n}=\frac{4m+2}{n}=\frac{n}{n}=1,$$

because in this case $l=m,\ |\beta_j|=1\ (j=\overline{1,l}),\ |\beta_0|=0$.

Remark. It should be especially noted that the estimate (7.40), generally speaking, is not true for dimensions $n>4m+2$. That's exactly the reason why we failed to prove the existence in large for the classical solution of problem (0.1)-(0.3) in cases $n>4m+2$.

2. Let $n = 4m$, $|\beta_0| = 0$. Then, using estimate (7.20) (for $s = 1$ and $|\alpha| = |\beta_0| = 0$), a priori estimate (7.18) (for $\alpha = \beta_j$ $(j = \overline{1,l})$ and $|\alpha| = 0$) and condition (7.12), we obtain that $\forall \tau \in [0,T]$:

$$\int_\Omega |u(\tau,x)|^{2\delta_{\beta,l}} \cdot \prod_{j=1}^{l} \left|D^{\beta_j} u(\tau,x)\right|^2 \cdot \left|D^{\beta_0} u_\tau(\tau,x)\right|^2 dx \le$$

$$\le \left\|D^{\beta_0} u_\tau(\tau,x)\right\|^2_{L_{2p_0}(\Omega)} \cdot \prod_{j=1}^{l} \left\|D^{\beta_j} u(\tau,x)\right\|^2_{L_{2p_j}(\Omega)} \cdot \|u(\tau,x)\|^{2\delta_{\beta,l}}_{L_{2\delta_{\beta,l}\cdot q}(\Omega)} \le$$

$$\le C \cdot \left\|D^{\beta_0} u_\tau(\tau,x)\right\|^2_{L_{2p_0}(\Omega)} \le \tilde{C} \cdot \|u\|^2_{B^{3,2,1}_{2,2,2,\tau}}, \qquad (7.41)$$

where

$$1 < p_0 < +\infty,\; p_j = \frac{n}{n-4m+2|\beta_j|} = \frac{n}{2|\beta_j|} \quad (j = \overline{0,l}),$$

$$1 < q < +\infty,\; \frac{1}{p_0} + \sum_{j=1}^{l} \frac{1}{p_j} + \frac{1}{q} = 1,$$

the possibility of choosing such numbers p_0 and q following from the relation below:

$$\sum_{j=1}^{l} \frac{1}{p_j} = \frac{2}{n} \cdot \sum_{j=1}^{l} |\beta_j| \le \frac{2m}{n} = \frac{2m}{4m} = \frac{1}{2} < 1.$$

3. Now we consider all other cases for integrals (7.28). It is evident that in these cases $n - 4m + 2|\beta_0| > 0$ and

$$n - \sum_{j=0}^{l} (n - 4m + 2|\beta_j|) = n - (n-4m)(l+1) - 2\sum_{j=0}^{l} |\beta_j| \ge$$

$$\ge n - (n-4m)(l+1) - 2(m - |\beta|) = 2m + 2|\beta| - (n-4m) \cdot l > 0.$$

Then, using estimate (7.20) (for $s=1, \alpha=\beta_0$), a priori estimate (7.18) (for $\alpha=\beta_j$ $(j=\overline{1,l})$ and $|\alpha|=0$) and condition (7.12), we obtain that $\forall \tau \in [0,T]$ estimate (7.41) holds with

$$p_j = \frac{n}{n-4m+2|\beta_j|} \quad (j=\overline{0,l}), \quad q = \frac{n}{n-\sum\limits_{j=0}^{l}(n-4m+2|\beta_j|)},$$

and it is evident that

$$\frac{1}{p_0} + \sum_{j=1}^{l}\frac{1}{p_j} + \frac{1}{q} = 1,$$

$$2\delta_{\beta,l} \cdot q \le 2 \cdot \frac{2m-(n-4m)\cdot l + 2|\beta|}{n-4m} \cdot \frac{n}{n-\sum\limits_{j=0}^{l}(n-4m+2|\beta_j|)} \le$$

$$\le 2 \cdot \frac{2m-(n-4m)\cdot l + 2|\beta|}{n-4m} \cdot \frac{n}{2m+2|\beta|-(n-4m)\cdot l} = \frac{2n}{n-4m}.$$

Thus, integrals (7.28) are also estimated in all cases, i.e. estimates (7.40) and (7.41) are true for them.

Now, using a priori estimates (7.21), (7.17) and, for integrals (7.25)-(7.28), a priori estimates (7.29), (7.30), (7.33), (7.34) and estimates (7.35)-(7.41), from (6.9) we obtain that $\forall t \in [0,T]$:

$$\|u\|^2_{B^{3,2,1}_{2,2,2,t}} \le C_1 + C_2 \cdot \int_0^t \|u\|^2_{B^{3,2,1}_{2,2,2,\tau}} d\tau, \tag{7.42}$$

where $C_1 > 0$ and $C_2 > 0$ are some constants not depending on u and t. On applying Bellman's inequality, from (7.42) we obtain

$$\|u\|^2_{B^{3,2,1}_{2,2,2,T}} \le C_1 \cdot \exp\{C_2 \cdot T\} \equiv C_3, \tag{7.43}$$

i.e. all the possible classical solutions $u(t,x)$ of problem (0.1)-(0.3), belonging to the space $B^{k,k-1,k-2}_{2,2,2,T}$, are a priori bounded in $B^{3,2,1}_{2,2,2,T}$. In view of the fact that for $n < 2m$ we have $k \equiv \left[\frac{n}{2m}\right] + 3 = 3$, this theorem is now proved for ***n<2m***. Therefore we will continue proving only for dimensions ***2m≤n≤4m+2***. Under the conditions of this theorem it is evident that the functions $u_s(t) = \int_\Omega u(t,x)v_s(x)dx$ satisfy not only systems (1.13) and (1.17), but also the system (1.19) with $l = 1$, $☞(u(\tau,x)) = F(\tau,x,u(\tau,x))$. Then, using Bessel's inequality, we obtain from system (1.19) (with $l = 1$) that $\forall t \in [0,T]$:

$$\|u\|^2_{B^{4,3,2}_{2,2,2,t}} \leq 2\|Z_4\|^2_{B^{4,3,2}_{2,2,2,T}} + C \cdot \int_0^t \left\| \mathsf{A}\left(\frac{\partial}{\partial\tau}\{\mathsf{F}\,(u(\tau,x))\}\right)\right\|^2_{L_2(\Omega)} d\tau \leq$$

$$\leq 2\|Z_4\|^2_{B^{4,3,2}_{2,2,2,T}} + \tilde{C} \cdot \int_0^t \left\|\frac{\partial}{\partial\tau}\{F(\tau,x,u(\tau,x))\}\right\|^2_{W_2^{2m}(\Omega)} d\tau, \tag{7.44}$$

where function $Z_4(t,x)$ is defined by (1.21), and $C > 0$, $\tilde{C} > 0$ are some numbers not depending on u and t.

Next, a priori estimate (7.43), in particular, implies the following a priori estimates:

$$\|u(t,x)\|_{W_2^{3m}(\Omega)} \leq C, \quad \|u_t(t,x)\|_{W_2^{2m}(\Omega)} \leq C \ \ \forall t \in [0,T]. \tag{7.45}$$

Due to the imbedding theorems of Sobolev and Pokhozhayev, a priori estimates (7.45), in turn, imply the following a priori estimates:

$$\left\|☞^{\alpha} u(t,x)\right\|_{C(\bar{Q}_T)} \leq C\begin{pmatrix} m \geq 2, & 2m \leq n \leq 4m+2; \\ m = 1, & 2m \leq n \leq 4m+1; \end{pmatrix} 0 \leq |\alpha| \leq k_0 \equiv 3m - \left[\frac{n}{2}\right] - 1 \Bigg), \tag{7.46}$$

$$\int_\Omega e^{|u(t,x)|^2} dx \leq C(m = 1,\ n = 4m + 2 = 6;\ t \in [0,T]), \tag{7.47}$$

$$\left\| D^{\alpha} u(t,x) \right\|_{L_q(\Omega)} \le C_q \begin{pmatrix} m \ge 1,\ 2m \le n \le 4m+2;\ k_0+1 \equiv 3m - \left[\frac{n}{2}\right] \le |\alpha| \le 3m; \\ 1 \le q \le \frac{2n}{n-6m+2|\alpha|},\ q < +\infty;\ t \in [0,T] \end{pmatrix}, \qquad (7.48)$$

where the constants $C > 0$ and $C_q > 0$ do not depend on u.

Besides, using the structure of the space $B^{4,3,2}_{2,2,2,T}$ and imbedding theorems of Sobolev, $\forall \tau \in [0,T]$ we have:

$$\left\| D^{\alpha} u_{\tau}(t,x) \right\|_{L_q(\Omega)} \le C_q \cdot \|u\|_{B^{4,3,2}_{2,2,2,\tau}} \begin{pmatrix} m=1,\ n=4m+2;\ 0 \le |\alpha| \le 2m; \\ 1 \le q \le \frac{2n}{n-6m+2|\alpha|},\ q < +\infty \end{pmatrix}. \qquad (7.49)$$

A priori estimate (7.46) (with $|\alpha| = 0$) implies the following a priori estimates:

$$\left\| \partial^{s} F(t,x,u(t,x)) / \partial t^{\alpha} \partial x^{\beta} \partial u^{\gamma} \right\|_{C(\overline{Q}_T)} \le C \begin{pmatrix} m \ge 2, \quad 2m \le n \le 4m+2 \\ \text{or } m=1, \quad 2m \le n \le 4m+1; \\ 0 \le \alpha + |\beta| + \gamma = s \le 2m+1; \\ \alpha \le 1; |\beta| \le m \end{pmatrix}. \qquad (7.50)$$

Next, it is evident that the expressions

$$\frac{\partial^{i}}{\partial x_1^{\alpha_1} \ldots \partial x_n^{\alpha_n}} \frac{\partial}{\partial \tau} \{F(\tau,x,u(\tau,x))\} \quad (i = \overline{0,2m}) \qquad (7.51)$$

consist of finite number of summands of the following form:

$$\frac{\partial^{1+|\alpha|+s} F(\tau,x,u(\tau,x))}{\partial \tau \partial x^{\alpha} \partial u^{s}} \cdot \prod_{i=1}^{s} D^{\alpha_i} u(\tau,x) \quad \left(0 \le s \le 2m;\ \sum_{i=1}^{s} |\alpha_i| + |\alpha| \le 2m\right), \qquad (7.52)$$

$$\frac{\partial^{1+|\beta|+l} F(\tau,x,u(\tau,x))}{\partial x^{\beta} \partial u^{1+l}} \cdot \prod_{j=1}^{l} D^{\beta_j} u(\tau,x) \cdot D^{\beta_0} u_{\tau}(\tau,x) \ \left(0 \le l \le 2m; \sum_{j=0}^{l} |\beta_i| + |\beta| \le 2m\right)$$

(7.53)

where $1 \le |\alpha_1| \le \ldots \le |\alpha_s| \le 2m;\ 1 \le |\beta_1| \le \ldots \le |\beta_l| \le 2m;$ besides, for $s = 0$ we assume that $\prod_{i=1}^{s} \ldots$ is 1, and for $l = 0$ we assume that $\prod_{j=1}^{l} \ldots$ is 1.

Therefore, due to a priori estimates (7.50), (7.46) and estimates (7.13), (7.14), in order to have an estimate ($\forall\, \tau \in [0,T]$) for the expression

$$\left\| \frac{\partial}{\partial \tau}\{F(\tau,x,u(\tau,x))\} \right\|^2_{W_2^{2m}(\Omega)}$$

it suffices to have an estimate ($\forall\, \tau \in [0,T]$) for the following integrals:

$$\int_\Omega \prod_{i=1}^{s} \left| D^{\alpha_i} u(\tau,x) \right|^2 dx \left(\begin{array}{l} m \ge 2, 2m \le n \le 4m+2;\ m=1,\ 2m \le n \le 4m+1; \\ 1 \le s \le 2m;\ \sum\limits_{i=1}^{s} |\alpha_i| \le 2m;\ 3m - \left[\frac{n}{2}\right] \le |\alpha_1| \le \ldots \le |\alpha_s| \le 2m \end{array} \right), \tag{7.54}$$

$$\int_\Omega \prod_{j=1}^{l} \left| D^{\beta_j} u(\tau,x) \right|^2 \cdot \left| D^{\beta_0} u_\tau(\tau,x) \right|^2 dx \left(\begin{array}{l} m \ge 2, 2m \le n \le 4m+2; \\ m=1,\ 2m \le n \le 4m+1; \\ 0 \le l \le 2m,\ \sum\limits_{j=0}^{l} |\beta_j| \le 2m, \\ 3m - \left[\frac{n}{2}\right] \le |\beta_1| \le \ldots \le |\beta_l| \le 2m \end{array} \right), \tag{7.55}$$

$$\int_\Omega e^{2|u(\tau,x)|^2} \cdot \prod_{i=1}^{s} \left| D^{\alpha_i} u(\tau,x) \right|^2 dx \left(\begin{array}{l} m=1,\ n=4m+2;\ 0 \le s \le 2m; \\ \sum\limits_{i=1}^{s} |\alpha_i| \le 2m;\ 3m - \left[\frac{n}{2}\right] \le |\alpha_1| \le \ldots \le |\alpha_s| \le 2m \end{array} \right), \tag{7.56}$$

$$\int_\Omega e^{2|u(\tau,x)|^2} \cdot \prod_{j=1}^{l} \left| D^{\beta_j} u(\tau,x) \right|^2 \cdot \left| D^{\beta_0} u_\tau(\tau,x) \right|^2 dx \left(\begin{array}{l} m=1,\ n=4m+2;\ 0 \le l \le 2m; \\ \sum\limits_{l=0}^{l} |\beta_j| \le 2m; \\ 3m - \left[\frac{n}{2}\right] \le |\beta_1| \le \ldots \le |\beta_l| \le 2m \end{array} \right)$$

(7.57)

Using a priori estimate (7.48), $\forall \tau \in [0,T]$ for integrals (7.54) we have:

$$\int_{\Omega}\prod_{i=1}^{s}\left|D^{\alpha_i}u(\tau,x)\right|^2 dx \le \prod_{i=1}^{s}\left\|D^{\alpha_i}u(\tau,x)\right\|_{L_{2p_i}(\Omega)}^2 \le C, \tag{7.58}$$

where $C > 0$ is some constant not depending on u and τ, and

$$1 \le p_i \le \frac{n}{n-6m+2|\alpha_i|},\ p_i < +\infty \quad (i=\overline{1,s});\ \sum_{i=1}^{s}\frac{1}{p_i}=1,$$

the possibility of last equality following from the fact that in case (7.54) we have $n - 6m < 0$ and

$$\sum_{i=1}^{s}\frac{n-6m+2|\alpha_i|}{n} = \frac{(n-6m)s+2\sum_{i=1}^{s}|\alpha_i|}{n} \le \frac{(n-6m)s+4m}{n} \le$$

$$\le \frac{n-6m+4m}{n} = \frac{n-2m}{n} < 1.$$

Next, using a priori estimates (7.47) and (7.48), $\forall \tau \in [0,T]$ for integrals (7.56) we have:

$$\int_{\Omega} e^{2|u(\tau,x)|^2}\cdot\prod_{i=1}^{s}\left|D^{\alpha_i}u(\tau,x)\right|^2 dx \le \left\{\int_{\Omega} e^{2q|u(\tau,x)|^2}dx\right\}^{\frac{1}{q}}\cdot\prod_{i=1}^{s}\left\|D^{\alpha_i}u(\tau,x)\right\|_{L_{2p_i}(\Omega)}^2 \le C, \tag{7.59}$$

where $C > 0$ is some constant not depending on u and τ, and

$$1 \le p_i \le \frac{n}{n-4m+2|\alpha_1|},\ p_i < +\infty \quad (i=\overline{1,s});\ 1 < q + \infty;\ \frac{1}{q}+\sum_{i=1}^{s}\frac{1}{p_i}=1,$$

the possibility of last equality following from the fact that in case (7.56) we have $n - 6m = 0$ and

$$\sum_{i=1}^{s}\frac{n-6m+2|\alpha_i|}{n}=\frac{2\sum_{i=1}^{s}|\alpha_i|}{n}\le\frac{4m}{n}=\frac{n-2}{n}<1\cdot$$

Now, using estimate (7.49) (for $\alpha=\beta_0$) and a priori estimate (7.48), $\forall\tau\in[0,T]$ for integrals (7.55) we have:

$$\int_{\Omega}\prod_{j=1}^{l}\left|\mathrm{D}^{\beta_j}u(\tau,x)\right|^2\cdot\left|\mathrm{D}^{\beta_0}u_\tau(\tau,x)\right|^2dx\le\left\|\mathrm{D}^{\beta_0}u_\tau(\tau,x)\right\|^2_{L_{2p_0}(\Omega)}\cdot$$

$$\cdot\prod_{j=1}^{l}\left\|\mathrm{D}^{\beta_j}u(\tau,x)\right\|^2_{L_{2p_i}(\Omega)}\le C\cdot\left\|\mathrm{D}^{\beta_0}u_\tau(\tau,x)\right\|^2_{L_{2p_0}(\Omega)}\le\widetilde{C}\cdot\|u\|^2_{B^{4,3,2}_{2,2,2,\tau}},\qquad(7.60)$$

where $C>0$ and $\widetilde{C}>0$ are some constants not depending on u and τ, and

$$1\le p_j\le\frac{n}{n-6m+2|\beta_j|},\ p_j<+\infty\ \ (j=\overline{0,l}),\ \sum_{j=0}^{l}\frac{1}{p_j}=1,$$

the possibility of last equality following from the fact that in case (7.55) we have $n-6m<0$ and

$$\sum_{j=0}^{l}\frac{n-6m+2|\beta_j|}{n}=\frac{(n-6m)(l+1)+2\sum_{j=0}^{l}|\beta_j|}{n}\le\frac{(n-6m)(l+1)+4m}{n}\le$$

$$\le\frac{n-6m+4m}{n}=\frac{n-2m}{n}<1.$$

And for integrals (7.57), using a priori estimates (7.47), (7.48) and estimate (7.49) (for $\alpha=\beta_0$), $\forall\tau\in[0,T]$ we have:

$$\int_{\Omega}e^{2|u(\tau,x)|^2}\cdot\prod_{j=1}^{l}\left|\mathrm{D}^{\beta_j}u(\tau,x)\right|^2\cdot\left|\mathrm{D}^{\beta_0}u_\tau(\tau,x)\right|^2dx\le$$

$$\le\left\|\mathrm{D}^{\beta_0}u_\tau(\tau,x)\right\|^2_{L_{2p_0}(\Omega)}\cdot\prod_{j=1}^{l}\left\|\mathrm{D}^{\beta_j}u_\tau(\tau,x)\right\|^2_{L_{2p_j}(\Omega)}\cdot\left\{\int_{\Omega}e^{2q|u(\tau,x)|^2}dx\right\}^{\frac{1}{q}}\le$$

$$\leq C \cdot \left\| \mathrm{D}^{\beta_0} u_\tau(\tau,x) \right\|^2_{L_{2p_0}(\Omega)} \leq \widetilde{C} \cdot \|u\|^2_{B^{4,3,2}_{2,2,2,\tau}}, \tag{7.61}$$

where $C > 0$ and $\widetilde{C} > 0$ are some constants not depending on u and τ, and

$$1 \leq p_j \leq \frac{n}{n - 6m + 2|\beta_j|}, \ p_j < +\infty \ \ (j = \overline{0,l})\,;\ 1 < q + \infty\,;\ \frac{1}{q} + \sum_{j=0}^{l} \frac{1}{p_j} = 1,$$

the possibility of last equality following from the fact that in case (7.57) we have $n - 6m = 0$ and

$$\sum_{j=0}^{l} \frac{n - 6m + 2|\beta_j|}{n} = \frac{2\sum_{j=0}^{l} |\beta_j|}{n} \leq \frac{4m}{n} < \frac{n-2}{n} < 1.$$

Now, using a priori estimates (7.50), (7.46) and, for integrals (7.54)-(7.57), estimates (7.58)-(7.61), we obtain from (7.44) that $\forall t \in [0,T]$:

$$\|u\|^2_{B^{4,3,2}_{2,2,2,t}} \leq C_1 + C_2 \cdot \int_0^t \|u\|^2_{B^{4,3,2}_{2,2,2,\tau}} d\tau, \tag{7.62}$$

where $C_1 > 0$ and $C_2 > 0$ are some constants not depending on u and t. On applying Bellman's inequality, from (7.62) we obtain:

$$\|u\|^2_{B^{4,3,2}_{2,2,2,T}} \leq C_1 \cdot \exp\{C_2 \cdot T\} \equiv C_3, \tag{7.63}$$

i.e. all the possible classical solutions $u(t,x)$ of problem (0.1)-(0.3), belonging to the space $B^{k,k-1,k-2}_{2,2,2,T}$, are a priori bounded in $B^{4,3,2}_{2,2,2,T}$. In view of the fact that for $2m \leq n < 4m$ we have $k \equiv \left[\frac{n}{2m}\right] + 3 = 4$, this theorem is now proved for ***2m≤n<4m***, too. Therefore we will continue proving only for dimensions ***4m≤n≤4m+2***. Under the conditions of this theorem it is evident that the functions $u_s(t) = \int_\Omega u(t,x) v_s(x) dx$ satisfy not only systems (1.13) and (1.17), but also the system (1.20) with $l = 1$, $\mathsf{F}(u(\tau,x)) = F(\tau,x,u(\tau,x))$. Then, using inequality (1.10) with $\mathsf{A}\left(\frac{\partial}{\partial\tau}\{\mathsf{F}(u(\tau,x))\}\right)$ instead of function $u(x)$, we obtain from system (1.20) (with $l=1$) that $\forall t\in[0,T]$:

$$\|u\|^2_{B^{5,4,3}_{2,2,2,t}} \le 2\|Z_5\|^2_{B^{5,4,3}_{2,2,2,T}} + C\cdot\int_0^t\left\|\mathsf{A}\left(\frac{\partial}{\partial\tau}\{\mathsf{F}(u(\tau,x))\}\right)\right\|^2_{W_2^m(\Omega)} d\tau \le$$

$$\le 2\|Z_5\|^2_{B^{5,4,3}_{2,2,2,T}} + \tilde{C}\cdot\int_0^t\left\|\frac{\partial}{\partial\tau}\{F(\tau,x,u(\tau,x))\}\right\|^2_{W_2^{3m}(\Omega)} d\tau, \tag{7.64}$$

where the function $Z_5(t,x)$ is defined by (1.21) with $l=1$, and $C>0$, $\tilde{C}>0$ are some constants not depending on u and t.

Next, a priori estimate (7.62), in particular, implies the following a priori estimates:

$$\|u(t,x)\|_{W_2^{4m}(\Omega)} \le C,\ \ \|u_t(t,x)\|_{W_2^{3m}(\Omega)} \le C\ \ \forall t\in[0,T]. \tag{7.65}$$

Due to Sobolev's imbedding theorems, a priori estimates (7.65), in turn, imply the following a priori estimates:

$$\left\|\partial_t^s\partial^\alpha(u(t,x))\right\|_{C(\overline{Q}_T)} \le C\left(4m\le n\le 4m+2;\ 0\le s\cdot m+|\alpha|\le k_0\equiv 4m-\left[\frac{n}{2}\right]-1;\ s\le 1\right) \tag{7.66}$$

$$\left\|\partial_t^s\partial^\alpha(u(t,x))\right\|_{L_q(\Omega)} \le C_q\left(\begin{array}{l} 4m\le n\le 4m+2;\ k_0+1=4m-\left[\frac{n}{2}\right]\le s\cdot m+|\alpha|\le 4m, s\le 1; \\ 1\le q\le\frac{2n}{n-8m+2(s\cdot m+|\alpha|)},\ \ q<+\infty;\ t\in[0,T]\end{array}\right) \tag{7.67}$$

where $C>0$ and $C_q>0$ are the constants not depending on u and t.

A priori estimate (7.66) (with $s=0$, $|\alpha|=0$) implies the following a priori estimates:

$$\left\|\partial^s F(t,x,u(t,x))/\partial t^\alpha \partial x^\beta \partial u^\gamma\right\|_{C(\bar{Q}_T)} \le C\begin{pmatrix} 4m \le n \le 4m+2; \\ 0 \le \alpha \cdot m + |\beta| + \gamma = s \le 3m+1; \\ \alpha \le 1; |\beta| \le 3m \end{pmatrix}. \tag{7.68}$$

Next, it is evident that the expressions

$$\frac{\partial^i}{\partial x_1^{\alpha_1} \ldots \partial x_n^{\alpha_n}} \frac{\partial}{\partial \tau}\{F(\tau,x,u(\tau,x))\} \quad (i=\overline{0,3m}) \tag{7.69}$$

consist of finite number of summands of the following form:

$$\frac{\partial^{1+|\alpha|+s} F(\tau,x,u(\tau,x))}{\partial \tau\, \partial x^\alpha \partial u^s} \cdot \prod_{i=1}^{s} \mathrm{D}^{\alpha_i} u(\tau,x) \quad (0 \le s \le 3m;\ |\alpha| + \sum_{i=1}^{s} |\alpha_i| \le 3m), \tag{7.70}$$

$$\frac{\partial^{1+|\beta|+l} F(\tau,x,u(\tau,x))}{\partial x^\beta \partial u^{1+l}} \cdot \prod_{j=1}^{l} \mathrm{D}^{\beta_j} u(\tau,x) \cdot \mathrm{D}^{\beta_0} u_\tau(\tau,x) \ (0 \le l \le 3m; |\beta| + \sum_{j=1}^{l} |\beta_j| \le 3m), \tag{7.71}$$

where $1 \le |\alpha_1| \le \ldots \le |\alpha_s| \le 3m;\ 1 \le |\beta_1| \le \ldots \le |\beta_l| \le 3m;$ besides, for $s = 0$ we assume that $\prod_{i=1}^{s} \ldots$ is 1, and for $l = 0$ we assume that $\prod_{j=1}^{l} \ldots$ is 1.

Therefore, due to a priori estimates (7.68) and (7.66), in order to have an estimate ($\forall \tau \in [0,T]$) for the expression

$$\left\|\frac{\partial}{\partial \tau}\{F(\tau,x,u(\tau,x))\}\right\|_{W_2^{3m}(\Omega)}^2,$$

it suffices to have an estimate ($\forall \tau \in [0,T]$) for the following integrals:

$$\int_\Omega \prod_{i=1}^{s} \left|\mathrm{D}^{\alpha_i} u(\tau,x)\right|^2 dx \begin{pmatrix} 4m \le n \le 4m+2;\ 1 \le s \le 2m; \\ \sum_{i=1}^{s} |\alpha_i| \le 3m;\ k_0 + 1 = 4m - \left[\frac{n}{2}\right] \le |\alpha_1| \le \ldots \le |\alpha_s| \le 3m \end{pmatrix}, \tag{7.72}$$

$$\int_\Omega \prod_{j=1}^{l} \left| D^{\beta_j} u(\tau,x) \right|^2 \cdot \left| D^{\beta_0} u_\tau(\tau,x) \right|^2 dx \left(\begin{array}{l} 4m \le n \le 4m+2;\ 0 \le l \le 3m;\ \sum_{j=0}^{l} |\beta_j| \le 3m; \\ 4m - \left[\frac{n}{2}\right] \le |\beta_1| \le \ldots \le |\beta_l| \le 3m; \\ 3m - \left[\frac{n}{2}\right] \le |\beta_0| \le 3m. \end{array} \right) \tag{7.73}$$

Using a priori estimates (7.67) (for $s = 0,\ \alpha = \alpha_i$), $\forall \tau \in [0,T]$ for integrals (7.72) we have:

$$\int_\Omega \prod_{i=1}^{s} \left| D^{\alpha_i} u(\tau,x) \right|^2 dx \le \prod_{i=1}^{s} \left\| D^{\alpha_i} u(\tau,x) \right\|^2_{L_{2p_i}(\Omega)} \le C, \tag{7.74}$$

where $C > 0$ is some constant not depending on u and τ, and

$$1 \le p_i \le \frac{n}{n - 8m + 2|\alpha_i|},\ p_i < +\infty \quad (i = \overline{1,s});\ \sum_{i=1}^{s} \frac{1}{p_i} = 1,$$

the possibility of last equality following from $n - 8m < 0$ and

$$\sum_{i=1}^{s} \frac{n - 8m + 2|\alpha_i|}{n} = \frac{(n-8m)\cdot s + 2\sum_{i=1}^{s} |\alpha_i|}{n} \le \frac{(n-8m)\cdot s + 6m}{n} \le$$

$$\le \frac{n - 8m + 6m}{n} = \frac{n-2m}{n} < 1.$$

Next, using a priori estimates (7.67) and (7.48) (for $s = 0,\ \alpha = \beta_j\ (j = \overline{1,l})$ and $s = 1,\ \alpha = \beta_0$), $\forall \tau \in [0,T]$ for integrals (7.73) we have:

$$\int_\Omega \prod_{j=1}^{l} \left| D^{\beta_j} u(\tau,x) \right|^2 \cdot \left| D^{\beta_0} u_\tau(\tau,x) \right|^2 dx \le$$

$$\leq \prod_{j=1}^{l} \left\| \mathrm{D}^{\beta_j} u(\tau,x) \right\|^2_{L_{2p_j}(\Omega)} \cdot \left\| \mathrm{D}^{\beta_0} u_\tau(\tau,x) \right\|^2_{L_{2p_0}(\Omega)} \leq C, \tag{7.75}$$

where $C > 0$ is a constant not depending on u and τ, and

$$1 \leq p_j \leq \frac{n}{n - 8m + 2|\beta_j|},\ p_j < +\infty \quad (j = \overline{1,l});$$

$$1 \leq p_0 \leq \frac{n}{n - 6m + 2|\beta_0|},\ p_0 < +\infty;\ \sum_{j=0}^{l} \frac{1}{p_j} = 1,$$

the possibility of last equality following from the fact that for $l = 0$ this equality is obvious, and for $l \geq 1$ (due to $n - 8m < 0$) we have

$$\frac{n - 6m + 2|\beta_0|}{n} + \sum_{j=1}^{l} \frac{n - 8m + 2|\beta_j|}{n} = \frac{n - 6m + (n-8m)\cdot l + 2\sum_{j=0}^{l}|\beta_j|}{n} \leq$$

$$\leq \frac{n - 6m + (n-8m)\cdot l + 6m}{n} = \frac{n + (n-8m)\cdot l}{n} < \frac{n}{n} = 1.$$

Thus, integrals (7.72) and (7.73) are a priori bounded.

Now, using a priori estimates (7.68), (7.66), (7.74) and (7.75), we obtain from (7.64) that $\forall t \in [0,T]$

$$\|u\|^2_{B^{5,4,3}_{2,2,2,t}} \leq C_0.$$

Consequently,

$$\|u\|^2_{B^{5,4,3}_{2,2,2,T}} \leq C_0, \tag{7.76}$$

i.e. all the possible classical solutions $u(t,x)$ of problem (0.1)-(0.3), belonging to the space $B^{k,k-1,k-2}_{2,2,2,T}$, are a priori bounded in $B^{5,4,3}_{2,2,2,T}$. In view of the fact that for

$$m \geq 2,\ 4m \leq n \leq 4m + 2;\ m = 1,\ 4m \leq n \leq 4m + 1 \tag{7.77}$$

we have $k \equiv \left[\dfrac{n}{2m}\right] + 3 = 5$, this theorem is now proved in cases (7.77), too. Therefore, to finish the proof of theorem, we only need to consider the case ***m=1, n=4m+2=6***. Under the conditions of this theorem it is evident that the functions $u_s(t) = \int\limits_{\Omega} u(t,x)v_s(x)dx$ satisfy not only systems (1.13) and (1.17), but also the system (1.19) with $l = 2$, ☞$(u(\tau,x)) = F(\tau,x,u(\tau,x))$. Then, using Bessel's inequality, we obtain from system (1.19) (with $l = 2$) that $\forall t \in [0,T]$:

$$\|u\|^2_{B^{6,5,4}_{2,2,2,t}} \leq 2\|Z_6\|^2_{B^{6,5,4}_{2,2,2,T}} + C \cdot \int_0^t \left\| \mathrm{A}^2 \left(\frac{\partial}{\partial \tau} \{\mathrm{F}\,(u(\tau,x))\} \right) \right\|^2_{L_2(\Omega)} d\tau \leq$$

$$\leq 2\|Z_6\|^2_{B^{6,5,4}_{2,2,2,T}} + \tilde{C} \cdot \int_0^t \left\| \frac{\partial}{\partial \tau} \{\mathrm{F}\,(u(\tau,x))\} \right\|^2_{W_2^{4m}(\Omega)} d\tau, \qquad (7.78)$$

where the function $Z_6(t,x)$ is defined by (1.21) with $l = 2$, and $C > 0$, $\tilde{C} > 0$ are some constants not depending on u and t.

Next, a priori estimate (7.76), in particular, implies the following a priori estimates:

$$\|u(t,x)\|_{W_2^{5m}(\Omega)} \leq C, \quad \|u_t(t,x)\|_{W_2^{4m}(\Omega)} \leq C \ \forall t \in [0,T]. \qquad (7.79)$$

Due to Sobolev's imbedding theorems, a priori estimates (7.79), in turn, imply the following a priori estimates:

$$\|u(t,x)\|_{C^1(\bar{Q}_T)} \leq C\,(m=1,\ n=6), \qquad (7.80)$$

$$\left\| ☞_t^s ☞^\alpha (u(t,x)) \right\|_{L_q(\Omega)} \leq C_q \begin{pmatrix} m=1,\ n=4m+2=6;\ 5m-\left[\dfrac{n}{2}\right]=2 \leq s\cdot m+|\alpha| \leq 5m=5; \\ s \leq 1;\ 1 \leq q \leq \dfrac{2n}{n-10m+2(s\cdot m+|\alpha|)},\quad q<+\infty;\ t\in[0,T] \end{pmatrix}, \qquad (7.81)$$

where $C > 0$ and $C_q > 0$ are the constants not depending on u and t.

A priori estimate (7.80) implies the following a priori estimates:

$$\left\| \partial^s F(t,x,u(t,x)) / \partial t^\alpha \partial x^\beta \partial u^\gamma \right\|_{C(\overline{Q}_T)} \le C \begin{pmatrix} m=1,\ n=4m+2=6; \\ 0 \le \alpha \cdot m + |\beta| + \gamma = s \le \\ \le 4m+1;\ \alpha \le 1; |\beta| \le 4m \end{pmatrix}. \tag{7.82}$$

Thus, due to a priori estimates (7.82) and (7.80), in order to have an estimate ($\forall \tau \in [0,T]$) for the expression

$$\left\| \frac{\partial}{\partial \tau} \{F(\tau,x,u(\tau,x))\} \right\|^2_{W_2^{4m}(\Omega)},$$

it suffices to have an estimate ($\forall \tau \in [0,T]$) for the following integrals:

$$\int_\Omega \prod_{i=1}^{s} \left| \mathrm{D}^{\alpha_i} u(\tau,x) \right|^2 dx \begin{pmatrix} m=1,\ n=4m+2=6;\ 1 \le s \le 4m; \\ \sum_{i=1}^{s} |\alpha_i| \le 4m;\ 2 \le |\alpha_1| \le \ldots \le |\alpha_s| \le 4m \end{pmatrix}, \tag{7.83}$$

$$\int_\Omega \prod_{j=1}^{l} \left| \mathrm{D}^{\beta_j} u(\tau,x) \right|^2 \cdot \left| \mathrm{D}^{\beta_0} u_\tau(\tau,x) \right|^2 dx \begin{pmatrix} m=1,\ n=4m+2;\ 0 \le l \le 4m;\ \sum_{j=0}^{l} |\beta_j| \le 4m; \\ 2 \le |\beta_1| \le \ldots \le |\beta_l| \le 4m;\ 1 \le |\beta_0| \le 4m. \end{pmatrix} \tag{7.84}$$

Now, using a priori estimates (7.81) (for $s=0,\ \alpha=\alpha_i$), for integrals (7.83) $\forall \tau \in [0,T]$ we have:

$$\int_\Omega \prod_{i=1}^{s} \left| \mathrm{D}^{\alpha_i} u(\tau,x) \right|^2 dx \le \prod_{i=1}^{s} \left\| \mathrm{D}^{\alpha_i} u(\tau,x) \right\|^2_{L_{2p_i}(\Omega)} \le C, \tag{7.85}$$

where $C > 0$ is a constant not depending on u and τ, and

$$1 \le p_i \le \frac{n}{n - 10m + 2|\alpha_i|},\ p_i < +\infty\ \ (i = \overline{1,s});\ \sum_{i=1}^{s} \frac{1}{p_i} = 1,$$

the possibility of last equality following from $n - 10m = -4 < 0$ and

$$\sum_{i=1}^{s}\frac{n-10m+2|\alpha_i|}{n}=\frac{(n-10m)\cdot s+2\sum_{i=1}^{s}|\alpha_i|}{n}\leq\frac{(n-10m)\cdot s+8m}{n}\leq\frac{n-10m+8m}{n}<1.$$

Next, using a priori estimates (7.81) (for $s = 0,\ \alpha=\beta_j\ \ (j=\overline{1,l})$ and $s = 1,\ \alpha=\beta_0$), for integrals (7.84) $\forall\tau\in[0,T]$ we have:

$$\int_{\Omega}\prod_{j=1}^{l}\left|D^{\beta_j}u(\tau,x)\right|^2\cdot\left|D^{\beta_0}u_\tau(\tau,x)\right|^2dx\leq$$

$$\leq\prod_{j=1}^{l}\left\|D^{\beta_j}u(\tau,x)\right\|^2_{L_{2p_j}(\Omega)}\cdot\left\|D^{\beta_0}u_\tau(\tau,x)\right\|^2_{L_{2p_0}(\Omega)}\leq C, \tag{7.86}$$

where $C > 0$ is a constant not depending on u and τ, and

$$1\leq p_j\leq\frac{n}{n-10m+2|\beta_j|},\ p_j<+\infty\quad(j=\overline{1,l});$$

$$1\leq p_0\leq\frac{n}{n-8m+2|\beta_0|},\ p_0<+\infty;\ \sum_{j=0}^{l}\frac{1}{p_j}=1,$$

the possibility of last equality following from the fact that for $l = 0$ this equality is obvious, and for $l \geq 1$ (due to $n - 10m < 0$) we have

$$\frac{n-8m+2|\beta_0|}{n}+\sum_{j=1}^{l}\frac{n-10m+2|\beta_j|}{n}=\frac{n-8m+(n-10m)\cdot l+2\sum_{j=0}^{l}|\beta_j|}{n}\leq$$

$$\leq\frac{n-8m+(n-10m)\cdot l+8m}{n}=\frac{n+(n-10m)\cdot l}{n}<\frac{n}{n}=1.$$

Thus, integrals (7.83) and (7.84) are a priori bounded.

Now, using a priori estimates (7.82), (7.80), (7.85) and (7.86), we obtain from (7.78):

$$\|u(t,x)\|_{B^{6,5,4}_{2,2,2,T}} \le C_0,$$

i.e. all the possible classical solutions $u(t,x)$ of problem (0.1)-(0.3), belonging to the space $B^{k,k-1,k-2}_{2,2,2,T}$, are a priori bounded in $B^{6,5,4}_{2,2,2,T}$. In view of the fact that for $m=1,\ n=4m+2$ we have $k \equiv \left[\dfrac{n}{2m}\right]+3=6$, the theorem is now proved in case ***m=1, n=4m+2***, too. Thus, theorem 8 is fully proved.

Remark 8. As in all Theorems 3, 4, 7 and 8 the found classical solutions $u(t,x)$ of problem (0.1)-(0.3) belong to the space $B^{\left[\frac{n}{2m}\right]+3,\left[\frac{n}{2m}\right]+2,\left[\frac{n}{2m}\right]+1}_{2,\,2,\,2,\,T}$, we have:

$$u(t,x)\in \bigcap_{i=0}^{2} C^{(i)}\left([0,T];\, W_2^{\left(\left[\frac{n}{2m}\right]+3-i\right)m}(\Omega)\right)\subset C^{2,2m}_{t,x}(\overline{Q}_T)\,.$$

REFERENCES

[1] Agmon S. The coerciveness problem for integro-differential forms // *J. Analyse. Math.*, 1958, 6, №2, pp.183-223.

[2] Schechter M. General boundary value problems for partial elliptic equations // Mathematics. *Periodical collection of translated foreign articles*, 1960, 4, №5, pp.93-122 *(in Russian).*

[3] Khudaverdiyev K.I. On the theory of multidimensional mixed problems for nonlinear hyperbolic equations // Doctoral thesis, Baku State University, 1973, 319 pages *(in Russian).*

[4] Khudaverdiyev F.K. Study of multidimensional mixed problem for nonlinear hyperbolic equations // Ph.D. thesis, Institue of Mathematics and Mechanics, Academy of Science of Azerbaijan, 1987, 159 pages *(in Russian).*

[5] Krasnoselski M.A. Topological methods in the theory of nonlinear integral equations // M.: GITTL, 1956, 329 pages *(in Russian).*

[6] Mathematical reference library. Functional analysis // M.: Nauka Publishers, 1964, 424 pages *(in Russian).*

[7] Khudaverdiyev K.I. On the mixed problem for a class of nonlinear equations of hyperbolic type // Proceedings of Azerbaijan State University, series of physical and mathematical sciences, 1960, №3, pp.17-25 *(in Russian).*

[8] Khudaverdiyev K.I. On the solution of mixed problem for a class of nonlinear equations of hyperbolic type // Proceedings of Azerbaijan State University, series of physical and mathematical sciences, 1960, №4, pp.45-55 *(in Russian).*

[9] Khudaverdiyev K.I. Solution of the mixed problem for some nonlinear equations of hyperbolic type // Proceedings of Azerbaijan State University, series of physical and mathematical sciences, 1961, №1, pp.21-27 *(in Russian).*

[10] Khudaverdiyev K.I. Study of mixed problem for a class of second order hyperbolic equations with nonlinear operator right side // Reports of the Academy of Science of Azerbaijan SSR, 1971, 27, №9, pp.3-7 *(in Russian).*

[11] Khudaverdiyev K.I. Study of weakly generalized solution of multidimensional mixed problem for a class of second order hyperbolic equations with nonlinear operator right side // Materials of the 6th scientific conference of the Faculty of Department of Physics and Mathematics of the Samarkand State Pedagogical Institute, 1972, pp.37-47 *(in Russian).*

[12] Khudaverdiyev K.I. Study of strongly generalized solution of multidimensional mixed problem for a class of second order hyperbolic equations with nonlinear operator right side // News of the Academy of Science of Azerbaijan SSR, series of physical and technical sciences, 1971, №5-6, pp.84-90 *(in Russian).*

[13] Khudaverdiyev K.I. Study of almost everywhere solution of multidimensional mixed problem for a class of second order hyperbolic equations with nonlinear operator right side // News of the Academy of Science of Azerbaijan SSR, series of physical and technical sciences, 1971, №5-6, pp.25-31 *(in Russian).*

[14] Khudaverdiyev K.I. Study of classical solution of multidimensional mixed problem for a class of second order hyperbolic equations with nonlinear operator right side by Fourier method // Reports of the Academy of Science of Azerbaijan SSR, 1971, 27, №10, pp.8-14 *(in Russian).*

[15] Khudaverdiyev K.I. Study of classical solution of multidimensional mixed problem for a class of second order quasilinear hyperbolic equations // Proceedings of Azerbaijan State University, series of physical and mathematical sciences, 1972, №1, pp.3-27 *(in Russian).*

[16] Khudaverdiyev F.K. Global solvability of multidimensional mixed problem for nonlinear hyperbolic equations // Theme Collection "Partial Differential Equations and Their Applications", Baku, Publishing House of the Azerbaijan State University, 1985, pp.61-68 *(in Russian).*

[17] Khudaverdiyev F.K. Study of almost everywhere solution of multidimensional mixed problem for nonlinear hyperbolic equations // Materials of the republican scientific conference of postgraduate students, Baku, Nauka Publishers, 1986, pp.23-25 *(in Russian).*

[18] Khudaverdiyev F.K. Study of almost everywhere solution of multidimensional mixed problem for nonlinear hyperbolic equations // Manuscript deposited in VINITI on 01.07.1986, №4744-B86, p.97 *(in Russian).*

[19] Khudaverdiyev F.K. Classical solvability in large of multidimensional mixed problem for nonlinear hyperbolic equations Theme Collection "Partial Differential Equations and Their Applications", Baku, Publishing House of the Azerbaijan State University, 1986, pp.86-102 *(in Russian).*

[20] Khudaverdiyev F.K. Study of classical solution of multidimensional mixed problem for nonlinear hyperbolic equations // Manuscript deposited in VINITI on 19.11.1986, №7868-B86, p.74 *(in Russian).*

[21] Halilov Z.I. On one vibration problem // News of the Azeri branch of the Academy of Science of USSR, 1942, №4, pp.168-169 *(in Russian).*
[22] Halilov Z.I. Solution of finite string vibration problem in medium with variable coefficient // Reports of the Academy of Science of Azerbaijan SSR, 1952, 8, №7, pp.333-337 *(in Russian).*
[23] Halilov Z.I. On the method of expansion in terms of eigenfunctions of principal part of equation when solving mixed problems // Reports of the Academy of Science of Azerbaijan SSR, 1954, 10, №3, pp.151-158 *(in Russian).*
[24] Halilov Z.I. On the method of expansion in terms of eigenfunctions of principal part of equation when solving mixed problems // Transactions of the Azerbaijan State University, series of physical and mathematical sciences, 1954, Volume 4, pp.5-33 *(in Russian).*
[25] Ladizhenskaya O.A. Mixed problem for hyperbolic equations // M.: Gostekhizdat, 1953, 279 pages *(in Russian).*
[26] Lopatinski Y.B. On some conditions of applicability of Fourier method // Transactions of the Azerbaijan State University, series of physical and mathematical sciences, 1952, Volume 3, pp.85-92 *(in Russian).*
[27] Kato T. Integration of the equation of evolution in a Banach space // *J. Math. Soc. Japan*, 1953, 5 №2, pp.208-243.
[28] Vishik M.I., Ladizhenskaya O.A. Boundary value problems for partial differential equations and some classes of operator equations // UMN, 1956, II, №6, pp.41-97 *(in Russian).*
[29] Korobeynik Y.F. Solving mixed problem by Fourier method for one integro-differential equation // *Reports of the Academy of Science of USSR*, 1957, 114, №1, pp.14-17 *(in Russian).*
[30] Levitan B.M. On substantiation of Fourier method for one-dimensional problems // Proceedings of Moscow Correspondence Pedagogical Institute, 1959, №3, pp.72-81 *(in Russian).*
[31] Iliin B.A. On solvability of mixed problems for hyperbolic and parabolic equations // UMN, 1960, 15, №2, pp.97-154 *(in Russian).*
[32] Brish N.I. On Fourier method for fourth order differential equation including second derivative in time // Reports of the Academy of Science of Belorussian SSR, 1962, 6, №1, pp.9-13 *(in Russian).*
[33] Brish N.I. On classical solution of mixed problems for non-stationary equations // Differential Equations, 1965, 1, №4, pp.523-528 *(in Russian).*
[34] Brish N.I. On mixed problems for some semihyperbolic equations // *Differential Equations*, 1970, 6, №4, pp.705-720 *(in Russian).*
[35] Brish N.I., Valeshkevich I.N. Fourier method for differential equations including second derivative in time // *Reports of the Academy of Science of USSR*, 1962, 146, №6, pp.1247-1250 *(in Russian).*
[36] Brish N.I., Valeshkevich I.N. Fourier method for non-stationary equations with general boundary conditions // *Differential Equations*, 1965, 1, №3, pp.393-399 *(in Russian).*
[37] Valeshkevich I.N. On some mixed problems for biharmonic wave equation // News of the Academy of Science of Belorussian SSR, series of physical and technical sciences, 1963, 3, pp.14-24 *(in Russian).*

[38] Kreiss H.O. Initial-boundary value problems for hyperbolic systems // *Uppsal Univ. Mag.*, 1969, 25 pages.

[39] Shirota Taira, Asano Kazuo. On certain mixed problem for hyperbolic equation of higher order. II. // *Proc. Japan. Acad.*, 1969, 45, №5, pp.383-393.

[40] Ikava Mitsuru. A mixed problem for hyperbolic equation of second order with a first order derivative boundary condition // *Publs. Res. Inst. Math. Sci.*, 1969, 5, №2, pp.119-147.

[41] Ikava Mitsuru. On the mixed problem for hyperbolic equations of second order with the Neumann boundary condition // *Osaka J. Math.*, 1970, 7, №1, pp.203-233.

[42] Reiko Sakamato. Mixed problems for hyperbolic equations. I. Energy inequalities // Mathematics. *Periodical collection of translated foreign articles*, 1972, 16, №1, pp.62-80 *(in Russian).*

[43] Reiko Sakamato. Mixed problems for hyperbolic equations. II. Existence theorems for zero initial conditions and energy inequalities for any initial conditions // Mathematics. *Periodical collection of translated foreign articles*, 1972, 16, №1, pp.81-89 *(in Russian).*

[44] von Wahl Wolf. Regular solutions of initial-boundary value problems for linear and nonlinear wave equations. I. // *Manuscr. Math.*, 1974, 13, №2, pp.187-206.

[45] Lichtenstein L. Zur Theorie partieller Differentialgleichungen zweiter Ordnung vom hyperbolischen Typus // *J. für die reine und angewandte Mathematik*, 1927, 158, №2, pp.80-91.

[46] Siddiqi M.R. Cauchy's problem in a nonlinear partial differential equation of the hyperbolic type // *Proc. Cambridge Phil. Soc.*, 1935, 31, pp.195-202.

[47] Siddiqi M.R. Boundary problems in a nonlinear partial differential equations // Lucknow University Studies. *Faculty of Science*, 1939, №11, pp.1-136.

[48] Bernstein S.N. On one class of functional partial equations // News *of the Academy of Science of USSR*, 1940, №4, pp.17-26 *(in Russian).*

[49] Barbuti U. Analisi esusterioli in problem di proposizione semi-lineari // *Ann. di Scuola Norm. Sup. di Pisa*, 1957, №3, pp.183-207.

[50] Chandirov G.I. Application of Fourier method to the study of some nonlinear problems // *Ph.D. thesis*, Baku, 1958.

[51] Khudaverdiyev K.I. On solving one-dimensional mixed problem for quasilinear second order hyperbolic equation and fourth order quasilinear equation by Fourier method // Ph.D. thesis, Baku, 1962 *(in Russian).*

[52] Huseynov A.I., Khudaverdiyev K.I. On solving one-dimensional mixed problem for quasilinear second order hyperbolic equations by Fourier method // *Reports of the Academy of Science of USSR*, 1963, 148, №3, pp.496-499 *(in Russian).*

[53] Kurzweil I. Exponentially stable integral manifolds averaging principle and continuous dependence on a parameter // *Czech. Math. J.,* 1966, 16 (91), №4, pp.463-492.

[54] Sobolevski P.E., Pogorelenko V.A. On solvability of mixed problems for one-dimensional quasilinear hyperbolic equations // *Ukrainian Mathematical Journal*, 1970, 22, №1, pp.114-121 *(in Russian).*

[55] Larkin N.A. On one class of quasilinear hyperbolic equations having solutions in large // *Reports of the Academy of Science of USSR*, 1979, 244, №1, pp.38-41 *(in Russian).*

[56] Schauder J. Gemischte Randwertaufgaben bei partiellen Differentialgleichungen vom hyperbolischen Typus // *Studia Math.*, 1936, 6, pp.190-198.

[57] Krzizanski M., Schauder J. Quasilineare Differentialgleichungen zweiter Ordnung vom hyperbolischen Typus // *Studia Math.*, 1936, 6, pp.162-189.
[58] Brauder F.E. On nonlinear wave equations // *Math. Z.*, 1962, 80, №3, pp.249-264.
[59] Lions J.-L., Strauss W.A. Sur certains problemes hyperboliques nonlineaires // *Comptes Rendus*, 1963, 257, №22, pp.3267-3270.
[60] Lions J.-L., Strauss W.A. Some nonlinear evolution equations // *Bull. Soc. Math. France*, 1965, 93, №1, pp.43-96.
[61] Lions J.-L. Ultra weak solutions of nonlinear partial differential equations // *An. Acad. brasil. cienc.*, 1980, 52, №1, pp.7-10.
[62] Segal I.E. The global Cauchy problem for a relativistic scalar field with power interaction // *Bull. Soc. Math. France*, 1963, 91, №2, pp.129-135.
[63] Sather J. The existence of a global classical solution of the initial-boundary value problem for $u + u^3 = f$ // *Arch. Pat. Mech. and Anal.*, 1966, 22, №4, pp.292-307.
[64] Sattinger D.H. On global solution of nonlinear hyperbolic equations // *Arch. Ration Mech. and Analysis*, 1968, 30, №2, pp.148-172.
[65] Walk Manfred. Verallgemeinerte Lösbarkeit des gemischten Problems quasilinearer Differentialgleichungen zweiter Ordnung vom hyperbolischen Typus // *Math. Nachr.*, 1969, 41, №4-6, pp.213-221.
[66] Strauss Walter. On weak solutions of semi-linear hyperbolic equations // *An. Acad. brasil. cienc.*, 1970, 42, №4, pp.645-651.
[67] Doktor Alexander. Mixed problem for semi-linear hyperbolic equation of second order with Dirichlet boundary condition // *Prelim. Commun. Comment. Math. Univ. Carol.*, 1972, 13, №1, pp.185-189.
[68] Doktor Alexander. Mixed problem for semi-linear hyperbolic equation of second order with the Dirichlet boundary condition // *Czech. Math. J.*, 1973, 23, №1, pp.95-122.
[69] Inone Atsushi. Sur $u + u^3 = f$ dans un domaine noncylindrique // *C. r. acad. Sci.*, 1972, 275, №14, A 659 – A 662.
[70] Inone Atsushi. Sur $u + u^3 = f$ dans un domaine noncylindrique // *J. Math. Anal. and Appl.*, 1974, 46, №3, pp.777-819.
[71] Kato Hiskao, Nanbu Tokumori. The existence of a local classical solution of the initial-boundary value problem for $D_t^2 u - \Delta u + (D_t u)^3 = 0$ // *Math. Repts. Coll. Gen. Educ. Kyushu Univ.*, 1973, 9, №1, pp.11-25.
[72] Ebihara Yukiyoshi. On the local classical solution of the mixed problem for nonlinear wave equation // *Math. Repts. Coll. Gen. Educ. Kyushu Univ.*, 1974, №2, pp.43-65.
[73] Ebihara Yukiyoshi, Nakao Mitsuhiro, Nanbu Tokumori. On the existence of global classical solution of initial-boundary value problem for $u + u^3 = f$ // *Pacif. J. Math.*, 1975, 60, №2, pp.63-70.
[74] Kalantarov V.K. On mixed problem for semilinear compound systems of parabolico-hyperbolic equations // Ph.D. thesis, Baku, 1974 *(in Russian)*.
[75] Caughey Thomas K., Ellison James. Existence, uniqueness and stability of solutions of a class of nonlinear partial differential equations // *J. Math. Anal. and Appl.*, 1975, 51, №1, pp.1-32.

[76] Nakao Mitsuhiro. Decay of classical solutions of a semilinear wave equation // *Math. Repts. Coll. Gen. Educ. Kyushu Univ.*, 1977, 11, №1, pp.39-45.
[77] Matsumura Akitaka. Global existence and asymptotics of the solutions of the second order quasilinear hyperbolic equations with the first order dissipation // *Publs. Res. Inst. Math. Sci.*, 1977, 13, №2, pp.349-379.
[78] Jerome Joseph W. The method of lines and nonlinear Klein-Gordon equation // *J. Different. Equat.*, 1978, 30, №1, pp.20-30.
[79] Clements John C. Existence theorems for some nonlinear equations of evolution // *Can. J. Math.*, 1970, 22, №4, pp.726-745.
[80] Yakubov S.Y. Solvability of Cauchy problem for abstract quasilinear second order hyperbolic equations and their applications // *Transactions of the Moscow Mathematical Society*, 1970, 23, pp.37-60 *(in Russian).*
[81] Pecher Hartmut. Die Existenz regularer Lösungen für Cauchy und Anfangs-Randwertprobleme für nichtlinearer Wellengleichungen // *Math. Z.*, 1974, 140, №3, pp.263-279.
[82] von Wahl Wolf. Regular solutions of initial-boundary value problems for linear and nonlinear wave equations. II. // *Math. Z.,* 1975, 142, №2, pp.121-130.
[83] Pokhozhayev S.I. On one class of quasilinear hyperbolic equations // *Mathematical Collection,* 1975, 96, №1, pp.152-166 *(in Russian).*
[84] Aliyev A.B. Solvability in large of Cauchy problem for quasilinear partial evolution and hyperbolic equations // Spectral Theory of Operators. Transactions of the 2nd all-USSR summer mathematical school on the spectral theory of operators. Baku, 1979, pp.6-18 *(in Russian).*
[85] Brenner Philip, von Wahl Wolf. Global classical solutions of nonlinear wave equations // *Math. Z.*, 1981, 176, №1, pp.87-121.
[86] Smirnov V.I. Course in higher mathematics. Volume 5 // Moscow, Nauka Publishers, 1959, 655 pages *(in Russian).*
[87] Beckenbach E., Bellman R. Inequalities // M., *Mir*, 1965, 276 pages *(in Russian).*
[88] Ladizhenskaya O.A., Uraltseva N.N. Linear and quasilinear equations of elliptic type // Moscow, Nauka Publishers, 1973, 576 pages *(in Russian).*
[89] Pokhozhayev S.I. On S.L.Sobolev's imbedding theorem in case $p \cdot l = n$ // Reports of scientific and technical conference in Moscow Energy Institute, Section of Mathematics, 1965, pp.158-170 *(in Russian).*

Reviewed by:
Prof. B.A.Iskenderov,
Corresponding Member of the National Academy of Science of Azerbaijan
Doctor of Sciences in Physics and Mathematics
Currently holding a position of the Head of the Department of Differential Equations in the Institute of Mathematics and Mechanics, National Academy of Science of Azerbaijan, Baku, Azerbaijan.

In: Evolution Equations
Editor: Arthur L. Claes

ISBN: 978-1-61209-652-0

Chapter 3

SOME APPEALING CLASSICAL AND STOCHASTIC EVOLUTION EQUATIONS

Shambhu N. Sharma*
Department of Electrical Engineering
National Institute of Technology, Surat, India

ABSTRACT

The dynamical system theory in combination with the notion of feedback formalizes the concept of mathematical control theory. The structure of dynamical systems is adjudged on the basis of dimensionality, time-invariance/time varying as well as driving input. The stochastic differential equation resulting from the phase space analysis of the fluctuation equation describing noisy dynamical systems has received remarkable success in different branches of sciences. Consider the stochastic differential system of the form

$$\dot{x}_t = f(x_t) + g(x_t)\xi_t,$$

where the terms $f(x_t)$, $g(x_t)\xi_t$ have interpretations as the system non-linearity and the random forcing term respectively. For the input noise process $\xi_t = \dot{B}_t$, where B_t is the Brownian motion, the resulting stochastic differential system would be the Itô stochastic differential system in which the state vector is Markovian. More over, the Ornstein-Uhlenbeck (OU) process-driven stochastic differential system can be recast as the Itô stochastic differential system by writing down the evolution equation of the OU process, i.e. $d\xi_t = -\beta\xi_t dt + \beta\, dB_t$, $\beta > 0$. The contribution to the stochastic evolution of the state vector comes from the system non-linearity as well as the dispersion matrix, which introduces slight modifications into the motion of the dynamical system. This book chapter encompasses *five* stochastic evolution equations as well as their special cases, i.e. classical evolution equations, which are regarded as the cornerstone formalism for analysing and accomplishing filtering for stochastic differential systems. The stochastic evolution equations of the chapter are about the

* Email: snsvolterra@gmail.com

filtering density evolution, the exact stochastic evolution of conditional moment as well as its approximate counterparts, the stochastic evolution of the scalar function of the multi-dimensional state vector. This chapter discusses evolution equations for the non-Markovian stochastic differential system as well, which are not widely available in literature.

The evolution equations of the chapter will be of interest to Researchers looking for understanding the theory of noisy dynamical systems in Markovian as well as non-Markovian setting. Rigorous mathematical treatments about stochastic differential systems can be found in authoritative books, i.e. Multi-dimensional diffusion processes authored by DS Strook and SRS Varadhan, as well as Brownian motion and stochastic calculus authored by I Karatzas and Steven E. Shreve. For a greater detail about stochastic filtering, the Author's International Journal of Control contribution (Sharma, S. N.; Parthasarathy, H. and Gupta, J.R.P. Third-order approximate Kushner filter for a non-linear dynamical system. *International Journal of Control*, 79(9), 1096-1106, 2006) can be consulted as well.

Keywords: Brownian motion, extended phase space analysis, fluctuation equations, stochastic evolution equations, master's equation.

1. Introduction

The dynamical system viewpoint has received considerable success for analysing the problems arising from control theory, satellite mechanics, financial mathematics, stochastic networks, and theoretical biology. Consider a system of two coupled second-order non-linear differential equations, i.e. $\ddot{x}_1 = F_1(t, x_1, \dot{x}_1, x_2, \dot{x}_2)$, $\ddot{x}_2 = F_2(t, x_1, \dot{x}_1, x_2, \dot{x}_2)$ (Arnold 1995), an appealing and attractive case, which has a central importance in satellite mechanics. After introducing the notion of phase space formulation of the coupled equation, the resulting equation is termed as finite-dimensional, time-varying, continuous-time, non-linear system. After accounting for random initial conditions and small stochastic accelerations, i.e. random perturbations, these lead to the concept of the noisy dynamical system involving the structure of stochastic differential equations. A nice exposition about the theory and applications of stochastic processes and stochastic differential equations in sciences can be found in celebrated books authored by Gihman and Skorohod (1972), Karatzas and Shreve (1991), Protter (1991). For a greater and rigorous understanding about stochastic processes and stochastic differential equations, the celebrated book authored by Strook and Varadhan (1979), which integrates the concepts of martingale and the theory of Markov processes, can be consulted. Furthermore, the Quantum Stochastic Differential Equation (QSDE) is utilized to describe the quantum stochastic differential system. The quantum Itô differential rules play the key role for analysing and filtering for the quantum stochastic differential system. Hudson and Parthasarathy pioneered the theory of the QSDE (Parthasarathy 1992). Subsequently, the Belavkin equation plays the key role to accomplish the quantum stochastic filtering theory, see equation (4.15) of Bouten (2004, p.78).

The linear stochastic differential system as well as the bilinear stochastic differential system offer simplified exposition in the sense that their closed form solutions are possible. The closed solution for the time-varying bilinear stochastic differential system can be found in standard books, e.g. Pugachev and Sinitsyn (1977), as well. After obtaining the closed form expressions for the solution vectors satisfying the linear as well as the bilinear stochastic differential systems respectively, we obtain the mean and variance. However, the closed form expression for the non-linear stochastic differential equation is not possible. For this reason, we introduce the concept of *evolution* equations for analysing stochastic differential systems as well as *stochastic evolution* equations for accomplishing their filtering theory. There are no publications on summarizing evolution equations and stochastic evolution equations for the Markovian as well as the non-Markovian stochastic differential systems. Here, this book chapter attempts to bridge the gap.

This chapter is chiefly devoted to summarizing evolution equations and stochastic evolution equations associated with the Brownian motion-driven stochastic differential system as well as the arbitrary process-driven. The theory of deriving the evolution equations for the stochastic differential system begins from deriving the evolution equation of the conditional probability density.

2. Five Stochastic Evolution Equations

The prediction theory of stochastic differential systems is formalized by deriving the evolution equation of conditional moment. On the other hand, the filtering theory of the stochastic differential system is accomplished by deriving the stochastic evolution of conditional moment. The starting point for arriving at the conditional moment evolution equation is to derive the expression for conditional probability density evolution equation. This section discusses five different stochastic evolution equations, i.e. (i) filtering density evolution equation (ii) the stochastic evolution of conditional moment (iii) the stochastic evolutions of conditional mean and variance, which are regarded as the exact filtering equation (iv)approximate filtering equations (v) the stochastic evolution of the scalar function of multi-dimensional vector. Subsequently, the stochastic equation, i.e. master's equation for the non-Markovian stochastic differential system, w'd be the subject of discussion. This section introduces a discussion about the evolution of conditional probability density for the OU process-driven stochastic differential system as well, where the OU process is a stationary, zero mean with relatively smaller correlation time.

The filtering density evolution equation for the Brownian motion-driven stochastic differential equation is a partial-integro stochastic differential equation, i.e.

$$dp = L(p)dt + (h - \hat{h})^T \varphi_n^{-1}(dz_t - \hat{h}dt)p, \tag{1}$$

where $L(.)$ is the Fokker-Planck operator, the filtering density $p = p(x, t|z_\tau, t_0 \leq \tau \leq t)$, $h(x_t, t)$ is the measurement non-linearity and the observation z_t is masked by the Brownian noise. Further more, the term 'φ_η' has an interpretation as the intensity of the noise introduced into observations. The result was developed by Harald J. Kushner, which involves the following: (i) the expression of the filtering density is derived and subsequently, the filtering density evolution equation is derived using the stochastic differential rule (Jazwinski

1970, p. 178). The above filtering density evolution equation is also termed as the Kushner-Stratonovich equation. Liptser-Shiryaev pioneered an alternative proof of the filtering density evolution equation, equation (1). The procedure involves the following: first derive the stochastic evolution $d\widehat{\phi(x_t)}$ of the conditional moment, where $\widehat{\phi(x_t)} = E(\phi(x_t)|z_\tau, t_0 \leq \tau \leq t)$, secondly, the stochastic evolution of the conditional characteristic function, a special case of the conditional moment evolution, thirdly, the definition of conditional expectation as well as integration by part formula lead to the filtering density evolution equation, see Liptser and Shiryaev (1977).

One can derive the stochastic evolution of conditional moment using the definition of conditional expectation, evolution of the filtering density of equation (1) and integration by part formula. Consider the double differentiable scalar function of n-dimensional state vector, $\phi : U \rightarrow R$ and the phase space $U \subset R^n$, we have

$$d\hat{\phi}(x_t) = \int \phi(x) dp(x, t|z_\tau, t_0 \leq \tau \leq t) dx$$

$$= (\int \phi(x) L(p(x, t|z_\tau, t_0 \leq \tau \leq t)) dx) dt$$

$$+ \int \phi(x)(h - \hat{h})^T \varphi_\eta^{-1} (dz_t - \hat{h} dt) p(x, t|z_\tau, t_0 \leq \tau \leq t) dx.$$

The definition of conditional expectation and the adjoint property of the Fokker-Planck operator imply that

$$d\hat{\phi}(x_t) = (\left\langle f^T \frac{\partial \phi(x_t)}{\partial x_t} \right\rangle + \frac{1}{2} \left\langle tr((G\psi_w G^T)(x_t, t) \frac{\partial^2 \phi(x_t)}{\partial x_t \partial x_t^T}) \right\rangle) dt$$

$$+ (\langle \phi h \rangle - \langle \phi \rangle \langle h \rangle)^T \varphi_\eta^{-1} (dz_t - \hat{h} dt), \qquad (2)$$

where the terms f and G are the system non-linearity and the process noise coefficient matrix respectively and the notation '$\langle \ \rangle$' of this chapter denotes the conditional expectation. Equation (2) can be regarded as the *second* stochastic evolution equation of this chapter. The stochastic filtering theory of the Brownian motion-driven stochastic differential system is grounded on equation (2).

Here, the Author explains the *third* stochastic evolution equation of this chapter. The stochastic evolutions of conditional mean and variance can be regarded as the special cases of the stochastic evolution of conditional moment, equation (2). Thanks to equation (2) of the this chapter, for the scalar functions $\phi(x_t) = x_i$ and $\phi(x_t) = x_i x_j$, we arrive at the i th component of the stochastic evolution of the conditional mean vector and (i, j) th component of the stochastic evolution of conditional variance matrix respectively, i.e.

$$d\hat{x}_i = \hat{f}_i(x_t, t)dt + (\widehat{x_i h} - \hat{x}_i \hat{h})^T \varphi_\eta^{-1}(t)(dz - \hat{h}dt),$$

$$(dP_t)_{ij} = (\widehat{x_i f_j} - \hat{x}_i \hat{f}_j + \widehat{f_i x_j} - \hat{f}_i \hat{x}_j + \widehat{(G\psi_w G^T)_{ij}(x_t, t)}$$

$$-(\widehat{x_i h} - \hat{x}_i \hat{h})^T \varphi_\eta^{-1}(t)(\widehat{h x_j} - \hat{h}\hat{x}_j))dt$$

$$+(\widehat{x_i x_j h} - \widehat{x_i x_j}\hat{h} - \hat{x}_i \widehat{x_j h} - \hat{x}_j \widehat{x_i h} + 2\hat{x}_i \hat{x}_j \hat{h})^T \varphi_\eta^{-1}(t)(dz - \hat{h}dt), \tag{3}$$

where $\hat{f}_i = E(f_i(x_t, t)|Y_t)$, $P_{ij} = E((x_i - E(x_i|Y_t))(x_j - E(x_j|Y_t))|Y_t)$. Note that equation (3), the system of conditional mean and conditional variance evolutions, describes the *filtering* evolution equation of the Brownian motion process-driven stochastic differential system.

The analytical and numerical solutions to equation (3), the above set of stochastic filtering evolution equations, are not possible because of infinite dimensionality. For this reason, approximate stochastic evolution equations are explored by replacing the global properties with their local counterparts. This can be accomplished by introducing the higher-order partials of system non-linearity, measurement non-linearity and diffusion coefficient into equation (3). As a result of this, we have

$$\hat{f}_i(x_t, t) = f_i(\hat{x}_t, t) + \frac{1}{2}\sum_{p,q} P_{pq} \frac{\partial^2 f_i(\hat{x}_t, t)}{\partial \hat{x}_p \partial \hat{x}_q}, \tag{4}$$

$$\widehat{x_i f_j} - \hat{x}_i \hat{f}_j = \sum_p P_{ip} \frac{\partial f_j(\hat{x}_t, t)}{\partial \hat{x}_p} + \frac{1}{2}\sum_{p,q,r} P_{ip} P_{qr} \frac{\partial^3 f_i(\hat{x}_t, t)}{\partial \hat{x}_p \partial \hat{x}_q \partial \hat{x}_r}, \tag{5}$$

$$\widehat{x_i h} - \hat{x}_i \hat{h} = \sum_p P_{ip} \frac{\partial h(\hat{x}_t, t)}{\partial \hat{x}_p} + \frac{1}{2}\sum_{p,q,r} P_{ip} P_{qr} \frac{\partial^3 h(\hat{x}_t, t)}{\partial \hat{x}_p \partial \hat{x}_q \partial \hat{x}_r}, \tag{6}$$

and

$$\widehat{x_i x_j h} - \widehat{x_i x_j}\hat{h} - \hat{x}_i \widehat{x_j h_k} - \hat{x}_j \widehat{x_i h_k} + 2\hat{x}_i \hat{x}_j \hat{h} = \sum_{p,q} P_{ip} P_{jq} \frac{\partial^2 h_k(\hat{x}_t, t)}{\partial \hat{x}_p \partial \hat{x}_q}. \tag{7}$$

After plugging equations (4)-(7) into equation (3), the set of exact stochastic evolution equations, the *approximate* stochastic evolution equations involving the third-order partials are

$$d\hat{x}_i = (f_i(\hat{x}_t,t) + \frac{1}{2}\sum_{p,q} P_{pq}\frac{\partial^2 f_i(\hat{x}_t,t)}{\partial\hat{x}_p\partial\hat{x}_q})dt$$

$$+(\sum_p P_{ip}\frac{\partial h^T(\hat{x}_t,t)}{\partial\hat{x}_p} + \frac{1}{2}\sum_{p,q,r} P_{ip}P_{qr}\frac{\partial^3 h^T(\hat{x}_t,t)}{\partial\hat{x}_p\partial\hat{x}_q\partial\hat{x}_r})\varphi_\eta^{-1}(dz - \hat{h}dt).$$

$$dP_{ij} = (\sum_p P_{ip}\frac{\partial f_j(\hat{x}_t,t)}{\partial\hat{x}_p} + \frac{1}{2}\sum_{p,q,r} P_{ip}P_{qr}\frac{\partial^3 f_i(\hat{x}_t,t)}{\partial\hat{x}_p\partial\hat{x}_q\partial\hat{x}_r} + \sum_p P_{jp}\frac{\partial f_i(\hat{x}_t,t)}{\partial\hat{x}_p}$$

$$+\frac{1}{2}\sum_{p,q,r} P_{jp}P_{qr}\frac{\partial^3 f_i(\hat{x}_t,t)}{\partial\hat{x}_p\partial\hat{x}_q\partial\hat{x}_r} + (G\psi_w G^T)_{ij}(\hat{x}_t,t) + \frac{1}{2}\sum_{p,q} P_{pq}\frac{\partial^2 (G\psi_w G^T)_{ij}(\hat{x}_t,t)}{\partial\hat{x}_p\partial\hat{x}_q}$$

$$+(\sum_p P_{ip}\frac{\partial h^T(\hat{x}_t,t)}{\partial\hat{x}_p} + \frac{1}{2}\sum_{p,q,r} P_{ip}P_{qr}\frac{\partial^3 h^T(\hat{x}_t,t)}{\partial\hat{x}_p\partial\hat{x}_q\partial\hat{x}_r})\varphi_\eta^{-1}(\sum_p P_{jp}\frac{\partial h(\hat{x}_t,t)}{\partial\hat{x}_p}$$

$$+\frac{1}{2}\sum_{p,q,r} P_{jp}P_{qr}\frac{\partial^3 h(\hat{x}_t,t)}{\partial\hat{x}_p\partial\hat{x}_q\partial\hat{x}_r}))dt + (\sum_{p,q} P_{ip}P_{jq}\frac{\partial^2 h^T(\hat{x}_t,t)}{\partial\hat{x}_p\partial\hat{x}_q})\varphi_\eta^{-1}(dz - \hat{h}dt). \quad (8)$$

Equation (8) describes the set of approximate filtering evolution equations and can be regarded as the *fourth* stochastic evolution equation. This structure of filtering equations received relatively *less* attention in literature. A detailed proof of the Theorem associated with equation (8), its effectiveness and application can be found in Sharma *et al.* (2006). Its special case received much attention and is widely available in celebrated books, e.g. Jazwinski (1970, p. 361), Pugachev and Sintsyn (1977, p. 452). Equation (3), the set of exact stochastic filtering equations, assumes the structure of stochastic differential equations. The Carleman linearization (Kowalski and Steeb 1991), which transforms finite dimensional non-linear system into a system of infinite dimensional linear systems, was successfully applied to approximate stochastic differential equations (Germani *et al. 2007*). Thus, the application of the Carleman linearization to derive the approximate stochastic filtering evolution equations can be explored.

Here, the Author states the *fifth* stochastic evolution equation, i.e. the stochastic evolution of scalar function of the multi-dimentional *random* state vector. In literature, it is also known as the multi-dimensional Itô differential rule. The multi-dimensional Itô differential rule is one of the most celebrated results of stochastic processes. It has found applications in deriving the expression of evolution of conditional moment, the Fokker-Planck equation, accomplishing the qualitative analysis of the stochastic differential system. The stochastic evolution $d\phi(x_t)$ of the scalar function of the n-dimensional random state vector using the stochastic differential rule can be stated as

$$d\phi(x_t) = \sum_i \frac{\partial \phi(x_t)}{\partial x_i} dx_i + \frac{1}{2}\sum_{i,j} dx_i dx_j \frac{\partial^2 \phi(x_t)}{\partial x_i \partial x_j}. \tag{9}$$

Note that the random state vector x_t, which satisfies the Brownian motion process-driven stochastic differential equation, is the subject of discussion. The scalar function $\phi(x_t)$ is twice continuously differentiable as well. After plugging the ith component of stochastic differential equation,

$$dx_i = f_i(x_t, t)dt + \sum_\gamma G_{i\gamma}(x_t, t)dB_\gamma,$$

into equation (9), we have

$$d\phi(x_t) = (\sum_i \frac{\partial \phi(x_t)}{\partial x_i} f_i(x_t, t) + \frac{1}{2}\sum_i (G\psi_w G^T)_{ii}(x_t, t)\frac{\partial^2 \phi(x_t)}{\partial x_i^2}$$

$$+ \sum_{i<j}(G\psi G^T)_{ij}(x_t, t)\frac{\partial^2 \phi(x_t)}{\partial x_i \partial x_j})dt + \sum_{1\le i\le n, 1\le \gamma \le r} \frac{\partial \phi(x_t)}{\partial x_i} G_{i\gamma}(x_t, t)dB_\gamma, \tag{10}$$

where the size of the vector Brownian motion process is r. Note that the contribution to the term $d\phi(x_t)$ coming from the second and third terms of the right-hand side of equation (10) is attributed to the Itô property $dB_\phi dB_\gamma = \delta_{\phi\gamma} dt$. The integral counterpart of equation (10) can be written as

$$\phi(x_t) = \phi(x_{t_0}) + (\sum_i \int_{t_0}^{t} \frac{\partial \phi(x_s)}{\partial x_i(s)} f_i(x_s, s)ds + \frac{1}{2}\sum_i \int_{t_0}^{t} (G\psi_w G^T)_{ii}(x_s, s)\frac{\partial^2 \phi(x_s)}{\partial x_i^2(s)} ds$$

$$+ \sum_{i<j} \int_{t_0}^{t} (G\psi_w G^T)_{ij}(x_s, s)\frac{\partial^2 \phi(x_s)}{\partial x_i(s)\partial x_j(s)} ds)$$

$$+ \sum_{1\le i\le n, 1\le \gamma\le r} \int_{t_0}^{t} \frac{\partial \phi(x_s)}{\partial x_i} G_{i\gamma}(x_s, s)\, dB_\gamma(s).$$

The Brownian motion process-driven stochastic differential equation can be regarded as a Markovian stochastic differential system. On the other hand, the coloured noise-driven stochastic differential system can be termed as the non-Markovian stochastic differential equation, since it generates the non-Markov process. The problem of analysing the non-Markovian stochastic differential system can be accomplished via stochastic equation assuming the structure of infinite series, which is a generalization of the Fokker-Planck equation. Here, we explain the stochastic equation in some detail. Consider the conditional probability density $p(x_1|x_2)$, where

$$p(x_1, x_2) = p(x_1|x_2)p(x_2). \tag{11}$$

After integrating over the variable x_2 and introducing the transition probability density, equation (11) becomes

$$p(x_1) = \int q_{t_1,t_2}(x_1, x_2)p(x_2)dx_2. \tag{12}$$

The relationship between the transition probability density $q_{t_1,t_2}(x_1, x_2)$ and the conditional characteristic function involves the inverse Fourier integral. The conditional characteristic function is the conditional moment generating function. As a result of these, equation (12) can be recast as

$$p(x_1) = \frac{1}{2\pi}\int e^{-iu(x_1-x_2)}(\sum_{0\le n}\frac{(iu)^n}{n!}\langle(x_1 - x_2)^n\rangle)p(x_2)dx_2du$$

$$= \sum_{0\le n}\int\frac{1}{n!}(\frac{1}{2\pi}\int(iu)^n e^{-iu(x_1-x_2)}du)\langle(x_1 - x_2)^n\rangle p(x_2)dx_2,$$

and

$$\frac{1}{2\pi}\int e^{-iu(x_1-x_2)}(iu)^n du = (-\frac{\partial}{\partial x_1})^n \delta(x_1 - x_2).$$

Thus,

$$p(x_1) = \sum_{0\le n}\int\frac{1}{n!}(-\frac{\partial}{\partial x_1})^n \delta(x_1 - x_2)\langle(x_1 - x_2)^n\rangle p(x_2)dx_2. \tag{13}$$

After introducing the notion of stochastic processes, taking $x_1 = x_\tau, x_2 = x$, equation (13) can be recast as

$$p(x_\tau) = \sum_{0\le n}\int\frac{1}{n!}(-\frac{\partial}{\partial x_\tau})^n \delta(x_\tau - x)\langle(x_\tau - x\ )^n\rangle p(x)dx$$

$$= \sum_{0\le n} \int \frac{1}{n!}(-\frac{\partial}{\partial x_\tau})^n \delta(x_\tau - x)k_n(x)\tau\, p(x)dx,$$

where $\left\langle \frac{(x_\tau - x)^n}{\tau} \right\rangle = k_n(x)$ and the time interval condition $\tau \to 0$ leads to

$\dot{p}(x) = \sum_{1\le n} \frac{1}{n!}(-\frac{\partial}{\partial x})^n k_n(x\ )p(x)$. (14) Equation (14) is popularly known as stochastic equation, *evolution* of conditional probability density for given initial states for the *non-Markovian* process x_t satisfying the stochastic differential equation $\dot{x}_t = F(x_t, \xi_t)$. Here the process ξ_t is an arbitrary process. The coefficient $k_n(x)$ can be explicitly stated for the stochastic differential equation

$$\dot{x}_t = f(x_t) + g(x_t)\xi_t,$$

see equation (4.180) of Stratonovich (1963). The evolution $d\hat{\phi}(x_t)$ of conditional moment for the non-Markovian stochastic differential system can be cast as

$$d\hat{\phi}(x_t) = \langle \sum_{1\le n} \frac{k_n}{n!}(x_t)\frac{\partial^n \phi(x_t)}{\partial x_t^n} \rangle dt. \tag{15}$$

One can arrive at equation (15) by using the definition of conditional expectation, equation (14), and integration by part formula. After a simple calculation and adopting the procedure stated for the *third* stochastic evolution equation, equation (3), of this book chapter, one can arrive at the expression for the evolutions $d\hat{x}_t$ and dP_x.

Here, the Author states and explains briefly about the evolution of conditional probability density for the OU process-driven stochastic differential system, where the OU process has finite, non-zero, smaller correlation time. The OU process is zero mean, stationary and Gauss-Markov process. The autocorrelation $R_{\xi\xi}(t-s)$ of the OU process ξ_t satisfying the stochastic differential equation $d\xi_t = \frac{-1}{\tau_{cor}}\xi_t dt + \frac{\sqrt{2D}}{\tau_{cor}} dB_t$ becomes

$$R_{\xi\xi}(t-s) = \frac{D}{\tau_{cor}} e^{-\frac{|t-s|}{\tau_{cor}}}. \tag{16}$$

The procedure is lengthy and details are omitted. Here, we simply state the steps associated with deriving the evolution equation, i.e. (i) derive the master's equation for the coloured noise-driven stochastic differential system using the functional calculus approach (ii) introduce the assumptions, i.e. zero mean, stationary, and Gaussianity, which lead to a quite simplified expression (iii) take the Taylor series expansion of the functional derivative $\frac{\delta x_t}{\delta \xi_s}$ in powers of $(s-t)$, where $s<t$, upto first-order expansion, since this corresponds to the relatively smaller correlation time (iv) replace the second-order cumulant term of the resulting equation with equation (16) because of the Gaussian assumption, we arrive at the *evolution equation*

$$\dot{p}(x) = -\frac{\partial}{\partial x} fp + D\frac{\partial}{\partial x}(g\frac{\partial}{\partial x} g(1+\tau_{cor} g(\frac{f}{g})')p). \tag{17}$$

For a greater detail about the proof of the above evolution equation, Hänggi (1995) can be consulted. Note that equations (1)-(3), (8), (10), (14), (15), (17) can be regarded as the main results of this topic. Some of the results can be found in celebrated books on stochastic processes and related fields. Some of them are not sufficiently well known in literature, especially equations (8), (15), (17) and merit further investigations.

CONCLUSION

This chapter can be considered as a result of a compilation of current knowledge for analysing the stochastic differential system in classical and stochastic evolution equations frameworks. This chapter encompasses the evolution of conditional probability density for given initial states for the non-Markov process. In this chapter, the evolution of conditional moment for the non-Markov process as well as the evolution of conditional probability density for the OU process-driven stochastic differential system are also explained that are not widely known in literature.

The Researchers can find celebrated books and seminal papers in the references encompassing works on stochastic differential systems in Markovian as well as non-Markovian setting.

REFERENCES

Arnold, V. I. Ordinary Differential Equations; The MIT Press: Cambridge and Massachusetts, 1995.

Bouten, L. Filtering and control in optics. *PhD Thesis Dissertation,* Nottingham University (UK), 2004.

Germani, A.; Manes, C. and Palumbo, P. Filtering of stochastic non-linear differential systems via a Carleman approximation approach. *IEEE Trans. Automat. Contrl,* 52, 2166-2172, 2007.

Gihman, I. I.; Skorohod, A.V. Stochastic Differential Equations; Springer-Verlag: Berlin, 1972.

Hänggi, P. The functional derivative and its use in the description of noisy dynamical systems. *Stochastic Processes Applied to Physics* (L. Pesquera and M. Rodriguez, eds.), World Scientific: Heyden, Philadelphia, 69–95, 1995.

Jazwinski, A. H. Stochastic Processes and Filtering Theory; Academic Press: New York and London, 1970.

Karatzas, I.; Shreve, S. E. Brownian Motion and Stochastic Calculus (graduate text in mathematics); Springer: New York, 1991.

Kowalski, K.; Steeb W-H. Non-Linear Dynamical Systems and Carleman Linearization; World Scientific: Singapore, New Jersy, 1991.

Liptser, R. S.; Shiryaev, A. N. Statistics of Random Processes (1); Springer: Berlin, 1977.

Parthasarathy, K. R. An Introduction to Quantum Stochastic Calculus; Birkhäuser: Boston, Berlin, 1992.

Protter, Philip E. Stochastic Integration and Differential Equations; Springer: Berlin, Heidelberg, New York, 1991.

Pugachev, V. S.; Sinitsyn, I. N. Stochastic Differential Systems (analysis and filtering); John-Wiley and Sons: Chichester and New York, 1977.

Sharma, S. N.; Parthasarathy, H. and Gupta, J.R.P. Third-order approximate Kushner filter for a non-linear dynamical system. *International Journal of Control*, 79(9), 1096-1106, 2006.

Stratonovich, R. L. Topics in the Theory of Random Noise (Vol 1and 2); Gordan and Breach: New York, 1963.

Strook, D. W.; Varadhan, S. R. S. Multidimensional Diffusion Processes (classics in mathematics); Springer: Berlin, Heidelberg, New York, 1979.

In: Evolution Equations
Editor: Arthur L. Claes

ISBN: 978-1-61209-652-0

Chapter 4

EXACT SOLUTION TO EINSTEIN'S EVOLUTION EQUATION FOR FLUID WITH EXPONENTIAL FREE PATHS DISTRIBUTION

G. L. Aranovich and M. D. Donohue
Department of Chemical and Biomolecular Engineering
The Johns Hopkins University, Baltimore, MD, U.S.A.

ABSTRACT

The evolution of fluid density in time and space is considered in the framework of Einstein's theory involving expansions of the density distribution in powers of displacement and time. By truncating these expansions in the evolution equation, Einstein derived the classical model of diffusion. Here, we consider Einstein's model without truncating the density expansions. This demonstrates limits to Einstein's truncations and demonstrates non-classical behavior for diffusion phenomenology in fluids with large mean-free paths, i.e. for fluids between the Knudsen and Fickian limits.

An exact solution to Einstein's evolution equation for a fluid is presented. This solution shows that the classical diffusion model is valid for fluids only in the limit of small mean-free paths, λ.

Here we demonstrate that, at large λ, Einstein's evolution equation predicts not only significant quantitative deviations from classical diffusion profiles, but also qualitative metamorphoses in diffusion mechanisms, such as co-existence of "laminar" and "turbulent" fluxes and "cooperative" transitions between molecular chaos and predominantly ballistic motion of molecules.

This has implications for fundamentals of fluids between the Knudsen and Fickian limits, and for a variety of fields where evolution of a system includes random, multi-scale displacement of particles, such as nanotechnology, vacuum techniques, biological systems, turbulence, and astrophysics.

EINSTEIN'S EVOLUTION EQUATION

To analyze diffusion in fluid, Einstein considered a single discrete time step, τ, and a distribution, $\phi(\Delta)$, of spatial displacements, Δ [1].

In the framework of this model, the temporally-evolved density distribution $n(x,t+\tau)$ can be expressed in terms of the average of the spatially-displaced density distribution $n(x+\Delta,t)$ over $\phi(\Delta)$ [1-3]:

$$n(x,t+\tau) = \int_{-\infty}^{\infty} n(x+\Delta,t)\phi(\Delta)d\Delta \tag{1}$$

For equation (1), Einstein derived a classical approximation by considering the expansions:

$$n(x,t+\tau) = n(x,t) + \tau\frac{\partial n(x,t)}{\partial t} + \ldots \tag{2}$$

and $n(x+\Delta,t) = n(x,t) + \Delta\frac{\partial n(x,t)}{\partial x} + \frac{\Delta^2}{2!}\frac{\partial^2 n(x,t)}{\partial x^2} + \ldots$ (3)

and plugged them into equation (1) to get

$$n + \tau\frac{\partial n}{\partial t} + \ldots = n\int_{-\infty}^{+\infty}\phi(\Delta)d\Delta + \frac{\partial^2 n}{\partial x^2}\int_{-\infty}^{+\infty}\frac{\Delta^2}{2!}\phi(\Delta)d\Delta + \frac{\partial^4 n}{\partial x^4}\int_{-\infty}^{+\infty}\frac{\Delta^4}{4!}\phi(\Delta)d\Delta + \ldots \tag{4}$$

Since $\phi(\Delta) = \phi(-\Delta)$, then $\int_{-\infty}^{+\infty}\Delta^i\phi(\Delta)d\Delta = 0$ for any odd i. This is why the right-hand side of equation (4) only has terms with even powers of Δ.

Einstein assumed that τ is small to neglect terms of τ^2 and higher in expansion (2) and Δ is small to neglect terms of Δ^3 and higher in expansion (3). Bearing in mind that

$$\int_{-\infty}^{+\infty}\phi(\Delta)d\Delta = 1 \tag{5}$$

and putting

$$\frac{1}{\tau}\int_{-\infty}^{+\infty}\frac{\Delta^2}{2}\phi(\Delta)d\Delta = D \tag{6}$$

Einstein [1] arrived at the classical (Fick's [4]) diffusion equation

$$\frac{\partial n}{\partial t} = D\frac{\partial^2 n}{\partial x^2} \quad (7)$$

constituting a basic phenomenological model for later developments, including modern modifications of diffusion equations [5 – 11]. Note that a key factor in the acceptance of Fick's model as the fundamental basis of diffusion phenomenology was Einstein's derivation of equation (7) from evolution equation (1).

Einstein's approximation given by equations (2) – (7) includes two assumptions: existence of infinite series (2) – (4) and validity of truncation assuming small Δ and τ. However, integration in equation (4) goes from $\Delta = -\infty$ to $\Delta = +\infty$; therefore, neglecting terms of Δ^3 and higher is at least questionable.

To avoid problems of expansions in series (2) – (4) and prevent losing information due to truncation of the series, in this paper we propose exact solution of equation (1) for exponential distribution function, $\phi(\Delta)$.

EXPONENTIAL FREE PATHS DISTRIBUTION

Consider free-path lengths, ξ, between subsequent collisions in a fluid with the classical distribution of ξ [12], $f(\xi) = A\exp(-\frac{\xi}{\lambda})$, where λ is the mean-free path. Then, $\xi = |\Delta|$ and $A = \frac{1}{2\lambda}$ (from equation (5)) which give

$$\phi(\Delta) = \frac{1}{2\lambda}\exp(-\frac{|\Delta|}{\lambda}) \quad (8)$$

Exponential distribution is not only a phenomenological ("practical") model, but it also has a theoretical basis for fluids [12] and it has been verified by various simulations, including molecular dynamics simulations for hard-sphere and square-well molecules [13].

With the exponential distribution of free path lengths (8), equation (1) can be written as

$$n(x,t+\tau) - n(x,t) = \frac{1}{2\lambda}\int_{-\infty}^{\infty} n(x+\Delta,t)\exp(-\frac{|\Delta|}{\lambda})d\Delta - n(x,t) \quad (9)$$

and presented in the following form

$$n(x,t+\tau) - n(x,t) = \frac{1}{2\lambda}\int_{0}^{\infty} n(x+\Delta,t)\exp(-\frac{\Delta}{\lambda})d\Delta + \frac{1}{2\lambda}\int_{0}^{\infty} n(x-\Delta,t)\exp(-\frac{\Delta}{\lambda})d\Delta - n(x,t) \quad (10)$$

The integrals in equation (10) can be transformed by integration in parts:

$$\frac{1}{2\lambda}\int_0^\infty n(x+\Delta,t)\exp(-\frac{\Delta}{\lambda})d\Delta = -\frac{1}{2}n(x+\Delta,t)\exp(-\frac{\Delta}{\lambda})\Big|_0^\infty + \frac{1}{2}\frac{\partial}{\partial x}\int_0^\infty n(x+\Delta,t)\exp(-\frac{\Delta}{\lambda})d\Delta =$$

$$= \frac{1}{2}n(x,t) + \frac{1}{2}\frac{\partial}{\partial x}\int_0^\infty n(x+\Delta,t)\exp(-\frac{\Delta}{\lambda})d\Delta \tag{11}$$

and

$$\frac{1}{2\lambda}\int_0^\infty n(x-\Delta,t)\exp(-\frac{\Delta}{\lambda})d\Delta = -\frac{1}{2}n(x-\Delta,t)\exp(-\frac{\Delta}{\lambda})\Big|_0^\infty + \frac{1}{2}\frac{\partial}{\partial x}\int_0^\infty n(x+\Delta,t)\exp(-\frac{\Delta}{\lambda})d\Delta =$$

$$= \frac{1}{2}n(x,t) - \frac{1}{2}\frac{\partial}{\partial x}\int_0^\infty n(x-\Delta,t)\exp(-\frac{\Delta}{\lambda})d\Delta \tag{12}$$

Combining equations (10) – (12) gives:

$$n(x,t+\tau) - n(x,t) = \frac{1}{2}\frac{\partial}{\partial x}\int_0^\infty [n(x+\Delta,t) - n(x-\Delta,t)]\exp(-\frac{\Delta}{\lambda})d\Delta \tag{13}$$

Note that equation (13) is equivalent to Einstein's evolution equation (1) with exponential distribution of free path lengths.

Exact Solutions to Equation (13) for System without Boundaries

Consider a one-dimensional system without boundaries for some initial profile which at time $t = 0$ begins diffusing into infinite space [2]. Assume that $\lambda = Const$; this is equivalent to $D = Const$, if D is defined by equations (6) and (8) at constant τ. Such a model is a "canonical" problem in the mathematical foundations of diffusion theory [14, 15]. For a delta-function profile initially, Fick's equation (7) gives well known (classical) solution

$$n(x,t) = \frac{1}{\sqrt{4\pi Dt}}\exp(-\frac{x^2}{4Dt}) \tag{14}$$

In this section, we solve this ("canonical") problem for equation (13) which gives a solution to Einstein's evolution equation (1) with exponential free paths distribution (8).

Separation of Variables

To derive a solution to equation (13), consider the method of separation of variables [2, 7, 16] with a partial solution (eigenfunctions) of equation (13) in the form:

$$n(x,t) = X(x)T(t) \tag{15}$$

Plugging this in equation (13) results in:

$$[T(t+\tau) - T(t)]X(x) = T(t)\frac{\partial}{\partial x}\int_0^{\infty}[X(x+\Delta) - X(x-\Delta)]\exp(-\frac{\Delta}{\lambda})d\Delta \tag{16}$$

which can be written as:

$$\frac{2}{D\tau}[\frac{T(t+\tau)}{T(t)} - 1] = \frac{1}{D\tau X(x)}\frac{\partial}{\partial x}\int_0^{\infty}[X(x+\xi) - X(x-\xi)]\exp(-\frac{\xi}{\lambda})d\xi \tag{17}$$

Since the left-hand side of equation (17) depends only on t and right-hand side depends only on x, both parts are constant [7, 16] which we denote $-\gamma^2$:

$$\frac{2}{D\tau}[\frac{T(t+\tau)}{T(t)} - 1] = \frac{1}{D\tau X(x)}\frac{\partial}{\partial x}\int_0^{\infty}[X(x+\xi) - X(x-\xi)]\exp(-\frac{\xi}{\lambda})d\xi = -\gamma^2 \tag{18}$$

This equation translates into the two separate equations:

$$T(t+\tau) + (\frac{1}{2}\gamma^2 D\tau - 1)T(t) = 0 \tag{19}$$

$$\frac{\partial}{\partial x}\int_0^{\infty}[X(x+\xi) - X(x-\xi)]\exp(-\frac{\xi}{\lambda})d\xi + D\tau\gamma^2 X(x) = 0 \tag{20}$$

Eigenfunctions for *X(x)*

The solutions of equation (20) can be represented as [16]

$$X_1(x) = C_1 \sin(\alpha\, x) \tag{21}$$

and

$$X_2(x) = C_2 \cos(\alpha x) \tag{22}$$

where C_1 and C_2 are constant. Plugging $X_1(x) = C_1 \sin(\alpha x)$ and $X_2(x) = C_2 \cos(\alpha x)$ in equation (20) gives correspondingly:

$$[\alpha \int_0^\infty \sin(\alpha\xi) \exp(-\frac{\xi}{\lambda}) d\xi - \frac{D\tau}{2}\gamma^2] \sin(\alpha x) = 0 \tag{23}$$

$$[\alpha \int_0^\infty \sin(\alpha\xi) \exp(-\frac{\xi}{\lambda}) d\xi - \frac{D\tau}{2}\gamma^2] \cos(\alpha x) = 0 \tag{24}$$

Since $\int_0^\infty \sin(\alpha\xi) \exp(-\frac{\xi}{\lambda}) d\xi = \frac{\alpha\lambda^2}{1+\alpha^2\lambda^2}$, equations (23) and (24) give after algebraic manipulations:

$$(\frac{\alpha^2\lambda^2}{1+\alpha^2\lambda^2} - \frac{D\tau}{2}\gamma^2) \sin(\alpha x) = 0 \tag{25}$$

$$(\frac{\alpha^2\lambda^2}{1+\alpha^2\lambda^2} - \frac{D\tau}{2}\gamma^2) \cos(\alpha x) = 0 \tag{26}$$

showing that eigenfunctions (21) and (22) are solutions of equation (20) if

$$\gamma^2 = \frac{2\alpha^2\lambda^2}{D\tau(1+\alpha^2\lambda^2)} \tag{27}$$

Therefore, solution of equation (20) can be written as

$$X(x) = C_1 \sin(\alpha x) + C_2 \cos(\alpha x) \tag{28}$$

with α satisfying equation (27).

Eigenfunctions for *T(t)*

The solution of equation (19) can be represented as:

$$T(t) = A \exp(\omega t) \tag{29}$$

where A is a constant. Plugging this function into equation (19) results in:

$$\exp(\omega\tau) = 1 - \frac{1}{2}D\tau\gamma^2 \tag{30}$$

Combining (30) with (27) gives:

$$\exp(\omega\tau) = \frac{1}{1+\alpha^2\lambda^2} \tag{31}$$

which results in

$$\omega = -\frac{1}{\tau}\ln(1+\alpha^2\lambda^2) \tag{32}$$

With equation (32), eigenfunction (29) for $T(t)$ can be written in the following form:

$$T(t) = A(1+\alpha^2\lambda^2)^{-\frac{t}{\tau}} \tag{33}$$

Solution for Arbitrary Initial Profile

Combining e igenfunctions (28) and (3 3), a par ticular solu tion of eq uation (13) for an arbitrary α is

$$\frac{C_1(\alpha)\sin(\alpha x) + C_2(\alpha)\cos(\alpha x)}{(1+\alpha^2\lambda^2)^{t/\tau}} \tag{34}$$

where $C_1(\alpha)$ and $C_2(\alpha)$ are constants depending on α. Therefore, general solution of equation (13) can be written as

$$n(x,t) = \int_0^\infty \frac{C_1(\alpha)\sin(\alpha x) + C_2(\alpha)\cos(\alpha x)}{(1+\alpha^2\lambda^2)^{t/\tau}} d\alpha \tag{35}$$

Requiring that initial distribution is

$$n(x,0) = \varphi(x) \tag{36}$$

gives from equation (35):

$$\varphi(x)=\int_0^\infty [C_1(\alpha)\sin(\alpha x)+C_2(\alpha)\cos(\alpha x)]d\alpha \tag{37}$$

Considering (37) as a Fourier integral for the function *φ(x)* [7], we obtain:

$$C_1(\alpha)=\frac{1}{\pi}\int_{-\infty}^{\infty}\varphi(\beta)\sin(\alpha\beta)d\beta \tag{38}$$

and

$$C_2(\alpha)=\frac{1}{\pi}\int_{-\infty}^{\infty}\varphi(\beta)\cos(\alpha\beta)d\beta \tag{39}$$

Plugging equations (38) and (39) into equation (35) gives the solution of equation (13) in the form of Fourier integral:

$$n(x,t)=\frac{1}{\pi}\int_0^\infty\int_{-\infty}^{\infty}\frac{\varphi(\beta)[\sin(\alpha\beta)\sin(\alpha x)+\cos(\alpha\beta)\cos(\alpha x)]}{(1+\alpha^2\lambda^2)^{t/\tau}}d\beta\, d\alpha = \\ =\frac{1}{\pi}\int_0^\infty\int_{-\infty}^{\infty}\frac{\varphi(\beta)\cos[\alpha(\beta-x)]}{(1+\alpha^2\lambda^2)^{t/\tau}}d\beta\, d\alpha \tag{40}$$

Formula (40) is an exact solution to equation (13). Since equation (13) is equivalent to Einstein's evolution equation (1) for an exponential distribution of free paths (8), formula (40) gives a solution to equation (1) for this case.

Solution for Delta Function Initial Profile

For a delta function initial profile, *φ(x)* = *δ(x)* and equation (40) turns into the following:

$$n(x,t)=\frac{1}{\pi}\int_0^\infty\frac{\cos(\alpha x)}{(1+\alpha^2\lambda^2)^{t/\tau}}d\alpha \tag{41}$$

Integration in equation (41) gives the following [17, page 376, #9.6.25], [18]:

$$n(x,t)=\frac{1}{\Gamma(t/\tau)\lambda\sqrt{\pi}}\left(\frac{|x|}{2\lambda}\right)^{-\nu}K_\nu\left(\frac{|x|}{\lambda}\right) \tag{42}$$

where $\nu = \frac{1}{2} - \frac{t}{\tau}$, Γ is the gamma function [17, 18], and K_ν is the modified Bessel function of the third kind [17, page 443]. Equation (42) is the exact solution to equation (13) for a delta function initially which at time $t = 0$ begins diffusing into infinite space.

Integration in equation (6) with $\phi(\Delta)$ defined by equation (8) gives $\tau = \lambda^2/D$. Using this in equation (42) allows presenting this solution in the form

$$n(x,t) = \frac{1}{\Gamma(Dt/\lambda^2)\lambda\sqrt{\pi}}\left(\frac{|x|}{2\lambda}\right)^{\frac{Dt}{\lambda^2}-\frac{1}{2}} K_{\frac{1}{2}-\frac{Dt}{\lambda^2}}\left(\frac{|x|}{\lambda}\right) \tag{43}$$

analogous to classical solution (14) for Fick's equation (7). It can be readily shown that, in the limit of small λ, equation (41) produces this classical function (14). Indeed, taking into account that

$$\lim_{\lambda\to 0} \frac{1}{(1+\alpha^2\lambda^2)^{t/\tau}} = \lim_{\lambda\to 0} \frac{1}{(1+\alpha^2\lambda^2)^{Dt/\lambda^2}} = \exp(-\alpha^2 Dt) \tag{44}$$

in the limit of small λ, equation (41) becomes

$$n(x,t) = \frac{1}{\pi}\int_0^\infty \cos(\alpha x)\exp(-\alpha^2 Dt)\,d\alpha \tag{45}$$

Integration in (45) results in equation (14) [7]. Note that applying equation (44) to solution (40) indicates that, in the limit of small λ, equation (40) turns into a classical solution to Fick's equation for an arbitrary initial profile [7].

PREDICTIONS OF EINSTEIN'S EVOLUTION EQUATION WITH EXPONENTIAL FREE PATH DISTRIBUTION

As we mentioned earlier, equation (13) is equivalent to Einstein's evolution equation (1) for an exponential distribution of free paths (8). For this case, equation (40) is an exact solution to equation (1) at an arbitrary initial profile and formula (43) is a particular case of equation (40) for delta function initial profile.

Analysis of equation (43) indicates that there are three stages of evolution where profiles are qualitatively different. These stages correspond to the following ranges of $\frac{Dt}{\lambda^2}$: (a) $0 < \frac{Dt}{\lambda^2} \le \frac{1}{2}$ (initial stage), (b) $\frac{1}{2} < \frac{Dt}{\lambda^2} < 1$ (intermediate stage), and (c) $\frac{Dt}{\lambda^2} \ge 1$ (transition to the classical diffusion profiles). Here we consider each of these stages.

(a) Initial Stage of $0 < \frac{Dt}{\lambda^2} \leq \frac{1}{2}$

Figure 1 shows profiles predicted by equation (43) for $\lambda = 2$ and relatively small Dt: *0.1* (1), *1.0* (2), and *2.0* (3).

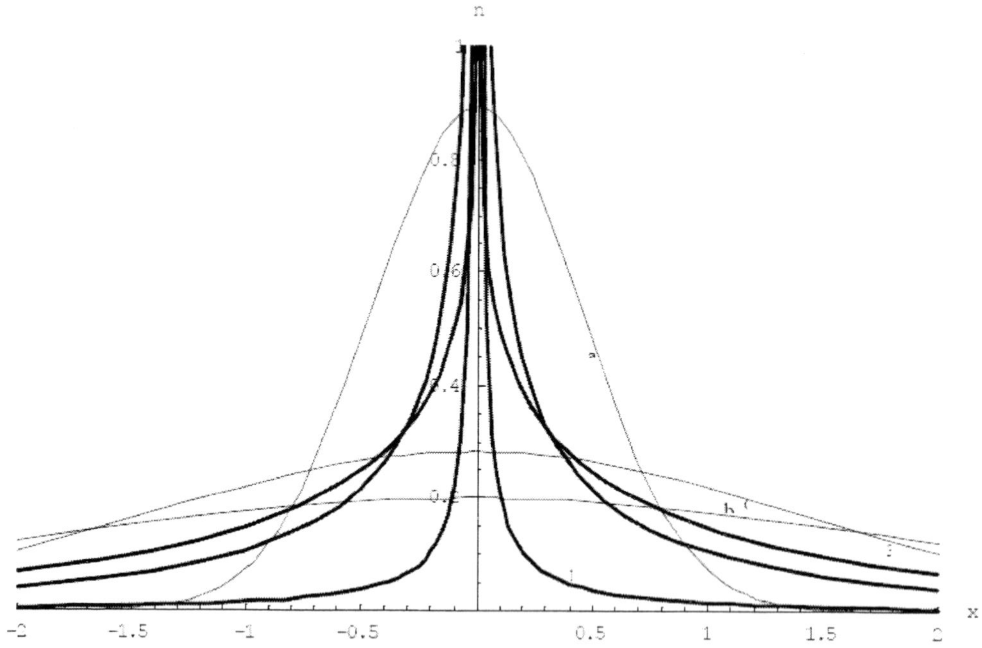

Figure 1. Profiles predicted by equation (43) (thick lines) for $\lambda = 2$ and $Dt = 0.1$ (1), *1.0* (2), and *2.0* (3). Thin lines show corresponding classical profiles (14) at $Dt = 0.1$ (a), *1.0* (b), and *2.0* (c).

As seen from Figure 1, $x = 0$ is still a singularity point in the profiles. At small $|x|/\lambda$ and $0 < \frac{Dt}{\lambda^2} < \frac{1}{2}$, $K_{\frac{1}{2}-\frac{Dt}{\lambda^2}}(\frac{|x|}{\lambda}) \approx \frac{1}{2}\Gamma(\frac{1}{2} - \frac{Dt}{\lambda^2})(\frac{|x|}{2\lambda})^{\frac{Dt}{\lambda^2}-\frac{1}{2}}$ (see [17, page 375, #9.6.9]). In this limit, equation (43) can be written as

$$n(x,t) \approx \frac{\Gamma(\frac{1}{2} - \frac{Dt}{\lambda^2})}{2\Gamma(Dt/\lambda^2)\lambda\sqrt{\pi}}(\frac{|x|}{2\lambda})^{\frac{2Dt}{\lambda^2}-1} \tag{46}$$

As seen from equation (46), in the range of $0 < \frac{Dt}{\lambda^2} < \frac{1}{2}$, there is an exponential singularity at $x = 0$ (curves 1 and 2 in Figure 1). Point $\frac{Dt}{\lambda^2} = \frac{1}{2}$ is the border of this range; at this point, for small $|x|/\lambda$, $K_{\frac{1}{2}-\frac{Dt}{\lambda^2}}(\frac{|x|}{\lambda}) = K_0(\frac{|x|}{\lambda}) \approx -\ln(\frac{|x|}{\lambda})$ (see [17, page 375, #9.6.8]).

Therefore, at $\frac{Dt}{\lambda^2} = \frac{1}{2}$ and small $|x|/\lambda$, equation (43) becomes

$$n(x,t) \approx -\frac{\ln(|x|/\lambda)}{\Gamma(Dt/\lambda^2)\lambda\sqrt{\pi}} \tag{47}$$

which indicates a logarithmic singularity at $x = 0$ (curve 3 in Figure 1).

(b) Intermediate Stage of $\frac{1}{2} < \frac{Dt}{\lambda^2} < 1$

In the range of $\frac{1}{2} < \frac{Dt}{\lambda^2} < 1$, the value of ν in the solution (42) becomes negative. Since $K_{-\nu}(z) = K_\nu(z)$ [17, page 375, #9.6.6], then for small $|x|/\lambda$ [17, page 375, #9.6.9]:

$$K_{\frac{1}{2}-\frac{Dt}{\lambda^2}}(|x|/\lambda) = K_{\frac{Dt}{\lambda^2}-\frac{1}{2}}(|x|/\lambda) \approx \frac{1}{2}\Gamma(\frac{Dt}{\lambda^2}-\frac{1}{2})(\frac{|x|}{2\lambda})^{\frac{1}{2}-\frac{Dt}{\lambda^2}} \tag{48}$$

From equations (43) and (48), at $\frac{1}{2} < \frac{Dt}{\lambda^2} < 1$ and small $|x|/\lambda$, we have:

$$n(x,t) \approx \frac{\Gamma(\frac{Dt}{\lambda^2}-\frac{1}{2})}{2\Gamma(\frac{Dt}{\lambda^2})\lambda\sqrt{\pi}} \tag{49}$$

Therefore, in the range of $\frac{1}{2} < \frac{Dt}{\lambda^2} < 1$, the value of $n(x, t)$ is finite as $x \to 0$.

To analyze $\frac{\partial n}{\partial x}$ at small $\frac{|x|}{\lambda}$, consider the following relation of the modified Bessel function of the third kind, $K_\nu(z)$, to the modified Bessel functions of the first, $I_\nu(z)$, and the second, $I_{-\nu}(z)$, kinds [17, page 375, #9.6.2] :

$$K_\nu(z) = \frac{1}{2}\pi\frac{I_{-\nu}(z) - I_\nu(z)}{\sin(\nu\pi)} \tag{50}$$

At small z [17, page 375, #9.7.6], $I_\nu(z) \approx \frac{1}{\Gamma(\nu+1)}(\frac{z}{2})^\nu$; then, equation (50) gives the following representation for $K_\nu(z)$:

$$K_\nu(z) \approx \frac{\pi}{2\sin(\nu\pi)}[\frac{(z/2)^{-\nu}}{\Gamma(-\nu+1)} - \frac{(z/2)^\nu}{\Gamma(\nu+1)}] \tag{51}$$

Taking into account that $K_{-\nu}(z) = K_\nu(z)$ [17, page 375, # 9.6.6] and using approximation (51), solution (43) can be written as

$$n(x,t) \approx \frac{1}{\Gamma(Dt/\lambda^2)\lambda\sqrt{\pi}}\frac{\pi}{2\sin(\nu\pi)}[\frac{1}{\Gamma(-\nu+1)} - \frac{(|x|/2\lambda)^{2\nu}}{\Gamma(\nu+1)}] \tag{52}$$

where $\nu = \frac{Dt}{\lambda^2} - \frac{1}{2}$. Equation (52) gives $\frac{\partial n}{\partial x} \approx Const(\frac{|x|}{2\lambda})^{2\nu-1}$ at small $\frac{|x|}{2\lambda}$. In the range of $\frac{1}{2} < \frac{Dt}{\lambda^2} < 1$, we have $0 < 2\nu < 1$; therefore, in this range, $\frac{\partial n}{\partial x}$ has singularity at $x = 0$ though n is finite at this point.

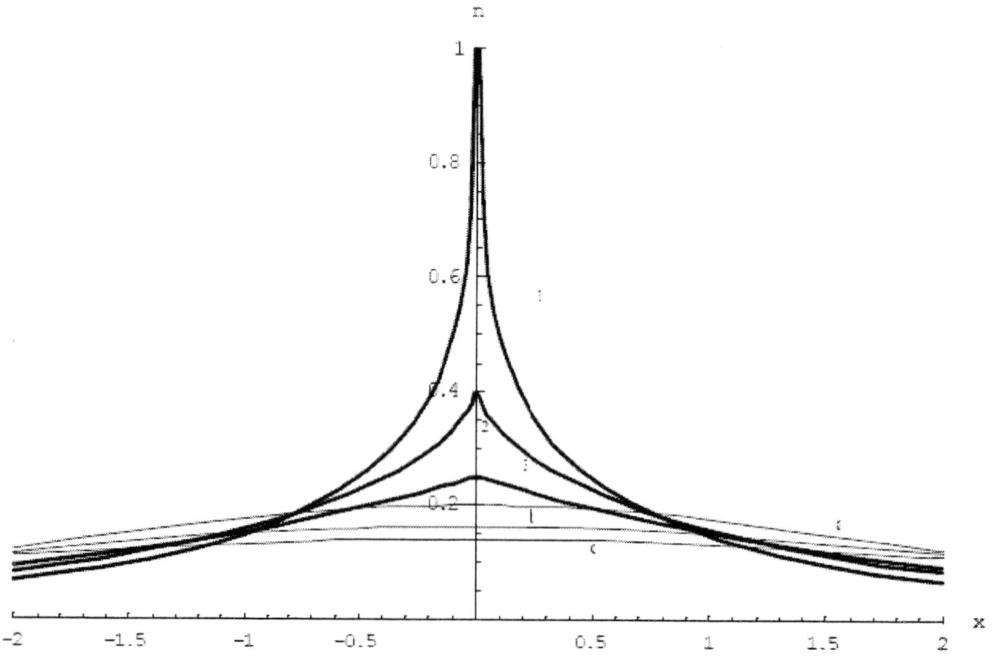

Figure 2. Profiles predicted by equation (43) (thick lines) for $\lambda = 2$ and $Dt = 2.0$ (1), 3.0 (2), and 4.0 (3). Thin lines show corresponding classical profiles (14) at $Dt = 2.0$ (a), 3.0 (b), and 4.0 (c).

Figure 2, curves 2 and 3, illustrate profiles predicted by equation (43) in the range of $\frac{1}{2} < \frac{Dt}{\lambda^2} < 1$ at $\lambda = 2$. Also shown (curve 1) is the profile on the border of this range at $\frac{Dt}{\lambda^2} = \frac{1}{2}$. As seen from Figure 2, at $Dt = 3.0$ and $Dt = 4.0$, there are no singularities at $x = 0$.

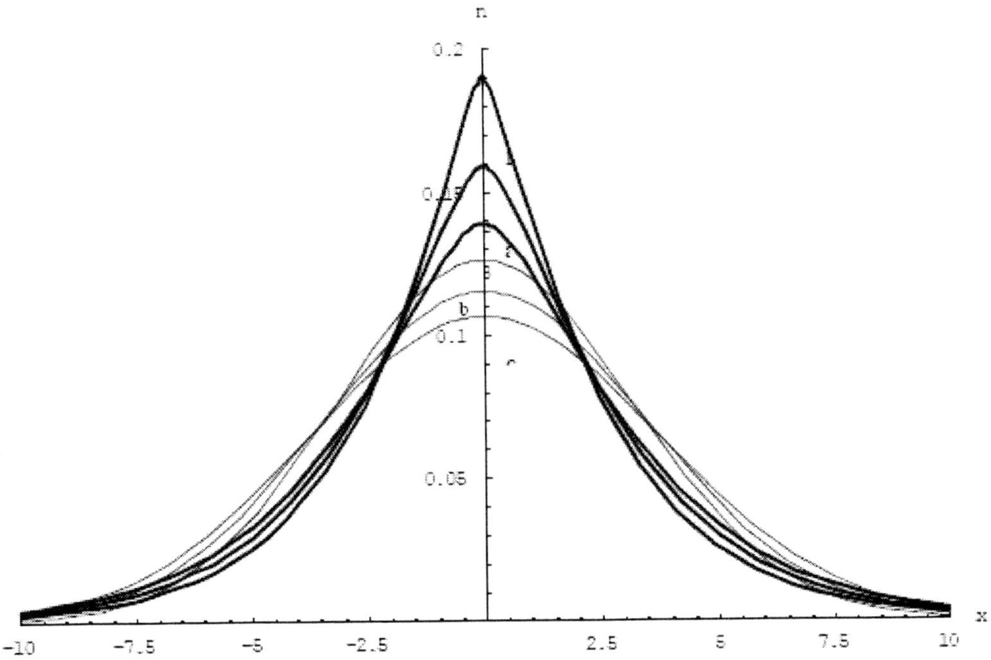

Figure 3. Profiles predicted by equation (43) (thick lines) for $\lambda = 2$ and $Dt = 5.0$ (1), 6.0 (2), and 7.0 (3). Thin lines show classical profiles (14) at $Dt = 5.0$ (a), 6.0 (b), and 7.0 (c).

(c) Stage of $\frac{Dt}{\lambda^2} \geq 1$

As seen from equation (52), singularity in the derivative, $\frac{\partial n}{\partial x}$, disappears as $\frac{Dt}{\lambda^2} \geq 1$. Figure 3 gives profiles predicted by equation (43) in this range of $\frac{Dt}{\lambda^2}$ for $\lambda = 2$. Thin line in this Figure shows a classical profile given by equation (14). As seen from Figure 3, predictions of equation (43) are significantly different from the classical profile.

Figure 4 gives profiles predicted by equation (43) for $\lambda = 2$ at relatively large $\frac{Dt}{\lambda^2}$ and a classical profile given by equation (14). As seen from Figure 4, at large Dt, predictions of equation (43) approach the classical profile.

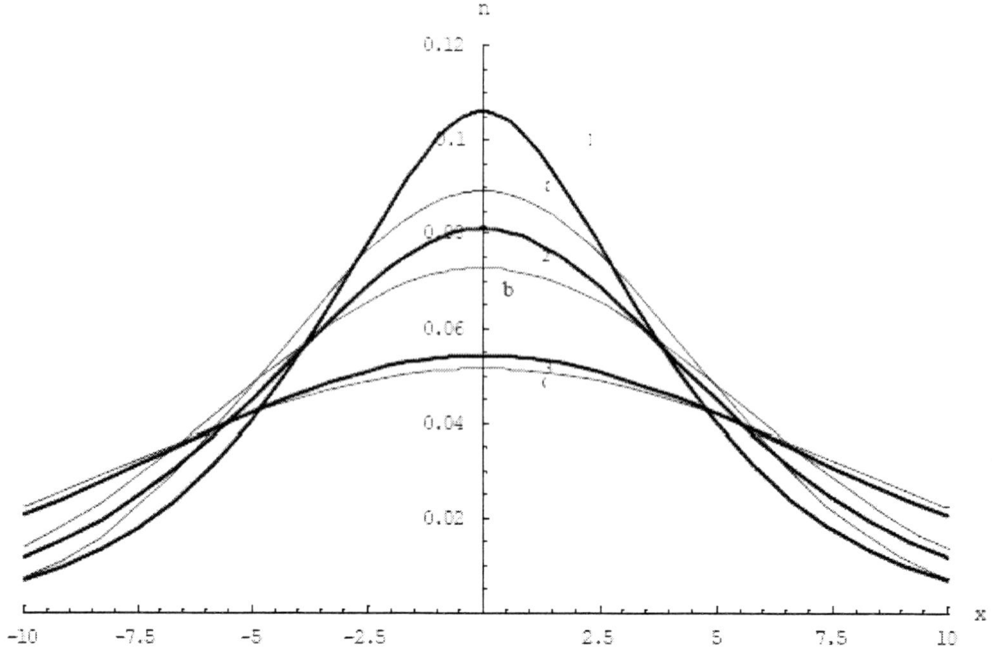

Figure 4. Profiles predicted by equation (43) (thick lines) for $\lambda = 2$ and $Dt = 10$ (1), 15 (2), and 30 (3). Thin lines show classical profiles (14) at $Dt = 10$ (a), 15 (b), and 30 (c).

FLOW TO DIFFUSION TRANSITION

As seen from Figures 1 – 4, equation (43) predicts a principle change in the profile during the (intermediate) stage (b) when $\frac{Dt}{\lambda^2}$ changes from $\frac{1}{2}$ to 1. Kinetic theory of gas gives a classical estimate, $D = \frac{1}{3}V\lambda$ [10], which results in $\frac{Dt}{\lambda^2} = \frac{1}{3}\frac{Vt}{\lambda}$ where V is a characteristic velocity. If L is some distance and V_f is the velocity characterizing a molecular flow, then $t = \frac{L}{V_f}$ and $\frac{Dt}{\lambda^2} = \frac{1}{3}\frac{Vt}{\lambda} = \frac{1}{3}\frac{VL}{V_f\lambda} = \frac{1}{6}\mathrm{Re}^*$ where $\mathrm{Re}^* = \frac{2VL}{V_f\lambda}$ is a kind of Reynolds number for the system [19, page 102] (defined semi-empirically as a ratio of "inertial force to the viscous force"). So, changing $\frac{Dt}{\lambda^2}$ from ½ to unity is equivalent of increasing Re* from 3 to 6. From this point of view, intermediate stage (b) is a transition from a laminar to a turbulent behavior. In terms of mass transfer, it is a transition from flow to diffusion.

This can be understood from the following consideration of evolution. At $t = 0$, delta function initial profile "ignites" flows off the point $x = 0$. While t is small, molecules do not have enough time for significant number of collisions. Therefore, evolution represents predominantly a (laminar) flow regime. Over time, at $\frac{Dt}{\lambda^2} > \frac{1}{2}$, growing number of collisions turns the system into a diffusion (turbulent) mode.

Figure 1 shows stages of evolution in the regime of a (laminar) flow where molecules move out of the point $x = 0$ with almost no collisions. Here, the value of density, n, remains mathematically infinite at $x = 0$ because at least one molecule sits on the infinitesimally small interval (similarly to the initial delta function profile). This occurs because there are molecules resting at $x = 0$ until a moment when collisions return previously flown away molecules and hit out (statistically spread out) those resting at $x = 0$.

When $\frac{Dt}{\lambda^2}$ crosses the point of $\frac{1}{2}$ (Figure 2), collisions do not let molecules to rest at $x = 0$. On this stage, collisions spread out molecules resting at $x = 0$ which results in vanishing singularity in profiles. The point of $\frac{Dt}{\lambda^2} = \frac{1}{2}$ can be considered as switching from flow to diffusion, though the whole range of $\frac{1}{2} < \frac{Dt}{\lambda^2} < 1$ is transitional from a predominantly laminar to a predominantly turbulent behavior. In a sense, diffusion and a (molecular level) turbulence are synonyms, defining "a chaotic or restless character or tendency" (Webster's quote [20, page 1188]).

Profiles in Figure 3 are smoothened by chaotic back-and-force movements due to collisions. This is already a diffusion (turbulent) mode, though the profiles are significantly different from the classical diffusion profile. On this stage, the system still "remembers" its predominantly laminar past, though laminar and turbulent flows coexist. In some sense, profiles in Figure 3 are transitional between "non-classical" profiles in Figure 2 and profiles approaching a classical diffusion profile as shown in Figure 4. On this last stage, the system "forgets" the predominantly laminar past and becomes a predominantly turbulent.

REFERENCES

[1] A. Einstein, Annalen der Physik 322 (1905) 549-560 (English transl. *Investigations on the Theory of the Brownian Movement*, Dover, NY, 1956).

[2] G.L. Aranovich and M.D. Donohue, Physica A 388 (2009) 3355-3370; Molecular Physics 105 (2007) 1085-1093; *Physica A* 373 (2007) 119-141.

[3] S. Abe and A.K. Rajagopal, *Physica A* 375 (2007) 129-132; S. Abe and S. Thurner, Physica A 356 (2005) 403-407; S.N. Majumdar, Current Science 89 (2005) 2076-2092.

[4] A. Fick, *Philosophical Magazine* 10 (1855) 30-39.

[5] I. Muller and T. Ruggeri, *Extended Thermodynamics*, in "Springer Tracts in Natural Philosophy", vol. 37 (Springer, NY, 1993); Seizo Ito, *Diffusion Equations* (American Mathematical Society, Providence, 1992).

[6] R.F. Bass, *Diffusions and Elliptic Operators* (Springer, NY, 1998).

[7] A.N. Tychonov and A.A. Samarski, *Partial Differential Equations of Mathematical Physics*, volume 1 (Holden-Day, Inc., San Francisco, 1964).

[8] R.E. Cunningham and R.J.J. Williams, *Diffusion in Gases and Porous Solids* (Plenum Press, New York, 1980).

[9] H.J.V. Tyrrell and K.R. Harris, *Diffusion in Liquids: Theoretical and Experimental Study* (Butterworth, Boston, 1984).

[10] E.L. Cussler, *Diffusion: Mass Transfer in Fluid Systems* (Cambridge University Press, Cambridge, 1997).

[11] R.B. Bird, W.E. Stewart and E.N. Lightfoot, *Transport Phenomena* (John Wiley, New York, 2002).

[12] R.D. Present, *Kinetic Theory of Gases* (McGraw-Hill, New York, 1958), section 3.1.; S. Chapman and T.G. Cowling, *The Mathematical Theory of Non-Uniform Gases* (University Press, Cambridge, 1939).

[13] T. Einwohner and B.J. Alder, *J. Chem. Phys.* 49 (1968) 1458-1473.

[14] L. Boltzmann, *Annalen der Physik und Chemie* 53 (1894) 959.

[15] G.E. Uhlenbeck and L.S. Ornstein, *Phys. Rev.* 36 (1930) 823.

[16] B.I. Boltaks, *Diffusion in Semiconductors* (Academic Press, NY, 1963).

[17] Handbook of Mathematical Functions, Ed. By M. Abramowitz and I.A. Stegun, National Bureau of Standards, Applied Mathematics Series 55, Tenth Printing, December 1972.

[18] A.P. Prudnikov, Yu.A. Brychkov and O.I. Marichev, *Integrals and Series* (Gordon and Breach Science Publishers, New York, 1986).

[19] P.D. McCormack and L. Crane, *Physical Fluid Dynamics* (Academic Press, NY, 1973).

[20] Webster's II, *New College Dictionary* (Houghton Mifflin Co., Boston, 2001).

In: Evolution Equations
Editor: Arthur L. Claes

ISBN: 978-1-61209-652-0

Chapter 5

SINGULARITY STRUCTURE ANALYSIS OF THE VECTOR NONLINEAR SCHRÖDINGER SYSTEM: MISCELLANEOUS EXCITATIONS AND THEIR INTERACTIONS

Victor K. Kuetche[1,2,*], ***Thomas B. Bouetou***[1,2,3,†]
and Timoleon C. Kofane[2,‡]
[1]National Advanced School of Engineering,
University of Yaounde I, Cameroon
[2]Department of Physics, Faculty of Science,
University of Yaounde I, Cameroon
[3]Department of Mathematics, Faculty of Science,
University of Yaounde I, Cameroon

Abstract

Throughout this chapter, we investigate the singularity structure analysis of the (2+1)-dimensional coupled nonlinear Schrödinger (CNLS) equations, and we show that these equations are Painlevé (P)-integrable. By means of the Weiss et al.'s methodology, we show the arbitrariness of the expansion cœfficients and the consistency of the truncation corresponding to a special Bäcklund transformation (BT) of these CNLS equations. In the wake of such transformation, following the Hirota's formalism, we derive a one-soliton solution. Besides, by using the Zakharov-Shabat (ZS) scheme which provides a general Lax-representation of an evolution system, we show that the (2+1)-dimensional CNLS system under interests is completely integrable. Furthermore, using the arbitrariness of the above cœfficients, we unearth and investigate a typical spectrum of localized and periodic coherent structures. As a result, we depict the elastic and nonelastic interactions amongst such structures and also a splitting phenomenon occurring during their motion.

PACS: 02.30.Ik, 05.45.Yv, 11.10.Lm, 42.81.Dp, 03.75.Lm, 42.50.Md, 42.65.Tg.

*E-mail address: vkuetche@yahoo.fr
†E-mail address: tbouetou@yahoo.fr
‡E-mail address: tckofane@yahoo.com

Keywords: (2+1)-dimensional coupled nonlinear Schrödinger, Bäcklund transformation, Hirota's formalism, Zakharov-Shabat scheme, coherent structures.

1. Introduction

Most natural systems are nonlinear and are therefore modelled by nonlinear systems of equations. In recent years, a certain class of higher-order nonlinear partial differential evolution (NLPDE) equations [1, 2] has become of special interest to mathematicians and theoretical physicists. Such equations possess a special type of elementary solution taking the form of localized disturbances which act somewhat like particles and are therefore known as solitons. The nonlinear Schrödinger (NLS) equation is a well-known example, describing the dynamical evolution of a slowly varying wave envelope under the influence of linear dispersion or diffraction and nonlinear self-phase modulation [3, 4, 5, 6, 7].

The (1+1)-dimensional NLS equation given in its simple form is

$$\imath\phi_t + P\phi_{xx} + Q\left|\phi\right|^2\phi = 0, \tag{1}$$

where ϕ is a physical observable, P is the linear dispersion cœfficient and Q is the nonlinear cœfficient, $\imath^2 = -1$ and subscripts denote partial derivatives. The NLS equation is a universal equation with applications in many different areas of applied mathematics and physics; e.g., in plasma physics, astrophysics, and nonlinear optics, just to name a few. The (1+1)-dimensional NLS equation belongs to the remarkable class of nonlinear evolution equations which can be solved by essentially linear methods. The analytical tool is the inverse scattering transform (IST) which can be viewed as a generalization of the well-known Fourier transform method used for linear problems [3, 4, 5, 6, 7]. Other extended forms of the (1+1)-dimensional NLS equation have been studied [2] amongst which the generalized derivative (1+1)-dimensional NLS equation [8], and they have been shown to be integrable.

In another way, one of the most important features of a soliton is its stability and many recent investigations have been devoted to the profile of soliton stability to various types of perturbations. In this way, the question of transverse stability of the plane soliton solutions to a (1+1)-dimensional NLS equation has been investigated by Anderson et al. [9] by means of a variational method. From this method, the following (2+1)-dimensional NLS equation [9] has been derived

$$\imath\phi_t + P_1\phi_{xx} + P_2\phi_{yy} + g\left(\left|\phi\right|^2\right)\phi = 0, \tag{2}$$

where ϕ is a physical observable, P_1 and P_2 are the linear dispersion cœfficients, and g is an arbitrary function of $|\phi|^2$. Besides, many other (2+1)-dimensional NLS equations have been investigated from a beam propagating in a Kerr material with a parabolic linear index of refraction [10], in intramolecular excitations nonlinearity coupled with lattice vibration of (2+1)-dimensional molecular systems [11], and while using Lie's method of infinitesimal transformation groups to similarity solutions to the (1+1)-and (2+1)-dimensional NLS [12] equations, just to name a few. These NLS equations can also be found by means of the similarity transformations with the effect of reducing a (3+1)-dimensional NLS equation of the form

$$\imath\phi_t + P_1\phi_{xx} + P_2\phi_{yy} + P_3\phi_{zz} + Q\left|\phi\right|^2\phi = 0, \tag{3}$$

where ϕ is a physical observable, P_1, P_2 and P_3 are the linear dispersion cœfficients, Q is the nonlinear cœfficient. Thus, the N-soliton solutions to the (3+1)-dimensional NLS equation [13] can be derived from those of lower dimensional NLS equations.

Moreover, knowing some essential properties of the (1+1)-dimensional NLS equation, some interests have recently been paid to a (1+1)-dimensional CNLS equation in terms of its integrability [14, 15]. There are different approaches to the analytical description of polarized (vector) soliton dynamics issued from the (1+1)-dimensional CNLS equations. One of the frequently used methods deals with various ansätzes which reduce the (1+1)-dimensional CNLS equations to a system of ordinary differential equations [16, 17, 18]. The variational approach has been successfully applied in refs. [19, 20, 21]. From this approach, conditions of resonance splitting of two-component soliton into two separating solitons of different polarization are obtained. By means of the Lie group analysis, some exact solutions to a (1+1)-dimensional CNLS equation with reference to the linear birefringence have been found [22]. Also, formulating the Kaup-Karpman-Maslov-type perturbation theory for solitons of the Manakov system [23], some essential properties of these structures have been derived [24]. At last, the singularity structure analysis combined to the Hirota's method have been intensively investigated in many physical systems, and followed with the construction of some exact soliton solutions ([25, 26, 27] and references therein). Based on this approach, Sahadevan et al. [14] have investigated the integrability of the following system

$$\imath\psi_t + P_1\psi_{xx} + \left(Q_1|\psi|^2 + Q_2|\phi|^2\right)\psi = 0, \tag{4a}$$

$$\imath\phi_t + P_2\phi_{xx} + \left(Q_1|\psi|^2 + Q_2|\phi|^2\right)\phi = 0, \tag{4b}$$

where ϕ and ψ are physical observables, P_1 and P_2 are the linear dispersion cœfficients, Q_1 and Q_2 are the nonlinear cœfficients. As a result, such equations have been shown to possess a N-soliton solution [14].

Due to the complexity of some natural phenomena such as waves on the surface of the ocean, folded protein in natural organisms, it is sometimes difficult to get a rough description of such problems by virtue only of (1+1)-dimensional evolution equations. Focusing particular interests to the (1+1)-dimensional CNLS equations, *the question logically arises whether the remarkable features of such system also survive in higher dimensions*. To obtain an answer to this question, we examine in this paper a (2+1)-dimensional CNLS equation recently investigated in ref. [28] as a model system of vector spatial solitons which components diffract in two transverse dimensions. These equations are expressed as follows [28]

$$\imath\psi_t + \Delta\psi + \sigma\left(|\psi|^2 + \alpha|\phi|^2\right)\psi = 0, \tag{5a}$$

$$\imath\phi_t + \Delta\phi + \sigma\left(\alpha|\psi|^2 + |\phi|^2\right)\phi = 0, \tag{5b}$$

where ψ and ϕ are complex-valued physical observables, functions of the independent variables t, x and y. The operator Δ stands for the Laplacian in two-dimensional space. Here, $\sigma = \pm 1$ is the parameter to distinguish between self-focusing or self-defocusing Kerr nonlinearity. The value of α varies over a wide range [29, 30].

As it is well known in the literature, the scalar NLS equation is not integrable in general (see [31, 32, 33] and references therein). However, Zakharov and Schulman (see [34] and references therein) have shown that the (1+1)-dimensional CNLS system (1) in the case $\alpha = 1$ possesses an infinite set of motion invariants and can be solved by the inverse scattering transform. Such results have been recovered by Sahadevan, Tamizhmani and Lakshmanan [14] through a systematic analysis of the P-integrability of the system. Motivated by this pioneering theoretical work of Zakharov and Schulman (see [34] and references therein) on integrability of the previous system, *we would like to investigate whether this complete integrability also remains sustainable, conditionally or not, when higher-dimensional manifold in which the system is embedded is regarded.*

In this chapter, we investigate the singularity structure analysis of Eq. (5) followed by the construction of some localized and periodic excitations amongst which the folded solitary waves (FSWs). This analysis which is based on Weiss-Tabor-Carnavale (WTC) scheme [35, 36], has been deeply studied in many physical systems [37, 38, 39].

In fact, the subject of integrability has considerably grown and progressed since the "early" integrability times of 70's ([40, 41, 42, 43] and references therein). The P-conjecture, as formulated by Ablowitz et al. [44, 45, 46], and Hastings and McLeod [47], asserts that every ordinary differential equation which arises as a similarity reduction of a NLPDE equation solvable by IST is of P-type, i.e. it has no movable singularities except poles perhaps after a transformation of variables. Ablowitz et al. [46], McLeod and Olver [48] have given proofs of the conjecture under certain restrictions. Subsequently, Weiss et al. [35, 36] have defined the P-property for nonlinear partial differential evolution (NLPDE) equations and developed a method for testing for a common particular type of movable singularity, without having to study any similarity reductions.

The organization of the chapter is settled as follows. In Sec. II, we present the physical motivation of the chapter. In Sec. III, through a systematic search of P-properties of Eq. (5), we establish that for the parametric choices $\sigma = -1$ and $\alpha = 1$, is the system free from movable critical singularity manifolds provided the expansion cœfficients are real-valued. We also derive the BT and construct the Hirota's bilinearization. Subsequently, we discuss a one-soliton solution derived from this bilinearization. In Sec. IV, we present the generalized Lax-representation of the CNLS system while proving its complete integrability. In Sec. V, using the arbitrariness of the expansion cœfficients, we construct some rich variety of localized excitations. In this section, we also investigate some periodic excitations and we depict the elastic interaction processes between such structures. Finally, the last section is devoted to a brief summary of our work.

2. Physical Motivation

Real matter in general does not have a linear shape and owing to its miscellaneous features, it occupies a specific volume in the space. In order to get a rough description of the dynamics of such media, it is useful to introduce independent spatio-temporal variables which span a multidimensional manifold in which the system is embedded.

Because of their multicomponent structure, vector solitons have richer propagation dynamics than their scalar, one-component counterparts. For example, a scalar soliton propagates with an unchanging profile by balancing selfphase modulation with dispersion or diffraction, while in the vector case one must also account for crossphase modulation between components. For collisions, scalar solitons are characterized by phase and position shifts, while vector solitons have the added dynamics of possible intensity redistributions between the component fields [23]. The motivation of the present work should be ascribed to the different applications of the higher-dimensional CNLS equations to real physical systems and the understanding of miscellaneous kinds of pattern formations of this coupled system.

Indeed, starting from wave-wave interactions [49, 50] to matter-wave interactions in systems [51] such as Bose-Einstein Condensates (BEC), a lot of water has flown over the bridge in the past decades and interests in CNLS systems continue to grow nowadays. The system (5) can be regarded as a higher-dimensional time-gated Manakov system [23] referring to prearranged sequences of soliton collisions that effect logical operators. Thus, the above system should play an important role in the construction of logic gates in three-dimensions such as NAND gate, COPY gate, FANOUT gate, NOT gate, the universal Turing equivalent computing machines and developing optical computing [52, 53, 54]. Solitons diffracting in two transverse spatial directions according to the previous model equations can be observed in the transverse effects in nonlinear optics [29, 55, 56, 57, 58, 59, 60, 61, 62], time-independent copropagation [63], but also in the dynamics of two incoherently interacting optical beam in bulk Kerr and saturable media [64].

In recent years, to enhance the capacity of high bit rate and long-distance optical communication systems [65], the emphasis was placed on the study of the propagation of ultrashort optical pulses with a subpicosecond or femtosecond width. In the optical soliton communication system, the transmission of information is usually in the form of a large number of soliton pulse series which inevitably interact mutually while affecting the quality and capacity of the system in terms of pulse distortion, transmission characteristic deterioration, transmission rate decrease and transmission distance shortening. However, such undesirable higher-order effects in the system, namely, the third-order dispersion, the self-steepening and the stimulated Raman scattering which can be used in energy conversion, collision-based amplification without any external amplification medium and pulse compression [66, 67, 68, 69, 70, 71, 72, 73]. The question can logically be formulated whether by taking the topological properties of the system into account, one would get further information about the soliton communication.

In addition to optical communication, in the context of biophysics, the system (5) can be extended to the study of the launching and propagation of solitons along the three spines of an alpha helix in protein [74] spanning a three-dimensional manifold. Similarly, such system is also of physical relevance in the theory of soliton wavelength division multiplexing [75], multichannel bit-parallel-wavelength optical networks [76], propagation of two self-trapped mutually incoherent wave packets in Kerr-like photorefractive media [77, 78] in which the observables ϕ and ψ stand for components of the beam. In the wake of recent investigations of Kanna and Lakshmanan [79] on shape changing collision and partially coherent solitons in CNLS equations, more light has been shed in the possible extension of the system (5) to AlGaAs planar waveguides [80, 81], photorefractive strontium barium niobate crystal using screening solitons [82], excitations by partially coherent light [83], and light bulb [84, 85, 86], for a better understanding of the properties of these substances.

In another hand, the CNLS equation is a special case of the Gross-Pitaevski system [87, 88] known to capture the ground-state properties of the BEC remarkably well, both in higher-and lower-dimensional configurations [89, 90]. In the presence of an optical lattice [91, 92], this picture changes drastically when quantum fluctuations become significant [93]. Thus, the system (5) can be extended to the study of the superfluidity density wave state of BEC in a three-dimensional spatial optical lattice [94] through a classical dynamical phase transition where modulational instability plays an important role. Besides, the frictionless dynamics of BEC under fast trap variations [95], and correlated hopping of bosonic atoms induced by optical lattices [96], atomic collisions on the spectrum of light scattered [97], three-body interactions of neutral bosons in optical lattices [98], quantum interference in the time-of-flight distribution [99], vortex splitting in subcritical NLS system [100], collapse of BEC with attractive interactions [101], Feshbach resonance in BEC [102], controllable scattering of vector BEC solitons [51], just to name a few, constitute other interesting systems to extend to higher-dimensional CNLS systems.

Owing to the wide variety of applications of the singular optical beam [103], ranging from optical tweezers to quantum information [104], optical vortices [105] attract growing attention nowadays and promise in carrying information through futurist optical photonic devices [29, 106]. Vortices in matter waves and stirred BEC provide a close link between self-focusing of light in nonlinear optics and the nonlinear dynamics of matter waves. Assigning the observables ϕ and ψ to the components of an optical field or the two-component matter waves in a BEC [107], the system (5) stands for a model system describing the dynamics of composite vortex solitons in photorefractive crystal and BEC.

3. P-analysis of the (2+1)-dimensional CNLS Equations

In order to investigate the integrability properties of Eq. (5), we rewrite this system in terms of four real-valued functions p, q, u and v defined by $\psi = p + \imath q$ and $\phi = u + \imath v$. Consequently, we derive the following equations

$$p_t + q_{xx} + q_{yy} + \sigma\left[p^2 + q^2 + \alpha\left(u^2 + v^2\right)\right] q = 0, \tag{6a}$$
$$-q_t + p_{xx} + p_{yy} + \sigma\left[p^2 + q^2 + \alpha\left(u^2 + v^2\right)\right] p = 0, \tag{6b}$$
$$u_t + v_{xx} + v_{yy} + \sigma\left[\alpha\left(p^2 + q^2\right) + u^2 + v^2\right] v = 0, \tag{6c}$$
$$-v_t + u_{xx} + u_{yy} + \sigma\left[\alpha\left(p^2 + q^2\right) + u^2 + v^2\right] u = 0. \tag{6d}$$

According to the P-analysis generalized to NLPDE equations by Weiss et al. [35, 36], a NLPDE equation is said to possess the P-property if the only singularities of the general integral which can live on arbitrary non-characteristic ('movable') hypersurfaces are poles. The P-test has proved to be a useful criterion for the identification of completely integrable NLPDE system. In order words, if the singular manifold is determined by

$$f(x, y, t) = 0, \quad f_{xy}(x, y, t) \neq 0, \tag{7}$$

and $u(x, y, t)$ is a solution to a NLPDE equation, it is required that

$$u = \sum_{k=0}^{\infty} u_k f^{k+s}, \tag{8}$$

where $u_0 \neq 0$, $f = f(x, y, t)$ and $u_k = u_k(x, y, t)$, ($k \in \mathbb{N}^{\star}$) are functions of (x, y, t) in a neighborhood of the manifold given by equation (7), and s is a negative integer.

Thus, in view of finding the powers at which some arbitrary functions could enter into the series, we consider the full Laurent series as follows

$$p = \sum_{k=0}^{\infty} p_k f^{k+\beta}, \quad q = \sum_{k=0}^{\infty} q_k f^{k+\gamma}, \quad u = \sum_{k=0}^{\infty} u_k f^{k+\kappa}, \quad v = \sum_{k=0}^{\infty} v_k f^{k+\lambda}. \tag{9}$$

Let us assume that the leading orders are of the forms

$$p \sim p_0 f^{\beta}, \quad q \sim q_0 f^{\gamma}, \quad u \sim u_0 f^{\kappa} \quad v \sim v_0 f^{\lambda}, \tag{10}$$

where β, γ, κ and λ are constants to be determined, and $f(x, y, t)$ is the singularity manifold. Using Eq. (10) into (6) and equating the most dominant terms, we obtain the unique choice

$$\beta = \gamma = \kappa = \lambda = -1, \tag{11}$$

and

$$\sigma\left[p_0^2 + q_0^2 + \alpha\left(u_0^2 + v_0^2\right)\right] + 2\left(f_x^2 + f_y^2\right) = 0, \tag{12a}$$
$$\sigma\left[\alpha\left(p_0^2 + q_0^2\right) + u_0^2 + v_0^2\right] + 2\left(f_x^2 + f_y^2\right) = 0. \tag{12b}$$

This implies that two of the four functions p_0, q_0, u_0 and v_0 are arbitrary without any parametric constraints. Indeed, the two equations of the system (12) are equivalent if and only if $\alpha = 1$. The other parameter σ takes arbitrary values. In a previous study, Zhang et al. [28] have shown that in the case of $\sigma = 1$ referring to the self-focusing Kerr nonlinearity, there are solutions with 'bright' background having elastic and nonelastic properties. In the next, we shall consider the following values $\sigma = -1$ and $\alpha = 1$, referring to a physical system with self-defocusing Kerr nonlinearity. We aim at showing that alongside the class of 'dark' solitons satisfying to the (1+1)-dimensional NLS equations [108] with self-defocusing Kerr nonlinearity, the (2+1)-dimensional counterparts enrich the above class of solutions with miscellaneous patterns.

Substituting Eq. (9) into (6) and vanishing all the cœfficients of the powers f^k, we can obtain the recursion relations to determine the functions p_k, q_k, u_k and v_k as follows

$$\mathcal{M}_k \mathcal{V}_k = \mathcal{T}_k, \tag{13}$$

where $\mathcal{M}_k$ is a square matrix, $\mathcal{V}_k = (p_k, q_k, u_k, v_k)^T$ and $\mathcal{T}_k = (P_k, Q_k, U_k, V_k)^T$ with

$$\begin{aligned} P_k &= p_{k-2,xx} + p_{k-2,yy} - q_{k-2,t} - (k-2)q_{k-1}f_t \\ &- 2(k-2)[2p_{k-1,x}f_x + 2p_{k-1,y}f_y + p_{k-1}(f_{xx} + f_{yy})] \\ &+ \sum_{j=1}^{k-1} p_j\left(p_0 p_{k-j} + q_0 q_{k-j} + u_0 u_{k-j} + v_0 v_{k-j}\right) \\ &+ \sum_{j=1}^{k-1}\sum_{i=0}^{j} p_i\left(p_{k-j}p_{j-i} + q_{k-j}q_{j-i} + u_{k-j}u_{j-i} + v_{k-j}v_{j-i}\right), \end{aligned} \tag{14}$$

$$\begin{aligned} Q_k &= q_{k-2,xx} + q_{k-2,yy} + p_{k-2,t} + (k-2)p_{k-1}f_t \\ &- 2(k-2)[2q_{k-1,x}f_x + 2q_{k-1,y}f_y + q_{k-1}(f_{xx}+f_{yy})] \\ &+ \sum_{j=1}^{k-1} q_j\left(p_0p_{k-j} + q_0q_{k-j} + u_0u_{k-j} + v_0v_{k-j}\right) \\ &+ \sum_{j=1}^{k-1}\sum_{i=0}^{j} q_i\left(p_{k-j}p_{j-i} + q_{k-j}q_{j-i} + u_{k-j}u_{j-i} + v_{k-j}v_{j-i}\right), \end{aligned} \tag{15}$$

$$U_k = P_k\Big|_{\substack{p_k \leftrightarrow u_k \\ q_k \leftrightarrow v_k}}, \quad V_k = Q_k\Big|_{\substack{p_k \leftrightarrow u_k \\ q_k \leftrightarrow v_k}}. \tag{16}$$

Eq. (16) merely tells that the couples (U_k, V_k) and (P_k, Q_k) are equivalent through the substitution of p_k by u_k, and q_k by v_k, and reciprocally.

The matrix $\mathcal{M}_k$ is given by

$$\mathcal{M}_k = \begin{bmatrix} A_{1k} & A_{2k} & A_{3k} & A_{4k} \\ A_{2k} & B_{1k} & B_{2k} & B_{3k} \\ A_{3k} & B_{2k} & C_{1k} & C_{2k} \\ A_{4k} & B_{3k} & C_{2k} & D_k \end{bmatrix}, \tag{17}$$

with

$$A_{1k} = 2p_0^2 - k(k-3)\left(f_x^2+f_y^2\right), \quad A_{2k} = 2p_0q_0, \quad A_{3k} = 2u_0p_0, \tag{18a}$$

$$B_{1k} = 2q_0^2 - k(k-3)\left(f_x^2+f_y^2\right), \quad B_{2k} = 2q_0u_0, \quad A_{4k} = 2v_0p_0, \tag{18b}$$

$$C_{1k} = 2u_0^2 - k(k-3)\left(f_x^2+f_y^2\right), \quad C_{2k} = 2u_0v_0, \quad B_{3k} = 2q_0v_0, \tag{18c}$$

$$D_k = 2v_0^2 - k(k-3)\left(f_x^2+f_y^2\right). \tag{18d}$$

Thus, the determinant Δ_k of the matrix $\mathcal{M}_k$ is given by

$$\Delta_k = -(k-4)(k-3)^3(k+1)k^3\left(f_x^2+f_y^2\right)^4. \tag{19}$$

The resonances are then found at

$$k = -1, 0, 0, 0, 3, 3, 3, 4. \tag{20}$$

The resonance at $k=-1$ corresponds to that of the singularity manifold f being arbitrary. If the model is P-integrable, it is required three resonance conditions at $k = 0, 3, 4$, which are satisfied identically such that the other three arbitrary functions among p_k, q_k, u_k and v_k can be introduced into the general series expansion given by Eq. (9).

From the leading order analysis, we know that the resonance at $k = 0$ is satisfied identically and three of p_0, q_0, u_0 and v_0 are arbitrary $(\sigma=-1, \alpha=1)$.

For $k=1$, we find that

$$p_1 = q_1 = u_1 = v_1 = 0, \tag{21}$$

provided that the system

$$p_0 f_t + q_0 (f_{xx} + f_{yy}) + 2 (q_{0x} f_x + q_{0y} f_y) = 0, \tag{22a}$$
$$q_0 f_t - p_0 (f_{xx} + f_{yy}) - 2 (p_{0x} f_x + p_{0y} f_y) = 0, \tag{22b}$$
$$u_0 f_t + v_0 (f_{xx} + f_{yy}) + 2 (v_{0x} f_x + v_{0y} f_y) = 0, \tag{22c}$$
$$v_0 f_t - u_0 (f_{xx} + f_{yy}) - 2 (u_{0x} f_x + u_{0y} f_y) = 0, \tag{22d}$$

holds.

We also see that the functions p_2, q_2, u_2 and v_2 vanish identically while p_0, q_0, u_0 and v_0 satisfy

$$p_{0t} + q_{0xx} + q_{0yy} = 0, \quad q_{0t} - p_{0xx} - p_{0yy} = 0, \tag{23a}$$
$$u_{0t} + v_{0xx} + v_{0yy} = 0, \quad v_{0t} - u_{0xx} - u_{0yy} = 0. \tag{23b}$$

On the other hand, solving the case $k = 3$, the following resonance condition is derived

$$p_0 p_3 + q_0 q_3 + u_0 u_3 + v_0 v_3 = 0, \tag{24}$$

so that three of the four functions p_3, q_3, u_3 and v_3 are arbitrary.

In the similar way, we easily check that any one of the functions p_4, q_4, u_4 and v_4 is arbitrary provided that the system

$$p_3 f_t + q_3 (f_{xx} + f_{yy}) + 2 (q_{3x} f_x + q_{3y} f_y) = 0, \tag{25a}$$
$$q_3 f_t - p_3 (f_{xx} + f_{yy}) - 2 (p_{3x} f_x + p_{3y} f_y) = 0, \tag{25b}$$
$$u_3 f_t + v_3 (f_{xx} + f_{yy}) + 2 (v_{3x} f_x + v_{3y} f_y) = 0, \tag{25c}$$
$$v_3 f_t - u_3 (f_{xx} + f_{yy}) - 2 (u_{3x} f_x + u_{3y} f_y) = 0, \tag{25d}$$

holds.

Eq. (6) admit sufficient number of arbitrary functions. Thus, this equation is P-integrable for the parametric restriction ($\sigma = -1, \alpha = 1$). It is well known that the P-analysis can also be used to obtain other interesting properties [35, 36]. In this chapter, we use the truncated P-expansion to obtain some interesting exact solutions.

Setting

$$p_k = q_k = u_k = v_k = 0, \quad k \geq 2, \tag{26}$$

Equation (9) becomes

$$p = p_0/f + p_1, \quad q = q_0/f + q_1, \quad u = u_0/f + u_1, \quad v = v_0/f + v_1, \tag{27}$$

where $\{p_1, q_1, u_1, v_1\}$ is also a solution of the (2+1)-dimensional CNLS equations. Thus, the truncated expansion given above is also a BT. Generally, in order to construct a typical family of solutions to Eq. (6) in a simple manner, it is useful to consider very simple expressions of p_1, q_1, u_1 and v_1. Thus, for convenience later, we select the seed solution

$$p_1 = q_1 = u_1 = v_1 = 0. \tag{28}$$

All the remainder equations to solve the functions p_0, q_0, u_0, v_0 and f are simplified to Eqs. (22) and (23) provided

$$p_0^2 + q_0^2 + u_0^2 + v_0^2 = 2\left(f_x^2 + f_y^2\right), \tag{29}$$

holds.

Using Eqs. (27) and (28) into (6), we derive

$$\left[\left(D_x^2 + D_y^2\right) f \cdot f + \mathcal{V}_0^T \mathcal{V}_0\right] \mathcal{V}_0 - f\left(\mathcal{A} D_t + D_x^2 + D_y^2\right) \mathcal{V}_0 \cdot f = 0, \tag{30}$$

where the operators D_x and D_t are defined by [109, 110, 111]

$$D_t^m D_x^n (G \cdot F) = (\partial_t - \partial_{t'})^m (\partial_x - \partial_{x'})^n G(x,t) F(x',t')|_{x=x';t=t'}. \tag{31}$$

The constant matrix $\mathcal{A}$ is given by

$$\mathcal{A} = \begin{bmatrix} 0 & -1 & 0 & 0 \\ 1 & 0 & 0 & 0 \\ 0 & 0 & 0 & -1 \\ 0 & 0 & 1 & 0 \end{bmatrix}, \tag{32}$$

such that $\mathcal{A}\mathcal{A} = -\mathcal{I}$, $\mathcal{I}$ being the identity matrix.

Eq. (30) can be decoupled to the following bilinear equations

$$\left(D_x^2 + D_y^2 - \mu f D_x D_y\right) f \cdot f + \mathcal{V}_0^T \mathcal{V}_0 = 0, \tag{33a}$$

$$\left(\mathcal{A} D_t + D_x^2 + D_y^2\right) \mathcal{V}_0 \cdot f - \mu \mathcal{V}_0 D_x D_y f \cdot f = 0, \tag{33b}$$

where μ may stand for a parameter.

In particular, from Eq. (33), we have

$$p^2 + q^2 + u^2 + v^2 = 2\left(\mu f \partial_{xy}^2 - \partial_{xx}^2 - \partial_{yy}^2\right) \ln(f). \tag{34}$$

Now, expanding the functions f and $\mathcal{V}_0$ as power series, and using them in Eq. (33), we can construct the N-soliton solution in the usual way. Nonetheless, the investigation of these N-soliton solution does not constitute the purpose of this paper. We shall briefly present the general scheme to derive such structures and we shall illustrate the simple case of one-soliton solution.

The P-test is actually useful in finding the BT and hence the Lax pairs such that the spectral problem and also the Hirota's bilinearization of the system can straightforwardly be written down. As illustration, even though the (1+1)-dimensional NLS equation has been solved by means of the IST method, its P-integrability has been established by Gibbon et al. [112] and its connection with the Hirota's bilinearization. In the same way, in a previous study, Sahadevan et al. [14] have investigated the (1+1)-dimensional CNLS equations. Avoiding to consider the problem in full generality, they have found conditions at which these equations are integrable, and the BT and Hirota's bilinearization have been derived. Thus, whether a system does pass the P-test or partially passes this test under some conditions, the result from the P-test can be used to find out other integrability properties such as BT, Lax pairs, just to name a few, so that the question of integrability should be clearly investigated.

Now, before exploiting the bilinear form of the (2+1)-dimensional CNLS equations to construct the one-soliton solution, it is useful to make an emphasis: in the wake of the P-analysis of this coupled system, without any truncation procedure, the general solutions $\mathcal{S} \equiv (p, q, u, v)^T$ are written as follows

$$\mathcal{S} = \left(\mathcal{V}_0 + \sum_{k=1}^{\infty} \mathcal{V}_k f^k\right) \Big/ f. \tag{35}$$

In the Hirota's bilinearization, we use this result to write the solution $\mathcal{S} = \mathcal{Y}/f$ where $\mathcal{Y}$ is an arbitrary matrix expressed as

$$\mathcal{Y} = \mathcal{V}_0 + \sum_{k=1}^{\infty} \mathcal{V}_k f^k, \tag{36}$$

satisfying the bilinear Eq. (33), provided to replace $\mathcal{V}_0$ by $\mathcal{Y}$. The form of the general solution $\mathcal{S} = \mathcal{Y}/f$ given by Eqs. (35) and (36) looks like we have truncated the P-expansion to $\mathcal{V}_k = 0, k \geq 1$. Looking forward avoiding any confusion, we shall consider the bilinear Eq. (33) to be expressed in terms of $\mathcal{Y}$ instead of $\mathcal{V}_0$. Thus, in order to find the soliton solution to this bilinear system by means of the Hirota's method [109, 110, 111], we consider the following expressions

$$\mathcal{Y} = \epsilon\mathcal{Y}_{(1)} + \epsilon^3\mathcal{Y}_{(3)} + \epsilon^5\mathcal{Y}_{(5)} + \cdots, \tag{37a}$$

$$f = 1 + \epsilon^2 f_{(2)} + \epsilon^4 f_{(4)} + \epsilon^6 f_{(6)} + \cdots, \tag{37b}$$

where $\mathcal{Y}_{(j)}$ and f_{j+1} $(j = 1, 3, 5)$ satisfy the following equations

1. order of ϵ,

$$\left(\partial^2_{xx} + \partial^2_{yy} + \mathcal{A}\partial_t\right)\mathcal{Y}_{(1)} = 0, \tag{38}$$

2. order of ϵ^2,

$$\left(\mu\partial^2_{xy} - \partial^2_{xx} - \partial^2_{yy}\right) f_{(2)} - \frac{1}{2}\mathcal{Y}^T_{(1)}\mathcal{Y}_{(1)} = 0, \tag{39}$$

3. order of ϵ^3,

$$\left(\partial^2_{xx} + \partial^2_{yy} + \mathcal{A}\partial_t\right)\mathcal{Y}_{(3)} + \left(\mathcal{A}D_t + D^2_x + D^2_y\right)\mathcal{Y}_{(1)} \cdot f_{(2)} - 2\mu\partial^2_{xy} f_{(2)} = 0, \tag{40}$$

4. order of ϵ^4,

$$\left(\mu\partial^2_{xy} - \partial^2_{xx} - \partial^2_{yy}\right) f_{(4)} + \frac{1}{2}\left(\mu D_x D_y - D^2_x - D^2_y\right) f_{(2)} \cdot f_{(2)} + \mu f_{(2)}\partial^2_{xy} f_{(2)} - \mathcal{Y}^T_{(1)}\mathcal{Y}_{(3)} = 0, \tag{41}$$

5. order of ϵ^5,

$$\left(\partial_{xx}^2+\partial_{yy}^2+\mathcal{A}\partial_t\right)\mathcal{Y}_{(5)}+\left(\mathcal{A}D_t+D_x^2+D_y^2\right)\left(\mathcal{Y}_{(1)}\cdot f_{(4)}+\mathcal{Y}_{(3)}\cdot f_{(2)}\right)$$
$$-\mu\mathcal{Y}_{(1)}\left(D_xD_yf_{(2)}\cdot f_{(2)}+2\partial_{xy}^2f_{(4)}\right)-2\mu\mathcal{Y}_{(3)}\partial_{xy}^2f_{(2)}=0, \tag{42}$$

6. order of ϵ^6,

$$\left(\mu\partial_{xy}^2-\partial_{xx}^2-\partial_{yy}^2\right)f_{(6)}+\left(\mu D_xD_y-D_x^2-D_y^2\right)f_{(2)}\cdot f_{(4)}$$
$$+\mu\left(f_{(2)}\partial_{xy}^2f_{(4)}+f_{(4)}\partial_{xy}^2f_{(2)}+\frac{1}{2}D_xD_yf_{(2)}\cdot f_{(2)}\right) \tag{43}$$
$$-\frac{1}{2}\left(\mathcal{Y}_{(3)}^T\mathcal{Y}_{(3)}+2\mathcal{Y}_{(1)}^T\mathcal{Y}_{(5)}\right)=0,$$

Now, in order to derive the one-soliton solution, it is useful to write $\mathcal{Y}_{(1)}$ as follows

$$\mathcal{Y}_{(1)}=\theta\exp(\eta)+\theta^\star\exp(\eta^\star), \tag{44}$$

where $\eta=k_1x+k_2y-\omega t+\eta_0$ and $\theta=(a,b,c,d)^T$, $k_1,k_2,\omega,\eta_0,a,b,c,d\in\mathbb{C}$. Thus, solving the Eq. (38) leads to the following system

$$\begin{bmatrix} k_1^2+k_2^2 & \omega \\ -\omega & k_1^2+k_2^2 \end{bmatrix}\begin{bmatrix} a \\ b \end{bmatrix}=\begin{bmatrix} 0 \\ 0 \end{bmatrix}, \tag{45a}$$

$$\begin{bmatrix} k_1^2+k_2^2 & \omega \\ -\omega & k_1^2+k_2^2 \end{bmatrix}\begin{bmatrix} c \\ d \end{bmatrix}=\begin{bmatrix} 0 \\ 0 \end{bmatrix}. \tag{45b}$$

Looking for nontrivial solutions to Eq. (45), the following dispersion equation is satisfied,

$$\left(k_1^2+k_2^2\right)^2+\omega^2=0. \tag{46}$$

The phase η now becomes

$$\eta=k_1x+k_2y-\varepsilon\imath\left(k_1^2+k_2^2\right)t+\eta_0, \tag{47}$$

with $\varepsilon=\pm1$. Besides, θ takes the following form

$$\theta=(a,-\varepsilon\imath a,c,-\varepsilon\imath c)^T. \tag{48}$$

Solving Eq. (39), it can easily be found that $f_{(2)}$ takes the following form,

$$f_{(2)}=\delta\exp(\eta+\eta^\star), \tag{49}$$

where

$$\delta=\frac{1}{2}\frac{|a|^2+|c|^2}{\mu k_{1R}k_{2R}-k_{1R}^2-k_{2R}^2}, \tag{50}$$

provided

$$\mu\neq\frac{k_{1R}^2+k_{2R}^2}{k_{1R}k_{2R}}. \tag{51}$$

We note that k_{1R} and k_{2R} stand for real parts of the complex-valued wave numbers k_1 and k_2, respectively. We pay interests to the quantity $\mathcal{W}=\sqrt{p^2+q^2+u^2+v^2}$. Two conditions are regarded as follows

1.

$$\mu < \frac{k_{1R}^2 + k_{2R}^2}{k_{1R}k_{2R}}, \tag{52}$$

$$\mathcal{W} = \frac{\sqrt{2\left(k_{1R}^2 + k_{2R}^2 - \mu k_{1R}k_{2R}\right)}}{\left|\sinh\left(\eta_R + \varrho\right)\right|}, \tag{53}$$

where

$$\varrho = \frac{1}{2}\ln\left[\frac{|a|^2 + |c|^2}{2\left(k_{1R}^2 + k_{2R}^2 - \mu k_{1R}k_{2R}\right)}\right]. \tag{54}$$

From Eq. (53), it is seen that the solution is undefined at some particular points meaning that it takes infinite values at these singular points.

2.

$$\mu > \frac{k_{1R}^2 + k_{2R}^2}{k_{1R}k_{2R}}, \tag{55}$$

$$\mathcal{W} = \sqrt{2\left(\mu k_{1R}k_{2R} - k_{1R}^2 - k_{2R}^2\right)}\, sech\left(\eta_R + \varpi\right), \tag{56}$$

with

$$\varpi = \frac{1}{2}\ln\left[\frac{|a|^2 + |c|^2}{2\left(\mu k_{1R}k_{2R} - k_{1R}^2 - k_{2R}^2\right)}\right]. \tag{57}$$

From Eq. (56), the solution has no singular points and takes the form of a hump. The amplitude of this soliton solution is given by $\sqrt{2\left(\mu k_{1R}k_{2R} - k_{1R}^2 - k_{2R}^2\right)}$. In order to get symmetry in the (x,y,t)-space, it is useful to take

$$\eta_{0R} = -\varpi, \tag{58}$$

such that η now reads

$$\eta_R = k_{1R}x + k_{2R}y - \omega_R t - \frac{1}{2}\ln\left[\frac{|a|^2 + |c|^2}{2\left(\mu k_{1R}k_{2R} - k_{1R}^2 - k_{2R}^2\right)}\right]. \tag{59}$$

In order to get more information about the interactions between such structures, the further subsequent equations shown above should be solved. Even N-soliton solutions ($N \geq 2$) can be investigated if full series instead of truncated ones are considered. Even though the localized soliton-like solutions displayed in Eqs. (53) and (56) can be obtained by symmetry reduction of the (2+1)-dimensional CNLS equations, it actually seems important to show that the Hirota's bilinearization of this system, which is found after a preliminary P-analysis of the system, holds very well such that the one-, two-, or N-soliton ($N \geq 3$) can be constructed. This constitutes another interesting investigation. However, in this chapter, we would like to focus our interests to other types of solutions that exist owing to the arbitrariness of some functions derived from the P-analysis.

4. General Lax-representation of the (2+1)-dimensional CNLS Equations

In other to check the complete integrability of the above coupled system (5), we now follow the ZS-scheme ([113, 114, 115]) which generalizes the Lax-method for higher-dimensional systems. Making use of the Matrix operators, we define two integral operators J_F and $J_\pm$ as follows

$$J_F = \int_{-\infty}^{\infty} dz F(x,z), \quad J_\pm = \int_{-\infty}^{\infty} dz K_\pm(x,z), \tag{60}$$

where the functions F and $K_\pm$ are $N \times N$ (N integer) matrices such that

$$K_+(x,z) = 0 \quad if \quad z < x, \tag{61a}$$
$$K_-(x,z) = 0 \quad if \quad z > x. \tag{61b}$$

The operators J_F and $J_\pm$ are related through the following identity

$$(I + J_+)(I + J_F) = (I + J_-), \tag{62}$$

where the quantity I stands for the unit matrix. Following the operator identity given above, the matrix Marchenko equations ([113, 114, 115]) for $K_\pm$ can be expressed in terms of the function F leading to the constructions of the solutions to the investigated system. However, reaching such a goal requires the introduction of two pairs of operators Δ_0 and Δ referred to as "undressed" and "dressed" operators, respectively. A typical choice is given by

$$\Delta_0^{(1)} = I\lambda\partial_t - M_0, \quad \Delta_0^{(2)} = I\mu\partial_y + L_0, \tag{63a}$$
$$\Delta^{(1)} = I\lambda\partial_t - M, \quad \Delta^{(2)} = I\mu\partial_y + L, \tag{63b}$$

where the quantities λ and μ are arbitrary constants, the matrices M_0, L_0, M and L are differential operators in x only. We note that $\partial_t \equiv \partial/\partial_t$ and $\partial_y \equiv \partial/\partial_y$. The undressed and dressed operators satisfy the following relations

$$[\Delta_0^{(i)}, J_F] = 0, \quad \Delta^{(i)}(I + J_+) = (I + J_+)\Delta_0^{(i)}, \tag{64}$$

in such a way that it easily comes

$$\lambda L_t + \mu M_y + [L, M] = 0. \tag{65}$$

This equation stands for the generalization of the Lax-equation, representing the system of nonlinear evolution equations. The procedure for solving the above system is now described. The variable cœfficients which arise in the dressed operators L and M constitute some functions satisfying the system of evolution equation. These functions are related to the kernel K_+, solution to the linear matrix Marchenko equations ([113, 114, 115]) where the function F satisfies to a pair of equations. In the wake of the above development, let us consider the following operators

$$\Delta_0^{(1)} = I(\imath 3\partial_t - \partial_{xx}^2), \quad \Delta_0^{(2)} = I\partial_y + \begin{pmatrix} 2 & 0 \\ 0 & 1 \end{pmatrix} \partial_x/3, \tag{66}$$

where $\imath^2 = -1$. Thus, following the commutation relations (see Eq. (64)) given above, the following equations are obtained

$$\imath 3F_t + F_{zz} - F_{xx} = 0, \quad 3F_y + \begin{pmatrix} 2 & 0 \\ 0 & 1 \end{pmatrix} F_x + F_z \begin{pmatrix} 2 & 0 \\ 0 & 1 \end{pmatrix} = 0. \tag{67}$$

Following the second equation of the system (64), it comes

$$\Delta^{(1)} = I(\imath 3\partial_t - \partial_{xx}^2 - W(x,y,t)) - 2d\hat{K}_+/dx, \tag{68a}$$

$$\Delta^{(2)} = I\partial_y + \begin{pmatrix} 2 & 0 \\ 0 & 1 \end{pmatrix} \partial_x/3 + V(x,t), \tag{68b}$$

where the matrices V and K_+ are expressed as

$$3V = \begin{pmatrix} 2 & 0 \\ 0 & 1 \end{pmatrix} \hat{K}_+ - \hat{K}_+ \begin{pmatrix} 2 & 0 \\ 0 & 1 \end{pmatrix}, \tag{69a}$$

$$\begin{pmatrix} 2 & 0 \\ 0 & 1 \end{pmatrix} K_{+x} + K_{+z} \begin{pmatrix} 2 & 0 \\ 0 & 1 \end{pmatrix} + 3VK_+ = 0, \tag{69b}$$

with $\hat{K}_+ = K(x, x; t)$. It is worth noting that the matrix $W(x, y, t)$ stands for an arbitrary differential operator which in general satisfies to some conditions in the derivation of an evolution equation solvable by the ZS-scheme. Let us express the matrix $\hat{K}_+$ as follows

$$\hat{K}_+ = \begin{pmatrix} A & B \\ C & D \end{pmatrix}. \tag{70}$$

Eq. (69) gives

$$3V = \begin{pmatrix} 0 & B \\ -C & 0 \end{pmatrix}, \quad 2A_x + BC = 0, \quad D_x - BC = 0. \tag{71}$$

By setting $B = \Phi$ and $C = \pm\Phi^\star$ (star referring to complex conjugation) where Φ stands for a polarized complex-valued observable which components ϕ and ψ span a two-dimensional manifold generated by the basis unit vectors $\mathbf{e}_\phi$ and $\mathbf{e}_\psi$ such that

$$\Phi = \phi\mathbf{e}_\phi + \psi\mathbf{e}_\psi. \tag{72}$$

The dressed operators are then expressed as follows

$$\Delta^{(1)} = I(\imath 3\partial_t - \partial_{xx}^2 - W(x,y,t)) - \begin{pmatrix} \mp|\Phi|^2 & 2\Phi_x \\ \pm 2\Phi_x^\star & 2|\Phi|^2 \end{pmatrix}, \tag{73a}$$

$$\Delta^{(2)} = I\partial_y + \begin{pmatrix} 2 & 0 \\ 0 & 1 \end{pmatrix} \partial_x/3 + \begin{pmatrix} 0 & \Phi/3 \\ \mp\Phi^\star/3 & 0 \end{pmatrix}. \tag{73b}$$

By setting

$$W = \begin{pmatrix} p & q \\ r & s \end{pmatrix}, \tag{74}$$

and using the generalized Lax system (65), we straightforwardly derive

$$\imath\Phi_t + \Delta\Phi \pm |\Phi|^2\Phi = 0, \tag{75}$$

where the differential operators p, q, r and s satisfy to the following relations

$$\begin{aligned}
q_y + 2q_x/3 + q\partial_x/3 + \Phi s/3 - p\Phi/3 + 2\Phi_{xy} - \Phi_{yy} &= 0, && \text{(76a)}\\
r_y + r_x/3 - r\partial_x/3 \pm s\Phi^\star/3 \mp \Phi^\star p/3 \pm 2\Phi^\star_{xy} - \Phi^\star_{yy} &= 0, && \text{(76b)}\\
p_y + 2p_x/3 + \Phi r/3 \mp (|\Phi|^2)_y &= 0, && \text{(76c)}\\
s_y + s_x/3 \mp \Phi^\star q \pm 2(|\Phi|^2)_y &= 0. && \text{(76d)}
\end{aligned}$$

The previous development of the ZS-scheme leading to the derivation of the CNLS system (75) shows that the system (5) under our study actually possesses a generalized Lax-representation in the particular case consisting of $\sigma = -1$ and $\alpha = 1$. Consequently, the system can be solved by means of the matrix Marchenko equations ([113, 114, 115]). But, it is not the main purpose of the present work. It is however useful to mention that following the present results, there actually exist some (2+1)-dimensional CNLS equations which are solvable, at least by the ZS-scheme and possess miscellaneous soliton solutions which stabilities against various types of perturbations can properly be discussed.

From the previous sections to the present one, we have shown that the system (5) is P-integrable and possesses both bilinear forms and generalized Lax-representation under the arbitrary choice $\sigma = -1$ and $\alpha = 1$. We can now conclude that the system is completely integrable. In the next, we now discuss physically the different pattern formations that such equations in (2+1)-dimensional space can support.

5. Localized and Periodic Excitations to the (2+1)-dimensional CNLS Equations

5.1. Localized Excitations to the (2+1)-dimensional CNLS Equations

Recently, great interests have been paid to many (2+1)-dimensional equations ([116, 117, 118] and references therein) and a rich diversity of coherent structures such as the FSWs have been found. These structures are more abundant than those of (1+1)-dimensional cases because some types of arbitrary functions have been included in the solution expressions [119, 120, 121].

We focus our interests either on the intensity of the waves which is proportional to the square of their amplitude as it is done in nonlinear optics propagation or on the density of atoms also proportional to the square of the amplitude of the wave as investigated in matter-wave interactions in systems such as BEC. However, in the different plots presented in the paper, we consider the square root of the previous observables. Thus, actually, there is no need to search for the explicit individual solutions necessarily owing to the fact that the sum of the square of their amplitudes are explicitly found.

We define a quantity $\mathcal{W} = \sqrt{|\phi|^2 + |\psi|^2}$ expressed as follows

$$\mathcal{W}^2(x, y, t) = \frac{\mathcal{V}_0^T \mathcal{V}_0}{f^2} = 2\frac{(f_x^2 + f_y^2)}{f^2}. \tag{77}$$

We note that Eq. (22) and (23) are rewritten in a compact form as follows

$$2\left(f_x\mathcal{V}_{0,x}+f_y\mathcal{V}_{0,y}\right)+\mathcal{B}\mathcal{V}_0=0,\quad \mathcal{V}_{0,t}-\mathcal{A}\left(\mathcal{V}_{0,xx}+\mathcal{V}_{0,yy}\right)=0, \tag{78}$$

where

$$\mathcal{B}=\left(f_{xx}+f_{yy}\right)\mathcal{I}+f_t\mathcal{A}. \tag{79}$$

Solving the first equation of the system (78) by using the characteristic method, it leads to the following set

$$\mathcal{V}_0=\mathcal{G}_0\left(y-\int\frac{f_y dx}{f_x}\right)\exp\left(\int\frac{dx}{2f_x\mathcal{B}^{-1}}\right), \tag{80}$$

where $\mathcal{G}_0$ is an arbitrary array-function of $\left(y-\int\frac{f_y}{f_x}dx\right)$ to be determined.

From equation (78), it can be shown that

$$f_t+\mathcal{E}(x,t)f_x+\mathcal{F}(y,t)f_y=0,\quad f_t+\mathcal{H}(x,t)f_x+\mathcal{K}(y,t)f_y=0, \tag{81}$$

where $\mathcal{E}$ and $\mathcal{H}$ are arbitrary functions of x and t, $\mathcal{F}$ and $\mathcal{K}$ being arbitrary functions of y and t, defined by

$$2\left(\frac{q_0}{p_0}\right)_x = \mathcal{E}(x,t)\left[1+\left(\frac{q_0}{p_0}\right)^2\right], \tag{82a}$$

$$2\left(\frac{q_0}{p_0}\right)_y = \mathcal{F}(y,t)\left[1+\left(\frac{q_0}{p_0}\right)^2\right], \tag{82b}$$

$$2\left(\frac{v_0}{u_0}\right)_x = \mathcal{H}(x,t)\left[1+\left(\frac{v_0}{u_0}\right)^2\right], \tag{82c}$$

$$2\left(\frac{v_0}{u_0}\right)_y = \mathcal{K}(y,t)\left[1+\left(\frac{v_0}{u_0}\right)^2\right], \tag{82d}$$

such that the compatibility equations $\partial^2_{xy}\left(\frac{q_0}{p_0}\right)=\partial^2_{yx}\left(\frac{q_0}{p_0}\right)$ and $\partial^2_{xy}\left(\frac{v_0}{u_0}\right)=\partial^2_{yx}\left(\frac{v_0}{u_0}\right)$ are satisfied.

Expressing $f(x,y,t)$ as follows,

$$f(x,y,t)=f_1(x,t)+f_2(y,t)+a_0, \tag{83}$$

where a_0 is an arbitrary constant, and solving Eq. (82), it easily comes that

$$f_1(x,t) = \mathcal{L}_1\left(y-\int\mathcal{E}(x,t)dt\right)-\int\varsigma(t)dt \tag{84a}$$

$$= \mathcal{M}_1\left(y-\int\mathcal{H}(x,t)dt\right)-\int\nu(t)dt, \tag{84b}$$

$$f_2(y,t) = \mathcal{L}_2\left(y-\int\mathcal{F}(y,t)dt\right)+\int\varsigma(t)dt \tag{84c}$$

$$= \mathcal{M}_2\left(y-\int\mathcal{K}(y,t)dt\right)+\int\nu(t)dt, \tag{84d}$$

where ς, υ, $\mathcal{L}_j$ and $\mathcal{M}_j$ $(j = 1, 2)$ stand for arbitrary functions. Thus, the functions f_1 and f_2 are arbitrary. For this reason, there exists a rich variety of structures due to the different expressions that these functions can take. It seems worth noting that the functions $\mathcal{E}$, $\mathcal{H}$, $\mathcal{F}$ and $\mathcal{K}$ are derived from Eq. (82) after solving the second equation of the system (78). Besides, solving Eq. (82), the following system is found

$$\frac{q_0}{p_0} = \tan\left(\frac{1}{2}\int \mathcal{E}(x,t)dx + \frac{1}{2}\int \mathcal{F}(y,t)dy + d_1(t)\right), \tag{85}$$

$$\frac{v_0}{u_0} = \tan\left(\frac{1}{2}\int \mathcal{H}(x,t)dx + \frac{1}{2}\int \mathcal{K}(y,t)dy + d_2(t)\right), \tag{86}$$

where d_1 and d_2 stand for arbitrary functions depending on time t. Equation (85) shows that for some arbitrary expressions of $\mathcal{E}$, $\mathcal{H}$, $\mathcal{F}$ and $\mathcal{K}$, $\frac{q_0}{p_0}$ and $\frac{v_0}{u_0}$ are found, provided the second equation of the system (78) be satisfied.

Multiple (2+1)-dimensional localized solitonic excitations such as FSWs with completely elastic interaction properties can be built-up from the (1+1)-dimensional multiple localized excitations. Let us introduce some localized functions $f_{1,x}$ and $f_{2,y}$, in the forms

$$f_{1,x} = \sum_{i=1}^{N} \kappa_i(\xi + c_i t), \quad x = \xi + \sum_{i=1}^{N} \varphi_i(\xi + c_i t), \tag{87a}$$

$$f_{2,y} = \sum_{i=1}^{M} \chi_i(\eta + v_i t), \quad y = \eta + \sum_{i=1}^{M} \vartheta_i(\eta + v_i t). \tag{87b}$$

The constants c_i $(i = 1, \ldots, N)$ and v_i $(i = 1, \ldots, M)$ stand for velocities. The quantities κ_i, φ_i $(i = 1, \ldots, N)$, χ_i and ϑ_i $(i = 1, \ldots, M)$ stand for localized excitations with the properties $\kappa_i(\pm\infty) = 0$, $\varphi_i(\pm\infty) = A_i^{\pm} = constant$ $(i = 1, \ldots, N)$, $\chi_i(\pm\infty) = 0$, $\vartheta_i(\pm\infty) = B_i^{\pm} = constant$ $(i = 1, \ldots, M)$. In some suitable regions, ξ (or η) is a multi-valued function of x (or y) by choosing some appropriate functions φ_i (or ϑ_i). Therefore, the function $f_{1,x}$ (or $f_{2,y}$) which is obviously an interaction solution of N (or M) localized excitations due to the property $\xi|_{x\to\infty} \to \infty$ (or $\eta|_{y\to\infty} \to \infty$), constitutes a multi-valued function of x (or y) in these areas, even though it is single-valued of ξ (or η).

As a result, at $t = 0$, we find four typical FSWs with the 'tent' shape Fig. 1(A), 'worm' shape of one type Fig. 1(B), 'worm' shape of another type Fig. 1(C), and 'worm'-'half-straight-line' shape Fig. 1(D). Another typical of FSWs looking bubble-like are shown in Figs. 2(A) and 2(B). We also derive a 'ghost'-like FSW usually referring to exponentially decaying ring soliton (see Fig. 2(C)); some typical peakons are obtained, but we depict in Fig. 2(D) only one type. The more detailed function selections of the figures are directly given in the figure caption. Amongst other localized solitary waves, there are 'singular' patterns which can be found in higher dimensional systems by selecting some appropriate lower dimensional piecewise smooth functions with 'singular' structures [122, 123]. In the wake of such interests, we depict a typical two 'singular' half-straight lines crossing at a right angle as presented in the column from left in Fig. 3 where particular expressions of f_1 and f_2 have been considered so that the derived solutions are exponentially localized in large scale and possess self-similar structure near the center of the excitation [124].

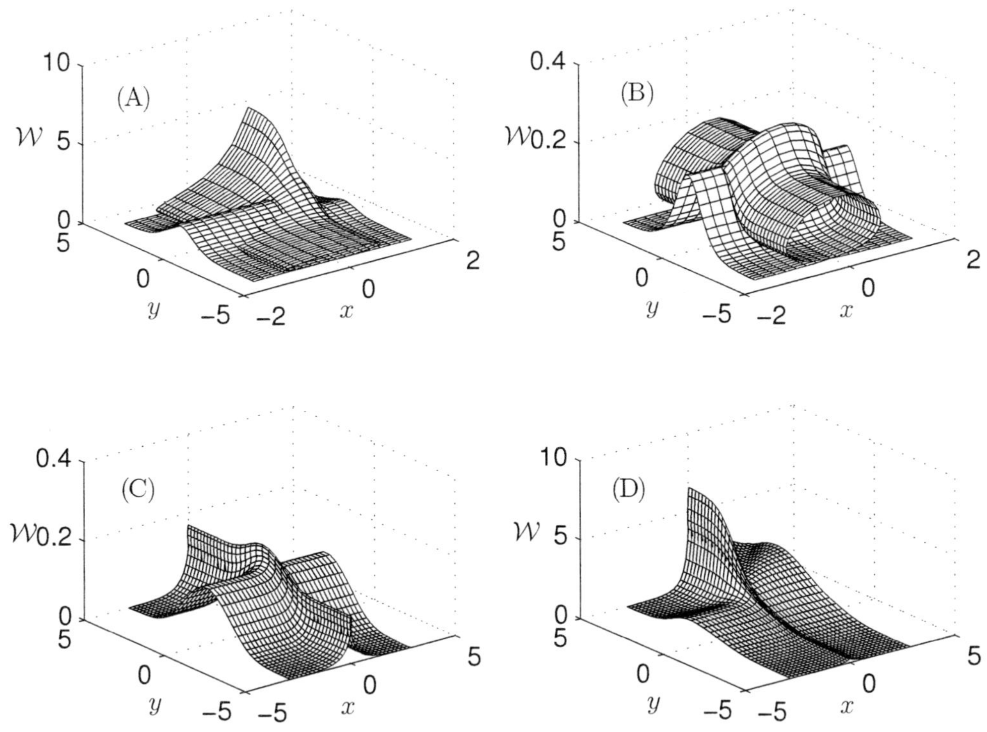

Figure 1. Four typical localized excitations for the physical observable $\mathcal{W}(x, y, t)$ at $t = 0$ with $f_{1,x} = -sech^2(\xi - t)$, $x = \xi + k_4 \tanh(\xi - t)$, $f_{2,y} = -sech^2(\eta)$, $y = \eta + l_3 \tanh(\eta)$: (A) 'tent'-like with $k_4 = -2.5$, $l_3 = 0$ and $a_0 = 1.9$; (B) 'worm'-like with $k_4 = -2.5$, $l_3 = 0$ and $a_0 = 8$; (C) 'worm'-like shape of another type with $k_4 = -1.15$, $l_3 = 1$ and $a_0 = 10$; (D) 'worm'-half-straight-line structure with $k_4 = -1.15$, $l_3 = 1$ and $a_0 = 1.9$.

Recently, a new dispersive shallow water equation known as Camassa-Holm (CH) equation has been derived by Camassa and Holm [125],

$$u_t - u_{xxt} + 3uu_x + 2ku_x = 2u_x u_{xx} + uu_{xxx}, \tag{88}$$

where u is an observable and k being a real-valued constant. They have shown that the solitary wave solution to the CH equation is expressed as follows [125]

$$u(x, t) = b \exp(-|x - bt|), \tag{89}$$

where b is an arbitrary constant. Due to the discontinuity of the first derivative at the wave peak, this solution has been called peakon. In the wake of these results, a great investigation has been recently paid to a wide class of (2+1)-dimensional models ([122, 124, 126] and references therein). It has been shown that if the functions f_1 and f_2 include some functions in space variables called peakons, there exists some quite rich (2+1)-dimensional peakons

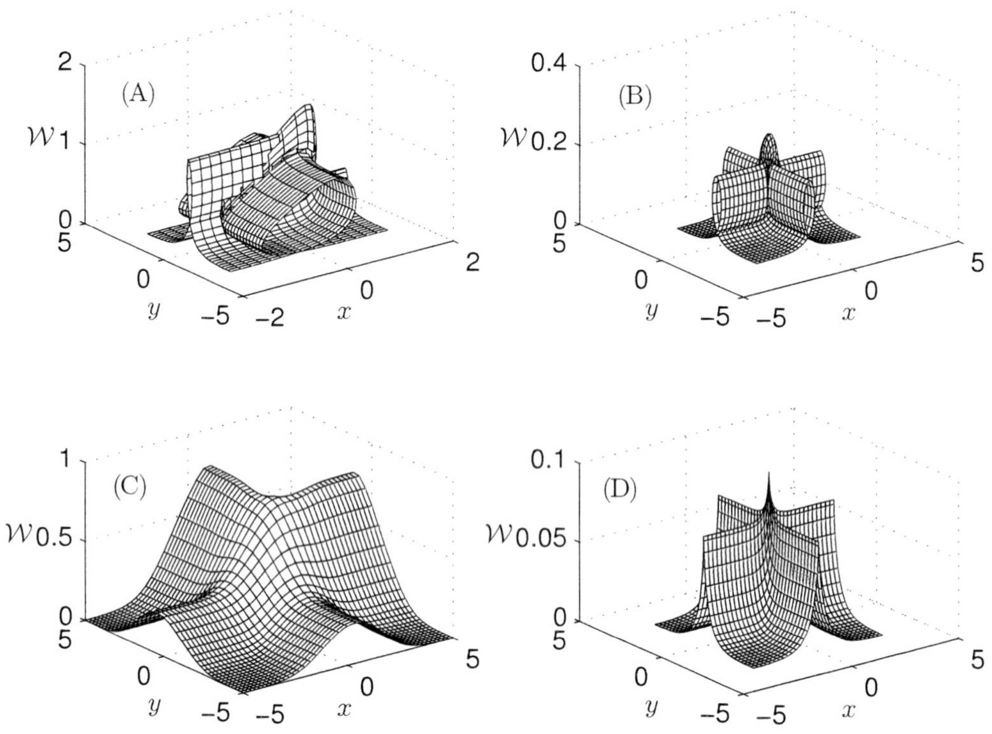

Figure 2. Bubble-like and single-valued localized excitations for the physical observable $\mathcal{W}(x, y, t)$ at $t = 0$ with $f_{1,x} = -sech^2(\xi - t)$, $x = \xi + k_4 \tanh(\xi - t)$, $f_{2,y} = -sech^2(\eta)$, $y = \eta + l_3 \tanh(\eta)$: (A) typical bubble of first kind with $k_4 = -2.5$, $l_3 = -1.4$ and $a_0 = 1.9$; (B) typical bubble of second kind with $k_4 = -1.6$, $l_3 = -1.6$ and $a_0 = 8$; (C) exponentially decaying ring soliton with $k_4 = 1$, $l_3 = 1$ and $a_0 = 4$; (D) peakon with $k_4 = -1$, $l_3 = -1$ and $a_0 = 20$.

from the lower dimensional peakons. Selecting f_1 and f_2 as piecewise functions, a typical class of peakon is given by [124]

$$f_1 = \sum_{i=1}^{M} \begin{cases} F_i(\theta_i), & \theta_i \leq 0, \\ 2F_i(0) - F_i(-\theta_i), & \theta_i > 0, \end{cases} \tag{90a}$$

$$f_2 = \sum_{i=1}^{N} \begin{cases} G_i(y), & y \leq 0, \\ 2G_i(0) - G_i(-y), & y > 0, \end{cases} \tag{90b}$$

where $\theta_i = x - c_i t$ $(i = 1, \cdots, M)$, c_i being an arbitrary constant standing for velocity. The quantity $F_i(\theta_i)$ $(i = 1, \cdots, M)$ and $G_i(y)$ $(i = 1, \cdots, N)$ are differentiable functions yielding many kinds of (2+1)-dimensional peakons. Alternatively, we make a selection of f_1 and f_2 as presented in the caption of Fig. 3. In addition to the peakon solutions, there

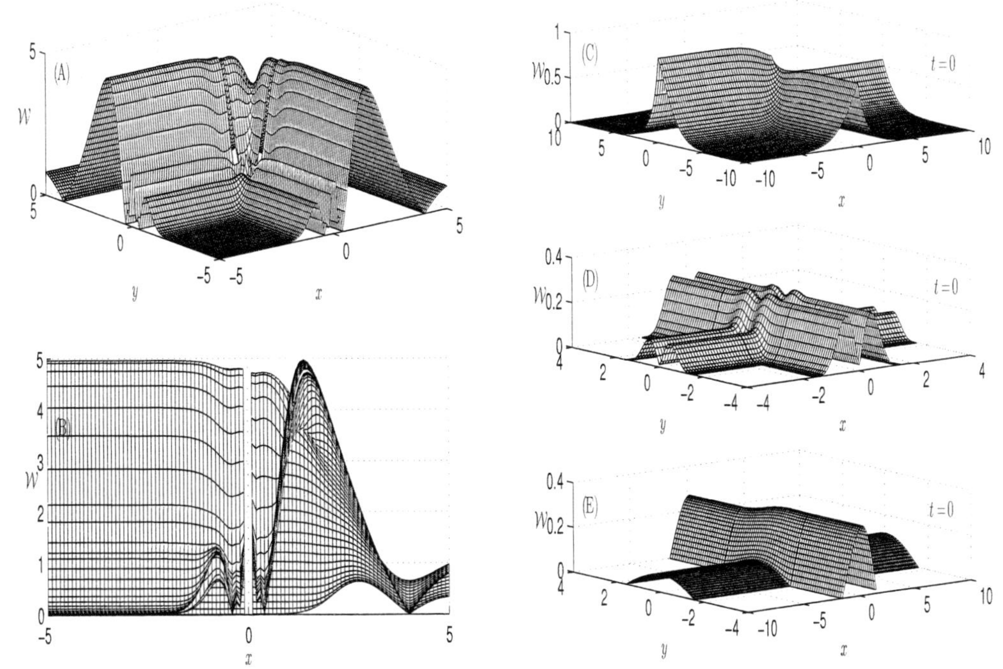

Figure 3. Typical 'singular' half-straight-line soliton $\mathcal{W}(x, y, t)$ (panels from left (A) and (B)) depicted under the following selections: $f_1(x) = \exp\{x[3/2 + \sin(\ln x^2)]\}$ and $f_2(y) = \exp\{y[3/2 + \sin(\ln y^2)]\}$ and $a_0 = 1$. Typical peakons $\mathcal{W}(x, y, t)$ (upper panel from right (C)) depicted at $t = 0$ under the following selections: $f_{1x} = \exp(|x - t|)$, $f_{2y} = \exp\left(\left|y - \frac{3}{2}t\right|\right)$ and $a_0 = 1$. A Type of structure with compact support $\mathcal{W}(x, y, t)$ (middle panel from right (D)) depicted at $t = 0$ under the following selections: $R_1^p(\varphi_1) = -2\cos^5(\varphi_1)$, $R_2^p(\varphi_2) = -\cos^5(\varphi_2)$ and $R^q(\phi) = \cos^5(\phi)$ with $\varphi_1 = x - t$, $\varphi_2 = x + 2t$, $\phi = y$, $\varphi_{11} = \varphi_{12} = \phi_1 = -\pi/2$, and $\varphi_{21} = \varphi_{22} = \phi_2 = \pi/2$. $a_0 = 17$. Another Type of structure with compact support $\mathcal{W}(x, y, t)$ (lower panel from right (E)) depicted at $t = 0$ under the following selections: $R_1^p(\varphi_1) = -2\sin(\varphi_1) - 2$, $R_2^p(\varphi_2) = -\sin(\varphi_2) - 1$ and $R^q(\phi) = \sin(\phi) + 1$ with $\varphi_1 = x - t$, $\varphi_2 = x + 2t$, $\phi = y$, $\varphi_{11} = \varphi_{12} = \phi_1 = -\pi/2$, and $\varphi_{21} = \varphi_{22} = \phi_2 = \pi/2$. $a_0 = 17$.

are some types of weak solutions with compact support known as compactons [124] with the following form

$$f_1 = \sum_{i=1}^{M} \begin{cases} 0, & \varphi_i \leq \varphi_{1i}, \\ R_i^p(\varphi_i) - R_i^p(\varphi_{1i}), & \varphi_{1i} < \varphi_{2i}, \\ R_i^p(\varphi_{2i}) - R_i^p(\varphi_{1i}), & \varphi_{2i} < \varphi_i, \end{cases} \tag{91a}$$

$$f_2 = \sum_{i=1}^{N} \begin{cases} 0, & \phi_i \leq \phi_{1i}, \\ R_i^q(\phi_i) - R_i^q(\phi_{1i}), & \phi_{1i} < \phi_{2i}, \\ R_i^q(\phi_{2i}) - R_i^q(\phi_{1i}), & \phi_{2i} < \phi_i, \end{cases} \tag{91b}$$

where $\varphi_i = x - c_i t$ $(i = 1, \cdots, M)$ and $\phi_i = y - v_i t$ $(i = 1, \cdots, M)$, c_i and v_i being arbitrary constants standing for velocities of waves. The quantities $R_i^p(\varphi_i)$ $(i = 1, \cdots, M)$ and $R_i^q(\phi_i)$ $(i = 1, \cdots, N)$ are differentiable functions yielding many kinds of (2+1)-dimensional waves with compact support. In the caption of Fig. 3, we make a selection of f_1 and f_2 for two typical structures with compact support as depicted in the figure.

It seems worth investigating the interaction processes between the previous structures to see whether they are elastic or not. Indeed, the scattering will be elastic if the essential properties of solitons such as amplitudes, velocities and wave-shapes do not change after the interaction. We discuss some novel coherent excitations, i.e. the 'bubbles'. If interactions amongst such FSWs are elastic, they are called foldons. Nonetheless, these elastic scattering processes are constructed if the asymptotic properties of Eqs. (77) and (87) are firstly discussed at $t \to \pm\infty$.

It is assumed that $F_i(\omega_i) \equiv F_i(\xi + c_i t) \equiv \int \kappa_i dx|_{\omega_i \to \pm\infty} \to F_i^{\pm}$ and ω_i invariant as $t \to \infty$. It is also assumed that f_2 is t-independent. At the ith excitation, the interaction properties amongst the localized excitations are described by the following equations

$$\mathcal{W}^2_{t \to \mp\infty} \to \sum_{i=1}^{N} 2 \left(\frac{\kappa_i(\omega_i) + f_{2,y}}{a_0 + F_i(\omega_i) + \Omega_i^{\mp} + f_2} \right)^2, \quad x|_{t \to \mp\infty} \to \xi + \Gamma_i^{\mp} + \varphi_i(\xi + c_i t), \tag{92}$$

where

$$\Omega_i^{\mp} = \sum_{j<i} F_j^{\mp} + \sum_{j>i} F_j^{\mp}, \quad \Gamma_i^{\mp} = \sum_{j<i} G_j^{\mp} + \sum_{j>i} G_j^{\mp}. \tag{93}$$

Thus, the total phase shift of the ith excitation is given by $\Gamma_i^{+} - \Gamma_i^{-}$, and it preserves its shape if $\Omega_i^{+} = \Omega_i^{-}$. Therefore, in order to construct completely elastic interaction properties, suitable localized functions F_i (and then κ_i) are selected such that $\Omega_i^{+} = \Omega_i^{-}$ $(i = 1, \ldots, N)$. To provide some illustration of the previous discussion, we have plotted in Fig. 4 two bubble-like FSWs which completely interact elastically. These FSWs are built-up from the (1+1)-dimensional localized multivalued functions generating loop solitons [127, 128, 129, 130, 131, 132, 133, 134, 135, 136]. Before interaction, the small and large foldons are initially located at $x = -3$ and $x = -1.5$, respectively. During interaction, they overlap three times, and after interaction, they shift to $x = +1.5$ and $x = +3$, respectively. Another illustration is provided by one typical peakon as depicted in Fig. 5. In this figure, there are two peaks with heights 0.6773 and 0.4163 located at time $t = -4$ at $(x = -4, y = 0)$ and $(x = 12, y = 0)$, respectively, which after interactions are elastically shifted to $(x = -12, y = 0)$ and $(x = 4, y = 0)$, respectively, at time $t = 4$. What seems remarkable here is that these two interacting structures do not pass through each other even though they overlap at time $t = 0$. Besides, amongst typical waves with compact support, there are elastic interactions, but on contrary to the foregoing peakons, the two interacting structures clearly pass through each other so that their initial shapes are not alter after the interacting process. This is clearly presented in Fig. 6. In general, all structures with compact support do not behave like solitons. In Fig. 7, we present a kind of inelastic interactions between two solitary waves with compact structures. It is important to mention here that if a NLPDE equation possesses soliton-like solutions, this property does not straightforwardly guarantee its integrability. Reciprocally, all integrable NLPDEs do not obviously possess solitonic structures even though, due to their integrability properties, they possess an infinite number of conserved quantities [124, 137, 138].

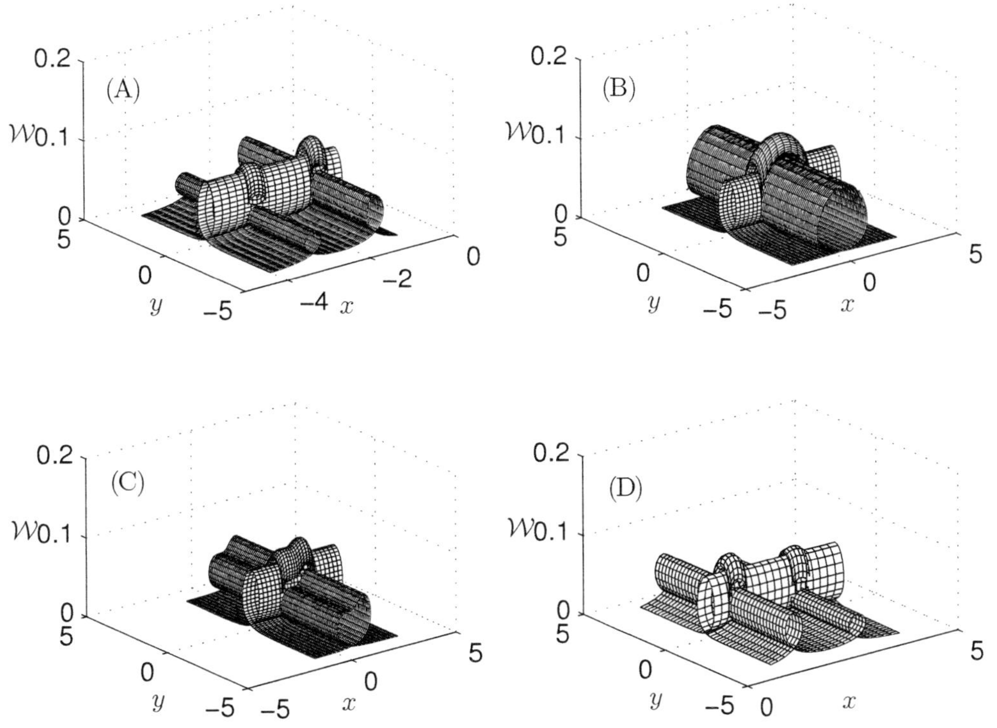

Figure 4. Evolution of two foldons for the physical observable $\mathcal{W}(x, y, t)$ with $f_{1,x} = 0.8sech^2(\xi) + 0.5sech^2(\xi - 0.25t)$, $x = \xi - 1.5\tanh(\xi) - 1.5\tanh(\xi - 0.25t)$, $f_{2,y} = sech^2(\eta)$, $y = \eta - 2\tanh(\eta)$ and $a_0 = 20$ at times, $(A)\ t = -18$, $(B)\ t = 3$, $(C)\ t = 7$, $(D)\ t = +18$.

More interestingly alongside the elastic interactions studied above are the fusion and the fission phenomena. In the fusion phenomenon, two or more solitary waves can fuse together to form a single solitary wave. However, in the fission phenomenon, it is an initial single solitary wave which splits into two or more other waves. In recent studies, these phenomena have been investigated both analytically and experimentally. For example, soliton fission and fusion have been found in Burgers, Sharma-Tasso-Olver systems [139], and generalized Sasa-Satsuma equation [140]. These phenomena are worth underlying due to their wide applications [141, 142, 143, 144]. In the wake of such interests, we have found some suitable selections for f_1 and f_2 such as to depict these phenomena. We consider the following system expressing f_1 and f_2 as follows

$$f_1 = \alpha \tanh(k_1 x + \omega_1 t) + \beta \tanh^3(k_2 x + \omega_2 t)/2, \quad f_2 = \gamma \tanh(k_3 y), \tag{94}$$

such that for $\alpha = 18$, $\beta = 1/2$, $\gamma = 1/3$, $k_1 = 1/2$, $k_2 = 1$, $k_3 = 1/3$, $\omega_1 = -2$, $\omega_2 = 2$ and $a_0 = 20$, the soliton fission phenomenon is observed. As depicted in Fig. 8, the

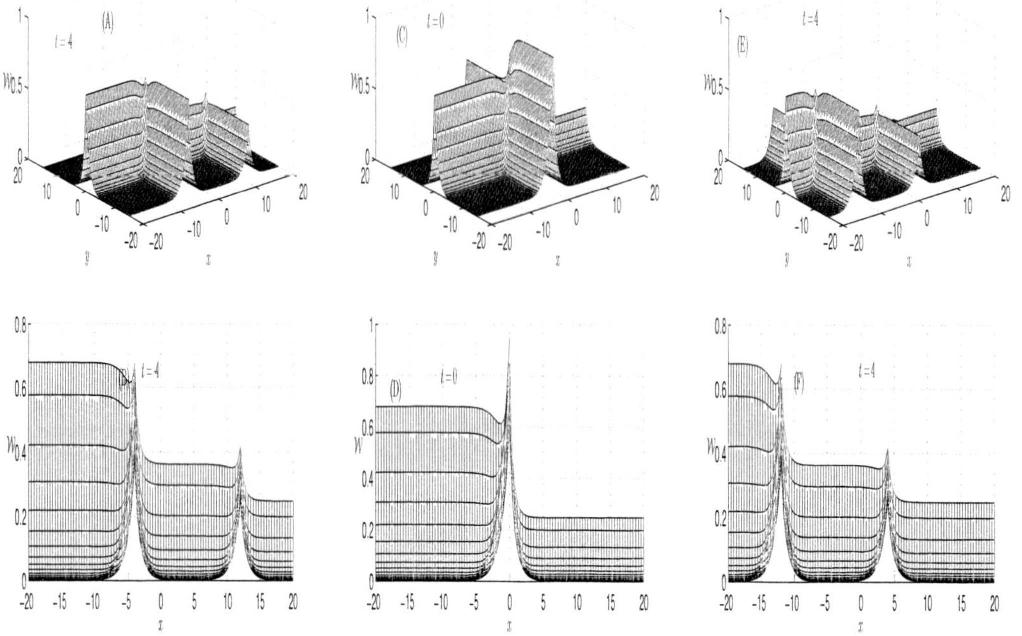

Figure 5. Elastic interaction amongst typical peakons $\mathcal{W}(x, y, t)$ depicted under the following selections: $f_{1x} = 1/\sinh(1 + |x - t|) + 1/\sinh(1 + |x + 3t|)$, $f_{2y} = 1/\sinh(1 + |y|)$ and $a_0 = 1$. The plots are depicted from left to right at times $t = -4$ (panels (A) and (B)), $t = 0$ (panels (C) and (D)) and $t = 4$ (panels (E) and (F)).

initial line soliton splits into two other line solitons which repulse each other in such a way that they move towards opposite directions. On contrary, while trying instead the following parameters: $\alpha = 1$, $\beta = 18$, $\gamma = 1$, $k_1 = 1$, $k_2 = 2$, $k_3 = 1/2$, $\omega_1 = 2$, and $\omega_2 = -2$ with $a_0 = 20$, a fusion phenomenon appears.

5.2. Periodic Excitations to the (2+1)-dimensional CNLS Equations

In this section, we express periodic excitations in terms of some selected Jacobi elliptic functions, namely sn-and cn-functions. It is worth noting that the Jacobi transformation implies that any solution found by one Jacobi elliptic function are transformed into equivalent one that can be obtained by another. First of all, we express the multi-periodic solutions as follows

$$f_1 = \sum_{k=1}^{k=N} A_k \mathrm{sn}(\xi_k, m_k), \quad f_2 = \sum_{k=1}^{k=M} B_k \mathrm{sn}(\eta_k, n_k), \tag{95}$$

where $\xi_k = \alpha_k x + \beta_k t$ and $\eta_k = \gamma_k y + \delta_k t$, and A_k, B_k, α_k, β_k, γ_k and δ_k being arbitrary constants. According to the definition of the Jacobi elliptic functions, the parameters m_k

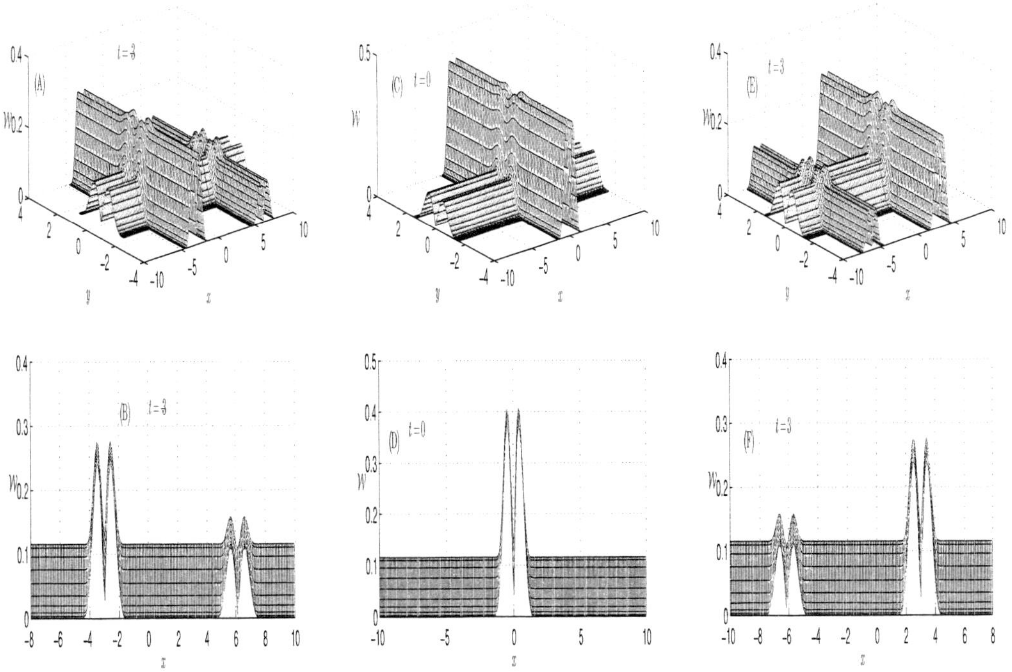

Figure 6. Elastic interaction amongst typical solitons with compact supports $\mathcal{W}(x, y, t)$ depicted under the following selections: $R_1^p(\varphi_1) = -2\cos^5(\varphi_1)$, $R_2^p(\varphi_2) = -\cos^5(\varphi_2)$ and $R^q(\phi) = \cos^5(\phi)$ with $\varphi_1 = x - t$, $\varphi_2 = x + 2t$, $\phi = y$, $\varphi_{11} = \varphi_{12} = \phi_1 = -\pi/2$, and $\varphi_{21} = \varphi_{22} = \phi_2 = \pi/2$. $a_0 = 17$. The plots are depicted from left to right at times $t = -3$ (panels (A) and (B)), $t = 0$ (panels (C) and (D)) and $t = 3$ (panels (E) and (F)).

and n_k should be less than unity.

Taking $N = 2$ and $M = 1$, we construct the following doubly periodic waves

$$f_1 = \mathtt{sn}(\xi_1, m_1) + \mathtt{sn}(\xi_2, m_2), \quad f_2 = \mathtt{sn}(\eta, n), \tag{96}$$

with $\xi_1 = x + t$, $\xi_2 = 2x + t$, $\eta = y + 4t$, $m_1 = 0.95$, $m_2 = 0.98$ and $n = 0.97$. This type of doubly periodic wave is depicted in Fig. 9(A) at $t = 0$ where we have taken $a_0 = 4$. For the other times $t \neq 0$ which are not reported here for some convenience, we can easily check that interactions amongst these waves are not roughly elastic. Due to elementary properties of the elliptic functions, we consider three asymptotic cases specified as follows, $m_k \rightarrow 1$ $(k = 1, 2)$ and $n \rightarrow 1$ corresponding to Fig. 9(B) which shows the interactions between half-straight-line solitons; $m_k \rightarrow 0$ $(k = 1, 2)$ and $n \rightarrow 1$ corresponding to Fig. 9(C) which represents an interaction between x-periodic half-straight-line soliton and exponentially decaying ring soliton; $m_k \rightarrow 1$ $(k = 1, 2)$ and $n \rightarrow 0$ corresponding to Fig. 9(D) which depicts an interaction between y-periodic half-straight-line soliton and exponentially decaying ring soliton, and finally $m_k \rightarrow 0$ $(k = 1, 2)$ and $n \rightarrow 0$ corresponding to Fig. 10(A) which shows x-and y-periodic exponentially decaying ring soliton. Fig.

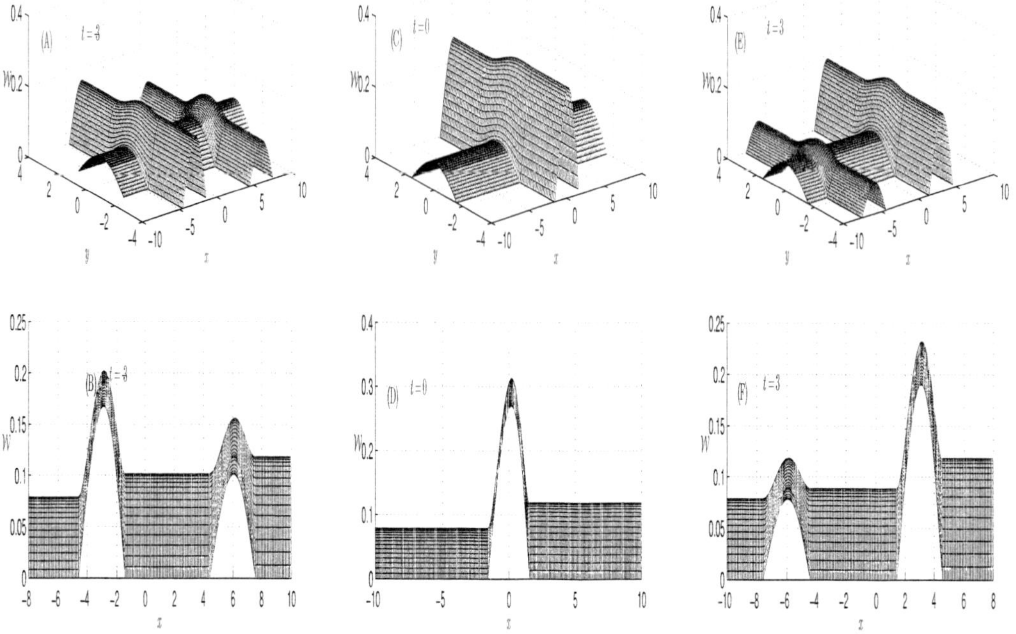

Figure 7. Nonelastic interaction amongst typical structures with compact support $\mathcal{W}(x, y, t)$ depicted under the following selections: $R_1^p(\varphi_1) = -2\sin(\varphi_1) - 2$, $R_2^p(\varphi_2) = -\sin(\varphi_2) - 1$ and $R^q(\phi) = \sin(\phi) + 1$ with $\varphi_1 = x - t$, $\varphi_2 = x + 2t$, $\phi = y$, $\varphi_{11} = \varphi_{12} = \phi_1 = -\pi/2$, and $\varphi_{21} = \varphi_{22} = \phi_2 = \pi/2$. $a_0 = 17$, $a_1 = 1$, $a_2 = 1$, $a_3 = 1/14$ and $\hat{\gamma}_2 = 0$. The plots are depicted from left to right at times $t = -3$ (panels (A) and (B)), $t = 0$ (panels (C) and (D)) and $t = 3$ (panels (E) and (F)).

10(B) represents a typical kink-kink interaction derived from the following expressions

$$f_1 = \exp(\xi_1) + \exp(\xi_2), \quad f_2 = \exp(\eta). \tag{97}$$

Besides, we can find some other interesting types of features such as periodic kink-half-straight-line soliton interaction depicted in Fig. 10(C) with the selections

$$f_1 = \mathtt{sn}(\xi_1, m_1) + \mathtt{sn}(\xi_2, m_2), \quad f_2 = \exp(\eta). \tag{98}$$

Moreover, it should be noted that all periodic waves cannot behave like solitons. There exist some of them which retain their amplitudes, velocities, shapes unchanged after interactions. They are namely periodic worm-like, periodic cusp-like and periodic hump-like waves which are all depicted at $t = 0$ in Figs. 11(A), 11(B) and 11(C), respectively, through the following selections

$$f_{1,x} = -sech(\xi - t)^2, \quad f_2 = \mathtt{sn}(\eta, n), \tag{99}$$

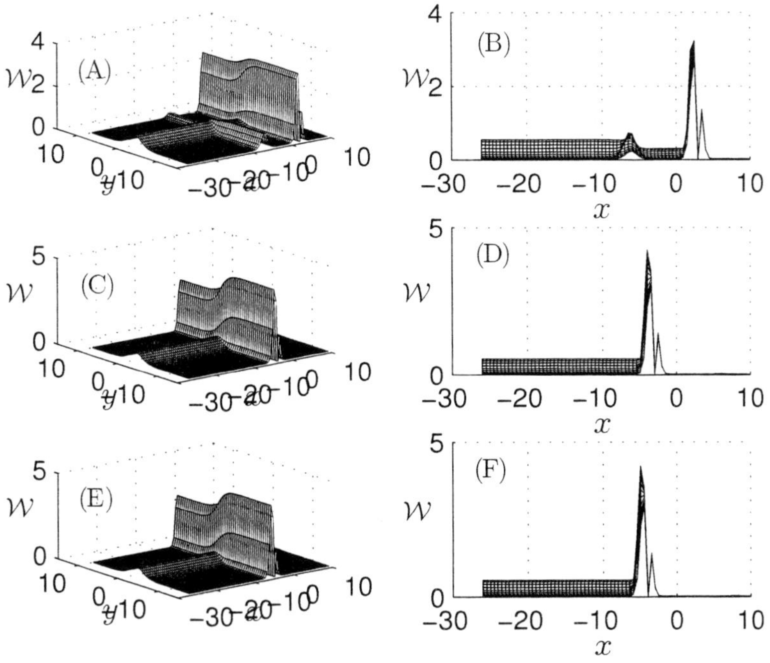

Figure 8. Fission phenomenon depicted by $\mathcal{W}(x, y, t)$ under the following selections: $\alpha = 18$, $\beta = 1/2$, $\gamma = 1/3$, $k_1 = 1/2$, $k_2 = 1$, $k_3 = 1/3$, $\omega_1 = -2$, $\omega_2 = 2$ and $a_0 = 20$. For upper panels, $t = -3$ (panels (A) and (B)), for middle panels, $t = 3$ (panels (C) and (D)) and for lower panels, $t = 4$ (panels (E) and (F)).

with

$$x = \xi + k_4 \tanh(\xi - t), \quad \eta = y + 4t. \tag{100}$$

In other to construct elastic interactions between such structures, we consider the following expressions

$$f_{1,x} = 1.2sech^2(\xi) + 0.8sech^2(\xi - 0.25t), \quad f_2 = \mathrm{sn}(\eta, n), \tag{101}$$

where

$$x = \xi - 1.5\tanh(\xi) - 1.5\tanh(\xi - 0.25t), \quad \eta = y + 4t. \tag{102}$$

Periodic worm-worm interactions are depicted in Fig. 12. During the scattering process, the periodic worms overlap three times and retain their initial shapes unchanged after interaction.

5.3. Discussion of the Physical Meanings of the Solutions

The profiles of Figs. 1 and 2 understand that when the transverse coordinates of a nonlinear medium is taken into account, for example when an optical beam is sent within a medium

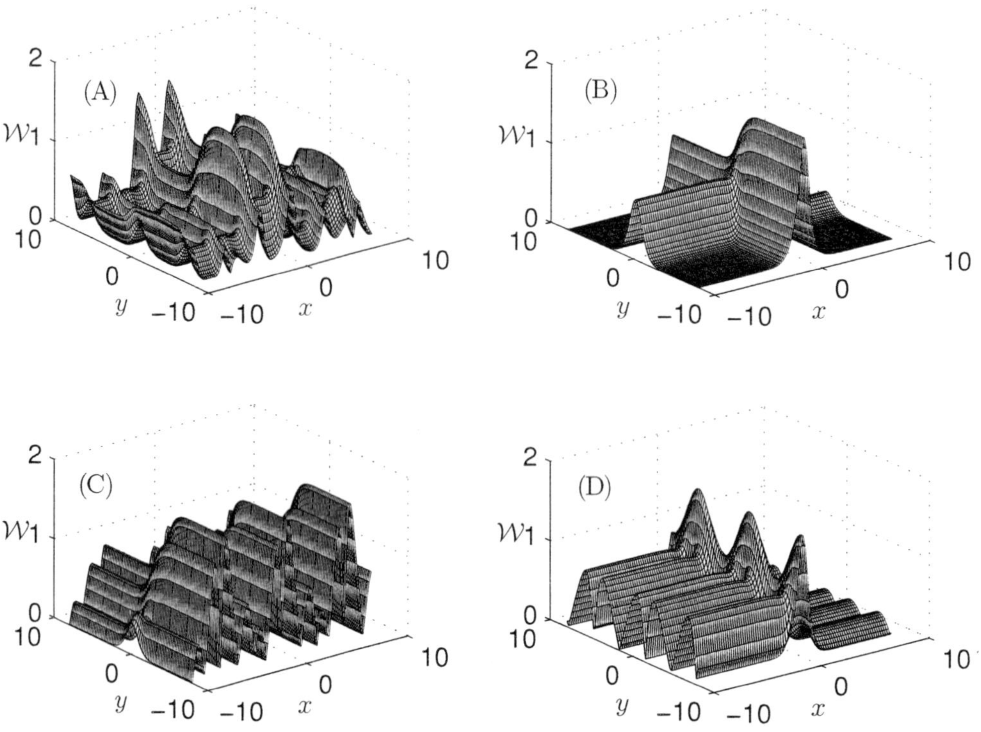

Figure 9. Four typical periodic excitations for the physical observable $\mathcal{W}(x, y, t)$ at $t = 0$ and $a_0 = 4$: (A) typical doubly periodic waves with $f_1 = sn(\xi_1|0.95) + sn(\xi_2|0.98)$, $f_2 = sn(\eta|0.97)$, $\xi_1 = x + t$, $\xi_2 = 2x + t$ and $\eta = y + 4t$; (B) interaction between half-straight-line solitons with $f_1 = \tanh(\xi_1) + \tanh(\xi_2)$, $f_2 = \tanh(\eta)$; (C) interaction between x-periodic half-straight-line soliton and exponentially decaying ring soliton with $f_1 = \sin(\xi_1) + \sin(\xi_2)$, $f_2 = \tanh(\eta)$; (D) interaction between y-periodic half-straight-line soliton and exponentially decaying ring soliton with $f_1 = \tanh(\xi_1) + \tanh(\xi_2)$, $f_2 = \sin(\eta)$.

where the width is not negligible with respect to the wavelength of the soliton, the intensity of the wave can either be singlevalued (see panels $1(A)$, $1(C)$ and $1(D)$, $2(C)$ and $2(D)$) or multivalued (see panels $1(B)$, $2(A)$ and $2(B)$) in a three-dimensional space. This is of particular importance since the results contrast with the well-known singlevalued patterns supported by the lower-dimensional systems. However, one important query would rely on the practical investigation of such patterns in optical spatial devices, but also in BEC systems [51] where the variations of atomic density are investigated. For a such system, during the matter-wave interaction, some kinds of defects can occur in the systems such as singularities in the propagation distance related to the fractal nature of the matter and consequently giving rise to fractal patterns (see panels $3(A)$ and $3(B)$). It is also of physical significance to note that amongst the singlevalued waves that can support an optical spatial device where a transverse spanning y-component is regarded, the intensity of the beam can

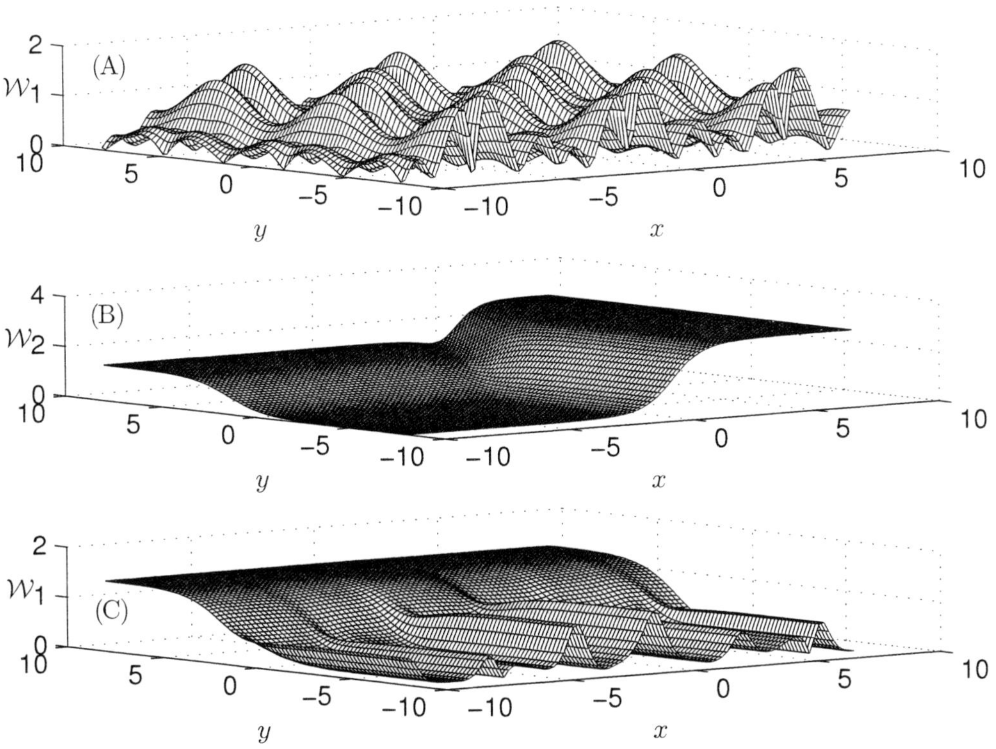

Figure 10. Three more typical periodic excitations for the physical observable $\mathcal{W}(x, y, t)$ at $t = 0$ and $a_0 = 4$: (A) x-and y-periodic exponentially decaying ring solitons with $f_1 = \sin(\xi_1) + \sin(\xi_2)$, $f_2 = \sin(\eta)$, $\xi_1 = x + t$, $\xi_2 = 2x + t$ and $\eta = y + 4t$; (B) kink-kink interaction with $f_1 = \exp(\xi_1) + \exp(\xi_2)$, $f_2 = \exp(\eta)$; (C) periodic kink-half-straight-line soliton scattering with $f_1 = sn(\xi_1|0.95) + sn(\xi_2|0.98)$ and $f_2 = \exp(\eta)$.

reach even relatively lower values when the waves present some peak (see panel $2(D)$) or compact supports (see panels $3(C)$, $3(D)$ and $3(E)$).

Discussing the Fig. 4, the presented patterns can represent the situation of an excitation of atoms of a given matter while interacting with a wave. The response of the matter to such excitation is reported on the variations of the density of the atoms with the space-time. This can be the case of trapped cold atoms in BEC [51] where the different excitations of the above multivalued case propagated in such medium would conserve its characteristics after collisions with others. However, what is of particular interest and can further be subjected to discussion is that at a located zone of the matter, the density of atoms can take many values showing for example that they are perhaps in thermic instability and that this instability conserves all its features in other locations as the time elapses. Apart from this elastic interaction, we also foresee the existence of some kinds of patterns with compact support supported by the above matter in a situation where the atoms are excited with very

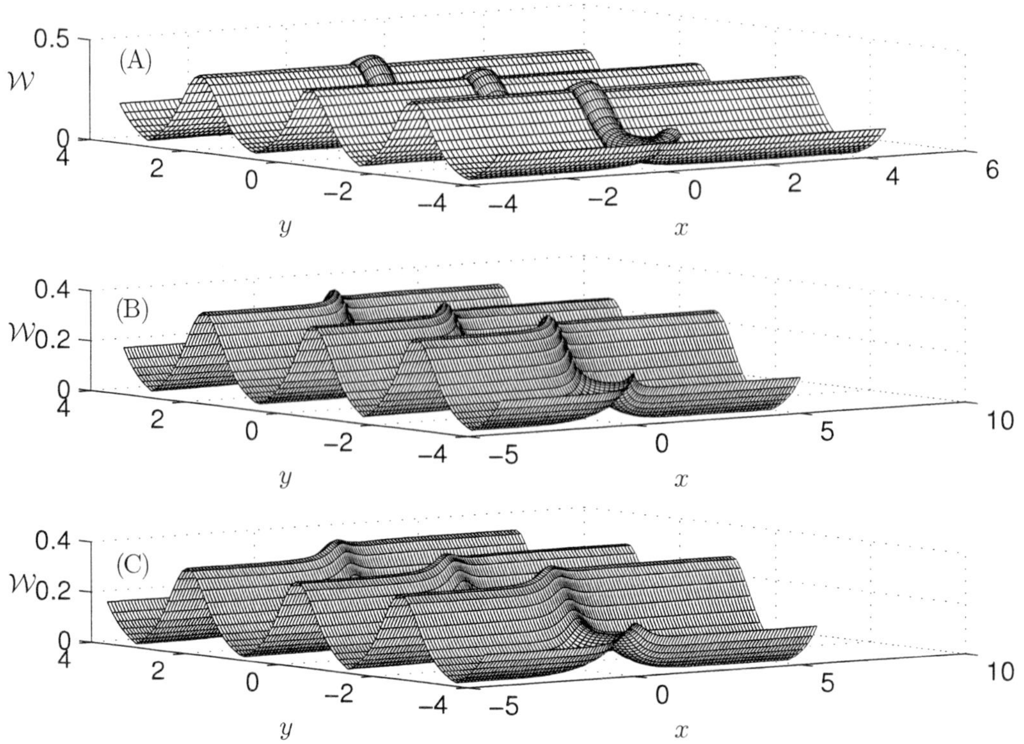

Figure 11. Periodic waves with $f_{1,x} = -sech^2(\xi - t)$, $f_2 = sn(\eta|0.8), x = \xi + k_4 \tanh(\xi - t), \eta = 2y + 4t$ and $a_0 = 8$ at $t = 0$: (A) 'worm'-pattern with $k_4 = -1.6$; (B) cusp-pattern with $k_4 = -1$; (C) hump pattern with $k_4 = -0.5$.

small and located perturbations such that the evolution of the density of atoms can break or retain its initial properties after elastic interactions (see Figs. 6 and 7). The nonelastic scattering observed in Fig. 7 physically shows that for some systems such as BEC [51] with atoms trapped in a particular thermal instability within a higher-dimensional manifold, the interaction of this matter with a wave can results to some induced undesirable effects such as wave distortion, transmission characteristic deterioration, transmission rate distortion, just to name a few. Nevertheless, the fundamental explanation, namely, inhomogeneity in the matter, line defects of the matter, should be investigated practically in order to explain concretely the existence of such defects.

Alongside the previous profiles, there is another one presented in Fig. 5 which can be found in nonlinear optics [29, 55, 56, 57, 58, 59, 60, 61, 62] through the dynamics of optical beam within a nonlinear medium embedded in a three-dimensional manifold where the intensity of the beam undergoes a singlevalued pattern with sharp shape. Although in some situation the scattering is not elastic at all, the interaction between the above peakons physically means that by sending many optical signals through a spatial fiber support, they can retain their initial shapes without alterations. However, while sending instead some signals with hump-like shape, they can breaks into baby solitons during the propagation

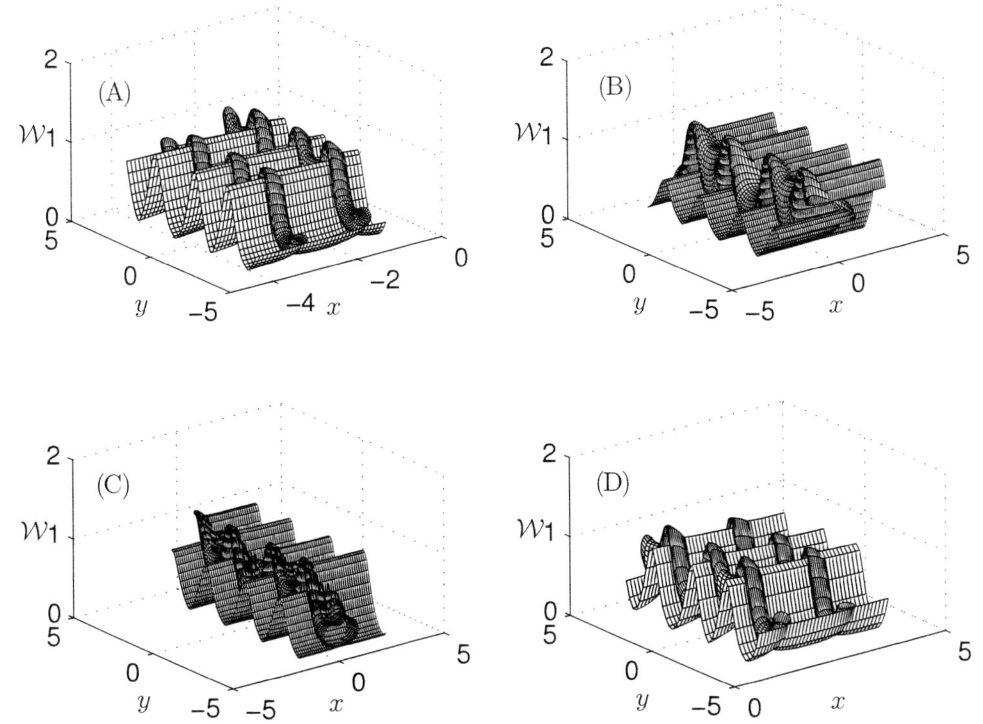

Figure 12. Evolution of two periodic-'worms' $f_{1,x} = 1.2sech^2(\xi) + 0.8sech^2(\xi - 0.25t)$, $f_2 = sn(\eta|0.8)$, $x = \xi - 1.5\tanh(\xi) - 1.5\tanh(\xi - 0.25t)$, $\eta = y + 4t$ and $a_0 = 6$: at times $t = -18$ (A), $t = 3$ (B), $t = 7$ (C) and $t = +18$ (D).

although conserving their total energy (See Fig. 8). Such kind of phenomenon is named fission and it depends on some parameters of the system.

All the above patterns can be created periodically in different supported systems namely the spatial optical fiber [29, 55, 56, 57, 58, 59, 60, 61, 62], the BEC [51], the alpha helix in protein [74], just to name a few. There are actually many ways to construct periodic patterns. For example, perturbing the atoms of a BEC [51] with excitations expressed in terms of sn-functions, the response of these atoms to the perturbations, expressed in terms of their dynamic behavior will be depicted through periodic patterns of their particle density (See Figs. 9, 10, 11 and 12).

6. Summary

In this chapter, we have systematically analyzed the P-properties of the (2+1)-dimensional CNLS equations, and shown that they are completely integrable according to the particular case $\sigma = -1$ and $\alpha = 1$ of self-defocusing Kerr-nonlinearity [29, 55, 56]. In this analysis, the associated BT and their connection with Hirota's bilinear formalism have been presented in such a way that even N-soliton solutions can properly be constructed. In order to check that the above system is completely integrable, we have constructed its general

Lax-representation following the ZS-scheme. It is important to note that there is actually no formal relations between the components of a vector NLS solutions. However, in some particular cases, the CNLS system can be reduced to a one-component NLS equation. Although it can appear that such one-component system is not integrable in general, one cannot conclude straightforwardly that the CNLS system is nonintegrable. Besides, the higher-dimensional systems has this peculiarity that it enriches the different classes of solutions of lower-dimensional counterparts with new ones no matter integrable or not are these lower-systems. This is due to the fact that the independent variables spanning the higher-dimensional manifold in which the system is embedded are not related to any equation. However, using a similarity reduction of the system to a lower-dimensional one, the obtained evolution equation is only a part of all possible reductions of the former system. Thus, the integrability properties of the reduced system cannot infer whether the higher-dimensional system is integrable or not.

Using the arbitrariness of the expansion cœfficients of the solutions expressed in the form of series, a wide range class of localized and periodic excitations amongst which the singlevalued and multivalued waves can be constructed, namely exponentially decaying ring solitons, 'saddle'-like ring solitons, peakons, bubbles, foldons, solitons with compact support, 'breather'-like solitons, half-straight-line solitons, worms, just to name a few [145, 146, 126, 147, 148, 149, 150, 151, 124, 122, 137, 152, 123]. In addition to many types of stable localized structures, there are many types of 'chaotic' or 'singular' patterns by selecting arbitrary functions as the 'chaotic' or 'singular' solutions [153, 154] to some lower dimensional nonintegrable models. Similar to the cases of (1+1)-dimensional integrable models [125], the weak solutions like the peakon [124, 122] solutions also exist in higher dimensions. Many kinds of peakons [124, 126, 122] have been constructed from single valued arbitrary piece-wise functions to multivalued arbitrary functions (say (1+1)-dimensional loop solitons).

For the existence of the abundant structures of (2+1)-dimensional CNLS equations, it is quite important but difficult to investigate the interaction properties for all the possible localized excitations. Concerning the multivalued waves, we have constructed some kinds of FSWs known as 'bubbles'. When behaving like particles in an elastic interaction, these 'bubbles' become solitons and are therefore named foldons. We have also investigated the scattering process amongst typical peakons and solitons with compact support. The interactions between peakons have been found to be elastic. However, during this scattering process, the two initial structures overlap at time $t = 0$ and subsequently repulse each other towards their initial positions. On contrary, we have constructed typical two-solitons with compact supports representing the attractive interaction between two single solitons with compact supports passing through each other during the process. The interactions between solitary waves with compact support are not always elastic. We have found a kind of such waves which do not retain their initial shapes after the scattering process. Usually, when two structures interact elastically, they separate after the interaction process. Nonetheless, we have unearthed soliton fusion and fission phenomena in the (2+1)-dimensional CNLS equations. These phenomena which are related to localized waves can also be investigated in periodic patterns.

There are many interesting features for the periodic interacting waves. These waves have been constructed by means of some Jacobi elliptic functions and we have found that they can be viewed as generalizations of exponentially decaying ring solitons, half-straight-line solitons, exponentially decaying ring-half-straight-line soliton and periodic solutions. All Jacobi elliptic wave solutions do not interact elastically during some scattering processes. We have however constructed some periodic waves such as periodic 'worm'-like structures which behave as solitons and present some peculiar features during the scattering processes. Due to the rich variety of solutions, we actually think that much is to be investigated in such structures and their scattering properties.

It is also important to mention here that the integrability properties of a given system do not necessarily suffer from the fact that some solutions to such system scatter through an inelastic process. In fact, when a system is shown to be integrable, that means it possesses an infinite set of conserved quantities. However, the conservation of some quantities such as the momentum, the amplitude, just to name a few, are not straightforward. Such results have recently been pointed out through many investigations. As illustration, recently, while investigating some (2+1)-dimensional integrable systems such as the Boiti-Leon-Pempinelli equation, the Nizhnik-Novikov-Veselov (NNV) system, dispersive long wave equation, the modified NNV system, just to name a few [116, 155, 156], nonelastic interactions have been depicted.

As previously shown, the WTC method stands for a powerful method amongst other techniques which prove the integrability of a given model, but also it is useful in finding some exact solutions. However, for some model systems, using this standard WTC's P-expansion method, it is not possible to find more physically significant non singular localized solutions. Thus, Conte [157] has developed an alternative P-analysis approach, the invariant P-analysis for such models. Subsequently, Fordy and Pickering [158, 159] have firstly used the nonstandard Conte's truncated expansions to find new exact solutions. A modification of the truncated Conte's expansion has been presented by Pickering [160, 161]. Lou [162, 163, 164] has generalized the P-expansion analysis approach and he has shown that the standard and nonstandard truncations of the generalized P-expansion lead to the construction of new explicit exact solutions including singular and nonsingular ones. We actually believe that, in order to further find more physically significant solutions to the (2+1)-dimensional CNLS equations, it is interesting to use the generalized WTC approach in a nonstandard truncation developed by Conte [157], Pickering [160, 161] and Lou [162, 163, 164] and look forward extending this method to the above system under investigation.

Acknowledgments

The authors would like to express their sincere thanks to Nova Science Publishers for their financial support in publication.

References

[1] F. Calogero and A. Degasperis, *Special Transform and Solitons* (North-Holland, Amsterdam, 1982).

[2] G. L. Jr. Lamb, *Elements of Soliton Theory*(Wiley, New York, 1980).

[3] V. E. Zakharov and P. B. Shabat, *Sov. Phys.–JETP* **62**, 34 (1974).

[4] J. Satsuma and N. Yajima, *Prog. Theor. Phys. Suppl.* **55**, 284 (1974).

[5] M. J. Ablowitz, D. J. Kaup, A. C. Newell, and Segur, *Stud. Appl. Math.* **53**, 249 (1974).

[6] M. J. Ablowitz and P. A. Clarkson, *Solitons, Nonlinear Evolution Equations and Inverse Scattering* (Cambridge Univ. Press, Cambridge, 1991).

[7] M. J. Ablowitz, B. Prinari, and A. D. Trubatch, *Discrete and Continuous Nonlinear Schrödinger Systems*, London Mathematical Society Lecture Note Series (Cambridge Univ. Press, Cambridge, 2004).

[8] P. A. Clarkson and C. M. Cosgrove, *J. Phys. A* **20**, 2003 (1987).

[9] D. Anderson, A. Boudeson, and M. Lisak, *J. Phys. A* **21**, 259 (1979).

[10] J. T. Manassah, *Opt. Lett.* **17**, 1259 (1992).

[11] E. Simo and T. C. Kofane, *Phys. Lett. A.* **192**, 233 (1994).

[12] M. Tajiri, *J. Phys. Soc. Jpn.* **52**, 1908 (1983).

[13] M. Tajiri and M. Hagiwara, *J. Phys. Soc. Jpn.* **53**, 1634 (1984).

[14] R. Sahadevan, K. M. Tamizhmani, and M. Lakshmanan, *J. Phys. A* **19**, 1783 (1986).

[15] C. R. Menyuk, *IEEE J. Quant. Elect.* **25**, 2674 (1988).

[16] D. N. Christodoulides and R. J. Joseph, *Opt. Lett.* **13**, 53 (1988).

[17] M. Haelterman and A. P. Sheppard, *Phys. Rev. E* **49**, 3384 (1994).

[18] Y. Chen and J. Atai, *Phys. Rev.* E **52**, 3102 (1995).

[19] T. Ueda and W. L. Kath, *Phys. Rev. A* **42**, 563 (1990).

[20] B. A. Malomed and N. F. Smyth, *Phys. Rev. E* **50**, 1535 (1994).

[21] M. Desaix, D. Anderson, and M. Lisak, *Phys. Rev. E* **50**, 2253 (1994).

[22] E. Alfinito, M. Leo, R. A. Leo, G. Soliani, and L. Solombrino, *Phys. Rev. E* **52**, 3159 (1995).

[23] S. V. Manakov, Sov. Phys. *JETP* **38**, 248 (1974).

[24] V. S. Shchesnovich and E. V. Doktorov, *Phys. Rev. E* **55**, 7626 (1997).

[25] T. Alagesan and K. Porsezian, *Chaos, Solitons Fractals* **8**, 1645 (1997).

[26] K. Porsezian and M. Lakshmanan, *J. Math. Phys.* **32**, 2923 (1991).

[27] S. Y. Lou, J. Lin, and X. Y. Tang, *Eur. Phys. J. B* **22**, 473 (2001).

[28] H. Q. Zhang, X. H. Meng, T. Xu, L. L. Li, and B. Tian, *Phys. Scripta* **75**, 537 (2007).

[29] Y. S. Kivshar and G. P. Agrawal, *Optical Solitons for Fibers to Photonic Crystals* (Academic, San Diego, 2003).

[30] R. W. Boyd, *Nonlinear Optics* (Academic, San Diego, 1992).

[31] V. E. Zakharov and A. B. Shabat, *Sov. Phys. -JETP* **34**, 62 (1972).

[32] V. E. Zakharov and A. B. Shabat, *Sov. Phys. -JETP* **37**, 823 (1973).

[33] K. Nakkeeran, *J. Phys. A: Math. Gen.* **33**, 3947 (2000).

[34] V. E. Zakharov and E. I. Schulman, *Physica D* **4**, 270 (1982).

[35] J. Weiss, M. Tabor, and G. Carnevale, *J. Math. Phys.* **24**, 522 (1983).

[36] J. Weiss, M. Tabor, and G. Carnevale, *J. Math. Phys.* **25**, 13 (1984).

[37] E. Yomba and T. C. Kofane, *J. Phys. Soc. Jpn.* **69**, 1027 (2000).

[38] E. Yomba and T. C. Kofane, *Chaos, Solitons Fractals* **17**, 847 (2003).

[39] S. Y. Lou and Q. X. Wu, *Phys. Lett. A* **262**, 344 (1999).

[40] E. Yomba, T.C. Kofane, and F. B. Pelap, *J. Phys. Soc. Jpn.* **65**, 2337 (1996).

[41] E. Yomba and T.C. Kofane, *Physica D* **125**, 105 (1999).

[42] E. Yomba and T.C. Kofane, *Phys. Scr.* **54**, 576 (1996).

[43] A. Ramani, B. Grammaticos, and T. Bountis, *Phys. Rep.* **180**, 159 (1989).

[44] M. J. Ablowitz and H. Segur, *Solitons and the Inverse Scattering Transform* (Society for Industrial and Applied Mathematics, Philadelphia, 1981).

[45] M. J. Ablowitz, A. Ramani, and H. Segur, *Lett. Nuov. Cimento* **23**, 333 (1978).

[46] M. J. Ablowitz, A. Ramani, and H. Segur, *J. Math. Phys.* **21**, 715 (1980).

[47] S. P. Hastings and J. B. McLeod, *Archive for Rational Mechanics and Analysis* **73**, 31 (1980).

[48] J. B. McLeod and P. J. Olver, *SIAM* **14**, 488 (1983).

[49] D. Rand, I. Glesk, C. S. Bres, D. A. Nolan, X. Chen, J. Koh, J. W. Fleischer, K. Steiglitz, and P. R. Prucnal, *Phys. Rev. Lett.* **98**, 053902 (2007).

[50] V. K. Kuetche, T. B. Bouetou, and T. C. Kofane, *Phys. Rev. E* **79**, 056605 (2009).

[51] J. Babarro, M. J. Paz-Alonso, H. Michinel, J. R. Salguiero, and D. N. Olivieri, Phys. Rev. A **71**, 043608 (2005).

[52] K. Steiglitz, *Phys. Rev. E* **63**, 016608 (2000).

[53] M. H. Jakubowski, K. Steiglitz, and R. Squier, *Phys. Rev. E* **58**, 6752 (1998).

[54] M. Lakshmanan and T. Kanna, *Pramana J. Phys.* **57**, 885 (2001).

[55] N. B. Abraham and W. J. Firth, *J. Opt. Soc. Am. B* **7**, 951 (1990).

[56] P. D. Maker and R. W. Terhume, *Phys. Rev. A* **137**, 801 (1965).

[57] Z. H. Musslimani, M. Segev, D. N. Christodoulides, and M. Soljačic, *Phys. Rev. Lett.* **84**, 1164 (2000).

[58] A. Hasewara, *Optical Solitons in Fibers* (Springer, Berlin, 1990).

[59] B. A. Malomed, H. E. Nistazakis, D. J. Frantzeskakis, and P. G. Kevrekidis, *Phys. Rev. A* **70**, 043616 (2004).

[60] M. I. Merhasin, B. A. Malomed, and R. Driben, *J. Phys. B: At. Mol. Opt. Phys.* **38**, 877 (2005).

[61] H. Sakaguchi and B. A. Malomed, *Phys. Rev. E* **73**, 026601 (2006).

[62] K. Porsezian and V. C. Kuriakose, *Optical Solitons: Theoretical and Experimental Challenges* (Springer, Berlin, 2003).

[63] L. Gagnon, *J. Phys. A: Math. Gen.* **25**, 2649 (1992).

[64] J. N. Malmberg, A. H. Carlsson, D. Anderson, E. A. Ostrovskaya, Y. S. Kivshar, and M. Lisak, *Opt. Lett.* **25**, 643 (2000).

[65] K. C. Chan and H. F. Liu, *IEEE J. Quantum Electron.* **31**, 2226 (1995).

[66] W. J. Liu, B. Tian, H. Q. Zhang, L. L. Li, and Y. S. Xue, *Phys. Rev. E* **77**, 066605 (2008).

[67] A. Janutka, *J. Phys. A: Math. Theor.* **41**, 285204 (2008).

[68] T. Kanna and M. Lakshmanan, *Phys. Rev. E* **67**, 046617 (2003).

[69] T. Kanna, M. Lakshmanan, P. T. Dinda, and N. Akhmediev, *Phys. Rev. E* **73**, 026604 (2006).

[70] T. Kanna, M. Vijayajayanthi, and M. Lakshmanan, *Phys. Rev. A* **76**, 013808 (2007).

[71] K. Porsezian, *Phys. Rev. E* **68**, 066607 (2003).

[72] X. G. He, D. Zhao, L. Li, and H. G. Luo, *Phys. Rev. E* **79**, 056610 (2009).

[73] H. Q. Zhang, B. Tian, X. Lu, H. Li, and X. H. Meng, *Phys. Lett. A* **373**, 4315 (2009).

[74] A. C. Scott, *Phys. Scr.* **29**, 279 (1984).

[75] S. Chakravarty, M. J. Ablowitz, J. R. Sauer, and R. B. Jenkis, *Opt. Lett.* **20**, 136 (1995).

[76] C. Yeh and L. Bergman, *Phys. Rev. E* **57**, 2398 (1998).

[77] A. Ankiewicz, W. Krolikowski, and N. N. Akhmediev, *Phys. Rev. E* **59**, 6079 (1999).

[78] N. N. Akhmediev, W. Krolikowski, and A. W. Snyder, Phys. Rev. Lett. **81**, 4632 (1998).

[79] T. Kanna and M. Lakshmanan, *Phys. Rev. Lett.* **86**, 5043 (2001).

[80] J. U. Kang, G. I. Stegeman, J. S. Aitchison, and N. N. Akhmediev, *Phys. Rev. Lett.* **76**, 3699 (1996).

[81] C. Anastassiou, M. Segev, K. Steiglitz, J. A. Giordmaine, M. Mitchell, M. Shih, S. Lan, and J. Martin, *Phys. Rev. Lett.* **83**, 2332 (1999).

[82] W. Krolikowski, N. N. Akhmediev, and B. Luther-Davies, *Phys. Rev. E* **59**, 4654 (1999).

[83] M. Mitchell, Z. Chen, M. F. Shih, and M. Segev, *Phys. Rev. Lett.* **77**, 490 (1996).

[84] M. Mitchell and M. Segev, *Nature (London)* **387**, 880 (1997).

[85] D. N. Christodoulides, T. H. Coskun, M. Mitchell and M. Segev, *Phys. Rev. Lett.* **80**, 2310 (1998).

[86] D. N. Christodoulides, T. H. Coskun, and R. I. Joseph, *Opt. Lett.* **22**, 1080 (1997).

[87] P. E. Gross, *J. Math. Phys.* **4**, 19 (1963).

[88] L. P. Pitaevskii, *Sov. Phys.-JETP* **13**, 451 (1961).

[89] C. J. Pethick and H. Smith, *Bose-Einstein Condensates in Dilute Gases* (Cambridge University Press, Cambridge, 1999).

[90] F. Dalfovo, S. Giorgini, L. P. Pitaevskii, and S. Stringari, *Rev. Mod. Phys.* **71**, 463 (1999).

[91] O. Morsch and M. Oberthaler, *Rev. Mod. Phys.* **78**, 179 (2006).

[92] I. Bloch, J. Dalibard, and W. Zwerger, *Rev. Mod. Phys.* **80**, 885 (2008).

[93] S. Sachdev, *Quantum Phase Transitions* (Cambridge University Press, Cambridge, 1999).

[94] P. Das, M. Vyas, and P. K. Panigrahi, *J. Phys. B: At. Mol. Opt. Phys.* **42**, 245304 (2009).

[95] J. G. Muga, X. Chen, A. Ruschhaupt, and D. G. Odelin, *J. Phys. B: At. Mol. Opt. Phys.* **42**, 241001 (2009).

[96] M. Eckholt and J. J. G. Ripoll, *New J. Phys.* **11**, 093028 (2009).

[97] Z. Haghshenasfard, M. H. Naderi, and M. Soltanolkotabi, *J. Phys. B: At. Mol. Opt. Phys.* **42**, 195501 (2009).

[98] P. R. Johnson, E. Tiesinga, J. V. Porto, and C. J. Williams, *New J. Phys.* **11**, 093022 (2009).

[99] M. M. Ali and H. S. Goan, *J. Phys. A: Math. Theor.* **42**, 385303 (2009).

[100] N. G. Berloff, *Fluid Dyn. Res.* **41**, 051403 (2009).

[101] B. J. D. Wuster, S. Wuster, and M. J. Davis, *New J. Phys.* **11**, 053017 (2009).

[102] D. Zhao, H. G. Luo, and H. Y. Chai, *Phys. Lett. A* **372**, 5644 (2008).

[103] J. F. Nye and M. V. Berry, *Proc. R. Soc. London A* **336**, 165 (1974).

[104] M. S. Soskin and M. V. Vasnetsov, *Prog. Opt.* **42**, 219 (2001).

[105] A. S. Desyatnikov, D. E. Pelinovsky, and J. Yang, *J. Math. Sci.* **151**, 3091 (2008).

[106] A. S. Desyatnikov, Yu. S. Kivshar, and L. Torner, *Prog. Opt.* **47**, 219 (2005).

[107] A. L. Fetter and A. A. Svidzinsky, *J. Phys.: Cond. Mat.* **13**, R135 (2001).

[108] X. G. He, D. Zhao, L. Li, and H.G. Luo, *Phys. Rev. E* **79**, 056610 (2009).

[109] R. Hirota, *Phys. Rev. Lett.* **27**, 1192 (1971).

[110] R. Hirota and J. Satsuma, *J. Phys. Soc. Jpn.* **40**, 611 (1980).

[111] R. Hirota, *Direct Methods in Soliton Theory* in Soliton, edited Bullough R K and Caudrey P J (Springer, Berlin, 1980).

[112] J. D. Gibbon, Radmore, M. Tabor, and D. Wood, *Stud. Appl. Math.* **72**, 39 (1985).

[113] A. B. Shabat, *Sov. Math. Dokl.* **14**, 1266 (1973).

[114] V. E. Zakharov and A. B. Shabat, *Funct. Anal. Appl.* **8**, 226 (1974).

[115] P. G. Drazin and R. S. Johnson, *Solitons: an Introduction* (Cambridge University Press, Cambridge, 1989).

[116] X. Y. Tang and S. Y. Lou, *J. Math. Phys.* **44**, 4000 (2003).

[117] X. Y. Tang and S. Y. Lou, *Comm. Theor. Phys.* **40**, 62 (2003).

[118] X. Y. Tang, J. Lin, X. M. Qian, and S. Y. Lou, *Int. J. Mod. Phys. B* **17**, 4343 (2003).

[119] C. L. Bai and H. Zhao, *J. Phys. A* **38**, 4375 (2005).

[120] C. L. Bai and H. Zhao, *J. Phys. A* **39**, 3283 (2006).

[121] C. L. Bai and H. Zhao, *Chin. J. Phys.* **44**, 94 (2006).

[122] J. Lin and H. M. Li, *Z. Naturforsch* **57**, 929 (2002).

[123] C. L. Bai, H. Zhao, and X. Y. Wang, Nonlinearity **19**, 1697 (2006).

[124] S. Y. Lou, X. Y. Tang, X. M. Qian, C. L. Chen, J. Lin, and S. L. Zhang, *Mod. Phys. Lett. B* **16**, 1075 (2002).

[125] R. Camassa and D. D. Holm, *Phys. Rev. Lett.* **71**, 1661 (1993).

[126] X. Y. Tang, S. Y. Lou, and Y. Zhang, *Phys. Rev. E* **66**, 046601 (2002).

[127] V. O. Vakhnenko, *J. Phys. A* **25**, 4181 (1992).

[128] V. K. Kuetche, T. B. Bouetou, and T. C. Kofane, *J. Phys. A* **39**, 12355 (2006).

[129] V. O. Vakhnenko and E. J. Parkes, *Nonlinearity* **11**, 1457 (1998).

[130] V. K. Kuetche, T. B. Bouetou, and T. C. Kofane, *J. Phys. Soc. Jpn.* **76**, 024004 (2007).

[131] A. J. Morrison, E. J. Parkes, and V. O. Vakhnenko, *Nonlinearity* **12**, 1427 (1999).

[132] V. K. Kuetche, T. B. Bouetou, and T. C. Kofane, *J. Phys. A* **40**, 5585 (2007).

[133] A. Sakovich and S. Sakovich, *J. Phys. A* **39**, L361 (2006).

[134] V. K. Kuetche, T. B. Bouetou, and T. C. Kofane, *J. Phys. Soc. Jpn.* **76**, 073001 (2007).

[135] H. Kakuhata and K. Konno, *J. Phys. Soc. Jpn.* **68**, 757 (1999).

[136] V. K. Kuetche, T. B. Bouetou, and T. C. Kofane, *Phys. Lett. A* **372**, 4891 (2008).

[137] S. Y. Lou, *J. Phys. A* **35**, 10619 (2002).

[138] C. Q. Dai and G. Q. Zhou, *Chin. Phys.* **16**, 1201 (2007).

[139] S. Wang, X. Y. Tang and S. Y. Lou, *Chaos, Solitons Fractals* **21**, 231 (2004).

[140] R. Radha and S. Y. Lou, *Phys. Scripta* **72**, 432 (2005).

[141] V. N. Serkin, V. M. Chapela, J. Percino, and T. L. Belyaeva, *Opt. Comm.* **192**, 237 (2001).

[142] W. Krolikowski and S. A. Holmstrom, *Opt. Comm.* **22**, 15 (2007).

[143] M. Hisakado, *Phys. Lett. A* **227**, 87 (1997).

[144] H. Ono and I. J. Nakata, *J. Phys. Soc. Jpn.* **63**, 40 (1994).

[145] S. Y. Lou, *Phys. Lett. A* **277**, 94 (2000).

[146] S. Y. Lou and H. Y. Ruan, *J. Phys. A* **34**, 305 (2001).

[147] S. Y. Lou, *J. Math. Phys.* **41**, 6509 (2000).

[148] B. G. Konopelchenko and C. Rogers, *J. Math. Phys.* **34**, 214 (1993).

[149] S. Y. Lou, *J. Phys. A* **28**, 7227 (1995).

[150] S. Y. Lou, *J. Phys. A* **29**, 5989 (1996).

[151] T. B. Bouetou, V. K. Kuetche, and T. C. Kofane, *J. Phys. A* **40**, 135208 (2008).

[152] C. L. Bai and H. Zhao, *Chaos, Solitons Fractals* **23**, 777 (2005).

[153] F. W. Lorenz, *J. Atmos. Sci.* **20**, 130 (1963).

[154] M. Clerc, P. Coullet, and E. Tirapegui, *Phys. Rev. Lett.* **83**, 3820 (1999).

[155] D.C. Qing and Z.G. Quan, *Chin. Phys.* **16**, 1201 (2007).

[156] C.Q. Dai and Y.Z. Ni, *Phys. Scr.* **74**, 584 (2006).

[157] R. Conte, *Phys. Lett. A* **140**, 383 (1989).

[158] A. P. Fordy and A. Pickering, *Phys. Lett. A* **160**, 347 (1991).

[159] A. Pickering, *J. Phys. A* **26**, 4395 (1993).

[160] A. Pickering, *J. Math. Phys.* **37**, 1894 (1996).

[161] R. Conte, M. Musette, and A. Pickering, *J. Phys. A* **28**, 179 (1995).

[162] S. Y. Lou, *Phys. Rev. Lett.* **80**, 5027 (1998).

[163] S. Y. Lou, *Z. Naturforsch* **53**, 251 (1998).

[164] S. Y. Lou and J. J. Xu,*J. Phys. A* **39**, 5364 (1998).

In: Evolution Equations
Editor: Arthur L. Claes
ISBN: 978-1-61209-652-0

Chapter 6

METHODS BASED ON PROJECTIVE EQUATIONS FOR SOLVING NONLINEAR EVOLUTION EQUATIONS

Alvaro H. Salas
FIZMAKO Research Group, Universidad de Caldas,
Department of Mathematics, Universidad Nacional de Colombia,
Nubia Campus, Manizales, Columbia

Abstract

In this chapter we present some ideas for solving nonlinear partial differential equations which are based on the idea of the projective Riccati equations. We show how to derive well known methods such as Conte's projective Riccati equations method, tanh-coth method, He's exp function method and a new method we called sn-ns method. We illustrate the effectiveness of the sn-ns method in the problem of findin new exact solutions to some relevant nonlinear pde's with the aid of Maple 14 and Mathematica 7.

Keywords: nonlinear PDE; traveling wave solution; projective Riccati equation method; Fan's tanh method; tanh–coth method; He's exp–method; Jacobi elliptic functions Duffin equation; quadratic Duffin equation; cubic-quintic Duffin equation;sn–ns method; Weierstrass elliptic functions method; Mathematica 7; Maple 14.

1. Introduction

The search of explicit solutions to nonlinear partial differential equations (NLPDE's) by using computational methods is one of the principal objectives in nonlinear science problems. Some powerful methods have been extensively used in the past decade to handle nonlinear PDE's. Some of them are : the tanh-method [1], the $\tanh - \coth$ method [2], the exp–function method [3], the projective Riccati equation method [4], the Jacobi elliptic functions method, the Weierstrass elliptic functions method and the G'/G method, which is equivalent to tanh method. Practically, there is no unifie method that could be used to handle all types of nonlinear problems.

The main purpose of this work consists on solving nonlinear polynomial ode's starting from the idea of projective Riccati equations.We show the way the projective Riccati equations may be successfully used to obtain powerful methods that allow us to solve many nonlinear differential equations (partial and ordinary) in closed form. We also derive exact solutions to following equations : Gardner equation, Sawada-Kotera equation, Duffin equation, cubic nonlinear Schrodinger equation, Klein-Gordon-Zakharov equations, quadratic Duffing equation, KdV equation, Boussinesq equation, symmetric regular long wave equation, generalized shallow water wave equation, Klein–Gordon equation with quadratic nonlinearity, Fitzhugh-Nagumo-Huxley equation and double sine-Gordon equation.

2. The Main Idea

In the search of the traveling wave solutions to nonlinear partial differential equation of the form

$$P(u, u_x, u_t, u_{xx}, u_{xt}, u_{tt}, \ldots) = 0, \tag{1}$$

the firs step consists on considering the wave transformation

$$u(x, t) = v(\phi(\xi)), \; \xi = x + \lambda t + \xi_0, \xi_0 = \text{arbitrary constant}, \tag{2}$$

for a suitable function $\phi = \phi(\xi)$, where λ is a constant. Usually, $\phi(\xi) = \xi$ (the identity function).
Using (2), equation (1) converts to an ordinary differential equation (ODE) with respect to the function $v(\xi)$

$$Q(v, v', v'', \ldots) = 0, \tag{3}$$

being Q a *polynomial* with respect to (shortly, w.r.t.) variables $v, v', v'', \ldots$.
To fin solutions to (3), we suppose that $v(\xi)$ can be expressed as

$$v(\xi) = H(f(\xi), g(\xi)), \tag{4}$$

where $H(f, g)$ is a *rational function* in the new variables $f = f(\xi), g = g(\xi)$ which satisfy the system

$$\begin{cases} f'(\xi) = rf(\xi)g(\xi), \\ g^2(\xi) = S(f(\xi)), \end{cases} \tag{5}$$

being $r \neq 0$ some constant to be determined and $S(f)$ a rational function in the variable $f = f(\xi)$. We show that the system (5) may be solved exactly in certain cases. In fact, taking

$$f(\xi) = \varphi^N(\xi), \tag{6}$$

where $\varphi(\xi) \neq 0$, and $N \neq 0$, system (5) reduces to

$$\varphi'(\xi) = \frac{r}{N}\varphi(\xi)g(\xi), \tag{7}$$

$$g^2(\xi) = S(\varphi^N(\xi)). \tag{8}$$

From (7)-(8) we get

$$(\varphi'(\xi))^2 = \frac{r^2}{N^2}\varphi^2(\xi)S(\varphi^N). \tag{9}$$

Equation (9) is of elliptic type. Choosing $S(f)$ and N adequately, we may obtain distinct methods to solve nonlinear PDEs. More exactly, suppose we have solved Eq. (9). Then, in view of (5) and (6) functions f and g may be computed by formulas

$$f(\xi) = \varphi^N(\xi) \text{ and } g(\xi) = \frac{f'(\xi)}{rf'(\xi)} = \frac{N}{r}\frac{\varphi'(\xi)}{\varphi(\xi)}. \tag{10}$$

To solve Eq. (3) we try one of the following ansätze :

$$v(\xi) = a_0 + \sum_{j=1}^{n} f^{j-1}(a_j f + b_j g). \tag{11}$$

$$v(\xi) = a_0 + \sum_{j=1}^{n} a_j f^j. \tag{12}$$

$$v(\xi) = a_0 + \sum_{j=1}^{n} \left(a_j f^j + b_j f^{-j}\right). \tag{13}$$

We substitute any of these ansätze into (3) and we obtain a polynomial equation either in the variables $f = f(\xi)$ and $g = g(\xi)$ or $f = f(\xi)$. We equate the coefficient of $f^i g^j$ $(i, j = 0, 1, 2, 3, ..)$ to zero and we obtain a system of polynomial equations in the variables a_j, b_i, λ, ...Solving this system with the aid of a symbolic computational package such as Mathematica or Maple, we obtain desired solutions. Sometimes, we replace ξ with $k\xi$ and then the corresponding system is regarded w.r.t. the variables $f = f(k\xi)$ and $g = g(k\xi)$, where k =const.

We also may solve coupled systems of nonlinear equations. Indeed, suppose that we have a coupled system of two equations in the form

$$\begin{cases} P(u, \widetilde{u}, u_x, \widetilde{u}_x, u_t, \widetilde{u}_t, u_{xx}, \widetilde{u}_{xx}, u_{xt}, \widetilde{u}_{xt}, u_{tt}, \widetilde{u}_{tt}, \ldots) = 0. \\ \widetilde{P}(u, \widetilde{u}, u_x, \widetilde{u}_x, u_t, \widetilde{u}_t, u_{xx}, \widetilde{u}_{xx}, u_{xt}, \widetilde{u}_{xt}, u_{tt}, \widetilde{u}_{tt}, \ldots) = 0. \end{cases} \tag{14}$$

We firs apply the wave transformation

$$\begin{cases} u(x, t) = v(\phi(\xi)), \ \widetilde{u}(x, t) = \widetilde{v}(\widetilde{\phi}(\xi)). \\ \xi = x + \lambda t + \xi_0, \xi_0 = \text{arbitrary constant.} \end{cases} \tag{15}$$

for a suitable pair of functions $\phi = \phi(\xi)$ and $\widetilde{\phi} = \widetilde{\phi}(\xi)$ in order to obtain a coupled system of two ode's in the form

$$\begin{cases} Q(v, \widetilde{v}, v', \widetilde{v}', v'', \widetilde{v}'', v''', \widetilde{v}''', \ldots) = 0. \\ \widetilde{Q}(v, \widetilde{v}, v', \widetilde{v}', v'', \widetilde{v}'', v''', \widetilde{v}''', \ldots) = 0. \end{cases}, \tag{16}$$

where Q and $\widetilde{Q}$ are *polynomials* w.r.t. the variables $v, \widetilde{v}, v', \widetilde{v}', v'', \widetilde{v}'', v''', \widetilde{v}''', \ldots$.
We seek solutions to system (16) in the forms

$$\begin{cases} v(\xi) = a_0 + \sum_{j=1}^{n} f^{j-1}(a_j f + b_j g). \\ \widetilde{v}(\xi) = \widetilde{a}_0 + \sum_{i=1}^{\widetilde{n}} f^{i-1}(\widetilde{a}_i f + \widetilde{b}_i g). \end{cases} \tag{17}$$

$$\begin{cases} v(\xi) = a_0 + \sum_{j=1}^{n} a_j f^j. \\ \widetilde{v}(\xi) = \widetilde{a}_0 + \sum_{i=1}^{\widetilde{n}} \widetilde{a}_i f^i. \end{cases} \tag{18}$$

$$\begin{cases} v(\xi) = a_0 + \sum_{j=1}^{n} \left(a_j f^j + b_j f^{-j}\right). \\ \widetilde{v}(\xi) = \widetilde{a}_0 + \sum_{i=1}^{\widetilde{n}} \left(\widetilde{a}_i f^i + \widetilde{b}_i f^{-i}\right). \end{cases} \tag{19}$$

The integers n and $\widetilde{n}$ are determined by the balancing method.

We substitute any of these ansätze (17), (18) or (19) into (16) and we obtain a system of two polynomial equations either w.r.t. the variables $f = f(\xi)$ and $g = g(\xi)$ or w.r.t. the variable $f = f(\xi)$. We equate the coefficient of $f^i g^j$ (respectively, the coefficient of f^i) $(i, j = 0, 1, 2, 3, ..)$ to zero and we obtain a system of polynomial equations w.r.t the variables $a_i, b_j, \widetilde{a}_i, \widetilde{b}_j, \lambda, \ldots$. Solving this system with the aid of a symbolic computational package such as Mathematica or Maple, we obtain desired solutions. Sometimes, with the aim to add an additional parameter k, we replace ξ with $k\xi$ and then the corresponding system is regarded w.r.t. the variables $f = f(k\xi)$ and $g = g(k\xi)$, where k =const.

The same technique is applied for solving systems of three or more equations.

3. The Methods

In this section we show how to derive distinct useful methods to fin solutions to nonlinear PDE's. These methods are based on the main idea of previous section.

3.1. Projective Riccati equations method by Conte

Let $N = -1, r = -1$ and $S(f) = R - 2mf + \frac{1}{R}(m^2 + \rho)f^2$, where $\rho = \pm 1$. In this case, system (5) reads

$$\begin{cases} f' = -fg \\ g^2 = R - 2mf + \frac{1}{R}(m^2 + \rho)f^2 \end{cases} \tag{20}$$

From system (20) we obtain following new version of the projective Riccati equations:

$$\begin{cases} f'(\xi) = -f(\xi)g(\xi) \\ g'(\xi) = R - g^2(\xi) + mf(\xi) \end{cases} \tag{21}$$

Eq. (9) may be written as

$$(\varphi')^2 = \varphi^2 S(\varphi^{-1}) = \frac{1}{R}(R^2\varphi^2 - 2mR\varphi + m^2 + \rho). \tag{22}$$

Following are solutions to Eq.(22):

For $\rho = -1$:

$$\varphi_1(\xi) = \frac{1}{R}\left(m + \cosh\left(\sqrt{R}\,\xi\right)\right) \tag{23}$$

For $\rho = 1$:

$$\varphi_2(\xi) = \frac{1}{R}\left(m + \sinh\left(\sqrt{R}\,\xi\right)\right) \tag{24}$$

Therefore, the following are solutions to system (21) :

$$f_1(\xi) = \frac{R}{m + \cosh\left(\sqrt{R}\,\xi\right)}, \tag{25}$$

$$g_1(\xi) = \frac{\sqrt{R}\sinh\left(\sqrt{R}\,\xi\right)}{m + \cosh\left(\sqrt{R}\,\xi\right)}. \tag{26}$$

$$f_2(\xi) = \frac{R}{m + \sinh\left(\sqrt{R}\,\xi\right)}, \tag{27}$$

$$g_2(\xi) = \frac{\sqrt{R}\cosh\left(\sqrt{R}\,\xi\right)}{m + \sinh\left(\sqrt{R}\,\xi\right)}. \tag{28}$$

If we eliminate the variable ξ in the (21), we get the differential equation

$$\frac{df}{dg} = \frac{fg}{g^2 + mf - R} \tag{29}$$

whose general solution is given by

$$g^2 = Cf^2 - 2mf + R,\ C = \text{const.} \tag{30}$$

This last equation is the firs integral to system (21). Its equilibrium points are

$$(0, -\sqrt{R}),\ (0, \sqrt{R}),\ (\tfrac{R}{m}, 0).$$

Some phase portraits of system (21) are shown in Figure 1.
To solve equation (3) we can try to fin solutions by using the expansion

$$v(\xi) = \alpha_0 + \sum_{j=1}^{n} f^{j-1}(\alpha_j f + \beta_j g), \tag{31}$$

The value of n is calculated by the balance procedure. Usually, $n = 2$. With this in mind and using the expressions given by solutions (25)-(28) we will use the following ansatz to solve equation (3)

$$v(\xi) = a_0 + \frac{a_1 + b_1\sinh\left(\sqrt{R}\xi\right)}{m + \cosh\left(\sqrt{R}\xi\right)} + \frac{a_2 + b_2\sinh\left(\sqrt{R}\xi\right)}{\left(m + \cosh\left(\sqrt{R}\xi\right)\right)^2}. \tag{32}$$

$$v(\xi) = c_0 + \frac{c_1 + d_1\cosh\left(\sqrt{R}\xi\right)}{m + \sinh\left(\sqrt{R}\xi\right)} + \frac{c_2 + d_2\cosh\left(\sqrt{R}\xi\right)}{\left(m + \sinh\left(\sqrt{R}\xi\right)\right)^2}. \tag{33}$$

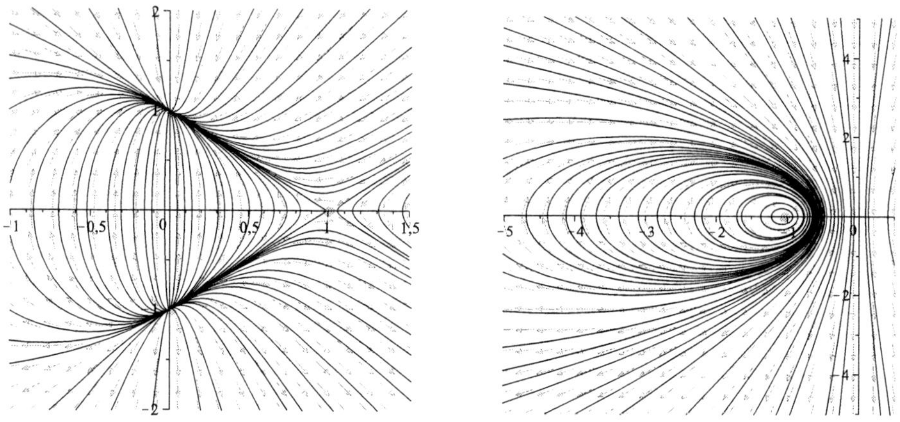

$R > 0$ $\qquad$ $R < 0$

Figure 1.

3.2. The tanh-coth method

Let $N = r = -1$ and $S(f) = (k^2 f - f^{-1})^2$. This choice gives us Eq. (9) in the form

$$(\varphi')^2 = \varphi^2 R(\varphi^{-1}) = \varphi^2 (k^2 \varphi^{-1} - \varphi)^2 = (k^2 - \varphi^2)^2. \tag{34}$$

This last equation gives

$$\varphi' = k^2 - \varphi^2. \tag{35}$$

The general solution to Riccati Eq. (35) is given by

$$\varphi(\xi) = k \coth(k\xi + C), \text{ where } k \neq 0 \text{ and } C = \text{arbitrary constant}. \tag{36}$$

$$\varphi(\xi) = \frac{1}{C - \xi}, \text{ where } k = 0 \text{ and } C = \text{arbitrary constant}. \tag{37}$$

Projective equations are

$$\begin{cases} f'(\xi) = -f(\xi) g(\xi). \\ g^2(\xi) = \left(k^2 f(\xi) - \dfrac{1}{f(\xi)}\right)^2. \end{cases} \tag{38}$$

Solutions to system (38) are given by

$$f(\xi) = \frac{1}{\varphi(\xi)} = \frac{1}{k} \tanh(k\xi) \text{ and } g(\xi) = -\frac{f'(\xi)}{f(\xi)} = -k \text{sech}(k\xi) \text{csch}(k\xi). \tag{39}$$

To solve equation (3) we can try to fin solutions using the expansion

$$v(\xi) = a_0 + \sum_{j=1}^{n} (a_j \tanh^j(k\xi) + b_j \coth^j(k\xi)) \tag{40}$$

or the expansion

$$v(\xi) = a_0 + \sum_{j=1}^{n} \left[a_j(C-\xi)^j + b_j(C-\xi)^{-j}\right]. \tag{41}$$

The value of n is calculated by the balance procedure. Usually, $n = 1, 2$. For this particular choices ansatz (40) and (41) take following forms

$$v(\xi) = a_0 + a_1 \tanh(k\xi) + b_1 \coth(k\xi). \tag{42}$$

$$v(\xi) = a_0 + a_1 \tanh(k\xi) + b_1 \coth(k\xi) + a_2 \tanh^2(k\xi) + b_2 \coth^2(k\xi). \tag{43}$$

$$v(\xi) = a_0 + a_1(C-\xi) + b_1(C-\xi)^{-1}. \tag{44}$$

$$v(\xi) = a_0 + a_1(C-\xi) + b_1(C-\xi)^{-1} + a_2(C-\xi)^2 + b_2(C-\xi)^{-2}. \tag{45}$$

Thus, we obtain the tanh-coth method [2].
If $b_1 = a_1$ the tanh-coth formula (42) reduces to

$$v(\xi) = a_0 + a_1 \tanh(k\xi) + b_1 \coth(k\xi) = a_0 + 2a_1 \tanh\left(2k\xi \pm \frac{i\pi}{2}\right) \tag{46}$$

since

$$\tanh x + \coth x = 2\tanh\left(2x \pm \frac{i\pi}{2}\right) \text{ for all } x. \tag{47}$$

In this case, the tanh-coth method does not give new solutions compared with the tanh method.
We also may try the expansion

$$v(\xi) = a_0 + \sum_{j=1}^{n} \tanh^{j-1}(k\xi)\left[a_j \tanh(k\xi) + b_j \text{sech}(k\xi)\text{csch}(k\xi)\right]. \tag{48}$$

In the case when $n = 1, 2$ this last ansatz takes the forms

$$v(\xi) = a_0 + a_1 \tanh(k\xi) + b_1 \text{sech}(k\xi)\text{csch}(k\xi). \tag{49}$$

$$v(\xi) = a_0 + a_1 \tanh(k\xi) + a_2 \tanh^2(k\xi) + b_1 \text{sech}(k\xi)\text{csch}(k\xi). \tag{50}$$

3.3. He's exp Function Method

Let $N = -1$, $r = 1$ and $S(f) = k^2 =$const. This choice gives us Eq. (9) in the form

$$(\varphi')^2 = \varphi^2 S(\varphi^{-1}) = \varphi^2 k^2 = (k\varphi)^2. \tag{51}$$

From this equation we get, in particular,

$$\varphi' = -k\varphi. \tag{52}$$

Eq. (52) has the particular solution

$$\varphi = \varphi(\xi) = e^{-k\xi}, \;\; f = \varphi^{-1} = e^{k\xi}.$$

Projective equations are

$$\begin{cases} f'(\xi) = f(\xi)g(\xi). \\ g^2(\xi) = k^2. \end{cases} \tag{53}$$

Solutions to (53) are

$$f(\xi) = e^{k\xi}, \ \ g(\xi) = \frac{f'(\xi)}{f(\xi)} = k. \tag{54}$$

He [3] suggested to look for solutions to Eq. (3) in the form

$$v(\xi) = \frac{\sum_{n=-c}^{d} a_n f^n}{\sum_{m=-p}^{q} b_m f^m} = \frac{\sum_{n=-c}^{d} a_n \exp(nk\xi)}{\sum_{m=-p}^{q} b_m \exp(mk\xi)}. \tag{55}$$

Most common variants of this ansatz are :

$$v(\xi) = \frac{a_{-1}\exp(-k\xi) + a_0 + a_1\exp(k\xi)}{1 + b_{-1}\exp(-k\xi) + b_1\exp(k\xi)}. \tag{56}$$

$$v(\xi) = \frac{a_{-1}\exp(-k\xi) + a_0 + a_1\exp(k\xi)}{\exp(-k\xi) + b_1\exp(k\xi)}. \tag{57}$$

$$v(\xi) = \frac{a_{-2}\exp(-2k\xi) + a_{-1}\exp(-k\xi) + a_0 + a_1\exp(k\xi) + a_2\exp(2k\xi)}{b_{-2}\exp(-2k\xi) + b_{-1}\exp(-k\xi) + b_0 + b_1\exp(k\xi) + b_2\exp(2k\xi)}. \tag{58}$$

3.4. New sn-ns Jacobi Elliptic Functions Method

Let $N = -1$, $r = 1$ and $S(f) = af^{-2} + b + cf^2$, where

$$\Delta = b^2 - 4ac > 0. \tag{59}$$

This choice gives us Eq. (9) in the form

$$(\varphi')^2 = a\varphi^4 + b\varphi^2 + c \tag{60}$$

We may express the general solution of this equation in terms of the Jacobi elliptic functions ns or nd as follows :

$$\varphi(\xi) = \pm\sqrt{\frac{-b+\sqrt{\Delta}}{2a}}\,\text{ns}\left(k(\xi + C) \mid m\right), \ C = \text{arbitrary constant.} \tag{61}$$

where

$$k = \sqrt{\frac{-b+\sqrt{\Delta}}{2}} \text{ and } m = \sqrt{\frac{b^2 - 2ac + b\sqrt{\Delta}}{2ac}} \tag{62}$$

Solution (61) is valid for $a > 0$, $b < 0$ and $0 < c \leq \frac{b^2}{4a}$.
On the other hand, function $\xi \rightarrow \sqrt{-1}\text{ns}\big(\sqrt{-1}k(\xi + C) \mid m\big)$ is real valued for any real k, m, ξ and C. We may verify that function

$$\phi(\xi) = \pm\sqrt{-1}\sqrt{\frac{-b+\sqrt{\Delta}}{2a}}\,\text{ns}\left(\sqrt{-1}k(\xi + \xi_0) \mid m\right), \tag{63}$$

where k and m are given by (62), is a solution to equation

$$(\varphi')^2 = a\varphi^4 - b\varphi^2 + c, \ a > 0, b < 0 \text{ and } 0 < c \leq \frac{b^2}{4a}. \tag{64}$$

Thus, we always may fin a solution to equation (60) when $a > 0$ and $0 < c \leq \frac{b^2}{4a}$ for any $b \neq 0$.
Now, let us assume that $a < 0$. It may be verifie that a solution to equation (60) is

$$\varphi(\xi) = \pm\sqrt{\frac{-b-\sqrt{\Delta}}{2a}}\text{nd}\left(k(\xi + C) \mid m\right), \ C = \text{arbitrary constant.} \tag{65}$$

where

$$k = \sqrt{\frac{b-\sqrt{\Delta}}{2}} \text{ and } m = \sqrt{-\frac{b^2 - 4ac + b\sqrt{\Delta}}{2ac}} \text{ for } b \neq 0 \text{ and } c > 0. \tag{66}$$

Observe that for any $A > 0$ function $\xi \to \text{nd}\left(\sqrt{-A}(\xi + C) \mid m\right)$ is real valued for any real values of m, ξ and C. We conclude that equation (60) has exact solutions for any $a \neq 0$ and $b \neq 0$.
Projective equations are

$$\begin{cases} f'(\xi) = f(\xi)g(\xi) \\ g^2(\xi) = \dfrac{a}{f^2(\xi)} + b + cf^2(\xi) \end{cases} \tag{67}$$

Taking $C = 0$ in (61) we see that solutions to system (67) are

$$f(\xi) = \varphi^{-1}(\xi) = \frac{\sqrt{a}}{k}\text{sn}\left(k\xi \mid m\right) \text{ and } g(\xi) = \frac{f'(\xi)}{f(\xi)} = k\text{sn}(k\xi \mid m)\text{cs}(k\xi \mid m)\text{ds}(k\xi \mid m). \tag{68}$$

This motivates us to seek solutions to equation (3) in the form

$$v(\xi) = a_0 + \sum_{j=1}^{n} \left[a_j \text{sn}^j(k\xi) + b_j \text{ns}^j(k\xi)\right]. \tag{69}$$

Usually, $n = 1, 2$ and then $v(\xi)$ has the forms

$$v(\xi) = a_0 + a_1\text{sn}\left(k\xi \mid m\right) + b_1\text{ns}\left(k\xi \mid m\right). \tag{70}$$

$$v(\xi) = a_0 + a_1\text{sn}\left(k\xi \mid m\right) + b_1\text{ns}\left(k\xi \mid m\right) + a_2\text{sn}^2\left(k\xi \mid m\right) + b_2\text{ns}^2\left(k\xi \mid m\right). \tag{71}$$

In the case when $m = 1$ this gives the tanh-coth method, since $\text{sn}\left(k\xi \mid 1\right) = \tanh(k\xi)$ and $\text{ns}(k\xi \mid 1) = \coth(k\xi)$.
Another possible anstatz is

$$v(\xi) = a_0 + \sum_{j=1}^{n} \text{sn}^{j-1}(k\xi)\left[a_j\text{sn}\left(k\xi \mid m\right) + b_j\text{sn}(k\xi \mid m)\text{cs}(k\xi \mid m)\text{ds}(k\xi \mid m)\right]. \tag{72}$$

If $n = 1, 2$, this ansatz reads

$$v(\xi) = a_0 + a_1 \text{sn}\,(k\xi \mid m) + b_1 \text{sn}(k\xi \mid m)\text{cs}(k\xi \mid m)\text{ds}(k\xi \mid m). \tag{73}$$

$$\begin{aligned} v(\xi) &= a_0 + a_1 \text{sn}\,(k\xi \mid m) + b_1 \text{sn}(k\xi \mid m)\text{cs}(k\xi \mid m)\text{ds}(k\xi \mid m) + a_2 \text{sn}^2\,(k\xi \mid m) \\ & b_2 \text{sn}^2\,(k\xi \mid m)\,\text{cs}(k\xi \mid m)\text{ds}(k\xi \mid m). \end{aligned} \tag{74}$$

We may consider similar ansätze by replacing sn by dn and ns by nd, respectively. We will call this the dn-nd method. Thus, two possible ansätze for this method are

$$v(\xi) = a_0 + a_1 \text{dn}\,(k\xi \mid m) + b_1 \text{nd}\,(k\xi \mid m)\,. \tag{75}$$

and

$$v(\xi) = a_0 + a_1 \text{dn}\,(k\xi \mid m) + b_1 \text{nd}\,(k\xi \mid m) + a_2 \text{dn}^2\,(k\xi \mid m) + b_2 \text{nd}^2\,(k\xi \mid m)\,. \tag{76}$$

Another useful ansatz to handle other equations of the form (3) is

$$v(\xi) = \frac{a_0 + a_1 \text{cn}(k\xi \mid m)}{1 + b_1 \text{cn}(k\xi \mid m)} \tag{77}$$

For example, this ansatz may be successfully applied to the cubic-quintic Duffin equation, which is define by

$$v''(\xi) + pv(\xi) + qv^3(\xi) + rv^5(\xi) = 0. \tag{78}$$

3.5. Weierstrass Elliptic Functions Method

Let $N = -1,\ r = -1$ and $S(f) = 4f^{-1} - g_2 f - g_3 f^2$ (g_2 =const and g_3 =const). In this case, Eq. (9) takes the form

$$(\varphi')^2 = 4\varphi^3 - g_2\varphi - g_3 \tag{79}$$

Solution to Eq. (79) is given by the Weierstrass $\wp$ function

$$\varphi = \wp = \wp(\xi) = \wp(\xi, g_2, g_3\}). \tag{80}$$

Projective equations are

$$\begin{cases} f'(\xi) = -f(\xi)g(\xi) \\ g^2(\xi) = \dfrac{4}{f(\xi)} - g_2 f(\xi) - g_3 f^2(\xi)\,. \end{cases} \tag{81}$$

Solutions to system (81) are

$$f(\xi) = \frac{1}{\varphi(\xi)} = \frac{1}{\wp(\xi)} \text{ and } g(\xi) = -\frac{f'(\xi)}{f(\xi)} = \frac{\wp'(\xi)}{\wp(\xi)}. \tag{82}$$

This motivates us to seek solutions to Eq. (3) in the form

$$v(\xi) = a_0 + \sum_{j=1}^{n} f^{j-1}(a_j f + b_j g). \tag{83}$$

Thus, if $n = 1$ ansatz (83) reduces to

$$v(\xi) = a_0 + \frac{a_1}{\wp(\xi)} + b_1 \frac{\wp'(\xi)}{\wp(\xi)}. \tag{84}$$

On the other hand, if $n = 2$ we seek solutions in the form

$$v(\xi) = a_0 + \frac{a_1 + b_1\wp'(\xi)}{\wp(\xi)} + \frac{a_2 + b_2\wp'(\xi)}{\wp^2(\xi)}, \tag{85}$$

We also may try following ansatz :

$$v(\xi) = a_0 + \sum_{j=1}^{n} \left[a_j\wp^j(\xi) + b_j\wp^{-j}(\xi)\right] \tag{86}$$

In the case when $n = 1, 2$ ansatz (86) takes the forms

$$v(\xi) = a_0 + a_1\wp(\xi) + \frac{b_1}{\wp(\xi)}. \tag{87}$$

$$v(\xi) = a_0 + a_1\wp(\xi) + \frac{b_1}{\wp(\xi)} + a_2\wp^2(\xi) + \frac{b_2}{\wp^2(\xi)}. \tag{88}$$

4. Examples

In this section we solve various important models related to nonlinear science by the methods described in previous sections.

4.1. Gardner Equation

This equation is also called KdV-mKdV equation and it reads

$$u_t + \alpha u u_x + \beta u^2 u_x + \gamma u_{xxx} = 0, \;\; \beta\gamma \neq 0. \tag{89}$$

Gardner equation is widely used in various branches of physics, such as plasma physics, flui physics, quantum fiel theory. It also describes a variety of wave phenomena in plasma and solid state. Studies of various physical structures of nonlinear equations had attracted much attention in connection with the important problems that arise in scientifi applications.

Transformation (2) converts this equation to

$$\lambda v' + \alpha v v' + \beta v^2 v' + \gamma v''' = 0. \tag{90}$$

We will solve Eq. (90) by the sn-ns method. Balancing gives $n = 1$. We seek solutions in the form

$$v(\xi) = a_0 + a_1 \text{sn}\,(k\xi \mid m) + b_1 \text{ns}\,(k\xi \mid m). \tag{91}$$

Substituting (91) into (90) and equating to zero the coefficient of the different powers of the Jacobi elliptic functions sn and ns yields following system :

$$\begin{cases} a_1^2 k\left(2a_0\beta+\alpha\right)=0. \\ a_1 k\left(a_1^2\beta+6k^2m^2\gamma\right)=0. \\ a_1\left(-a_1b_1k\beta-a_0k\alpha-a_0^2k\beta+k^3m^2\gamma+k^3\gamma+\lambda\right)=0. \\ b_1^2 k\left(2a_0\beta+\alpha\right)=0. \\ b_1\left(-a_1b_1k\beta-a_0k\alpha-a_0^2k\beta+k^3m^2\gamma+k^3\gamma+\lambda\right)=0. \\ b_1 k\left(b_1^2\beta+6k^2\gamma\right)=0. \end{cases} \tag{92}$$

Solving this system gives

- $a_0=-\frac{\alpha}{2\beta}$, $a_1=\pm\frac{km\sqrt{-6\beta\gamma}}{\beta}$, $b_1=0$, $\lambda=\frac{\alpha^2+4\beta\gamma\left(m^2+1\right)k^2}{4\beta}$:

$$\begin{aligned} u(x,t) &= -\frac{\alpha}{2\beta}\pm\frac{km\sqrt{-6\beta\gamma}}{\beta}\text{sn}\left(k\xi|m\right), \\ \xi &= x+\frac{\alpha^2+4\beta\gamma\left(m^2+1\right)k^2}{4\beta}t+\xi_0. \end{aligned} \tag{93}$$

- $a_0=-\frac{\alpha}{2\beta}$, $a_1=0$, $b_1=\pm\frac{k\sqrt{-6\beta\gamma}}{\beta}$, $\lambda=\frac{\alpha^2+4\beta\gamma\left(m^2+1\right)k^2}{4\beta}$:

$$\begin{aligned} u(x,t) &= -\frac{\alpha}{2\beta}\pm\frac{k\sqrt{-6\beta\gamma}}{\beta}\text{ns}\left(k\xi|m\right), \\ \xi &= x+\frac{\alpha^2+4\beta\gamma\left(m^2+1\right)k^2}{4\beta}t+\xi_0. \end{aligned} \tag{94}$$

- $a_0=-\frac{\alpha}{2\beta}$, $a_1=\frac{km\sqrt{-6\beta\gamma}}{\beta}$, $b_1=\pm\frac{k\sqrt{-6\beta\gamma}}{\beta}$, $\lambda=\frac{\alpha^2+4\beta\gamma\left(m^2+6m+1\right)k^2}{4\beta}$:

$$\begin{aligned} u(x,t) &= -\frac{\alpha}{2\beta}+\frac{km\sqrt{-6\beta\gamma}}{\beta}\text{sn}\left(k\xi|m\right)\pm\frac{k\sqrt{-6\beta\gamma}}{\beta}\text{ns}\left(k\xi|m\right), \\ \xi &= x+\frac{\alpha^2+4\beta\gamma\left(m^2+6m+1\right)k^2}{4\beta}t+\xi_0. \end{aligned} \tag{95}$$

- $a_0=-\frac{\alpha}{2\beta}$, $a_1=-\frac{km\sqrt{-6\beta\gamma}}{\beta}$, $b_1=\pm\frac{k\sqrt{-6\beta\gamma}}{\beta}$, $\lambda=\frac{\alpha^2+4\beta\gamma\left(m^2\mp 6m+1\right)k^2}{4\beta}$:

$$\begin{aligned} u(x,t) &= -\frac{\alpha}{2\beta}-\frac{km\sqrt{-6\beta\gamma}}{\beta}\text{sn}\left(k\xi|m\right)\pm\frac{k\sqrt{-6\beta\gamma}}{\beta}\text{ns}\left(k\xi|m\right), \\ \xi &= x+\frac{\alpha^2+4\beta\gamma\left(m^2\mp 6m+1\right)k^2}{4\beta}t+\xi_0. \end{aligned} \tag{96}$$

Observe that. since $\text{sn}(\xi|0)=\sec\xi$, $\text{ns}(x|0)=\csc(x)$, $\text{sn}(\xi|1)=\tanh(\xi)$ and $\text{ns}(\xi|1)=\coth(\xi)$, we obtain trigonometric and hyperbolic solutions when $m\to 0$ and $m\to 1$, respectively. They are :

$$u(x,t)=-\frac{\alpha}{2\beta}\pm\frac{k\sqrt{-6\beta\gamma}}{\beta}\csc(k\xi),\ \xi=x+\frac{\alpha^2+4\beta\gamma k^2}{4\beta}t+\xi_0. \tag{97}$$

$$u(x,t) = -\frac{\alpha}{2\beta} \pm \frac{k\sqrt{-6\beta\gamma}}{\beta} \sec(k\xi), \xi = x + \frac{\alpha^2 + 4\beta\gamma k^2}{4\beta} t + \xi_0. \quad (98)$$

$$u(x,t) = -\frac{\alpha}{2\beta} \pm \frac{k\sqrt{-6\beta\gamma}}{\beta} \tanh(k\xi), \xi = x + \frac{\alpha^2 + 8\beta\gamma k^2}{4\beta} t + \xi_0. \quad (99)$$

$$u(x,t) = -\frac{\alpha}{2\beta} \pm \frac{k\sqrt{-6\beta\gamma}}{\beta} \coth(k\xi), \xi = x + \frac{\alpha^2 + 8\beta\gamma k^2}{4\beta} t + \xi_0.$$

$$\begin{aligned} u(x,t) &= -\frac{\alpha}{2\beta} + \frac{k\sqrt{-6\beta\gamma}}{\beta} \tanh(k\xi) \pm \frac{k\sqrt{-6\beta\gamma}}{\beta} \coth(k\xi), \\ \xi &= x + \frac{\alpha^2 + 32\beta\gamma k^2}{4\beta} t + \xi_0. \end{aligned} \quad (100)$$

$$\begin{aligned} u(x,t) &= -\frac{\alpha}{2\beta} - \frac{k\sqrt{-6\beta\gamma}}{\beta} \tanh(k\xi) + \frac{k\sqrt{-6\beta\gamma}}{\beta} \coth(k\xi), \\ \xi &= x + \frac{\alpha^2 - 16\beta\gamma k^2}{4\beta} t + \xi_0. \end{aligned} \quad (101)$$

$$\begin{aligned} u(x,t) &= -\frac{\alpha}{2\beta} - \frac{k\sqrt{-6\beta\gamma}}{\beta} \tanh(k\xi) - \frac{k\sqrt{-6\beta\gamma}}{\beta} \coth(k\xi), \\ \xi &= x + \frac{\alpha^2 + 32\beta\gamma k^2}{4\beta} t + \xi_0. \end{aligned} \quad (102)$$

Solutions (94)- (102) are valid for $\beta\gamma < 0$. If $\beta\gamma > 0$ we seek solutions by the dn-nd method, i.e. we look for solutions in the form (75). Two of these solutions are

$$\begin{aligned} u(x,t) &= -\tfrac{\alpha}{2\beta} - k\sqrt{\frac{6\gamma}{\beta}} \mathrm{dn}(k\xi|m), \\ \xi &= x + \left(\frac{\alpha^2}{4\beta} + k^2\gamma(m^2-2)\right) t + \xi_0. \end{aligned} \quad (103)$$

$$\begin{aligned} u(x,t) &= -\tfrac{\alpha}{2\beta} + k\sqrt{\frac{6\gamma(1-m^2)}{\beta}} \mathrm{nd}(k\xi|m), \\ \xi &= x + \left(\frac{\alpha^2}{4\beta} + k^2\gamma(m^2-2)\right) t + \xi_0. \end{aligned} \quad (104)$$

Letting $m \to 1$ we see that solution (103) degenerates into

$$\begin{aligned} u(x,t) &= -\frac{\alpha}{2\beta} - k\sqrt{\frac{6\gamma}{\beta}} \mathrm{sech}(k\xi), \\ \xi &= x + \left(\frac{\alpha^2}{4\beta} - k^2\gamma\right) t + \xi_0. \end{aligned} \quad (105)$$

Letting $\alpha = 0$ we obtain solutions to mKdV equation, which reads

$$u_t + \beta u^2 u_x + \gamma u_{xxx} = 0, \ \beta\gamma \neq 0. \quad (106)$$

We could think that Gardner equation is more general than mKdV equation. However, these equations are in essence the same. Indeed, let us apply the similarity transformations

$$\widetilde{t}=\gamma t,\quad \widetilde{x}=x+\frac{\alpha^2}{4\beta}t,\quad \widetilde{u}=\sqrt{\frac{|\beta|}{|\gamma|}}\left(u+\frac{\alpha}{2\beta}\right) \tag{107}$$

to Gardner equation (89). We obtain the mKdV equation

$$\widetilde{u}_{\widetilde{t}}+\varepsilon\widetilde{u}_{\widetilde{x}}^2\widetilde{u}+\gamma\widetilde{u}_{\widetilde{x}\widetilde{x}\widetilde{x}}=0, \tag{108}$$

where ε =sign$(\beta\gamma)=\pm1$.
This means that if $\widetilde{u}=\widetilde{u}(\widetilde{t},\widetilde{x})$ is a solution to mKdV equation, then function define by

$$u=u(t,x)=\sqrt{\frac{|\gamma|}{|\beta|}}\widetilde{u}\left(\gamma t,x+\frac{\alpha^2}{4\beta}t\right)-\frac{\alpha}{2\beta} \tag{109}$$

is a solution to Gardner equation (89). For other solutions to MkdV equation, see [13].

4.2. Sawada-Kotera Equation

This equation is also called Caudrey-Dodd-Gibbon equation and reads [34].

$$u_t+u_{xxxxx}+5uu_{xxx}+5u_xu_{xx}+5u^2u_x=0. \tag{110}$$

Transformation (2) gives this equation in the ode form

$$\lambda u'(\xi)+5u^2(\xi)u'(\xi)+5k^2u'(\xi)u''(\xi)+5k^2u(\xi)u'''(\xi)+k^4u'''''(\xi)=0. \tag{111}$$

Balancing gives $n=2$. We seek solutions to this equation in the ansatz form (71), i.e.

$$v(\xi)=a_0+a_1\text{sn}\,(k\xi\mid m)+b_1\text{ns}\,(k\xi\mid m)+a_2\text{sn}^2\,(k\xi\mid m)+b_2\text{ns}^2\,(k\xi\mid m)\,. \tag{112}$$

Inserting (112) into (111) and equating to zero the coefficient of the different powers of the Jacobi elliptic functions sn and ns yields a large algebraic system. Solving it we obtain following solutions :

- $\lambda=-16m^4k^4-44m^2k^4-16k^4+20m^2a_0k^2+20a_0k^2-5a_0^2$, $a_1=0$, $a_2=-6k^2m^2$, $b_1=0$, $b_2=0$:

$$u(t,x)=a_0-6k^2m^2\text{sn}^2\left(k\left(x+\lambda t+\xi_0\right)|m\right). \tag{113}$$

- $\lambda=-16m^4k^4-44m^2k^4-16k^4+20m^2a_0k^2+20a_0k^2-5a_0^2$, $a_1=0$, $a_2=0$, $b_1=0$, $b_2=-6k^2$:

$$u(t,x)=a_0-6k^2\text{ns}^2\left(k\left(x+\lambda t+\xi_0\right)|m\right). \tag{114}$$

- $\lambda=-16k^4\left(m^4-m^2+1\right)$, $a_0=4k^2\left(m^2+1\right)$, $a_1=0$, $a_2=-12k^2m^2$, $b_1=0$, $b_2=0$:

$$u(t,x)=4k^2\left(m^2+1\right)-12k^2m^2\text{sn}^2\left(k\left(x+\lambda t+\xi_0\right)|m\right). \tag{115}$$

- $\lambda = -16k^4(m^4 - m^2 + 1)$, $a_0 = 4k^2(m^2+1)$, $a_1 = 0$, $a_2 = 0$, $b_1 = 0$, $b_2 = -12k^2$:

$$u(t,x) = 4k^2(m^2+1) - 12k^2\text{ns}^2(k(x+\lambda t+\xi_0)|m). \quad (116)$$

- $\lambda = -16k^4(m^4 + 14m^2 + 1)$, $a_0 = 4k^2(m^2+1)$, $a_1 = 0$, $a_2 = -12k^2m^2$, $b_1 = 0$, $b_2 = -12k^2$:

$$u(t,x) = 4k^2(m^2+1) - 12k^2m^2\text{sn}^2(k(x+\lambda t+\xi_0)|m) - 12k^2\text{ns}^2(k(x+\lambda t+\xi_0)|m). \quad (117)$$

- $\lambda = -16m^4k^4 + 16m^2k^4 - 16k^4 + 20m^2a_0k^2 + 20a_0k^2 - 5a_0^2$, $a_1 = 0$, $a_2 = -6k^2m^2$, $b_1 = 0$, $b_2 = -6k^2$:

$$u(t,x) = a_0 - 6k^2m^2\text{sn}^2(k(x+\lambda t+\xi_0)|m) - 6k^2\text{ns}^2(k(x+\lambda t+\xi_0)|m). \quad (118)$$

Letting $m \to 1$ we obtain hyperbolic solutions in terms on $\tanh$ and $\coth$ functions. They are :

- $\lambda = 40a_0k^2 - 5a_0^2 - 76k^4, a_1 = 0, a_2 = -6k^2, b_1 = 0, b_2 = 0$:

$$u(t,x) = a_0 - 6k^2\tanh^2\left(k\left(x + \left(40a_0k^2 - 5a_0^2 - 76k^4\right)t + \xi_0\right)\right). \quad (119)$$

- $\lambda = -16k^4, a_0 = 8k^2, a_1 = 0, a_2 = -12k^2, b_1 = 0, b_2 = 0$:

$$u(t,x) = 8k^2 - 12k^2\tanh^2\left(k\left(x - 16k^4t\right)\right). \quad (120)$$

- $\lambda = -16k^4, a_0 = 8k^2, a_1 = 0, a_2 = 0, b_1 = 0, b_2 = -12k^2$:

$$u(t,x) = 8k^2 - 12k^2\coth^2\left(k\left(x - 16k^4t\right)\right). \quad (121)$$

- $\lambda = 40a_0k^2 - 5a_0^2 - 16k^4, a_1 = 0, a_2 = -6k^2, b_1 = 0, b_2 = -6k^2$:

$$u(t,x) = a_0 - 6k^2\left[\tanh^2(k(x+\lambda t+\xi_0)) + \coth^2(k(x+\lambda t+\xi_0))\right]. \quad (122)$$

- $\lambda = -256k^4, a_0 = 8k^2, a_1 = 0, a_2 = -12k^2, b_1 = 0, b_2 = -12k^2$:

$$u(t,x) = 8k^2 - 12k^2\left[\tanh^2\left(k\left(x - 256k^4t + \xi_0\right)\right) + \coth^2\left(k\left(x - 256k^4t + \xi_0\right)\right)\right]. \quad (123)$$

We also obtain other solutions by using ansatz (74). This gives following solutions :

- $\lambda = -(m^4 - m^2 + 1)k^4 + 5(m^2+1)k^2a_0 - 5a_0^2$, $a_1 = 0$, $a_2 = -3k^2m^2$, $b_1 = 0$, $b_2 = \pm 3k^2m$:

$$u(t,x) = a_0 \pm 3k^2m\,\text{cn}(\xi|m)\text{dn}(\xi|m) - 3k^2m^2\text{sn}^2(\xi|m), \xi = k(x+\lambda t+\xi_0). \quad (124)$$

- $\lambda = -k^4\left(m^4+14m^2+1\right)$, $a_0 = k^2\left(m^2+1\right)$, $a_1 = 0, a_2 = -6k^2m^2$, $b_1 = 0, b_2 = \pm 6k^2m$:

$$u(t,x) = k^2\left(m^2+1\right) \pm 6k^2m\mathbf{cn}(\xi|m)\mathbf{dn}(\xi|m) - 6k^2m^2\mathbf{sn}^2(\xi|m),\ \xi = k\left(x+\lambda t+\xi_0\right). \tag{125}$$

- $\lambda = -k^4\left(61m^4+119m^2+61\right)$, $a_0 = 4k^2\left(m^2+1\right)$, $a_1 = 0$, $a_2 = -3k^2m^2$, $b_1 = 0$, $b_2 = \pm 3k^2m$:

$$u(t,x) = 4k^2\left(m^2+1\right) \pm 3k^2m\mathbf{cn}(\xi|m)\mathbf{dn}(\xi|m) - 3k^2m^2\mathbf{sn}^2(\xi|m), \xi = k\left(x+\lambda t+\xi_0\right). \tag{126}$$

4.3. Duffing Equation $v''(\xi)+pv(\xi)+qv^3(\xi)=0$.

Let us consider the equation

$$v''(\xi)+pv(\xi)+qv^3(\xi)=0, \tag{127}$$

where p and q are non-zero constants. This equation is very important, since some relevant physical models described by a nonlinear pdes may be studied once this equation is solved. Two of them are related to cubic nonlinear Schrodinger equation and Klein-Gordon-Zakharov equations.

To fin solutions we multiply equation (127) by $v'(\xi)$ and we integrate it w.r.t. ξ. The resulting equation is

$$\left(\frac{dv}{d\xi}\right)^2 = -\frac{q}{2}v^4(\xi) - pv^2(\xi) - 2C, \tag{128}$$

where C is the constant of integration. This equation has form (60) and we already know that there exists an exact solution to it for any p and q. Instead, we may apply directly the sn-ns method (respectively, the dn-nd method) to it. Balancing gives $n=1$. We seek solutions in the form

$$v(\xi) = a_0 + a_1\mathbf{sn}\left(k\xi \mid m\right) + b_1\mathbf{ns}\left(k\xi \mid m\right) \tag{129}$$

Inserting (129) into (127) we obtain a polynomial equation w.r.t. the variable ζ =sn$(k\xi \mid m)$. Equating to zero the coefficient of ζ^j $(j = 0,1,2,..)$ yields following algebraic system :

$$\begin{cases} 3a_0a_1^2q = 0. \\ a_1\left(a_1^2q+2k^2m^2\right) = 0. \\ a_1b_1\left(3a_1b_1q+3a_0^2q-k^2m^2-k^2+p\right) = 0. \\ a_0\left(6a_1b_1q+a_0^2q+p\right) = 0. \\ b_1\left(b_1^2q+2k^2\right) = 0. \end{cases} \tag{130}$$

Solving system (130) gives solutions as follows :

- $a_0 = 0, a_1 = \frac{\sqrt{-2pm}}{\sqrt{\left(m^2+6m+1\right)q}}$, $b_1 = \frac{\sqrt{-2p}}{\sqrt{\left(m^2+6m+1\right)q}}$, $k = \frac{\sqrt{p}}{\sqrt{m^2+6m+1}}$:

$$v(\xi) = \pm\frac{\sqrt{-2p}}{\sqrt{\left(m^2+6m+1\right)q}}\left(m\ \mathbf{sn}\left(\frac{\sqrt{p}}{\sqrt{m^2+6m+1}}\xi \mid m\right) + \mathbf{ns}\left(\frac{\sqrt{p}}{\sqrt{m^2+6m+1}}\xi \mid m\right)\right). \tag{131}$$

- $a_0 = 0, a_1 = \frac{\sqrt{-2pm}}{\sqrt{(m^2-6m+1)q}}, b_1 = -\frac{\sqrt{-2p}}{\sqrt{(m^2-6m+1)q}}, k = \frac{\sqrt{p}}{\sqrt{m^2-6m+1}}$:

$$v(\xi) = \pm\frac{\sqrt{-2p}}{\sqrt{(m^2-6m+1)\,q}}\left(m\,\mathrm{sn}\left(\frac{\sqrt{p}}{\sqrt{m^2-6m+1}}\xi \mid m\right) - \mathrm{ns}\left(\frac{\sqrt{p}}{\sqrt{m^2-6m+1}}\xi \mid m\right)\right). \tag{132}$$

- $a_0 = 0, a_1 = \frac{m\sqrt{-2p}}{\sqrt{q(m^2+1)}}, b_1 = 0, k = \frac{\sqrt{p}}{\sqrt{m^2+1}}$:

$$v(\xi) = \pm m\sqrt{-\frac{2p}{q(m^2+1)}}\,\mathrm{sn}\left(\sqrt{\frac{p}{m^2+1}}\xi \mid m\right). \tag{133}$$

- $a_0 = 0,\ a_1 = 0,\ b_1 = \frac{\sqrt{-2p}}{\sqrt{q(m^2+1)}}, k = \frac{\sqrt{p}}{\sqrt{m^2+1}}$:

$$v(\xi) = \pm\frac{\sqrt{-2p}}{\sqrt{q(m^2+1)}}\mathrm{ns}\left(\frac{\sqrt{p}}{\sqrt{m^2+1}}\xi \mid m\right). \tag{134}$$

Letting $m \to 1$ we obtain trigonometric and hyperbolic solutions :

$$v(\xi) = \sqrt{-\frac{p}{2q}}\tanh\left(\frac{\sqrt{p}}{2\sqrt{2}}\xi\right) + \sqrt{-\frac{p}{2q}}\coth\left(\frac{\sqrt{p}}{2\sqrt{2}}\xi\right). \tag{135}$$

$$v(\xi) = -\sqrt{-\frac{p}{2q}}\tan\left(\frac{\sqrt{p}}{2}\xi\right) - \sqrt{-\frac{p}{2q}}\cot\left(\frac{\sqrt{p}}{2}\xi\right) = \sqrt{-\frac{2p}{q}}\csc(\sqrt{p}\xi). \tag{136}$$

$$v(\xi) = \pm\sqrt{-\frac{p}{q}}\coth\left(\sqrt{\frac{p}{2}}\xi\right), \tag{137}$$

$$v(\xi) = \pm\sqrt{-\frac{p}{q}}\tanh\left(\sqrt{\frac{p}{2}}\xi\right). \tag{138}$$

4.4. Cubic Nonlinear Schrodinger Equation

This equation reads

$$iu_t + u_{xx} + \mu\,|u|^2\,u = 0, \tag{139}$$

where $u = u(x,t)$ is a complex-valued function of two real variables x and t and μ is a non-zero real parameter and $i = \sqrt{-1}$. The physical model of the cubic nonlinear Schrodinger equation (shortly, NLS equation) equation (139) and its generalized variants occur in various areas of physics such as nonlinear optics, water waves, plasma physics, quantum mechanics, superconductivity and Bose–Einstein condensate theory. It also has applications in optics since it models many nonlinearity effects in a fibe , including but not limited to self-phase modulation, four-wave mixing, second harmonic generation, stimulated Raman scattering, etc. For water waves, the NLS equation (139) describes the evolution of the envelope of modulated nonlinear wave groups. All these physical phenomena can be better understood with the help of exact solutions for a given μ. When $\mu > 0$, equation is said to be attractive. If $\mu < 0$ we say that it is repulsive. Recently, Ma and Chen [6] obtained some solutions to equation (139).

We seek solutions to Eq. (139) in the form

$$u(x,t) = v(\xi)\exp(i(\alpha x + \gamma t)),\ \xi = x - 2\alpha t + \xi_0 \tag{140}$$

where α and γ are some real constants to be determined.. Insering (140) into (139) and simplifying we obtain

$$v''(\xi) - (\gamma + \alpha^2)v(\xi) + \mu v^3(\xi) = 0 \tag{141}$$

This last equation has the form (127) with

$$p = -(\gamma + \alpha^2) \text{ and } q = \mu. \tag{142}$$

Thus, making use of solutions (131)- (138) for the choices given by (142) we obtain exact solutions to Schrodinger equation (139) in the form (140).

4.5. Klein-Gordon-Zakharov Equations

These equations are define by the following system of coupled nonlinear pde's [7] :

$$\begin{cases} u_{tt} - c_0^2\nabla^2 u + f_0^2 u + \delta u w = 0. \\ w_{tt} - c_0^2\nabla^2 w - b\nabla^2|u|^2 = 0. \end{cases}, \tag{143}$$

where $u(x,y,z,t)$ and $w(x,y,z,t)$ are differentiable functions, x, y, z, t are independent variables and c_0, f_0, δ, b are arbitrary parameters. In order to solve (143), we firs consider the wave transformation

$$\begin{cases} u(x,y,z,t) = v(\xi)e^{i(x+y+z+\sigma t)} \\ w(x,y,z,t) = w(\xi) \\ \xi = x + y + z + \lambda t, \end{cases} \tag{144}$$

where $v(\xi)$ and $w(\xi)$ are real-valued functions. Inserting (144) into (143) and choosing

$$\sigma = \frac{3c_0^2}{\lambda}, \tag{145}$$

the system of Klein-Gordon-Zakharov equations may be written in the form

$$\begin{cases} v''(\xi) + Av(\xi) + Bw(\xi)v(\xi) = 0. \\ w''(\xi) + C(v^2(\xi))'' = 0. \end{cases}, \tag{146}$$

where

$$\begin{cases} A = \dfrac{\lambda^2 f_0^2 + 3c_0^2\lambda^2 - 9c_0^4}{\lambda^2(\lambda^2 - 3c_0^2)} \\ B = \dfrac{\delta}{\lambda^2 - 3c_0^2} \\ C = -\dfrac{3b}{\lambda^2 - 3c_0^2} \\ \lambda^2 - 3c_0^2 \neq 0. \end{cases} \tag{147}$$

Integrating the second equation in (146) once with respect to ξ and setting the constant of integration equal to zero gives

$$w'(\xi) + C(v^2(\xi))' = 0. \tag{148}$$

Integrating once more we obtain

$$w(\xi) = D - Cv^2(\xi), \tag{149}$$

where D is the constant of integration.
Substituting (149) into firs equation in (146), using (147) and after some simplification we obtain the elliptic-like equation

$$v''(\xi) + \frac{3c_0^2(\lambda^2 - 3c_0^2) + \lambda^2(f_0^2 + D\delta)}{\lambda^2\left(\lambda^2 - 3c_0^2\right)}v(\xi) + \frac{3b\delta\lambda^2}{\lambda^2\left(\lambda^2 - 3c_0^2\right)^2}v(\xi)^3 = 0. \tag{150}$$

This last equation has the form (127) with

$$p = \frac{3c_0^2(\lambda^2 - 3c_0^2) + \lambda^2(f_0^2 + D\delta)}{\lambda^2\left(\lambda^2 - 3c_0^2\right)} \quad \text{and} \quad q = \frac{3b\delta\lambda^2}{\lambda^2\left(\lambda^2 - 3c_0^2\right)^2}. \tag{151}$$

In consequence, making use of solutions (131)- (138) for the choices given by (151) we obtain analytic solutions to Klein-Gordon-Zakharov equations (146) from (144), (145) and (149).

4.6. Quadratic Duffing Equation $v''(\xi) = pv^2(\xi) + qv(\xi) + r$

Let us consider following second order and second degree nonlinear ode :

$$v''(\xi) = pv^2(\xi) + qv(\xi) + r, \tag{152}$$

where p, q and r are constants and $p \neq 0$. Solutions to this equation may be used to study some important physical models whose associated pde's may be solved after making the traveling wave transformation (2). As we will show in next subsections some examples of nonlinear partial differential equations where this equation arises are :

- KdV equation : $u_t + 6uu_x + u_{xxx} = 0$.
- Boussinesq equation : $u_{tt} + \alpha uu_{xx} + \alpha u_x^2 + \beta u_{xxxx} = 0$.
- Symmetric regular long wave equation : $u_{tt} + u_{xx} + uu_{xt} + u_x u_t + u_{xxtt} = 0$.
- Generalized shallow water wave equation : $u_{xxtt} + \alpha u_x u_{xt} + \beta u_t u_{xx} - u_{xt} - u_{xx} = 0$.
- Klein–Gordon equation with quadratic nonlinearity : $u_{tt} - \alpha^2 u_{xx} + \beta u - \gamma u^2 = 0$.

Balancing gives $n = 2$. We seek solutions to Eq. (152) in the form (71), i.e.

$$v(\xi) = a_0 + a_1\text{sn}\,(k\xi \mid m) + b_1\text{ns}\,(k\xi \mid m) + a_2\text{sn}^2\,(k\xi \mid m) + b_2\text{ns}^2\,(k\xi \mid m)\,. \tag{153}$$

Inserting this ansatz into (152) gives following algebraic system :

$$\begin{cases} a_1\left(k^2m^2-a_2p\right)=0. \\ a_2\left(6k^2m^2-a_2p\right)=0. \\ 4a_2k^2m^2+4a_2k^2+a_1^2p+2a_0a_2p+a_2q=0. \\ 2a_2b_1p+a_1k^2m^2+a_1k^2+2a_0a_1p+a_1q=0. \\ b_1\left(k^2-b_2p\right)=0, \\ b_2\left(6k^2-b_2p\right)=0. \\ 2a_0b_2p+4b_2k^2m^2+4b_2k^2+b_1^2p+b_2q=0. \\ 2a_0b_1p+2a_1b_2p+b_1k^2m^2+b_1k^2+b_1q=0. \\ 2a_1b_1p+2a_2b_2p-2a_2k^2+a_0^2p+a_0q-2b_2k^2m^2+r=0. \end{cases} \tag{154}$$

Solving this system we obtain following solutions :

- $k=\frac{1}{2}\sqrt[4]{\frac{q^2-4pr}{m^4-m^2+1}}, a_0=\frac{q}{2p}-\frac{1}{2p}\left(m^2+1\right)\sqrt{\frac{q^2-4pr}{m^4-m^2+1}}, a_1=0, a_2=\frac{3m^2}{2p}\sqrt{\frac{q^2-4pr}{m^4-m^2+1}},$
  $b_1=0, b_2=0:$

$$v(\xi)=a_0+\frac{3m^2}{2p}\sqrt{\frac{q^2-4pr}{m^4-m^2+1}}\text{sn}^2\left(\frac{1}{2}\sqrt[4]{\frac{q^2-4pr}{m^4-m^2+1}}\xi\mid m\right). \tag{155}$$

- $k=\frac{1}{2}\sqrt[4]{\frac{q^2-4pr}{m^4-m^2+1}}, a_0=-\frac{q}{2p}-\frac{1}{2p}\left(m^2+1\right)\sqrt{\frac{q^2-4pr}{m^4-m^2+1}}, a_1=0, a_2=0,$
  $b_1=0, b_2=\frac{3}{2p}\sqrt{\frac{q^2-4pr}{m^4-m^2+1}}:$

$$v(\xi)=a_0+\frac{3}{2p}\sqrt{\frac{q^2-4pr}{m^4-m^2+1}}\text{ns}^2\left(\frac{1}{2}\sqrt[4]{\frac{q^2-4pr}{m^4-m^2+1}}\xi\mid m\right). \tag{156}$$

- $k=\frac{1}{2}\sqrt[4]{\frac{q^2-4pr}{m^4+14m^2+1}}, a_0=-\frac{q}{2p}-\frac{1}{2p}\left(m^2+1\right)\sqrt{\frac{q^2-4pr}{m^4-m^2+1}}, a_1=0, a_2=\frac{3m^2}{2p}\sqrt{\frac{q^2-4pr}{m^4+14m^2+1}},$
  $b_1=0, b_2=\frac{3}{2p}\sqrt{\frac{q^2-4pr}{m^4+14m^2+1}}:$

$$\begin{aligned} v(\xi) &= a_0+\frac{3}{2p}\sqrt{\frac{q^2-4pr}{m^4+14m^2+1}}(3m^2\text{sn}^2\left(\frac{1}{2}\sqrt[4]{\frac{q^2-4pr}{m^4+14m^2+1}}\xi\mid m\right)+ \\ &\quad \text{ns}^2\left(\frac{1}{2}\sqrt[4]{\frac{q^2-4pr}{m^4+14m^2+1}}\xi\mid m\right)). \end{aligned} \tag{157}$$

- $k=\frac{\sqrt{-1}}{2}\sqrt[4]{\frac{q^2-4pr}{m^4-m^2+1}}, a_0=\frac{1}{2p}\left[\left(m^2+1\right)\sqrt{\frac{q^2-4pr}{m^4-m^2+1}}-q\right], a_1=0, a_2=0,$
  $b_1=0, b_2=-\frac{3}{2p}\sqrt{\frac{q^2-4pr}{m^4-m^2+1}}:$

$$v(\xi)=a_0-\frac{3}{2p}\sqrt{\frac{q^2-4pr}{m^4-m^2+1}}\text{ns}^2\left(\frac{\sqrt{-1}}{2}\sqrt[4]{\frac{q^2-4pr}{m^4-m^2+1}}\xi\mid m\right). \tag{158}$$

- $k = \frac{\sqrt{-1}}{2}\sqrt[4]{\frac{q^2-4pr}{m^4-m^2+1}}, a_0 = \frac{q}{2p} - \frac{1}{2p}\left(m^2+1\right)\sqrt{\frac{q^2-4pr}{m^4-m^2+1}}, a_1 = 0, a_2 = -\frac{3m^2}{2p}\sqrt{\frac{q^2-4pr}{m^4-m^2+1}}$,
  $b_1 = 0, b_2 = 0$:

$$v(\xi) = a_0 - \frac{3m^2}{2p}\sqrt{\frac{q^2-4pr}{m^4-m^2+1}}\text{sn}^2\left(\frac{\sqrt{-1}}{2}\sqrt[4]{\frac{q^2-4pr}{m^4-m^2+1}}\xi \mid m\right). \tag{159}$$

- $k = \frac{\sqrt{-1}}{2}\sqrt[4]{\frac{q^2-4pr}{m^4+14m^2+1}}, a_0 = \frac{q}{2p} - \frac{1}{2p}\left(m^2+1\right)\sqrt{\frac{q^2-4pr}{m^4+14m^2+1}}, a_1 = 0, a_2 = -\frac{3m^2}{2p}\sqrt{\frac{q^2-4pr}{m^4+14m^2+1}}$,
  $b_1 = 0, b_2 = -\frac{3}{2p}\sqrt{\frac{q^2-4pr}{m^4+14m^2+1}}$:

$$\begin{aligned} v(\xi) &= a_0 - \frac{3}{2p}\sqrt{\frac{q^2-4pr}{m^4+14m^2+1}}(3m^2\text{sn}^2\left(\frac{\sqrt{-1}}{2}\sqrt[4]{\frac{q^2-4pr}{m^4+14m^2+1}}\xi \mid m\right) + \\ &\text{ns}^2\left(\frac{\sqrt{-1}}{2}\sqrt[4]{\frac{q^2-4pr}{m^4+14m^2+1}}\xi \mid m\right)). \end{aligned} \tag{160}$$

Letting $m \to 1$ we obtain trigonometric and hyperbolic solutions :

$$v(\xi) = a_0 + \frac{3}{2p}\sqrt{q^2-4pr}\tanh^2\left(\frac{1}{2}\sqrt[4]{q^2-4pr}\xi\right). \tag{161}$$

$$v(\xi) = a_0 + \frac{3}{2p}\sqrt{q^2-4pr}\coth^2\left(\frac{1}{2}\sqrt[4]{q^2-4pr}\xi\right). \tag{162}$$

$$v(\xi) = a_0 + \frac{3}{4p}\sqrt{q^2-4pr}(3\tanh^2\left(\frac{1}{4}\sqrt[4]{q^2-4pr}\xi\right) + \coth^2\left(\frac{1}{4}\sqrt[4]{q^2-4pr}\xi\right)). \tag{163}$$

Now, let us seek solutions in the ansatz form (74), i.e.

$$\begin{aligned} v(\xi) &= a_0 + a_1\text{sn}\left(k\xi \mid m\right) + b_1\text{ns}(k\xi \mid m)\text{cn}(k\xi \mid m)\text{dn}(k\xi \mid m) + a_2\text{sn}^2\left(k\xi \mid m\right) + \\ &b_2\text{cn}(k\xi \mid m)\text{dn}(k\xi \mid m). \end{aligned} \tag{164}$$

Inserting this ansatz into (152) and solving the resulting algebraic system yields following solutions to Eq.(152) :

- $k = \sqrt{-1}\sqrt[4]{\frac{q^2-4pr}{m^4+14m^2+1}}, a_0 = -\frac{q}{2p} + \frac{1}{2p}\left(m^2+1\right)\sqrt{\frac{q^2-4pr}{m^4+14m^2+1}}, a_1 = 0, a_2 = -\frac{3m^2}{p}\sqrt{\frac{q^2-4pr}{m^4+14m^2+1}}$,
  $b_1 = 0, b_2 = \pm\frac{3m}{p}\sqrt{\frac{q^2-4pr}{m^4+14m^2+1}}$:

$$\begin{aligned} v(\xi) &= -\frac{q}{2p} + \frac{1}{2p}\left(m^2+1\right)\sqrt{\frac{q^2-4pr}{m^4+14m^2+1}} - \\ &\frac{3m^2}{p}\sqrt{\frac{q^2-4pr}{m^4+14m^2+1}}\text{sn}^2\left(\sqrt{-1}\sqrt[4]{\frac{q^2-4pr}{m^4+14m^2+1}}\xi \mid m\right) \pm \\ &\frac{3m}{p}\sqrt{\frac{q^2-4pr}{m^4+14m^2+1}}\text{cn}(\sqrt{-1}\sqrt[4]{\frac{q^2-4pr}{m^4+14m^2+1}}\xi \mid m)\text{dn}(\sqrt{-1}\sqrt[4]{\frac{q^2-4pr}{m^4+14m^2+1}}\xi \mid m). \end{aligned} \tag{165}$$

- $k = \sqrt[4]{\frac{q^2-4pr}{m^4+14m^2+1}}, a_0 = -\frac{q}{2p} - \frac{1}{2p}\left(m^2+1\right)\sqrt{\frac{q^2-4pr}{m^4+14m^2+1}}, a_1 = 0, a_2 = \frac{3m^2}{p}\sqrt{\frac{q^2-4pr}{m^4+14m^2+1}},$

$b_1 = 0, b_2 = \pm\frac{3m}{p}\sqrt{\frac{q^2-4pr}{m^4+14m^2+1}}$:

$$\begin{aligned} v(\xi) &= -\frac{q}{2p} - \frac{1}{2p}\left(m^2+1\right)\sqrt{\frac{q^2-4pr}{m^4+14m^2+1}} - \\ &\frac{3m^2}{p}\sqrt{\frac{q^2-4pr}{m^4+14m^2+1}}\,\mathrm{sn}^2\left(\sqrt[4]{\frac{q^2-4pr}{m^4+14m^2+1}}\xi \mid m\right) \pm \\ &\frac{3m}{p}\sqrt{\frac{q^2-4pr}{m^4+14m^2+1}}\,\mathrm{cn}(\sqrt[4]{\frac{q^2-4pr}{m^4+14m^2+1}}\xi \mid m)\mathrm{dn}(\sqrt[4]{\frac{q^2-4pr}{m^4+14m^2+1}}\xi \mid m). \end{aligned} \tag{166}$$

We may obtain other solutions by making use of the dn-nd method.

4.7. KdV Equation

This is the equation

$$u_t + 6uu_x + u_{xxx} = 0. \tag{167}$$

If $u = v(\xi), \xi = x + \lambda t + \xi_0$ this equation takes the form

$$\lambda v'(\xi) + 6v(\xi)v'(\xi) + v'''(\xi) = 0. \tag{168}$$

Integrating this equation w.r.t. ξ yields

$$v''(\xi) = -3v^2(\xi) + \lambda v(\xi) + C, \tag{169}$$

where C is the constant of integration. Equation (169) has the form (152) with

$$p = -3, q = \lambda \text{ and } r = C. \tag{170}$$

Exact solutions to KdV equation may be derived from (155)-(164) and (170).

The KdV equation may also be solved by the Weierstrass elliptic functions method.

4.8. Boussinesq Equation

This equation reads

$$u_{tt} + \alpha uu_{xx} + \alpha u_x^2 + \beta u_{xxxx} = 0, \ \ \alpha\beta \neq 0. \tag{171}$$

After the traveling wave transformation $u = v(\xi), \xi = x + \lambda t + \xi_0$ and integrating twice w.r.t ξ we obtain following ode :

$$v''(\xi) = -\frac{\alpha}{2\beta}v^2(\xi) - \frac{\lambda^2}{\beta}v(\xi) - \frac{D\xi + C}{\beta}, \tag{172}$$

where C and D are the constants of integration. Setting $D = 0$ we obtain an equation of the form (152) with

$$p = -\frac{\alpha}{2\beta}, q = -\frac{\lambda^2}{\beta} \text{ and } r = -\frac{C}{\beta}. \tag{173}$$

As we can see, exact solutions to Boussinesq equation are calculated from (155)-(164) and (173).

4.9. Symmetric Regular Long Wave Equation

This equation is define by

$$u_{tt} + u_{xx} + uu_{xt} + u_x u_t + u_{xxtt} = 0. \tag{174}$$

Let $u = v(\xi), \xi = x + \lambda t + \xi_0$. Applying this transformation and integrating twice w.r.t. ξ we obtain the ode

$$v''(\xi) = -\frac{1}{2\lambda}v^2(\xi) - \frac{\lambda^2+1}{\lambda^2}v(\xi) - \frac{D\xi + C}{\lambda^2}, \tag{175}$$

where C and D are the constants of integration. Setting $D = 0$ we obtain an equation of the form (152) with

$$p = -\frac{1}{2\lambda}, q = --\frac{\lambda^2+1}{\lambda^2} \text{ and } r = -\frac{C}{\lambda^2}. \tag{176}$$

Thus, exact solutions to symmetric regular long wave equation are easily found from (155)-(164) taking into account (176).

4.10. Generalized Shallow Water Wave Equation

This equation is given by

$$u_{xxtt} + \alpha u_x u_{xt} + \beta u_t u_{xx} - u_{xt} - u_{xx} = 0. \tag{177}$$

Let

$$u = u(x,t) = V(\xi), V(\xi) = \int v(\xi)d\zeta, \xi = x + \lambda t. \tag{178}$$

Substituting (178) into (177) and integrating once w.r.t. ξ we obtain :

$$v''(\xi) = -\frac{\alpha+\beta}{2\lambda}v^2(\xi) + \frac{\lambda+1}{\lambda^2}v(\xi) + \frac{C}{\lambda^2}, \tag{179}$$

where C is the constant of integration. Eq. (152) has the form (152) with

$$p = -\frac{\alpha+\beta}{2\lambda}, q = \frac{\lambda+1}{\lambda^2} \text{ and } r = \frac{C}{\lambda^2}. \tag{180}$$

It is evident that we may fin exact solutions to generalized shallow water wave equation (177) from (155)-(164) for the choices given by (180).

4.11. Klein–Gordon Equation with Quadratic Nonlinearity

The general Klein-Gordon equation has the form

$$u_{tt} - \alpha^2 u_{xx} + \beta u + f(u) = 0. \tag{181}$$

In the case when $f(u) = -\gamma u^2$ we obtain the so called Klein–Gordon equation with quadratic nonlinearity :

$$u_{tt} - \alpha^2 u_{xx} + \beta u - \gamma u^2 = 0. \tag{182}$$

Let $u = v(\xi), \xi = x + \lambda t + \xi_0$. After this traveling wave transformation, Eq. (182) reduces to

$$v''(\xi) = -\frac{\gamma}{\alpha^2 - \lambda^2} v^2(\xi) + \frac{\beta}{\alpha^2 - \lambda^2} v(\xi),\ \alpha^2 \neq \lambda^2, \tag{183}$$

which is an equation of the form (152) with

$$p = -\frac{\gamma}{\alpha^2 - \lambda^2}, q = \frac{\beta}{\alpha^2 - \lambda^2} \text{ and } r = 0. \tag{184}$$

Again, exact solutions to Eq. (182) are obtained from (155)-(164) taking into account (184).

4.12. Fitzhugh-Nagumo-Huxley Equation

This equation reads

$$u_t - u_{xx} + u(1-u)(\alpha - u) = 0,\ \alpha = \text{const.} \tag{185}$$

This equation is an important model in the study of neuron axon [9]. Let

$$u = u(x,t) = v(\xi),\ \ \xi = x + \lambda t + \xi_0. \tag{186}$$

The corresponding ODE is

$$v''(\xi) - \lambda v'(\xi) - v(\xi)(v(\xi) - 1)(v(\xi) - \alpha) = 0. \tag{187}$$

Application of the sn-ns method gives only trivial solutions, since we get $\lambda = 0$. Instead, we may use other methods. If we apply the tanh-coth method or the exp method, we obtain non-trivial solutions. Indeed, balancing gives $n = 1$. Following the tanh-coth method, we try ansatz (42) :

$$v(\xi) = a_0 + a_1 \tanh(k\xi) + b_1 \coth(k\xi). \tag{188}$$

Inserting (188) into (185) and solving the corresponding algebraic system gives following solutions to Eq. (185) :

$$u(x,t) = \frac{1}{2}\left(1 + \tanh\left[\frac{1}{2\sqrt{2}}(x - \frac{2\alpha - 1}{\sqrt{2}}t + \xi_0)\right]\right). \tag{189}$$

$$u(x,t) = \frac{1}{2}\left(1 + \coth\left[\frac{1}{2\sqrt{2}}(x - \frac{2\alpha - 1}{\sqrt{2}}t + \xi_0)\right]\right). \tag{190}$$

$$u(x,t) = \frac{1}{4}\left(2 + \tanh\left[\frac{1}{4\sqrt{2}}(x - \frac{2\alpha - 1}{\sqrt{2}}t + \xi_0)\right] + \coth\left[\frac{1}{4\sqrt{2}}(x - \frac{2\alpha - 1}{\sqrt{2}}t + \xi_0)\right]\right). \tag{191}$$

$$u(x,t) = \frac{\alpha}{2}\left(1 + \tanh\left[\frac{\alpha}{2\sqrt{2}}(x - \frac{2 - \alpha}{\sqrt{2}}t + \xi_0)\right]\right). \tag{192}$$

$$u(x,t) = \frac{\alpha}{2}\left(1 + \coth\left[\frac{\alpha}{2\sqrt{2}}(x - \frac{2 - \alpha}{\sqrt{2}}t + \xi_0)\right]\right). \tag{193}$$

$$u(x,t) = \frac{\alpha}{4}\left(2 + \tanh\left[\frac{\alpha}{4\sqrt{2}}(x - \frac{2\alpha - 1}{\sqrt{2}}t + \xi_0)\right] + \coth\left[\frac{\alpha}{4\sqrt{2}}(x - \frac{2\alpha - 1}{\sqrt{2}}t + \xi_0)\right]\right). \tag{194}$$

4.13. Double sine-Gordon Equation

Our last example deals with the double sine-Gordon equation. This equation in a normalized form reads

$$u_{tt} - u_{xx} + \sin(u) - \frac{1}{2}\sin(2u) = 0. \tag{195}$$

This equation is an important model in the study of the DNA molecule [10].
The application of the tanh-coth method gives only the trivial solution $u = 0$. If we apply the sn-ns method, we obtain non-trivial solutions. Indeed, let

$$u = 2\arctan(v), v = v(\xi),\ \xi = x + \lambda t + \xi_0. \tag{196}$$

Inserting ansatz (196) into (195) gives the ODE

$$2v^3(\xi) - 2(\lambda^2 - 1)v(\xi)(v'(\xi))^2 + (\lambda^2 - 1)v''(\xi) + (\lambda^2 - 1)v^2(\xi)v''(\xi) = 0. \tag{197}$$

Let

$$v(\xi) = a_0 + a_1\mathbf{sn}(k\xi \mid m) + b_1\mathbf{ns}(k\xi \mid m). \tag{198}$$

Substituting (198) into (197) and solving the corresponding algebraic system gives following solutions to Eq. (195):

- $a_0 = 0, a_1 = \frac{\sqrt{2}}{2}\sqrt{m^2+1}\sqrt{-1}, b_1 = 0, \lambda = \frac{\sqrt{k^2(m^2-1)^2-2(m^2+1)}}{k(m^2-1)}$:

$$v(\xi) = \frac{\sqrt{2}}{2}\sqrt{m^2+1}\sqrt{-1}\mathbf{sn}(k\left(x + \frac{\sqrt{k^2\left(m^2-1\right)^2 - 2\left(m^2+1\right)}}{k\left(m^2-1\right)}t + \xi_0\right) \mid m). \tag{199}$$

- $a_0 = 0, a_1 = 0, b_1 = \frac{\sqrt{2}}{2m}\sqrt{m^2+1}\sqrt{-1}, \lambda = \frac{\sqrt{k^2(m^2-1)^2-2(m^2+1)}}{k(m^2-1)}$:

$$v(\xi) = \frac{\sqrt{2}}{2m}\sqrt{m^2+1}\sqrt{-1}\mathbf{ns}(k\left(x + \frac{\sqrt{k^2\left(m^2-1\right)^2 - 2\left(m^2+1\right)}}{k\left(m^2-1\right)}t + \xi_0\right) \mid m). \tag{200}$$

- $a_0 = 0,\ a_1 = \frac{\sqrt{m(m^2-6m+1)}}{2\sqrt{2}(m-1)},\ b_1 = -\frac{\sqrt{m(m^2-6m+1)}}{2\sqrt{2}m(m-1)},\ \lambda = -\frac{\sqrt{k^2(m+1)^4-2(m^2-6m+1)}}{k(m+1)^2}$:

$$\begin{aligned} v(\xi) &= \frac{\sqrt{m\left(m^2-6m+1\right)}}{2\sqrt{2}(m-1)}\mathbf{sn}(k\xi \mid m) - \frac{\sqrt{m\left(m^2-6m+1\right)}}{2\sqrt{2}m\left(m-1\right)}\mathbf{ns}(k\xi \mid m), \\ \xi &= x - \frac{\sqrt{k^2(m+1)^4 - 2(m^2-6m+1)}}{k(m+1)^2}t + \xi_0 \end{aligned} \tag{201}$$

5. Conclusions

We successfully obtained exact solutions for some important physical models by techniques based on projective equations. Mainly, we have used the sn-ns method. In our opinion, this is the most appropriate of all methods we have studied, since it provides elliptic function solutions and trigonometric and hyperbolic solutions. However, in the cases when the sn-ns method does not work (this occurs for Fitzhugh–Nagumo–Huxley equation), we may try other methods, such as the tanh-coth method as well as the exp method. On the other hand, there are some equations for which the tanh-coth method gives only trivial solutions(this is the case of the double sine-Gordon equation).

We think that some of the results we obtained are new in the open literature. Other results concerning exact solutions of nonlinear pde's may be found in [11]–[48]

References

[1] E. Fan & Y.C. Hon, *Generalized tanh Method Extended to Special Types of Nonlinear Equations*, Z. Naturforsch. A, **57**(2002), no. 8, 692-700.

[2] A. M Wazwaz, *The extended tanh method for new solitons solutions for many forms of the fifth-order KdV equations*, Appl. Math. Comput, **84-2** (2007), 1002-1014.

[3] He-Huan & Li-Na Zhang, *Generalized solitary solutions and compacton-like solution of the Jaulent-Miodek equations using the Exp-function method*, Phy. Lett A., **372**(2008), 1044-1047.

[4] R. Conte & M. Musette, *Link betwen solitary waves and projective Riccati equations*, J. Phys. A Math. **25** (1992), 5609-5623.

[5] A. Wazwaz, *The variable separated ODE and the tanh methods for solving the combined and the double combined sinh–cosh-Gordon equations*, Applied Mathematics and Computation **177** (2006) 745–754.

[6] W. Ma & M. Chen, *Direct search for exact solutions to the nonlinear Schrödinger equation, Applied Mathematics and Computation* **215** (2009) 2835–2842.

[7] M.J Ablowitz, P.A Clarkson, Soliton, Nonlinear Evolution Equations and Inverse Scattering. Cambridge University Press, New York (1991).

[8] C.A. Gómez & A. H. Salas, *New exact solutions for the combined sinh-cosh-Gordon equation, Lecturas Matematicas*, special issue(2006), 87-93.

[9] A. Scott, *Neuroscience-A Mathematical Primer*, Springer-Verlag New York, Inc. (2002).

[10] L.V. Yakushevich, *Nonlinear Physics of DNA*, Chichester and NewYork, Wiley (1998).

[11] A. H. Salas, C. A. Gómez, *Computing Exact Solutions to a Generalized Lax-Sawada-Kotera-Ito Seventh-Order KdV Equation*, Mathematical Problems in Engineering, vol. 2010 (2010), Article ID 524567.

[12] A. H. Salas,C. A. Gómez , *Exact solutions for a Third-Order KdV Equation with Variable Coefficients and Forcing Term*, Mathematical Problems in Engineering, vol. 2009(2009), Article ID 737928, 13 pages, doi:10.1155/2009/737928.

[13] C. A. Gómez, A. H. Salas, *Exact Solutions to KdV6 Equation by Using a New Approach of the Projective Riccati Equation Method*, Mathematical Problems in Engineering, vol. 2010 (2010), Article ID 797084.

[14] A. H. Salas, C. A. Gómez, *Application of the Cole–Hopf transformation for finding exact solutions for several forms of the seventh-order KdV equation* (KdV7), Mathematical Problems in Engineering, vol. 2010 (2010), Article ID 194329.

[15] C. A. Gómez, A. H. Salas, *Exact solutions for the generalized BBM equation with variable coefficients*, Mathematical Problems in Engineering, vol. 2010 (2010), Article ID 498249.

[16] A. H. Salas, C. A. Gómez, J. G. Escobar, *Exact solutions for the general fifth–order KdV equation by the extended tanh method*, Journal of Mathematical Sciences : Advances and Applications, vol. 1, No. 2(2008), pp. 305–310.

[17] C. A. Gómez, A. H. Salas *Exact solutions for a new integrable system* (KdV6), Journal of Mathematical Sciences : Advances and Applications, vol. 1, No. 2(2008), pp. 401–413.

[18] A. H. Salas, C. A. Gómez, *A practical approach to solve coupled systems of nonlinear PDE's*, Journal of Mathematical Sciences : Advances and Applications, vol. 3, No. 1(2009), pp. 101–107.

[19] C. A. Gómez, A. H. Salas *Solutions for a class of fifth–order nonlinear partial differential system*, Journal of Mathematical Sciences : Advances and Applications, vol 3, No. 1(2009), pp. 121–128.

[20] A. H. Salas, *Symbolic Computation of Solutions for a forced Burgers equation* , Applied Mathematics and Computation 216 (2010) 18–26.

[21] A. H. Salas, *Exact solutions for the general fifth KdV equation by the exp function method*, Applied Mathematics and Computation 205 (2008) 291–297.

[22] C. A. Gómez, A. H. Salas, *The Cole–Hopf transformation and improved tanh–coth method applied to new integrable system* (KdV6), Applied Mathematics and Computation 204 (2008) 957–962.

[23] A. H. Salas, C. A. Gómez & J. E. Castillo, *New abundant solutions for the Burgers equation*, Computers and Mathematics with Applications 58 (2009) 514 520.

[24] C. A. Gómez, A. H. Salas, *The generalized tanh–coth method to special types of the fifth–order KdV equation*, Applied Mathematics and Computation 203 (2008) 873–880.

[25] A. H. Salas, *New solutions to Korteweg–De Vries (kdV) equation by the Riccati equation expansion method*, International Journal of Applied Mathematics (IJAM) **22**(2009), pp. 1169-1177.

[26] C. A. Gómez, A. H. Salas, *Special forms of the fifth-order KdV equation with new periodic and soliton solutions*, Applied Mathematics and Computation, **189** (2007), pp. 1066-1077.

[27] C. A. Gómez, A. H. Salas, *Exact solutions for a reaction–diffusion equation by using the generalized tanh method*, Scientia et Technica, Universidad Tecnológica de Pereira, Risaralda–Colombia Vol 13, No. 035(2007) pp. 409–410.

[28] C. A. Gómez, A. H. Salas, *The variational iteration method combined with improved generalized tanh–coth method applied to Sawada–Kotera equation,* Applied Mathematics and Computation 217 (2010) 1408–1414.

[29] C. A. Gómez, A. H. Salas, *Special symmetries to standard Riccati equations and applications,* Applied Mathematics and Computation 216 (2010) 3089–3096.

[30] C. A. Gómez, A. H. Salas, *New periodic and soliton solutions for the Generalized BBM and Burgers–BBM equations,* Applied Mathematics and Computation 217 (2010) 1430–1434.

[31] A. H. Salas, J.E. Castillo, J. G. Escobar, *About the seventh–order Kaup–Kupershmidt equation and its solutions*, http://arxiv.org, arXiv:0809.2870(2008).

[32] A. H. Salas, C. A. Gómez and J.E. Castillo, *New solutions for the modified generalized Degasperis–Procesi equation*, Applied Mathematics and Computation 215 (2009) 2608–2615.

[33] A. H. Salas, *Some exact solutions for the Caudrey–Dodd–Gibbon equation*, http://arxiv.org, arXiv:0805.2969(2008).

[34] A. Salas, *Some solutions for a type of generalized Sawada–Kotera equation* , Applied Mathematics and Computation 196 (2008) 812–817.

[35] A. H. Salas, C. A. Gómez, *El software Mathematica en la búsqueda de soluciones exactas de ecuaciones diferenciales no lineales en derivadas parciales mediante el uso de la ecuación de Riccati*. Memorias del Primer Seminario Internacional de Tecnologias en Educación Matemática, Universidad Pedagógica Nacional, Santafé de Bogotá, Colombia 1 (2005), pp. 379-387.

[36] A. H. Salas, C. A. Gómez and L.L. Palomá, *Travelling wave solutions to the reduced Ostrovsky equation*, International Journal of Applied Mathematics (IJAM) **23**(2010), pp. 249-256.

[37] A. H. Salas, *Symbolic Computation of exact solutions to KdV equation* , Canadian Applied Mathematics Quarterly, vol. 16, No. 4(2008).

[38] A. H. Salas, C. A. Gómez, J.E. Castillo, *Symbolic Computation of Solutions for the General Fifth-order KdV Equation*, International Journal of Nonlinear Science, vol.9, No.4(2010), pp. 394-401.

[39] C. A. Gómez, A. H. Salas, *A generalized Riccati method to solve NLPDEs,* Far East Journal of Applied Mathematics, vol. 40, No. 2(2010), pp. 89 - 175

[40] J.E. Castillo H, A.H. Salas, J.G. Escobar, *Exact solutions for a nonlinear model*, Applied Mathematics and Computation 217 (2010) 1646–1651.

[41] A. H. Salas, *Exact solutions of coupled sine Gordon equations* , Nonlinear Analysis: Real World Applications 11 (2010) 3930 3935.

[42] A. H. Salas, *Computing solutions to a forced KdV equation,* Nonlinear Analysis: Real World Applications 12 (2011) 1314–1320.

[43] A. H. Salas, C.A. Gómez, *Computing exact solutions for some fifth KdV equations with forcing term,* Applied Mathematics and Computation 204 (2008) 257–260.

[44] A. H. Salas, *Exact solutions to mKdV equation with variable coefficients,* Applied Mathematics and Computation 216 (2010) 2792–2798.

[45] A. H. Salas, *Computing exact solutions to a generalized Lax seventh-order forced KdV equation (KdV7)*, Applied Mathematics and Computation 216 (2010) 2333–2338.

[46] A.H. Salas, J.E. Castillo, *New exact solutions to sinh–cosh-Gordon equation by using techniques based on projective Riccati equations* , Computers and Mathematics with Applications 61 (2011) 470–481.

[47] A. H. Salas, C. A. Gómez, *A note on the G'/G method and the improved tanh-coth method,* Far East Journal of Applied Mathematics, Volume 48, Issue 1, Pages 1 - 73 (November 2010)

[48] A. H. Salas, C.A. Gómez, *Exact Solutions for the KdV6 and mKdV6 equations via tanh, coth and sech methods*, Applications and Applied Mathematics: An International Journal (AAM), Vol. 5, Issue 10 (December 2010).

In: Evolution Equations
Editor: Arthur L. Claes

ISBN: 978-1-61209-652-0

Chapter 7

HOMOGENEOUS AND NONLINEAR GENERALIZED MASTER EQUATIONS: A NEW TYPE OF EVOLUTION EQUATIONS

Victor F. Los
Institute of Magnetism, Nat. Acad. of Sci. of Ukraine,
Kiev, Ukraine

Abstract

Deriving the kinetic (irreversible) equations from the reversible microscopic dynamics of the many-body systems remains one of the principal tasks of statistical physics. One of the key assumptions used to deriving the Boltzmann equation is the Bogoliubov principle of weakening of initial correlations which implies that on a sufficientl large time scale all initial correlations (existing at the initial instant) are damped. This approach leads to the evolution equations that are not valid on all time scales and do not allow considering all the stages of the system evolution, in particular, the short-term and transient regimes, which are of scientifi and practical interest for considering the ultrafast relaxation and non-Markovian processes, decoherence phenomena and onset of irreversibility. To take initial correlations into account, a method, based on the conventional time-independent projection operator technique, that allows converting the conventional linear inhomogeneous (containing a source caused by initial correlations) time-convolution generalized master equation (TC-GME) and time-convolutionless GME (TCL-GME) into the homogeneous form exactly, is proposed. This approach results in the exact linear time-convolution and time-convolutionless homogeneous generalized master equations (TC-HGME and TCL-HGME) which take the dynamics of initial correlations into account via modifie memory kernels governing the evolution of the relevant part of a distribution function of a many-particle system. These equations describe the evolution of the relevant part of a distribution function on all time scales including the initial stage when the initial correlations matter. The obtained TC-HGME is applied to the spatially homogeneous dilute gas of classical and quantum particles. However, to derive the desired nonlinear equations (the Boltzmann equation in particular) from this actually linear equation, we should make an additional approximation neglecting the time-retardation of a one-particle distribution function which restricts the time scale by times much smaller than the relaxation time for a one-particle distribution function. To obtain the actually nonlinear evolution equations and avoid any restrictions on the time scales, we develop a new method, based on using a time-dependent operator (generally not a projection operator) converting a distribution function of a total system into the relevant form, that allows deriving the new exact nonlinear generalized master equations. The derived inhomogeneous nonlinear GME is a generalization of the linear Nakajima-Zwanzig TC-GME and can be viewed as an alternative to the BBGKY chain. To include the initial correlations into consideration, we convert the obtained inhomogeneous nonlinear GME into the homogeneous form by the method which we used for conventional linear GMEs. The obtained exact homogeneous nonlinear GME describes all evolution stages of the system of interest and treats initial correlations on an equal footing with collisions by means of the modifie memory kernel. As an application, we obtain a new homogeneous nonlinear equation retaining initial correlations for a one-particle distribution function of the spatially inhomogeneous nonideal gas of classical particles. We show that on the kinetic time scale, the time-reversible terms resulting from initial correlations vanish (if the particle dynamics have the ergodic mixing property) and this equation can be converted into the Vlasov-Landau and Boltzmann equations without any additional commonly used approximations.

1. Introduction

While the effects of equilibrium statistical mechanics seem to be well understood and described, nonequilibrium statistical physics does not enjoy such an opinion. One of the basic tasks of the latter remains deriving the appropriate evolution equations for the measurable values (statistical expectations) characterizing a nonequilibrium state of a many-particle system. It is expected that these equations should generally be evolution equations converting into kinetic (or other irreversible) equations on some time scale. The principal questions are how to derive such irreversible equations rigorously from underlying microscopic (reversible) classical or quantum dynamical equations and what is the reason for the observable irreversibility. Several approaches are usually used to address these problems, commonly starting with the linear Liouville-von-Neumann equation for a distribution function (statistical operator) of the N-particle ($N \gg 1$) system under consideration.

One approach leads to the chain of coupled first-orde differential equations for s-particle distribution functions $F_s(x_1, \ldots, x_s, t)$ ($1 \leq s \leq N$) known as the BBGKY hierarchy [1]. The most consistent approach to decoupling the BBGKY hierarchy was developed by Bogoliubov [1]. He suggested the principle of weakening of initial correlations, which implies that at a sufficientl large time $t - t_0 \gg t_{cor}$ (t_{cor} is the correlation time due to interparticle interaction), all initial correlations (existing at the initial instant t_0) are damped and the time-dependence of multiparticle distribution functions is consequently determined completely by the time-dependence of a one-particle distribution function (Bogoliubov's ansatz). This assumes the existence of the time interval (time scale)

$$t_{cor} \ll t - t_0 \ll t_{rel} \tag{1.1}$$

(where t_{rel} is the relaxation time for a one-particle distribution function) and leads to an approximate conversion (valid only on the large time scale indicated above) of the inhomogeneous equation (including two-particle correlations) for a one-particle distribution function (of the BBGKY hierarchy) into a homogeneous equation. The idea behind this is that the perturbation expansions of the homogeneous equations are much more effective than that for the inhomogeneous ones (like the equations of the BBGKY chain), where the expansions of the functions, rather than equations, are involved. The Bogoliubov ansatz allows thus to avoid the so-called "secular terms", i.e the terms growing with time, which emerge after integration of the BBGKY chain inhomogeneous equation for s-particle distribution function $F_s(x_1, \ldots, x_s, t)$ $(1 \leq s < N)$ over time and make the perturbation expansion of these functions ineffective on the time scale of interest in kinetic theory. The Bogoliubov principle of weakening of initial correlations provides a solid mathematical foundation to the Boltzmann hypothesis of the molecular chaos (Stosszablansatz) which leads to the coarse-grained (smoothed) over the time interval (1.1) evolution of the one-particle distribution function (described by the famous Boltzmann kinetic equation).

Using this approach, Bogoliubov successfully derived the kinetic (Markovian) equations for a one-particle distribution function, the classical and quantum Boltzmann equations in particular, which describe the time-evolution with the characteristic time t_{rel} but is unsuitable for taking the initial evolution stage $t_0 \leqslant t \leqslant t_{cor}$ into account.

The Bogoliubov principle of weakening of initial correlations is introduced as the boundary (initial) condition for the BBGKY chain. For example, in classical physics, it can be written as

$$\lim_{t-t_0 \to \infty} U(t, t_0) \left[F_N(x_1, \ldots, x_N, t_0) - f_r(x_1, \ldots, x_N, t_0)\right] = 0,$$

$$f_r(x_1, \ldots, x_N, t) = \prod_{i=1}^{N} F_1(x_i, t), \qquad (1.2)$$

where $F_N(x_1, \ldots, x_N, t_0)$ is an N-particle distribution function at $t = t_0$, $x_i = (\mathbf{x}_i, \mathbf{p}_i)$ is the coordinate of the ith particle in the phase space, $f_r(x_1, \ldots, x_N, t)$ is the relevant distribution function with no correlations (the general definitio of the relevant distribution function will be given later), and $U(t, t_0)$ is the operator determining the evolution of all coordinates x_i in time. The firs inequality in (1.1) implies that all correlations between the particles vanish for $t - t_0 \gg t_{cor}$, and the second inequality in (1.1), $t - t_0 \ll t_{rel}$, allows neglecting the time retardation and replacing $U(t, t_0) f_r(t_0)$ with $U(t, t_0) f_r(t)$ (in the firs approximation in the small density of particles n). Thus, decoupling the BBGKY chain is nonlinear procedure (1.2), which results in a closed nonlinear equation (like the Boltzmann equation) for a one-particle distribution function $F_1(x_i, t)$ and introduces a nonlinearity into the process originally described by the linear reversible Liouville-von-Neumann equation. It is important to note, that shifting t_0 to $-\infty$ in the Bogoliubov limiting procedure (1.2) (in order to get rid of dealing with a singled out instant of time, when a distribution function $F_N(t)$ is replaced with its relevant part) leads to an irreversibility of the relaxation process.

But we note that only the small-scale initial correlations with $t_{cor} \ll t_{rel}$ can in fact be neglected on the kinetic time scale. Trying to obtain the collision integral in the kinetic equation with collisions of three particles (in the two-dimensional case) or more particles (the terms of the second and higher orders in n), we should handle the divergencies, in particular, resulting from the large-scale initial correlations with $t_{cor} \gtrsim t_{rel}$ (see, e.g., [2], [3]). Hence, the correct expansion of the collision integral in powers of n and consideration of kinetic fluctuation are only possible if we account for initial correlations with $t_{cor} \gtrsim t_{rel}$ and $l_{cor} \gtrsim l$, where $l_{cor} = \overline{v}t_{cor}$, $l = \overline{v}t_{rel}$, $\overline{v}$ is the characteristic (mean) particle velocity, and l is the particle mean-free path [4]. Moreover, there can be correlations which do not damp, e.g. due the conservation laws, such as the collective plasma excitations; also, the initial quantum correlations, caused by the particle statistics, do not damp with time.

Attempts to obtain the Boltzmann equation rigorously (especially in quantum physics) continue. This is important from both the conceptual and practical standpoints. For example, following [5], where the classical Boltzmann equation was rigorously obtained for a system of hard spheres at very low densities and for short times, the authors of [6] analogously (but not rigorously) obtained the quantum Boltzmann equation but using the factorized initial condition, i.e., the absence of initial correlations or random phase approximation (RPA) at $t = t_0$, which is doubtful in principle [7].

On the other hand, we note that, as to the best of our knowledge, there is not yet a way to take initial correlations into account exactly when dealing with the BBGKY chain.

In another approach, leading to the so called generalized master equations (GMEs), an N-particle distribution function (or statistical operator) is divided into the essential (relevant) and inessential (irrelevant) parts using time-independent projection operators P and Q

$$\begin{aligned} F_N(x_1,\ldots,x_N,t) &= f_r(x_1,\ldots,x_N,t) + f_i(x_1,\ldots,x_N,t), \\ f_r(x_1,\ldots,x_N,t) &= PF_N(x_1,\ldots,x_N,t), f_i(x_1,\ldots,x_N,t) = QF_N(x_1,\ldots,x_N,t), \\ P + Q &= 1. \end{aligned} \tag{1.3}$$

We note, that the relevant and irrelevant parts depend on coordinates (and/or momenta) of all N particles in contrast to the reduced distribution functions (like $F_1(x_i)$). The relevant part, which is usually of main interest, is, as a rule, a vacuum (slowly changing) part of the distribution function (statistical operator), i.e., the part with no correlations (like $f_r(x_1,\ldots,x_N,t)$ in (1.2)). Applying the projectors P and Q to the Liouville-von-Neumann equation, we can obtain the time-convolution (non-Markovian) GME (TC-GME) [8], [9], [10] and time-convolutionless (time-local) GME (TCL-GME) [11], [12], which are the exact inhomogeneous equations for the relevant part of a distribution function with a source (irrelevant part) containing all many-particle correlations at the initial instant t_0. We stress that the time-independent (linear) projection operators P and Q used in this approach commute with the time-derivative operator in the Liouville-von-Neumann equation for the distribution function (statistical operator) of the whole system. Therefore, procedure (1.3) is linear and transforms the linear Liouville-von-Neumann equation into the linear GMEs. Bogoliubov's principle of weakening of initial correlations or simply the factorized initial conditions (RPA) are commonly used to exclude initial many-particle correlations (a source), which results in an approximate closed linear homogeneous GMEs for the relevant part of the distribution function (statistical operator). The latters are then used to obtain the linear evolution (master) equations for an interesting reduced distribution function (especially like that for an open subsystem interacting with a thermal bath).

There is also an approach based on the Kadanoff-Baym equations for the two-time correlations functions [13]. The original Kadanoff-Baym equations do not contain initial correlations and follow from the exact equations for correlation functions under the assumption of Bogoliubov's principle of weakening of initial correlations. In the framework of this approach, several attempts were made to include into account the initial correlations. A review of these works can be found in [14].

All kinetic equations which follow from that principle are unable to describe the initial stage of the evolution, $t_0 \leqslant t \leqslant t_{cor}$. Therefore, the influenc of initial correlations, which can be important for, e.g., the ultrafast relaxation processes, is not described by these equations. Moreover, the time-scale $t \lesssim t_{cor}$ may be of significan interest, because we can expect that on this time-scale the transition from the short-term "non-kinetic" evolution of a multi-particle system to the long-term "kinetic" behavior occurs.

In particular, what we have said above means that the proper handling of the initial correlations is crucial in obtaining the closed evolution equations. Neglecting the initial correlations results in the evolution equations which are not valid on all time scales (but, e.g., only on the time scale (1.1)) and do not allow consecutive considering the all stages of the system of interest evolution, in particular, the stage when a system switches from the initial short-term (reversible) evolution (when initial correlations play a role) to the long-term (irreversible) kinetic regime. The outlined problem is common for both principal methods of deriving the kinetic equations: the BBGKY hierarchy and the GMEs. Moreover, the initial correlations can be important for considering the ultrafast relaxation and non-Markovian processes and also for describing the decoherence phenomenon (see, e.g., [15]).

A progress achieved in the research on dynamics of the many-particle systems makes it possible, in principle, to study the entire evolution process of systems in statistical mechanics on any time scale and to determine the conditions under which a system evolves towards the equilibrium state. To do that, the dynamics of initial correlations should be included into consideration, and our goal is to obtain new evolution equations exactly accounting the initial correlations.

To take the initial correlations into account, we firs suggest a method, based on the conventional time-independent projection operator technique (specifie by (1.3)), that allows converting the linear inhomogeneous TC-GME and TCL-GME into the homogeneous form exactly [16, 17]. This approach leads to the exact linear time-convolution and time-convolutionless homogeneous GMEs (TC-HGME and TCL-HGME), which take the dynamics of initial correlations into account via modifie memory kernels governing the evolution of the relevant part of the distribution function (statistical operator) of a many-particle system. These equations describe the evolution (influence by initial correlations) of the relevant part of a distribution function on all time scales including the initial stage $t - t_0 \lesssim t_{cor}$, when the transition from the reversible evolution to the irreversible kinetic behavior (resulting from the stochastic instability of the system dynamics) is expected to occur.

We analyze how the obtained linear TC-HGME works for a spatially homogeneous dilute gas of classical and quantum particles. But to obtain the desired nonlinear equations (the Boltzmann equation in particular) describing the evolution of a classical or quantum gas of particles from the linear TC-HGME, we should make an additional approximation neglecting the time retardation of the one-particle distribution function (statistical operator). Hence, in this approach, we are not restricted by the firs inequality in (1.1), $t_{cor} \ll t - t_0$, but still need the second inequality. The obtained (in the linear approximation in the gas density) new homogeneous equations for a one-particle distribution function (statistical operator) include initial correlations caused by interaction between particles (as well as the quantum initial correlations in the quantum case) and describe the evolution process on the time scale which includes the initial stage $t - t_0 \lesssim t_{cor}$. We show that on the appropriate time scale these equations convert to the classical and quantum Boltzmann equations if all initial correlations (due to interparticle interactions) vanish on this time scale.

We note, that the obtained linear homogeneous GMEs, when applied to an open quantum system interacting with a thermal bath, are not subject to the time interval restriction $t-t_0 \ll t_{rel}$ (here t_{rel} is the relaxation time due to the system-bath interaction) and describe all stages of a system of interest evolution including the initial stage $t - t_0 \lesssim t_{cor}$.

We see, that, strictly speaking, it is impossible to obtain a nonlinear master equation (like the Boltzmann equation) in the framework of the linear time-independent projection operator formalism without additional approximations (like discussed above neglection of time retardation which is justifie by the second inequality in (1.1)). Thus, a nonlinear equation for a density matrix is often simply postulated in the form of a Lindblad-type master equation whose generator depends parametrically on this density matrix (see, e.g., [18]).

To overcome the mentioned restriction, we suggest an approach allowing to obtain a new class of evolution equations for the relevant part of a distribution function (statistical operator) that suitable for deriving both the nonlinear and linear evolution equations for the reduced distribution functions (statistical operators) of interest and that can hold on any time scale [19]. The class contains exact nonlinear inhomogeneous and homogeneous GMEs deduced from the Liouville-von-Neumann equation using a (generally) nonlinear time-dependent operator $P(t)$ that converts the many-particle distribution function (statistical operator) into the appropriate relevant form. The time-dependent operator $P(t)$ is not generally a projection operator. The nonlinear inhomogeneous GME can be considered as an alternative to the BBGKY chain. In order to include the initial correlations into consideration, this equation is converted into nonlinear homogeneous GME by the method used for the linear GMEs. In the case of a time-independent linear projection operator P (1.3), the obtained equations reduce to the conventional linear GME and obtained TC-HGME. In contrast to these linear equations, the obtained nonlinear equations are more general and therefore more convenient, in particular, for obtaining Boltzmann-like equations and studying the spatially inhomogeneous nonideal gas of particles for which the spatial nonlocality and time retardation are essential (see, e.g., [4]).

We apply the derived nonlinear GMEs to the spatially inhomogeneous nonideal diluted gas of classical particles. In particular, from the obtained homogeneous nonlinear GME in the linear approximation in the density parameter, we obtain a new homogeneous nonlinear equation for a one-particle distribution function that holds on all time scales (no restrictions like (1.1) are needed) and takes the influenc of the dynamics of initial correlations into account in all stages of the evolution process. In this approximation (linear in n), the evolution of collisions and initial correlations (in the memory kernel) is governed by the exact two-particle propagator, and the obtained equation is closed in this sense. If the gas dynamics have the necessary property (ergodic mixing fl w in the phase space), then all initial correlations vanish at the kinetic time scale, and this equation can be converted into the irreversible nonlinear Boltzmann equation with a self-consistent nonlinear Vlasov term. No additional commonly used assumption, like special initial conditions (factorized ones) or a principle (weakening of initial correlations), is needed.

2. Time-Convolution Homogeneous GME

We start with the Liouville-von-Neumann equation for a distribution function (statistical operator) $F_N(t)$ of N ($N \gg 1$) particles

$$\frac{\partial}{\partial t} F_N(t) = LF_N(t). \tag{2.1}$$

Here, $F_N(t)$ is a symmetric function of N variables $x_i = (\mathbf{r}_i, \mathbf{p}_i)$ $(i = 1, ..., N)$ representing the coordinates and momenta of the particles (classical physics) or a statistical operator (quantum mechanics), which satisfy the normalization conditions

$$\int dx_1 \ldots \int dx_N F_N(x_1, \ldots x_N, t) = 1, Tr F_N = 1, \tag{2.2}$$

L is the Liouville operator acting on $F_N(t)$ in the case of classical physics as

$$LF_N(t) = \{H, F_N(t)\}_P = \sum_{i=1}^{N} \{ \frac{\partial H}{\partial \mathbf{r}_i} \frac{\partial F_N(t)}{\partial \mathbf{p}_i} - \frac{\partial H}{\partial \mathbf{p}_i} \frac{\partial F_N(t)}{\partial \mathbf{r}_i} \}, \tag{2.3}$$

where $\{H, F_N(t)\}_P$ is the Poisson bracket and H is the Hamilton function for the system under consideration, while in the quantum physics case L is the superoperator acting on a statistical operator as

$$LF_N(t) = \frac{1}{i\hbar} [H, F_N(t)], \tag{2.4}$$

where $[,]$ is a commutator and H is the Hamiltonian. Note, that for simplicity we consider the case of a closed isolated system when H does not depend on time t.

The formal solution to Eq. (2.1) is

$$F_N(t) = e^{L(t-t_0)} F_N(t_0), \tag{2.5}$$

where $F_N(t_0)$ is a distribution function (statistical operator) at an initial instant t_0, when the initial condition for the Liouville-von-Neumann equation (2.1) should be set.

Employing the projection operator technique [8], [9], [10], and applying the projection operators $P = P^2$ and $Q = Q^2 = 1 - P$ to Eq. (2.1), it is easy to obtain the equations for the relevant $f_r(t) = PF_N(t)$ and irrelevant $f_i(t) = QF_N(t)$ (see (1.3)) parts of $F_N(t)$

$$\frac{\partial}{\partial t} f_r(t) = PL[f_r(t) + f_i(t)], \tag{2.6}$$

$$\frac{\partial}{\partial t} f_i(t) = QL[f_r(t) + f_i(t)]. \tag{2.7}$$

A formal solution to Eq. (2.7) has the form

$$f_i(t) = \int_{t_0}^{t} \exp[QL(t-t_1)]QLf_r(t_1)dt_1 + \exp[QL(t-t_0)]f_i(t_0). \qquad (2.8)$$

Inserting this solution into (2.6), we obtain the conventional exact time-convolution generalized master equation (TC-GME) known as the Nakajima-Zwanzig equation for the relevant part of the distribution function (statistical operator)

$$\frac{\partial}{\partial t}f_r(t) = PLf_r(t) + \int_{t_0}^{t} PL\exp[QL(t-t_1)]QLf_r(t_1)dt_1 + PL\exp[QL(t-t_0)]f_i(t_0). \qquad (2.9)$$

It is important to stress that $f_r(t)$ and $f_i(t_0)$ are the basic quantities we are dealing with in Eq. (2.9). All functions of dynamical variables, the average values of which we can calculate with the help of $f_r(t)$ by multiplying equation (2.9) with the corresponding functions (operators) from the right and integrating over the relevant variables (taking a trace), are dependent only on the variables which are not projected off by P (P integrates off all excessive variables in $F_N(t)$). Therefore, if we represent $f_r(t)$ and $f_i(t_0)$ in (2.9) as $f_r(t) = PF_N(t)$ and $f_i(t_0) = QF_N(t_0)$, correspondingly, then the projection operators P and Q in these expressions act only on $F_N(t)$ but not on the functions of dynamical variables of interest to the right of them. This is the essence of the reduced description method, when, in order to calculate the average values of the functions dependent on a much smaller number of variables than the whole distribution function $F_N(t)$, we actually need only the reduced distribution function (density matrix) like the one-particle distribution functions determining the relevant distribution function $f_r(t)$ in (1.2).

Serving as a basis for many applications, Eq. (2.9), nevertheless, contains the undesirable and in general non-negligible inhomogeneous term (the last term in the right hand side of (2.9)), which depends (via $f_i(t_0)$) on the same large number of variables as the distribution function $F_N(t_0)$ at the initial instant ($f_i(t_0)$ is not composed of the reduced distribution functions). Therefore, Eq. (2.9) does not provide for a complete reduced description of a multiparticle system in terms of relevant (reduced) distribution function. Applying Bogoliubov's principle of weakening of initial correlations (allowing to eliminate the influenc of $f_i(t_0)$ on the large enough time scale $t - t_0 \gg t_{cor}$) or using a factorized initial condition (RPA),when $f_i(t_0) = QF_N(t_0) = 0$, one can achieve the above-mentioned goal and obtain the homogeneous GME for $f_r(t)$, i.e. Eq. (2.9) with no initial condition term. However, obtained in such a way homogeneous GME is either approximate and valid only on a large enough time scale (when all initial correlations vanish) or applicable only for a rather artifi cial (actually unreal, as pointed in [7]) initial conditions (no correlations at an initial instant of time).

In order to obtain a homogeneous equation for the relevant part of the distribution function (statistical operator), we try to transfer the inhomogeneous initial correlations (irrelevant) term in the right hand side of Eq. (2.9) to the (super)operator (kernel) acting on the relevant part $f_r(t)$. To achieve this goal we suggest to present the initial (irrelevant) term

$f_i(t_0) = QF_N(t_0)$ as a following exact identity

$$f_i(t_0) = [QF_N(t_0)]F_N^{-1}(t_0)e^{-L(t-t_0)}(P+Q)e^{L(t-t_0)}F_N(t_0)$$
$$= C_0 \exp[-L(t-t_0)][f_r(t) + f_i(t)]. \quad (2.10)$$

We assume here that the inverse operator $F_N^{-1}(t_0) = [f_r(t_0) + f_i(t_0)]^{-1}$ exists (see below), use that $P + Q = 1$ and that the backward propagator of the system is $\exp[-L(t-t_0)]$. In (2.10) the initial correlation parameter C_0 is introduced

$$C_0 = [QF_N(t_0)]F_N{}^{-1}(t_0) = f_i(t_0)[f_r(t_0) + f_i(t_0)]^{-1}$$
$$= f_i(t_0)f_r^{-1}(t_0)[1 + f_i(t_0)f_r^{-1}(t_0)]^{-1}$$
$$= (1 - C_0)f_i(t_0)f_r^{-1}(t_0), \quad (2.11)$$

where the projection operator Q acts only on $F_N(t_0)$, which is indicated by enclosing $QF_N(t_0)$ in brackets.

Thus, the additional identity (2.10) has been obtained by multiplying the irrelevant part by the unity $F_N^{-1}(t_0)F_N(t_0)$ (which implies the existence of $F_N^{-1}(t_0)$) and inserting the unities $\exp[-L(t-t_0)]\exp[L(t-t_0)] = 1$ and $P + Q = 1$. Therefore, neither divergency (due to possible vanishing of $F_N(t_0)$) nor indetermination of the $0/0$ type (behaviors of the numerator and denominator in $F_N(t_0)/F_N(t_0) = 1$ are similar) can happen. This holds over all further (identical) manipulations (see below). As it is seen from (2.11), the correlation parameter is a series in $f_i(t_0)f_r^{-1}(t_0)$ and, therefore, one may only need a formal existence of the function (operator) $f_r^{-1}(t_0)$ (see also (6.23)), which is invert to the relevant distribution function (statistical operator) chosen with the help of the appropriate projection operator P (generally, it can provide some restriction on the class of appropriate projectors). The relevant part, which is mainly of interest, is, as a rule, a vacuum (relatively slowly changing) part of a distribution function (statistical operator), i.e. the part with no correlations (e.g., a product of the one-particle distribution functions as in (1.2)). It seems plausible that the invert relevant part of the distribution function (statistical operator) define in a pointed above sense (uncorrelated part) can always be constructed.The examples of construction of the appropriate invert distribution function (operator) $f_r^{-1}(t_0)$ will be given below.

As a result of introducing the additional identity (2.10), we have two equations, (2.8) and (2.10), connecting $f_i(t)$ with $f_i(t_0)$. Finding $f_i(t_0)$ from these equations and inserting it into (2.9), we obtain the following exact equation for the relevant part of a distribution function (statistical operator):

$$\frac{\partial}{\partial t} f_r(t) = PLR(t-t_0)f_r(t) + \int_{t_0}^{t} PLR(t-t_0)\exp[QL(t-t_1)]QLf_r(t_1)dt_1, \quad (2.12)$$

where function $R(t-t_0)$ is define as

$$R(t-t_0) = 1 + C(t-t_0) = \frac{1}{1 - C_0(t-t_0)},$$
$$C(t-t_0) = e^{QL(t-t_0)} \frac{1}{1 - C_0 e^{-L(t-t_0)} e^{QL(t-t_0)}} C_0 e^{-L(t-t_0)},$$
$$C_0(t-t_0) = e^{QL(t-t_0)} C_0 e^{-L(t-t_0)}. \quad (2.13)$$

We have derived the time-convolution homogeneous generalized master equation (TC-HGME) (2.12) for the relevant part of the distribution function (statistical operator). This equation holds in both the classical and quantum physics cases if the proper redefinitio of the symbols is done and all (super)operators exist (we will address the latter problem below). We have not removed any information while deriving equation (2.12), and, therefore, it is exact integro-differential equation which takes into account the initial correlations and their dynamics via the modificatio of the (super)operator (memory kernel) in GME (2.9) acting on the relevant part of the distribution function (statistical operator) $f_r(t)$. The obtained exact kernel of TC-HGME (2.12) can serve as a starting point for consecutive perturbation expansions. In many cases such expansions of the homogeneous equations (like (2.12)) have much broader range of validity than that for the inhomogeneous ones (like (2.9)).

The initial correlations are entering the TC-HGME (2.12) through the function $R(t)$ (2.13). This function represents a sum of the infinit series of expansion in the properly define time dependent parameter $C_0(t-t_0)$ which describes the influenc of initial correlations in time. Presence of the function $R(t) \neq 1$ in Eq. (2.12) has led to the modificatio of the firs term in the right-hand side of the GME (2.9) as well as the second (collision) term. This modificatio reflect the influenc of initial correlations on the corresponding processes (fl w and relaxation) with time. The TC-HGME also allows treating the correlations arising from collisions and initial correlations on an equal footing by expanding the memory kernel in series in some small parameter.

However, the problem of the existence (convergence) of $R(t-t_0)$ can be raised. The function $R(t-t_0)$ behaves properly at all times. Moreover, the expansion of the kernel in (2.12) can result in canceling the pole in function $R(t-t_0)$ as we will see below. In such cases there is no problem with the existence of $R(t-t_0)$.

3. Time-Convolutionless Homogeneous GME

Now, let us turn to the case of the so called time-convolutionless (time-local) GME (TCL-GME). It is believed that such TCL-GME can be more easily solved and be more convenient for studying the non-Markovian processes than the integral TC-GME (2.9) (see [18], [11], [12]).

Using the identity $F_N(\tau) = \exp[-L(t-\tau)][f_r(t) + f_i(t)]$, we turn Eq. (2.8) into the form

$$f_i(t) = [1 + A(t-t_0)]^{-1}\left[-A(t-t_0)f_r(t) + e^{QL(t-t_0)}f_i(t_0)\right],$$

$$A(t-t_0) = -\int_0^{t-t_0} e^{QL\tau}QLPe^{-L\tau}d\tau = e^{QL(t-t_0)}Qe^{-L(t-t_0)} - Q. \tag{3.1}$$

Inserting Eq. (3.1) for $f_i(t)$ into equation (2.6), we obtain the TCL-GME [11], [12]

$$\frac{\partial}{\partial t}f_r(t) = PL\,[1 + A(t-t_0)]^{-1}\left[f_r(t) + e^{QL(t-t_0)}f_i(t_0)\right], \tag{3.2}$$

which contains the undesirable inhomogeneous initial condition term $\propto f_i(t_0)$.

Solving Eqs. (3.1) and (2.10), we get for $f_i(t_0)$

$$f_i(t_0) = \chi^{-1} C_0 e^{-L(t-t_0)} \left(1 + A(t-t_0)\right)^{-1} f_r(t), \tag{3.3}$$

where

$$\chi^{-1} = \left[1 - C_0 e^{-L(t-t_0)} \left(1 + A(t-t_0)\right)^{-1} e^{QL(t-t_0)}\right]^{-1}. \tag{3.4}$$

The inverse operator (3.4), needed for obtaining $f_i(t_0)$ from Eq. (3.3), is define by expanding it into the series in the term $\propto C_0$. At $t = t_0$, $A(t-t_0) = 0$, and there is no singularity at the initial moment of time if $|C_0| < 1$. From the existence of $(1 + A(t-t_0))^{-1}$[11], [12], it follows that χ^{-1} exists at any time if $|C_0| < 1$. As for the function $R(t-t_0)$ (2.13) in TC-HGME (2.12), an expansion in some small parameter can result in canceling the pole in (3.4) and, therefore, the condition $|C_0| < 1$ can be not necessary.

Inserting Eq. (3.3) into Eq. (3.2) and using (3.4), we obtain a desired time-convolutionless (time-local) homogeneous generalized master equation (TCL-HGME) for the relevant part of the distribution function (statistical operator) $f_r(t)$

$$\frac{\partial}{\partial t} f_r(t) = PL\left[1 + A(t-t_0)\right]^{-1} \left\{1 - C_0(t-t_0)\left[1 + A(t-t_0)\right]^{-1}\right\}^{-1} f_r(t). \tag{3.5}$$

Based on the above mentioned arguments, one can expect that all operators entering equation (3.5) exist.

Thus, using the introduced identity (2.10), we have transferred the inhomogeneous initial correlations term of equation (3.2) into the operator acting on the relevant part of the distribution function (statistical operator). Equation (3.5) is exact homogeneous time-local equation for the relevant part of a distribution function (statistical operator) which accounts for initial correlations via the time-dependent parameter of initial correlations $C_0(t-t_0)$ define by (2.13). This equation is expected to work on any time scale and to describe the entire evolution process. Equation (3.5) can also appear more convenient for studying the non-Markovian processes than the TC-HGME (2.12).

4. Evolution Equation Retaining Initial Correlations for a Homogeneous Dilute Gas of Classical Particles

In this section we consider a gas of N ($N \gg 1$) interacting identical classical particles. It is supposed that the Hamiltonian of the system can be split into two parts

$$H = H^0 + H',$$
$$H^0 = \sum_{i=1}^{N} H_i^0, \; H_i^0 = \frac{\mathbf{p}_i^2}{2m},$$
$$H' = \sum_{i<j=1}^{N} V_{ij}, \; V_{ij} = V(|\mathbf{x}_i - \mathbf{x}_j|). \tag{4.1}$$

Here, H^0describes an ideal gas of noninteracting particles having the momenta $\mathbf{p}_i$ and mass m, and the Hamiltonian of interaction between particles H' is assumed to be a sum of potentials V_{ij} dependent only on the difference of coordinates of two particles. We do not suppose here that the interaction is weak and assume that all usual necessary requirements to the properties of forces, by which particles interact with each other, are met. In particular, we suppose that the bound states are not formed (an interaction is repulsive).

To obtain the evolution equation in this case, we presume that the density of particles $n = N/V$ (V is the volume of the system) is small enough and use a perturbation theory with the small parameter

$$\gamma = r_o^3 n \ll 1, \tag{4.2}$$

where r_0 is the effective radius of the interparticle interaction. This small parameter guarantees that the collisions between particles are the well separated events and define the time hierarchy, because $t_{cor}/t_{rel} \sim \gamma \ll 1$.

First of all, we have to defin the projection operator P selecting the relevant part $f_r(x_1, \ldots, x_N, t)$ of the $N-$particle distribution function. It is convenient to deal with the $N-$particle distribution function define as $f_N(x_1, \ldots, x_N, t) = V^N F_N(x_1, \ldots, x_N, t)$, which satisfie the same Liouville equation (2.1). Then, the required projection operator is

$$P = \left[\prod_{i=2}^{N} f_1(x_i) \right] \frac{1}{V^{N-1}} \int dx_2 \cdots \int dx_N \, . \tag{4.3}$$

Here, $f_1(x_i)$ is a one-particle distribution function

$$f_1(x_i, t) = V \int dx_1 \cdots \int dx_{i-1} \int dx_{i+1} \cdots \int dx_N F_N(x_{1,\ldots,} x_N, t) \tag{4.4}$$

taken at the initial instant, $f_1(x_i) = f_1(x_i, t_0)$. In the definitio (4.4), a symmetry of a distribution function relatively to a permutation of space-phase coordinates $x_i = (\mathbf{x}_i, \mathbf{p}_i)$ is taken into account.

In general, the s-particle ($s \leq N$) distribution function is define as

$$f_s(x_1, \ldots, x_s, t) = V^s \int dx_{s+1} \cdots \int dx_N F_N(x_1, \ldots, x_N, t) \, . \tag{4.5}$$

From (2.2) we have the normalization conditions for the reduced distribution functions f_s

$$\int dx_1 \cdots \int dx_s f_s(x_1, \ldots, x_s, t) = V^s \, . \tag{4.6}$$

Applying the projection operator (4.3) to the N-particle distribution function $f_N(t) = V^N F_N(t)$, we obtain the following relevant part f_r

$$f_r = P f_N(t) = \left[\prod_{i=2}^{N} f_1(x_i) \right] f_1(x_1, t) \, . \tag{4.7}$$

We suppressed in (4.7) the arguments of the $f_r(x_1, \ldots, x_N, t)$ for brevity.

The irrelevant part comprising initial correlations, $f_i(t_0) = (1 - P) f_N(t_0) = f_N(t_0) - f_r(t_0)$, can always be represented by the following cluster expansion in terms of multiparticle correlation functions (see, for example, [20])

$$f_i(t_0) = \sum_{i,j=1,i<j}^{N} g_2(x_i,x_j) \prod_{k=1,k\neq i,j}^{N-2} f_1(x_k) + \sum_{i,j,k=1,i<j<k}^{N} g_3(x_i,x_j,x_k) \prod_{l=1,l\neq i,j,k}^{N-3} f_1(x_l) + \ldots, \tag{4.8}$$

where $\prod_{k=1,k\neq i,j}^{N-2} f_1(x_k)$ and $\prod_{l=1,l\neq i,j,k}^{N-3} f_1(x_l)$ stand for the products of $N-2$ and $N-3$ one-particle distribution functions with $k \neq i,j$ and $l \neq i,j,k$, respectively, whereas $g_2(x_i,x_j)$ and $g_3(x_i,x_j,x_k)$ are the irreducible two-particle and three-particle correlation functions (further terms in (4.8) are define in the same way).

In order to fin out the rules of acting of the Liouvillian L and the functions of the Liouvillian, which enter Eq. (2.12), on the relevant distribution function (4.7), note, that in accordance with the Hamiltonian (4.1), the Liouvillian L can be presented as

$$L = L^0 + L^{'} ,$$

$$L^0 = \sum_{i=1}^{N} L_i^0, L_i^0 = -\mathbf{v}_i \cdot \nabla_i, \mathbf{v}_i = \frac{\mathbf{p}_i}{m}, \nabla_i = \frac{\partial}{\partial \mathbf{x}_i},$$

$$L^{'} = \sum_{i<j=1}^{N} L_{ij}^{'}, L_{ij}^{'} = (\nabla_i V_{ij}) \cdot (\frac{\partial}{\partial \mathbf{p}_i} - \frac{\partial}{\partial \mathbf{p}_j}) . \tag{4.9}$$

We suppose, as usual, that all functions $\Phi(x_1,\ldots,x_N)$, define on the phase space, and their derivatives vanish at the boundaries of the configurationa space and at $\mathbf{p}_i = \pm\infty$. These properties of distribution functions and the form of the Liouville operators (4.9) lead to the following relations

$$\int dx_i L_i^0 \Phi(x_i,\ldots,x_N,t) = 0,$$

$$\int dx_i \int dx_j L_{ij}^{'} \Phi(x_1,\ldots,x_N,t) = 0. \tag{4.10}$$

These equations also follow, in general, from the Liouville-von-Neumann equation (2.1) and conditions (2.2).

Let us consider the firs term in the right-hand side of Eq. (2.12). Applying operators (4.9) to the relevant distribution function (4.7) and taking into account (4.10), we have

$$PL^0 f_r = \left[\prod_{i=2}^{N} f_1(x_i)\right] L_1^0 f_1(x_1,t),$$

$$PL^{'} f_r = \left[\prod_{i=2}^{N} f_1(x_i)\right] n \int dx_2 L_{12}^{'} f_1(x_2) f_1(x_1,t) . \tag{4.11}$$

To calculate the additional term (hereinafter we put $t_0 = 0$ because Eq. (2.12) is valid for any initial instant t_0)

$$C(t) = P(L^0 + L^{'})e^{Q(L^0+L^{'})t} \frac{1}{1 - C_0 e^{-(L^0+L^{'})t} e^{Q(L^0+L^{'})t}} C_0 e^{-(L^0+L^{'})t} f_r(t) , \tag{4.12}$$

arising due to initial correlations, we use the following relations

$$e^{(A+B)t} = e^{At} + \int_0^t d\theta e^{A(t-\theta)} B e^{(A+B)\theta}, \tag{4.13}$$

$$e^{L^0 t}\Phi(x_1, \ldots, x_N, t) = \left[\prod_{i=1}^{N} e^{L_i^0 t}\right] \Phi(x_1, \ldots, x_N, t)$$
$$= \Phi(\mathbf{x}_1 - \mathbf{v}_1 t, \mathbf{p}_1, \ldots, \mathbf{x}_N - \mathbf{v}_N t, \mathbf{p}_N, t), \tag{4.14}$$

where A and B are the arbitrary operators. Equation (4.14) follows from the definitio (4.9) of the Liouvillian L^0 of the free propagating particles. We also note, that the parameter of initial correlations (2.11) $C_0 = f_i(0)/[f_r(0) + f_i(0)]$ can be expanded into the power series in $f_i(0)/f_r(0)$. Using (4.7) and (4.8), we have

$$\frac{f_i(0)}{f_r(0)} = \sum_{i<j=1}^{N} \frac{g_2(x_i, x_j)}{f_1(x_i) f_1(x_j)} + \sum_{i<j<k} \frac{g_3(x_i, x_j, x_k)}{f_1(x_i) f_1(x_j) f_1(x_k)} + \ldots \tag{4.15}$$

If we use (4.13) and expand (4.12) in PL^0, $L^{'}$, $QL^{'}$ and $f_i(0)/f_r(0)$, then, it is easy to see that (4.12) can be presented as a sum of products of operators $PL^0, L^{'}, PL^{'}, \exp(L^0\theta)$, one-particle distribution functions and correlations functions (4.15). Applying the projection operator (4.3) to such products and accounting for (4.7), (4.9), (4.10), (4.14) and (4.15), we fin that (4.12) can be presented as a series expansion in the density parameter γ (4.2).

In what follows, we restrict ourselves to the linear approximation in the density parameter γ (4.2). Also, in order to simplify formulae, a spatially homogeneous case will be considered. In this case a one-particle distribution function does not depend on a particle coordinate, $f_1(x_j, t) = f_1(\mathbf{p}_j, t)$ ($\int f(\mathbf{p}, t)d\mathbf{p} = 1$), and the multiparticle correlation functions depend only on the differences of coordinates, e.g., $g_2(x_i, x_j) = g_2(\mathbf{x}_i - \mathbf{x}_j, \mathbf{p}_i, \mathbf{p}_j)$. A spatial homogeneity allows us to cancel all terms with PL^0(such as the firs equation (4.11)), because, according to (4.9) and (4.10), all these terms contain a derivative with respect to a variable $\mathbf{x}_1$, which is not under the integration. The second equation (4.11) also vanishes in a space homogeneous case, because we consider a potential V_{ij} (4.1) dependent of the particle coordinates difference. Taking also into account, that the terms containing $PL^{'}$ result in the expressions proportional at least to the firs power of n, we can replace operators $Q = 1 - P$ in (4.12) by the unity and simplify it to

$$C(t) = PL^{'} e^{(L^0 + L^{'})t} \frac{1}{1 - C_0} C_0 e^{-(L^0 + L^{'})t} f_r(t). \tag{4.16}$$

Expanding formally $\frac{1}{1-C_0}$ and C_0 in the power series in f_i/f_r, restricting ourselves to the firs order in n and then converting the series, we obtain the following expression for the contribution of initial correlations (4.12)

$$C(t) = \left[\prod_{i=2}^{N} f_1(\mathbf{p}_i)\right] n \int dx_2 L_{12}^{'} e^{L_{12}t} \frac{g_2(x_1, x_2)}{f_1(\mathbf{p}_1) f_1(\mathbf{p}_2)} e^{-L_{12}t} f_1(\mathbf{p}_2) f_1(\mathbf{p}_1, t), \tag{4.17}$$

where $L_{12} = L^0_{12} + L'_{12}$, $L^0_{12} = L^0_1 + L^0_2$, i.e. L_{12} is the two-particle Liouvillian (4.9), and $f_1(\mathbf{p}_i) = f_1(\mathbf{p}_i, t_0)$. The parameter of initial correlations $C_0(x_1, \ldots, x_N)$ in (4.16) under the accepted linear approximation in n (binary collisions) reduces to the parameter $C^{12}_0(x_1, x_2)$, which contains only a two-particle correlation function $g_2(x_1, x_2) = g_2(\mathbf{x}_1 - \mathbf{x}_2, \mathbf{p}_1, \mathbf{p}_2)$:

$$\frac{1}{1 - C^{12}_0(x_1, x_2)} C^{12}_0(x_1, x_2) = \frac{g_2(x_1, x_2)}{f_1(\mathbf{p}_1) f_1(\mathbf{p}_2)} \,. \tag{4.18}$$

Deriving (4.17), we have also taken into account that each additional integration over $x_3, \ldots$ adds an additional power of n and, therefore, in the linear approximation in n, all formulae can contain no more than one integration over the phase space.

Consider now the second (collision) term of the TC-HGME (2.12). Initial correlations also contribute to this term through the second term of function $R(t) = 1 + [R(t) - 1]$ (2.13). We start with the conventional collision term of the TC-GME taking $R(t) = 1$. To simplify this term, we note, that the following relations hold in general

$$\begin{aligned} PL^0 e^{L^0 t} Q &= 0, \\ e^{QL^0 t} Q &= e^{L^0 t} Q \,. \end{aligned} \tag{4.19}$$

The second equation (4.19) follows from the firs one, which in its turn follows from (4.3), (4.6), (4.9), (4.10), (4.13) and (4.14). The term in the considered collision integral, which is proportional to $QL^0 f_r(t)$, gives zero contribution in the case of spatial homogeneity. Thus, the kinetic term under consideration acquires the form

$$K(t) = \int_0^t dt_1 PL' e^{(L^0 + QL')t_1} QL' f_r(t - t_1) \,. \tag{4.20}$$

Using the procedure, described above, we have the collision term (4.20) in the accepted linear approximation in the density parameter γ as

$$K(t) = \left[\prod_{i=2}^{N} f_1(\mathbf{p}_i)\right] n \int_0^t dt_1 \int dx_2 L'_{12} e^{L_{12} t_1} L'_{12} f_1(\mathbf{p}_2) f_1(\mathbf{p}_1, t - t_1) \,. \tag{4.21}$$

In the same way we can consider the contribution of initial correlations to the collision integral. Accounting for the second term in $R(t)$ (2.13) in the kinetic term of equation (2.12) leads to the modificatio of the collision integral (4.21). In the linear approximation in γ we have for the kinetic term of equation (2.12) (instead of (4.21))

$$\begin{aligned} \tilde{K}(t) = \prod_{i=2}^{N} f_1(\mathbf{p}_i) n \int_0^t dt_1 \int dx_2 L'_{12} R_{12}(t) e^{L_{12} t_1} L'_{12} \\ \times f_1(\mathbf{p}_2) f_1(\mathbf{p}_1, t - t_1) \,, \end{aligned} \tag{4.22}$$

$$R_{12}(t) = 1 + e^{L_{12} t} \frac{g_2(x_1, x_2)}{f_1(\mathbf{p}_1) f_1(\mathbf{p}_2)} e^{-L_{12} t} \,. \tag{4.23}$$

Now, collecting all obtained formulae (4.17), (4.22) and (4.23) and accounting for the definitio of the relevant distribution function (4.7), we obtain from the TC-HGME (2.12) the following equation for a one-particle momentum distribution function

$$\frac{\partial}{\partial t} f_1(\mathbf{p}_1, t) = n \int dx_2 L'_{12} G_{12}(t) f_1(\mathbf{p}_2) f_1(\mathbf{p}_1, t)$$
$$+ n \int dx_2 \int_0^t dt_1 L'_{12}[1 + G_{12}(t)] e^{L_{12}t_1} L'_{12} f_1(\mathbf{p}_2) f_1(\mathbf{p}_1, t - t_1) , \quad (4.24)$$

where

$$G_{12}(t) = R_{12} - 1$$
$$= e^{L_{12}t} \frac{g_2(x_1, x_2)}{f_1(\mathbf{p}_1) f_1(\mathbf{p}_2)} e^{-L_{12}t} \quad (4.25)$$

is the parameter of initial correlations in the linear in γ approximation which is determined by the time evolution of a two-particle correlation function. It is interesting to note that the expression like (4.25) appears as an additional term in the (correlation) entropy caused by two-particle correlation function and obtained by means of quasi-equilibrium statistical operator (see, e.g. [21]). This additional contribution to entropy is essential when a two-particle correlation function damps slowly.

Equation (4.24) is a new equation for a one-particle distribution function obtained from TC-HGME (2.12) in the linear approximation in n and which accounts for both initial correlations and collisions in this approximation exactly. It has been obtained not using the Bogoliubov principle of weakening of initial correlations or any other hypothesis. First term in the right-hand side of this equation (linear in L'_{12}) is exclusively define by initial correlations evolving with time. Initial correlations also modify the collision integral (second term in the right-hand side of this equation). It contains, except the term related to an interaction between particles (determined by the Liouvillian L'_{12}), also time-dependent initial correlations determined by the two-particle correlation function (4.25)). Evolution in time of initial correlations as well as of the collision term is determined by the exact two-particle propagator $\exp(L_{12}t)$ only, and in this sense Eq. (4.24) is closed. As it follows from (4.13), the propagator $\exp(L_{12}t)$ satisfie the integral equation

$$\exp(L_{12}t) = \exp(L^0_{12}t) + \int_0^t dt_1 \exp[L^0_{12}(t - t_1)] L'_{12} \exp(L_{12}t_1) , \quad (4.26)$$

where $\exp(L^0_{12}t)$ is the propagator for noninteracting particles. In the case of a weak interparticle interaction, we can iterate the integral equation (4.26) in L'_{12} and obtain equation (4.24) in the desired approximation in the small interaction parameter (note, that the correlation function $g_2(x_1, x_2)$ also depends on an interaction between particles).

The obtained homogeneous integro-differential equation (4.24) differs from that which could be found from the conventional GME (2.9) using the projection operator (4.3) and the principle of weakening of initial correlations or RPA. In the latter case we would have equation (4.24) in which the time-dependent correlation function factor $G_{12}(t)$ determining the dynamics of initial correlations would be set zero. Moreover, such an equation would be valid only at at the time scale $t \gg t_{cor}$, whereas equation (4.24) is valid on any time scale.

It should be noted, that there can be the large-scale correlations associated with, e.g., the quantities that are conserved. These correlations can only vanish on the time scale of the order of the relaxation time t_{rel} of a distribution function $f_1(\mathbf{p}_i, t)$. If at $t \gg t_{cor}$ $(t_{cor} \ll t_{rel})$ the correlation function $G_{12}(t)$ was set zero, this would signify that all initial correlations vanish on this time scale, i.e. the large-scale correlations do not matter.

Equation (4.24) describes the entire evolution process and is expected to switch from initial (reversible) regime into the kinetic (irreversible) one, automatically. It will be shown below, that Eq. (4.24) converts on the appropriate time scale into the conventional Boltzmann equation if all correlations (initial and caused by collisions) damp with time.

5. Connection to the Classical Boltzmann Equation

Obtained closed equation (4.24) for a one-particle distribution function is not a kinetic equation in the conventional sense. It does not describe generally an irreversible evolution with time because it is time-reversible if the parameter of initial correlations (4.18) remains unchanged when all the particle velocities are reversed (see below). Thus, a time-asymmetric behaviour can be secured by the special choice of an initial condition. If we suppose that $C_0^{12}(\mathbf{x}_1 - \mathbf{x}_2, \mathbf{p}_1, \mathbf{p}_2) = C_0^{12}(\mathbf{x}_1 - \mathbf{x}_2, -\mathbf{p}_1, -\mathbf{p}_2)$, i.e., $g_2(\mathbf{x}_1 - \mathbf{x}_2, \mathbf{p}_1, \mathbf{p}_2) = g_2(\mathbf{x}_1 - \mathbf{x}_2, -\mathbf{p}_1, -\mathbf{p}_2)$ and $f_1(\mathbf{p}_i) = f_1(-\mathbf{p}_i)$, then equation (4.24) is invariant under the transformation $t \to -t$, because such a transformation results in changing the sign of all Liouvillians L_i^0 and $L_{ij}^{'}$. Such reversibility is understandable on the very early stage of the system evolution, when $t \lesssim t_{cor}$ and memory of the initial state has not been lost. On the time scale $t \gg t_{cor}$ the situation can change.

It can be seen from (4.24) that, in order to enter the kinetic (irreversible) stage of the evolution, the reversible terms connected with initial correlations should vanish on some time scale. Moreover, irreversibility of collision integral may be realized if it is possible to extend the integral over t_1 to infinit . Thus, we have to consider the behaviour of initial correlations and the integral over t_1 in Eq. (4.24) in time.

It should be noted that the irreversible evolution can go "forwards" or "backwards". The question as to which of these evolutions should be chosen cannot be answered on the basis of the present consideration of the time-reversible Liouville equation only, and we consider the evolution in the future $(t > 0)$.

In the accepted approximation of the binary collisions, the initial correlations are described by a correlation function $g_2(x_1, x_2)$ depending on coordinates and momenta of only two particles. The time evolution of this binary correlation function is governed by the two-particle propagator (4.26). In the zero approximation in the interaction $L_{12}^{'}$, the action of the propagator $\exp(L_{12}^0 t)$ on a correlation function $g_2(x_1, x_2)$ (and on all functions of coordinates) results in shifting the coordinates $\mathbf{x}_1$and $\mathbf{x}_2$ with time according to (4.14)

$$\exp(L_{12}^0 t) g_2(x_1, x_2) = g_2(\mathbf{r} - \mathbf{g}t, \mathbf{p}_1, \mathbf{p}_2) \,, \tag{5.1}$$

where $\mathbf{r} = \mathbf{x}_1 - \mathbf{x}_2$ is the initial distance between particles 1 and 2 and $\mathbf{g} = \mathbf{v}_1 - \mathbf{v}_2$ is the relative velocity. Therefore, in this case a binary correlation function changes with time according to a change of a distance between particles $|\mathbf{v}_1 - \mathbf{v}_2|t$, which is linear in time. In general, an evolution of a two-particle correlation function $g_2(x_1, x_2)$ under the action of the exact two-particle propagator $\exp(L_{12}t)$ can be found from the well-known solution to the two-body problem.

It is important to know how a correlation function behaves as the time passes. If there is a big fraction of a "parallel motion" and a distance between particles (or, more generally, a distance between points in the phase space) does not change significantl , a correlation function does not vanish, i.e. the initial correlations do not disentangle with time. It can be so even in the case of the ergodic fl w in the phase space. If a distance between particles grows with time and the radius of correlation between particles $r_{cor} \backsim r_0$ is limited, the correlation function $g_2(x_1, x_2)$ fades with time. But for existence of a limited relaxation time, i.e. a finit time for vanishing a correlation function, this function should diminish with time exponentially. As it is well known, such an exponential disentangling of correlations with time can be guaranteed by a mixing ergodic fl w in the phase space. The mixing ergodic fl w is closely connected with the property of the dynamical systems known as the local (stochastic) instability (see,e.g.,[22]). If we assume that the dynamics of the system of classical particles under consideration have the property of mixing fl w (and this is usually the case), then the initial correlations weak with time and disappear after a short correlation time t_{cor}. In this case, at $t \gg t_{cor}$, we can neglect the contribution of initial correlations to equation (4.24) and put $G_{12}(t) = 0$.

This cannot, however, be done if there are the correlations associated with the large-scale fluctuation such as the collective excitations, for example, the plasma oscillations. A relaxation time for such fluctuation can be of the order of the kinetic relaxation time $t_{rel} \gg t_{cor}$, and, therefore, they are essential on the kinetic and other large time scale regimes. The conventional kinetic equations do not take into account such large-scale correlations and do not describe the fluctuation of the distribution functions. A kinetic theory of fluctuation can be found in, e.g., [4].

Dependence of the initial correlation function $g_2(x_1, x_2)$ and a potential of an interparticle interaction $V_{ij} = V(|\mathbf{x}_i - \mathbf{x}_j|)$ on a distance between particles is essential for the behaviour of integrals over $\mathbf{x}_2$ in (4.24) with time. Let us consider an example, when this dependence is define as

$$g_2(x_1, x_2) = g_0 \exp(-\frac{r^2}{r_{cor}^2})\phi(\mathbf{p}_1, \mathbf{p}_2)\,,$$

$$V(r) = V_0 \exp(-\frac{r^2}{r_0^2}), \tag{5.2}$$

where g_0 and V_0 are the constant parameters and $\phi(\mathbf{p}_1, \mathbf{p}_2)$ is a properly normalized function of the particle momenta. We will estimate the time dependence of the terms in Eq. (4.24), determined by initial correlations, in the case of a weak interparticle interaction, when the time evolution is governed by the "free" propagator $\exp(L_{12}^0 t)$. The correlation function (5.2) under the action of $\exp(L_{12}^0 t)$ transforms according to (5.1) as

$$\exp(L_{12}^0 t) g_2(x_1, x_2) = g_2(\mathbf{r} - \mathbf{g}t, \mathbf{p}_1, \mathbf{p}_2) = g_0 \exp(-\frac{|\mathbf{r} - \mathbf{g}t|^2}{r_{cor}^2})\phi(\mathbf{p}_1, \mathbf{p}_2)\,. \tag{5.3}$$

The function (5.3) goes to zero at $t \to \infty$ and any fi ed distance $\mathbf{r}$ and velocity $\mathbf{g}$.

Using (5.2) and (5.3), the firs right-hand side term of Eq. (4.24) can be presented as

$$n \int d\mathbf{p}_2 \int d\mathbf{r} \left[\frac{\partial}{\partial \mathbf{r}} V_0 \exp(-\frac{r^2}{r_0^2}) \right] (\frac{\partial}{\partial \mathbf{p}_1} - \frac{\partial}{\partial \mathbf{p}_2}) \frac{g_2(\mathbf{r} - \mathbf{g}t, \mathbf{p}_1, \mathbf{p}_2)}{f_1(\mathbf{p}_1) f_1(\mathbf{p}_2)} f_1(\mathbf{p}_2) f_1(\mathbf{p}_1, t) , \tag{5.4}$$

where $g_2(\mathbf{r} - \mathbf{g}t, \mathbf{p}_1, \mathbf{p}_2)$ is define by (5.3).

It is easy to see, that the integral over $\mathbf{r}$ (5.4) is not equal to zero only if $t < t_{cor}$, where $t_{cor} \backsim r_{cor}/\bar{v} \sim r_0/\bar{v}$ and $\bar{v}$ is a typical (mean) particle velocity. At $t \gg t_{cor}$ the integral (5.4) practically vanishes because of the finit range r_0 of an interparticle interaction. Of course, such a behaviour is possible if a contribution of the "parallel motion" with small g to (5.4) is negligible. The term, related to initial correlations and contributing to the second right-hand side term of Eq. (4.24) (collision integral), displays the same behaviour with time. Thus, in the considered example, the terms with initial correlations in Eq. (4.24) vanish at $t \gg t_{cor}$ if the particles dynamics are characterized by the ergodic mixing fl w in the phase space.

If the effective interaction between particles vanishes at a distance between particles $r \gg r_{cor}$, then at $t_1 \gg t_{cor}$ the action of $\exp(L_{12}^0 t_1)$ on the interaction Liouvillian $L_{12}^{'}$ (4.9) under the integral over t_1 in (4.24) results in increasing the distance between particles beyond the radius of inter-particle interaction $r_0 \backsim r_{cor}$ and, therefore, in vanishing the integrand. As follows from Eq. (4.24), the initial moment $t = 0$ refers to the time just before the collision. Given that the time between collisions t_{rel}, which determines the time of an essential change of the one-particle distribution function $f_1(\mathbf{p}_i, t)$, exceeds considerably t_{cor} due to the condition (4.2), $f_1(\mathbf{p}_1, t - t_1)$ does not practically change within the interval $0 \leqslant t_1 \leqslant t_{cor}$. Thus, $f_1(\mathbf{p}_1, t - t_1)$ under the integral in (4.24) can be substituted by $f_1(\mathbf{p}_1, t)$ with the accuracy of $t_{cor}/t_{rel} \backsim \gamma \ll 1$. Moreover, at $t \gg t_{cor}$ the upper limit of integral over t_1 can be extended to infinit . After such a transformation the collision integral becomes Markovian and irreversible.

Therefore, the transition from the microscopic time scale $0 \leqslant t \leqslant t_{cor}$ to the macroscopic one $t \gg t_{cor}$ is essential for obtaining an irreversible Markovian kinetic equation. On the macroscopic time scale the reversible-in-time terms of Eq. (4.24) caused by initial correlations may vanish (if the particle dynamics have the necessary properties) and the collision term acquires the desired form. The extension of the upper limit of the integral in Eq. (4.24) to infinit ($t \to \infty$) and the existence of this limit are also essential for obtaining the irreversible equation from reversible Eq. (4.24). Existence of the limiting value of such an integral over time (at $t \to \infty$) was proved in [23] in the thermodynamic limit ($V \to \infty$, $N \to \infty$, $n = V/N$ is finite and when the time scales of collisions and of relaxation are widely separated ($t_{cor} \ll t_{rel}$).

In order to obtain the desired nonlinear equation for the one-particle distribution function $f_1(\mathbf{p}_1, t)$, we should make an additional approximation neglecting the time retardation of the one-particle distribution function which is possible on the time scale $t \ll t_{rel}$. In this case, we can approximate the function $f_1(\mathbf{p}_2) = f_1(\mathbf{p}_2, 0)$ in Eq. (4.24) by $f_1(\mathbf{p}_2, t)$ with the accuracy of $\gamma \ll 1$ (for more details see below).

Summing up, we can rewrite Eq. (4.24) on the considered macroscopic time scale (1.1) as

$$\frac{\partial}{\partial t} f_1(\mathbf{p}_1, t) = n \int dx_2 \int_0^\infty dt_1 L'_{12} \exp(L_{12}t) L'_{12} f_1(\mathbf{p}_1, t) f_1(\mathbf{p}_2, t) \,. \tag{5.5}$$

This equation is Markovian, time-irreversible and describes the relaxation of the system with the characteristic time t_{rel}. Using the two-particle scattering theory, it can be shown that Eq. (5.5) is equivalent to the conventional Boltzmann equation (see, e.g., [20]). In the case of a classical gas with a weak interparticle interaction, the exact two-particle propagator $\exp(L_{12}t)$ in (5.5) can be replaced by the "free" propagator $\exp(L^0_{12}t)$ according to (4.26). In this case (5.5) is equivalent to the Landau equation (see, [20]).

It is worth noting, that we have managed to obtain the Boltzmann equation (5.5), which is nonlinear as it should be, using the linear theory of projection operators. Actually, Eq. (4.24), which has been used for deriving the Boltzmann equation (5.5), is linear with respect to the distribution function $f_1(\mathbf{p}_1, t)$ but contains a one-particle distribution function $f_1(\mathbf{p}_2)$ at the initial moment $t = 0$ (this follows from using the time-independent projection operator (4.3) resulting in the relevant distribution function (4.7)). Change in time of the momentum distribution function can be presented as $f_1(\mathbf{p}_i, t) = f_1(\mathbf{p}_i) + (\partial f_1/\partial t)_0 t + \ldots$, where the derivative can be estimated as $\partial f_1/\partial t \sim f_1/t_{rel}$ and the relaxation time t_{rel} is determined by the collision integral in (4.24). On the time scale (1.1), there is no difference in Eq. (4.24) between $f_1(\mathbf{p}_2)$ and $f_1(\mathbf{p}_2, t)$ and that makes this equation nonlinear. Thus, the second inequality in (1.1), $t \ll t_{rel}$, is essential for obtaining a nonlinear evolution equation from the linear equation (2.12), which accounts for initial correlations, i.e. derived with no restriction on the time scale define by the firs inequality in (1.1), $t \gg t_{cor}$.

The described procedure of obtaining the time-irreversible Eq. (5.5) from Eq. (4.24) clearly indicates that irreversibility emerges on the macroscopic time scale (1.1) as a result of the loss with time of the information about both the initial correlations and correlations that emerge due to collisions.

6. Evolution Equation Retaining Initial Correlations for a Homogeneous Dilute Gas of Quantum Particles

Let us now test the general scheme outlined in Section 2 by applying it to the system of interacting quantum particles. In the quantum physics case the derived HGMEs hold in the forms obtained above but the Liouville operator L in Eq. (2.1) (and elsewhere) as well as the operator $\exp(Lt)$ are now the superoperators acting according to (2.4) on any operator $A(t)$ (particularly on the statistical operator $F_N(t)$ of N interacting quantum particles) as

$$LA(t) = \frac{1}{i\hbar}[H, A(t)], e^{Lt_1}A(t) = e^{\frac{1}{i\hbar}Ht_1}A(t)e^{-\frac{1}{i\hbar}Ht_1}, \tag{6.1}$$

respectively. Accordingly, Eqs. (2.12) and (3.5) for the relevant part of the statistical operator $f_r(t) = PF_N(t)$ hold with the appropriate redefinition of the symbols and operations used in these equations. The order of operators in these equations also matters.

In the space representation, the system of N quantum particles is described by the density matrix $F_N(\mathbf{r}_1,\ldots,\mathbf{r}_N,\mathbf{r}_1^{'},\ldots,\mathbf{r}_N^{'},t)$. This matrix should satisfy the symmetry conditions which reflec the statistics of the particles under consideration, i.e.

$$\begin{aligned} P_{ij}F_N = F_N P_{ij} = \theta F_N, \\ \theta = \pm 1, \end{aligned} \tag{6.2}$$

where P_{ij} is the operator of transmutation of any two variables $\mathbf{r}_i$and $\mathbf{r}_j$, when it stands to the left of F_N, or of $\mathbf{r}_i^{'}$and $\mathbf{r}_j^{'}$, when it acts from the write side of F_N (this rule holds for any operators acting on the matrixes). The sign plus is applicable for bosons and minus should be used for fermions.

We focus on the derivation of the evolution equation for a one-particle density matrix $f_1(\mathbf{r}_1,\mathbf{r}_1^{\prime},t)$, which is define according to the following definitio of the $s-$particle $(s \leq N)$ density matrix

$$f_s(\mathbf{r}_1,\ldots,\mathbf{r}_s,\mathbf{r}_1^{'},\ldots,\mathbf{r}_s^{'},t) = V^s Tr_{(s+1,\ldots,N)} F_N(\mathbf{r}_1,\ldots,\mathbf{r}_N,\mathbf{r}_1^{'},\ldots,\mathbf{r}_N^{'},t), \tag{6.3}$$

where $Tr_{(s+1,\ldots,N)}$ denotes the trace taken over the coordinates of $N-s$ particles $(s+1,\ldots,N)$. From the normalization condition (2.2) $TrF_N = 1$ it follows that $\frac{1}{V}Trf_1 = 1$.

It is convenient to introduce the following projection operator (compare with the defi nition (4.3))

$$P = \left[\prod_{i=2}^{N} f_1(\mathbf{r}_i,\mathbf{r}_i^{\prime})\right]\frac{1}{V^{N-1}}Tr_{(2,\ldots,N)}, \tag{6.4}$$

where $f_1(r_i,r_i^{\prime}) = f_1(r_i,r_i^{\prime},t_0)$ is the one-particle density matrix taken at the initial moment of time t_0. Therefore, the chosen relevant density matrix under consideration is

$$Pf_N(t) = f_r(t) = \left[\prod_{i=2}^{N} f_1(\mathbf{r}_i,\mathbf{r}_i^{\prime})\right] f_1(\mathbf{r}_1,\mathbf{r}_1^{\prime},t) \tag{6.5}$$

in accordance with the definitio (6.3) for the N-particle density matrix $f_N = V^N F_N$. As it is seen from equation (6.5), the selected relevant part of the N -particle density matrix does not contain any correlations. In order to defin the irrelevant part of the density matrix $f_i(t_0) = (1-P)f_N(t_0)$, which enters equations (2.12) and (3.5) and comprise all initial correlations, we have also to take into account the quantum correlations arising due to the symmetry condition (6.2). These correlations exist independently of the interaction between particles, which we introduce by means of the following Hamiltonian

$$\begin{aligned} H &= H^0 + H^{'}, \\ H^0 &= \sum_{i=1}^{N} K_i,\ H^{'} = \sum_{i<j=1}^{N} \Phi_{i,j}, \\ K_i &= -\frac{\hbar^2}{2m}\nabla^2_{\mathbf{r}_i},\ \Phi_{i,j} = \Phi(|\mathbf{r}_i - \mathbf{r}_j|). \end{aligned} \tag{6.6}$$

Thus, $f_N(t_0) = f_r(t_0) + f_i(t_0)$ can be presented in the form of the cluster expansion with

$$f_r(t_0) = \prod_{i=1}^{N} f_1(i) \tag{6.7}$$

and

$$f_i(t_0) = \sum_{i,j=1,i<j=1}^{N} \widetilde{g}_2(i,j) \prod_{k=1,k\neq i,j}^{N-2} f_1(k) + \sum_{i.j.k=1,i<j<k=1}^{N} \widetilde{g}_3(i,j,k) \prod_{l=1,l\neq i,j,k}^{N-3} f_1(l) + \ldots, \tag{6.8}$$

where we have used the short notation $f_1(i) = f_1(\mathbf{r}_i, \mathbf{r}_i^{'})$ (i.e. $i = \mathbf{r}_i, \mathbf{r}_i^{'}, i = 1, \ldots, N$) and, e.g., $\prod_{k=1,k\neq i,j}^{N-2} f_1(k)$ stands for the product of $N-2$ one-particle functions $f_1(k)$ with k taking values $1, \ldots, N$ but $k \neq i, j$ (compare with (4.8)). A two-particle $\widetilde{g}_2(i,j)$ and a three-particle $\widetilde{g}_3(i,j,k)$ correlation functions are define as (the correlation functions for more particles can be written down in the same way)

$$\begin{aligned}
\widetilde{g}_2(i,j) &= \theta P_{ij} f_1(i) f_1(j) + g_2(i,j), \\
\widetilde{g}_3(i,j,k) &= (P_{ij}P_{jk} + P_{ij}P_{ik}) f_1(i) f_1(j) f_1(k) + (\theta P_{ik} + \theta P_{jk}) g_2(i,j) f_1(k) \\
&+ (\theta P_{ij} + \theta P_{ik}) g_2(j,k) f_1(i) + (\theta P_{ij} + \theta P_{jk}) g_2(i,k) f_1(j) \\
&+ g_3(i,j,k).
\end{aligned} \tag{6.9}$$

The terms in the right-hand sides of Eqs. (6.9) with one, two, etc. permutation operators P_{ij} represent a two-, three-particle, etc. quantum correlations emerging due to the proper symmetry properties of $f_N(t_0)$ guaranteed by the symmetrization operators, e.g.,

$$f_2(i,j) = \theta P_{ij} f_2(\mathbf{r}_i, \mathbf{r}_j, \mathbf{r}_i^{'}, \mathbf{r}_j^{'}). \tag{6.10}$$

The irreducible two-, three-, etc. correlation functions, $g_2(i,j)$, $g_3(i,j,k)$, etc., are due to the interaction between particles $\Phi_{i,j}$ and are proportional to an interaction parameter ε in some power. Note, that in contrast to the classical physics case (see (4.8)), each correlation function in (6.9) contains correlations of all possible orders in ε, e.g. $\widetilde{g}_2(i,j)$ includes two-particle quantum correlations (existing even in the absence of interaction, when $\varepsilon = 0$) and two-particle correlations of the firs order in ε ($g_2(i,j)$). The same is valid for $\widetilde{g}_3(i,j,k)$, which contains a three-particle correlations of the zero, firs and second ($g_3(i,j,k)$) orders in ε.

The following useful relations hold for the permutation operators

$$\begin{aligned}
P_{ij}P_{jk} &= P_{ik}P_{ij}, \\
P_{jk}P_{ij} &= P_{ij}P_{ik}.
\end{aligned} \tag{6.11}$$

Let us now derive the evolution equation for a gas of interacting quantum particles in the lowest (first order in the small density of particles parameter n. The corresponding dimensionless small parameter is again (see (4.2))

$$\gamma = r_0^3 n \sim t_{cor}/t_{rel} \ll 1. \tag{6.12}$$

We will use the TC-HGME given by Eqs.(2.12) and (2.13) with appropriate redefinition of the symbols, as it was mentioned above.

The Liouville superoperator $L = L^0 + L^{'}$, corresponding to the Hamiltonian (6.6), is define by equation (6.1) as

$$L^0 = \sum_{i=1}^{N} L_i^0, L_i^0 = \frac{1}{i\hbar}(-\frac{\hbar^2}{2m})(\nabla^2_{\mathbf{r}_i} - \nabla^2_{\mathbf{r}_i^{'}}),$$
$$L^{'} = \sum_{i<j=1}^{N} L^{'}_{ij}, L^{'}_{ij} = \frac{1}{i\hbar}\left[\Phi(|\mathbf{r}_i - \mathbf{r}_j|) - \Phi(|\mathbf{r}_i^{'} - \mathbf{r}_j^{'}|)\right], \tag{6.13}$$

where we have used the abovementioned rule, that an operator standing on the right side (in the commutator) acts on the primed variables.

Using the definition (6.13), one can easily show that the following relations (similar to the classical physics case) hold

$$Tr_{(i)} L_i^0 \varphi_N(1, \ldots, i, \ldots, N) = 0,$$
$$Tr_{(i,j)} L^{'}_{ij} \varphi_N(1, \ldots, i, \ldots, j, \ldots, N) = 0, \tag{6.14}$$

where $\varphi_N(1, \ldots, i, \ldots, N)$ is some matrix, define for the $N-$particle system, which in the space representation takes the form $\varphi_N(1, \ldots, i, \ldots, N) = \varphi_N(\mathbf{r}_1, \ldots, \mathbf{r}_i, \ldots, \mathbf{r}_N, \mathbf{r}_1^{'}, \ldots, \mathbf{r}_i^{'}, \ldots, \mathbf{r}_N^{'})$ and satisfie the necessary boundary conditions.

Thus, using the projection operator (6.4) and the relevant density matrix (6.5), the firs two terms in TC-HGME (2.12) $P(L^0 + L^{'})f_r(t)$ (which coincide with those in TC-GME (2.9)) can be presented as

$$PL^0 f_r(t) = \left[\prod_{i=2}^{N} f_1(i)\right] \frac{1}{i\hbar}[-\frac{\hbar^2}{2m}\left(\nabla^2_{\mathbf{r}_1} - \nabla^2_{\mathbf{r}_1^{'}}\right)] f_1(\mathbf{r}_1, \mathbf{r}_1^{'}, t),$$
$$PL^{'} f_r(t) = \left[\prod_{i=2}^{N} f_1(i)\right] \frac{n}{i\hbar} \int d\mathbf{r}_2 [\Phi(|\mathbf{r}_1 - \mathbf{r}_2|)$$
$$- \Phi(|\mathbf{r}_1^{'} - \mathbf{r}_2|)] f_1(\mathbf{r}_2, \mathbf{r}_2) f_1(\mathbf{r}_1, \mathbf{r}_1^{'}, t)\}. \tag{6.15}$$

To simplify the derivation, we restrict ourselves to the spatially homogeneous case, when all matrixes are invariant under translations and, in particular, the matrix $f_1(\mathbf{r}_i, \mathbf{r}_i^{'})$ should be of the form

$$f_1(\mathbf{r}_i, \mathbf{r}_i^{'}, t) = f_1(\mathbf{r}_i - \mathbf{r}_i^{'}, t). \tag{6.16}$$

Then, it is easy to see that for functions (6.16), H^0 commutes with $f_r(t)$ and Eqs. (6.15) vanish, i.e.

$$L^0 f_r(t) = 0, PL^{'} f_r(t) = 0. \tag{6.17}$$

Also, in this spatially homogeneous case, for any matrix $\varphi_N(1, \ldots, N)$

$$PL^0 \varphi_N(1, \ldots, N) = 0. \tag{6.18}$$

Let us now consider the terms in TC-HGME (2.12) caused by initial correlations. We note, that in the considered case of identical interacting quantum particles, it is impossible to simply disregard the inhomogeneous term in the exact TC-GME (2.9) (assuming that correlations damp with the distance between particles) due to always existing (at any distances) quantum initial correlations (see (6.9)). Bogoliubov [1] tackled this problem by introducing a special boundary (initial) condition to the BBGKY chain at $t_0 \to -\infty$ implying that at this limiting moment of time the particles are located at distances beyond the correlation radius r_0, which means that correlation functions in (6.9), $g_2(i,j)$, $g_3(i,j,k)$, etc., caused by interaction between particles (and proportional to some power of ε), vanish, and

$$\lim_{t_0 \to -\infty} f_i(t_0) = \gamma_N \prod_{i=1}^{N} f_1(i), \tag{6.19}$$

where

$$\gamma_N = (1 + \theta P_{1N} + \ldots + \theta P_{N-1,N}) \ldots (1 + \theta P_{12}) \tag{6.20}$$

is the symmetrization operator for N particles. Therefore, in the Bogoliubov approach, the multi-particle density matrix at $t_0 \to -\infty$ is the properly symmetrized product of the one-particle density matrixes (compare with (6.7)-(6.9)). As in the classical physics case, this boundary condition introduces irreversibility into the system evolution and thus allows obtaining the quantum kinetic equations [24]. Also, as in the case of a dilute gas of classical particles, in order to obtain the closed kinetic equation for a one-particle density matrix, Bogoliubov used the specifi form of time-dependence of the multi-particle density matrix (valid only at $t - t_0 \gg t_{cor}$), i.e. he assumed that on this time scale the dependence on time of the multi-particle density matrices is determined by the time-dependence of a one-particle density matrix.

Here, we do not use any of the abovementioned Bogoliubov's assumptions. Thus, let us consider the (super)operator (see (2.13))

$$C(t) = R(t) - 1 = e^{QLt} \frac{1}{1 - C_0 e^{-Lt} e^{QLt}} C_0 e^{-Lt} \tag{6.21}$$

determining the influenc of initial correlations in Eq. (2.12) on the evolution process (hereinafter we put $t_0 = 0$, because the HGME is valid for any initial moment t_0). In order to expand exponential (super)operators in the power series, we use the relation (4.13) which is also valid for any (super)operators A and B.

Using (4.13) and expanding (6.21) in PL^0 and QL', one can see that all terms with PL^0 vanish due to Eq. (6.18) and, therefore, all QL in (6.21) can be replaced by $L^0 + QL'$. Taking into account that the terms containing PL' result in the expressions proportional to at least the firs power of n (see, e.g.,(6.15)) and looking for the lowest (first order approximation in n of Eq. (2.12), we can substitute $Q = 1 - P$ in (6.21) with the unity and simplify this expression to

$$C(t) = e^{Lt} \frac{1}{1 - C_0} C_0 e^{-Lt}. \tag{6.22}$$

Using expression (2.11) for C_0, we have

$$\frac{1}{1 - C_0} C_0 = f_i(0) f_r^{-1}(0). \tag{6.23}$$

This is a remarkable result showing that in this approximation there is no problem with the existence (convergence/invertability) of $R(t)$ (2.13) (the same has been shown in Sec. 4 for the classical physics case). In connection with that it is worth reminding, that if in some part of the definitio space $f_r(0)$ goes to zero, so should do $f_i(0)$ (at the same rate or more promptly than the relevant part). It follows from the possibility of breaking the statistical operator into the uncorrelated (f_r) and correlated (f_i) parts (see (6.7)-(6.9)), which in turn caused by the fact that if an interaction between particles is switched off ($\varepsilon = 0$), the statistical operator should be given by symmetrized product of one-particle operators. Moreover, in the equations for the relevant part of statistical operator obtained below, as a result of approximation (6.23), the factor $f_i(0)f_r^{-1}(0)$ is always multiplied by $f_r(0)$ (see (6.41) below and take into account that in the approximation under consideration the relevant part changes slowly). This fact additionally guarantees an absence of any divergencies (in other words, any manipulations with initial condition term should result in terms behaving like $f_i(0)$ in accordance with the identity (2.10)).

Now, it is necessary to defin $f_r^{-1}(0)$. In the spatially homogeneous case under consideration, it is convenient to work in the momentum representation. The relation between the space and momentum representations for a one-particle density matrix of interest is

$$f_1(\mathbf{r}, \mathbf{r}', t) = \frac{1}{(2\pi\hbar)^3} \int f_1(\mathbf{p}, \mathbf{p}', t) e^{i\mathbf{pr}/\hbar} e^{-i\mathbf{p}'\mathbf{r}'/\hbar} d\mathbf{p} d\mathbf{p}',$$
$$f_1(\mathbf{p}, \mathbf{p}', t) = (2\pi\hbar)^3 w(\mathbf{p}, \mathbf{t}) \delta(\mathbf{p} - \mathbf{p}'), \tag{6.24}$$

where a function $w(\mathbf{p})$ (with the dimensionality $[p^{-3}]$) is normalized to the unity and corresponds to a classical momentum distribution function. According to the definitio

$$\int f_1(\mathbf{p}, \mathbf{p}'', t) f_1^{-1}(\mathbf{p}'', \mathbf{p}', t) d\mathbf{p}'' = \delta(\mathbf{p} - \mathbf{p}'), \tag{6.25}$$

we can take for $f_1^{-1}(\mathbf{p}, \mathbf{p}')$ the following matrix

$$f_1^{-1}(\mathbf{p}, \mathbf{p}', t) = (2\pi\hbar)^{-3} w^{-1}(\mathbf{p}, \mathbf{t}) \delta(\mathbf{p} - \mathbf{p}'). \tag{6.26}$$

Note, that the δ-functions in (6.24)-(6.26) (related to the fact that a one-particle distribution function is diagonal in the spatially homogeneous case) always stay under the integrals or are divided out.

In accordance with (6.24) and (6.26), an invert one-particle matrix in the space representation may be define as

$$f_1^{-1}(\mathbf{r}, \mathbf{r}', t) = \frac{1}{(2\pi\hbar)^3} \int f_1^{-1}(\mathbf{p}, \mathbf{p}', t) e^{i\mathbf{pr}/\hbar} e^{-i\mathbf{p}'\mathbf{r}'/\hbar} d\mathbf{p} d\mathbf{p}'$$
$$= \frac{1}{(2\pi\hbar)^6} \int w^{-1}(\mathbf{p}, \mathbf{t}) e^{i\mathbf{p}(\mathbf{r}-\mathbf{r}')} d\mathbf{p}. \tag{6.27}$$

It satisfie the necessary relation

$$\int f_1(\mathbf{r}, \mathbf{r}'', t) f_1^{-1}(\mathbf{r}'', \mathbf{r}', t) d\mathbf{r}'' = \delta(\mathbf{r} - \mathbf{r}'). \tag{6.28}$$

Now,we can defin an operator, invert to the relevant operator (6.7), as

$$f_r^{-1}(0) = \prod_{i=1}^{N} f_1^{-1}(i), \tag{6.29}$$

because, due to (6.25) and (6.28),

$$[f_r(0)f_r^{-1}(0)] = \prod_{i=1}^{N} I(i). \tag{6.30}$$

Here and further, (i) stands for $(\mathbf{r}_i, \mathbf{r}_i^{'})$ or $(\mathbf{p}_i, \mathbf{p}_i^{'})$ depending on the representation used, and $I(i)$ is the unity matrix in the space of i-particle

$$I(i) = I(\mathbf{r}_i, \mathbf{r}_i^{'}) = \delta(\mathbf{r}_i - \mathbf{r}_i^{'}),$$
$$I(i) = I(\mathbf{p}_i, \mathbf{p}_i^{'}) = \delta(\mathbf{p}_i - \mathbf{p}_i^{'}) \tag{6.31}$$

in the space and momentum representations, respectively.

Applying the introduced definition for invert matrixes and using Eq. (6.8) for the irrelevant density matrix, the correlation parameter (6.23) can be rewritten in the following way

$$f_i(0)f_r^{-1}(0) = \sum_{i,j=1,i<j}^{N} \tilde{g}(i,j)f_1^{-1}(i)f_1^{-1}(j) \prod_{k=1,k\neq i,j}^{N-2} I(k)$$
$$+ \sum_{i,j,k=1,i<j<k}^{N} \tilde{g}(i,j,k)f_1^{-1}(i)f_1^{-1}(j)f_1^{-1}(k) \prod_{l=1,l\neq i,j,k}^{N-3} I(l) + \ldots, \tag{6.32}$$

where, e.g., the two-particle part of (6.32), according to the expressions for correlation functions (6.9), is

$$\tilde{g}(i,j)f_1^{-1}(i)f_1^{-1}(j) = \theta P_{ij} + g(i,j)f_1^{-1}(i)f_1^{-1}(j), \tag{6.33}$$

and the matrix elements of the permutation operator P_{ij} are define in the space and momentum representations as

$$P_{ij}(\mathbf{r}_i, \mathbf{r}_j, \mathbf{r}_i^{'}, \mathbf{r}_j^{'}) = \delta(\mathbf{r}_i - \mathbf{r}_j^{'})\delta(\mathbf{r}_j - \mathbf{r}_i^{'}),$$
$$P_{ij}(\mathbf{p}_i, \mathbf{p}_j, \mathbf{p}_i^{'}, \mathbf{p}_j^{'}) = \delta(\mathbf{p}_i - \mathbf{p}_j^{'})\delta(\mathbf{p}_j - \mathbf{p}_i^{'}). \tag{6.34}$$

Let us consider the additional (to the fl w) term in TC-HGME (2.12), $F(t) = PL^{'}C(t)f_r(t)$, caused by initial correlations, in the linear approximation in n. Using (6.22), (6.23) and (6.32), (6.33), one can obtain in this approximation

$$F(t) = PL^{'}e^{Lt}\frac{1}{1-C_0}C_0e^{-Lt}f_r(t)$$
$$= \left[\prod_{i=2}^{N} f_1(i)\right] \frac{n}{i\hbar}Tr_{(2)}L_{12}^{'}e^{-\frac{i}{\hbar}(H_{12}^0+H_{12}^{'})t}[\theta P_{12}$$
$$+ g(1,2)f_1^{-1}(1)f_1^{-1}(2)]e^{\frac{i}{\hbar}(H_{12}^0+H_{12}^{'})t}f_1(2)f_1(1,t). \tag{6.35}$$

Deriving (6.35), we have taken into consideration that each additional integration over the coordinate or momentum space adds an additional power of n and, therefore, in the linear approximation in n, all formulae can contain no more than one integration (as in (6.35)). Thus, Eq. (6.35) is determined only by the two-particle dynamics described by the Hamiltonian $H_{12} = H_{12}^0 + H_{12}'$ (and the corresponding Liouvillian $L_{12} = L_{12}^0 + L_{12}'$). We note, that (6.35) is different from the corresponding additional (to fl w) term (caused by initial correlations) obtained in Sec. 4 for the classical physics case: except a two-particle correlation function, Eq. (6.35) contains a symmetrization operator P_{12} responsible for quantum correlations.

Taking into account (6.34) and invariance of the Hamiltonian H_{12} with regard to renaming the particles, i.e. $H_{12}(\mathbf{p}_1, \mathbf{p}_2, \mathbf{p}_1', \mathbf{p}_2') = H_{12}(\mathbf{p}_2, \mathbf{p}_1, \mathbf{p}_2', \mathbf{p}_1')$, it can be shown that in the operator form we have (see, e.g. [25])

$$P_{12}H_{12}P_{12}^{-1} = H_{12} \tag{6.36}$$

(i.e. a symmetrization operator commutes with the Hamiltonian) and, therefore, $e^{-\frac{i}{\hbar}(H_{12}^0+H_{12}')t}P_{12}e^{\frac{i}{\hbar}(H_{12}^0+H_{12}')t} = P_{12}$. Now, it is easy to show that the firs term in (6.35), caused by the symmetrization operator P_{12}, vanishes in the spatially homogeneous case (for non-symmetrized $f_r(t)$ it is demonstrated by Eq. (6.17)). Thus,

$$PL_{12}'P_{12}f_r(t) = n[\prod_{i=2}^{N} f_1(i)] \int d\mathbf{r}_2 \frac{1}{i\hbar}[\Phi(|\mathbf{r}_1 - \mathbf{r}_2|)$$

$$-\Phi(\left|\mathbf{r}_1' - \mathbf{r}_2\right|)]f_1(\mathbf{r}_1 - \mathbf{r}_2)f_1(\mathbf{r}_2 - \mathbf{r}_1', t) = 0. \tag{6.37}$$

This can be easily proven, if we substitute the integration variable $\mathbf{r}_2$ in (6.37) by $\rho = \mathbf{r}_1 - \mathbf{r}_2$ and $\rho' = \mathbf{r}_2 - \mathbf{r}_1'$ in the firs and second integrals, respectively, and take into consideration, that in the adopted approximation (firs order in n) the time-dependence of a one-particle density matrix can be disregarded due to the fact that the change with time is described by the (super)operator which is already of the firs order in n for the spatially homogeneous system.

Hence, the contribution of initial correlations to the fl w term of Eq. (2.12) in the firs order in n and for the spatially homogeneous system is

$$PL'C(t)f_r(t) = \left[\prod_{i=2}^{N} f_1(i)\right] nTr_{(2)}L_{12}'G_{12}^q(t)f_1(2)f_1(1,t), \tag{6.38}$$

where

$$G_{12}^q(t) = e^{-\frac{i}{\hbar}H_{12}t}g(1,2)f_1^{-1}(1)f_1^{-1}(2)e^{\frac{i}{\hbar}H_{12}t} \tag{6.39}$$

is a time-dependent quantum two-particle correlation function. This result is similar to that obtained for the classical physics case (see (4.24) and (4.25)): the initial correlations lead to the appearance of the additional, linear in interaction between particles (H_{12}'), term which is determined by the evolution in time of a two-particle correlation function. Moreover, this result is closed in the sense that the time-evolution is also completely determined by the two-particle dynamics with the Hamiltonian H_{12}.

The collision term of Eq. (2.12) in the adopted approximation can be obtained in the same way. Accounting for equations (6.17), (6.18), (4.13), (6.22), (6.23), (6.32), (6.33),

(6.36) and remembering that any additional power of PL' and any additional integration (taking a trace) over the coordinate or momentum space adds an additional power of density n, it is not difficul to show that in the linear approximation in n the collision integral in (2.12) reduces to

$$K^q(t) = \left[\prod_{i=2}^{N} f_1(i)\right] n \int_0^t dt_1 Tr_{(2)} L'_{12} \left[1 + \theta P_{12} + G^q_{12}(t)\right] e^{L_{12}t_1} \times L'_{12} f_1(2) f_1(1, t - t_1). \quad (6.40)$$

This result is different from that obtained for the classical physics case, $K(t) + \widetilde{K}(t)$, given by Eqs. (4.21), (4.22), due to the presence of quantum correlations stipulated by a symmetrization operator P_{12}.

Collecting the obtained results (formulae (6.38) and (6.40)) and accounting for the definition of the relevant statistical operator (6.5), we obtain from the TC-HGME (2.12) the following equation for a one-particle density matrix in the linear approximation in n for the spatially homogeneous case

$$\frac{\partial}{\partial t} f_1(1, t) = n Tr_{(2)} L'_{12} G^q_{12}(t) f_1(2) f_1(1, t) + n \int_0^t dt_1 Tr_{(2)} L'_{12} \left[1 + \theta P_{12} + G^q_{12}(t)\right] e^{L_{12}t_1} L'_{12} f_1(2) f_1(1, t - t_1). \quad (6.41)$$

We have obtained a new closed homogeneous integro-differential equation for a one-particle density matrix retaining initial correlations which is valid on any time scale (i.e. we can use equation (6.41) for considering in detail all stages of evolution of a one-particle density matrix). No approximation like the Bogoliubov principle of weakening of initial correlations (valid only on a large time scale $t \gg t_{cor}$) has been used for deriving equation (6.41). This equation is exact in the linear approximation in n and, therefore, accounts for initial correlations and collisions in this approximation exactly. The firs term in the right-hand side of (6.41), linear in interaction L'_{12}, is determined by initial correlations caused by interaction between particles (this term is absent if we apply the boundary condition (6.19)). Initial correlations, both quantum and due to interaction, also modify the second (collision) term of (6.41). Note, that the symmetrization operator θP_{12} (accounting for the particle statistics) appears in this term as a result of a regular procedure outlined above but not as a result of using the boundary condition (6.19), as it is the case when one applies the Bogoliubov principle of weakening of initial correlations [1]. Also, when this principle is applied, the correlation function $G^q_{12}(t)$ does not show up in the second term of the right-hand side of (6.41)). Evolution with time in (6.41) is completely determined by the exact two-particle dynamics, described by the Hamiltonian H_{12} (which is quite natural for the considered linear in n approximation), and in this sense the obtained equation is closed.

7. Connection to the Quantum Boltzmann Equation

Equation (6.41) is not a kinetic equation in the conventional sense because it is time reversible (if the correlation function $g_2(i, j) f_1^{-1}(i) f_1^{-1}(j)$ does not change when the particle velocities are reversed). Thus, one of the possibilities to secure a time-asymmetric behaviour can be the special choice of an initial condition (e.g., the factorized one, like (6.19)).

It is instructive to consider the transition from Eq. (6.41) to an irreversible kinetic equation describing the evolution, say, into the future ($t > 0$). As it is seen from (6.41), in order to enter the kinetic (irreversible) stage of the evolution, the reversible terms caused by initial correlation function, $G^q_{12}(t)$, should vanish on some time scale. Let us suppose that this is the case, on the time scale $t \gg t_{cor}$, i.e.

$$G^q_{12}(t) = 0, t \gg t_{cor}. \tag{7.1}$$

Here, the correlation time can be estimated as $t_{cor} \sim r_0/\overline{v}$, where r_0 is the radius of inter-particle interaction and $\overline{v}$ is a typical mean particle velocity. Thus, if the dynamics of the system of particles under consideration is such (e.g., a mixing ergodic fl w in the phase space) that all initial correlations caused by inter-particle interaction vanish on the time scale $t \gg t_{cor}$, as it is supposed by condition (7.1), then Eq. (6.41) reduces to

$$\frac{\partial}{\partial t} f_1(1,t) = n \int_0^t dt_1 Tr_{(2)} L'_{12} e^{L_{12} t_1} L'_{12} (1 + \theta P_{12}) f_r^{12}(t - t_1),$$
$$f_r^{12}(t - t_1) = f_1(2) f_1(1, t - t_1), t \gg t_{cor}, \tag{7.2}$$

where (6.36) is also taken into account. Making use of the definition (6.1), this equation can be rewritten as

$$\frac{\partial f_1(1,t)}{\partial t} = -\frac{n}{\hbar^2} \int_{-\infty}^{\infty} dE \int_{-\infty}^{\infty} dE' \int_0^t dt_1 e^{\frac{1}{i\hbar}(E - E')t_1} Tr_{(2)} \Phi_{12}(E, E', t - t_1), t \gg t_{cor},$$
$$\Phi_{12}(E, E', t - t_1) = \left[H'_{12}, \delta(E - H_{12}) \left[H'_{12}, (1 + \theta P_{12}) f_r^{12}(t - t_1)\right] \delta(E' - H_{12})\right]. \tag{7.3}$$

In the adopted firs approximation in n, we can consider $f_r^{12}(t - t_1)$ under the integral in the zero approximation in n, in which it does not change with time. Thus, one can approximate the two-particle relevant part of a density matrix under the integral in (7.3) as $f_r^{12}(t - t_1) = f_1(1,t) f_1(2,t)$. This approximation is valid on the time scale $t \ll t_{rel}$ (see inequality (1.1 at $t_0 = 0$)) allowing the neglection of time retardation and allows to convert the linear equation for $f_1(1,t)$ (6.41) (or (7.3)) into the nonlinear one. Now, the integral in (7.3) over t_1 can be calculated. Note, that for finit t equation (7.3) is time reversible and integral over t_1 is a periodic function of t for a finit volume of a system V. This equation becomes time irreversible if the upper limit of integral can be extended to infinit , $t \rightarrow \infty$, and this limiting value of the integral over t_1 exists. However, this limiting value of the integral can evidently exist only for a large enough volume of the system under consideration. In other words, in order to exclude the long Poincare's cycles, the thermodynamic limiting procedure ($V \rightarrow \infty, N \rightarrow \infty, n = \frac{N}{V}$ is finite is needed before calculating the integral. The existing of this integral in the thermodynamic limit was proved in [23]. If the mentioned limiting value of the integral exists, then the integral can be considered in the Abel's sense, i.e.

$$\lim_{t\to\infty}\int\limits_0^t dt_1 e^{\frac{1}{i\hbar}(E-E^{'})t_1}$$

$$=\lim_{\varepsilon\to 0^+}\lim_{t\to\infty}\int\limits_0^t dt_1 e^{-\varepsilon t_1}e^{\frac{1}{i\hbar}(E-E^{'})t_1}=-i\hbar\frac{\Pr}{E-E^{'}}+\pi\hbar\delta(E-E^{'}), \tag{7.4}$$

where the order of taking limits matters and $\Pr$ is a symbol of taking a principal value of an integral.

Thus, we consider the system evolution on the time interval, which satisfie two inequalities: $t\gg t_{cor}$ (to exclude the influenc of initial correlations) and $t\ll t_{rel}$ (to go from linear Eq. (6.41) to nonlinear equation). This is the interval (1.1) which existence is guaranteed by the condition (6.12). Using (7.4), Eq. (7.3) can be rewritten as

$$\frac{\partial f_1(1,t)}{\partial t}=-\frac{\pi n}{\hbar}\int dETr_{(2)}\Phi_{12}(E,t),$$
$$\Phi_{12}(E,t)=\left[H_{12}^{'},\delta(E-H_{12})\left[H_{12}^{'},(1+\theta P_{12})f_1(1,t)f_1(2,t)\right]\delta(E-H_{12})\right]. \tag{7.5}$$

Obtaining (7.5), we have taken into account that Eq. (7.3) contains only the diagonal matrix element $\Phi_{12}(E,E^{'},\mathbf{p}_1,\mathbf{p}_2,\mathbf{p}_1,\mathbf{p}_2,t)$ as it should be because of taking a trace over $\mathbf{p}_2$ and because the left hand side of (7.3) is diagonal relatively to $\mathbf{p}$ and $\mathbf{p}^{'}$(see (6.24)). The latter property of $Tr_{(2)}\Phi_{12}(E,t)$ can be also proven explicitly. This diagonal element has the symmetry property $\Phi_{12}(E,E^{'},\mathbf{p}_1,\mathbf{p}_2,\mathbf{p}_1,\mathbf{p}_2,t)=\Phi_{12}(E^{'},E,\mathbf{p}_1,\mathbf{p}_2,\mathbf{p}_1,\mathbf{p}_2,t)$, and, therefore, the imaginary (principal) part of the integral over E and $E^{'}$ vanishes, as it follows from (7.3) and (7.4). The mentioned symmetry of $\Phi_{12}(E,E^{'},\mathbf{p}_1,\mathbf{p}_2,\mathbf{p}_1,\mathbf{p}_2,t)$ (7.3) is a consequence of the symmetry of matrix elements of H_{12}^0 and $H_{12}^{'}$ (and, therefore, of H_{12}). The latter symmetry follows from the definitio (6.6) and means that $H_{12}^0(\mathbf{p}_1,\mathbf{p}_2,\mathbf{p}_1^{'},\mathbf{p}_2^{'})=H_{12}^0(\mathbf{p}_1^{'},\mathbf{p}_2^{'},\mathbf{p}_1,\mathbf{p}_2)$ and $H_{12}^{'}(\mathbf{p}_1,\mathbf{p}_2,\mathbf{p}_1^{'},\mathbf{p}_2^{'})=H_{12}^{'}(\mathbf{p}_1^{'},\mathbf{p}_2^{'},\mathbf{p}_1,\mathbf{p}_2)$, where

$$H_{12}^0(\mathbf{p}_1,\mathbf{p}_2,\mathbf{p}_1^{'},\mathbf{p}_2^{'})=\left[K(\mathbf{p}_1)+K(\mathbf{p}_2)\right]\delta(\mathbf{p}_1-\mathbf{p}_1^{'})\delta(\mathbf{p}_2-\mathbf{p}_2^{'}),\, K(\mathbf{p}_i)=\frac{\mathbf{p}_i^2}{2m},$$
$$H_{12}^{'}(\mathbf{p}_1,\mathbf{p}_2,\mathbf{p}_1^{'},\mathbf{p}_2^{'})=\frac{1}{(2\pi\hbar)^3}\nu(\mathbf{p}_1-\mathbf{p}_1^{'})\delta(\mathbf{p}_1+\mathbf{p}_2-\mathbf{p}_1^{'}-\mathbf{p}_2^{'}),$$
$$\nu(\mathbf{p})=\int\Phi(|\mathbf{r}|)\exp(-\frac{i}{\hbar}\mathbf{p}\mathbf{r})d\mathbf{r},\nu(\mathbf{p})=\nu(-\mathbf{p}). \tag{7.6}$$

The (operator) δ-function of the Hamiltonian H (entering equation (7.5) with a two-particle Hamiltonian H_{12}) can be expressed via the imaginary part of Green's function $G(E)$:

$$\delta(E-H)=\mp\frac{1}{\pi}\operatorname{Im}G(E^{\pm}),G(E^{\pm})=\frac{1}{E^{\pm}-H},$$
$$E^{\pm}=E\pm i\varepsilon,\varepsilon=0^+. \tag{7.7}$$

On the other hand, there is the so called optical theorem relating $\delta(E-H)$ (the imaginary part of Green's function $G(E^{\pm})$ for the full Hamiltonian $H=H^0+H^{'}$) and the interaction

Hamiltonian $H^{'}$ to $\delta(E - H^0)$ (the imaginary part of Green's function $G^0(E^{\pm}) = (E^{\pm} - H^0)^{-1}$) and the T-matrix (see, e.g., [21]):

$$H'\delta(E - H)H^{'} = T^{+}\delta(E - H^0)T^{-} = T^{-}\delta(E - H^0)T^{+}, \quad (7.8)$$

where the T-matrix is define as

$$T^{\pm} = T(E^{\pm}) = H^{'}G(E^{\pm})G^{0^{-1}}(E^{\pm}) \quad (7.9)$$

and satisfie the equations

$$T^{\pm} = H^{'} + H^{'}G^0(E^{\pm})T^{\pm} = H^{'} + T^{\pm}G^0(E^{\pm})H^{'}. \quad (7.10)$$

The relations (7.7) - (7.9) allow to convert $\Phi_{12}(E,t)$ (7.5) into the following form

$$\Phi_{12}(E,t) = \left[H_{12}^{'-1}, T_{12}^{+}\delta(E - H_{12}^0)T_{12}^{-}\left[H_{12}^{'-1}, (1+\theta P_{12})f_1(1,t)f_1(2,t)\right]T_{12}^{-}\delta(E - H_{12}^0)T_{12}^{+}\right], \quad (7.11)$$

where $T_{12}^{\pm}$ refers to the T-matrix (7.9) define for the two-particle Hamiltonian $H_{12} = H_{12}^0 + H_{12}^{'}$. Using (7.10), it is not difficul to show that

$$T_{12}^{-}\left[H_{12}^{'-1}, (1+\theta P_{12})f_1(1,t)f_1(2,t)\right]T_{12}^{-} = -\left[T_{12}^{-}, (1+\theta P_{12})f_1(1,t)f_1(2,t)\right], \quad (7.12)$$

where we have also taken into account that $G_{12}^0(E^{\pm}) = (E^{\pm} - H_{12}^0)^{-1}$ commutes with $(1+\theta P_{12})f_1(1,t)f_1(2,t)$ for the spatially homogeneous case (see also (6.17)).

Using (7.12), (7.11), (7.10), and diagonality of $G^0(E^{\pm})$ and of both sides of Eq. (7.5) in the momentum representation, we can transform this equation into the following form

$$\frac{\partial f_1(1,t)}{\partial t} = -\frac{\pi n}{\hbar}\int dE Tr_{(2)}\left[T_{12}^{+}, \delta(E - H^0)[T_{12}^{-}, (1+\theta P_{12})f_1(1,t)f_1(2,t)]\delta(E - H^0)\right]. \quad (7.13)$$

To calculate the trace (integral over $\mathbf{p}_2$) in (7.13), one need to have the matrix elements of T-matrix in the $\mathbf{p}$-representation. Iterating equation (7.10) and calculating the matrix elements in the momentum representation, one can show that the T_{12} matrix (describing the particle collisions) guarantees the momentum conservation, i.e. $T_{12}^{\pm}(\mathbf{p}_1, \mathbf{p}_2, \mathbf{p}_1^{'}, \mathbf{p}_2^{'}) \sim \delta(\mathbf{p}_1 + \mathbf{p}_2 - \mathbf{p}_1^{'} - \mathbf{p}_2^{'})$. For example, in the firs approximation in $H_{12}^{'}$, we have $T_{12}^{\pm(1)}(\mathbf{p}_1, \mathbf{p}_2, \mathbf{p}_1^{'}, \mathbf{p}_2^{'}) = H_{12}^{'}(\mathbf{p}_1, \mathbf{p}_2, \mathbf{p}_1^{'}, \mathbf{p}_2^{'})$, where the matrix element of the interaction Hamiltonian is given by (7.6). In general, it is not difficul to obtain from (7.10) and (7.6) the following expression for the matrix element of T_{12} in the momentum representation:

$$\begin{aligned} T_{12}^{\pm}(\mathbf{p}_1, \mathbf{p}_2, \mathbf{p}_1^{'}, \mathbf{p}_2^{'}) &= \frac{1}{(2\pi\hbar)^3}t^{\pm}(\mathbf{p}_1, \mathbf{p}_2, \mathbf{p}_1^{'}, \mathbf{p}_2^{'})\delta(\mathbf{p}_1 + \mathbf{p}_2 - \mathbf{p}_1^{'} - \mathbf{p}_2^{'}), \\ t^{\pm}(\mathbf{p}_1, \mathbf{p}_2, \mathbf{p}_1^{'}, \mathbf{p}_2^{'}) &= \nu(\mathbf{p}_1 - \mathbf{p}_1^{'}) \\ &+ \frac{1}{(2\pi\hbar)^3}\int d\mathbf{p}_1^{''}\int d\mathbf{p}_2^{''}\frac{\nu(\mathbf{p}_1 - \mathbf{p}_1^{''})\nu(\mathbf{p}_1^{''} - \mathbf{p}_1^{'})}{E^{\pm} - K(\mathbf{p}_1^{''}) - K(\mathbf{p}_2^{''})}\delta(\mathbf{p}_1 + \mathbf{p}_2 - \mathbf{p}_1^{''} - \mathbf{p}_2^{''}) \\ &+ \ldots \end{aligned} \quad (7.14)$$

This T-matrix obeys the following symmetry conditions

$$T_{12}^{\pm}(\mathbf{p}_1, \mathbf{p}_2, \mathbf{p}_1^{'}, \mathbf{p}_2^{'}) = T_{12}^{\pm}(\mathbf{p}_2, \mathbf{p}_1, \mathbf{p}_2^{'}, \mathbf{p}_1^{'}),$$
$$T_{12}^{\pm}(\mathbf{p}_1, \mathbf{p}_2, \mathbf{p}_1^{'}, \mathbf{p}_2^{'}) = T_{12}^{\pm}(\mathbf{p}_1^{'}, \mathbf{p}_2^{'}, \mathbf{p}_1, \mathbf{p}_2). \qquad (7.15)$$

Calculating the trace (integral over $\mathbf{p}_2$) in (7.13) with the use of (7.14), (7.15), (6.24) and (6.34), we obtain the following equation for the momentum distribution function

$$\frac{\partial w(\mathbf{p}_1, t)}{\partial t} = \frac{\pi n}{(2\pi\hbar)^3\hbar} \int d\mathbf{p}_2 d\mathbf{p}_1^{'} d\mathbf{p}_2^{'} \delta\left[K(\mathbf{p}_1) + K(\mathbf{p}_2) - K(\mathbf{p}_1^{'}) - K(\mathbf{p}_2^{'})\right]$$
$$\times\, \delta(\mathbf{p}_1 + \mathbf{p}_2 - \mathbf{p}_1^{'} - \mathbf{p}_2^{'}) \left|t^{+}(\mathbf{p}_1, \mathbf{p}_2, \mathbf{p}_1^{'}, \mathbf{p}_2^{'}) + \theta t^{+}(\mathbf{p}_1, \mathbf{p}_2, \mathbf{p}_2^{'}, \mathbf{p}_1^{'})\right|^2$$
$$\times \left[w(\mathbf{p}_1^{'}, t) w(\mathbf{p}_2^{'}, t) - w(\mathbf{p}_1, t) w(\mathbf{p}_2, t)\right]. \qquad (7.16)$$

Equation (7.16) is the quantum Boltzmann equation in the linear approximation in the particle density n (only the binary collisions have been accounted for). It includes (as it should be) the quantum mechanical processes of particle exchange at scattering (the term proportional to θ), which are due to the quantum statistics of particles. This irreversible kinetic equation (7.16) has been obtained with no use of a factorized initial condition (which introduces irreversibility) for the density matrix of the system under consideration. In the suggested here approach, the kinetic equation (7.16) follows from the evolution equation (6.41), which is exact in the linear approximation in n and describes an evolution process on any time scale treating the initial correlations and correlations due to collisions on the equal footing. The described procedure of obtaining the time-irreversible Eq. (7.16) from Eq. (6.41) clearly indicates that irreversibility emerges on the macroscopic time scale (1.1) as a result of the damping of both the initial correlations caused by interaction and correlations caused by collisions (see also [26]). The latter can be secured by the appropriate properties of the system's dynamics (e.g., the ergodic mixing fl w in the phase space). One can, therefore, expect that Eq. (6.41) would switch automatically from a reversible behaviour to an irreversible one (described by the Boltzmann equation (7.16)) if all correlations vanish with time while going from microscopic to a large enough time scale. Thus, the influenc of initial correlations on an evolution process can be revealed.

Before going to the next Section, we would again like to stress, that the obtained above TC-HGME and TCL-HGME are linear equations for the relevant (reduced) statistical operator because they have been derived from the linear Liouville-von-Neumann equation by the conventional linear time-independent projection operator technique. These equations, when applied, e.g., to the (sub)system interacting with a thermal bath, describe the (sub)system's evolution on all time scales (no restrictions like those determined by (1.1)). But we have seen, that in order to obtain from these equations the nonlinear evolution equations accounting for initial correlations, the second inequality in (1.1) ($t - t_0 \ll t_{rel}$) should still be satisfied Therefore, we have to turn to a different approach in order to obtain from the Liouville-von-Neumann equation the nonlinear evolution equations valid on all time scales, i.e. to lift all restrictions imposed by inequalities (1.1).

8. Nonlinear Inhomogeneous GME

We are considering the evolution equations for many-particle systems. It is natural to expect that these equations should generally be nonlinear evolution equations on any time scale due to interaction between particles. Strictly speaking, it is impossible to obtain a nonlinear evolution equation (like the Boltzmann equation) in the framework of the linear time-independent projection operator formalism without additional approximations (as we have seen above). The problem is that the conventional time-independent projection operator (4.3) (or (6.4)) leads to the relevant part of an N-particle distribution function (4.7) (or (6.5)) which is linear in the time-dependent one-particle distribution function of interest. That is why, a nonlinear equation for a density matrix is often simply postulated in the form of a Lindblad-type master equation whose generator depends parametrically on this density matrix (see, e.g., [18]).

Therefore, we need a new approach which would allow to obtain, e.g., a nonlinear relevant distribution function (statistical operator) like $f_r(x_1,\ldots,x_N,t) = \prod\limits_{i=1}^{N} F_1(x_i,t)$ in (1.2).

We start again with the Liouville-von-Neumann equation (2.1) describing the evolution of a distribution function (statistical operator) $F_N(t)$ of a system of N $(N \gg 1)$ classical or quantum particles but in more general case, when the Hamiltonian and the corresponding Liouville (evolution) (super)operator $L(t)$, acting on $F_N(t)$ according to Eqs. (2.3), (2.4) may depend on time

$$\frac{\partial}{\partial t} F_N(t) = L(t) F_N(t). \tag{8.1}$$

The distribution function (or statistical operator) is subject to the normalization condition (2.2).

The formal solution to equation (8.1) can be written as

$$F_N(t) = U(t,t_0)F_N(t_0), U(t,t_0) = T\exp\left[\int\limits_{t_0}^{t} ds L(s)\right], \tag{8.2}$$

where T denotes the chronological time-ordering operator, which orders the product of time-dependent operators such that their time arguments increase from right to left and $F_N(t_0)$ is the value of a distribution function (statistical operator) at some initial time t_0.

We introduce the generally time-dependent operator $P(t)$ converting the distribution function $F_N(t)$ into a relevant distribution function $f_r(t)$, which as a rule is a vacuum (without correlations) many-particle distribution function, i.e., a product of the reduced distribution functions that are necessary and sufficien for describing the evolution of the measurable values (statistical expectations). The problem is to obtain the closed evolution equation for such a reduced distribution function in view of the existence of correlations between the elements of the considered system including initial correlations at the initial time t_0.

We therefore defin the relevant distribution function (statistical operator) as

$$f_r(t) = P(t)F_N(t) = P(t)U(t,t_0)F_N(t_0). \tag{8.3}$$

We note that the operator $P(t)$ is generally not a projection operator. Moreover, the action of $P(t)$ on $F_N(t)$ is no longer a linear operation (in contrast to applying the conventional time-independent projection operator [9]) because $P(t)$ can depend on the time-dependent distribution function of interest (see below). Therefore, by applying $P(t)$ to the linear Liouville equation (8.1), we generally obtain a nonlinear equation. The operator $P(t)$ does not commute with the derivative $\frac{\partial}{\partial t}$ (the time-independent linear projection operators commute with $\frac{\partial}{\partial t}$) and $P(t^{'})F_N(t) \neq f_r(t)$ (for $t^{'} \neq t$).

Using Liouville-von-Neumann equation (8.1), the equation of motion for $f_r(t)$ (8.3) may be written as

$$\frac{\partial f_r(t)}{\partial t} = \left[\frac{\partial P(t)}{\partial t} + P(t)L(t)\right][f_r(t) + f_i(t)], \tag{8.4}$$

where $f_i(t) = [F_N(t) - f_r(t)]$ and $\frac{\partial P(t)}{\partial t}$ is the operator obtained by taking the derivative of the operator $P(t)$, i.e., we assume that the operator $\frac{\partial P(t)}{\partial t}$ exists. To make sense of $\frac{\partial P(t)}{\partial t}$, we can defin $P(t)$ as

$$P(t) = C(t)D, \tag{8.5}$$

where $C(t)$ is a well define (operator) function of time and D is a time-independent (super)operator acting on $F_N(t)$ and integrating over all unnecessary variables in the spirit of the reduced description method (more details of the specificatio of $P(t)$ are given below).

As it follows from (8.4), we can obtain the equation for the irrelevant part of a distribution function $f_i(t)$ by using Eqs. (8.1) and (8.4) and splitting $F_N(t)$ into the relevant and irrelevant parts. Thus,

$$\frac{\partial f_i(t)}{\partial t} = \left[Q(t)L(t) - \frac{\partial P(t)}{\partial t}\right][f_i(t) + f_r(t)], \tag{8.6}$$

where $Q(t) = 1 - P(t)$ and therefore $f_i(t) = Q(t)F_N(t)$.

The formal solution of equation (8.6) is

$$f_i(t) = \int_{t_0}^{t} dt^{'} S(t,t^{'})\left[Q(t^{'})L(t^{'}) - \frac{\partial P(t^{'})}{\partial t^{'}}\right] f_r(t^{'}) + S(t,t_0)f_i(t_0), \tag{8.7}$$

where

$$S(t,t_0) = T\exp\left\{\int_{t_0}^{t} ds\left[Q(s)L(s) - \frac{\partial P(s)}{\partial s}\right]\right\}.. \tag{8.8}$$

We can represent the latter operator as the series

$$\begin{aligned} S(t,t_0) = 1 &+ \int_{t_0}^{t} dt_1\left[Q(t_1)L(t_1) - \frac{\partial P(t_1)}{\partial t_1}\right] \\ &+ \int_{t_0}^{t} dt_1 \int_{t_1}^{t} dt_2\left[Q(t_2)L(t_2) - \frac{\partial P(t_2)}{\partial t_2}\right] \\ &\times \left[Q(t_1)L(t_1) - \frac{\partial P(t_1)}{\partial t_1}\right] + \ldots . \end{aligned} \tag{8.9}$$

We note that if the operators P and L are independent of time, then $S(t,t_0) = \exp\left[QL(t-t_0)\right]$.

Substituting (8.7) in (8.4), we obtain

$$\frac{\partial f_r(t)}{\partial t} = \left[P(t)L(t) + \frac{\partial P(t)}{\partial t}\right]\{f_r(t) \\ + \int_{t_0}^{t} dt' S(t,t')\left[Q(t')L(t') - \frac{\partial P(t')}{\partial t'}\right] f_r(t') \\ + S(t,t_0) f_i(t_0)\}. \tag{8.10}$$

Equation (8.10) is a generalization of the Nakajima-Zwanzig time-convolution generalized master equation (TC-GME) (see [8, 9, 10]) for the relevant part of the distribution function (statistical operator) to the case of time-dependent operators P and Q. As previously noted, $P(t)$ and $Q(t)$ are generally not projection operators in the usual sense, and in deriving (8.10) we do not use such a property of the projection operators as $P^2 = P$, $Q^2 = Q$ (for $P(t)P(t')$ and $Q(t)Q(t')$ such a property holds only for $t = t'$, as is shown below). The obtained equation, like the conventional TC-GME, is an exact inhomogeneous integro-differential equation for the relevant part of the distribution function (statistical operator) containing the irrelevant part of the distribution function (a source) at the initial instant $f_i(t_0) = F_N(t_0) - f_r(t_0)$. But, Eq. (8.10) is generally equivalent to the nonlinear equation for the reduced distribution function of interest (e.g., for the one-particle distribution function considered in the next section) and is therefore also convenient for studying the evolution of many-particle systems described by nonlinear Boltzmann-type equations (in the kinetic regime) where the nonlinearity results from particle collisions.

For time-independent P and Q, Eq. (8.10) reduces to the conventional linear TC-GME. This type of equation is more suitable, for example, for studying the evolution of a subsystem interacting with a large system in thermal equilibrium (a thermal bath). Although, it is possible to obtain from the linear TC-GME the nonlinear evolution equation but on the time scale $t - t_0 \ll t_{rel}$, as we have seen in the previous sections.

To simplify Eq. (8.10) and make sense of the derivative $\frac{\partial P(t)}{\partial t}$, we consider an operator $P(t)$ of form (8.5). This representation of $P(t)$ is suitable for studying many cases of interest (see Sec. 9). Also, splitting a distribution function (statistical operator) into the relevant $P(t)F_N(t)$ and irrelevant $Q(t)F_N(t)$ parts makes sense if the operator $P(t)$ satisfie the relation

$$P(t)P(t) = P(t) \tag{8.11}$$

for any t. In this case, if $P(t) = C(t)D$, then Eq. (8.11) implies

$$DC(t) = 1 \tag{8.12}$$

at any instant t. Equation (8.12) generally represents the normalization condition for the distribution function(s) making up $C(t)$ (see (2.2) and (4.6)) and implies that $C(t)$ depends on the variables that are removed from $F_N(t)$ by the operator D ($f_r(t) = P(t)F_N(t)$ depends on the complete set of variables of the distribution function $F_N(t)$ as indicated by

(1.3)). From condition (8.12) we have

$$P(t)P(t') = P(t), Q(t)Q(t') = Q(t'), P(t)Q(t') = 0,$$
$$Q(t')P(t) = P(t) - P(t'). \tag{8.13}$$

Condition (8.12) also leads to the relations

$$D\frac{\partial P(t)}{\partial t} = D\frac{\partial C(t)}{\partial t}D = 0, \tag{8.14}$$

for the time-derivative of $P(t)$, where we assume that the operators D and $\partial/\partial t$ commute (D is independent of time).

We also introduce the reduced distribution function (statistical operator)

$$f_{red}(t) = DF_N(t) = Df_r(t), \tag{8.15}$$

which is actually needed for calculating the expectation values of interest in a nonequilibrium state. By (8.12), we have

$$D\frac{\partial f_r(t)}{\partial t} = \frac{\partial f_{red}(t)}{\partial t}, Df_i(t) = DQ(t)F_N(t) = 0. \tag{8.16}$$

Hence, applying the operator D to equation (8.10) from the left, we obtain

$$\frac{\partial f_{red}(t)}{\partial t} = DL(t)\{f_r(t) + \int_{t_0}^{t} dt' S(t,t') \left[Q(t')L(t') - \frac{\partial P(t')}{\partial t'}\right] f_r(t')$$
$$+ S(t, t_0)f_i(t_0)\}, \tag{8.17}$$

where we use (8.12), (8.14), and (8.16).

Without loss of generality, we can split the Hamiltonian or the Liouvillian of a system into two terms as $H(t) = H^0(t) + H'(t)$ or $L(t) = L^0(t) + L'(t)$, where H^0 or L^0 is related to the energy of the noninteracting particles (subsystems) and H' or L' describes the interaction between particles (subsystems). Then, in addition to satisfying condition (8.12), the operator D, which selects the reduced distribution function, should commute with L^0 as

$$DL^0(t) = L^0_{red}(t)D, \tag{8.18}$$

where $L^0_{red}(t)$ is the reduced Liouvillian L^0 depending only on the variables of the reduced distribution function (statistical operator) $f_{red}(t)$. Expression (8.18) means that both sides of this relation act on the arbitrary function (operator) define on the phase (Hilbert) space. Taking into account that $C(t')L^0_{red}(t) = L^0_{red}(t)C(t')$ ($C(t')$ depends on the set of variables that are removed from $F(t)$ by D, and $L^0_{red}(t)$ is independent of them), we can write Eq. (8.18) in the more general form

$$P(t')L^0(t) = L^0_{red}(t)P(t'). \tag{8.19}$$

Commutation relation (8.18) follows from a self-consistency argument because if $L'(t) = 0$, then the evolution of a "free" subsystem (not interacting with the rest of the system) should be described in Eq. (8.17) only by $L^0_{red}(t)$ ($DL^0(t)f_r(t) = L^0_{red}(t)f_{red}(t) = L^0_{red}(t)Df_r(t)$; also see [1]). It can be seen from the relations

$$DL^0(t)S(t,t_1)Q(t_2) = 0,$$
$$DL^0S(t,t_1)\frac{\partial P(t_2)}{\partial t_2} = 0, \tag{8.20}$$

which follow from (8.18), (8.8), (8.14), and (8.16), that this is the case. Hence, all terms in (8.17) are proportional to $L^{'}(t)$ except the firs term, $L^0_{red}(t)f_{red}(t)$.

Analogously, using (8.18), (8.8), and relations that follow from (8.12), we can prove the equations

$$S(t,t_1)Q(t_2) = \overline{U}(t,t_1)Q(t_2),$$
$$S(t,t_1)\frac{\partial P(t_2)}{\partial t_2} = \overline{U}(t,t_1)\frac{\partial P(t_2)}{\partial t_2}, \tag{8.21}$$

where

$$\overline{U}(t,t_1) = T\exp\left\{\int_{t_1}^{t} ds\left[L^0(s) + Q(s)L^{'}\right]\right\}. \tag{8.22}$$

Hence, using (8.20) and (8.21) in (8.17), we can finall rewrite Eq. (8.10) as the equation for the relevant part of the distribution function (statistical operator)

$$\begin{aligned}\frac{\partial Df_r(t)}{\partial t} &= DL(t)f_r(t)\\ &+ DL^{'}(t)\int_{t_0}^{t} dt^{'}\overline{U}(t,t^{'})\left[Q(t^{'})L(t^{'}) - \frac{\partial P(t^{'})}{\partial t^{'}}\right]f_r(t^{'})\\ &+ DL^{'}(t)\overline{U}(t,t_0)f_i(t_0),\end{aligned} \tag{8.23}$$

where $Df_r(t) = f_{red}(t)$ and (8.18) holds.

We demonstrate that basic conditions (8.12) and (8.18) are satisfie for a gas of interacting classical particles in the next section.

9. Nonlinear Inhomogeneous Evolution Equation for a Nonideal Inhomogeneous Dilute Gas of Classical Particles

In this section, we apply Eq. (8.23) to the case of a gas of N ($N \gg 1$) identical classical particles. The Liouville operator L for such a system can be represented in the form (4.9), where L^0 corresponds to the kinetic energy $H^0 = \sum_{i=1}^{N} \mathbf{p}_i^2/2m$ of particles with the momenta $\mathbf{p}_i$ and mass m and $L^{'}$ corresponds to the particle interaction $H^{'} = \sum_{i<j=1}^{N} V_{ij}$ with the pair potential $V_{ij} = V(|\mathbf{x}_i - \mathbf{x}_j|)$. Hence, L is independent of time in the considered case. We do not assume here that the interaction is weak, but we do assume that all necessary requirements for the properties of forces by which the particles interact are satisfie (in particular,

bound states do not form). As usual, we also assume that all functions $\Phi(x_1, \ldots, x_N)$ define on the phase space and their derivatives vanish at the boundaries of the configuratio space and at $\mathbf{p}_i = \pm\infty$. These boundary conditions and the explicit form of the Liouville operators (4.9) lead to relations (4.10) (see Sec. 4).

We seek an evolution equation for a one-particle distribution function (4.4)

$$f_1(x_i, t) = V \int dx_1 \cdots \int dx_{i-1} \int dx_{i+1} \cdots \int dx_N F_N(x_1, \ldots, x_N, t). \tag{9.1}$$

For this, it is convenient to defin the operator $P(t)$ as

$$P(t) = C(t)D, C(t) = \left[\prod_{i=2}^{N} f_1(x_i, t)\right], D = \frac{1}{V^{N-1}} \int dx_2 \cdots \int dx_N \tag{9.2}$$

(compare with (4.3)).

We apply the projection operator (9.2) to $f_N(t) = V^N F_N(t)$, which satisfie the Liouville-von- Neumann equation (8.1), and obtain the following relevant part of the distribution function

$$f_r(t) = P(t) f_N(t) = \prod_{i=1}^{N} f_1(x_i, t). \tag{9.3}$$

From definitio (9.3) we can see an advantage of the time-dependent operator $P(t)$ (9.2): applying it to an N-particle distribution function $f_N(t)$, we can defin relevant part (9.3) of the distribution function, which seems more natural and symmetric than (4.7) used in Sec. 4 and obtained using the time-independent projection operator (4.3) (we had to use the time-independent projection operator because the TC-HGME considered in Sec. 2 holds only for such operators). Moreover, relevant part (9.3) has a clear physical meaning because it represents a part of an N-particle distribution function without correlations which is often of most interest.

Hence, the irrelevant part

$$f_i(t) = Q(t) f_N(t) = f_N(t) - \prod_{i=1}^{N} f_1(x_i, t) \tag{9.4}$$

describes all interparticle correlations, in particular, the initial ones (existing at the initial instant t_0), which compose the inhomogeneous term in (8.23). The irrelevant part of the N-particle distribution function at t_0, $f_i(t_0) = f_N(t_0) - f_r(t_0)$, can always be represented (and also at any other time t) by the cluster expansion (4.8).

Using normalization condition (4.6) for a one-particle distribution function $\int f_1(x_i, t) dx_i = V$, we can easily see that by applying the integral operator D to $C(t)$ (both are define by (9.2)), we obtain $DC(t) = 1$, i.e., condition (8.12) and its consequences given by (8.13) and (8.14) are satisfied Also, in this case,

$$\begin{aligned} f_{red}(t) &= D f_r(t) = f_1(x_1, t), \\ DQ(t) &= 0, D f_i(t) = 0 \end{aligned} \tag{9.5}$$

in accordance with (8.15) and (8.16).

Commutation rules (8.18) and (8.19) also follow from (4.9), (4.10), and (9.2), because

$$DL^0 = L_1^0 D$$
$$P(t)L^0 = L_1^0 P(t), \tag{9.6}$$

i.e., $L^0_{red} = L_1^0$ in the case under consideration. Therefore, the identities (8.20) - (8.22) are also satisfied where L^0 and $L^{'}$ are independent of time in the considered case.

Using (4.10), (9.5), and (9.6), we can represent Eq. (8.23) in the case under consideration as

$$\frac{\partial f_1(x_1,t)}{\partial t} = L_1^0 f_1(x_1,t) + n\int dx_2 L^{'}_{12} f_1(x_2,t) f_1(x_1,t)$$
$$+ DL^{'}\int_{t_0}^{t} dt^{'}\overline{U}(t,t^{'})\left[(L^0 - L_1^0) + Q(t^{'})L^{'} - \frac{\partial P(t^{'})}{\partial t^{'}}\right] f_r(t^{'})$$
$$+ DL^{'}\overline{U}(t,t_0) f_i(t_0), \tag{9.7}$$

where $n = N/V$ is the density of particles and D, $f_r(t)$, and $f_i(t_0)$ are respectively define in (9.2), (9.3), and (4.8). Evolution operator (8.22) can be represented by the series similar to (8.9)

$$\overline{U}(t,t^{'}) = 1 + \int_{t^{'}}^{t} dt_1[L^0 + Q(t_1)L^{'}]$$
$$+ \int_{t^{'}}^{t} dt_1 \int_{t_1}^{t} dt_2 [L^0 + Q(t_2)L^{'}][L^0 + Q(t_1)L^{'}] + \ldots. \tag{9.8}$$

We consider equation (9.7) in the firs approximation in the particle density n. The corresponding dimensionless small parameter of the perturbation expansion is given by (4.2). From the second equation in (4.10), we can easily see that all terms in expansion (9.8) containing $P(t)L^{'}$result in expressions proportional to at least the firs power of n (like the second term in the right-hand side of (9.7)). Therefore, in the firs approximation in n, all terms of (9.8) with $P(t)L^{'}$ can be neglected because the terms of Eq. (9.7) containing $\overline{U}(t,t^{'})$ are already of the firs order in n (applying $DL^{'}$ leads to an expression of at least the firs order in n). In this approximation, all terms under the integrands in (9.8) are independent of time, and the operator $\overline{U}(t,t^{'})$ reduces to

$$\overline{U}(t,t^{'}) = \exp[(L^0 + L^{'})(t - t^{'})]. \tag{9.9}$$

It is now convenient to use the expansion for the exponential operator (9.9) (see (4.13))

$$e^{(L^0+L^{'})(t-t^{'})} = e^{L^0(t-t^{'})} + \int_{t^{'}}^{t} d\theta e^{L^0(t-\theta)} L^{'} e^{(L^0+L^{'})\theta}. \tag{9.10}$$

Using this expansion and relations (4.10), we obtain the following evolution equation for a one-particle distribution function in the linear approximation in the density parameter (4.2) from (9.7):

$$\frac{\partial f_1(x_1,t)}{\partial t} = L_1^0 f_1(x_1,t) + n\int dx_2 L'_{12} f_1(x_2,t) f_1(x_1,t)$$
$$+ n\int dx_2 L'_{12} \int_0^{t-t_0} dt_1 e^{(L_{12}^0+L'_{12})t_1}[(L_2^0 + L'_{12}) f_1(x_2, t-t_1) f_1(x_1, t-t_1)$$
$$+ \frac{\partial f_1(x_2,t-t_1)}{\partial t_1} f_1(x_1,t-t_1)]$$
$$+ n\int dx_2 L'_{12} e^{(L_{12}^0+L'_{12})(t-t_0)} g_2(x_1,x_2), \quad (9.11)$$

where $L_{12}^0 = L_1^0 + L_2^0$. In obtaining (9.11), we also use the definitio (4.8) for the irrelevant part of the distribution function and take into account that each additional integration over $x_3, \ldots$ adds an additional power of n (therefore, in the linear approximation in n, all formulae contain no more than one integration over the phase space).

Equation (9.11) is the main result in this section. We must stress that (9.11) is a nonlinear inhomogeneous time-convolution (non-Markovian) master equation for a one-particle distribution function containing initial correlations (a source). We note, that this new equation is rigorously derived here (the desired nonlinear master equation is usually postulated [18]). Equation (9.11) is exact in the linear approximation in small density parameter (4.2) and holds on any time scale and for any spatial inhomogeneity of the system under consideration. The firs term in the right-hand side is a conventional fl w term. The second term, a nonlinear Vlasov term, represents the self-consistent fiel acting on the given (first particle and determined by all particles of the system. We note that it is impossible (without additional approximations) to obtain this term (appearing only in a spatially inhomogeneous case) from the GME (or HGME) containing time-independent projection operator (4.3) (see Sec. 4). The term quadratic in L'_{12} describes particle collisions and can lead to dissipation in the system. The terms with L_2^0 (the space derivative) and with the time derivative take the change in space and time of the one-particle distribution function on the respective microscopic scales (also see below) of the orders of r_0 and $t_{cor} \sim r_0/\overline{v}$ into account. These two terms determine the contribution of the pair collisions to the nondissipative characteristics (thermodynamic functions) of the nonideal gas (see, e.g., [4]) and are usually absent in the standard derivation (from the BBGKY chain) of the kinetic equations in the approximation linear in n. Taking these terms into account poses a certain problem because they usually appear in the next (second) approximation in n (see, e.g., [27]), which is inconsistent with the dissipative three-particle collision term appearing in this approximation. The last (irrelevant) term in (9.11) takes initial correlations (at $t = t_0$) into account and is given by the two-particle correlation function $g_2(x_1, x_2)$ in the linear approximation in n.

We examine the terms in (9.11) more closely, especially their behavior in time. The evolution in time is governed by the exact two-particle propagator $G_{12}(t) = \exp(L_{12}t)$ ($L_{12} = L_{12}^0 + L'_{12}$) (this is natural for the considered dilute gas in the lowest approximation in density), which satisfie the integral equation (4.26).The action of "free" propagator $G^0(t) = \exp(L^0 t)$ on any function define on the phase space is given by Eq. (4.14).

If the particle dynamics have the necessary properties such as a mixing ergodic fl w in the phase space resulting from the local stochastic instability, then the distances between particles rapidly (exponentially) increase with time under the action of $G_{12}(t)$ (the contribution of a "parallel motion" is negligible). It was proved that such behavior is the case, for example, for the Boltzmann dilute gas of hard spheres and is now viewed as common for the most real systems considered in statistical physics (see, e.g., [4]). The mixing ergodic fl w in the phase space results in the exponential damping of correlations and is the reason for irreversibility. Hence, if the effective particle interaction $V(|\mathbf{r}_i - \mathbf{r}_j|)$ vanishes at a distance $|\mathbf{r}_i - \mathbf{r}_j| > r_0$, then the integrand of the integral over t_1 in (9.11) vanishes for $t - t_0 > t_{cor}$ and the same holds (under the action of $G_{12}(t - t_0)$) for the initial correlation term determined by the function $g_2(x_1, x_2)$, which also depends on the particle interaction (we discuss this in more detail below, and it was considered in the spatially homogeneous case in Sec. 5).

Condition (4.2) implies the existence of the time hierarchy

$$t_{cor} \ll t_{rel}, \tag{9.12}$$

where $t_{rel} \sim \gamma^{-1} t_{cor}$ is the relaxation time for a one-particle distribution function $f_1(x_i, t)$ ($1/nr_0^2$ is the particle mean free path). In the initial evolution stage $t_0 \leqslant t \leqslant t_{cor}$, which is very interesting and essential for understanding the irreversibility problem and for studying the non-Markovian processes (the memory effects), decoherence phenomena, and the ultrafast relaxation effects, the initial correlations can be important. But if we consider the time scale $t - t_0 \gg t_{cor}$, then the initial correlations are damped (as a result of mixing fl w), and the source in Eq. (9.11) containing $g_2(x_1, x_2)$ can be neglected. But as Bogoliubov noted in [1], this can be done only on time interval (1.1), which exists because of (9.12). The latter conclusion follows because the corrections to the solution of (9.11) due to initial correlations turn out to be proportional to $|t - t_0|$ (secular terms) and are small only for $|t - t_0| \ll t_{rel}$. Therefore, the expansion of the irrelevant initial condition term in n (as in passing from (9.7) to (9.11)) is ineffective on the time scale of interest in the kinetic theory $t - t_0 \gtrsim t_{rel}$.

To get rid of the undesirable initial condition terms, it is often assumed that $f_i(t_0) = 0$, which corresponds to RPA and is incorrect in principle (see [7]). Therefore (as mentioned in Sec. 1), to study the kinetic evolution stage $t \gtrsim t_{rel}$, a more sophisticated approach like that based on the principle of weakening of initial correlations (Bogoliubov's ansatz) [1] is needed. But the Bogoliubov approach does not allow considering the initial evolution stage $t_0 \leqslant t \leqslant t_{cor}$ and the situations where the large-scale correlations with $t_{cor} \gtrsim t_{rel}$ play an essential role.

If $f_1(x_i, t)$ changes significantl on the macroscopic (hydrodynamic) length l_h with the corresponding time scale $t_h = l_h/\overline{v}$ (for the spatially homogeneous gas, there are only the characteristic scales l and t_{rel} on which $f_1(x_i, t)$ changes), then the terms with L_2^0 and $\frac{\partial F_1(x_2, t-t')}{\partial t'}$ in Eq. (9.11) can be neglected because they are both at least of the order of γ, i.e., their contribution is of the second order in n (but these terms should not be omitted if we wish to take the nondissipative characteristics of the nonideal gas into account self-consistently). On the kinetic time scale $t - t_0 \gtrsim t_{rel}$, the upper limit of integration over t_1 can be extended to infinit ($t_0 \to -\infty$) because of the damping of correlations resulting from collisions (as noted above). If we also accept the RPA (or principle of weakening of initial correlations) for $t_0 \to -\infty$, i.e., neglect the initial condition term with $g_2(x_1, x_2)$, then Eq. (9.11) reduces to

$$\frac{\partial f_1(x_1,t)}{\partial t} = L_1^0 f_1(x_1,t) + n\int dx_2 L_{12}^{'} f_1(x_2,t) f_1(x_1,t)$$
$$+ n\int dx_2 L_{12}^{'} \int_0^{\infty} dt_1 e^{(L_{12}^0 + L_{12}^{'})t_1} L_{12}^{'} f_1(x_2,t) f_1(x_1,t), \qquad (9.13)$$

where we take into account that for $t \gtrsim t_{rel} \gg t_{cor}$ in the adopted approximation, we can replace $F_1(x_i, t - t_1)$ with $F_1(x_i, t)$ (the integration over t_1 in (9.11) gives an essential contribution only up to $t_1 \sim t_{cor}$). In the last (collision) term in (9.13), the distribution functions of the colliding particles can be taken at the same space point $\mathbf{r}_1 = \mathbf{r}_2$ with the adopted accuracy $\sim r_0/r_h \ll 1$ (or $r_0/l \ll 1$).

If the last integral in Eq. (9.13) exists (see [23]), then Eq. (9.13) represents the nonlinear Markovian kinetic equation irreversible in time. In the case of a weak particle interaction, this equation in the firs approximation in the small interaction parameter ε gives the reversible Vlasov equation (Eq. (9.13) with neglect of the third collision term in the right-hand side, which is of the second order in ε). In the second approximation in ε, it coincides with the Vlasov-Landau kinetic equation (see, e.g., [20]). In the spatially homogeneous case, where a one-particle distribution function is independent of the particle coordinate, $f_1(x_j,t) = f_1(\mathbf{p}_j,t)$ ($\int f_1(\mathbf{p},t)d\mathbf{p} = 1$), the second (Vlasov) term in (9.13) vanishes because we consider a potential V_{ij} given by (4.9) that depends on the particle coordinate difference. In this case, Eq. (9.13) is equivalent to the nonlinear Boltzmann equation (see, e.g., [20]). We stress that the nonlinearity of (9.13) appears quite naturally in this approach (based on the time-dependent $P(t)$ given by (9.2)) in contrast to the time-independent projection operator approach presented in Secs. 2-7. In the latter case, we must approximate $f_1(x_2,t_0)$ by $f_1(x_2,t)$ (which is possible on the time scale given by (1.1)) in the evolution equation in order to obtain a nonlinear equation from the linear GME or HGME (in the case of the BBGKY chain approach, the nonlinearity is introduced by principle of weakening of initial correlations (1.2)).

10. Nonlinear Homogeneous GME

In Sec. 8, we obtained the nonlinear GME for the relevant part of a distribution function (statistical operator). But Eq. (8.10) is not closed in the sense that it is inhomogeneous and contains an irrelevant term generally containing all multiparticle correlations. As noted above, to eliminate this irrelevant part and obtain a homogeneous equation (like (9.13)), for example, for a one-particle distribution function, the Bogoliubov principle of weakening of initial correlations or the RPA is usually used.

To take initial correlations into account and to obtain the evolution equations applicable on all time scales, we use here a method for converting conventional linear inhomogeneous GMEs into the homogeneous GMEs presented in Secs. 2 and 3. We now apply this approach to equation (8.23). Our goal is to convert Eq. (8.23) into a homogeneous equation containing the irrelevant part of the distribution function (statistical operator) $f_i(t_0)$ in the "mass" (super)operator acting on the relevant part of the distribution function (statistical operator) $f_r(t)$. Such an equation allows treating the initial correlations (contained in $f_i(t_0)$)

on an equal footing with all other correlations (collisions) and is applicable on any time scale including the initial evolution stage when the initial correlations are certainly essential. This is so because there is no need for restrictions given by inequalities (1.1) that allow neglecting the initial correlations and avoiding the "secular terms" in the solution of the evolution equation (the perturbation expansion of the "mass" (super)operator does not lead to such terms on the large (kinetic) time scale). In contrast to the HGME obtained in Sec. 2, the equation derived below is nonlinear and is also applicable to the nonconservative systems (with time-dependent Hamiltonians).

Following Sec. 2, we identically represent the irrelevant part of the distribution function (statistical operator) in the form

$$\begin{aligned} f_i(t_0) &= F_N(t_0) - f_r(t_0) = Q(t_0)F_N(t_0) \\ &= [Q(t_0)F_N(t_0)]\, F_N^{-1}(t_0)U^{-1}(t,t_0)[P(t)+Q(t)]U(t,t_0)F_N(t_0) \\ &= C_0U^{-1}(t,t_0)[f_r(t)+f_i(t)], \\ C_0 &= [Q(t_0)F_N(t_0)]\, F_N^{-1}(t_0), U^{-1}(t,t_0) = T_- \exp[-\int_{t_0}^{t} dsL(s)], \end{aligned} \tag{10.1}$$

where $U^{-1}(t,t_0)$ is the backward-in-time evolution operator for the density matrix $F_N(t)$ (compare with (8.2)), $U^{-1}(t,t_0)U(t,t_0) = 1$, T_- is the antichronological time-ordering operator arranging the time-dependent operators $L(s)$ such that the time arguments increase from left to right, $F_N^{-1}(t_0)$ is the inverse of $F_N(t_0)$, $F_N^{-1}(t_0)F_N(t_0) = 1$, $P(t) + Q(t) = 1$. Hence, additional identity (10.1) is obtained by multiplying the irrelevant part by the unity $F_N^{-1}(t_0)F_N(t_0)$ (we assume that $F_N^{-1}(t_0)$ exists) and inserting the unities $U^{-1}(t,t_0)U(t,t_0) = 1$ and $P(t) + Q(t) = 1$. Therefore, neither a divergence (due to $F_N(t_0)$ possible vanishing) nor an indeterminacy of the $0/0$ type (the behaviors of the numerator and denominator in $F_N(t_0)/F_N(t_0) = 1$ are similar) can occur. This holds for all further (identical) manipulations (see below).

In (10.1) we introduced the parameter of initial correlations

$$\begin{aligned} C_0 &= [Q(t_0)F_N(t_0)]F_N^{-1}(t_0) = f_i(t_0)[f_r(t_0)+f_i(t_0)]^{-1} \\ &= f_i(t_0)f_r^{-1}(t_0)[1+f_i(t_0)f_r^{-1}(t_0)]^{-1} \\ &= (1-C_0)f_i(t_0)f_r^{-1}(t_0). \end{aligned} \tag{10.2}$$

It is important that operator $Q(t_0)$ acts only on $F_N(t_0)$ in (10.1) and (10.2), which is indicated by enclosing $Q(t_0)F_N(t_0)$ in brackets. This follows because $f_r(t)$ and $f_i(t_0)$ are the basic quantities we are dealing with in Eq. (8.23). All functions of dynamical variables, whose average values we can calculate using $f_r(t)$ (or $f_{red}(t)$) by multiplying equation (8.23) by the corresponding functions (operators) from the right and taking an average value (a trace), depend only on the variables that are not integrated off by D (D integrates over all excess variables in $F_N(t)$). Therefore, if we correspondingly represent $f_r(t)$ and $f_i(t_0)$ in (8.23) as $f_r(t) = P(t)F_N(t)$ and $f_i(t_0) = Q(t_0)F_N(t_0)$, then the projection operators $P(t) = C(t)D$ and $Q(t) = 1 - P(t)$ in these expressions act only on $F_N(t)$ but not on the functions (if any) to the right of them. This is the essence of the reduced description method: to calculate the average values of the functions that depend on a much smaller number of variables than the whole distribution function $F_N(t)$ does, we in fact need only the reduced distribution function (density matrix) $f_{red}(t)$.

As can be seen from (10.2), the correlation parameter is a series in $f_i(t_0)f_r^{-1}(t_0)$, and we may therefore only need the formal existence of the function (operator) $f_r^{-1}(t_0)$, which is the inverse of the relevant distribution function (statistical operator) chosen using the appropriate operator $P(t_0)$. It seems plausible that the inverse of the relevant part of the distribution function (statistical operator) define in the sense noted above (uncorrelated part) can be constructed (see Secs. 4,6, and the next section).

We now have two equations, (9.1) and (10.1), relating $f_i(t)$ and $f_i(t_0)$. Using relations (8.21), findin $f_i(t_0)$ from the indicated equations as a function of $f_r(t)$, and substituting it in (8.23), we obtain the equation

$$\frac{\partial Df_r(t)}{\partial t} = DL^0(t)f_r(t) + DL'(t)R(t,t_0)f_r(t)$$
$$+ DL'(t)R(t,t_0)\int_{t_0}^{t} dt'\overline{U}(t,t')\left[Q(t')L(t') - \frac{\partial P(t')}{\partial t'}\right] f_r(t'), \tag{10.3}$$

where the operator $R(t,t_0)$ is define as

$$R(t,t_0) = 1 + C(t,t_0),$$
$$C(t,t_0) = \overline{U}(t,t_0)\left[1 - C_0(t,t_0)\right]^{-1} C_0U^{-1}(t,t_0),$$
$$C_0(t,t_0) = C_0U^{-1}(t,t_0)\overline{U}(t,t_0), \tag{10.4}$$

$Df_r(t) = f_{red}(t)$ and relation (8.18) holds.

Equation (10.3) is the central result of this section. We have derived the desired homogeneous generalized evolution equation for the relevant part of the distribution function (statistical operator). This equation differs from the linear TC-HGME obtained in Sec. 2 from the conventional Nakajima-Zwanzig TC-GME using the time-independent projection operator technique. Equation (10.3) is generally nonlinear and holds for systems with a time-dependent Hamiltonian. It holds in both the classical and quantum physics cases if the symbols are properly redefine and all (super)operators exist (we address the latter question below). We have not made any approximations in deriving Eq. (10.3), and it is therefore an exact integro-differential equation that takes the initial correlations and their dynamics into account via the modifie (super)operator (memory kernel) of (8.23) acting on the relevant part of the distribution function (statistical operator) $f_r(t)$.

The obtained exact kernel of (10.3) can serve as a starting point for effective perturbation expansions. In many cases, such expansions of homogeneous equations (like (10.3)) have a much broader applicability domain than those of the inhomogeneous equations (like (8.23)) when the expansions of the functions (f_r, f_i), rather than equations, are involved (see also [1]). Particularly, we can expect that the divergencies appearing in the higher-order terms in the gas density expansion [2] are tackled more conveniently using the "mass" (super)operator expansion. As can be seen from (10.3), the time evolution of initial correlations ($C(t,t_0)$) influence the collision integral (and also the second term) and therefore leads to an additional dependence on the density n (when the expansion in n is used). It is known, that initial correlations in the higher orders in n become more long-lived and are damped only at $t - t_0 \sim t_{rel}$ because of successive pair collisions (see [2, 3]). While the initial correlations in the lowest approximation in n (pair collision approximation) are typically the short-scale ones with the characteristic $t_{cor} \sim r_0/\overline{v} \ll t_{rel}$ and $l_{cor} \sim \overline{v}t_{cor} \ll l$, they are large-scale ones in the next approximations in n and therefore cannot be neglected on the kinetic time scale. Hence, the initial correlations in the higher orders terms (than linear) of the expansion of Eq. (4.20) in n may result in the cut-off of the discussed divergencies of the kinetic equation at $t \sim t_{rel}$ (this was noted in [28]). The foregoing means that the expansion of the "mass" (super)operator governing the evolution of $f_r(t)$ in (10.3) provides a new possibility for expanding the kinetic coefficient in n.

Equation (10.3) reduces to the linear TC-HGME obtained in Sec. 2 if the (projection) operator P and the Hamiltonian H are independent of time and relations (8.18) and (8.19) hold.

As in Sec. 2, the problem of the existence (convergence) of $R(t, t_0)$ can be raised. The function $R(t, t_0)$ behaves properly at all times. Moreover, expanding the kernel of (10.3) may result in canceling the pole in the function $R(t, t_0)$. For the linear TC-HGME, this correspondingly shown in Secs. 4 and 6 in the linear approximation in the small density for a spatially homogeneous dilute gas of classical and quantum particles. In such cases there is no problem with the existence of $R(t, t_0)$. Examining (10.2) (from which the relation $(1 - C_0)^{-1}C_0 = f_i(t_0)/f_r(t_0)$ follows), we can expect that the same is true for $R(t, t_0)$ given by (10.4) (see below).

11. Nonlinear Evolution Equation for a Nonideal Inhomogeneous Dilute Gas of Classical Particles with Initial Correlations Taken into Account

We now apply Eq. (10.3) to a system of N ($N \gg 1$) identical interacting classical particles described by the Liouvillian (4.9) and considered in Sec. 4. To obtain an equation for the one-particle distribution function $f_1(x_i, t)$ $(i = 1, \ldots, N)$, we use the operator $P(t)$ given by (9.2). Hence, taking (4.10), (9.5), and (9.6) into account, we obtain the equation for the one-particle distribution function from (10.3):

$$\frac{\partial f_1(x_1,t)}{\partial t} = L_1^0 f_1(x_1,t) + n\int dx_2 L_{12}^{'} f_1(x_2,t) f_1(x_1,t) + DL^{'}C(t,t_0)f_r(t)$$
$$+ DL^{'}\int_{t_0}^{t} dt^{'} R(t,t_0)\overline{U}(t,t^{'}) \left[(L^0 - L_1^0) + Q(t^{'})L^{'} - \frac{\partial P(t^{'})}{\partial t^{'}}\right] f_r(t^{'}). \quad (11.1)$$

We consider Eq. (11.1) in the linear approximation in density parameter (4.2). Given that the action of $P(t)$ or D results in expressions proportional to at least the firs power in the density n and that all terms on the right-hand side of (11.1) (except the firs fl w term) are already proportional to n (or D), in the firs approximation in n, we can neglect all terms with $P(t)$ in (11.1) including such terms in $\overline{U}(t,t^{'})$. Then $\overline{U}(t,t^{'}) = U(t,t^{'})$ in this approximation and is given by (9.9). Therefore, we can approximate $U^{-1}(t,t_0)\overline{U}(t,t_0)$ in (10.4) by unity and $C_0(t,t_0)$ by C_0. Hence, in the adopted approximation, the correlation parameter is

$$C(t,t_0) = U(t,t_0)[f_i(t_0)f_r^{-1}(t_0)]U^{-1}(t,t_0), \quad (11.2)$$

as follows from (10.2) and (10.4). To calculate contribution (11.2) of initial correlations to evolution equation (11.1), we should use definition (9.3) and (4.8) of $f_r(t_0)$ and $f_i(t_0)$.

Proceeding as in Sec. 9, i.e., using (4.10), (9.5), (9.10), and the fact (noted above) that each additional integration over the phase space leads to an additional power of n, we can represent Eq. (11.1) in the firs approximation in n as

$$\frac{\partial f_1(x_1,t)}{\partial t} = L_1^0 f_1(x_1,t) + n\int dx_2 L_{12}^{'}[1+C_{12}(t-t_0)]f_1(x_2,t)f_1(x_1,t)$$
$$+n\int dx_2 L_{12}^{'}[1+C_{12}(t-t_0)]\int\limits_0^{t-t_0} dt_1 e^{L_{12}t_1}[(L_2^0+L_{12}^{'})f_1(x_2,t-t_1)f_1(x_1,t-t_1)$$
$$+\frac{\partial f_1(x_2,t-t_1)}{\partial t_1}f_1(x_1,t-t_1)], \tag{11.3}$$

where

$$C_{12}(t-t_0) = e^{L_{12}(t-t_0)}\frac{g_2(x_1,x_2)}{f_1(x_1)f_1(x_2)}e^{-L_{12}(t-t_0)}, \tag{11.4}$$

$L_{12} = L_{12}^0 + L_{12}^{'}$, $L_{12}^0 = L_1^0 + L_2^0$ and $f_1(x_i) = f_1(x_i,t_0)$.

Homogeneous nonlinear equation (11.3) is the central result of this section. This is a new equation that (in the linear approximation in n) exactly describes the evolution of a one-particle distribution function of a nonideal classical gas of particles with arbitrary spatial inhomogeneity and on any time scale. This equation takes initial correlations into account exactly (in the adopted approximation in n) and treats them on an equal footing with the particle collisions and other processes. The initial correlations enter the memory kernel (mass operator) governing the evolution of the one-particle distribution function and do not result in any undesirable inhomogeneous term in evolution equation (11.3) (cf. (9.11)). The contribution of initial correlations to the evolution process is given by time-dependent correlation parameter (11.4), which describes the evolution in time of the normalized two-particle correlation function $\frac{g_2(x_1,x_2)}{f_1(x_1)f_1(x_2)}$. The time-evolution of all terms in (11.3) is determined by the exact two-particle propagator $G_{12}(t) = \exp(L_{12}t)$.

We consider the terms in the right-hand side of (11.3) and their evolution in time in more detail. The firs term is a conventional fl w term resulting from the gas inhomogeneity. The second term represents the Vlasov self-consistent fiel modifie by initial correlations. If $C_{12}(t-t_0)$ vanishes (under the action of $G_{12}(t-t_0)$) with time for $t-t_0 \gtrsim t_{cor}$ (see below), then we obtain the conventional Vlasov term. The third term in the right-hand side of (11.3) describes the particle collisions and the influenc on the collisions of the change in the one-particle distribution function on the microscopic scales in space and time. All these processes are modifie by initial correlations given by $C_{12}(t-t_0)$. Equation (11.3) thus takes the influenc of all pair collisions and correlations on the dissipative and nondissipative characteristics of the nonideal spatially inhomogeneous gas of particles into account (this is not the case in the conventional approaches [28]).

Noting that the term with L_2^0 is of the order of r_0/r_h and the term with the time derivative of $f_1(x_2,t)$ is $\sim t_{cor}/t_{rel}$ ($\sim \gamma$), we can neglect these terms (if we are not interested in the nondissipative corrections to the nonideal gas dynamics) in the adopted approximation (linear in n). Equation (11.3) then takes the simpler form

$$\frac{\partial f_1(x_1,t)}{\partial t} = L_1^0 f_1(x_1,t) + n\int dx_2 L_{12}'[1+C_{12}(t-t_0)]f_1(x_2,t)f_1(x_1,t)$$
$$+n\int dx_2 L_{12}'[1+C_{12}(t-t_0)]$$
$$\times \int_0^{t-t_0} dt_1 e^{L_{12}t_1} L_{12}' f_1(x_2,t-t_1)f_1(x_1,t-t_1), \qquad (11.5)$$

where the distribution functions of the colliding particles in the last (collision) term can be taken at the same space point $\mathbf{r}_1 = \mathbf{r}_2$ with the adopted accuracy $\sim r_0/r_h \ll 1$ (in the case of plasma, it is not good to do this in the second term because of the long-range Coulomb interaction).

As an example, we consider the evolution in time of the terms in (11.5) originating from the initial correlations (terms with $C_{12}(t-t_0)$) when the dependence of the initial correlation function and the interparticle interaction on the distance between the particles is given by

$$g_2(x_1,x_2) = g_0\exp(-\frac{r^2}{r_{cor}^2})\phi(x_1,x_2)\,,$$
$$V(r) = V_0\exp(-\frac{r^2}{r_0^2}), \qquad (11.6)$$

where $\mathbf{r} = \mathbf{x}_1 - \mathbf{x}_2$ is the distance between particles 1 and 2, $r_{cor} \sim r_0$, g_0 and V_0 are constant parameters, and $\phi(x_1,x_2)$ is a properly normalized function of x_1 and x_2. We estimate the time dependence of the terms in (11.5) determined by initial correlations in the case of a weak interparticle interaction when the time evolution is governed by the "free" propagator $G_{12}^0(t) = \exp(L_{12}^0 t)$. Then we have (see (4.14))

$$G_{12}^0(t)\frac{g_2(x_1,x_2)}{f_1(x_1)f_1(x_2)} = g_0\exp(-\frac{|\mathbf{r}-\mathbf{g}t|^2}{r_{cor}^2})\frac{\phi(\mathbf{x}_1-\mathbf{v}_1t,\mathbf{p}_1,\mathbf{x}_2-\mathbf{v}_2t,\mathbf{p}_2)}{f_1(\mathbf{x}_1-\mathbf{v}_1t,\mathbf{p}_1)f_1(\mathbf{x}_2-\mathbf{v}_2t,\mathbf{p}_2)}, \qquad (11.7)$$

where $\mathbf{g} = \mathbf{v}_1 - \mathbf{v}_2$ is the relative velocity. Function (11.7) tends to zero at $t \to \infty$ with any fi ed distance $\mathbf{r}$ and velocity $\mathbf{g}$. Using (4.9), (11.6), and (11.7), we can represent the term determining the contribution of initial correlation to the second term in the right-hand side of (11.5) as

$$n\int d\mathbf{p}_2\int d\mathbf{r}\left[\frac{\partial}{\partial\mathbf{r}}V_0\exp(-\frac{r^2}{r_0^2})\right](\frac{\partial}{\partial\mathbf{p}_1}-\frac{\partial}{\partial\mathbf{p}_2})g_0\exp(-\frac{|\mathbf{r}-\mathbf{g}t|^2}{r_{cor}^2})$$
$$\times\frac{\phi(\mathbf{x}_1-\mathbf{v}_1t,\mathbf{p}_1,\mathbf{x}_1-\mathbf{r}-\mathbf{v}_2t,\mathbf{p}_2)}{f_1(\mathbf{x}_1-\mathbf{v}_1t,\mathbf{p}_1)f_1(\mathbf{x}_1-\mathbf{r}-\mathbf{v}_2t,\mathbf{p}_2)}f_1(\mathbf{x}_1-\mathbf{r},\mathbf{p}_2,t)f_1(\mathbf{x}_1,\mathbf{p}_1,t) \qquad (11.8)$$

(where we set $t_0 = 0$ for brevity). It is easy to see that the integral over $\mathbf{r}$ in (11.8) is nonzero only if $t < t_{cor}$, where $t_{cor} \backsim r_{cor}/\bar{v} \sim r_0/\bar{v}$. For $t \gg t_{cor}$, the integral (11.8) practically vanishes because of the finit interparticle interaction range r_0. Physically, this means that the initial correlations propagate outward from the region of interest determined by r_0 and do not return in the approximation linear in n under consideration (also see [20]).

Of course, such behavior is possible if the contribution to (11.8) of the "parallel motion" with small g is negligible. Therefore, the mixing fl w in the phase space is necessary to ensure that the initial correlation term vanishes. The same behavior of the initial correlation term in the master equation was found in [29, 30] in the case of a quantum mechanical system interacting with a thermal bath with the mixing property in the van Hove limit. The term related to initial correlations and contributing to the third term in right-hand side of Eq. (11.5) (collision integral) displays the same behavior with time. In the considered example, the terms with initial correlations in Eq. (11.5) thus vanish for $t \gg t_{cor}$ (in the considered firs approximation in n).

If we pass to the kinetic time scale $t - t_0 \gtrsim t_{rel} \gg t_{cor}$ in Eq. (11.5), then the initial correlation terms determined by the two-particle correlation function $C_{12}(t - t_0)$ given by (11.4) vanish on this time scale because of the mixing fl w in the phase space (as in the example considered above), and we can rewrite Eq. (11.5) in the form of the irreversible Markovian kinetic equation

$$\frac{\partial f_1(x_1,t)}{\partial t} = L_1^0 f_1(x_1,t) + n\int dx_2 L_{12}^{'} f_1(x_2,t) f_1(x_1,t) + n\int dx_2 L_{12}^{'} \int_0^\infty dt_1 e^{L_{12}t_1} L_{12}^{'} f_1(x_2,t) f_1(x_1,t). \quad (11.9)$$

For an interparticle interaction with the small parameter $\varepsilon \ll 1$, the two-particle propagator $\exp(L_{12}t)$ in the collision term in (11.9) can be replaced with the "free" propagator $\exp(L_{12}^0 t)$ (see (4.26)), and Eq. (11.9) reduces to the Vlasov-Landau equation in the second approximation in ε (see, e.g., [20]). In the spatially homogeneous case, Eq. (11.9) coincides with the nonlinear Boltzmann equation for the momentum distribution function $f_1(\mathbf{p}_1, t)$.

We note, that all stages of the evolution of the system under consideration can be followed (in the adopted approximation) using Eq. (11.3) beginning from the initial reversible regime $t_0 \leqslant t \leqslant t_{cor}$. If the reversible terms in (11.3), related to the initial correlations $C_{12}(t - t_0)$, vanish with time and the last integral $\int_0^\infty$ in (11.9) exists (see Secs. 5 and 7), then the evolution of the system can switch automatically from the reversible to the irreversible stage described by kinetic equation (11.9).

12. Conclusion

In the preceding sections we have introduced a new type of evolution equations: the homogeneous generalized master equations (HGMEs) and nonlinear GMEs. The idea behind this is that we tried to obtain an evolution equations which are capable of describing all stages of the (sub)system of interest evolution in time beginning from the initial stage dependent of the initial system state (initial correlations). Although the BBGKY chain or conventional Nakajima-Zwanzig TC-GME and TCL-GME are formally valid for all $t > t_0$, the actual consideration of the transient (from reversible initial $0 < t - t_0 < t_{cor}$ to irreversible kinetic $t_{cor} < (t - t_0) \sim t_{rel}$) regime and of the influenc of initial correlations on the evolution process poses essential difficultie (see, e.g., [14] and references therein).

In the framework of the conventional time-independent projection operator approach, we have derived the exact time-convolution homogeneous generalized master equation

(TC-HGME) (2.12) and the exact time-convolutionless generalized master equation (TCL-HGME) (3.5) for the relevant part of a distribution function (statistical operator). In the derivation we have not used any approximation (like a factorizing initial condition or RPA) or principle (like the Bogoliubov principle of weakening of initial correlations). These equations have several advantages as compared to the conventional generalized master equations (GMEs). The HGMEs contain the parameter of initial correlations depending on time in the "mass" (super)operator acting on the relevant part of a distribution function (statistical operator). These equations allow treating the initial correlations consistently and on an equal footing with the collision integral by expanding the "mass" (super)operator into the series in the appropriate small parameter. The obtained equations are valid on any time scale, in particular on the initial stage of evolution $t_0 \leqslant t - t_0 \leqslant t_{cor}$, which can be important for studying the irreversibility problem, the non-Markovian and the ultrafast relaxation processes. The HGMEs enable the consideration of the entire evolution process of the relevant part of the distribution function (statistical operator) and of the influenc of initial correlations on this process. These equations can automatically switch, in principle, from the initial (reversible) regime into the kinetic (irreversible) one if the particle dynamics are characterized by the ergodic mixing fl w in the phase space leading to the damping of initial correlations and correlations caused by collisions.

By appropriate selection of the projection operator, one can rewrite the obtained HGMEs for the system under consideration. We have considered a spatially homogeneous dilute gas of classical and quantum particles with an arbitrary repulsive inter-particle interaction and obtained in the linear approximation in the small density parameter (4.2) the closed equation for a one-particle distribution function (4.24) and a one-particle density matrix (6.41) retaining initial correlations. In this approximation, the evolution equation for a one-particle momentum distribution function contains only binary collisions and a two-particle time-dependent (via only two-particle dynamics) correlation function in the parameter accounting for initial correlations. We have shown as on the macroscopic time-scale (1.1) these equations may become equivalent to the Boltzmann classical and quantum equations if all correlations caused by inter-particle interaction vanish on this time scale.

We have also derived (not postulated as it is usually done) the nonlinear GMEs describing the evolution of the relevant part of a distribution function (statistical operator). The new approach leading to these equations is based on using the nonlinear time-dependent operator $P(t)$ determining the relevant part of the distribution function (statistical operator). This operator is generally not a projection operator. The obtained inhomogeneous nonlinear GME (8.10) (and (8.23)) can be viewed as a generalization of the Nakajima-Zwanzig linear TC-GME. In the case of time-independent projection operators P and Q, the obtained equation reduces to the conventional TC-GME. This inhomogeneous nonlinear GME is equally useful for deriving both the nonlinear and linear evolution equations for the reduced distribution function (statistical operator) of interest in contrast to the linear TC-GME, which is naturally more convenient for deriving linear evolution equations, for example, for a subsystem interacting with a thermal bath. Using the obtained equation, we derived new inhomogeneous nonlinear equation (9.11) for a one-particle distribution function in the linear approximation in the small density parameter of a gas of classical particles. This equation, which holds for an arbitrary spatial inhomogeneity, contains the space and time changes of the one-particle distribution function on the microscopic time scale (contributing to the nondissipative characteristics of a nonideal gas and thus eliminating the inconsistency in the standard approach) and an irrelevant part (a source) given by the initial two-particle correlation function. Although the obtained inhomogeneous nonlinear GME allows deriving the closed nonlinear equation for a one particle distribution function, an additional assumption like the RPA or the Bogoliubov principle of weakening of initial correlations is still needed in order to obtain either the Vlasov-Landau or the Boltzmann kinetic equation.

To take initial correlations into account, we used the method of Sec. 2 for converting the inhomogeneous nonlinear GME into the homogeneous form. The obtained exact homogeneous nonlinear GME (10.3) for the relevant part of the distribution function (statistical operator) describes all evolution stages of the (sub)system of interest including the initial stage where the initial correlations play a role. In deriving this equation, which can be used to obtain both the nonlinear and linear evolution equations for the reduced distribution functions, we used no approximation like the RPA or the principle of weakening of initial correlations. The initial correlations are treated in this equation on an equal footing with collisions via the modifie memory kernel, which is a starting point for effective perturbation expansions. To test this equation, we used it for a spatially inhomogeneous nonideal gas of classical particles and obtained new homogeneous nonlinear evolution equation (11.3) for a one-particle distribution function retaining initial correlations in the memory kernel in the approximation linear in the density. Equation (11.3) describes all evolution stages, takes all two-particle correlations (collisions) into account, and converts into the nonlinear Boltzmann or Vlasov-Landau equation on the appropriate time scale when all initial correlations (and the ones due to collisions) vanish. It is important to note, that in order to obtain the nonlinear evolution (kinetic) equation, we are not now restricted by the second inequality (1.1) ($t - t_0 \ll t_{rel}$) as it was in the case of linear TC-HGME. The irreversible Boltzmann and Vlasov-Landau equations can be thus obtained from the Liouville equation without conventional additional approximations (like the RPA or principle of weakening of initial correlations) if the system dynamics have the necessary properties (such as an ergodic mixing fl w in the phase space).

References

[1] Bogoliubov, N. N. Problems of Dynamical Theory in Statistical Physics (in Russian); Gostekhizdat: Moscow, 1946. English transl; *Stud. Statist. Mech.,* Vol. 1; North-Holland: Amsterdam, 1962.

[2] Dorfman, J. R.; Cohen, E. G. D. *J Math Phys* 1967, 8, 282-297.

[3] Ferziger, J. H.; Kaper, H. G. *Mathematical Theory of Transport Processes in Gases*; North - Holland: Amsterdam, 1972.

[4] Klimontovich, Yu. L. *Statistical Physics* (in Russian); Nauka: Moscow, 1982. English. transl; Harwood Academic: Chur, 1986.

[5] Lanford III, O. E. Time Evolution of Large Classical Systems; Moser, E. J.; *Dynamical Systems, Theory and Applications*; Lect. Notes Phys.; Springer: Berlin, 1975; Vol. 38, pp. 1-111.

[6] Benedetto, D.; Castella, F.; Esposito, R.; Pulvirenti, M. *J Stat Phys* 2004, 116, 381-410.

[7] Van Kampen, N. G. *J Stat Phys* 2004, **115**, 1057-1072.

[8] Nakajima, S. *Progr Theor Phys* 1958, 20, 948-959.

[9] Zwanzig, R. *J Chem Phys* 1960, 33, 1338-1341.

[10] Prigogine, I. Non-Equilibrium Statistical Mechanics; *Monogr. Statist. Phys. and Thermodynamics*, Vol. 1; Interscience: New York, 1962.

[11] Shibata, F.; Takahashi, Y.; Hashitsume, N. *J Stat Phys* 1977, 17, 171-187.

[12] Shibata, F.; Arimitsu, T. *J Phys Soc Jap* 1980, 49, 891-897.

[13] Kadanoff, L. P.; Baym, G. *Quantum Statistical Mechanics*; Benjamin: New York, 1962.

[14] Zubarev, D. N.; Morozov, V. G.; Röpke, G. *Statistical Mechanics of Nonequilibrium Processes*; Akademie Verlag: Berlin, 1997, Vol. 2.

[15] Bellomo, B.; Compagno, G.; Petruccione, F. *J Phys A* 2005, 38, 10203-10216.

[16] Los, V. F. *J Phys A* 2001, 34, 6389-6403.

[17] Los, V. F. J Stat Phys 2005, 119, 241-271.

[18] Breuer, H.-P.; Petruccione, F. The Theory of Open Quantum Systems; Oxford Univ. Press: Oxford, 2002.

[19] Los, V. F. *Theor Math Phys* 2009, 160, 1124-1143.

[20] Balescu, *R.* Equilibrium and Nonequilibrium Statistical Mechanics; Wiley-Interscience: New York, 1975.

[21] Röpke, G. Statistiche Mechanik für dus Nichtgleichgewicht; VEB Deutscher Verlag der Wissenchaften: Berlin, 1987.

[22] Zaslavsky, G. M. *Stochasticity of Dynamical Systems;* Science: Moscow, 1984.

[23] Prigogine, I.; Resibois, P. *Physica* 1961, 27, 629-646.

[24] Bogoliubov, N. N.; Gurov, K. P. *J Exper Theor Phys* 1947, 17, 614-628.

[25] Bogoliubov, N. N.; Bogoliubov (Jr.), N. N. *Introduction into Quantum Statistical Mechanics*; Science: Moscow, 1984.

[26] Lebowitz, J. L. *Physica A* 1999, **263**, 516-527.

[27] Uhlenbeck, G. E.; Ford, G. W. *Lectures in Statistical Mechanics;* Amer. Math. Soc.: Providence, R I, 1963, Vol. 1.

[28] Klimontovich, Yu. L. Kinetic Theory of Nonideal Gases and Nonideal Plasmas (in Russian); Nauka: Moscow, 1975; English transl.; *Intern. Ser. Natural Philosophy,* Vol. 105; Pergamon: Oxford, 1982.

[29] Tasaki, S.; Yuasa, K.; Facchi, P.; Kimura, G.; Nakasato, H.; Ohba, I.; Pascazio, S. *Ann Phys* 2007, **332**, 631-656.

[30] Yuasa, K.; Tasaki, S.; Facchi, P.; Kimura, G.; Nakasato, H.; Ohba, I.; Pascazio, S. *Ann Phys* 2007, **332**, 656-676.

Reviewed by Professor Viktor G. Baryakhtar
Institute of Magnetism, 36-b Vernadsky Blvd., 03142 Kiev, Ukraine,
National Technical University "Kiev Polytechnic Institute",
37 Peremogy Ave., 03056 Kiev, Ukraine,
victor.baryakhtar@gmail.com

In: Evolution Equations
Editor: Arthur L. Claes

ISBN: 978-1-61209-652-0

Chapter 8

THE COSMOLOGICAL LINEAR AND NONLINEAR GROWTH IN THE ACCELERATING UNIVERSE

Seokcheon Lee*
Institute of Physics, Academia Sinica,
Taipei, Taiwan 11529, R.O.C.

Abstract

The present cosmological observations have confirme that the present Universe is expanding at an accelerating rate. One of the most popular solutions to this conundrum is introducing the so called "dark energy" which is the dominant energy contribution to the present energy of the Universe with its equation of state, $\omega_{\rm de} \simeq 1$. The effects of a dark energy component can be obtained mainly from both the cosmic microwave background (CMB) observations and the large scale structure formation ones. Even though early time properties of the dark energy dynamics can be extracted from the acoustic feature of the CMB anisotropy, the relevant characters of it have to be proved by investigating physical processes occurring at low redshifts when it starts dominating the cosmic expansion. Thus, the dark energy consequences on the large scale structure formation is the one of the most powerful tools. In this chapter, we investigate the cosmological implications of dark energy on both the sub-horizon scale growth of the matter density perturbation and the clustering by using the so called "spherical collapse model". We provide both the exact and approximate solutions for linear and non-linear perturbations. We provide the differences on the large scale structure formations between different dark energy models. We also investigate the cluster number calculations based on this specifi model.

PACS 95.36.+x, 98.62.Py, 98.65.-r.

Keywords: Large scale structure, dark energy, growth factor, galaxy clusters, cluster counting.

AMS Subject Classification: 85-01.

*E-mail address: skylee@phys.sinica.edu.tw

1. Introduction

The present accelerated expanding Universe scenario has been being confirme by the contemporary cosmological observations. In order to explain this puzzle, one needs to introduce either the extraordinary form of cosmological component or the deviation of our gravity theory in cosmological scale from the general relativity theory.

Both linear and non-linear density perturbations have been studied for several decades mainly for the so called standard cold dark matter (SCDM) (*i.e.* the matter component dominate the entire energy density of the Universe) picture. However, we now need to investigate those quantities with the coexistence of the dark energy component. And we focus on these issue in this chapter.

In the next section section, we obtain the exact analytic solution of subhorizon scale matter density perturbation (*i.e.* growth factor) for the general dark energy models with its equation of state $\omega_{\rm de}$ being constant. We also analyze its properties for the different choice of cosmological parameter. We show how this exact analytic solution helps the interpretation of Large Scale Structure (LSS) survey data.

In section III, the exact solution for the evolution of the background scale factor in a fla universe with the dark energy is given as a function of time. We obtain the exact and approximate solutions of the evolution of the overdensity radius with high accuracy. From the virial theorem, we obtain the exact value of the virial epoch for given models. From this, the value of the non-linear overdensity is obtained., which are lower than the known values. We show that clusters are formed earlier than those from the conventional model.

In section IV, we obtain the dark energy dependence of the mass fluctuatio which is consistent with the CMB normalization. We use the correct critical density threshold contrast δ_c to calculate the comoving number density of clusters for both PS and ST formalism. We show that PS formalism with this correct value of δ_c can predict the consistent cluster number with both the simulation and the observed data at the high mass region.

2. Growth Factor

Gravitational instability is believed to be a basic mechanism for the large scale structure of our universe. As times evolves, matter accumulates in initially overdense regions and is eventually attracted to the region to form structure. Mass in an overdense region is attracted to the center by gravity but repelled by pressure. The evolution equations for the dark matter overdensity are the full set of Boltzmann equations from the linear theory of scalar gravitational perturbation theory. Even though we will consider the cold dark matter with the dark energy like the quintessence fiel (QCDM) in a spatially fla universe, we do not need this full set of equations because the large scale structure (LSS) measurements probe scales 100 kpc $\sim$ 100 Mpc (subhorizon scale) where all the matter density perturbation modes $\delta_{\rm m}(\vec{k},a)$ grow uniformly. In other words, other cosmological components like dark energy and radiation do not contribute to the gravitational potential. Thus, the effect of the existence of dark energy appears only through the Hubble parameter, $H(a)$ and one can use the linear growth factor $D_g(a)$, define by $\delta_{\rm m}(\vec{k},a) \propto \delta_H(\vec{k})D_g(a)$.

We use the fla Friedmann-Robertson-Walker universe to probe the subhorizon scale linear density perturbations of matter δ_m in the matter dominated epoch,

$$H^2 \equiv \left(\frac{\dot{a}}{a}\right)^2 = \frac{8\pi G}{3}(\rho_m + \rho_{de}) = \frac{8\pi G}{3}\rho_{cr}\,, \tag{1}$$

$$2\frac{\ddot{a}}{a} + \left(\frac{\dot{a}}{a}\right)^2 = -8\pi G \omega_{de}\rho_{de}\,, \tag{2}$$

where ω_{de} is the equation of state (EOS) of dark energy, ρ_{cr} is the critical energy density, ρ_m and ρ_{de} are the energy densities of the matter and the dark energy, respectively. We consider the constant ω_{de} and set the present scale factor $a_0 = 1$. At subhorizon scales ($k \gg aH$), all interesting modes of the matter density perturbation $\delta_m(\vec{k}, a)$ grow uniformly as long as the dark energy do not cluster [1–3]. It means that we only consider the matter perturbation in Poisson equation in this scale. Thus, the growth factor $D_g(a)$ is define as

$$\delta_m(\vec{k}, a) \propto \delta_H(k) D_g(a)\,, \tag{3}$$

where $\delta_H(k)$ is the scalar amplitude at the horizon crossing generated during the cosmic inflation The alternative definitio of the D_g are also commonly used (see for example, [4])

$$\delta_m(\vec{k}, a) \equiv \delta_0(k) D_g(a)\,, \tag{4}$$

where $\delta_0(k)$ is the present density contrast. Then we obtain the evolution equation of $D(a)$ from the linear density perturbation equations [5, 6],

$$\frac{d^2D}{da^2} + \left(\frac{d\ln H}{da} + \frac{3}{a}\right)\frac{dD}{da} - \frac{4\pi G\rho_m}{(aH)^2} D = 0\,. \tag{5}$$

We use D in Eq. (5) instead of D_g because the general solution of Eq. (5) does not guarantee that D is the growing mode solution. We are able to fin the exact analytic solution of $D(a)$ for any value of the constant ω_{de} [7–9]. After replacing new parameters $Y = Qa^{3\omega_{de}}$ and $Q = \frac{\Omega_m^0}{\Omega_{de}^0}$ in Eq. (5), we obtain

$$Y\frac{d^2D}{dY^2} + \left[1 + \frac{1}{6\omega_{de}} - \frac{1}{2(Y+1)}\right]\frac{dD}{dY} - \left[\frac{1}{6\omega_{de}^2 Y} - \frac{1}{6\omega_{de}^2 Y(Y+1)}\right] D = 0\,. \tag{6}$$

Here $\Omega_m^0 = \frac{\rho_m^0}{\rho_{cr}^0}$ and $\Omega_{de}^0 = 1 - \Omega_m^0$ are the present energy density contrast of the matter and the dark energy, respectively. We replace a trial solution $D(Y) = cY^\alpha B(Y)$ into Eq (6) to get

$$Y(1+Y)\frac{d^2B}{dY^2} + \left[\frac{3}{2} - \frac{1}{6\omega_{de}} + \left(2 - \frac{1}{6\omega_{de}}\right)Y\right]\frac{dB}{dY} + \left[\frac{(3\omega_{de}+2)(\omega_{de}-1)}{12\omega_{de}^2}\right] B = 0\,,$$

$$\text{when } \alpha = \frac{1}{2} - \frac{1}{6\omega_{de}}\,. \tag{7}$$

The above equation becomes the so called "hypergeometric" equation when we replace $X = -Y$, which has the complete solution [10]

$$B(Y) = c_1 F\left[\frac{1}{2} - \frac{1}{2\omega_{de}}, \frac{1}{2} + \frac{1}{3\omega_{de}}, \frac{3}{2} - \frac{1}{6\omega_{de}}, -Y\right] + c_2 Y^{\frac{1-3\omega_{de}}{6\omega_{de}}} F\left[-\frac{1}{3\omega_{de}}, \frac{1}{2\omega_{de}}, \frac{1}{2} + \frac{1}{6\omega_{de}}, -Y\right], \tag{8}$$

where F is the hypergeometric function. Thus, the exact analytic solution of the above Eq. (5) is

$$D(a) = c_1 \left(\frac{\Omega_{\rm m}^0}{\Omega_{\rm de}^0}\right)^{\frac{3\omega_{\rm de}-1}{6\omega_{\rm de}}} a^{\frac{3\omega_{\rm de}-1}{2}} F\left[\frac{1}{2}-\frac{1}{2\omega_{\rm de}}, \frac{1}{2}+\frac{1}{3\omega_{\rm de}}, \frac{3}{2}-\frac{1}{6\omega_{\rm de}}, -\frac{\Omega_{\rm m}^0}{\Omega_{\rm de}^0}a^{3\omega_{\rm de}}\right]$$
$$+c_2 F\left[-\frac{1}{3\omega_{\rm de}}, \frac{1}{2\omega_{\rm de}}, \frac{1}{2}+\frac{1}{6\omega_{\rm de}}, -\frac{\Omega_{\rm m}^0}{\Omega_{\rm de}^0}a^{3\omega_{\rm de}}\right]. \tag{9}$$

$D(a)$ in Eq. (9) is just the general solution of the second order differential equation (5). Thus it does not have any physical meaning yet. It may represent the growing mode, the decaying mode or none of them before we choose the integral constants c_1 and c_2. If we want to have the correct growing mode solution from the above analytic solution, then this solution should follow the behavior of growing mode solution at an early epoch, say $a_i \simeq 0.1$. In other words, the coefficient of the general solution should be fi ed by using the initial conditions of the growth factor,

$$D_g(a_i) \simeq a_i \quad \text{and} \quad \left.\frac{dD_g}{da}\right|_{a_i} \simeq 1. \tag{10}$$

After we fi the coefficient from the initial conditions, we are able to determine the growth factor $D_g(a)$ from the general form of solution $D(a)$ in Eq. (9). If one wants to obtain the decaying mode solution $D_d(a)$ from Eq. (9), then one needs to adopt the decaying mode initial conditions to obtain the correct coefficient

$$D_d(a_i) \simeq a_i^{-\frac{3}{2}} \quad \text{and} \quad \left.\frac{dD_d}{da}\right|_{a_i} \simeq -\frac{3}{2}a_i^{-\frac{5}{2}}. \tag{11}$$

In literature, it is often used to normalize the growth mode with respect to the present value of it. We can obtain $\frac{\delta(Y)}{\delta(Y_0)}$ where $Y_0 = Q$ means the Y value at the present. In this case we only need to know the ratio of $\frac{c_1}{c_2}$ to determine the growth mode from Eq. (10). We obtain this relation from Eqs. (9) and (10) with simple algebra.

$$\frac{c_1}{c_2}\left(a_i, Q, \omega_{\rm de}\right) = 2a_i^{\frac{1-3\omega_{\rm de}}{2}} Q^{\frac{1}{6\omega_{\rm de}}-\frac{1}{2}}(9\omega_{\rm de}-1)\Bigg(-(1+3\omega_{\rm de})F \times$$
$$\left[-\frac{1}{3\omega_{\rm de}}, \frac{1}{2\omega_{\rm de}}, \frac{1}{2}+\frac{1}{6\omega_{\rm de}}, -a_i^{3\omega_{\rm de}}Q\right] + 3a_i^{3\omega_{\rm de}}QF\left[1-\frac{1}{3\omega_{\rm de}}, 1+\frac{1}{2\omega_{\rm de}}, \frac{3}{2}+\frac{1}{6\omega_{\rm de}}, -a_i^{3\omega_{\rm de}}Q\right]\Bigg)$$
$$\Bigg/ 3(3\omega_{\rm de}+1)(\omega_{\rm de}-1)\Bigg(a_i^{3\omega_{\rm de}}Q(3\omega_{\rm de}+2)F\left[\frac{3}{2}-\frac{1}{2\omega_{\rm de}}, \frac{3}{2}+\frac{1}{3\omega_{\rm de}}, \frac{5}{2}-\frac{1}{6\omega_{\rm de}}, -a_i^{3\omega_{\rm de}}Q\right]$$
$$+(1-9\omega_{\rm de})F\left[-\frac{1}{2\omega_{\rm de}}+\frac{1}{2}, \frac{1}{2}+\frac{1}{3\omega_{\rm de}}, \frac{3}{2}-\frac{1}{6\omega_{\rm de}}, -a_i^{3\omega_{\rm de}}Q\right]\Bigg) \tag{12}$$

Even though $\frac{c_1}{c_2}$ is a function of three quantities a_i, Q (*i.e.* Ω_m^0), and $\omega_{\rm de}$, its dependence on a_i and Q is quite small. When we vary both a_i and Ω_m^0 from 0.01 to 0.1 and from 0.2 to 0.35, respectively the changes on $\frac{c_1}{c_2}$ are about 10^{-4} % for both cases.

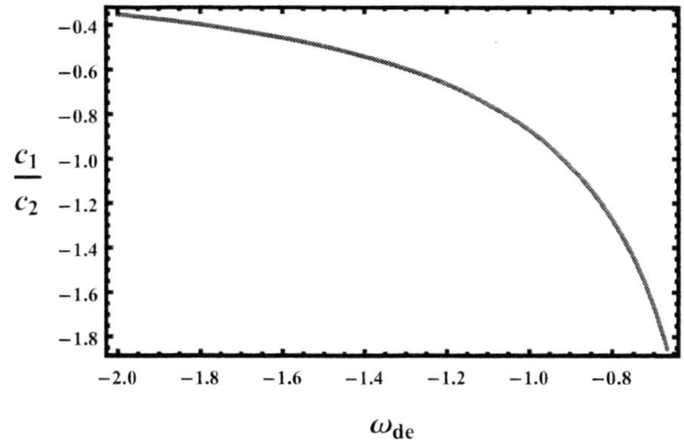

Figure 1. Behavior of $\frac{c_1}{c_2}$ as a function of $\omega_{\rm de}$ when $\Omega_{\rm m}^0 = 0.3$ and $a_i = 0.1$.

In Fig. 1, the behavior of the ratio of two coefficient in the growth mode solution are depicted as a function of $\omega_{\rm de}$. As $\omega_{\rm de}$ decreases, the magnitude of the ratio $\frac{c_1}{c_2}$ also decreases. This can be better understood with Fig. 2.

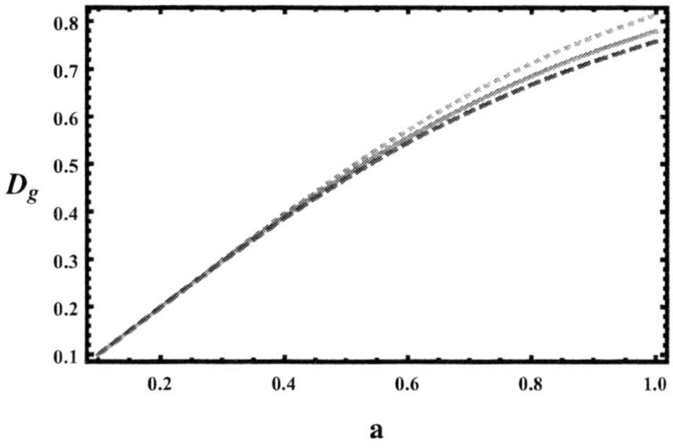

Figure 2. Evolutions of $D_g(a)$ when $\Omega_{\rm m}^0 = 0.3$ for $\omega_{\rm de} = -1.2, -1.0$, and -0.9 (from top to bottom).

In Fig. 2, we show the behaviors of the growth factor D_g for the different dark energy models (*i.e.* for the different values of $\omega_{\rm de}$) when $\Omega_{\rm m}^0 = 0.3$. The dotted, solid, and dashed lines correspond to $\omega_{de} = -1.2, -1.0$ and -0.9, respectively. As $\omega_{\rm de}$ decreases, D_g maintains the linear growth factor proportional to a for a longer time. This is easy to understand. If there is the same amount of the matter at the present epoch for the different models, then there will be more matter component in the past for the smaller value of $\omega_{\rm de}$. Thus, dark energy model with the smaller value of $\omega_{\rm de}$ maintains the longer linear growth behavior.

We are also able to obtain the exact value of the growth index $f(a) = \frac{d\ln D_g}{d\ln a}$ in any epoch. We also get the exact value of the growth index parameter $\gamma(a) = \frac{\ln f}{\ln \Omega_{\rm m}(a)}$. We investigate the behaviors of $f(a)$ and $\gamma(a)$ based on this exact analytic growth factor for the different dark energy models as shown in Fig. 2.

In Fig. 3a, we show the cosmological evolution of $f(a)$ for the different dark energy models when $\Omega_m^0 = 0.3$. Dashed, solid, and dotted lines correspond to $\omega_{de} = -0.9, -1.0$ and -1.2 dark energy models, respectively. As we show in Fig. 3a, the smaller value of ω_{de} gives the larger value of the present growth index $f(a=1)$. The present values of the growth index are $0.511, 0.513$, and 0.516 for corresponding models, respectively. Thus we may not see any differences between different models from the present growth index. If we investigate the f values at $z \simeq 0.15$ (*i.e.* $a \simeq 0.87$), then the growth index values will be $f(z=0.15) = 0.588, 0.598$, and 0.618. The current 2dF observation value is $0.36 \leq f(z=0.15) \leq 0.66$ [11, 12].

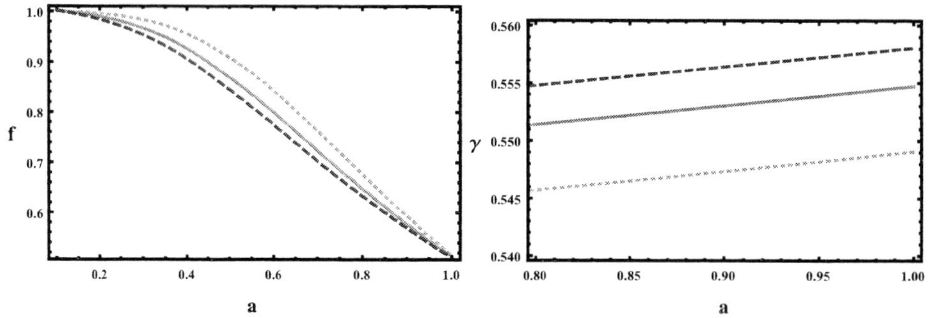

Figure 3. a) Evolution of $f(a)$ when $\Omega_m^0 = 0.3$. b) $\gamma(a)$ evolution with the same Ω_m^0. Dashed, solid, and dotted lines correspond to $\omega_{de} = -0.9, -1.0$, and -1.2, respectively.

In Fig. 3a, the behavior of $f(a)$ shows stronger model dependence around $0.25 \leq z \leq 1.5$. Thus, future observations in this redshift range should give us better observational constraints on the growth index. In Fig. 3b, we show the evolution of the growth index parameter $\gamma(a)$ for corresponding models with the same notation as Fig. 3a. The present values $\gamma(a=1) = 0.5581, 0.5547$, and 0.5491 for $\omega_{de} = -0.9, -1.0$, and -1.2, respectively. There is only 2% deviation from the different models. We are able to obtain the γ values at $z = 0.15$ (equally $a \simeq 0.87$). We get $\gamma(z=0.15) = 0.5559, 0.5525$, and 0.5468 for corresponding models. This shows not much changing in γ between the models. However, γ values at high z show very strong Ω_m^0 dependence and vary dramatically [7]. Thus, we had better to limit $z < 2 \sim 3$ for the investigation of γ. We summarize the results in Table 1. We also indicate the coefficient of the growth factor in this table, which are obtained from the growth factor initial conditions Eq. (10). In the last column, we compare the present growth index parameter γ_{ws}^0 obtained from Ref. [13]. Those values are quite close to the ones obtained from the exact analytic solution. Thus, the approximation of γ given in Ref. [13] is a good one as long as one probes γ near the present.

One interesting feature is that we can obtain the exact analytic solution of f and γ without using c_1 and c_2 for $\omega_{de} = -\frac{1}{3}$ and -1 cases [7]. The textbook solution for the

Table 1. ω_{de} is the eos of the dark energy. $f_{(z=0)}$ and $f_{(z=0.15)}$ are the values of the growth index at the present and $z = 0.15$, respectively. $\gamma_{(z=0)}$ and $\gamma_{(z=0.15)}$ are growth index parameters at the corresponding epoches. c_1 and c_2 are the coefficients of the growth factor D_g obtained from the initial conditions in Eq. (10). γ^0_{ws} are the values of the growth index parameters obtained from Ref. [13]

ω_{de}	$f_{(z=0)}$	$\gamma_{(z=0)}$	$f_{(z=0.15)}$	$\gamma_{(z=0.15)}$	c_1	c_2	γ^0_{ws}
-0.9	0.511	0.5581	0.588	0.5559	-1.12344	1.08851	0.556
-1.0	0.513	0.5547	0.598	0.5525	-0.943314	1.08464	0.553
-1.2	0.516	0.5491	0.618	0.5468	-0.716894	1.07822	0.548

growing mode solution of Eq. (5) for $\omega_{de} = -1$ or $-1/3$ is [14–16]

$$D_g(a) = \frac{5\Omega^0_{\rm m}}{2}\frac{H(a)}{H_0}\int_0^a X^{-3/2}(a')da' , \tag{13}$$

where $X(a) = (aH/H_0)^2 = \Omega^0_{\rm m} a^{-1} + \Omega^0_{\rm de} a^{-1-3\omega_{\rm de}}$. The growth index f is expressed from Eq. (13) as

$$f(\Omega^0_m, \omega_{de}, a) = -\frac{3}{2} - \frac{3\omega_{\rm de}(1-\Omega^0_{\rm m})a^{-1-3\omega_{\rm de}}}{2X} + \frac{aX^{-3/2}(a)}{\int_0^a X^{-3/2}(a')da'} . \tag{14}$$

We obtain the exact analytic form of the growth index for $\omega_{\rm de} = -1$ or $-1/3$ at the present epoch ($a = 1, X(a = 1) = 1$) as

$$f(\Omega^0_{\rm m}, \omega_{\rm de}, 1) = -\frac{3}{2} - \frac{3}{2}\omega_{\rm de}(1-\Omega^0_{\rm m}) - 3\omega_{\rm de}(\Omega^0_{\rm m})^{\frac{3}{2}} \frac{\Gamma[1-\frac{5}{6\omega_{\rm de}}]/(\Gamma[\frac{-5}{6\omega_{\rm de}}]\Gamma[1])}{F[\frac{3}{2},\frac{-5}{6\omega_{\rm de}},1-\frac{5}{6\omega_{\rm de}},-\frac{\Omega^0_{\rm de}}{\Omega^0_{\rm m}}]} , \tag{15}$$

where Γ is the gamma function and F is the hypergeometric function. We emphasize that this formula is correct only for $\omega_{\rm de} = -1/3$ or -1. For $\Omega^0_{\rm m} = 1$ one get $f = 1$ for all a, which is consistent with $D_g \propto a$. For $\omega_{\rm de} = -\frac{1}{3}$, f is matched with a non-fla without the cosmological constant $f(a=1) = -1 - \Omega^0_{\rm m}/2 + 5/2(\Omega^0_{\rm m})^{3/2}/F[3/2, 5/2, 7/2, 1-(\Omega^0_{\rm m})^{-1}]$ as shown in Ref. [17]. For the cosmological constant, the above formula becomes

$$f(\Omega^0_{\rm m}, -1, 1) \equiv f^0_L = -\frac{3}{2}\Omega^0_{\rm m} + 3(\Omega^0_{\rm m})^{\frac{3}{2}} \frac{\Gamma[\frac{11}{6}]/\Gamma[\frac{5}{6}]}{F[\frac{3}{2},\frac{5}{6},\frac{11}{6},-\frac{\Omega^0_{\rm m}}{\Omega^0_{\rm m}}]} , \tag{16}$$

where we use $\Gamma[1] = 1$. From the exact form of the growth index f^0_L, we are able to obtain the exact present value of the growth index parameter γ^0_L,

$$\gamma^0_L \equiv \frac{\ln f^0_L}{\ln \Omega^0_{\rm m}} = \frac{\ln\left[-\frac{3}{2}\Omega^0_{\rm m} + (\Omega^0_{\rm m})^{\frac{3}{2}} \frac{\Gamma[\frac{11}{6}]/\Gamma[\frac{5}{6}]}{F[\frac{3}{2},\frac{5}{6},\frac{11}{6},-\Omega^0_{\rm de}/\Omega^0_{\rm m}]}\right]}{\ln \Omega^0_{\rm m}} . \tag{17}$$

The variation of the present growth index for different values of $\Omega^0_{\rm m}$ is from 0.4073 to 0.6028 for $0.2 \leq \Omega^0_{\rm m} \leq 0.4$. Thus, there is up to 32% differences in f^0_L for the viable ranges of $\Omega_{\rm m}^0$. However, this $\Omega_{\rm m}^0$ dependence is decreased by using the parameter γ_L^0. γ_L^0 is changed from 0.5580 to 0.5524 for the same ranges of $\Omega_{\rm m}^0$. We obtain $\gamma_L^0 = 0.5547$ for $\Omega_{\rm m}^0 = 0.3$ to be consistent with the previous result.

3. Spherical Collapse Model

In this section, we investigate a spherical overdensity model for the non-clustering dark energy with the constant $\omega_{\rm de}$. The spherical collapse model (SCM) is a simple and power tool for investigating the evolution of nonlinear structure and bound systems in the Universe as it behaves in exactly the same way as a closed sub-universe [5, 18]. The collapse will never occur due to the dissipative physics eventually intervene and convert the kinetic energy of collapse into random motion which is called as the virial theorem.

The background evolution equations of a fla FRW universe are already given in Eqs. (1) and (2). The matter and the DE evolves as

$$\dot{\rho}_{\rm m} + 3\left(\frac{\dot{a}}{a}\right)\rho_{\rm m} = 0\,, \tag{18}$$

$$\dot{\rho}_{\rm de} + 3(1+\omega_{\rm de})\left(\frac{\dot{a}}{a}\right)\rho_{\rm de} = 0\,, \tag{19}$$

Now we investigate a spherical perturbation in the matter density. $\rho_{\rm cluster}$ is the matter density within the spherical overdensity radius R. The flatnes condition is not held because of the perturbation in the matter. Thus, we have another set of equations governing the dynamics of the spherical perturbation [19] :

$$\frac{\ddot{R}}{R} = -\frac{4\pi G}{3}\left[\rho_{\rm cluster} + (1+3\omega_{\rm de})\rho_{\rm dec}\right]\,, \tag{20}$$

$$\dot{\rho}_{\rm cluster} + 3\left(\frac{\dot{R}}{R}\right)\rho_{\rm cluster} = 0\,, \tag{21}$$

$$\begin{aligned} &\dot{\rho}_{\rm dec} + 3(1+\omega_{\rm de})\left(\frac{\dot{R}}{R}\right)\rho_{\rm dec} = \alpha\Gamma\,, \\ &\text{where } \Gamma = 3(1+\omega_{\rm de})\left(\frac{\dot{R}}{R} - \frac{\dot{a}}{a}\right)\rho_{\rm dec} \quad \text{with } 0 \leq \alpha \leq 1\,, \end{aligned} \tag{22}$$

where $\rho_{\rm dec}$ is the DE density within the cluster. $\alpha = 0$ gives the clustering behavior of DE and $\alpha = 1$ means the homogenous DE satisfying Eq. (19). For $\alpha = 1$ case, the system does not conserve energy as it collapses from the turnaround to the virialized state. We limit our consideration to only the non-clustering DE.

Now we adopt the notations in Ref. [13] to investigate the evolutions of a and R,

$$x = \frac{a}{a_{\rm ta}}\,, \tag{23}$$

$$y = \frac{R}{R_{\rm ta}}\,. \tag{24}$$

Then from Eqs. (1) and (20), we obtain

$$\frac{dx}{d\tau} = \frac{1}{\sqrt{x\Omega_{\rm m}(x)}} = \sqrt{x^{-1} + \frac{1}{Q_{\rm ta}} x^{-3\omega_{\rm de}-1}}, \tag{25}$$

$$\frac{d^2y}{d\tau^2} = -\frac{1}{2}\left[\zeta y^{-2} + (1+3\omega_{\rm de})\frac{1}{Q_{\rm cta}} x^{-3(1+\omega_{\rm de})} y\right], \tag{26}$$

where $d\tau \equiv H_{\rm ta}\sqrt{\Omega_{\rm m}(x_{\rm ta})}dt$, $\zeta \equiv \frac{\rho_{\rm cluster}}{\rho_{\rm m}}|_{z_{\rm ta}}$, $Q_{\rm ta} \equiv \frac{\rho_{\rm m}}{\rho_{\rm de}}|_{z_{\rm ta}} = \frac{\Omega_{\rm m}^0}{\Omega_{\rm de}^0}(1+z_{\rm ta})^{-3\omega_{\rm de}}$.

We are able to obtain the analytic solution of Eq. (25)

$$\int_0^x dx'\sqrt{x'\Omega_{\rm m}(x')} = \int_0^\tau d\tau' \Rightarrow \frac{2}{3}x^{\frac{3}{2}}F\left[\frac{1}{2}, -\frac{1}{2\omega_{\rm de}}, 1-\frac{1}{2\omega_{\rm de}}, -\frac{x^{-3\omega_{\rm de}}}{Q_{\rm ta}}\right] = \tau, \tag{27}$$

where F is the hypergeometric function and we use the boundary condition $x = 0$ when $\tau = 0$ (see Appendix for details). This solution is the same as the one in Ref. [20, 21] when $\omega_{\rm de} = -\frac{1}{3}$. The above equation gives the well known relation $a(t) \propto \sinh^{\frac{2}{3}}\left[\frac{3}{2}\sqrt{\Omega_{\rm de}^0}{\rm H_0 t}\right]$ when $\omega_{\rm de} = -1$. From this equation we are able to fin the exact turnaround time $\tau_{\rm ta}$ for general $\omega_{\rm de}$ models,

$$\tau_{\rm ta} = \frac{2}{3}F\left[\frac{1}{2}, -\frac{1}{2\omega_{\rm de}}, 1-\frac{1}{2\omega_{\rm de}}, -Q_{\rm ta}^{-1}\right], \tag{28}$$

where we use the fact that $x_{\rm ta} = 1$. It is useful to have the exact analytic form of the turnaround time to investigate the accuracies of both numerical and analytical calculations.

As expected, $\tau_{\rm ta}$ depends on $\omega_{\rm de}$, $\Omega_{\rm m}^0$, and $z_{\rm ta}$ as given in Eq. (28). We show these properties of $\tau_{\rm ta}$ in Fig. 4. In the left panel of Fig. 4, we show $\tau_{\rm ta}$ as a dependence on the values of $\Omega_{\rm m}^0$ for the different DE models. We choose $z_{\rm ta} = 1$ in this figure The dot-dashed, solid, dotted, and dashed lines (from top to bottom) correspond to $\omega_{\rm de} = -1.2, -1.0, -0.8$, and $-1/3$, respectively. If we have the same $\Omega_{\rm m}^0$ for the different models, then it takes longer time to the turnaround for the smaller values of $\omega_{\rm de}$. The smaller value of $\omega_{\rm de}$ means the higher pressure against the gravity. Thus, it takes longer time to reach to the maximum radius of the overdensity for the higher pressure universe when we have the same amount of $\Omega_{\rm m}^0$. We also need more matter to overcome the higher pressure. We also show the $z_{\rm ta}$ dependence of $\tau_{\rm ta}$ for the different DE models when we choose $\Omega_{\rm m}^0 = 0.3$ in the right panel of Fig. 4. Again the smaller values of $\omega_{\rm de}$ give the larger $\tau_{\rm ta}$. From the definitio of $Q_{\rm ta}$, we can fin that as $z_{\rm ta}$ increases so does $Q_{\rm ta}$. For large $z_{\rm ta}$, $Q_{\rm ta}^{-1}$ approaches to 0 and thus $\tau_{\rm ta}$ approaches to $\frac{2}{3}F\left[\frac{1}{2}, -\frac{1}{2\omega_{\rm de}}, 1-\frac{1}{2\omega_{\rm de}}, 0\right] = \frac{2}{3} \sim 0.67$ independent of $\omega_{\rm de}$.

Now we work on the evolution of y in Eq. (26). For general $\omega_{\rm de}$, there is no analytic solution of this equation. However, we are able to obtain the exact analytic solutions for $\omega_{\rm de} = -1/3$ and -1. First, we consider the case when $\omega_{\rm de} = -\frac{1}{3}$. In this case, the second term in the right hand side of Eq. (26) disappears and the equation becomes [22]

$$\frac{dy^2}{d\tau^2} = -\frac{1}{2}\zeta y^{-2} \equiv f(y) \Rightarrow \int_0^y \frac{dy'}{\sqrt{-c_1 + 2\int^{y'} f(y'')dy''}} = c_2 \pm \tau. \tag{29}$$

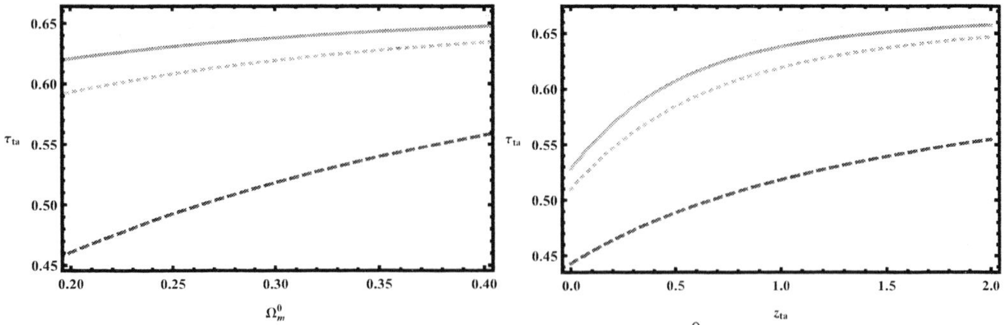

Figure 4. $\tau_{\rm ta}$ for the different values of $\omega_{\rm de}$. a) $\tau_{\rm ta}$ v.s. $\Omega_{\rm m}^0$ for the different values of $\omega_{\rm de} = -1.2, -1.0, -0.8$, and $-1/3$ (from top to bottom) when $z_{\rm ta} = 1.0$. b) $\tau_{\rm ta}$ v.s. $z_{\rm ta}$ for the same models when we choose $\Omega_{\rm m}^0 = 0.3$.

The solutions of the above equation are given by (see Ref. [20])

$$\text{ArcSin}[\sqrt{\text{y}}] - \sqrt{\text{y}(1-\text{y})} = \sqrt{\zeta_{13}}\tau\,, \text{when } \tau \leq \tau_{\rm ta13}\,, \tag{30}$$

$$\sqrt{y(1-y)} - \text{ArcSin}[\sqrt{\text{y}}] + \frac{\pi}{2} = \sqrt{\zeta_{13}}(\tau - \tau_{\rm ta13})\,, \text{when} \tau \geq \tau_{\rm ta13}\,, \tag{31}$$

where ζ_{13} and $\tau_{\rm ta13}$ denote respectively ζ and $\tau_{\rm ta}$ for $\omega_{\rm de} = -\frac{1}{3}$. Even though the above solutions given in Eqs. (30) and (31) are only true for $\omega_{\rm de} = -\frac{1}{3}$, these solutions can be good approximate solutions for general $\omega_{\rm de}$ when we consider the high-redshift clusters. It is because the high redshift clusters which are formed at the high $z_{\rm vir}$ require the high $z_{\rm ta}$ in each model. As $z_{\rm ta}$ increases, so does $Q_{\rm ta}$ for the fi ed value of $\Omega_{\rm m}^0 \sim 0.3$. Thus, the second term in Eq. (26) is negligible compared to the firs one and this equation approximately equals to Eq. (29). Also these solutions are the same as the ones in the Einstein de Sitter (EdS) universe ($\Omega_{\rm m}^0 = 1$ and $\Omega_{\rm de}^0 = 0$). It is because in the EdS universe, $Q_{\rm ta}^{-1} = 0$ and the evolution equation of y is the same as in Eq. (29). Also the solution given in Eq. (30) is a good approximate solution for general $\omega_{\rm de}$ when we consider the evolution of y at early epoch. As $y \rightarrow 0$, the second term in Eq. (29) is negligible again.

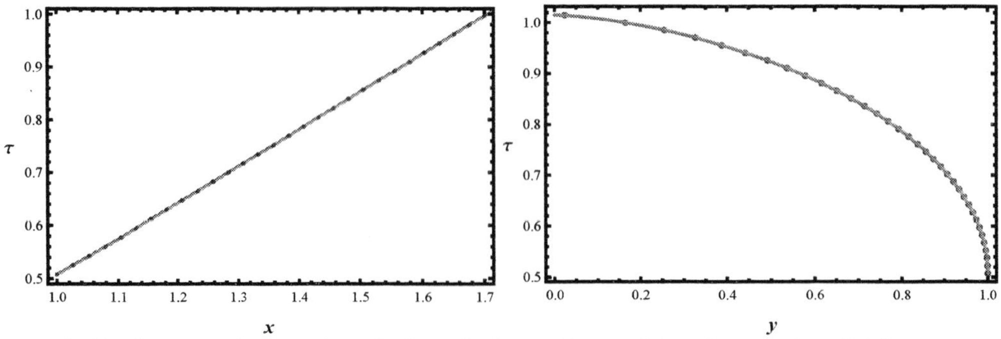

Figure 5. Both numerical and analytic solutions, denoted by dots and solid lines respectively. a) Cosmological evolutions of τ for $\omega_{\rm de} = -\frac{1}{3}$ as a function of x when we choose $z_{\rm ta} = 0.8$ and $\Omega_{\rm m}^0 = 0.3$. b) Evolutions of τ as a function of y for the same values of $z_{\rm ta}$ and $\Omega_{\rm m}^0$ as in a).

As long as we know the exact virialized epoch $z_{\rm vir}$, Eq. (31) provides the exact value of ζ for the different values of $\Omega_{\rm m}^0$ and $z_{\rm ta}$ when $\omega_{\rm de} = -\frac{1}{3}$. We will investigate $z_{\rm vir}$ based on the virial theorem later. However, we firs consider the simpler estimation to obtain ζ values for the general $\omega_{\rm de}$. If we defin z_c as the redshift at which the cluster formally collapses to $R = 0$ according to a spherical solution, then this collapse time is twice the turnaround time $t_c = 2t_{\rm ta}$ (*i.e.* $\tau(z_c) = 2\tau(z_{\rm ta})$) [13]. If we adopt this fact, then we are able to obtain ζ from the above analytic solution in Eq. (31)

$$\zeta_{13} = \left(\frac{\pi/2}{\frac{2}{3}F\left[\frac{1}{2},\frac{3}{2},\frac{5}{2},\frac{-1}{Q_{\rm ta13}}\right]} \right)^2 = \left(\frac{3\pi}{4}\right)^2 \left(F\left[\frac{1}{2},\frac{3}{2},\frac{5}{2},-\frac{(1-\Omega_{\rm m}^0)}{\Omega_{\rm m}^0}(1+z_{\rm ta})^{-1}\right] \right)^{-2}. \quad (32)$$

When $\Omega_{\rm m}^0 = 1$, $F\left[\frac{1}{2},\frac{3}{2},\frac{5}{2},0\right] = 1$. Thus, $\zeta_{13} = (\frac{3\pi}{4})^2$ when $\Omega_{\rm m}^0 = 1$. This factor $(\frac{3\pi}{4})^2$ is the well-known value of ζ for the EdS universe [5,23]. The general value of ζ for $\omega_{\rm de} = -\frac{1}{3}$ when $\Omega_{\rm m}^0 \neq 1$ is given by Eq. (32). After we obtain the value of ζ_{13} given in Eq. (32), we are able to fin the values of x and y at any $\tau_{\rm ta} \leq \tau \leq \tau_c$ from Eqs. (27) and (31). We constrain $\tau \geq \tau_{\rm ta}$ to show the evolution of x after the overdensity radius reaches its maximum. In the left panel of Fig. 5, we show the evolution of τ as a function of x when $z_{\rm ta} = 0.8$ and $\Omega_{\rm m}^0 = 0.3$. The numerical solution (denoted by dots) is definitel the same as the analytic solution (solid line). In the right panel of Fig. 5, we show the evolutions of both analytic and numerical solutions of $y(\tau)$ for the same values of $z_{\rm ta}$ and $\Omega_{\rm m}^0$ as in the left panel of Fig. 5.

Now we consider Eq. (26) when $\omega_{\rm de} = -1$. The differential equation of y for $\omega_{\rm de} = -1$ becomes

$$\frac{d^2y}{d\tau^2} = -\frac{1}{2}\left[\zeta y^{-2} - \frac{2}{Q_{\rm ta}}y\right] \Rightarrow \int^y \frac{dy'}{\sqrt{-c_1 + \frac{\zeta}{y'} + \frac{y'^2}{Q_{\rm ta}}}} = c_2 \pm \tau. \quad (33)$$

We obtain $c_1 = \zeta + Q_{\rm ta}^{-1}$ from Eq. (33) by using the initial condition $\frac{dy}{d\tau}|_{z_{\rm ta}} = 0$. The solution of the above equation is given by (see Appendix for details)

$$c_F\left(\text{EllipticF}\left[\text{ArcSin}[\text{B}],\text{C}\right] + c_P\text{EllipticPi}\left[\text{A},\text{ArcSin}[\text{B}],\text{C}\right]\right) = c_2 - \tau \text{ , when } \tau \geq \tau_{\rm ta}, \quad (34)$$

where EllipticF and EllipticPi correspond to the elliptic integral of the firs kind and the incompelete elliptic integral of the third kind, respectively [24]. Also $A = \frac{2\sqrt{1+4Q_{\rm ta}\zeta}}{3+\sqrt{1+4Q_{\rm ta}\zeta}}$, $B = \frac{\sqrt{3+\frac{(2Q_{\rm ta}\zeta-4)}{(1-y)}+\sqrt{1+4Q_{\rm ta}\zeta}}}{\sqrt{2(1+4Q_{\rm ta}\zeta)}}$, $C = \frac{2\sqrt{1+4Q_{\rm ta}\zeta}}{-1+2Q_{\rm ta}\zeta+\sqrt{1+4Q_{\rm ta}\zeta}}$, $c_F = -\frac{2\sqrt{2Q_{\rm ta}}}{\sqrt{2Q_{\rm ta}\zeta-1+\sqrt{1+4Q_{\rm ta}\zeta}}}$, and $c_P = -\frac{2(Q_{\rm ta}\zeta-2)}{(3+\sqrt{1+4Q_{\rm ta}\zeta})}$. The integral constant c_2 is able to be obtained when we use the boundary condition $y(\tau_{ta}) = 1$. Even though this is the exact analytic solution of y for $\omega_{\rm de} = -1$, this formula has some problems. First, this is inconvenient to use. Second, this formula does not produce the correct value when $B > 1$. However, it is useful to have this analytic solution when we derive the form of fittin formulae for ζ and y for general $\omega_{\rm de}$ cases.

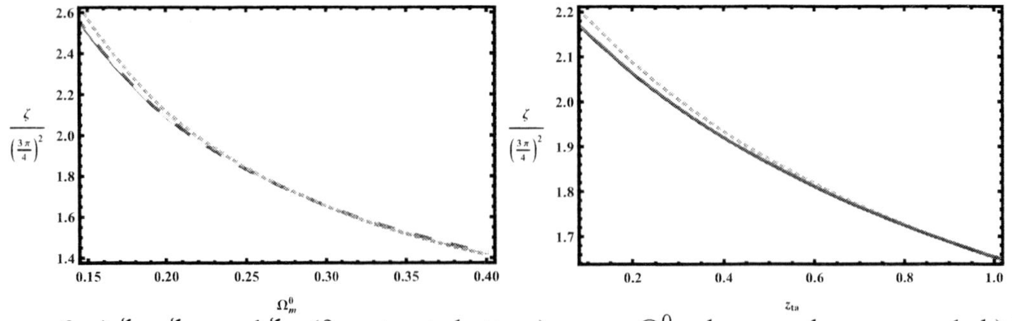

Figure 6. a) $\zeta_{\rm ws}$, ζ_{13}, and $\zeta_{\rm sk}$ (from top to bottom) versus $\Omega_{\rm m}^0$ when we choose $z_{\rm ta} = 1$. b) ζs versus $z_{\rm ta}$ when we choose $\Omega_{\rm m}^0 = 0.3$.

ζ_{13} given in Eq. (32) is exact only for $\omega_{\rm de} = -\frac{1}{3}$. Thus, we need to fin ζ for general $\omega_{\rm de}$. As we show in both $\omega_{\rm de} = -\frac{1}{3}$ and -1 cases, ζ is a function of both $\omega_{\rm de}$ and $Q_{\rm ta}$. Thus, we are able to derive a reasonable fittin formula for ζ as a function of $\omega_{\rm de}$ and $Q_{\rm ta}$. When we investigate the numerical solutions of Eqs. (26) for general $\omega_{\rm de}$, we obtain the fittin formula of ζ

$$\zeta_{\rm sk} = \left(\frac{3\pi}{4}\right)^2 \Omega_{\rm mta}^{-0.724+0.157\Omega_{\rm mta}+(1+\omega_{\rm de})(1+3\omega_{\rm de})(0.064-0.368\Omega_{\rm mta})}, \tag{35}$$

where $\Omega_{\rm mta} = \Omega_{\rm m}(x_{\rm ta}) = \frac{1}{1+Q_{\rm ta}^{-1}}$ and it becomes 1 when $\Omega_{\rm m}^0 = 1$. Thus, this choice of the fittin formula is consistent with the fact that $\zeta = (\frac{3\pi}{4})^2$ in the EdS universe. Also $\Omega_{\rm mta}$ is a function of $\omega_{\rm de}$, $z_{\rm ta}$, and $\Omega_{\rm m}^0$, and thus it is a function of all the variables in the consideration. The above fittin formula has less than 3 % errors when we limit our consideration to $0.2 \leq \Omega_{\rm m}^0 \leq 0.34$, $z_{\rm ta} \geq 0.7$, and $-1.1 \leq \omega_{\rm de} \leq -1/3$. The special choice of this fittin form of ζ is independent of α when $\omega_{\rm de} = -\frac{1}{3}$ or -1 in order to be consistent with real models. In Ref. [13], ζ is given by another form for the fittin formula,

$$\zeta_{\rm ws} = \left(\frac{3\pi}{4}\right)^2 \Omega_{\rm mta}^{-0.79+0.26\Omega_{\rm mta}-0.06\omega_{\rm de}}. \tag{36}$$

In Fig. 6, we compare the values of $\zeta_{\rm sk}$ and $\zeta_{\rm ws}$ for $\omega_{\rm de} = -\frac{1}{3}$ with the exact ζ_{13} given in Eq.(32) for the different values of $\Omega_{\rm m}^0$ and $z_{\rm ta}$. We use $z_{\rm ta} = 1$ and normalize ζs with $(\frac{3\pi}{4})^2$ in the left panel of Fig. 6. The dashed line depicts $\zeta_{\rm ws}$ and the solid and dot-dashed lines describe ζ_{13} and $\zeta_{\rm sk}$, respectively. ζ_{13} and $\zeta_{\rm sk}$ are too close to each other and look like a single line. We show ζs as a dependence on $z_{\rm ta}$ when we choose $\Omega_{\rm m}^0 = 0.3$ in the right panel of Fig. 6. Again both ζ_{13} and $\zeta_{\rm sk}$ are very close to each other in most of the $z_{\rm ta}$ range.

As given in Eq. (26), y should be a function of $\omega_{\rm de}$, $Q_{\rm ta}$, and ζ. Even though the function of y for the general value of $\omega_{\rm de}$ should not be the same as Eq. (31), we modify this equation to obtain the approximate analytic solution of y as

$$\begin{aligned}
&\sqrt{y(1-y)} - \text{ArcSin}[\sqrt{y}] + \frac{\pi}{2} + \left(-\frac{(1+3\omega_{\rm de})}{3}\Omega_{\rm mta}\right)^{2.7+0.1\Omega_{\rm mta}-3.4(1+\omega_{\rm de})-0.02\zeta_{\rm sk}+3(z_{\rm ta}-0.6)} \\
&= \sqrt{\zeta_{\rm sk}}(\tau - \tau_{\rm ta}) \\
&= \frac{2}{3}\sqrt{\zeta_{\rm sk}}\left(x^{\frac{3}{2}} F\left[-\frac{1}{2\omega_{\rm de}}, \frac{1}{2}, 1 - \frac{1}{2\omega_{\rm de}}, -\frac{x^{-3\omega_{\rm de}}}{Q_{\rm ta}}\right] - F\left[-\frac{1}{2\omega_{\rm de}}, \frac{1}{2}, 1 - \frac{1}{2\omega_{\rm de}}, -\frac{1}{Q_{\rm ta}}\right]\right).
\end{aligned} \tag{37}$$

This approximate solution is obtained from interpolating y values for the wide ranges of the cosmological parameters and one can further work on the similar fittin form by changing interested ranges of the cosmological parameters. There are several reasons for the specifi choice of the correction term $\left(-\frac{(1+3\omega_{\rm de})}{3}\Omega_{\rm mta}\right)^{2.7+0.1\Omega_{\rm mta}-3.4(1+\omega_{\rm de})-0.02\zeta_{\rm sk}+3(z_{\rm ta}-0.6)}$ in the above equation. First, we know that the firs two terms in Eq. (37) are exact when $\omega_{\rm de}=-\frac{1}{3}$. Thus, the correction term should not appear for $\omega_{\rm de}=-\frac{1}{3}$ and thus we put $(1+3\omega_{\rm de})$ in this correction. Also as $z_{\rm ta}$ increases, the contribution of the correction term decreases. Thus, in the approximate solution we have the power $3\,(z_{\rm ta}-0.6)$ term. Also we fin that $z_{\rm ta}\sim 0.6$ in order to have $z_{\rm vir}\sim 0$ in the concordance model.

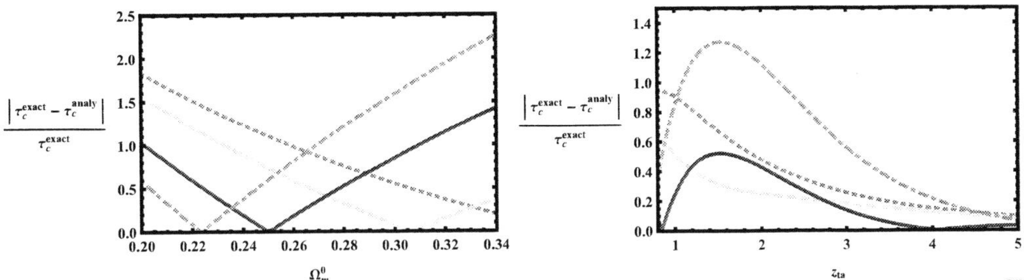

Figure 7. Error (%) of the analytic solution of τ_c obtained from Eq. (37). a) Error of $\tau_c^{\rm analy}$ against $\Omega_{\rm m}^0$ when $z_{\rm ta}=1.1(z_{\rm vir}\sim 0.3)$ for the different $\omega_{\rm de}=-0.8,-0.9,-1.0,$ and -1.05 models, denoted by the dotted, dashed, solid, and dot-dashed lines, respectively. b) Error of $\tau_c^{\rm analy}$ against $z_{\rm ta}$ when $\Omega_{\rm m}^0=0.27$ for the different $\omega_{\rm de}$ models.

Since we know the exact value of τ_c from Eq. (27), we are able to estimate the goodness of the approximate solution given in Eq. (37). The errors of τ_c for the different models obtained from Eq. (37) are shown in Fig. 7. We show the errors of $\tau_c^{\rm analy}$ against $\Omega_{\rm m}^0$ with $z_{\rm ta}=1.1$ for the different $\omega_{\rm de}$ models in the left panel of Fig. 7. The dotted, dashed, solid, and dot-dashed lines correspond to $\omega_{\rm de}=-0.8,-0.9,-1.0,$ and -1.05, respectively. If we limit $0.24\leq\Omega_{\rm m}^0\leq 0.3$, then the errors are less than 1.6 % for all the models. We also depict the errors of $\tau_c^{\rm analy}$ against $z_{\rm ta}$ with $\Omega_{\rm m}^0=0.27$ for the different $\omega_{\rm de}$ models in the right panel of Fig. 7. Again, the dotted, dashed, solid, and dot-dashed lines correspond to $\omega_{\rm de}=-0.8$, -0.9, -1.0, and -1.05, respectively. The errors of $\tau_c^{\rm analy}$ for $\omega_{\rm de}\geq-1.0$ are less than 1 % for the entire range of $z_{\rm ta}$.

If only the matter virializes, then we are not able to use the energy conservation in the virial theorem because the dark energy does not virialize with the dark matter [25–28].

$$U_{\rm cluster}(z_{\rm ta})+U_{\rm de}(z_{\rm ta})\geq\frac{1}{2}U_{\rm cluster}(z_{\rm vir})+2U_{\rm de}(z_{\rm vir})\,, \tag{38}$$

where $U_{\rm cluster}$ and $U_{\rm de}$ are the potential energies associated with the spherical mass overdensity and with the DE, respectively. Thus, in the presence of the smooth DE, the dark matter can reach a quasi-equilibrium state where virialization holds instantaneously. The potential energies in the above Eq. (38) are given by $U_i=\frac{1}{2}\int\rho_i\Phi_i dV$ where the gravitational potential Φ_i in a spherical homogeneous configuratio is given by $\Phi_i(r) = -2\pi G(1+3\omega_i)\rho_i(R^2 - r^2/3)$. To estimate the virial radius $y_{\rm vir} = \frac{R_{\rm vir}}{R_{\rm ta}}$ we use the assumption $\rho_{\rm de}(z_{\rm vir}) = \rho_{\rm de}(z_{\rm ta})$ as in the reference [26]. With this approximation, we are able to obtain the equation of $y_{\rm vir} = \frac{R_{\rm vir}}{R_{\rm ta}}$ for any value of $\omega_{\rm de}$ by using Eq. (38) :

$$4(1+3\omega_{\rm de})y_{\rm vir}^3 - 2\Big[\zeta Q_{\rm ta} + (1+3\omega_{\rm de})\Big]y_{\rm vir} + \zeta Q_{\rm ta} = 0\,. \tag{39}$$

An interesting feature in this equation is that we obtain $y_{\rm vir} = \frac{1}{2}$ like that in the EdS universe when $\omega_{\rm de} = -\frac{1}{3}$ for any given values of $\Omega_{\rm m}^0$ and $z_{\rm ta}$ [20]. Thus, we are able to obtain $x_{\rm vir}$ of the $\omega_{\rm de} = -\frac{1}{3}$ model for the different values of $\Omega_{\rm m}^0$ and $z_{\rm ta}$ from Eq. (31). With simple algebra by using Eqs. (27), (28), (31), and (32), we obtain

$$\begin{aligned} x_{\rm vir}^{\frac{3}{2}} F\left[\frac{1}{2},\frac{3}{2},\frac{5}{2},-\frac{x_{\rm vir}}{Q_{\rm ta13}}\right] &= \left(\frac{1}{\pi}+\frac{3}{2}\right)F\left[\frac{1}{2},\frac{3}{2},\frac{5}{2},-\frac{1}{Q_{\rm ta13}}\right] \\ &= \frac{3}{2}Q_{\rm ta13}^{\frac{3}{2}}\left(\sqrt{\frac{x_{\rm vir}}{Q_{\rm ta13}}}\left(1+\frac{x_{\rm vir}}{Q_{\rm ta13}}\right) - \ln\left[\sqrt{\frac{x_{\rm vir}}{Q_{\rm ta13}}} + \sqrt{1+\frac{x_{\rm vir}}{Q_{\rm ta13}}}\right]\right), \end{aligned} \tag{40}$$

where we use the relations $\frac{x_{\rm vir}}{Q_{\rm ta}}F\left[\frac{1}{2},\frac{3}{2},\frac{5}{2},-\frac{x_{\rm vir}}{Q_{\rm ta}}\right] = \frac{3}{2}\left(1+\frac{x_{\rm vir}}{Q_{\rm ta}}\right)F\left[\frac{1}{2},\frac{3}{2},\frac{3}{2},-\frac{x_{\rm vir}}{Q_{\rm ta}}\right] - \frac{3}{2}F\left[\frac{1}{2},\frac{1}{2},\frac{3}{2},-\frac{x_{\rm vir}}{Q_{\rm ta}}\right]$, $F\left[\frac{1}{2},\frac{3}{2},\frac{3}{2},-\frac{x_{\rm vir}}{Q_{\rm ta}}\right] = \left(1+\frac{x_{\rm vir}}{Q_{\rm ta}}\right)^{-\frac{1}{2}}$, and $F\left[\frac{1}{2},\frac{1}{2},\frac{3}{2},-\frac{x_{\rm vir}}{Q_{\rm ta}}\right] = \left(\sqrt{\frac{x_{\rm vir}}{Q_{\rm ta}}}\right)^{-1}\ln\left[\sqrt{\frac{x_{\rm vir}}{Q_{\rm ta}}}+\sqrt{1+\frac{x_{\rm vir}}{Q_{\rm ta}}}\right]$ in the second equality [24]. Even though we obtain the analytic expression of $x_{\rm vir}$ in Eq. (40), generally this equation can be solved in a non-algebraic way. However, $x_{\rm vir}$ can be represented by the approximate analytic form for the wide ranges of the cosmological parameters ($\Omega_{\rm m}^0$ and $z_{\rm ta}$) if one analyzes the form by interpolating $x_{\rm vir}$ values.

For the EdS universe, Eqs. (27), (31), and (32) become

$$\frac{2}{3}x^{\frac{3}{2}} = \tau\,, \tag{41}$$

$$\sqrt{y(1-y)} - \mathrm{ArcSin}[\sqrt{y}] + \frac{\pi}{2} = \sqrt{\zeta}(\tau - \frac{2}{3}) = \frac{\pi}{2}\left(x^{\frac{3}{2}} - 1\right)\,, \tag{42}$$

$$\zeta = (\frac{3\pi}{4})^2\,, \tag{43}$$

where we use $F[\frac{1}{2},-\frac{1}{2\omega_{\rm de}},1-\frac{1}{2\omega_{\rm de}},0] = 1$ and $\tau_{\rm ta} = \frac{2}{3}$. The above equations can be generalized to any value of $\omega_{\rm de}$ when $\Omega_{\rm m}^0$ approaches to 1 or $z_{\rm ta}$ goes to a very high value (for most of the parameter spaces $z_{\rm ta} \geq 5$). In the EdS universe, $y_{\rm vir} = \frac{1}{2}$ from Eq. (39). The commonly used assumption to obtain the nonlinear overdensity Δ_c in the EdS universe is that $\tau_c = 2\tau_{\rm ta} = \frac{4}{3}$ which is the collapsing time for $y_c = 0$ even though one uses $y_{\rm vir} = \frac{1}{2}$ to obtain Δ_c. By using this assumption, one obtain $x_c = 2^{\frac{2}{3}}$ from Eq. (41). Thus, one obtains

$$\Delta_c^{\rm EdS} = \zeta\left(\frac{x_c}{y_{\rm vir}}\right)^3 = \left(\frac{3\pi}{4}\right)^2\left(\frac{2^{\frac{2}{3}}}{2^{-1}}\right)^3 = 18\pi^2 \simeq 178\,. \tag{44}$$

However, we can obtain the correct value of $\Delta_{\rm vir}$ for the EdS universe because we know the exact relations between τ, x, and y. After replacing $y_{\rm vir} = \frac{1}{2}$ into Eq. (42), we obtain the virialization time $\tau_{\rm vir} = \frac{2}{3\pi} + 1 < \tau_c$. From this correct value of $\tau_{\rm vir}$, we obtain $x_{\rm vir} = (\frac{1}{\pi} + \frac{3}{2})^{\frac{2}{3}}$ by using Eq. (41). Thus, the correct value of the non-linear overdensity $\Delta_{\rm vir}$ for the EdS universe is [20]

$$\Delta_{\rm vir}^{\rm EdS} = \zeta \left(\frac{x_{\rm vir}}{y_{\rm vir}} \right)^3 = \left(\frac{3\pi}{4} \right)^2 \left(\frac{\left(\frac{1}{\pi} + \frac{3}{2}\right)^{\frac{2}{3}}}{2^{-1}} \right)^3 = 18\pi^2 \left(\frac{1}{2\pi} + \frac{3}{4} \right)^2 \simeq 147 \,. \tag{45}$$

As $\omega_{\rm de}$ becomes large, the structure is formed earlier and $\Delta_{\rm vir}$ becomes large. There is only gravity involved in the structure formation in the EdS universe and thus it requires less overdensity than the one with the pressure. Thus, the minimum overdensity for general $\omega_{\rm de}$ models is about 147 instead of 178. We show this in Tables 2 and 3. As $z_{\rm ta}$ increases, $\Delta_{\rm vir}$ approaches to 147 instead of 178.

The ratio of cluster to background density at the virialized epoch $z_{\rm vir}$ for $\omega_{\rm de} = -\frac{1}{3}$ becomes

$$\Delta_{\rm vir} \equiv \Delta(z_{\rm vir}) = \frac{\rho_{\rm cluster}}{\rho_{\rm m}} \bigg|_{z_{\rm vir}} = \zeta \left(\frac{x_{\rm vir}}{y_{\rm vir}} \right)^3 = 18\pi^2 \frac{x_{\rm vir}^3}{4(F\left[\frac{1}{2}, \frac{3}{2}, \frac{5}{2}, -\frac{x_{\rm vir}}{Q_{\rm ta}}\right])^2} \,, \tag{46}$$

where we use Eqs. (18), (21), and (32). We also use the fact that $y_{\rm vir} = \frac{1}{2}$ independent of $Q_{\rm ta}$. We are able to obtain $\Delta_{\rm vir}$ for the given cosmological parameters by using Eqs. (40) and (46). It is also useful to express $z_{\rm vir}$ by using $z_{\rm ta}$ and $x_{\rm vir}$ from Eq. (23)

$$z_{\rm vir} = \frac{1 + z_{\rm ta}}{x_{\rm vir}} - 1 \,. \tag{47}$$

Now we are able to extend the previous consideration to general values of $\omega_{\rm de}$ by using Eqs. (27), (35), (37), and (39). We show several quantities for the different models in Tables 2 and 3. We need to clarify several things in these tables. First, structures start to form earlier for lager $\omega_{\rm de}$ and thus $\Delta_{\rm vir}$ rises with increasing $\omega_{\rm de}$. Thus, we are able to distinguish between the models from the observed cluster at a given $z_{\rm vir}$. The other important feature is that as $\Omega_{\rm m}^0$ and $z_{\rm ta}$ increase, the model dependence of the observed quantity $\Delta_{\rm vir}$ decreases. Thus, the low-redshift clusters are the better objects to be used for studying the properties of DE. Again as $z_{\rm ta}$ increases, $\Delta_{\rm vir}$ approaches to 147 instead of $18\pi^2 \simeq 178$.

From Eq. (46), we are able to investigate the linear perturbation at early epoch $\Delta \stackrel{\tau \to 0}{\longrightarrow} 1 + \delta_{\rm lin}$ [20]

$$\Delta \equiv 1 + \delta_{\rm NL} \stackrel{\tau \to 0}{\longrightarrow} 1 + \delta_{\rm lin} = 1 + \frac{3}{5} \left(\sqrt{\zeta} \frac{3}{2} \tau \right)^{\frac{2}{3}} = 1 + \frac{3}{5} \left(\sqrt{\zeta} \sqrt{\Omega_{\rm mta}} \frac{t}{t_{\rm ta}} \right)^{\frac{2}{3}} \,, \tag{48}$$

Table 2. ζ_{sk}, x_{vir}, y_{vir}, and Δ_{vir} with the given values of the cosmological parameters for the different DE (ω_{de}) models. As ω_{de} becomes large, the structure is formed earlier and Δ_{vir} becomes large

ω_{de}	$z_{ta} = 0.7$								$z_{ta} = 1.1$							
	$\Omega_m^0 = 0.24(z_{vir} \sim 0.02)$				$\Omega_m^0 = 0.3(z_{vir} \sim 0.04)$				$\Omega_m^0 = 0.24(z_{vir} \sim 0.31)$				$\Omega_m^0 = 0.3(z_{vir} \sim 0.33)$			
	ζ_{sk}	x_{vir}	y_{vir}	Δ_{vir}	ζ_{sk}	x_{vir}	y_{vir}	Δ_{vir}	ζ_{sk}	x_{vir}	y_{vir}	Δ_{vir}	ζ_{sk}	x_{vir}	y_{vir}	Δ_{vir}
$-\frac{1}{3}$	11.2	1.62	0.50	384	9.8	1.60	0.50	322	10.2	1.61	0.50	339	9.0	1.59	0.50	289
−0.6	9.1	1.65	0.53	282	8.2	1.63	0.52	250	8.0	1.61	0.52	236	7.3	1.59	0.52	214
−0.8	8.1	1.66	0.53	244	7.4	1.64	0.53	222	7.1	1.60	0.52	201	6.7	1.58	0.52	188
−0.9	7.8	1.66	0.54	231	7.2	1.64	0.53	213	6.8	1.60	0.53	193	6.5	1.58	0.52	182
−1.0	7.6	1.66	0.54	222	7.1	1.65	0.53	212	6.6	1.60	0.53	189	6.4	1.59	0.52	181

Table 3. ζ_{sk}, x_{vir}, y_{vir}, and Δ_{vir} with the given values of the cosmological parameters for the different DE (ω_{de}) models. Independent of model parameters, Δ_{vir} converges to 147 as z_{ta} increases

ω_{de}	$z_{ta} = 1.4$								$z_{ta} = 5.0$							
	$\Omega_m^0 = 0.24(z_{vir} \sim 0.53)$				$\Omega_m^0 = 0.3(z_{vir} \sim 0.54)$				$\Omega_m^0 = 0.24(z_{vir} \sim 3.0)$				$\Omega_m^0 = 0.3(z_{vir} \sim 3.0)$			
	ζ_{sk}	x_{vir}	y_{vir}	Δ_{vir}	ζ_{sk}	x_{vir}	y_{vir}	Δ_{vir}	ζ_{sk}	x_{vir}	y_{vir}	Δ_{vir}	ζ_{sk}	x_{vir}	y_{vir}	Δ_{vir}
$-\frac{1}{3}$	9.6	1.60	0.50	316	8.6	1.58	0.50	271	7.2	1.55	0.50	215	6.8	1.50	0.50	183
−0.6	7.4	1.59	0.52	216	6.9	1.57	0.51	198	5.9	1.52	0.50	161	5.8	1.50	0.50	154
−0.8	6.7	1.57	0.52	184	6.4	1.56	0.52	176	5.7	1.50	0.50	151	5.6	1.50	0.50	150
−0.9	6.4	1.57	0.52	178	6.2	1.55	0.51	171	5.6	1.50	0.50	150	5.6	1.49	0.50	147
−1.0	6.3	1.57	0.52	175	6.1	1.56	0.51	170	5.6	1.50	0.50	150	5.6	1.49	0.50	147

where we use $H_{\rm ta} = \frac{2}{3} t_{\rm ta}^{-1}$ in the early (matter-dominated) epoch. This is equal to the famous result for the EdS universe ($\sqrt{\zeta} = \frac{3\pi}{4}$ and $\Omega_{\rm mta} = 1$), $\delta_{\rm lin} = \frac{3}{5}\left(\frac{3\pi}{4}\frac{t}{t_{\rm ta}}\right)^{\frac{2}{3}}$ [5, 20]. As τ approaches to zero, the exact solutions of x and y given in Eqs. (27) and (30) become the good approximate solutions for general values of $\omega_{\rm de}$. In this limit, $y_{\rm vir} \simeq \frac{1}{2}$ and thus $\tau_{\rm vir} \simeq \left(\frac{3}{2} + \frac{1}{\pi}\right)\tau_{\rm ta}$ from Eq. (30). Thus,

$$\delta_{\rm lin}(z_{\rm vir}) = \frac{3}{5}(\sqrt{\zeta})^{\frac{2}{3}}\left(\left(\frac{3}{4} + \frac{9\pi}{8}\right)\frac{1}{\sqrt{\zeta}}\right)^{\frac{2}{3}} = \frac{3}{20}(6+9\pi)^{\frac{2}{3}} \simeq 1.58\,, \tag{49}$$

where we use $\sqrt{\zeta} = \frac{\pi}{2\tau_{\rm ta}}$. The above result given in Eq. (49) is true for any value of $\omega_{\rm de}$ because $\omega_{\rm de}$ dependence of x and y is disappeared in the $\tau \to 0$ limit. From this critical density threshold $\delta_{\rm lin}(z_{\rm vir})$, we are able to obtain $\delta_{\rm lin}$ at any epoch by using the relation

$$\delta_{\rm lin}(z) = \frac{D_g(z)}{D_g(z_{\rm vir})}\delta_{\rm lin}(z_{\rm vir})\,, \tag{50}$$

where D_g is the linear growth factor. There is the exact analytic form of D_g for the DE model with the constant eos [7–9].

4. Cluster Number Counts

The formation of the large scale dark matter (DM) potential wells of clusters is solely determined by gravitational physics irrelevant to the gas dynamical processes, star formation, and feedback. Also clusters are the largest virialized objects in the Universe with their abundance and evolution simply related to the linear matter power spectrum. Thus, the abundance of clusters and their distribution in redshift should be determined purely by the geometry of the Universe and the power spectrum of initial density fluctuations As a result, the clusters of galaxies provide a useful probe of the fundamental cosmological parameters including the investigation into the dark energy equation of state $\omega_{\rm de}$, because the linear growth factor D_g, the cosmological volume element, as well as the primordial scalar amplitude at horizon crossing δ_H depend on $\omega_{\rm de}$ [3, 29].

Abundance of rich clusters has been commonly used to constrain the matter power spectrum because it is sensitive to the normalization of the power spectrum on cluster scales [13, 30]. The amplitude of matter perturbations is sensitive to the presence of dark energy, through the normalization of the matter power spectrum to the raw large scales probed by the cosmic microwave background (CMB) anisotropies [31]. The normalizations of the matter power spectrum obtained from both methods can be represented by the *rms* mass fluctuatio on $8h^{-1}$Mpc scales. However, there exists discrepancy in σ_8 values resulted from two different normalization methods [32]. While σ_8 is almost constant in the former method, it drops rapidly for larger $\omega_{\rm de}$ because of the increasingly strong integrated Sachs-Wolfe (ISW) effect in the latter.

The linear power spectrum for δ in QCDM models is given by

$$P(k,a) = A_Q k^{n_s} T_Q^2(k)\left(\frac{D_g(a)}{D_g(a_0)}\right)^2, \tag{51}$$

where A_Q is a normalization, n_s is the spectral index of the primordial adiabatic density perturbations, $T_Q(k)$ is the transfer function, D_g is the linear growth factor, and the present scale factor normalized as $a_0 = 1$. Even though the transfer function $T_Q(k)$ does depend on $\omega_{\rm de}$, its change happens only on large scale $k \leq 0.01$ hMpc^{-1}. Thus, this correction in the transfer function does not affect the value of the *rms* mass fluctuation The main effect of the quintessence fiel is on the normalization A_Q which can be written as $A_Q = 2\pi^2\delta_H^2(c/H_0)^{n_s+3}$, where [3]

$$\begin{aligned}
\delta_H &= 2.05\times10^{-5}\alpha_0^{-1}(\Omega_{\rm m}(a))^{c_1+c_2\ln\Omega_{\rm m}(a)}\exp[c_3(n_s-1)+c_4(n_s-1)^2]\,,\text{with} \quad (52)\\
c_1 &= -0.789|\omega_{\rm de}|^{0.0754-0.211\ln|\omega_{\rm de}|}\,, c_2=-0.118-0.0727\omega_{\rm de}\,, c_3=-1.037\,, c_4=-0.138\,,\\
\alpha &= (-\omega_{\rm de})^s \text{ with s} = (0.012-0.036\omega_{\rm de}-0.017\omega_{\rm de}^{-1})\Big(1-\Omega_{\rm m}(\rm a)\Big)\\
&\quad +(0.098+0.029\omega_{\rm de}-0.085\omega_{\rm de}^{-1})\ln\Omega_{\rm m}(a)\,. \quad (53)
\end{aligned}$$

Note that the notation $\alpha_0 = \alpha(a_0)$ and δ_H is the amplitude at horizon crossing. We use $\delta_H = 2.05\times10^{-5}$ to be consistent with WMAP 7 year data, and the cosmological parameters $\Omega_{\rm m}^0 = 0.272$, $h = 0.7$, $n_s = 0.963$, $\Omega_{\rm b}^0 = 0.0456$, and $\sigma_8 = 0.809$ from WMAP + BAO + H_0 measurements given in Ref. [33]. The *rms* linear mass fluctuatio at top-hat smoothing scale R is given by

$$\sigma_R^2(a) \equiv \left\langle \left|\frac{\delta M}{M(R,a)}\right|^2 \right\rangle = \frac{1}{2\pi^2}\int_0^\infty k^2P(k,a)\Big|W(kR)\Big|^2 dk, \quad (54)$$

where the filterin radius R is the Lagrangian radius of a halo of mass M at the present time, $R = (\frac{3M}{4\pi\rho_m(a_0)})^{1/3}$, and $W(kR) = \frac{3}{(kR)^3}\Big(\sin[kR]-(kR)\cos[kR]\Big)$ is the top-hat window function. Traditionally, the cluster abundance is used to put a constraint on the dispersion of the density contrast at the scale 8 h^{-1}Mpc, denoted σ_8. We also denote the mass inside a sphere of radius $R_8 = 8$ h^{-1}Mpc as $M_8 = 5.95\times10^{14}\Omega_{\rm m}^0 h^{-1}{\rm M}_\odot$ where ${\rm M}_\odot$ means the solar mass and we use the present critical energy density $\rho_{\rm crit}^0 = 2.775\times10^{11}{\rm M}_\odot h^2{\rm Mpc}^{-3}$. Now we show the present matter power spectra and σ_8 values for different values of $\omega_{\rm de}$ in Fig. 8. In the left panel of Fig. 8, the dot-dashed, solid, and dashed lines correspond to the matter power spectrum for $\omega_{\rm de} = -1.1$, -1.0, and -0.8, respectively. The differences in the power spectra of $\omega_{\rm de} = -1.1$ and -0.8 from -1.0 are about 7 and 17 %, respectively. The amplitude of the power spectra are comparable with the one in Ref. [34]. In the right panel of Fig. 8, we also compare the COBE normalized σ_8 (solid line) with the one from Ref. [13] (dashed one). There exists the discrepancy in the dependence of σ_8 on $\omega_{\rm de}$ between the two different normalizations other than the magnitudes. It is easy to understand this as shown in Ref. [32]. For the fi ed value of $\Omega_{\rm m}^0$, the dark energy dominates the cosmic expansion earlier as $\omega_{\rm de}$ increases and thus enhancing the dynamics of the gravitational potential which results in an increasing integrated Sachs-Wolfe (ISW) effect on large scales. This effect is properly shown in the COBE normalized σ_8 only. Also σ_8 obtained from the X-ray cluster population as in Ref. [13] suffers from the uncertainties due to the uncertainties in modelling. Thus, we use the COBE normalized σ_8 in our cluster abundance calculation.

Even though the COBE normalized σ_8 is the proper and the accurate one, we need to express it as a function of M in the cluster abundance calculation. When $\omega_{\rm de} = -1$,

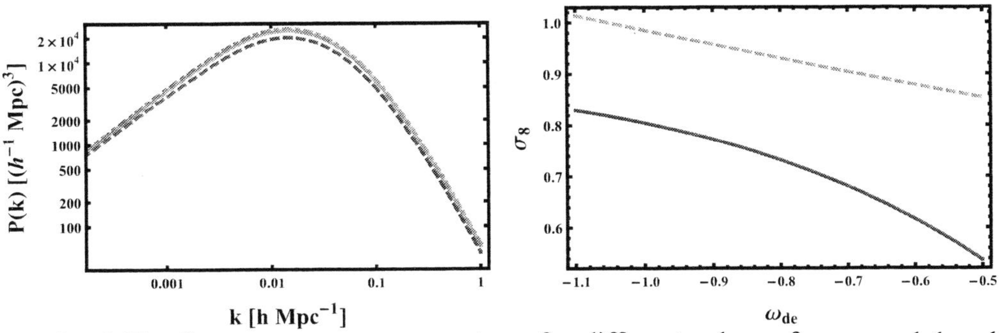

Figure 8. a) The linear matter power spectrum for different values of $\omega_{\rm de} = -1.1$, -1.0, and -0.8 (from top to bottom). b) σ_8 values from the COBE normalization and the Wang and Steinhardt's approximation (from bottom to top).

σ_8 is well known [35] and we can parameterize $\omega_{\rm de}$ dependence of σ_8 by $\sigma_8(\omega_{\rm de}) = (-\omega_{\rm de})^{0.72+0.36\omega_{\rm de}}\sigma_8(\omega_{\rm de} = -1)$ with less than 2 % errors for $-1.1 \leq \omega_{\rm de} \leq -0.6$. For the cosmological parameters adopted from the WMAP, we can express

$$\sigma(M,z) \simeq (-\omega_{\rm de})^{0.72+0.36\omega_{\rm de}} \left(3.90 - 0.215 \log\left[\frac{M}{h^{-1}M_{\odot}}\right] \right) \left(\frac{D_g(z)}{D_g(z_0)} \right) . \tag{55}$$

We show $\sigma(M, z = 0)$ for different values of $\omega_{\rm de}$ in Fig. 9. The dot-dashed, solid, and dashed lines correspond to $\omega_{\rm de} = -1.1$, -1.0, and -0.8, respectively.

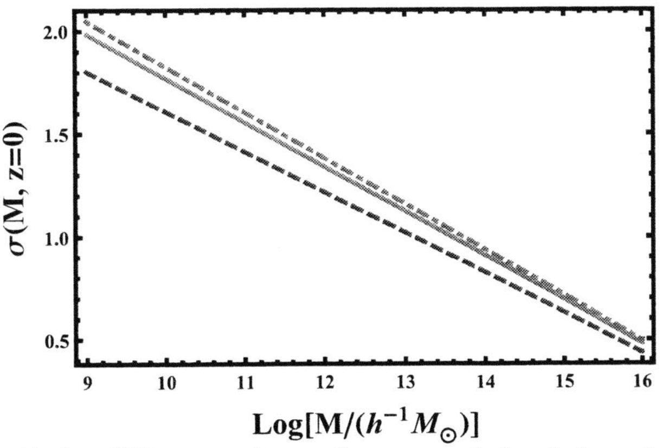

Figure 9. $\sigma(M, z = 0)$ for different values of $\omega_{\rm de} = -1.1, -1.0$, and -0.8 (from top to bottom).

Friedmann's equation for the QCDM is written as in Eq. (1). Thus, the cosmic volume per unit redshift is also given by

$$V(z) = \int_0^z 4\pi d_A^2(z') \left| \frac{cdt}{dz} \right| (z') dz' , \tag{56}$$

where $d_A = \frac{c}{1+z}\int_0^z \frac{dz'}{H(z')}$ is the angular diameter distance between redshifts 0 and z. $V(z)$ is the proper volume of a sphere of radius z around the observer. We show the dependence of $V(z)$ on $\omega_{\rm de}$ in the left panel of Fig. 10. As $\omega_{\rm de}$ increases, $V(z)$ decreases.

The exact analytic growing mode solution D_g of δ for any value of the constant $\omega_{\rm de}$ is introduced in the previous section. We just repeat the behaviors of the growth factor D_g for different dark energy models (*i.e.* for different values of $\omega_{\rm de}$) when $\Omega_{\rm m}^0 = 0.272$ in the right panel of Fig. 10. The dot-dashed, solid, and dashed lines correspond to $\omega_{\rm de} = -1.1, -1.0$, and -0.8, respectively. As $\omega_{\rm de}$ decreases, D_g maintains the linear growth factor proportional to a for a longer time.

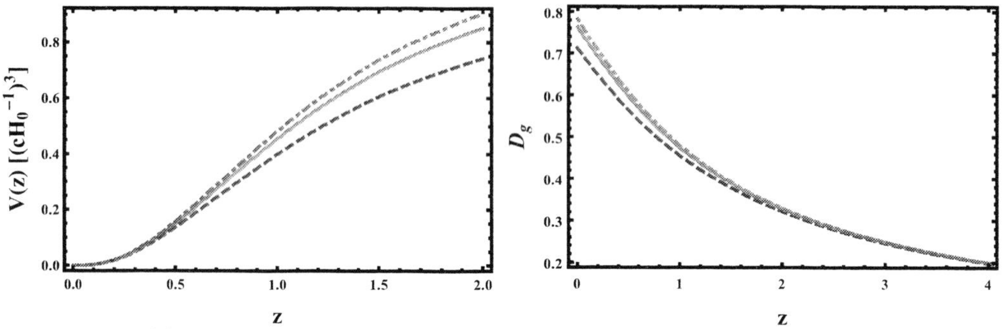

Figure 10. a) $V(z)$ for different values of $\omega_{\rm de} = -1.1, -1.0$, and -0.8 (from top to bottom). b) $D_g(z)$ for the same values of $\omega_{\rm Q}$ as in the left panel.

The mass function $f(M)$ from the Press-Schechter (PS) formalism is related to the comoving number density dn of objects in the range dM as $dn = (\rho_m^0/M)|d\ln\sigma/dM|f(M)dM$ [36,37]. PS theory which relates the comoving number density of virialized objects to their mass is given by

$$dn(M,z) = \sqrt{\frac{2}{\pi}}\frac{\rho_m^0}{M^2}\left|\frac{d\ln\sigma}{d\ln M}\right|\frac{\delta_c}{\sigma}\exp\left[-\frac{\delta_c^2}{2\sigma^2}\right]dM, \tag{57}$$

where the critical density threshold $\delta_c = \frac{\rho_{\rm linear}}{\rho_m}$ is predicted for a spherical overdensity of radius $R = (3M/4\pi\rho_m)^{1/3}$ and mass M according to the linear theory. However, the PS mass function is known to predict too many low mass clusters and too few high mass clusters, as well as too few clusters at high z [38]. One remark is that the correct value of $\delta_c = 1.58$ was recently obtained independent of the value of $\omega_{\rm de}$ instead of 1.69 [20, 21, 39]. If we use this correct value of δ_c, then the PS formalism shows the improved predictions for the number densities of both low and high mass clusters. We show this in Fig. 11. In the left panel of Fig. 11, we show the comoving number density of clusters n of mass great than M for $\omega_{\rm de} = -1$. The solid, dashed, dot-dashed, and long dashed lines correspond to $z = 0$, 0.5, 1.0, and 2.0, respectively. The circular ($z \simeq 0$) and triangular ($0.18 \le z \le 0.85$) dots represent the data from Ref. [40]. We also show the changes of n at high and low masses at different redshifts in the right panel of Fig. 11. At the low mass $M = 10^9[h^{-1}M_\odot]$, there is about 2 % decrease in n at present. At the high mass $M = 10^{16}[h^{-1}M_\odot]$, n increases about 58 % today. Also at high z, n increases about 75 and 114 % at $z = 0.5$ and 1.0, respectively.

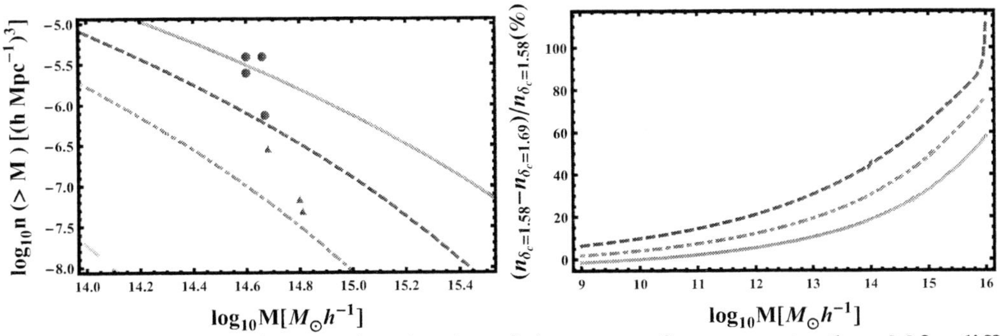

Figure 11. a) The comoving number density of clusters n of mass greater than M for different values of $z = 0$, 0.5, 1.0, and 2.0 (from top to bottom) when $\omega_Q = -1$ and $\delta_c = 1.58$. The circular ($z \simeq 0$) and triangular ($0.18 \leq z \leq 0.85$) dots represent the data from Ref. [40]. b) Errors of n when we use the correct threshold density contrast $\delta_c = 1.58$ instead of 1.69 for different values of $z = 0$, 0.5, and 1.0 (from bottom to top).

Using the correct value of δ_c, we can predict more massive clusters and more clusters at high z. However, we may still have too many low mass clusters even by using correct value of δ_c from the PS formalism. This is shown is Fig. 12. If we strictly limit the mass of clusters as bigger than $10^{14} M_\odot$, then PS formalism might be good enough for the cluster abundance calculation. Also there might be other mechanisms than only gravity for the low mass cluster formations. In any case, PS formalism shows the deviation from the simulation at low mass region. Thus, we need to consider another popular numerical fi for the differential mass function given by Sheth and Tormen (ST) [41,42],

$$f_{\rm ST}(\sigma) = A\sqrt{\frac{2b}{\pi}} \exp\left[-\frac{b\delta_c^2}{2\sigma^2}\right]\left[1+\left(\frac{\sigma^2}{b\delta_c^2}\right)^p\right]\frac{\delta_c}{\sigma}, \tag{58}$$

where $A = 0.3222$, $b = 0.75$, and $p = 0.3$ are three parameters tuned to fi with numerical simulations. A is fi ed by the normalization that all dark matter particles reside in halos. However, ST mass function deviates from the simulation results by as much as 40 % at the high mass end [43]. This problem can be cured when we use the correct δ_c. Thus, we introduce a simple but quite similar to the results in Ref. [43] by using the original ST formalism [44] :

$$f_{\rm LN}(\sigma,z) = 0.32\sqrt{\frac{2(0.67)}{\pi}} \exp\left[-\frac{0.67\delta_c^2}{2\sigma^2}\right]\left[1+\left(\frac{\sigma^2}{0.67\delta_c^2}\right)^{0.32}\right]\frac{\delta_c}{\sigma}. \tag{59}$$

We show the comoving number density of the clusters for $\omega_{\rm de} = -1.0$ at different z in the left panel of Fig. 13. The solid, dashed, dot-dashed, and large dashed lines correspond to $z = 0$, 0.5, 1.0, and 2.0, respectively. In the right panel of Fig. 13, we show the relative errors of the comoving number densities for two different models at two different redshifts. For example, the solid line represents the relative errors of n between $\omega_{\rm de} = -1.1$ and -1.0

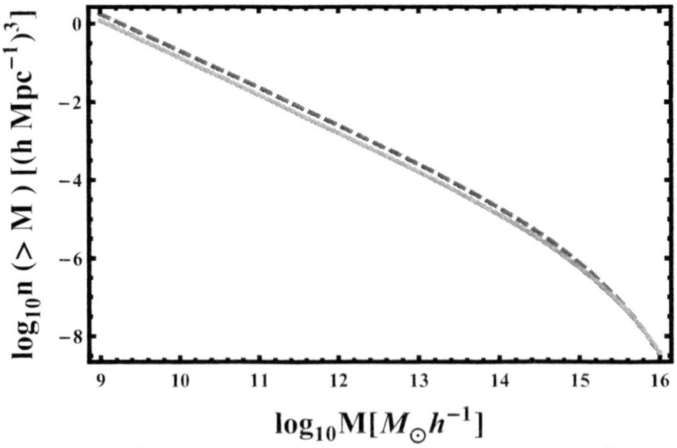

Figure 12. $n(>M)$ from both PS formalism with correct value of $\delta_c = 1.58$ (dashed line) and ST formalism with $\delta_c = 1.69$ (solid line).

at z = 0, $|n(>M)_{\omega_{de}=-1.1,z=0} - n(>M)_{\omega_{de}=-1.0,z=0}|/n(>M)_{\omega_{de}=-1.0,z=0} \times 100$ (%). At low mass $M = 10^9 h^{-1} M_\odot$, the differences of n between two different models ($\omega_{de} = -1.1$ and -1.0) are only 0.5 and 1 % at $z = 0$ and 1, respectively. At this mass, the differences of n between $\omega_{de} = -0.8$ and -1.0 are also only 1 and 4 % at $z = 0$ and 1, respectively. However, the differences of n between $\omega_{de} = -1.1$ and -1.0 at high mass $M = 10^{16} h^{-1} M_\odot$ are 28 and 48 % at $z = 0$ and 1, respectively. For the two models $\omega_{de} = -0.8$ and -1.0, the differences become 58 and 80 % at two different $z = 0$ and 1, respectively.

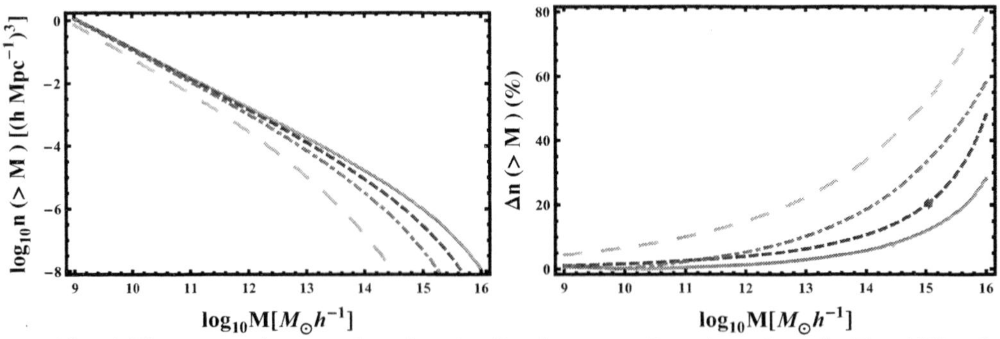

Figure 13. a) The comoving number density for the mass function given in Eq. (59) when $\omega_{de} = -1$ at $z = 0$, 0.5, 1, and 2 (from top to bottom). b) The relative errors of comoving number density for the two different dark energy models between $\omega_{de} = -1.1$ and -1.0 at two different redshifts $z = 0$ (solid) and $z = 1$ (dashed) and for models between $\omega_{de} = -0.8$ and -1.0 at $z = 0$ (dot-dashed) and 1 (long dashed).

5. Conclusion

In section II, we have obtained the exact analytic solution of subhorizon scale matter density perturbation (*i.e.* growth factor) for the general dark energy models with its equation of state ω_{de} being constant. We have also analyzed its properties for the different choice of cosmological parameter.

Even though we have obtained the exact analytic solution of D_g for the general DE models, this solution is limited to the constant ω_{de} models. Thus, the applications of this solution to the real observations are very limited. However, we can apply this solution to the more general cases like the time-varying ω_{de} by interpolating between models with constant ω_{de}. If we are able to obtain a good constraint on ω_{de} from the cosmological background evolution observations, then we will be able to constrain D_g and f very accurately. Thus, the exact analytic solution of D_g can be used as the very useful tool for the interpretation of LSS survey data.

In section III, the exact solution for the evolution of the background scale factor in a fla universe with the dark energy is given as a function of time. This provides the exact value of collapsing time and makes it possible to test the accuracies of both the analytic and the numerical solutions. We obtain the exact and approximate solutions of the evolution of the overdensity radius with high accuracy. From these solutions we are able to estimate the non-linear overdensity at any epoch.

Even though the analytic forms of y and ζ are obtained for the constant ω_{de} models, they can be generalized to the slowly varying $\omega_{de} \geq -\frac{1}{3}$ because they provide the very accurate values of them for the wide range of cosmological parameters.

The virial theorem provides the exact value of the virial epoch for given models. From this, the value of the non-linear overdensity is obtained. In the high matter density and/or the high-redshift clusters, the model dependence of the observed quantities becomes weaker. We are able to categorize the physical quantities of the low and high redshift clusters for the given models.

From the correct value of the virial radius of an EdS universe, we obtain the correct linear and non-linear overdensities for an EdS universe, which are lower than the known values. We show that clusters are formed earlier than those from the conventional model. Thus, the mass of the cluster is larger than one obtained from the conventional model. This will affect to the prediction for the different models for the cluster abundances, the weak gravitational lensing, etc.

In section IV, We obtain the dark energy dependence of the mass fluctuatio which is consistent with the CMB normalization. Even though we limit our analysis for the constant ω_{de}, we can extend our analysis for the slowly time varying ω_{de} or use the interpolation of the constant ω_{de} to study the general time varying ω_{de}.

We use the correct critical density threshold contrast δ_c to calculate the comoving number density of clusters for both PS and ST formalism. We show that PS formalism with this correct value of δ_c can predict the consistent cluster number with both the simulation and the observed data at the high mass region. Even though, the PS formalism with this correct value of δ_c still predicts too many low mass clusters, this might be due to other mechanisms for the cluster formation at the low mass region in addition to gravity. Thus, PS formalism might be good enough to explain the cluster abundance.

We obtain that the dark energy dependence of the comoving number densities for the high mass $M = 10^{16}h^{-1}M_{\odot}$ is about 25 and 40 % at two different redshifts $z = 0$ and 1. Thus, observation will need to be as accurate as this level to probe the property of dark energy from the cluster physics.

A. Appendix

We derive the exact solution of τ as a function of x given in Eq. (27). After replacing the variables $Z = -\frac{x^{-3\omega_{\rm de}}}{Q_{\rm ta}}$ and $T = (\frac{x'}{x})^{-3\omega_{\rm de}}$, Eq. (27) becomes [24]

$$\int_0^x \frac{dx'}{\sqrt{x'^{-1} + \frac{x'^{-3\omega_{\rm de}-1}}{Q_{\rm ta}}}} = -\frac{x^{\frac{3}{2}}}{3\omega_{\rm de}} \int_0^1 T^{-1-\frac{1}{2\omega_{\rm de}}} (1 - ZT)^{-\frac{1}{2}} dT$$
$$= \frac{2}{3} x^{\frac{3}{2}} F\left[\frac{1}{2}, -\frac{1}{2\omega_{\rm de}}, 1 - \frac{1}{2\omega_{\rm de}}, -\frac{x^{-3\omega_{\rm de}}}{Q_{\rm ta}}\right], \quad (1)$$

where F is the hypergeometric function and we use the gamma function relation $\Gamma[1+b] = b\Gamma[b]$.

The evolution equation of τ as a function of y when $\omega_{\rm de} = -\frac{1}{3}$ is given in Eq. (29). The exact analytic solution of this equation is known [20] :

$$\text{ArcSin}[\sqrt{y}] - \sqrt{y(1-y)} = \sqrt{\zeta_{13}}\tau \text{ , when } \tau \leq \tau_{\rm ta13} \text{ ,} \quad (2)$$
$$\sqrt{y(1-y)} - \text{ArcSin}[\sqrt{y}] + \frac{\pi}{2} = \sqrt{\zeta_{13}}(\tau - \tau_{\rm ta13}) \text{ , when } \tau \geq \tau_{\rm ta13} \text{ ,} \quad (3)$$

where ζ_{13} is given in Eq. (32) and $\tau_{\rm ta13}$ is $\tau_{\rm ta}$ for $\omega_{\rm de} = -\frac{1}{3}$. $\tau_{\rm ta13}$ is obtained from Eq. (28) as

$$\tau_{\rm ta13} = \frac{2}{3} F\left[\frac{1}{2}, \frac{3}{2}, \frac{5}{2}, -\frac{1}{Q_{\rm ta13}}\right], \quad (4)$$

where $Q_{\rm ta13}$ is $Q_{\rm ta}$ for $\omega_{\rm de} = -\frac{1}{3}$.

The equation governing y for $\omega_{\rm de} = -1$ is given by Eq. (33) with $c_1 = \zeta + Q_{\rm ta}^{-1}$.

$$\int^y \frac{dy'}{\sqrt{\zeta(\frac{1}{y} - 1) + \frac{1}{Q_{\rm ta}}(y^2 - 1)}} = c_2 \pm \tau. \quad (5)$$

The above equation (5) is solved analytically by [24]

$$\int^y \frac{dy'}{\sqrt{\zeta(\frac{1-y}{y}) + Q_{\rm ta}^{-1}(y^2 - 1)}} = -\frac{2\sqrt{2Q_{\rm ta}}}{\sqrt{2Q_{\rm ta}\zeta - 1 + \sqrt{1 + 4Q_{\rm ta}\zeta}}} \Bigg(\text{EllipticF}\Big[\text{ArcSin}[\text{B}], \text{C}\Big]$$
$$- \frac{2(Q_{\rm ta}\zeta - 2)}{(3 + \sqrt{1 + 4Q_{\rm ta}\zeta})} \text{EllipticPi}\Big[\text{A}, \text{ArcSin}[\text{B}], \text{C}\Big] \Bigg) \quad (6)$$

$$\text{where A} = \frac{2\sqrt{1 + 4Q_{\rm ta}\zeta}}{3 + \sqrt{1 + 4Q_{\rm ta}\zeta}},$$
$$B = \frac{\sqrt{3 + \frac{(2Q_{\rm ta}\zeta - 4)}{(1-y)} + \sqrt{1 + 4Q_{\rm ta}\zeta}}}{\sqrt{2(1 + 4Q_{\rm ta}\zeta)}}$$
$$C = \frac{2\sqrt{1 + 4Q_{\rm ta}\zeta}}{-1 + 2Q_{\rm ta}\zeta + \sqrt{1 + 4Q_{\rm ta}\zeta}},$$

where EllipticF and EllipticPi represent the elliptic integral of the firs kind and the incompelete elliptic integral of the third kind, respectively.

We also check the linear perturbation at early epoch, $\Delta \stackrel{\tau\to 0}{\longrightarrow} 1+\delta_{\rm lin}$. We use Eq. (2) because we consider $\tau \to 0$. Even though this solution is exact only for $\omega_{\rm de} = -\frac{1}{3}$, it is safe for us to use it for the general values of $\omega_{\rm de}$ because the second term in Eq. (26) becomes zero in $y \to 0$ limit. Thus, $\delta_{\rm lin}$ obtained from this limit is independent of the value of $\omega_{\rm de}$. $\frac{2}{3}x^{\frac{3}{2}} \simeq \tau$ and $\frac{2}{3}y^{\frac{3}{2}} + \frac{1}{5}y^{\frac{5}{2}} \simeq \sqrt{\zeta}\tau$ as $\tau \to 0$. Thus we obtain

$$\Delta \equiv 1+\delta_{\rm NL} = \zeta\left(\frac{x}{y}\right)^3 \stackrel{\tau\to 0}{\longrightarrow} \zeta\left(\frac{1}{\sqrt{\zeta}}\left(1+\frac{3}{10}y\right)\right)^2 \simeq 1+\frac{3}{5}y \Rightarrow \delta_{\rm NL} \to \delta_{\rm lin} = \frac{3}{5}\zeta^{\frac{1}{3}}\left(\frac{3}{2}\tau\right)^{\frac{2}{3}}. \tag{7}$$

References

[1] R. Dave, R. R. Caldwell, and P. J. Steinhardt, *Phys. Rev. D* **66**, 023516 (2002) [arXiv:astro-ph/0206372].

[2] R. R. Caldwell, R. Dave, and P. J. Steinhardt, *Phys. Rev. Lett.* **80**, 1582 (1998) [arXiv:astro-ph/9708069].

[3] C.-P. Ma, R. R. Caldwell, P. Bode, and L. Wang, *Astrophys. J.* **521**, L1 (1999) [arXiv:astro-ph/9906174].

[4] M. Bartelmann, *Rev. Mod. Phys.* **82**, 331 (2010) [arXiv:0906.5036].

[5] P. J. E. Peebles, *Large-Scale Structure of the Universe* (Princeton University Press, 1980).

[6] W. B. Bonnor, *Mon. Not. R. Astr. Soc.* **117**, 104 (1957).

[7] S. Lee and K.-W. Ng, [arXiv:0905.1522].

[8] S. Lee and K.-W. Ng, *Phys. Lett. B* **688**, 1 (2010) [arXiv:0906.1643].

[9] S. Lee and K.-W. Ng, *Phys. Rev. D* **82**, 043004 (2010) [arXiv:0907.2108].

[10] P. M. Morse and H. Feshbach, *Methods of Theoretical Physics, Part I* (McGraw-Hill Science, New York, 1953).

[11] L. Verde et al., *Mon. Not. R. Astron. Soc.* **335**, 432 (2002) [astro-ph/0112161].

[12] E. Hawkins et al., *Mon. Not. R. Astron. Soc.* **346**, 78 (2003) [astro-ph/0212375].

[13] L. Wang and P. J. Steinhardt, *Astrophys. J.* **508**, 483 (1998) [astro-ph/9804015].

[14] D. J. Heath, *Mon. Not. R. Astron. Soc.* **179**, 351 (1977).

[15] V. Sahni and P. Coles, *Phys. Rep.* **262**, 1 (1995) [arXiv:astro-ph/9505005]; U. Alam, V. Sahni, and A. A. Starobinsky, [arXiv:0812.2846].

[16] S. Dodelson, *Modern Cosmology* (Academic Press, San Diego, 2002).

[17] O. Lahav, P. B. Lilje, J. R. Primack, and M. J. Rees, *Mon. Not. R. Astron. Soc.* **251**, 128 (1991).

[18] J. E. Gunn and J. R. Gott III, *Astrophys. J.* **176**, 1, 1972.

[19] D. F. Mota and C. van de Bruck, *Astron. Astrophys.* **421**, 71 (2004) [arXiv:astro-ph/0401504].

[20] S. Lee, *Phys. Lett. B.* **685**, 110 (2010) [arXiv:0909.0826].

[21] S. Lee and K.-W. Ng, *JCAP* **1010**, 028 (2010) [arXiv:0910.0126].

[22] A. D. Polyanin and V. F. Zaitsev, *Handbook of Exact Solutions for Ordinary Differential Equations, 2nd Edition* (Chapman and Hall/CRC, Boca Raton, 2003).

[23] T. Kihara, *Publ. Astron. Soc. Japan.* **20**, 220 (1968).

[24] M. Abramowitz and I. A. Stegun, *Handbook of Mathematical Functions with Formulas, Graphs, and Mathematical Tables* (Dover Publications, New York, 1970).

[25] I. Maor and O. Lahav, *JCAP* **0507**, 003 (2005) [arXiv:astro-ph/0505308].

[26] P. Wang, *Astrophys. J.* **640**, 18 (2006) [arXiv:astro-ph/0507195].

[27] R. Caimmi, *New. Astron.* **12**, 327 (2007) [arXiv:gr-qc/0608030].

[28] S. Basilakos and N. Voglis, *Mon. Not. Roy. Astron. Soc.* **374**, 269 (2007) [arXiv:astro-ph/0610184].

[29] Z. Haiman, J. J. Mohr, and G. P. Holder, *Astrophys. J.* **553**, 545 (2000) [arXiv:astro-ph/0002336].

[30] P. T. P. Viana and A. R. Liddle, [arXiv:astro-ph/9511007].

[31] E. F. Bunn and M. White, *Astrophys. J.* **480**, 6 (1997) [arXiv:astro-ph/9607060].

[32] M. Bartelmann, F. Perrotta, and C. Baccigalupi, *Astron. Astrophys.* **396**, 21 (2002) [arXiv:astro-ph/0206507].

[33] E. Komatsu *et al*., [arXiv:1001.4538].

[34] M. Tegmark and M. Zaldarriaga, *Phys. Rev. D* **66**, 103508 (2002) [arXiv:astro-ph/0207047]

[35] T. Kitayama and Y. Suto, *Astrophys. J.* **469**, 480 (1996) [arXiv:astro-ph/9604141].

[36] W. H. Press and P. Schechter, *Astrophys. J.* **187**, 425 (1974).

[37] J. R. Bond, S. Cole, G. Efstathiou, and N. Kaiser, *Astrophys. J.* **379**, 440 (1991).

[38] C. Horellou and J. Berge, *Mon. Not. R. Astron. Soc.* **360**, 1393 (2005) [arXiv:astro-ph/0504465].

[39] S. Lee, *Mod. Phys. Lett. A* **25**, 874 (2010).

[40] R. G. Carlberg, S. L. Morris, H. K. C. Yee, and E. Ellingson, *Astrophys. J.* **479**, L19 (1997) [arXiv:astro-ph/9612169].

[41] R. K. Sheth and G. Tormen, *Mon. Not. R. Astron. Soc.* **308**, 119 (1999).

[42] R. K. Sheth and G. Tormen, *Mon. Not. R. Astron. Soc.* **329**, 61 (2002).

[43] S. Bhattachavya, K. Heitmann, M. White, Z. Lukić, C. Wagner, and S. Habib, [arXiv:1005.2239].

[44] S. Lee and K.-W. Ng, [arXiv:1010.2291].

In: Evolution Equations
Editor: Arthur L. Claes
ISBN: 978-1-61209-652-0

Chapter 9

NONLINEAR EVOLUTION EQUATIONS AND SEMI-SIMPLE LIE ALGEBRAS

***R. Zhdanov*[1,]* *, V. Lahno*[2] *and C. Z. Qu*[3]**
[1]BIO-key International, Eagan, MN, USA
[2]State Pedagogical University, Poltava, Ukraine
[3]Department of Mathematics, Northwest University, People's Republic of China

Abstract

We present the detailed account of our recent results on group classificatio of nonlinear evolution equations in one spatial variable. In this publication we focus on nonlinear evolution equations that admit semi-simple Lie algebras. We give exhaustive description of second-order evolution equations whose symmetry algebras contain semi-simple subalgebras. The same problem is solved for the class of third-order quasi-linear (KdV type) evolution equations. We present the exhaustive classificatio of fourth-order quasi-linear evolution equations invariant under semi-simple Lie algebras.

Introduction

Utilization of group properties of partial differential equations (PDEs) has already become a universal and convenient analytical tool for solving these equations. However for this technique to work equations under study should possess nontrivial symmetry properties. Within this viewpoint the class of differential equations splits into two subclasses of equations with nontrivial symmetry, $\mathcal{S}$, and without any symmetries.

In what follows we restrict our analysis to equations from $\mathcal{S}$. The basic facts and all the necessary information about group analysis of differential equations can be found in [5, 6, 33, 37].

The chapter is organized as follows. In the firs section we give the detailed description of our classificatio approach. Applying this approach to the general second-order evolution

*E-mail address: renat.zhdanov@bio-key.com (Corresponding author)

equation

$$u_t = F(t, x, u, u_x, u_{xx}) \tag{0.1}$$

we derive complete classificatio of Eqs.(0.1), whose invariance algebras contain a semi-simple Lie algebra. In the second section we solve the same problem for the class of KdV-type nonlinear evolution equations

$$u_t = u_{xxx} + F(t, x, u, u_x, u_{xx}). \tag{0.2}$$

Section 3 is devoted to group analysis of the general quasi-linear fourth-order evolution equation

$$u_t = F(t, x, u, u_x, u_{xx}, u_{xxx})u_{xxxx} + G(t, x, u, u_x, u_{xx}, u_{xxx}). \tag{0.3}$$

We give the complete account of inequivalent equations of the form (0.3) whose invariance algebras are semi-simple.

In the above formulas F and G are arbitrary smooth functions and $F \neq 0$. Hereafter we adopt the notations $u = u(t, x)$, $u_t = \partial u/\partial t$, $u_x = \partial u/\partial x$, $u_{xx} = \partial^2 u/\partial x^2$, $u_{xxx} = \partial^3 u/\partial x^3$, $u_{xxxx} = \partial^4 u/\partial x^4$.

Note that the complete calculation details are given for the case of the second-order evolution equations only. For the remaining cases we present the fina results omitting the intermediate calculations.

1. Group Analysis of Second-Order Equations

The basic question of the group analysis of the general class of second-order evolution equations (0.1) sounds like this: Is it possible to describe all possible functions F such that Eq.(0.1) admits nontrivial Lie transformation group? By non-trivial symmetry group we mean a Lie transformation group which is at least one-parameter. The firs paper on group classificatio of a subclass of linear equations from the class (0.1) was published by the creator of the theory of transformation groups Sophus Lie as early as in 1881 [26]. However, the real boom of interest in group classificatio of differential equations was initiated by the Ovsyannikov's paper [36]. It was followed by numerous publications [3, 8, 9, 14, 17, 18, 21, 22, 24, 35, 38, 41, 46, 47, 49] analyzing various specifi subclasses of the general class of evolution equations (0.1). A detailed account of group properties of the equations considered in the above papers can be found in [3, 21, 40].

Surprisingly, there is still no ultimate solution of the group classificatio problem for the general evolution equation (0.1). The main reason is that the class of equations (0.1) is too general for the traditional Ovsyannikov's classificatio method to be practical. This method is not very efficien when the class of equations under study involves functions of several variables.

Recently, we developed an efficien approach to solving group classificatio problem for low-dimensional partial differential equations. It enabled us to classify the broad classes of heat conductivity [3, 46], Schrödinger [48], third-order evolution [18] and wave [24] equations admitting nontrivial Lie symmetry. Note that some elements of this approach were utilized earlier by Fushchych & Serov [10], Gagnon & Winternitz [13] and Zhdanov et al

[45] in order to perform symmetry classificatio of the nonlinear d'Alembert, Schrödinger and multi-component wave equations, correspondingly.

In what follows we apply the method of [3] to obtain exhaustive classificatio of evolution equations (0.1) that admit n-parameter Lie transformation group containing a semi-simple group for all possible values of $n >= 1$. As group classificatio of linear PDEs of the form (0.1) has already been performed in [26], we consider essentially nonlinear evolution equations only. By 'essentially nonlinear' we mean PDEs (0.1) that cannot be linearized by point transformations of the space of the variables t, x, u.

1.1. Preliminary Group Analysis of Eq.(0.1)

It is a common knowledge, that the most general (in Lie's sense) transformation group admitted by (0.1) is generated by the infinitesima operators

$$\mathrm{v} = \tau\partial_t + \xi\partial_x + \eta\partial_u. \tag{1.1}$$

Here $\tau = \tau(t, x, u), \xi = \xi(t, x, u), \eta = \eta(t, x, u)$ are arbitrary smooth functions define in the space of $\mathrm{V} = \mathbb{R}^2 \times \mathbb{R}^1$ of two independent $\mathbb{R}^2 = \langle t, x\rangle$ and one dependent $\mathbb{R}^1 = \langle u\rangle$ variables.

Constructing the second prolongation of the infinitesima operator v we get

$$\widetilde{\mathrm{v}} = \mathrm{v} + \eta^t\frac{\partial}{\partial u_t} + \eta^x\frac{\partial}{\partial u_x} + \eta^{tt}\frac{\partial}{\partial u_{tt}} + \eta^{tx}\frac{\partial}{\partial u_{tx}} + \eta^{xx}\frac{\partial}{\partial u_{xx}},$$

where

$$\begin{aligned}
\eta^t &= D_t(\eta) - u_t D_t(\tau) - u_x D_t(\xi),\\
\eta^x &= D_x(\eta) - u_t D_x(\tau) - u_x D_x(\xi),\\
\eta^{xx} &= D_x(\eta^x) - u_{tx} D_x(\tau) - u_{xx} D_x(\xi),\\
D_t &= \frac{\partial}{\partial t} + u_t\frac{\partial}{\partial u} + u_{tt}\frac{\partial}{\partial u_t} + u_{tx}\frac{\partial}{\partial u_x} + \dots,\\
D_x &= \frac{\partial}{\partial x} + u_x\frac{\partial}{\partial u} + u_{tx}\frac{\partial}{\partial u_t} + u_{xx}\frac{\partial}{\partial u_x} + \dots.
\end{aligned}$$

We do not give the formulas for the coefficient η^{tt}, η^{tx} since they are not used in the sequel.

Acting by $\widetilde{\mathrm{v}}$ on Eq.(0.1) we arrive at the following invariance criterion:

$$\left(\eta^t - \tau F_t - \xi F_x - \eta F_u - \eta^x F_{u_x} - \eta^{xx} F_{u_{xx}}\right)\big|_{u_t \to F(t,x,u,u_x,u_{xx})} = 0. \tag{1.2}$$

The subscript formula in (1.2) means that one needs to replace u_{tx} with D_xF, u_{tt} with D_tF and u_t with F in the expression within the parentheses.

If we construct the general solution of (1.2), then we obtain the most general (local) Lie transformation group admitted by Eq.(0.1). Note that this group is also called the classical Lie symmetry of nonlinear PDE (0.1).

Analyzing relations (1.2) we prove the following technical assertion.

Assertion 1.1 *The most general invariance group of Eq.(0.1) is generated by the infinitesimal operators*

$$\mathrm{v} = \tau(t)\partial_t + \xi(t,x,u)\partial_x + \eta(t,x,u)\partial_u, \tag{1.3}$$

the functions τ, ξ, η *and* F *satisfying the following equation:*

$$\begin{aligned}\eta_t - u_x\xi_t + (\eta_u - \tau_t - u_x\xi_u)F = [\eta_x + u_x(\eta_u - \xi_x) - u_x^2\xi_u]F_{u_x}\\ +[\eta_{xx} + u_x(2\eta_{xu} - \xi_{xx}) + u_x^2(\eta_{uu} - 2\xi_{xu}) - u_x^3\xi_{uu}\\ +u_{xx}(\eta_u - 2\xi_x) - 3u_xu_{xx}\xi_u]F_{u_{xx}} + \tau F_t + \xi F_x + \eta F_u.\end{aligned} \tag{1.4}$$

As the forms of unknown functions τ, ξ, η depend essentially on the function F, it is customary to call (1.4) the classifying equation.

Now the problem of group classificatio of Eq.(0.1) becomes entirely algorithmic. It reduces to constructing all possible solutions of a single partial differential equation (1.4). It is that simple. The difficult , however, is that we have to deal with the under-determined system of partial differential equations that consists of one equation for three unknown functions τ, ξ, η. To make things even more complicated, this equation contains unknown function F of variables t, x, u, u_x, u_{xx}, which is to be determined as well.

To proceed any further we need extra information, either on the form of the function F or on the form of the functions τ, ξ, η which would narrow down the class of invariant equations or the set of possible symmetries. Say, we may put $F = f(u)u_{xx} + \frac{df(u)}{du}u_x^2$, split the resulting equation by the variables u_x, u_{xx} and obtain an over-determined system of partial differential equations for τ, ξ, η. Solving the latter yields the classical Ovsyannikov's classificatio result [36].

An alternative approach would be fixin *a priori* the symmetry group of Eq.(0.1) and solving (1.4) for this specifi choice of coefficient of the infinitesima group operator, τ, ξ, η. To this end one can also use the popular nowadays technique called conventionally "the method of moving frames". The latter can be used efficientl , if one succeeds in obtaining the explicit form of the finit group transformations of the infinitesima symmetry group whose differential invariants are to be constructed (see, e.g., [34] and the references therein).

The problem with the above approaches is that some functions F providing for extension of symmetry group of Eq.(0.1) might be lost. To prevent this from happening, the classifying equation should be the only source of constraints on the form of unknown functions τ, ξ, η, F. Another difficult is that the method of moving frames is inefficien when the group under consideration is infinite-paramete and this is exactly the case we are dealing with.

In [45] we suggest the efficien way to approach this problem based on the fact that the general solution (τ, ξ, η) of PDE (1.4) can always be represented as a linear combination of its basis solutions, $\mathrm{v}_a = (\tau_a, \xi_a, \eta_a)$, $a = 1, \ldots, n$, forming a Lie algebra ℓ_n. Consequently, if we succeed in describing all possible subalgebras of the infinite-dimensiona Lie algebra ℓ_∞ generated by operators (1.3) and solve for each of them the classifying equation (1.4), then the problem of group classificatio of initial Eq.(0.1) will be completely solved.

So that we can re-formulate the problem of group classificatio of Eq.(0.1) in a purely algebraic way. Namely, to solve the classificatio problem we need to

- construct all subalgebras of the infinite-dimensiona Lie algebra ℓ_∞, and

- select those subalgebras whose basis elements satisfy the classifying equation (1.4).

Saying it another way, we can replace (1.2) with the (possibly infinite set of systems of PDEs

$$\begin{cases} \text{Equation } (1.4)|_{\mathbf{v}\to\mathbf{v}_a}, \quad a = 1,\dots,n, \\ Q_i\tau_j - Q_j\tau_i = \sum_{k=1}^{n} C_{ij}^{k}\,\tau_k, \\ Q_i\xi_j - Q_j\xi_i = \sum_{k=1}^{n} C_{ij}^{k}\,\xi_k, \\ Q_i\eta_j - Q_j\eta_i = \sum_{k=1}^{n} C_{ij}^{k}\,\eta_k, \end{cases} \tag{1.5}$$

where $Q_i = \tau_i\partial_t + \xi_i\partial_x + \eta_i\partial_u$, C_{ij}^k are structure constants of a Lie algebra ℓ_n, $i, j = 1,\dots,n$ and $n = 1, 2, 3, \dots$.

If we solve over-determined system of PDEs (1.5) for *all* possible dimensions $n \geq 1$ of *all* admissible Lie algebras ℓ_n, then the problem of group classificatio of Eq.(0.1) is completely solved. Consequently, group classificatio of the general evolution equation (0.1) reduces to integrating over-determined systems of PDEs (1.5) for all $n = 1, 2, \dots, n_{\max}$, where $n_{\max}$ is the maximal dimension of the Lie algebra admitted by the equation under study. Note that $n_{\max}$ may be equal to ∞.

According to the Magadeev theorem [29], we have either $n_{\max} <= 7$ or $n_{\max} = \infty$. And what is more, in the latter case the corresponding invariant equation is mapped into a linear PDE by a contact transformation. So to describe all essentially nonlinear evolution equations (i.e., those inequivalent to linear ones) one needs, in fact, to consider all possible Lie algebras of the dimension up to 7. However, combining our Lie algebraic classificatio approach with the Ovsyannikov method it suffice to consider Lie algebras of the dimension $n < 5$ only. At this point, the classical Mubarakzyanov results [30, 31] come into play. He described all inequivalent abstract Lie algebras of the dimension $n < 6$. This means that the structure constants C_{ij}^k in (1.5) are already known.

Let us mention here the Reid's procedure for calculating the Lie algebra admitted by PDE without integrating determining equations [42]. The basic idea of his approach is investigating compatibility of systems (1.5) thus deriving the admissible forms of the structure constants C_{ij}^k. In our mind a more natural approach is actually to integrate equations (1.5) so that the compatibility conditions come as a by-product.

Summarizing we conclude that if we

1) construct all realizations of all subalgebras of ℓ_∞ by operators, coefficient of which satisfy Eq.(1.5), and
2) prove that these are maximal invariance algebras of (0.1),

then the problem of group classificatio of Eq.(0.1) is completely solved.

Thus it is clear *what* to do to achieve complete group classificatio of the generic class of PDEs (0.1). However, there is still a question *how* to do this. Saying it another way, what is the practical approach to classify all inequivalent subalgebras of the algebra ℓ_∞? Before introducing the details of our approach to solving this problem we give a brief account of necessary notions and facts from the general theory of Lie algebras.

Let the symbol L stand for a Lie algebra. Denote by the symbol $[L, L]$ the Lie algebra spanned by all the possible commutators of basis elements of L. Then the Lie algebra N is called a subalgebra of L provided $[N, N] \subset N$. If the relation $[L, N] \subset N$ holds true, then the Lie algebra N is called the ideal of L.

Given two ideals, N_1 and N_2 of the algebra L, the Lie algebra $[N_1, N_2]$ is also the ideal of L. Consider the following sequence of ideals:

$$L^{[0]} = L, \quad L^{[1]} = [L^{[0]}, L], \ldots, L^{[n]} = [L^{[n-1]}, L], \ldots$$

If $L^{[n]} = 0$ for some $n > 1$, then the Lie algebra L is called nilpotent. If, in particular, $L^{[1]} = 0$, then L is commutative or Abelian Lie algebra.

Now consider another sequence of ideals of L (the composition series of the Lie algebra L)

$$L^{(0)} = L, \quad L^{(1)} = [L^{(0)}, L^{(0)}], \ldots, L^{(n)} = [L^{(n-1)}, L^{(n-1)}], \ldots$$

The algebra L is called solvable if there is $n > 0$ such that $L^{(n)} = 0$. The Lie algebra R is called the radical of L, provided it is the maximal solvable ideal of L containing any other solvable ideal of L.

The Lie algebra L is called semi-simple, provided it does not contain non-zero solvable ideals. The algebra L is called simple if it contains no ideals different from 0 and L and $L^{[1]} \neq 0$. Evidently, a simple algebra is semi-simple. On the other hand, every semi-simple Lie algebra can be decomposed into a direct sum of simple Lie algebras.

The fundamental Levi-Maltsev theorem (see, e.g., [2, 3]) says that for any Lie algebra L with radical R there exists a semi-simple Lie algebra S such that

$$L = S \Subset R.$$

This relation is called the Levi decomposition of the algebra L, and what is more, the algebra S is called the Levi factor of L. The further details can be found, for example, in [2, 3].

Due to the Levi-Maltsev theorem the problem of classificatio of subalgebras of the algebra ℓ_∞ can be divided into three subproblems:

1. Classificatio of semi-simple subalgebras.
2. Classificatio of solvable subalgebras.
3. Classificatio of subalgebras which are semi-direct sums of semi-simple and solvable Lie algebras.

Remarkably, it is possible to complete all of the above classification for the case of the very general class of PDEs (0.1) starting with one-dimensional subalgebras and increasing step-by-step the dimension of the algebras involved.

Our algorithm for group classificatio in its present form has been suggested in [3, 46]. As the firs step, we compute the maximal equivalence group, $\mathcal{E}$, of the class of PDEs (0.1). The equivalence group consists of all of the one-to-one transformations of the space V

$$\bar{t} = \alpha(t, x, u), \quad \bar{x} = \beta(t, x, u), \quad v = \gamma(t, x, u), \quad \frac{D(\alpha, \beta, \gamma)}{D(t, x, u)} \neq 0, \tag{1.6}$$

which preserve the differential structure of equation (0.1). Note that we use a concept of equivalence group which is different from that employed by Ovsyannikov. Indeed, the group $\mathcal{E}$ does not involve the function F as an extra variable, as it is customary within the framework of the Ovsyannikov's classificatio scheme [37].

Making change of variables (1.6) in Eq.(0.1) and requiring for the transformed equation to belong to the same class of PDEs as the initial one, namely,

$$v_{\bar{t}} = \Phi(\bar{t}, \bar{x}, v, v_{\bar{x}}, v_{\bar{x}\bar{x}}),$$

one gets the following result (see, e.g., [23])

Assertion 1.2 *The maximal equivalence group of the class of equations (0.1) is formed by the transformations*

$$\bar{t} = T(t), \quad \bar{x} = X(t, x, u), \quad v = U(t, x, u), \; T' = \frac{dT}{dt} \neq 0, \; \frac{D(X, U)}{D(x, u)} \neq 0. \quad (1.7)$$

Note that the above group contains arbitrary functions, which means that it is the infinite-paramete Lie transformation group.

The equivalence group splits the class of equations (0.1) into conjugacy subclasses, each of them being uniquely characterized by a single representative. So it suffice to provide a representative of each conjugacy class in order to get the complete description of invariant equations. This property is used to simplify the form of infinitesima operators forming the basis of admissible realizations of Lie algebras. Classificatio of these realizations is thus reduced to constructing inequivalent realizations of abstract Lie algebras by differential operators in three real variables of the form (1.3). Note that the systematic analysis of realizations of Lie algebras by differential operators in one and two variables was performed by Sophus Lie itself [25]–[28].

Consider, as an example, one-dimensional Lie algebras generated by operators (1.3). Change of variables (1.7) reduces (1.3) to the form

$$\tilde{\mathrm{v}} = \tau T' \partial_{\bar{t}} + (\tau X_t + \xi X_x + \eta X_u)\partial_{\bar{x}} + (\tau U_t + \xi U_x + \eta U_u)\partial_v. \quad (1.8)$$

If $\tau \neq 0$, then choosing a solution of equation $\tau T' = 1$ as T and taking as X, U the fundamental solution of the system of equations

$$\tau X_t + \xi X_x + \eta X_u = 0, \quad \tau U_t + \xi U_x + \eta U_u = 0, \quad \frac{D(X, U)}{D(x, u)} \neq 0,$$

we get $\tilde{\mathrm{v}} = \partial_{\bar{t}}$.

Now if $\tau = 0$, then $\xi \neq 0$ or $\eta \neq 0$ (otherwise, operator (1.3) vanishes identically). In the case when $\xi \neq 0, \eta = 0$, making the change of variables

$$\bar{t} = t, \quad \bar{x} = u, \quad v = x,$$

which belongs to the group $\mathcal{E}$, we reduce (1.3) to the form $\tilde{\mathrm{v}} = \xi(\bar{t}, \bar{x}, v)\partial_v$. Consequently, we can suppose that $\eta \neq 0$ without any loss of generality. Choosing as X and U non-vanishing identically solutions of the system of PDEs

$$\xi X_x + \eta X_u = 0, \quad \xi U_x + \eta U_u = 1,$$

we transform (1.8) to become $\tilde{\mathrm{v}} = \partial_v$. We summarize the above reasonings in the form of lemma.

Lemma 1.1 *Operator (1.3) is equivalent to one of the canonical operators*

$$\mathrm{v}_1 = \partial_t, \quad \mathrm{v}_2 = \partial_u.$$

Thus there are two $\mathcal{E}$-inequivalent classes of realizations of one-dimensional Lie algebras by operators (1.3). Consequently, there are only two inequivalent classes of PDEs (0.1) admitting one-dimensional Lie algebras. They are easily obtained by integration of the determining equations for the corresponding infinitesima operators ∂_t, ∂_u.

Theorem 1.1 *There are two inequivalent classes of invariant equations of the form (0.1) admitting one-parameter Lie groups. Below we give the representatives of these classes and the corresponding one-dimensional invariance Lie algebras* A_1

$$\begin{aligned} u_t &= F(x, u, u_x, u_{xx}) : A_1^1 = \langle \partial_t \rangle, \\ u_x &= F(t, x, u_x, u_{xx}) : A_1^2 = \langle \partial_u \rangle. \end{aligned}$$

What is more, if the function F *is arbitrary, then the algebras* A_1^1 *and* A_1^2 *are maximal in Lie's sense invariance algebras admitted by the corresponding PDEs.*

As we noted in [47] the set of invariant equations (0.1) is naturally split into two classes, C_1 and C_2. Let $A_k = \langle \mathrm{v}_1, \mathrm{v}_2, \ldots, \mathrm{v}_k \rangle$, where

$$\mathrm{v}_i = \tau^i \partial_t + \xi^i \partial_x + \eta^i \partial_u, \quad i = 1, 2, \ldots, k \qquad (1.9)$$

be the maximal invariance algebra of Eq.(0.1). We say that this equation belongs to the firs class, C_1, if the functions τ^i $(i = 1, 2, \ldots, k)$ are linearly independent. Otherwise, it belongs to C_2.

According to [47] any nonlinear PDE from the second class C_2 can be mapped into quasi-linear evolution equations by a non-point transformation.

So if one constructs all $\mathcal{E}$-inequivalent equations belonging to the classes C_1, C_2, then the problem of group classificatio of the general evolution equation is solved. We provide full calculation details for the firs class C_1. In the case of class C_2 we will give outlines of the proofs of the corresponding theorems together with the lists of invariant equations and their maximal invariance algebras.

In the sequel we classify invariant equations from the class C_1 and those equations from C_2 whose invariance algebras are either semi-simple or semi-direct sums of semi-simple and solvable Lie algebras. We obtain exhaustive classificatio of these equations.

1.2. Group Classification of Equations from C_1

It is the direct consequence of Lemma 1.1 that one of the basis operators of the invariance algebra of an equation from C_1 can always be chosen in the from ∂_t. Also according to Theorem 1.1 the most general equation from the class C_1 that admits one-dimensional Lie algebra is equivalent to PDE

$$u_t = F(x, u, u_x, u_{xx}), \qquad F_{u_{xx}} \neq 0. \qquad (1.10)$$

The maximal invariance algebra of the above equation reads as $A_1^1 = \langle \partial_t \rangle$.

Now we turn to equations admitting two-dimensional Lie algebras. It so happens that equations from C_1 cannot admit two-dimensional commutative algebras.

Lemma 1.2 *Class C_1 does not contain PDEs invariant under two-dimensional commutative Lie algebras.*

Proof. Suppose that the assertion of the lemma does not hold. Then one of the basis operators of the symmetry algebra can be reduced to the form $v_1 = \partial_t$, while the second operator, v_2, is of the generic form (1.3). Inserting these operators into the commutation relation for the two-dimensional commutative Lie algebra, $[v_1, v_2] = 0$, yields that $\tau_t = 0$, $\xi_t = 0$, $\eta_t = 0$. Hence it follows that the equation under study belongs to the class C_2. We arrive at the contradiction proving the lemma.

It follows from the Levi-Maltsev theorem (see, e.g., [2]) that the set of finite dimensional real Lie algebras consists of solvable algebras and Lie algebras having non-trivial Levi ideal. That is why we proceed now to studying equations from C_1 that admit solvable and semi-simple Lie algebras of symmetry operators (1.8).

1.2.1. Invariance under Solvable Lie Algebras

There are two inequivalent two-dimensional solvable Lie algebras

$$A_{2.1} = \langle e_1, e_2 \rangle : [e_1, e_2] = 0,$$
$$A_{2.2} = \langle e_1, e_2 \rangle : [e_1, e_2] = e_2.$$

The case of commutative algebra $A_{2.1}$ has already been considered. Turn to the algebra $A_{2.2}$. Without loss of generality we may choose one of the basis operators, say, e_2 to be equal to ∂_t (Lemma 1.1).

Taking as e_1 an arbitrary operator of the form (1.3) with $\tau \neq 0$ and inserting e_1 into (1.11) yields $\tau_t = -1, \xi_t = \eta_t = 0$. Consequently, the operator e_1 has the form

$$e_1 = -t\partial_t + \xi(x, u)\partial_x + \eta(x, u)\partial_u, \tag{1.11}$$

where ξ, η are arbitrary smooth functions. If $\xi = \eta = 0$, then we get the realization $\langle -t\partial_t, \partial_t \rangle$. Next, provided $|\xi| + |\eta| \neq 0$, there is a transformation from the group $\mathcal{E}$ which does not alter e_2 and reduce e_1 (1.11) to the form $\tilde{e}_1 = -\bar{t}\partial_{\bar{t}} - v\partial_v$.

Thus there exist two inequivalent realizations of the Lie algebra $A_{2.2}$

$$\langle -t\partial_t, \partial_t \rangle, \quad \langle -t\partial_t - u\partial_u, \partial_t \rangle$$

that can be admitted by equations from C_1.

However, if we insert coefficient of the firs basis operator of the firs realization into the classifying equation (1.4), we get $F = 0$. Consequently, C_1 contains no equations invariant under the firs realization of $A_{2.2}$.

A similar analysis of the second realization yields the following class of $A_{2.2}$-invariant PDEs:

$$u_t = F(x, \omega, \mathrm{w}), \quad \omega = u^{-1}u_x, \quad \mathrm{w} = u^{-1}u_{xx}.$$

Consider now three-dimensional solvable Lie algebras. It is a common knowledge that any solvable three-dimensional Lie algebra, $A_3 = \langle e_1, e_2, e_3 \rangle$, is either equivalent to one of the two decomposable

$$A_{3.1} = A_{2.1} \oplus A_1 : [e_i, e_j] = 0, \quad i, j = 1, 2, 3;$$
$$A_{3.2} = A_{2.2} \oplus A_1 : [e_1, e_2] = e_2, \quad [e_1, e_3] = [e_2, e_3] = 0;$$

or equivalent to one of the seven non-decomposable algebras

$$
\begin{aligned}
A_{3.3} &: [e_2, e_3] = e_1, \quad [e_1, e_2] = [e_1, e_3] = 0;\\
A_{3.4} &: [e_1, e_3] = e_1, \quad [e_2, e_3] = e_1 + e_2, \quad [e_1, e_2] = 0;\\
A_{3.5} &: [e_1, e_3] = e_1, \quad [e_2, e_3] = e_2, \quad [e_1, e_2] = 0;\\
A_{3.6} &: [e_1, e_3] = e_1, \quad [e_2, e_3] = -e_2, \quad [e_1, e_2] = 0;\\
A_{3.7} &: [e_1, e_3] = e_1, \quad [e_2, e_3] = qe_2, \quad [e_1, e_2] = 0 \ (0 < |q| < 1);\\
A_{3.8} &: [e_1, e_3] = -e_2, \quad [e_2, e_3] = e_1, \quad [e_1, e_2] = 0;\\
A_{3.9} &: [e_1, e_3] = qe_1 - e_2, \quad [e_2, e_3] = e_1 + qe_2, \quad [e_1, e_2] = 0 \ (q > 0)
\end{aligned}
$$

(see, e.g., [30]).

Each of the above algebras contains a two-dimensional commutative subalgebra. By force of Lemma 1.2 there are no equations from C_1 which admit one of the above three-dimensional algebras. Since any n-dimensional solvable Lie algebra contains $(n-1)$-dimensional solvable subalgebra, we arrive at the following assertion.

Lemma 1.3 *Any equation from C_1 invariant under solvable Lie algebra of the dimension $n >= 2$ is equivalent to PDE*

$$u_t = F(x, \omega, \mathrm{w}), \quad \omega = u^{-1} u_x, \quad \mathrm{w} = u^{-1} u_{xx}.$$

If F is arbitrary, then the maximal invariance algebra of the above equation is the two-dimensional Lie algebra $\langle -t\partial_t - u\partial_u, \partial_t \rangle$.

1.2.2. Invariance under Semi-simple Lie Algebras

The semi-simple Lie algebras of the lowest dimension read as [2]

$$
\begin{aligned}
so(3) &= \langle e_2, e_2, e_3 \rangle : [e_1, e_2] = e_3, \quad [e_1, e_3] = -e_2, \quad [e_2, e_3] = e_1;\\
sl(2, \mathbb{R}) &= \langle e_1, e_2, e_3 \rangle : [e_1, e_2] = 2e_2, \quad [e_1, e_3] = -2e_3, \quad [e_2, e_3] = e_1.
\end{aligned}
$$

The above algebras do not contain the commutative algebra $A_{2.1}$. However, semi-simple algebras of higher dimension $n > 3$ do contain subalgebras equivalent to $A_{2.1}$ (see, e.g., [2]) and, consequently, cannot be invariance algebras of PDEs from the class C_1. Thus the only semi-simple Lie algebras that may lead to new invariant equations are the algebras $so(3)$ and $sl(2, \mathbb{R})$.

Consider firs the algebra $so(3)$. Inserting $e_1 = \partial_t$, e_2, e_3, where the latter two are of generic form (1.3), into the firs two commutation relations of $so(3)$ and integrating the equations obtained yield

$$e_2 = C \cos t\partial_t + (\alpha \cos t + \beta \sin t)\partial_x + (\gamma \cos t + \theta \sin t)\partial_u, \quad e_3 = [\partial_t, e_2].$$

Here C is an arbitrary non-zero constant, $\alpha = \alpha(x, u)$, $\beta = \beta(x, u)$, $\gamma = \gamma(x, u)$ and $\theta = \theta(x, u)$ are arbitrary real-valued functions. Inserting the obtained expressions for e_2, e_3 into the remaining commutation relation we get $C^2 = -1$. Since this equation has no real solutions, the class C_1 contains no $so(3)$-invariant equations.

Turn now to the algebra $sl(2,\mathbb{R})$. Let the operator e_3 be of the form ∂_t and the operators e_1, e_2 be of generic form (1.3). Inserting the expressions for e_1, e_2, e_3 into the commutation relations and solving the obtained equations we get four $\mathcal{E}$-inequivalent realizations of $sl(2,\mathbb{R})$:

$$\langle 2t\partial_t, -t^2\partial_t, \partial_t\rangle, \quad \langle 2t\partial_t + x\partial_x, -t^2\partial_t - tx\partial_x, \partial_t\rangle,$$
$$\langle 2t\partial_t + x\partial_x, -t^2\partial_t - tx\partial_x + x^2\partial_u, \partial_t\rangle, \quad \langle 2t\partial_t + x\partial_x, -t^2\partial_t + x(x^2-t)\partial_x, \partial_t\rangle.$$

The firs realization cannot be invariance algebra of an equation of the form (0.1).

Requiring invariance of (0.1) under the second realization yields the following system of equations for $F = F(x,u,u_x,u_{xx})$:

$$2F = -u_xF_{u_x} - 2u_{xx}F_{u_{xx}} + xF_x, \quad xu_x - 2tF = tu_xF_{u_x} + 2tu_{xx}F_{u_{xx}} - txF_x.$$

As the function F is independent of t, we have $xu_x = 0$. Consequently, this realization cannot be admitted by an equation of the form (0.1).

The third realization gives rise to the following system of equations for the function F:

$$2u_{xx}F_{u_{xx}} + u_xF_{u_x} - xF_x = 2F, \quad 2F_{u_{xx}} + 2xF_{u_x} + x^2F_u = xu_x.$$

This system is compatible. Its general solution is

$$F = x^{-1}uu_x - x^{-2}u^2 + x^{-2}\tilde{F}(\omega, \mathrm{w}), \quad \omega = x^2u_{xx} - 2u, \quad \mathrm{w} = 2u - xu_x.$$

Finally, the last realization yields the following system of determining equations for $F = F(x,u,u_x,u_{xx})$:

$$2u_{xx}F_{u_{xx}} + u_xF_{u_x} - xF_x = 2F, \quad 6x(u_x + xu_{xx})F_{u_{xx}} + 3x^2u_xF_{u_x} - x^3F_x = -xu_x,$$

whence

$$F = -\frac{1}{4}x^{-1}u_x + x^{-3}u_x^{-1}\tilde{F}(u,\omega), \quad \omega = u_x^{-2}u_{xx} + 3x^{-1}u_x^{-1}.$$

Summarizing the above reasonings, we conclude that the class $\mathcal{C}_1$ contains only two inequivalent classes of PDEs which are invariant under semi-simple Lie algebras. In both cases the maximal invariance algebras are isomorphic to $sl(2,\mathbb{R})$.

1.2.3. Finalizing Group Classification of Equations from $\mathcal{C}_1$

To complete group classificatio we have to consider equations from $\mathcal{C}_1$ invariant under the Lie algebras ℓ having nontrivial Levi factor and non-zero radical. It is a common knowledge that these Lie algebras should be decomposable into semi-direct sums of semi-simple ℓ_1 and solvable ℓ_2 Lie algebras. Structure of Lie algebras having the Levi factor $sl(2,\mathbb{R})$ is studied in [44]. Since the solvable algebra, ℓ_2, is either one-dimensional or isomorphic to $A_{2.2}$, it follows from the results of [44] that there are no algebras ℓ that are invariance algebras of equations (0.1) belonging to the firs class $\mathcal{C}_1$.

Theorem 1.2 *There are, at most, four inequivalent classes of PDEs from C_1 that admit nontrivial invariance algebras. The representative of the first class is (1.10), its maximal symmetry algebra is $\langle \partial_t \rangle$. The representative of the second class is given in Lemma 1.3, its maximal invariance algebra being isomorphic to $A_{2.2}$. Two other classes of PDEs are presented below*

$$\begin{aligned} u_t &= x^{-1}uu_x - x^{-2}u^2 + x^{-2}\tilde{F}(\omega, \mathrm{w}), \quad \omega = x^2u_{xx} - 2u, \quad \mathrm{w} = 2u - xu_x : \\ & sl^1(2,\mathbb{R}) = \langle 2t\partial_t + x\partial_x, -t^2\partial_t - tx\partial_x + x^2\partial_u, \partial_t \rangle; \\ u_t &= -\frac{1}{4}x^{-1}u_x + x^{-3}u_x^{-1}\tilde{F}(u,\omega), \quad \omega = u_x^{-2}u_{xx} + 3x^{-1}u_x^{-1} : \\ & sl^2(2,\mathbb{R}) = \langle 2t\partial_t + x\partial_x, -t^2\partial_t + x(x^2 - t)\partial_x, \partial_t \rangle. \end{aligned}$$

The algebras $sl^1(2,\mathbb{R})$ and $sl^2(2,\mathbb{R})$ are maximal invariance algebras of the corresponding PDEs, provided the functions $\tilde{F}$ are arbitrary.

1.3. Invariance of Equations from C_2 under Lie Algebras Having Nontrivial Levi Factor

As we mentioned above, to describe equations (0.1) invariant under Lie algebras having nontrivial Levi factor we need to construct equations, which are invariant under semi-simple Lie algebras. That is why we begin group classificatio of equations from C_2 by considering realizations of semi-simple Lie algebras by operators (1.3).

1.3.1. Invariance under Semi-simple Lie Algebras

It follows from the definitio of class C_2 that one of the basis operators of an equation from C_2 can be chosen as ∂_u. Consequently, any invariant equation from C_2 can always be transformed to

$$u_t = F(t, x, u_x, u_{xx}), \quad F_{u_{xx}} \neq 0. \tag{1.12}$$

Consider firs the semi-simple algebras of the lowest dimension, namely, $so(3)$ and $sl(2,\mathbb{R})$.

Algebra $so(3)$. Let $e_1 = \partial_u$ and let the operators e_2, e_3 be of the form (1.3). Inserting these operators into the firs two commutation relations of $so(3)$ and solving the PDEs obtained we get within the equivalence relation $\mathcal{E}$ the following formulas for e_2 and e_3:

$$e_2 = \alpha(t,x)\cos u\partial_x + \Big(\beta(t,x)\cos u + \gamma(t,x)\sin u\Big)\partial_u, \quad e_3 = [\partial_t, e_2].$$

The remaining commutation relation yields

$$\alpha\beta = 0, \quad \alpha\gamma_x - \beta^2 - \gamma^2 = 1. \tag{1.13}$$

If $\alpha = 0$, then $\beta^2 + \gamma^2 = -1$ and system (1.13) has no real solutions. In the case $\alpha \neq 0$ we have $\beta = 0$. Making the change of variables from $\mathcal{E}$

$$\bar{t} = t, \quad \bar{x} = X(t,x) \ (\alpha X_x = 1), \quad v = u$$

we get

$$e_2 = \cos u\partial_x + \gamma(t,x)\sin u\partial_u, \quad e_3 = [\partial_t, e_2],$$

where the function $\gamma = \gamma(t,x)$ is a solution of PDE $\gamma_x = 1 + \gamma^2$, i.e., $\gamma = \tan(x + \varphi(t))$. Making the equivalence transformation

$$\bar{t} = t, \quad \bar{x} = x + \varphi(t), \quad v = u$$

we reduce γ to the form $\tan x$. Consequently, there is only one realization of the algebra $so(3)$ by operators (1.3), namely,

$$so^1(3) = \langle \partial_u, \cos u\partial_x + \tan x \sin u\partial_u, -\sin u\partial_x + \tan x\cos u\partial_u \rangle.$$

Inserting the coefficient of the above operators into the classifying equation (1.4) yields the following system of PDEs for the function $F = F(t, x, u_x, u_{xx})$:

$$\begin{aligned} &u_xF - (\sec^2 x + u_x^2)F_{u_x} - (2\tan x\sec^2 x - u_x^2\tan x + 3u_xu_{xx})F_{u_{xx}} = 0,\\ &u_x\tan xF_{u_x} + (2u_x\sec^2 x + u_x^3 + u_{xx}\tan x)F_{u_{xx}} + F_x - \tan xF = 0. \end{aligned}$$

The general solution of the above system reads as

$$\begin{aligned} F &= \sqrt{\sec^2 x + u_x^2}\tilde{F}(t,\omega), \\ \omega &= \Big(u_{xx}\cos x - (2 + u_x^2\cos^2 x)u_x\sin x\Big)(1 + u_x^2\cos^2 x)^{-3/2}. \end{aligned} \tag{1.14}$$

What is more, the algebra $so(3)$ is the maximal invariance algebra of the corresponding equation, provided $\tilde{F}$ is arbitrary.

Lemma 1.4 *Any $so(3)$-invariant equation (1.12) from $\mathcal{C}_2$ is equivalent to PDE*

$$u_t = \sqrt{\sec^2 x + u_x^2}\,\tilde{F}(t,\omega), \quad \omega = \Big(u_{xx}\cos x - (2+u_x^2\cos^2 x)u_x\sin x\Big)(1+u_x^2\cos^2 x)^{-3/2}.$$

Its maximal symmetry algebra is

$$so^1(3) = \langle \partial_u, \cos u\partial_x + \tan x\sin u\partial_u, -\sin u\partial_x + \tan x\cos u\partial_u \rangle.$$

Algebra $sl(2,\mathbb{R})$. Let $e_3 = \partial_u$ and e_1, e_2 be of the form (1.3). In order to satisfy the commutation relations of the algebra $sl(2,\mathbb{R})$ the operators e_1, e_2, e_3 have to be of the form

$$e_2 = (\alpha u + \beta)\partial_x + (-u^2 + \gamma u + \theta)\partial_u, \quad e_1 = -[\partial_u, e_2], \quad e_3 = \partial_u,$$

where $\alpha = \alpha(t,x)$, $\beta = \beta(t,x)$, $\gamma = \gamma(t,x)$, $\theta = \theta(t,x)$ are solutions of the system of PDEs

$$2\beta = -\alpha\beta_x - \alpha\gamma + \beta\alpha_x, \quad 4\theta = -\alpha\theta_x - \gamma^2 + \beta\gamma_x. \tag{1.15}$$

If the function α does not vanish identically, then using the transformation

$$\bar{t} = t, \quad \bar{x} = X(t,x), \quad v = u + U(t,x),$$

where X, U are solutions of PDEs $\alpha X_x = X, XU = \beta X_x$, we simplify e_1, e_2

$$e_2 = xu\partial_x + (-u^2 + \gamma u + \theta)\partial_u, \quad e_1 = -[\partial_u, e_2], \quad e_3 = \partial_u.$$

In this case, system (1.15) takes the form $x\gamma = 0, \quad 4\theta = -x\theta_x - \gamma^2$, whence it follows that $\gamma = 0$, $\theta = \mu(t)x^{-4}$. Provided $\mu = 0$, we have the following realization of the algebra $sl(2, \mathbb{R})$:

$$\langle 2u\partial_u - x\partial_x, -u^2\partial_u + xu\partial_x, \partial_u\rangle.$$

If $\mu \neq 0$, then making the transformation

$$\bar{t} = t, \quad \bar{x} = |\mu|^{-\frac{1}{4}}x, \quad v = u,$$

we arrive at the two realizations of the algebra $sl(2, \mathbb{R})$:

$$\langle 2u\partial_u - x\partial_x, (x^{-4} - u^2)\partial_u + xu\partial_x, \partial_u\rangle, \quad \langle 2u\partial_u - x\partial_x, -(x^{-4} + u^2) + xu\partial_x, \partial_u\rangle.$$

Given the condition $\alpha = 0$, we get from (1.15) the following realization of $sl(2, \mathbb{R})$:

$$\langle 2u\partial_u, -u^2\partial_u, \partial_u\rangle.$$

However, this realization cannot be invariance algebra of an equation of the form (1.12), while the preceding ones give rise to the three classes of $sl(2, \mathbb{R})$-invariant equations.

Lemma 1.5 *There are only three inequivalent classes of equations from C_2 whose invariance algebras are isomorphic to $sl(2, \mathbb{R})$. Below we present the canonical forms of these equations together with their maximal invariance algebras.*

$$\begin{aligned} u_t &= xu_x\tilde{F}(t, \omega), \quad \omega = x^{-5}u_x^{-3}u_{xx} + 2x^{-6}u_x^{-2} : \\ &\quad sl^3(2, \mathbb{R}) = \langle 2u\partial_u - x\partial_x, -u^2\partial_u + xu\partial_x, \partial_u\rangle; \\ u_t &= x^{-2}\sqrt{4 + x^6u_x^2}\,\tilde{F}(t, \omega), \quad \omega = (4 + x^6u_x^2)^{-3/2}\left(x^4u_{xx} + 5x^3u_x + \frac{1}{2}x^9u_x^3\right) : \\ &\quad sl^4(2, \mathbb{R}) = \langle 2u\partial_u - x\partial_x, (x^{-4} - u^2)\partial_u + xu\partial_x, \partial_u\rangle; \\ u_t &= x^{-2}\sqrt{|x^6u_x^2 - 4|}\,\tilde{F}(t, \omega), \quad \omega = (x^6u_x^2 - 4)^{-3/2}\left(x^4u_{xx} + 5x^3u_x - \frac{1}{2}x^9u_x^3\right) : \\ &\quad sl^5(2, \mathbb{R}) = \langle 2u\partial_u - x\partial_x, -(x^{-4} + u^2)\partial_u + xu\partial_x, \partial_u\rangle. \end{aligned}$$

Semi-simple algebras of the dimension higher than 3. According to the general classificatio of semi-simple Lie algebras [2] the next possible dimension of a semi-simple Lie algebra is six. There are four non-isomorphic semi-simple algebras of the dimension six, $so(4)$, $so(3, 1)$, $so(2, 2)$ and $so^*(4)$.

It is well-known that $so(4) = so(3) \oplus so(3)$, $so^*(4) \sim so(3) \oplus sl(2, \mathbb{R})$. Consequently, we can utilize the results of classificatio of realizations of $so(3)$ and $sl(2, \mathbb{R})$.

To classify $so(4)$-invariant equations (0.1) we need to construct all realizations of $so(3)$ by operators (1.3), which commute with the basis operators of the realization $so^1(3)$. It is straightforward to verify that such realizations do not exist. Similarly, to classify $so^*(4)$-invariant equations (0.1) we have to construct all realizations of $sl(2, \mathbb{R})$ by operators (1.3), which commute with the basis operators of the realization $so^1(3)$. The only realization of $sl(2,\mathbb{R})$ obeying the above constraint is equivalent to $\langle 2t\partial_t, -t^2\partial_t, \partial_t\rangle$. The latter cannot be invariance algebra of an equation from $\mathcal{C}_2$. Consequently, there are no equations belonging to the class $\mathcal{C}_2$, which are invariant under the algebras isomorphic to $so(4)$ and $so^*(4)$.

The same assertion holds for the algebra $so(3,1)$. Indeed, the algebra $so(3,1)$ has the Cartan decomposition $\langle e_1, e_2, e_3\rangle \dotplus \langle N_1, N_2, N_3\rangle$, where $\langle e_1, e_2, e_3\rangle = so(3)$, $[e_i, N_j] = \sum_{l=1}^{3} \varepsilon_{ijl} N_l$, $[N_i, N_j] = -\sum_{l=1}^{3} \epsilon_{ijl} e_l$, $i,j,l = 1,2,3$, and ε_{ijl} is the anti-symmetric third-order tensor with $\varepsilon_{123} = 1$. Taking $so(3) = so^1(3)$ after simple calculations we get the forms of operators N_1, N_2, N_3

$$N_1 = \cos u\partial_u, \quad N_2 = -\sec u \cos x\partial_x + \sin u \sin x\partial_u, \quad N_3 = \sec u \sin x\partial_x + \sin u \cos x\partial_u.$$

Inserting the coefficient of the operator N_1 into the classifying equation (1.4) yields $F_{u_{xx}} = 0$, whence it follows that the realization obtained cannot be invariance algebra of PDE (0.1).

In order to classify realizations of the algebra $so(2,2)$, we make use of the fact that $so(2,2) \sim sl(2,\mathbb{R}) \oplus sl(2,\mathbb{R})$. So that we can choose $\langle e_i, \bar{e}_i \,|i = 1,2,3\rangle$ as the basis of $so(2,2)$. Here $\langle e_1, e_2, e_3\rangle = sl(2,\mathbb{R})$, $\langle \bar{e}_1, \bar{e}_2, \bar{e}_3\rangle = sl(2,\mathbb{R})$ with $[e_i, \bar{e}_j] = 0$, $(i,j = 1,2,3)$. Choosing as e_i $(i = 1,2,3)$ the basis operators of one of the realizations $sl^k(2,\mathbb{R})$ $(k = 1,2,\ldots,5)$ and analyzing the commutation relations of $so(2,2)$ we come to the conclusion that the class of operators (1.3) does not contain realizations of the algebra $so(2,2)$ which are admitted by a nonlinear equation of the form (0.1).

The same assertion holds true for the semi-simple algebras $sl(3,\mathbb{R})$, $su(3)$ and $su(2,1)$, which are eight dimensional (there are no seven-dimensional semi-simple algebras). Actually, this is true for any semi-simple algebra of the dimension $n > 3$. Indeed, there are four basic types of classical simple Lie algebras over the fiel of real numbers

- Type $A_{n-1}(n > 1)$ contains four real forms of the algebra $sl(n,\mathbb{C}) : su(n), sl(n,\mathbb{R})$, $su(p,q)$ $(p+q = n, p \geq q)$, $su^*(2n)$.

- Type $D_n(n > 1)$ contains three real forms of the algebra $so(2n,\mathbb{C})$: $so(2n)$, $so(p,q)$ $(p+q = 2n, p \geq q)$, $so^*(2n)$.

- Type $B_n(n > 1)$ contains two real forms of the algebra $so(2n+1,\mathbb{C})$: $so(2n+1)$, $so(p,q)$ $(p+q = 2n+1, p > q)$.

- Type $C_n(n > 1)$ contains three real forms of the algebra $sp(n,\mathbb{C})$: $sp(n)$, $sp(n,\mathbb{R})$, $sp(p,q)$ $(p+q = n, p > q)$.

As $su^*(4) \sim so(5,1)$, and furthermore the algebra $so(5,1)$ contains $so(4)$ as a subalgebra, the class of operators (1.3) does not contain realizations of the algebras A_{n-1} $(n > 1)$ and D_n $(n > 1)$, which differ from $sl^k(2,\mathbb{R})$ $(k = 1,2,\ldots,5)$.

The same assertion holds for the algebras B_n $(n > 1)$ and C_n $(n \geq 1)$, as well. Indeed, algebras of the type B_n with $n >= 2$ contains $so(4)$ and $so(3,1)$. What is more, $sp(2,\mathbb{R}) \sim so(3,2)$ and $so(3,1)$ is the subalgebra of $so(3,2)$, $sp(1,1) \sim so(4,1)$ and $so(3,1)$ is the subalgebra of $so(4,1)$, $sp(2) \sim so(5)$ and $so(4)$ is the subalgebra of $so(5)$.

It remains to analyze the exceptional semi-simple Lie algebras $G_1, G_2, F_4, E_6, E_7, E_8$. We consider the case of the algebra G_2 only, the remaining algebras are handled in the same way.

A Lie algebra of the type G_2 contains the compact real form g_2 and the non-compact real form g_2'. Since $g_2 \cap g_2' \sim su(2) \oplus su(2) \sim so(4)$, the class of operators (1.3) does not contain realizations of the algebras g_2 and g_2', which would be invariance algebras of Eq.(0.1). The same assertion holds true for the remaining exceptional semi-simple Lie algebras G_1, F_4, E_6, E_7, E_8.

Let us summarize the above reasonings in the form of theorem.

Theorem 1.3 *Any equation from the class C_2, whose invariance algebra is semi-simple, is equivalent to one of the equations given in Lemmas 1.4, 1.5.*

1.3.2. Invariance under the Algebras Having Nontrivial Levi Factor

Now we utilize the results of classificatio of inequivalent equations (0.1) admitting semi-simple symmetry algebras to describe PDEs (0.1) whose symmetry algebras admit Levi decomposition. The class of Lie algebras that admit Levi decomposition splits into two non-intersecting subclasses

- subclass of Lie algebras which are decomposable into direct sums of semi-simple and solvable Lie algebras, and
- subclass of Lie algebras which are semi-direct sums of a Levi factor and non-zero radical.

1.3.2.1. Invariance under Direct Sum of Semi-simple and Solvable Lie Algebras

To describe equations whose symmetry algebras are direct sums semi-simple and solvable Lie algebras we can use the explicit forms of realizations of semi-simple algebras constructed in the previous subsections. What is more, we need to consider realizations of semi-simple algebras belonging both to the class C_1 and to the class C_2.

Consider the case of the $sl^1(2, \mathbb{R})$-invariant equation. We look for possible extensions of the realization $sl^1(2, \mathbb{R})$ by operators (1.3) which commute with its basis operators. Analysis of the commutativity conditions yields the general form of the additional symmetry operators

$$\mathrm{v} = C_1 x\partial_x + (C_2 + 2C_1 u)\partial_u, \tag{1.16}$$

where C_1, C_2 are arbitrary constants. So we need to describe all possible solvable Lie algebras that have basis operators (1.16). Skipping intermediate calculations we formulate the fina result. The list of isomorphic solvable Lie algebras realized by operators (1.16) is formed by the two one-dimensional algebras $\langle \partial_u \rangle$, $\langle x\partial_x + 2u\partial_u \rangle$ and one two-dimensional algebra $L_2 = \langle \partial_u, x\partial_x + 2u\partial_u \rangle$. Note that the latter is isomorphic to $A_{2.2}$.

Inserting the coefficient of the above operators into the classifying equation and solving the resulting equations yield the forms of the unknown functions F in the corresponding invariant equations

1. Algebra $sl^1(2,\mathbb{R}) \oplus \langle \partial_u \rangle$

$$u_t = \frac{1}{4}u_x^2 + x^{-2}\tilde{F}(\omega), \quad \omega = x^2 u_{xx} - x u_x;$$

2. Algebra $sl^1(2,\mathbb{R}) \oplus \langle x\partial_x + 2u\partial_u \rangle$

$$u_t = x^{-1}uu_x - x^{-2}u^2 + x^{-2}(2u - xu_x)^2\tilde{F}(\omega), \quad \omega = (x^2u_{xx} - 2u)(2u - xu_x)^{-1};$$

3. Algebra $sl^1(2,\mathbb{R}) \oplus \langle \partial_u, x\partial_x + 2u\partial_u \rangle$

$$u_t = mx^2u_{xx}^2 - 2mxu_xu_{xx} + \left(m + \frac{1}{4}\right)u_x^2, \quad m \neq 0.$$

Under arbitrary $\tilde{F}$ the given algebras are maximal in Lie's sense invariance algebras of the corresponding equations.

Analogously we prove that the basis operators of a solvable Lie algebra, which commute with the basis operators of $sl^2(2,\mathbb{R})$, are necessarily of the form

$$\mathrm{v} = \eta(u)\partial_u. \tag{1.17}$$

Further computations show that the maximal dimension of a solvable Lie algebra having basis operators (1.17) is two. There are two inequivalent realizations, one-dimensional $L_1 = \langle \partial_u \rangle$ and two-dimensional $L_2 = \langle -u\partial_u, \partial_u \rangle$ with $L_2 \sim A_{2.2}$. Solving the corresponding classifying equations yields the forms of the right-hand sides of invariant equations (0.1)

1. Algebra $sl^2(2,\mathbb{R}) \oplus \langle \partial_u \rangle$

$$u_t = -\frac{1}{4}x^{-1}u_x + x^{-3}u_x^{-1}\tilde{F}(\omega), \quad \omega = u_x^{-2}u_{xx} + 3x^{-1}u_x^{-1}.$$

2. Algebra $sl^2(2,\mathbb{R}) \oplus \langle -u\partial_u, \partial_u \rangle$

$$u_t = -\frac{1}{4}x^{-1}u_x + mx^{-3}u_x^{-1}(u_x^{-2}u_{xx} + 3x^{-1}u_x^{-1})^{-2}, \quad m \neq 0.$$

Given an arbitrary $\tilde{F}$ and m the given algebras are maximal in Lie's sense invariance algebras of the corresponding equations.

The most general form of operators (1.3) commuting with the basis operators of $sl^3(2,\mathbb{R})$ is

$$\mathrm{v} = \tau(t)\partial_t + \xi(t)x\partial_x. \tag{1.18}$$

Our analysis shows that there exist only four $\mathcal{E}$-inequivalent realizations of solvable Lie algebras by operators (1.18), which are invariance algebras of equations of the form (0.1). Namely, there are two one-dimensional realizations, $\langle \partial_t \rangle$, $\langle tx\partial_x \rangle$, and two two-dimensional realizations, $\langle -t\partial_t - mx\partial_x, \partial_t \rangle (m \in \mathbb{R})$, $\langle t\partial_t, tx\partial_x \rangle$. Below we list the corresponding invariant equations.

1. Algebra $sl^3(2,\mathbb{R}) \oplus \langle \partial_t \rangle$

$$u_t = x u_x \tilde{F}(\omega), \quad \omega = x^{-5} u_x^{-3} u_{xx} + 2x^{-6} u_x^{-2}.$$

2. Algebra $sl^3(2,\mathbb{R}) \oplus \langle tx\partial_x \rangle$

$$u_t = \frac{x u_x}{4t} \ln\left(x^{-5} u_x^{-3} u_{xx} + 2x^{-6} u_x^{-2}\right) + x u_x \tilde{F}(t).$$

3. Algebra $sl^3(2,\mathbb{R}) \oplus \langle -t\partial_t - m x\partial_x, \partial_t \rangle$

$$u_t = \lambda x u_x (x^{-5} u_x^{-3} u_{xx} + 2x^{-6} u_x^{-2})^{1/4m}, \quad \lambda \neq 0, \quad m \neq 0, \quad \pm\frac{3}{4}.$$

4. Algebra $sl^3(2,\mathbb{R}) \oplus \langle t\partial_t, tx\partial_x \rangle$

$$u_t = \frac{x u_x}{4t} \ln(x^{-5} u_x^{-3} u_{xx} + 2x^{-6} u_x^{-2}) + \frac{\lambda x u_x}{t}, \quad \lambda \in \mathbb{R}.$$

Under arbitrary $\tilde{F}$, m and λ the given algebras are maximal in Lie's sense invariance algebras of the corresponding equations.

A similar analysis of extensions of the realizations algebras $sl^4(2,\mathbb{R})$, $sl^5(2,\mathbb{R})$ and $so^1(3)$ yields three more invariant equations. Below we give the right-hand sides of invariant equations and their maximal invariance algebras.

$$\begin{aligned}
& sl^4(2,\mathbb{R}) \oplus \langle \partial_t \rangle : u_t = x^{-2}\sqrt{4 + x^6 u_x^2}\,\tilde{F}(\omega), \\
& \quad \omega = (4 + x^6 u_x^2)^{-\frac{3}{2}} \left(x^4 u_{xx} + 5x^3 u_x + \frac{1}{2} x^9 u_x^3 \right); \\
& sl^5(2,\mathbb{R}) \oplus \langle \partial_t \rangle : u_t = x^{-2}\sqrt{|x^6 u_x^2 - 4|}\,\tilde{F}(\omega), \\
& \quad \omega = |x^6 u_x^2 - 4|^{-\frac{3}{2}} \left(x^4 u_{xx} + 5x^3 u_x - \frac{1}{2} x^9 u_x^3 \right); \\
& so^1(3) \oplus \langle \partial_t \rangle : u_t = \sqrt{\sec^2 x + u_x^2}\,\tilde{F}(\omega), \\
& \quad \omega = \left(u_{xx} \cos x - (2 + u_x^2 \cos^2 x) u_x \sin x \right) (1 + u_x^2 \cos^2 x)^{-3/2}.
\end{aligned}$$

1.3.2.2. Invariance under Semi-direct Sum of Semi-simple and Solvable Lie Algebras

To perform classificatio of equations from C_2 whose invariance algebras are isomorphic to semi-direct sum of semi-simple and solvable Lie algebras we need to apply a more sophisticated strategy. It is based on the well-known fact of the group analysis of differential equations, which is, the higher dimension of the invariance algebra admitted by PDE (0.1) the less arbitrary is the function F. So at some point, instead of an arbitrary function of f ve variables t, x, u, u_x, u_{xx} equation (0.1) would contain an arbitrary function of one variable or even arbitrary constants. When this is the case, we apply the Ovsyannikov classificatio method [37], since the determining equations split into over-determined systems of PDEs that can be effectively integrated.

So our approach to classificatio of equations invariant under semi-direct sum of semi-simple and solvable Lie algebras consist of the two major steps. Firstly, utilizing the results of classificatio of lower dimensional Lie algebras that can be decomposed into semi-direct sum of Levi factor and solvable radical [44] we describe all invariant equations containing arbitrary functions of four, three, two, one arguments and/or arbitrary constants. The second step is utilizing the Ovsyannikov's approach to classify those equations that have either arbitrary functions of one variable or arbitrary parameters. This will provide complete classificatio of Eq.(0.1) within the considered class of Lie algebras.

We prove that in order to implement the firs step of our method it suffice to consider

1. Algebras which are semi-direct sums of semi-simple and solvable Lie algebras of the dimension $n \leq 6$, if the Levi factor is isomorphic to $so(3)$.

2. Algebras which are semi-direct sums of semi-simple and solvable Lie algebras of the dimension $n \leq 5$, if the Levi factor is isomorphic to $sl(2,\mathbb{R})$.

Further analysis shows that without any loss of generality we can restrict our considerations to the Lie algebras $sl(2,\mathbb{R}) \subset\!\!\!\!+ A_{2.1}$, $so(3) \subset\!\!\!\!+ A_{3.1}$.

After completing the firs step, we apply the Ovsyannikov method to finaliz the classification In addition, we utilize this method to complete group classificatio of invariant equations containing arbitrary functions of one variable or arbitrary constants obtained in the previous subsection.

In a sequel, we use the list of non-isomorphic four-dimensional solvable Lie algebras obtained in [30]. This list is formed by ten decomposable algebras $A_{3.i} \oplus A_1$ $(i = 1, 2, \ldots, 9)$, $2A_{2.2} = A_{2.2} \oplus A_{2.2}$ and ten non-decomposable algebras $A_{4.i} = \langle e_1, e_2, e_3, e_4 \rangle$ $(i = 1, 2, \ldots, 10)$ (we give non-zero commutation relations only)

$$
\begin{array}{lll}
A_{4.1} & : & [e_2, e_4] = e_1, \quad [e_3, e_4] = e_2; \\
A_{4.2} & : & [e_1, e_4] = qe_1, \quad [e_2, e_4] = e_2, \quad [e_3, e_4] = e_2 + e_3, \quad q \neq 0; \\
A_{4.3} & : & [e_1, e_4] = e_1, \quad [e_3, e_4] = e_2; \\
A_{4.4} & : & [e_1, e_4] = e_1, \quad [e_2, e_4] = e_1 + e_2, \quad [e_3, e_4] = e_2 + e_3; \\
A_{4.5} & : & [e_1, e_4] = e_1, \quad [e_2, e_4] = qe_2, \quad [e_3, e_4] = pe_3, \quad -1 \leq p \leq q \leq 1, \ pq \neq 0; \\
A_{4.6} & : & [e_1, e_4] = qe_1, \quad [e_2, e_4] = pe_2 - e_3, \quad [e_3, e_4] = e_2 + pe_3, \quad q \neq 0, \ p \geq 0; \\
A_{4.7} & : & [e_2, e_3] = e_1, \quad [e_1, e_4] = 2e_1, \quad [e_2, e_4] = e_2, \quad [e_3, e_4] = e_2 + e_3; \\
A_{4.8} & : & [e_2, e_3] = e_1, \quad [e_1, e_4] = (1 + q)e_1, \quad [e_2, e_4] = e_2, \quad [e_3, e_4] = qe_3, \quad |q| \leq 1; \\
A_{4.9} & : & [e_2, e_3] = e_1, \quad [e_1, e_4] = 2qe_1, \quad [e_2, e_4] = qe_2 - e_3, \quad [e_3, e_4] = e_2 + qe_3, \quad q \geq 0; \\
A_{4.10} & : & [e_1, e_3] = e_1, \quad [e_2, e_3] = e_2, \quad [e_1, e_4] = -e_2, \quad [e_2, e_4] = e_1.
\end{array}
$$

We provide full calculation details for the case of the algebra $sl^1(2,\mathbb{R}) \subset\!\!\!\!+ A_{2.1}$. The algebra $so(3) \subset\!\!\!\!+ A_{3.1}$ is handled in the same way.

Let $sl(2,\mathbb{R}) = \langle e_1, e_2, e_3 \rangle$, $A_{2.1} = \langle e_4, e_5 \rangle$. Then the basis elements of $sl^1(2,\mathbb{R})$ are $\mathcal{E}$-equivalent to $e_1 = 2t\partial_t + x\partial_x$, $e_2 = -t^2\partial_t - tx\partial_x + x^2\partial_u$, $e_3 = \partial_t$. The remaining non-zero commutation relations of the algebra $sl^1(2,\mathbb{R}) \subset\!\!\!\!+ A_{2.1}$ read

$$[e_1, e_4] = e_4, \quad [e_1, e_5] = -e_5, \quad [e_2, e_5] = e_4, \quad [e_3, e_4] = e_5. \tag{1.19}$$

Inserting operators e_4, e_5 of the form (1.3) into (1.19) and solving the resulting equations we obtain the four inequivalent realizations of the algebra $sl^1(2,\mathbb{R}) \oplus A_{2.1}$, basis operators of $A_{2.1}$ having the form

$$\begin{array}{ll} 1) & e_4 = t\partial_x + 2tx^{-1}u\partial_u, \quad e_5 = \partial_x + 2x^{-1}u\partial_u; \\ 2) & e_4 = t\partial_x + (tx^{-1}u - x)\partial_u, \quad e_5 = \partial_x + x^{-1}u\partial_u; \\ 3) & e_4 = tx^{-1}\partial_u, \quad e_5 = x^{-1}\partial_u; \\ 4) & e_4 = (tu + x^2)\partial_x + (2ux + 2tx^{-1}u^2)\partial_u, \quad e_5 = u\partial_x + 2x^{-1}u^2\partial_u. \end{array}$$

However, only the second and the third realization give rise to the symmetry algebras of equations of the form (0.1). The corresponding invariant equations are

$$\begin{array}{ll} sl^1(2,\mathbb{R}) \oplus & \langle t\partial_x + (tx^{-1}u - x)\partial_u, \partial_x + x^{-1}u\partial_u\rangle : \\ & u_t = \lambda u_{xx} + 2\lambda x^{-2}u - 2\lambda x^{-1}u_x + x^{-1}uu_x - x^{-2}u^2, \ \lambda = 0; \\ sl^1(2,\mathbb{R}) \oplus & \langle (tu + x^2)\partial_x + 2(xu + tx^{-1}u^2)\partial_u, u\partial_x + 2x^{-1}u^2\partial_u\rangle : \\ & u_t = x^{-1}uu_x - x^{-2}u^2 + \lambda x^{-2}(2u - xu_x)(x^2u_{xx} + 2u - 2xu_x)^{-1}, \ \lambda = 0. \end{array}$$

Note that the f ve-dimensional Lie algebras presented above are maximal in Lie's sense.

Analysis of the realizations $sl^2(2,\mathbb{R})$, $sl^4(2,\mathbb{R})$, $sl^5(2,\mathbb{R})$ shows that they do cannot be extended up to a realization of the algebra $sl(2,\mathbb{R}) \oplus A_{2.1}$. Similarly, the realization $so^1(3)$ cannot be extended up to a realization of the algebra $so^1(3) \oplus A_{3.1}$.

The realization $sl^3(2,\mathbb{R})$ do yield new realizations of the algebra $sl(2,\mathbb{R}) \oplus A_{2.1}$. We give these below together with the corresponding invariant equations.

$$sl^3(2,\mathbb{R}) \oplus \langle -\partial_x + x^{-1}u\partial_u, x^{-1}\partial_u\rangle : u_t = x^{-1}(xu_{xx} + 2u_x)^{1/3}F(t). \quad (1.20)$$

$$sl^3(2,\mathbb{R}) \oplus \langle x^2u\partial_x, x^2\partial_x\rangle : u_t = x^3u_x^2(xu_{xx} + 2u_x)^{-1/3}F(t). \quad (1.21)$$

However these f ve-dimensional Lie algebras are not maximal. To fin the most extensive symmetry algebras we apply the infinitesima Lie algorithm directly. First of all, using the fact that $F(t) \neq 0$ we can make the change of variables

$$\bar{t} = \int F(t)dt, \quad \bar{x} = x, \quad v = u$$

and get $F \equiv 1$. Next, utilizing the Lie's algorithm we obtain that the maximal invariance algebra of PDE (1.20) with $F = 1$ is the seven-dimensional Lie algebra

$$sl^3(2,\mathbb{R}) \oplus \langle \partial_t, x\partial_x + \frac{4}{3}t\partial_t, x^{-1}\partial_u, -\partial_x + x^{-1}u\partial_u\rangle.$$

This algebra is isomorphic to the Lie algebra $sl(2,\mathbb{R}) \oplus A_{4.5}$ with $q = 1, p = \frac{4}{3}$. The maximal invariance algebra admitted by (1.21) with $F = 1$ is the seven-dimensional Lie algebra

$$sl^3(2,\mathbb{R}) \oplus \langle \partial_t, \frac{4}{3}t\partial_t - x\partial_x, x^2u\partial_x, x^2\partial_x\rangle.$$

It is isomorphic to the algebra $sl(2,\mathbb{R}) \oplus A_{4.5}$ with $q = 1, p = \frac{4}{3}$.

Now we proceed to finalizin group classificatio of the equations obtained in the previous subsection. Since these equations contain arbitrary functions of at most one variable, we can apply directly the Ovsyannikov classificatio method.

Consider, as an example, the case of realization $sl^1(2,\mathbb{R}) \oplus \langle\partial_u\rangle$. Inserting the function

$$F = \frac{1}{4}u_x^2 + x^{-2}\tilde{F}(\omega), \quad \omega = x^2 u_{xx} - x u_x$$

into the classifying equation (1.4) and splitting the obtained relation by the powers of the independent variable u_x yields the following over-determined system of PDEs for τ, ξ, η and $\tilde{F}$:

$$\begin{gathered}\left(x^{-2}(2x^{-1}\xi + \eta_u - 2\xi_x)\omega - x^{-1}\eta_x + \eta_{xx}\right)\tilde{F}_\omega = x^{-2}(\eta_u - \tau_t + 2x^{-1}\xi)\tilde{F},\\ (3x^{-2}\xi_u\omega + \xi_{xx} + x^{-1}\xi_x - x^{-2}\xi - 2\eta_{xu})\tilde{F}_\omega = x^{-2}\xi_u\tilde{F} + \frac{1}{2}\eta_x + \xi_t, \qquad (1.22)\\ (2\xi_{xu} - \eta_{uu} + 2x^{-1}\xi_u)\tilde{F}_\omega = \frac{1}{4}\eta_u + \frac{1}{4}\tau_t - \frac{1}{2}\xi_x, \quad \xi_{uu}\tilde{F}_\omega = -\frac{1}{4}\xi_u.\end{gathered}$$

If the function $\tilde{F}$ is arbitrary, then the realization $sl^1(2,\mathbb{R}) \oplus \langle\partial_u\rangle$ is the maximal symmetry algebra of the corresponding equation.

It follows from the last two equations from (1.22) that either $\tilde{F}$ is linear function of ω or

$$\xi_u = \eta_{uu} = 0, \quad 2\xi_x - \eta_u - \tau_t = 0. \qquad (1.23)$$

Provided $\tilde{F} = \lambda\omega + C$, $\lambda \neq 0$, $C \in \mathbb{R}$, then the maximal invariance algebra is infinite dimensional. In the case when $C \neq 3\lambda$ it is formed by the basis elements of the realization $sl^1(2,\mathbb{R}) \oplus \langle\partial_u\rangle$ and by the infinit set of operators

$$\mathrm{v}_\infty = \alpha(t,x)\exp\left(\frac{u}{4\lambda}\right)\partial_u,$$

where $\alpha = \alpha(t,x)$ is an arbitrary solution of the equation

$$\alpha_t = \lambda\alpha_{xx} - \lambda x^{-1}\alpha_x - \frac{C}{4\lambda}x^{-2}\alpha.$$

In the case when $C = 3\lambda$, the corresponding PDE admits two more symmetry operators $t\partial_x + 2(\lambda x^{-1}t - x)\partial_u$ and $\partial_x + 2\lambda x^{-1}\partial_u$.

However, the change of variables from $\mathcal{E}$

$$\bar{t} = t, \quad \bar{x} = x, \quad u = 4\lambda \ln|v|, \quad v = v(\bar{t},\bar{x})$$

reduce the equation under study to a linear heat conductivity equation

$$v_{\bar{t}} = \lambda v_{\bar{x}\bar{x}} - \lambda x^{-1}v_{\bar{x}} + \frac{C}{4\lambda}\bar{x}^{-2}v.$$

Consequently, the equation in question is equivalent to linear PDE and therefore is excluded from further consideration.

If the function $\tilde{F}$ is a non-linear function of ω, i.e., $\tilde{F}_{\omega\omega} \neq 0$, then taking into account (1.23) we derive from (1.22) the system of two equations for the functions $\tau, \xi = \xi(t,x)$, $\eta = (2\xi_x - \tau_x)u + \theta(t,x)$ and $\tilde{F}$

$$\left(x^{-2}(2x^{-1}\xi + \eta_u - 2\xi_x)\omega - x^{-1}\eta_x + \eta_{xx}\right)\tilde{F}_\omega = x^{-2}(\eta_u - \tau_t + 2x^{-1}\xi)\tilde{F} + \eta_t,$$
$$(2\eta_{xu} - \xi_{xx} - x^{-1}\xi_x + x^{-2}\xi)\tilde{F}_\omega = -\tfrac{1}{2}\eta_x - \xi_t. \qquad (1.24)$$

Analysis of the firs equation yields the following admissible forms of the function $\tilde{F}$

$$\begin{aligned}
\tilde{F} &= \lambda\exp(p\omega) + m, \quad \lambda p \neq 0, \quad m \in \mathbb{R};\\
\tilde{F} &= \lambda\ln|\omega + b| + m, \quad \lambda \neq 0, \quad b, m \in \mathbb{R};\\
\tilde{F} &= \lambda|\omega + b|^p + m, \quad \lambda p \neq 0, \quad p \neq 1, \quad b, m \in \mathbb{R}.
\end{aligned}$$

Inserting these expressions into (1.24) shows that extension of symmetry algebra is only possible when $\tilde{F} = \lambda\omega^2$. However, the maximal invariance algebra of this equation has already been obtained earlier in this subsection.

A similar analysis of the remaining invariant equations from Subsection 1.3.2.1 yields the following results:

- The only extension of the realization $sl^2(2,\mathbb{R}) \oplus \langle\partial_u\rangle$, which is invariance algebra of an equation of the form (0.1), is the realization $sl^2(2,\mathbb{R}) \oplus \langle -u\partial_u, \partial_u\rangle$.

- The only extension of the algebra $sl^3(2,\mathbb{R}) \oplus \langle tx\partial_x\rangle$, which is invariance algebra of an equation of the form (0.1), is the realization $sl^3(2,\mathbb{R}) \oplus \langle t\partial_t, tx\partial_x\rangle$.

- The list of possible extensions of the algebra $sl^3(2,\mathbb{R}) \oplus \langle\partial_t\rangle$, which are invariance algebras of equations of the form (0.1), is exhausted by the following algebras:

  1) $sl^3(2,\mathbb{R}) \oplus \langle -t\partial_t - mx\partial_x, \partial_t\rangle$,

  2) $sl^3(2,\mathbb{R}) \oplus \langle t\partial_t, tx\partial_x\rangle$,

  3) $sl^3(2,\mathbb{R}) \Subset \langle\partial_t, x\partial_x + \frac{4}{3}t\partial_t, x^{-1}\partial_u, -\partial_x + x^{-1}u\partial_u\rangle$,

  4) $sl^3(2,\mathbb{R}) \Subset \langle\partial_t, \frac{4}{3}t\partial_t - x\partial_x, x^2u\partial_x, x^2\partial_x\rangle$.

- The realizations $sl^4(2,\mathbb{R}) \oplus \langle\partial_t\rangle$, $sl^5(2,\mathbb{R}) \oplus \langle\partial_t\rangle$ and $so^1(3) \oplus \langle\partial_t\rangle$ do not admit extensions to realizations admitted by equations of the form (0.1).

Summarizing the above results we give the fina list of inequivalent equations from the class $\mathcal{C}_2$ that have nontrivial Levi factor.

$$
\begin{aligned}
sl^1(2,\mathbb{R}) &\oplus \langle \partial_u \rangle : u_t = \frac{1}{4}u_x^2 + x^{-2}F(\omega),\\
&\quad \omega = x^2 u_{xx} - x u_x;\\
sl^1(2,\mathbb{R}) &\oplus \langle x\partial_x + 2u\partial_u \rangle : u_t = x^{-1}uu_x - x^{-2}u^2 + x^{-2}(2u - xu_x)^2 F(\omega),\\
&\quad \omega = (x^2 u_{xx} - 2u)(2u - xu_x)^{-1};\\
sl^2(2,\mathbb{R}) &\oplus \langle \partial_u \rangle : u_t = -\frac{1}{4}x^{-1}u_x + x^{-3}u_x^{-1}F(\omega),\\
&\quad \omega = u_x^{-2}u_{xx} + 3x^{-1}u_x^{-1};\\
sl^3(2,\mathbb{R}) &\oplus \langle \partial_t \rangle : u_t = xu_x F(\omega),\\
&\quad \omega = x^{-5}u_x^{-3}u_{xx} + 2x^{-6}u_x^{-2};\\
sl^3(2,\mathbb{R}) &\oplus \langle tx\partial_x \rangle : u_t = \frac{xu_x}{4t}\ln\left(x^{-5}u_x^{-3}u_{xx} + 2x^{-6}u_x^{-2}\right) + xu_x F(t),\\
sl^4(2,\mathbb{R}) &\oplus \langle \partial_t \rangle : u_t = x^{-2}\sqrt{4 + x^6 u_x^2}\,F(\omega),\\
&\quad \omega = (4 + x^6 u_x^2)^{-3/2}(x^4 u_{xx} + 5x^3 u_x + \frac{1}{2}x^9 u_x^3);\\
sl^5(2,\mathbb{R}) &\oplus \langle \partial_t \rangle : u_t = x^{-2}\sqrt{x^6 u_x^2 - 4}\,F(\omega),\\
&\quad \omega = (x^6 u_x^2 - 4)^{-3/2}(x^4 u_{xx} + 5x^3 u_x - \frac{1}{2}x^9 u_x^3);\\
so^1(3) &\oplus \langle \partial_t \rangle : u_t = \sqrt{\sec^2 x + u_x^2}\,F(\omega),\\
&\quad \omega = (1 + u_x^2\cos^2 x)^{-3/2}\ \ u_{xx}\cos x \quad (2 + u_x^2\cos^2 x)u_x \sin x\ ;\\
sl^1(2,\mathbb{R}) &\oplus \langle \partial_u, x\partial_x + 2u\partial_u \rangle : u_t = \lambda x^2 u_{xx}^2 - 2\lambda x u_x u_{xx} + \left(\lambda + \frac{1}{4}\right)u_x^2,\ \ \lambda = 0;\\
sl^2(2,\mathbb{R}) &\oplus \langle -u\partial_u, \partial_u \rangle : u_t = -\frac{1}{4}x^{-1}u_x + \lambda x^{-3}u_x^{-1}(u_x^{-2}u_{xx} + 3x^{-1}u_x^{-1})^{-2},\ \lambda = 0;\\
sl^3(2,\mathbb{R}) &\oplus \langle -t\partial_t - mx\partial_x, \partial_t \rangle : u_t = \lambda x u_x |x^{-5}u_x^{-3}u_{xx} + 2x^{-6}u_x^{-2}|^{1/4m},\ \ \lambda = 0,\\
&\quad m = 0, \pm\frac{3}{4};\\
sl^3(2,\mathbb{R}) &\oplus \langle t\partial_t, tx\partial_x \rangle : u_t = \frac{xu_x}{4t}\ln\left(x^{-5}u_x^{-3}u_{xx} + 2x^{-6}u_x^{-2}\right) + \frac{\lambda x u_x}{t},\ \ \lambda \in \mathbb{R};\\
sl^1(2,\mathbb{R}) &\in\!\!\!\!+ \langle t\partial_x + (tx^{-1} - x)\partial_u, \partial_x + x^{-1}u\partial_u \rangle :\\
&\quad u_t = \lambda u_{xx} + 2\lambda x^{-2}u - 2\lambda x^{-1}u_x + x^{-1}uu_x - x^{-2}u^2,\ \ \lambda = 0;\\
sl^1(2,\mathbb{R}) &\in\!\!\!\!+ \langle (tu + x^2)\partial_x + 2(xu + tx^{-1}u^2)\partial_u, u\partial_x + 2x^{-1}u^2\partial_u \rangle :\\
&\quad u_t = x^{-1}uu_x - x^{-2}u^2 + \lambda x^{-2}(2u - xu_x)(x^2 u_{xx} + 2u - 2xu_x)^{-1},\ \ \lambda = 0;\\
sl^3(2,\mathbb{R}) &\in\!\!\!\!+ \langle \partial_t, x\partial_x + \frac{4}{3}t\partial_t, x^{-1}\partial_u, -\partial_x + x^{-1}u\partial_u \rangle : u_t = x^{-1}(xu_{xx} + 2u_x)^{1/3};\\
sl^3(2,\mathbb{R}) &\in\!\!\!\!+ \langle \partial_t, \frac{4}{3}t\partial_t - x\partial_x, x^2u\partial_x, x^2\partial_x \rangle : u_t = x^3 u_x^2 (xu_{xx} + 2u_x)^{-1/3}.
\end{aligned}
$$

This completes classificatio of second-order nonlinear evolution equations invariant whose invariance algebra contains a semi-simple Lie algebra. The completeness of classificatio ensures that any equation of the form (0.3) that admit a Lie algebra with nontrivial

Levi factor can be reduced to one of the invariant equations above by a suitable change of variables (1.6).

2. Group Analysis of Third-order Evolution Equations

Now we turn to group analysis of third-order nonlinear evolution equations (0.2). Note that Lie and higher Lie symmetries of a number of particular equations of the form (0.2) have been studied in [15, 16, 43, 18] (see, also, the relevant citations therein).

We begin by applying the infinitesima Lie method in order to derive the determining equations for the coefficient of the most general symmetry operator that can be admitted by (0.2). Skipping the intermediate calculations we present the fina result in the following assertion.

Theorem 2.1 *The most general symmetry group of nonlinear equation* (0.2) *for an arbitrary function* F *is generated by the vector field*

$$X = \tau(t)\partial_t + \left(\frac{\dot{\tau}}{3}x + \rho(t)\right)\partial_x + \phi(x,t,u)\partial_u, \tag{2.1}$$

where $\tau(t)$, $\rho(t)$ *and* $\phi(x,t,u)$ *are arbitrary solutions of the determining equation*

$$\begin{aligned}
&-3\,u_x\,\dot{\rho} - x\,u_x\,\ddot{\tau} - 9\,u_x\,u_{xx}\,\phi_{uu} - 3\,{u_x}^3\,\phi_{uuu}\\
&+3\,\phi_t - 9\,u_{xx}\,\phi_{xu} - 9\,{u_x}^2\,\phi_{xuu} - 9\,u_x\,\phi_{xxu} - 3\,\phi_{xxx} + 3\,(\phi_u - \dot{\tau})\,F\\
&+\left(2\,u_{xx}\,\dot{\tau} - 3\,u_{xx}\,\phi_u - 3\,{u_x}^2\,\phi_{uu} - 6\,u_x\,\phi_{xu} - 3\,\phi_{xx}\right)\,F_{u_{xx}}\\
&+\left(u_x\,\dot{\tau} - 3\,u_x\,\phi_u - 3\,\phi_x\right)\,F_{u_x} - 3\,\phi\,F_u - 3\,\tau\,F_t - (3\,\rho + x\,\dot{\tau})\,F_x = 0.
\end{aligned} \tag{2.2}$$

Here the dot over a symbol stands for the time derivative.

Therefore to describe all possible Lie symmetries of PDE (0.2) we need to integrate equation (2.2). However (2.2) is a single PDE for four unknown functions, τ, ρ, ϕ and F, which means that it is under-determined. Let us remind that the effectiveness of the infinitesima Lie approach is based on the fact that the determining equations for symmetries of a fi ed differential equation form an over-determined system of PDEs. Appending the compatibility conditions to the system of determining equations enables to construct its general solution.

As we explained in the previous section, additional constraints on the form of solutions of (2.2) come from the fact that the general solution, $\{\tau, \rho, \phi\}$, of Eq.(2.2) after being inserted into (2.1) gives rise to a Lie algebra. Consequently, with explicit forms of the commutation relations of the low-dimensional Lie algebras, $A_1, \ldots, A_N$, in hand we can integrate (2.2) for each algebra A_i, $i = 1, \ldots, N$. This yields the list of most general equations of the form (0.2) invariant under the Lie algebras $A_1, \ldots, A_N$.

In this section we restrict our considerations to the case of symmetry algebras that contain a subalgebra which is semi-simple. What is more, we exclude from considerations invariant PDEs which are locally equivalent to the linear differential equations. Note that the exhaustive group classificatio of third-order linear evolution equations admitting nontrivial Lie symmetries is obtained in [18].

According to [18] any nonlinear equation of the form (0.2) invariant under a semi-simple Lie algebra is locally equivalent to one of the equations listed in Theorem 2.1.

Theorem 2.2 *The class of PDEs* (0.2) *contains two inequivalent equations whose invariance algebra are semi-simple*

$$\begin{aligned} u_t &= u_{xxx} - \frac{3}{2}u_x^{-1}u_{xx}^2 + u_x G(x,t), \\ u_t &= -x^{-3}\Big(2xu_x + \frac{1}{9}x^2u_x^2 - G(3u - xu_x, 6u - x^2u_{xx})\Big). \end{aligned}$$

The maximal invariance algebras of the above equations under arbitrary G *read as*

$$\begin{aligned} \mathrm{sl}^1(2,\mathbb{R}) &= \{\partial_u, u\partial_u, -u^2\partial_u\}, \\ \mathrm{sl}^2(2,\mathbb{R}) &= \{\partial_t, t\partial_t + \frac{1}{3}x\partial_x, -t^2\partial_t - \frac{2}{3}tx\partial_x - x^3\partial_u\}. \end{aligned}$$

We also prove in [18] that a nonlinear equation of the form (0.2), whose invariance algebra is isomorphic to a Lie algebra having non-trivial Levi decomposition, is equivalent to one of the following equations:

$$u_t = u_{xxx} - \frac{3}{2}u_x^{-1}u_{xx}^2 + u_x\tilde{G}(x), \quad \mathrm{sl}^1(2,R) \oplus \{\partial_t\}, \tag{2.3}$$

$$u_t = u_{xxx} - \frac{3}{2}u_x^{-1}u_{xx}^2 + \lambda x^{-2}u_x, \quad \lambda \neq 0, \quad \mathrm{sl}^1(2,\mathbb{R}) \oplus \{\partial_t, t\partial_t + \frac{1}{3}x\partial_x\}, \tag{2.4}$$

$$u_t = u_{xxx} - \frac{3}{2}u_x^{-1}u_{xx}^2, \quad \mathrm{sl}^1(2,\mathbb{R}) \oplus \{\partial_t, \partial_x, t\partial_t + \frac{1}{3}x\partial_x\}, \tag{2.5}$$

$$u_t = u_{xxx} - 2x^{-2}u_x - \frac{1}{9}x^{-1}u_x^2 + x^{-3}\tilde{G}(\sigma), \; \sigma = x^2u_{xx} - 2xu_x, \quad \mathrm{sl}^2(2,\mathbb{R}) \oplus \{\partial_u\}, \tag{2.6}$$

where $\tilde{G}$ is an arbitrary function of x or σ. Moreover, the associated symmetry algebras are maximal.

This completes description of third-order evolution equations (0.2) invariant under semi-simple algebras.

3. Group Analysis of Fourth-order Evolution Equations

Evolution equations of the form (0.3) have the wide range of applications in physics. In particular, the Kuramoto-Sivashinsky equation

$$u_t = -u_{xxxx} - u_{xx} - \frac{1}{2}u_x^2, \tag{3.1}$$

the extended Fisher-Kolmogorov equation

$$u_t = -u_{xxxx} + u_{xx} - u^3 + u, \tag{3.2}$$

and the Swift-Hohenberg equation

$$u_t = -u_{xxxx} - 2u_{xx} - u^3 + (\kappa - 1)u, \quad \kappa \in \mathbf{R} \tag{3.3}$$

are obtained from Eq.(0.3) by specifying appropriately the arbitrary functions F and G.

In addition, class (0.3) includes the equation describing thin fil fl ws

$$u_t = -\Big(u^3 u_{xxx} + f(u, u_x, u_{xx})\Big)_x.$$

The latter is used for modeling of flui fl ws in physical situations such as coating, draining of foams, and movement of contact lenses [32].

One more important particular case of Eq.(0.3) reads as

$$u_t = -(f(u)u_{xxx})_x + (g(u)u_x)_x. \tag{3.4}$$

This equation describes motion of thin viscous film (for further details, see, [4] and the references therein). Provided $f(u) = u^n$, $g(u) \equiv 0$, Eq.(3.4) turns into the generalized lubrication equation. Depending on the choice of the parameter n, the general lubrication equation describes [7] (see, also, [21])

- capillary-driven fl w under $n = 3$,
- slip models in the vicinity of $u \to 0$ under $n < 3$,
- the thickness of a thin bridge between two masses of flui in a Hele-Shaw cell under $n = 1$.

Recently, using the classificatio approach of [3] we performed in [19, 20] group classificatio of equations of the form (0.3). Below we present a short overview of the results of [19, 20] that describe equations (0.3) invariant under semi-simple algebras.

According to [20] the most general symmetry group of (0.3) is generated by the infinitesima operator

$$V = \tau(t)\partial_t + \xi(t, x, u)\partial_x + \eta(t, x, u)\partial_u, \tag{3.5}$$

where τ, ξ and η are real-valued functions satisfying the classifying equation

$$\begin{aligned}
&(4\xi_u u_x + 4\xi_x - \dot{\tau})F - \tau F_t - \xi F_x - \eta F_u + (u_x\xi_x - u_x\eta_u + u_x^2\xi_u - \eta_x)F_{u_x}\\
&\quad(-u_{xx}\eta_u - \eta_{xx} + u_x^3\xi_{uu} - u_x^2\eta_{uu} - 2u_x\eta_{xu} + 2u_{xx}\xi_x + u_x\xi_{xx} + 2u_x^2\xi_{xu}\\
&\quad + 3u_x u_{xx}\xi_u)F_{u_{xx}} + (-3u_x^2\eta_{x,u,u} - \eta_{xxx} + u_x^4\xi_{uuu} - 3u_x\eta_{xxu} - 3u_{xx}\eta_{xu}\\
&\quad + 3u_{xxx}\xi_x + 3u_{xx}^2\xi_u + 3u_{xx}\xi_{xx} + 3u_x^3\xi_{xuu} + u_x\xi_{xxx} + 6u_x^2u_{xx}\xi_{uu} - u_x^3\eta_{uuu}\\
&\quad + 9u_x u_{xx}\xi_{xu} - u_{xxx}\eta_u + 4u_x u_{xxx}\xi_u + 3u_x^2\xi_{xxu} - 3u_x u_{xx}\eta_{uu})F_{u_{xxx}} = 0,
\end{aligned}$$

$$(-12u_xu_{xx}\eta_{xuu} - u_x^4\eta_{uuuu} + u_x\xi_{xxxx} + u_x^5\xi_{uuuu} + 4u_x^4\xi_{xuuu} - 6u_x^2\eta_{xxuu}$$
$$+ 6u_x^3\xi_{xxuu} - 4u_{xxx}\eta_{xu} - 6u_{xx}\eta_{xxu} - 4u_x\eta_{xxxu} + 12u_{xx}^2\xi_{xu} - 4u_x^3\eta_{xuuu}$$
$$- 3u_{xx}^2\eta_{uu} + 4u_x^2\xi_{xxxu} + 4u_{xx}\xi_{xxx} + 6u_{xxx}\xi_{xx} - \eta_{xxxx} - 6u_x^2u_{xx}\eta_{uuu}$$
$$16u_xu_{xxx}\xi_{xu} - 4u_xu_{xxx}\eta_{uu} + 10u_x^2u_{xxx}\xi_{uu} + 15u_xu_{xx}^2\xi_{uu} + 10u_x^3u_{xx}\xi_{uuu}$$
$$+ 10u_{xx}u_{xxx}\xi_u + 24u_x^2u_{xx}\xi_{xuu} + 18u_xu_{xx}\xi_{xxu})F + (\eta_u - \dot{\tau} - u_x\xi_u)G$$
$$- \tau G_t - \xi G_x - \eta G_u + (u_x\xi_x - u_x\eta_u + u_x^2\xi_u - \eta_x)G_{u_x} + (-u_{xx}\eta_u - \eta_{xx}$$
$$+ u_x^3\xi_{uu} - u_x^2\eta_{uu} - 2u_x\eta_{xu} + 2u_{xx}\xi_x + u_x\xi_{xx} + 2u_x^2\xi_{xu} + 3u_xu_{xx}\xi_u)G_{u_{xx}}$$
$$(-3u_x^2\eta_{xuu} - \eta_{xxx} + u_x^4\xi_{uuu} - 3u_x\eta_{xxu} - 3u_{xx}\eta_{xu} + 3u_{xxx}\xi_x + 3u_{xx}^2\xi_u$$
$$+ 3u_{xx}\xi_{xx} + 3u_x^3\xi_{xuu} + u_x\xi_{xxx} + 6u_x^2u_{xx}\xi_{uu} - u_x^3\eta_{uuu} + 9u_xu_{xx}\xi_{xu}$$
$$- u_{xxx}\eta_u + 4u_xu_{xxx}\xi_u + 3u_x^2\xi_{xxu} - 3u_xu_{xx}\eta_{uu})G_{u_{xxx}} - u_x\xi_t + \eta_t = 0, \qquad (3.6)$$

Here the dot over a symbol stands for differentiation with respect to t.

Now in order to describe all the possible equations (0.3) we need to integrate under-determined system of PDEs (3.6). To able to do this we need additional information on the possible forms of unknown functions τ, ξ, η and F, G. This information comes from the fact that inserting the general solution $\{\tau, \xi, \eta\}$ of (3.6) into (3.5) gives rise to a Lie algebra. Based on this simple idea is our approach to integration of (3.6) for the case of semi-simple Lie algebras developed in [19]. Applying this approach we prove that the list of all possible inequivalent equations (0.3), whose symmetry algebras are semi-simple, is exhausted by the six $sl(2,\mathbb{R})$-invariant PDEs. The corresponding results are given below in Theorems 3.1, 3.2.

Theorem 3.1 *There are at most six inequivalent realizations of* $sl(2,\mathbb{R})$ *by operators* (3.5) *which can be admitted by Eq.* (0.3). *These realizations and corresponding forms of the functions* F, G *are presented below*

$$sl^1(2,\mathbb{R}) = \langle 2t\partial_t + x\partial_x, -t^2\partial_t - tx\partial_x + x^2\partial_u, \partial_t\rangle :$$
$$F = x^2\tilde{F}(\omega_1,\omega_2,\omega_3), \quad G = \frac{\tilde{G}(\omega_1,\omega_2,\omega_3)}{x^2} + \frac{xuu_x - u^2}{x^2}, \quad \omega_1 = xu_x - 2u,$$
$$\omega_2 = x^2u_{xx} - 2u, \quad \omega_3 = x^3u_{xxx};$$

$$sl^2(2,\mathbb{R}) = \langle 2t\partial_t + x\partial_x, -t^2\partial_t + (x^3 - tx)\partial_x, \partial_t\rangle :$$
$$F = \frac{\tilde{F}(u,\omega_1,\omega_2)}{x^3u_x^5}, \quad G = \frac{18x^2u_{xxx} + 87xu_{xx} + 105u_x}{x^6u_x^5}\tilde{F}(u,\omega_1,\omega_2)$$
$$+\frac{\tilde{G}(u,\omega_1,\omega_2)}{x^3u_x} - \frac{u_x}{4x}, \quad \omega_1 = \frac{xu_{xx} + 3u_x}{xu_x^2}, \quad \omega_2 = \frac{x^2u_{xxx} + 9xu_{xx} + 15u_x}{x^2u_x^3};$$

$$sl^3(2,\mathbb{R}) = \langle 2x\partial_x - u\partial_u, -x^2\partial_x + xu\partial_u, \partial_x\rangle :$$
$$F = \frac{\tilde{F}(t,\omega_1,\omega_2)}{u^8}, \quad G = -8\frac{2u_{xxx}u^2 - 9u_xu_{xx}u + 9u_x^3}{u^{11}}\tilde{F}(t,\omega_1,\omega_2)$$
$$+u\tilde{G}(t,\omega_1,\omega_2), \quad \omega_1 = \frac{uu_{xx} - 2u_x^2}{u^6}, \quad \omega_2 = \frac{u_{xxx}u^2 - 9u_xu_{xx}u + 12u_x^3}{u^9};$$

$$sl^4(2,\mathbb{R}) = \langle 2x\partial_x, -x^2\partial_x, \partial_x\rangle :$$
$$F = \frac{\tilde{F}(t,u,\omega)}{u_x^4}, \quad G = 6\frac{u_{xx}}{u_x^6}(u_{xx}^2 - u_xu_{xxx})\tilde{F}(t,u,\omega) + u\tilde{G}(t,u,\omega),$$
$$\omega = \frac{3u_{xx}^2 - 2u_xu_{xxx}}{u_x^4};$$

$$sl^5(2,\mathbb{R}) = \langle 2x\partial_x - u\partial_u, (\frac{1}{u^4} - x^2)\partial_x + xu\partial_u, \partial_x\rangle :$$
$$F = \frac{u^4}{(u^6+4u_x^2)^2}\tilde{F}(t,\omega_1,\omega_2), \quad G = -\frac{u}{4(u^6+4u_x^2)^4}[146880u_x^8$$
$$+192u(641u^5 - 167u_{xx})u_x^6 - 896u^2u_{xxx}u_x^5 + 96u^2(417u^{10} - 208u_{xx}u^5$$
$$+28u_{xx}^2)u_x^4 + 32u^3u_{xxx}(u^5 + 20u_{xx})u_x^3 + 6u^3(855u^{15} - 1092u_{xx}u^{10} - 176u_{xx}^2u^5$$
$$-160u_{xx}^3)u_x^2 + 32u^9u_{xxx}(2u^5 + 5u_{xx})u_x + 3u^{14}(77u^{10} - 162u_{xx}u^5 + 36u_{xx}^2)]$$

$$\times\tilde{F}(t,\omega_1,\omega_2) + \frac{(u^6+4u_x^2)^{\frac{1}{2}}}{u^2}\tilde{G}(t,\omega_1,\omega_2), \quad \omega_1 = \frac{u^3}{(u^6+4u_x^2)^{\frac{3}{2}}}(u^6 - 2uu_{xx}$$
$$+10u_x^2), \quad \omega_2 = \frac{u^3}{(u^6+4u_x^2)^3}[u_{xxx}u^8 - 9u_xu_{xx}u^7 + 12u_x^3u^6 + 4u_x(u_xu_{xxx}$$
$$-3u_{xx}^2)u^2 + 36u_x^3u_{xx}u - 60u_x^5];$$

$$sl^6(2,\mathbb{R}) = \langle 2x\partial_x - u\partial_u, -(x^2 + \frac{1}{u^4})\partial_x + xu\partial_u, \partial_x\rangle :$$
$$F = \frac{u^4}{(u^6-4u_x^2)^2}\tilde{F}(t,\omega_1,\omega_2), \quad G = -\frac{u}{4(u^6-4u_x^2)^4}[146880u_x^8 - 192u$$
$$\times(641u^5 + 167u_{xx})u_x^6 - 896u^2u_{xxx}u_x^5 + 96u^2(417u^{10} + 208u_{xx}u^5 + 28u_{xx}^2)u_x^4$$
$$-32u^3u_{xxx}(u^5 - 20u_{xx})u_x^3 - 6u^3(855u^{15} + 1092u_{xx}u^{10} - 176u_{xx}^2u^5 + 160u_{xx}^3)$$
$$\times u_x^2 + 32u^9u_{xxx}(2u^5 - 5u_{xx})u_x + 3u^{14}(77u^{10} + 162u_{xx}u^5 + 36u_{xx}^2)]$$
$$\times\tilde{F}(t,\omega_1,\omega_2) + \frac{(4u_x^2-u^6)^{\frac{1}{2}}}{u^2}\tilde{G}(t,\omega_1,\omega_2), \quad \omega_1 = \frac{u^3}{(4u_x^2-u^6)^{\frac{3}{2}}}(u^6 + 2uu_{xx}$$
$$-10u_x^2), \quad \omega_2 = \frac{u^3}{(u^6-4u_x^2)^3}[u_{xxx}u^8 - 9u_xu_{xx}u^7 + 12u_x^3u^6 + 4u_x(3u_{xx}^2$$
$$-u_xu_{xxx})u^2 - 36u_x^3u_{xx}u + 60u_x^5].$$

The algebras $sl^i(2,\mathbb{R})$, $(i = 1,2,\ldots,6)$ are maximal invariance algebras of the corresponding PDEs provided the functions $\tilde{F}$ and $\tilde{G}$ are arbitrary.

Theorem 3.2 *The invariant equations listed in the Theorem 3.1 exhaust the list of all possible inequivalent PDEs* (0.3) *whose invariance algebras are semi-simple.*

4. Concluding Remarks

We have completely solved the problem of describing nonlinear second-order evolution equations whose invariance algebra contains semi-simple Lie algebra. Surprisingly, the list of inequivalent equations is not too large. If we waive the requirement that the invariance algebra contains a semi-simple subalgebra and allow for solvable invariance algebras the number of the corresponding invariant equations would increase dramatically.

The same is true for the cases of the third- and fourth order nonlinear evolution equations considered in the current paper. This fact fit perfectly into an intuitive idea that the collection of nonlinear PDEs used to model the various processes in nature should not be too extensive. On the other hand, any PDE that is supposed to model a process in nature should reflec the symmetry of the latter and thus should admit a nontrivial symmetry group. By this reason the invariant equations constructed in this paper are formidable candidates for the roles of models of evolution processes in nature.

One of the most important applications of the results obtained is construction of exact solutions of nonlinear evolution equations by means of the symmetry reduction procedure (see, e.g., [11, 12, 21] and the references therein).

Another interesting application of the group classificatio results is that they can be used to classify nonlocal symmetries of nonlinear evolution equations [50]-[52]. Recently, we developed the efficien approach enabling to construct broad families of nonlinear evolution equations admitting nonlocal symmetries which cannot be obtained within the classical Lie infinitesima approach [20, 50, 51, 52].

References

[1] Akhatov I.S., Gazizov R.K. and Ibragimov N.K., Non-local symmetries: Heuristic approach, *Journal of Soviet Mathematics*, 1991, 55, 1401-1450.

[2] Barut A.O. and Raczka R., *Theory of Group Representations and Applications*, Warszawa, PWN-Polish Scientifi Publishers, 1977.

[3] Basarab-Horwath P., Lahno V. and Zhdanov R., The structure of Lie algebras and the classificatio problem for partial differential equations, *Acta Appl. Math.*, 2001, 69, 43–94.

[4] Bertozzi A.L., The mathematics of moving contact lines in thin liquid films *Notices Amer. Math. Soc.*, 1998, 45, 689–697.

[5] Bluman G.W. and Cole J.D., *Similarity Methods for Differential Equations*, Berlin, Springer, 1974.

[6] Bluman G. and Kumei S., *Symmetries and Differential Equations*, New York, Springer, 1989.

[7] Constantin P., Dupont T.F., Goldstein R.E., Kadanoff L.P., Shelley M.J. and Zhou S.-M., Droplet breakup in a model of the Hele-Shaw cell, *Phys. Rev. E*, 1993, 47, 4169–4181.

[8] Dorodnitsyn V.A., On the invariant solutions of nonlinear heat transfer equation with a source, *Zhurn. Vych. Matem. i Matem. Fiz.*, 1982, 22, 1393–1400.

[9] Edwards M.P., Classical symmetry reductions of nonlinear diffusion-convection equations, *Phys. Lett. A*, 1994, 190, 149–154.

[10] Fushchych W.I., *Algebraic-Theoretical Studies in Mathematical Physics* , Kyiv, Institute of Mathematics, 6–28.

[11] Fushchych W.I., Shtelen W.M. and Serov M.I., *Symmetry Analysis and Exact Solutions of the Equations of Mathematical Physics* , Dordrecht, Kluwer, 1993.

[12] Fushchych W. and Zhdanov R., *Symmetry and Exact Solutions of Nonlinear Dirac Equations*, Kyiv, Naukova Ukraina Publishing, 1997.

[13] Gagnon L. and Winternitz P., Symmetry classes of variable coefficien nonlinear Schrödinger equations, *J. Phys. A: Math. Gen.*, 1993, 26, 7061–7076.

[14] Gandarias M.L., Classical point symmetries of a porous medium equation, *J. Phys. A: Math. Gen.*, 1996, 29, 607–633.

[15] Gazeau J.P. and Winternitz P., Symmetries of variable coefficien Korteweg-de Vries equations, *J. Math. Phys.*, 1992, 33, 4087–4102.

[16] Güngör F., Sanielevici M. and Winternitz P., On the Integrability properties of variable coefficien Korteweg-de Vries equations, *Can. J. Phys.*, 1996, 74, 676–684.

[17] Güngör F., Group classificatio and exact solutions of a radially symmetric porous-medium equation, *Int J. of Non-Linear Mech.*, 2002, 37, 245–255.

[18] Güngör F., Lagno V.I. and Zhdanov R.Z., Symmetry classificatio of KdV-type nonlinear evolution equations, *J. Math. Phys.*, 2004, 45, 2280–2813.

[19] Huang Q., Lahno V., Qu C.Z. and Zhdanov R., Preliminary group classificatio of a class of fourth-order evolution equations, *J. Math. Phys.*, 2009, 50, 023503.

[20] Huang Q., Qu C.Z. and Zhdanov R., Classificatio of classical and non-local symmetries of fourth-order nonlinear evolution equations, 2009, ArXiV:0905.2033.

[21] Ibragimov N.H. (Ed.), *CRC Handbook of Lie Group to Differential Equations* , Boca Raton, CRC Press, 1994–1996, 1–3.

[22] Katkov V.L., Group classificatio of solutions of the Hopf equation, *Zhurn, Pkikl. Mekh. i Tekh. Fiz.*, 1965, no.6, 105–106.

[23] Kingston J.G. and Sophocleous C., On form-preserving point transformations of partial differential equations, *J. Phys. A.: Math. Gen.*, 1998, 31, 1597–1620.

[24] Lahno V. and Zhdanov R., Group classificatio of nonlinear wave equations, *J. Math. Phys.*, 2005, 46, 053301.

[25] Lie S., Theorie der Transformationgruppen, *Math. Ann.*, 1880, 16, 441–528.

[26] Lie S., Über die Integration durch bestimmte Integrale von einer Classe linearer partieller Differentialgleichungen, *Arch. Math.*, 1881, 6, 328–368.

[27] Lie S., *Theorie der Transformationsgruppen*, Leipzig, B.G.Teubner, 1983, 3.

[28] Lie S. *Gesammetle Abhandlungen*, Leipzig, B.G.Teubner, 1924, 5.

[29] Magadeev B.A., On group classificatio of nonlinear evolution equations, *Algebra i Analiz*, 1993, 5, 141–156.

[30] Mubarakzyanov G.M., On solvable Lie algebras, *Izv. Vys. Ucheb. Zaved. Mat.*, 1963, 32, 114–123.

[31] Mubarakzjanov G.M., Classificatio of real structures of Lie algebras of fift order, *Izv. Vys. Ucheb. Zaved. Mat.*, 1963, 34, 99–106.

[32] Myers T.G., Thin film with high surface tension, *SIAM Rev.*, 1998, 40, 441–462.

[33] Olver P., *Applications of Lie Groups to Differential Equations*, New York, Springer, 1986.

[34] Olver P.J., Moving frames, *J. Symb. Comp.*, 2006, 36, 501–512.

[35] Oron A. and Rosenau Ph., Some symmetries of the nonlinear heat and wave equations, *Phys. Lett. A*, 1986, 118, 172–176.

[36] Ovsyannikov L.V., Group properties of the equation of heat conductivity, *Dokl. AN SSSR*, 1959, 125, 492–494.

[37] Ovsyannikov L.V., *Group Analysis of Differential Equations*, New York, Academic Press, 1982.

[38] Pallikaros C. and Sophocleous C., On point transformations of generalized nonlinear diffusion equations, *J. Phys. A: Math. Gen.*, 1995, 28, 6459–6465.

[39] Patera J. and Winternitz P., Continuous subgroups of the fundamental groups of physics. III. The de Sitter groups, *J. Math. Phys.*, 1977, 18, 1449–1456.

[40] Polyanin A.D. and Zaitsev V.F., *Handbook of Nonlinear Partial Differential Equations*, Boca Ranton, CRC Press, 2004.

[41] Popovych R.O. and Ivanova N.M., New results on group classificatio of nonlinear diffusion-convection equations, *J. Phys. A: Math. Gen.*, 2004, 37, 7547–7565.

[42] Reid G.J., Finding abstract Lie symmetry algebras of differential equations without integrating determining equations, *Eur. J. Appl. Math.*, 1991, 2, 319–340.

[43] Sokolov V.V. and Shabat A.B., Classificatio of integrable evolution equations, *Sov. Sci. Rev. C*, 1984, 4, 221–280.

[44] Turkowski P., Low-dimensional real Lie algebras, *J. Math. Phys.* 1988,, 29, 2139–2144.

[45] Zhdanov R.Z., Fushchych W.I. and Marko P.V., New scale-invariant nonlinear differential equations for a complex scalar field *Physica D*, 1996, 95, 158–162.

[46] Zhdanov R.Z. and Lahno V.I., Group classificatio of heat conductivity equations with a nonlinear source, *J. Phys. A.: Math. Gen.*, 1999, 32, 7405–7418.

[47] Zhdanov R.Z. and Lahno V.I., Group classificatio of the general evolution equation: local and quasilocal symmetries, *SIGMA*, 1995, 1.

[48] Zhdanov R.Z. and Roman O.V., On preliminary symmetry classificatio of nonlinear Schroedinger equations with some applications to Doebner-Goldin model, *Rep. Math. Phys.*, 2000, 45, 273–291.

[49] Yung C.M., Verburg K. and Baveye P., Group classificatio and symmetry reductions of the non-linear diffusion-convection equation $u_t = (D(u)u_x)_x - K'(u)u_x$, *Int. J. Non-Linear Mech.*, 1994, 29, 273–278.

[50] Zhdanov R., On relation between potential and contact symmetries of evolution equations, *J. Math. Phys.*, 2009, 50, 053522.

[51] Zhdanov R., Nonlocal symmetries of evolution equations, *Nonlinear Dynamics*, 2010, 60, 403–411.

[52] Zhdanov R., On Lie group characterization of quasi-local symmetries of nonlinear evolution equations, *Journal of Lie Theory*, 2010, 20, 375–392.

In: Evolution Equations
Editor: Arthur L. Claes

ISBN: 978-1-61209-652-0

Chapter 10

A Variational Alternative to the Use of Condensed Energy in Crystal Plasticity Based on the Evolution of Microstructure History

***D. M. Kochmann*[1*] *and* *K. Hackl*[2]**

[1] Graduate Aerospace Laboratories,
California Institute of Technology, Pasadena, CA, USA
[2] Lehrstuhl für Mechanik – Materialtheorie,
Ruhr-Universität Bochum, Germany

Abstract

The study of the origin and evolution of microstructures in crystalline solids has gained much interest, virtue of the essential influence of such microscale structures on the macroscopic mechanical properties of the material. The formation of regular, fine-scale patterns on the material's microlevel has been reasoned to stem from non-(quasi)convex energy potentials, which result in a lack of homogeneous solution states to the underlying thermodynamic extremum principles of minimum potential energy and maximum dissipation. The theory of relaxation has provided beneficial tools for understanding and analyzing the origin and subsequent development of microstructures in finite-strain plasticity models, whose prediction and simulation is commonly based on condensed potentials to overcome the non-(quasi)convexity and to render the problem well-posed while considering all admissible microfluctuations, i.e., microstructures. We outline an incremental variational alternative to the application of condensed potentials, which allows to account for more physical insight and which greatly expands the regime of applicability of material models. Also, we illustrate the incremental approach by its application to an incompressible Neo-Hookean solid with one and multiple active slip systems, but the method can readily be generalized to problems of microstructure evolution.

*E-mail address: kochmann@caltech.edu

PACS 61.72.-y, 62.20.F-, 81.16.Rf.

Keywords: Microstructure, Finite-Strain Plasticity, Relaxation, Variational Methods.

1. Introduction

The origin and evolution of microstructures in crystal plasticity is of enormous importance due to their considerable influence on the effective, macroscopic mechanical properties of engineering materials. Besides the habitual appearance of a vast amount of defects in the periodic crystal lattice (such as, e.g., dislocations or vacancies, grain boundaries, or interfaces in multiphase materials), experiments confirm that various engineering metals tend to form complex microstructural patterns of a remarkably high degree of regularity, such as laminate or labyrinth patterns of dislocation walls separating ordered domains of rather homogeneous deformation [7]. The formation of these microstructures in models of crystal plasticity has been theorized to stem from the non-(quasi)convexity of the governing energetic potentials, which give rise to small-scale fluctuations as infimizing sequences, to be observed as material microstructures.

Energetic models for crystal plasticity are based on the concepts of energy and dissipation and describe the deformed body in terms of the deformation gradient $\boldsymbol{F} = \boldsymbol{\nabla}\phi$ and a set of internal variables $\boldsymbol{z}$. For a specific Helmholtz free energy density $\Psi(\boldsymbol{\nabla}\phi, \boldsymbol{z})$ the total energy of the deformed body Ω at time t is given by

$$\mathcal{I}(t, \phi, \boldsymbol{z}) = \int_\Omega \Psi(\boldsymbol{\nabla}\phi, \boldsymbol{z})\,\mathrm{d}V - \ell(t, \phi) \tag{1}$$

where $\ell(t, \phi)$ represents the potential of external forces. Thermodynamically conjugate stresses are derived from

$$\boldsymbol{P} = \frac{\partial \Psi}{\partial \boldsymbol{F}}, \qquad \boldsymbol{Q} = -\frac{\partial \Psi}{\partial \boldsymbol{z}}. \tag{2}$$

Dissipation arises as a consumption of energy due to irreversible processes (e.g., dislocation gliding or climbing). The evolution of the internal variables $\boldsymbol{z}$ is then governed by either a so-called inelastic potential $J(\boldsymbol{z}, \boldsymbol{Q})$ or its Legendre transform, the dissipation functional [3] (where the dot denotes differentiation with respect to time)

$$\Delta(\boldsymbol{z}, \dot{\boldsymbol{z}}) = \sup\left\{ \dot{\boldsymbol{z}} \cdot \boldsymbol{Q} - J(\boldsymbol{z}, \boldsymbol{Q}) \,\middle|\, \boldsymbol{Q} \right\}. \tag{3}$$

Evolution equations for the internal variables then follow as

$$\dot{\boldsymbol{z}} \in \frac{\partial J}{\partial \boldsymbol{Q}}, \qquad \text{or} \qquad \boldsymbol{Q} \in \frac{\partial \Delta}{\partial \dot{\boldsymbol{z}}}. \tag{4}$$

The variational principles of thermodynamics can be employed to describe the time-continuous evolution of microstructures in a deformed solid. The principle of minimum potential energy yields the actual displacement field for given essential boundary conditions ϕ_0 on the body's surface subset Γ via

$$\phi = \operatorname{argmin}\left\{ \mathcal{I}(t, \phi, \boldsymbol{z}) \,\middle|\, \phi = \phi_0 \text{ on } \Gamma \right\}, \tag{5}$$

while the internal variables follow from the principle of maximum dissipation $\mathcal{D}$,

$$\dot{\boldsymbol{z}} = \operatorname{argmax}\left\{ \mathcal{D}(\boldsymbol{Q}, \dot{\boldsymbol{z}}) \,\middle|\, \boldsymbol{Q} \,:\, \mathcal{D} = \boldsymbol{Q} \cdot \dot{\boldsymbol{z}} \right\}. \tag{6}$$

Alternatively the internal variables can be computed from the principle of minimum dissipation potential [3, 10],

$$\dot{\boldsymbol{z}} = \operatorname{argmin}\left\{ \dot{\Psi}(\boldsymbol{\nabla}\boldsymbol{\phi}, \boldsymbol{z}) + \Delta(\boldsymbol{z}, \dot{\boldsymbol{z}}) \,\middle|\, \dot{\boldsymbol{z}} \right\}. \tag{7}$$

When the energetic potentials lack quasiconvexity, no homogeneous minimizer of the above minimization problem may exist [1]. As a consequence, the material does not respond by means of a homogeneous deformation state but breaks up into multiple compatible domains at local energy minima [8], which can be described by employing the quasiconvex hull of the free energy (which accounts for all admissible micro-fluctuations $\boldsymbol{\varphi}$):

$$Q\Psi(\boldsymbol{F}) = \inf\left\{ \frac{1}{|\omega|} \int_{\omega} \Psi(\boldsymbol{F} + \boldsymbol{\nabla}\boldsymbol{\varphi})\, \mathrm{d}V \,\middle|\, \boldsymbol{\varphi} \,:\, \boldsymbol{\varphi} = \boldsymbol{0} \text{ on } \partial\omega \right\} \tag{8}$$

in the minimum principles. By considering the associated potentials in a time-incremental setting, the initiation of microstructures has been studied by using condensed energy functionals [2, 4, 6, 17, 15, 18] as follows. For rate-independent materials, principle (7) can be integrated to account for instantaneous changes of the value of $\boldsymbol{z}$, which yields the balance law

$$\Psi(\boldsymbol{\nabla}\boldsymbol{\phi}, \boldsymbol{z}_1) - \Psi(\boldsymbol{\nabla}\boldsymbol{\phi}, \boldsymbol{z}_0) = -D(\boldsymbol{z}_0, \boldsymbol{z}_1), \tag{9}$$

where $D(\boldsymbol{z}_0, \boldsymbol{z}_1)$ represents the dissipation distance [16]. When applied to a finite time-increment $[t_n, t_{n+1}]$, this allows for an approximate formulation. The new state (i.e., $\boldsymbol{\phi}_{n+1}$ and $\boldsymbol{z}_{n+1}$ at time t_{n+1}) is obtained for given boundary conditions at time t_{n+1} and for the known initial value $\boldsymbol{z}_n = \boldsymbol{z}(t_n)$ via the following principle [3, 16]:

$$\{\boldsymbol{\phi}_{n+1}, \boldsymbol{z}_{n+1}\} = \operatorname{argmin}\left\{ \int_{\Omega} [\Psi(\boldsymbol{\nabla}\boldsymbol{\phi}, \boldsymbol{z}) + D(\boldsymbol{z}_n, \boldsymbol{z})]\, \mathrm{d}V - \ell(t_{n+1}, \boldsymbol{\phi}) \,\middle|\, \boldsymbol{\phi}, \boldsymbol{z} \right\}. \tag{10}$$

Carrying out the minimization with respect to $\boldsymbol{z}$ in (10) beforehand gives the so-called condensed energy

$$\Psi_n^{\mathrm{cond}}(\boldsymbol{F}) = \inf\left\{ \Psi(\boldsymbol{F}, \boldsymbol{z}) + D(\boldsymbol{z}_n, \boldsymbol{z}) \,\middle|\, \boldsymbol{z} \right\}, \tag{11}$$

which has been applied to numerous problems in the scientific literature to model the initiation of microstructures in solids. When possible, the computation and use of the condensed potentials provide a convenient and efficient way to account for microstructures in a multiscale manner. In particular, when applied to a single time step, analytical solutions are feasible and provide important insight, see e.g. [5]. However, the condensed potentials often lack some physical meaning as they do not account for the actual microstructural changes during a time step. Here, we outline an alternative approach based on the incremental solution of the variational principles, which allows to account for the exact changes of the microstructure during a time increment, which, in turn, opens up the opportunity to include more physical information about changes of the microstructure (e.g., regarding the change of hardening variables or laminate orientation) and to better simulate the mechanical behavior during non-monotonic deformation paths (e.g., during cyclic loading).

The incremental variational approach is based on solving the evolution equations (7) in combination with semi-relaxed potentials incrementally during each time step. This requires the a-priori computation of a relaxed energy density (i.e., the quasiconvex hull $Q\Psi$) or an approximation (here, e.g., the rank-one-convex hull) of the potentials involved. Note that we only carry out a partial relaxation to arrive at semi-relaxed potentials, by minimizing with respect to the elastic variables only, whereas the internal (dissipative) variables follow from the principle of minimum dissipation potential. For problems of crystal plasticity, it is convenient to make use of an approximation of the rank-one-convex hull since it admits a physical interpretation in terms of a lamellar microstructure, where each lamella represents a domain of constant deformation state and internal variables, separated by planar interfaces (observable as dislocation walls in micrographs). The computed relaxed potentials of energy density and dissipation potential are combined in the principle of minimum dissipation potential (7), which is then solved iteratively to compute the complete evolution of the deformed solid in time. If an analytical computation of the relaxed energy density is not feasible, it can, at least in principle, be determined on-the-fly during the solution process of the evolution equations. This is required, in particular, when considering higher-order laminates, where an analytical relaxation is not available. In the following section we briefly outline the application of this method to the model problem of finite-strain crystal plasticity in an incompressible Neo-Hookean solid.

2. Finite-strain crystal plasticity

In finite-strain plasticity models the deformation is characterized by the deformation gradient $\boldsymbol{F} = \nabla \boldsymbol{\phi}$, which splits multiplicatively into its elastic and plastic parts (i.e., $\boldsymbol{F} = \boldsymbol{F}_{\mathrm{e}} \boldsymbol{F}_{\mathrm{p}}$), and by a set of internal variables $\boldsymbol{z}$ to describe the microstructure at the material point level. For n active slip systems $(\boldsymbol{s}_i, \boldsymbol{m}_i)$ (with slip directions $\boldsymbol{s}_i$ and slip plane unit normals $\boldsymbol{m}_i$), the flow rule for the plastic deformation with slip rates $\dot{\gamma}_i$ reads

$$\dot{\boldsymbol{F}}_{\mathrm{p}} \boldsymbol{F}_{\mathrm{p}}^{-1} = \sum_{i}^{n} \dot{\gamma}_i \, \boldsymbol{s}_i \otimes \boldsymbol{m}_i. \tag{12}$$

Among the various models available, let us choose an incompressible Neo-Hookean solid for simplicity, whose energy density is given by

$$\Psi(\boldsymbol{F}_{\mathrm{e}}, \boldsymbol{p}) = \frac{\mu}{2} \left(\operatorname{tr} \boldsymbol{F}_{\mathrm{e}}^T \boldsymbol{F}_{\mathrm{e}} - 3 \right) + \sum_{i=1}^{n} \sum_{j=1}^{n} \kappa_{ij} \, |p_i \, p_j|^{\alpha/2}, \qquad \det \boldsymbol{F} = 1, \tag{13}$$

where μ denotes the material's shear modulus, p_i are internal hardening variables (also termed history variables) obeying the flow rule $\dot{p}_i = |\dot{\gamma}_i|$. $\boldsymbol{\kappa}$ is a tensor of hardening moduli, and α is 2 or 4 (for a linear or higher-order hardening law, respectively). Note that incompressibility requires that not only $\det \boldsymbol{F}_{\mathrm{p}} = 1$ but also $\det \boldsymbol{F}_{\mathrm{e}} = \det \boldsymbol{F} = 1$. This assumption is made to allow for an analytical relaxation of the non-quasiconvex energy density.

For a flow rule of Tresca-type, the dissipation potential can be formulated as

$$\Delta(\dot{\gamma}) = r \sum_{i}^{n} |\dot{\gamma}_i|, \tag{14}$$

where r is a material constant, the critical resolved shear stress. For the given model problem, the condensed energy functional for a homogeneously deformed body (i.e., without microstructure) takes the form

$$\Psi^{\text{cond}}(\boldsymbol{F}) = \inf \left\{ \Psi(\boldsymbol{F}\boldsymbol{F}_{\text{p}}^{-1}(\boldsymbol{\gamma}), \boldsymbol{p}) + D(\boldsymbol{\gamma}) \,\middle|\, \boldsymbol{\gamma} : p_i = |\gamma_i| \quad \forall \quad i = 1, \ldots, n \right\} \tag{15}$$

where we collect variables by defining $\boldsymbol{\gamma} = \{\gamma_1, \ldots, \gamma_n\}$ and $\boldsymbol{p} = \{p_1, \ldots, p_n\}$. Investigating the course of $\Psi^{\text{cond}}(\boldsymbol{F})$ for given active slip systems along a known deformation path $\boldsymbol{F}(t)$ conveys the nonconvex nature of the condensed energy potential [14], which hence results in the formation of microstructures.

One particular type of microstructures is a laminate of first order, which is characterized by N domains of volume fractions λ_i, separated by parallel interfaces with common unit normal $\boldsymbol{n}$. Within each laminate domain, we find constant values $\boldsymbol{z}_i$ of the internal variables and $\boldsymbol{F}_i$ of the deformation gradient. The latter can be written as

$$\boldsymbol{F}_i = \boldsymbol{F}(\mathbf{I} + \boldsymbol{a}_i \otimes \boldsymbol{n}) \tag{16}$$

to ensure compatibility across interfaces (i.e., the laminate is restricted to deformations that identically satisfy the Hadamard compatibility condition of rank-one-connectivity). Following the above framework, one can analytically determine the relaxed energy [14] for the special case of a laminate of first order with all slip systems within each domain being contained within a common slip plane (i.e., each domain is allowed to arbitrarily activate one (generally three-dimensional) slip system, but within every domain plastic slip is restricted to occur only in a single slip plane of the lattice). Besides, one needs to determine the corresponding relaxed dissipation potential, where we can account for dissipation not only due to changes of the plastic slip in each domain but also due to changes of the volume fractions [11, 14]. For example, for a two-domain laminate ($N = 2$) with a single active slip system ($n = 1$), where λ denotes the volume fraction of domain 1, we have introduced the dissipation potential

$$\Delta^*(\lambda, \gamma_i, \dot{\lambda}, \dot{\gamma}_i) = r\left((1-\lambda)\,|\dot{\gamma}_1| + \lambda\,|\dot{\gamma}_2| + \left|\dot{\lambda}(\gamma_1 - \gamma_2)\right|\right), \tag{17}$$

where the first two terms represent dissipation due to changes of the plastic slip within domains, whereas the final term accounts for dissipation due to volume fraction changes. For details about the relaxed dissipation potentials see e.g. [14]. Furthermore, dissipation due to a rotational rearrangements of the laminate microstructure can be accounted for by inclusion of the additional amount of dissipation required to rearrange the microstructure upon changing $\boldsymbol{n}$:

$$D\ (\boldsymbol{\lambda}, \boldsymbol{z}) = \sum_{i,j=1}^{N} \lambda_i \lambda_j D(\boldsymbol{z}_i, \boldsymbol{z}_j). \tag{18}$$

Via the principle given by Eq. (7) we now arrive at evolution equations for the volume fractions $\boldsymbol{\lambda}$ and the set of plastic slips γ, i.e., we incrementally solve the stationarity conditions from minimizing the total power now involving the relaxed energy density $\Psi^{\mathrm{rel}}(\boldsymbol{F}, \boldsymbol{\lambda}, \gamma, \boldsymbol{p})$ and the corresponding dissipation potential $\Delta^*(\boldsymbol{\lambda}, \dot{\boldsymbol{\lambda}}, \gamma, \dot{\gamma})$. This results in evolution equations of the form

$$0 \in \frac{\partial \Psi^{\mathrm{rel}}}{\partial \boldsymbol{\lambda}} + \frac{\partial \Delta^*}{\partial \dot{\boldsymbol{\lambda}}}, \tag{19}$$

$$0 \in \frac{\partial \Psi^{\mathrm{rel}}}{\partial \gamma} + \frac{\partial \Psi^{\mathrm{rel}}}{\partial \boldsymbol{p}} \operatorname{sign} \dot{\gamma} + \frac{\partial \Delta^*}{\partial \dot{\gamma}}. \tag{20}$$

Note that the aforementioned evolution equations allow for a variety of modifications to material models as compared to the application of condensed energy potentials. In particular, the incremental numerical solution of the stationarity conditions allows to include additional flow rules for internal variables. For example, we have included an updating procedure for the hardening variables, which accommodates the change of hardening within each domain upon changes of the volume fractions [11].

3. Results

The outlined incremental variational framework has been applied to the evolution of laminate microstructures in finite-strain crystal plasticity with a single and multiple active slip systems. Figure 1 illustrates the path of the energy at the material-point level during a plane-strain simple-shear test with a single active slip systems. Clearly, the unrelaxed energy (corresponding to a homogeneous deformation state) reveals its nonconvexity, which gives rise to the formation of microstructure. In a direct comparison, the present approach with energy path $\Psi^{\mathrm{rel},1}$ reproduces the solution Ψ^{cond} from using condensed potentials up to a minimal deviation (see the comparison of those two curves in the graphics). Therefore, for monotonic loading paths the present approach agrees well with the condensed formulation. However, as we include the update procedure for the internal hardening variables, we obtain the solution $\Psi^{\mathrm{rel},2}$, which indicates a significant reduction of energy as compared to the condensed energy path. The inclusion of additional microstructural mechanisms can hence result in a reduction of the energy considerably below that obtained from condensation of the potentials involved.

The difference between the condensed solution and the one from the present approach is more pronounced for a tension-compression test as illustrated in Figure 2, where the present incremental formulation yields an essential reduction in energy and consequently also a considerably different stress-strain response when accounting for hardening updates (the graphics illustrate both the path of energy and of the Cauchy shear stress during a plane-strain tension-compression test with a single slip system). This energy reduction is due to the relaxation of some of the mathematical constraints tacitly implied when carrying out the analysis based on condensed potentials [12].

The presented method hence provides elegant numerical tools to accurately compute the evolution of microstructures in crystal plasticity. A further advantage lies in its applicability to arbitrary deformation paths including cyclic loading, which is of high interest for engi-

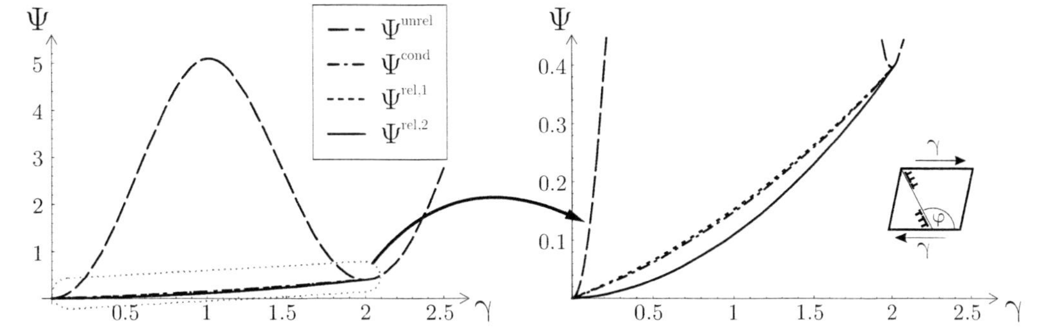

Figure 1. Comparison of the nonrelaxed energy (homogeneous deformation state), condensed energy, and relaxed energy from the present approach with and without hardening updates, for a plane-strain simple-shear test. The right graphic shows a magnification of the left.

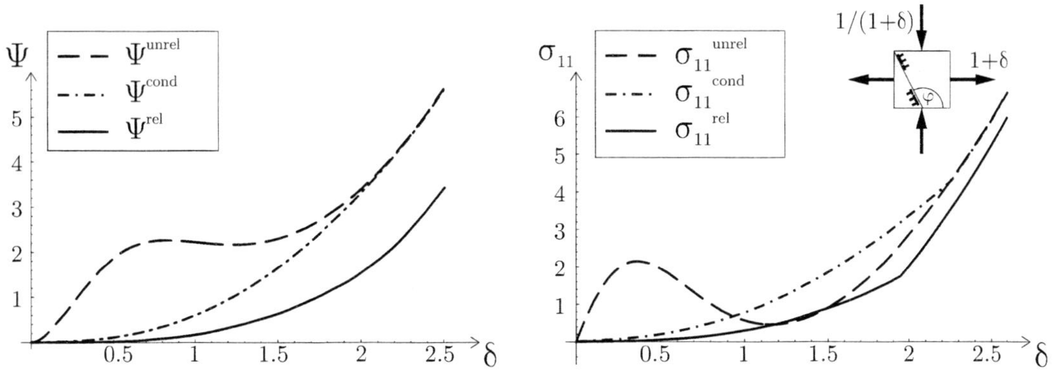

Figure 2. Comparison of nonrelaxed energy (homogeneous deformation state), condensed energy, and relaxed energy from the present approach, for a plane-strain tension-compression test, and corresponding shear stress curves, with hardening updates

neering materials and applications. Application of the outlined procedure to a cyclic plane-strain simple-shear test results in the stress-strain hysteresis loops shown in Figure 3, where the two curves correspond to zero hardening ($\kappa = 0$) and moderate hardening ($\kappa = 0.05\mu$) of the material. The gradual transformation of the hysteresis during cyclic loading with hardening leads to an interesting observation on the microscopic level (not shown here for conciseness) [13]. During subsequent deformation cycles, a steady-state laminate forms which concentrates plastic deformation in slip bands of low volume fraction (having typical characteristics of experimentally observed persistent slip band structures). The present incremental variational approach can conveniently be employed to model the origin and evolution of such band structures during cyclic straining.

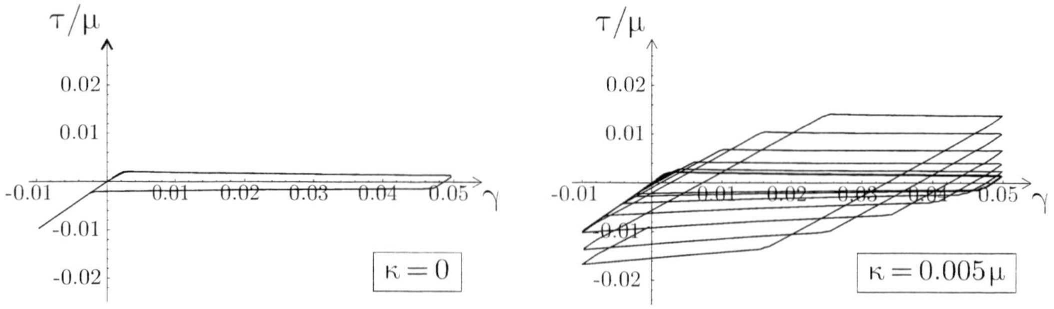

Figure 3. Stress-strain hysteresis for a cyclic plane-strain simple-shear test with a single active slip system, without ($\kappa = 0$) and with hardening ($\kappa = 0.05\mu$).

4. Conclusion

An incremental variational approach has been presented for the simulation of microstructures in crystal plasticity during their evolution in time. In contrast to many approaches in the scientific literature, we do not employ condensed energy potentials to include the formation of microstructure into the energetic constitutive framework for the nonconvex minimization problem. Instead, we incrementally solve the stationarity conditions of the principle of minimum dissipation potential for the updates of the internal variables, where we make use of a semi-relaxed energy potential. This requires the (semi-)relaxation of the potentials involved a-priori with respect to the elastic variables (to allow for their inclusion in the thermodynamic minimum principles) or the concurrent relaxation of the potentials on-the-fly while solving the evolution equations for the updates of all internal variables (e.g., for higher-order laminates). The benefits of this method result in a model that can be designed closer to physical reality, since the exact microstructural changes during each time step can be accounted for at will. Also, the present formulation can be applied to arbitrary deformation histories including, e.g., cyclic loading. The applicability of the general framework has been demonstrated by a model problem in finite-strain crystal plasticity, where the origin and evolution of laminate microstructures is modeled based on the present incremental variational approach. The inclusion of additional flow rules for the update procedure of the hardening variables upon volume fraction changes, for the rotational rearrangement of the laminate, as well as the application to cyclic loading have become possible and yield promising and interesting results.

References

[1] Ball, J. M. Convexity conditions and existence theorems in nonlinear elasticity. Arch. Rat. Mech. Anal. **63**, 337-403 (1977).

[2] Bartels, S., Carstensen, C., Hackl, K., Hoppe, U.: Effective relaxation for microstructure simulation: algorithms and applications. Comp. Meth. Appl. Mech. Eng. **193**,

5143–5175 (2004).

[3] Carstensen, C., Hackl, K., Mielke, A.: Non-convex potentials and microstructures in finite-strain plasticity. Proc. R. Soc. Lond. A **458**, 299–317 (2002).

[4] Carstensen, C., Conti, S., Orlando, A.: Mixed analytical-numerical relaxation in finite single-slip crystal plasticity. Continuum Mech. Thermodyn. **20**, 275–301 (2008).

[5] Conti, S., Dolzmann, G., Klust, C.: Relaxation of a class of variational models in crystal plasticity, Proc. Royal Soc. London **465**, 1735–1742 (2009).

[6] Conti, S., Theil, F. Single-slip elastoplastic microstructures. Arch. Rat. Mech. Anal. **178**, 125–148 (2005).

[7] Dmitrieva, O., Dondl, P., Müller, S., Raabe, D., 2009. Lamination microstructure in shear deformed copper single crystals. Acta Mater. 57, 3439-3449.

[8] Govindjee, S., Mielke, A., Hall, G.J. The free energy of mixing for n-variant martensitic phase transformations using quasi-convex analysis. J. Mech. Phys. Solids **51**, 1–26 (2003).

[9] Hackl, K. Relaxed potentials and evolution equations. In: Gumbsch, P. (Ed.), Third International Conference on Multiscale Materials Modeling. Fraunhofer IRB Verlag (2006).

[10] Hackl, K., Fischer, F. D. On the relation between the principle of maximum dissipation and inelastic evolution given by dissipation potentials. Proc. R. Soc. Lond. A **464**, 117-132 (2007).

[11] Hackl, K., Kochmann, D. M. Relaxed potentials and evolution equations for inelastic microstructures. In: Daya Reddy, B. (ed.), Theoretical, Computational and Modelling Aspects of Inelastic Media, IUTAM Bookseries, Springer, 27–39 (2008).

[12] Hackl, K., Kochmann, D. M. An incremental strategy for modeling laminate microstructures in finite plasticity – energy reduction, laminate orientation and cyclic behavior, in: R. de Borst, E. Ramm (Eds.), Multiscale Methods in Computational Mechanics, vol. 17, Springer, pp. 117-134 (2010).

[13] Kochmann, D. M., Hackl, K. Influence of hardening on the cyclic behavior of laminate microstructures in finite crystal plasticity. Techn. Mech. **30**, 387–400 (2010).

[14] Kochmann, D. M., Hackl, K. The evolution of laminates in finite crystal plasticity: a variational approach. Continuum Mech. Thermodyn., prepublished online (2010), doi: 10.1007/s00161-010-0174-5.

[15] Miehe, C., Lambrecht, E., Gürses, E. Analysis of material instabilities in inelastic solids by incremental energy minimization and relaxation methods: evolving deformation microstructures in finite plasticity. J. Mech. Phys. Solids **52**, 2725–2769 (2004).

[16] Mielke, A. Finite elastoplasticity, Lie groups and geodesics on SL(d). In: Newton, P., Weinstein, A., Holmes, P. (Eds.), Geometry, Dynamics, and Mechanics. Springer, Berlin (2002).

[17] Mielke, A. Deriving new evolution equations for microstructures via relaxation of variational incremental problems. Comp. Meth. Appl. Mech. Eng. **193**, 5095–5127 (2004).

[18] Ortiz, M., Repetto, E. A.: Nonconvex energy minimization and dislocation structures in ductile single crystals. J. Mech. Phys. Solids **47**, 397–462 (1999).

In: Evolution Equations
Editor: Arthur L. Claes

ISBN: 978-1-61209-652-0

Chapter 11

GEOMETRIC METHOD OF RECONSTRUCTING EVOLUTION EQUATIONS FROM EXPERIMENTAL DATA

E. Nikulchev*

Moscow State University of Printing Arts, Moscow, Russia

1. INTRODUCTION

One of the most important tasks in solving problems of control technical systems is the evolutionary behavior formalization of the object. These models are required for the formation control algorithms. The engineers operate only with the basic linear models, leaving the solution of problems of regulation of nonlinear phenomena in industrial controllers. As a result, there are actually functioning control systems that work not only optimal, but, even in manual mode set up control processes.

We can distinguish three classical approaches to modeling of nonlinear phenomena in technical systems. The first is to use stochastic models; it is assumed that the nonlinear oscillations are realization of a random process whose characteristics are to be found. Obviously, in this case can not be rigorously justified assumptions, and assumptions that are not explicitly probabilistic nature, and the observed behavior is possible for a whole class of nonlinear objects. Assessing the adequacy of the construction of stochastic models is to test the hypotheses match the original assumptions. The second approach - the construction of models based on the physical properties of all occurring phenomena in technical systems and the withdrawal of evolution equations. Despite the evidence of such a decision, it is almost impossible for technical systems, since closed and structural complexity of structures defines a knowledge-intensive and cumbersome nature of the physical simulation. Obtaining necessary to control the evolution equations of the models is a separate complex challenge. The third - for the use of an approach to modeling is to choose the kind of mathematical model as an evolution equation and the subsequent identification of parameters or

* Moscow State University of Printing Arts, Prynishnikova St, 2A, Moscow, Russia, 127550, e-mail: nikulchev@mail.ru

nonparametric model identification. Model is considered adequate if the evaluation of the adequacy of a given criterion, calculated as the residual model dependence of the experimental data is within acceptable limit.

Modeling of dynamical system consists of a selection of three components: i) the definition of phase space in terms of restrictions; ii) selection of discrete or continuous time, and iii) the law of evolution — the mapping of any given point in phase space and any time value in one uniquely a certain state of the system. For the discrete time evolution law is sought in the forms:

$$x(t+1) = \psi(x,t,u),$$
$$y = h(x(t)), \qquad t \in \mathbb{Z},$$

for continuous time:

$$\dot{x} = f(x,u,t),$$
$$y = \tilde{h}(x(t)), \qquad t \in \mathbb{R}^1,$$

where x — state vector; y — measured processes (hereinafter, if it is not defined features of the problem, we assume that monitoring is available, all the states); t — time; f, ψ, h, $\tilde{h}$ — continuous and smooth vector functions.

The simulation method presented in this chapter is to construct, based on studies of experimental data, equations, whose solution has an adequate dynamic behavior modeling in this qualitative dynamic geometry of the process.

The first work on the reconstruction of the strange attractor from the time series has been publishing the results on hydrodynamics [1]. The article shows that you can get a satisfactory picture of the strange attractor of the geometric dimensions of a small, if the variables x, appearing in the equations of the dynamical system $dx/dt = F(x)$, use the m-dimensional vectors, derived from the elements of time series on the same principle, that in the problems of autoregression. That same year, F. Takens reported on his theorem, which was published a year later [2]. That it is the basis of all algorithms for time series analysis methods of nonlinear dynamics. The problem of determining the form of a dynamical system from its one-dimensional realization belongs to a class of incorrect problems. Unlike the problem of analyzing this issue is ambiguous, since there are infinitely many dynamical systems of various kinds can play the existing signal with a given degree of accuracy.

The method of the global reconstruction of a dynamic system of equations for its one-dimensional realization was proposed in [3, 4]. The algorithm is as follows. One-dimensional realization of the process in a system, which is considered a "black box" recovered phase portrait on the Takens theorem, topologically equivalent to the attractor of the original system. According to a priori given equations, is the method of least squares a set of unknown coefficients

Now there is considerable amount of publications, developing and constantly improving the proposed the method [5–11]. For example, in R. Brown and others [12] to reconstruction dynamic equations on the experimental time series with a broadband continuous spectrum use additional information about the dynamic and statistical properties of the original system contained in the implementation. In obtaining equation takes into account the values of Lyapunov exponents and the probability density, calculated from the original time series. However, the resulting evolution equations have a very cumbersome, inconvenient to use. In [13] used the hidden variables to write model equation. In [12] describes a method for synchronizing the model with the original data. In a number of O.L. Anosov proposed reconstruction algorithm scalar differential equations for systems with delay.

However, the feature of many studies is that the proposed methods are illustrated with examples of simple low-dimensional model systems when we know in advance what should be the result of global reconstruction. It does not show substantial benefits given by the modifications of the basic method [3]. Described in the publications of the algorithms are tested on a number of known model systems with small dimension and a simple form right sides. The efficiency of the method is demonstrated by the example of time series generated by the real "black boxes".

The challenge is the need to work with noisy data in the processing of experimental time series. On the one hand, more desirable is the use of sequential differentiation to restore the phase trajectory, because it can get a model that contains, in general, approximately *n* times smaller than the coefficients of the various non-linear than when using the method of delays. But differentiation will inevitably lead to increased noise components of high order. Without pre-filtering the time dependence of the second derivative can be noise-like process. In addition, methods of attachment are obvious flaws in the analysis significantly heterogeneous implementations, i.e., signals in which areas with fast motion alternating with areas of slow motions.

Arbitrary choice of nonlinearities, as a rule, does not allow for a successful reconstruction of the dynamic equations for real systems. In particular, in [14] indicated the presence of three typical cases:

1. Reconstruction of locally describe the phase trajectory of the original system. In this case, the reconstructed model is unstable in the sense that the solution of these equations reproduces the signal under investigation only in a short period of time.
2. There is poor local predictability of the phase trajectory, but there is visual similarity of the phase portraits. Reconstruction of solution the stable in the sense of Poisson. In this case, the attractor of the reconstructed model has metric characteristics similar to those of the original attractor.
3. There is a good local predictability of the phase trajectory of any point in time values exceeding the characteristic correlation time. The phase portrait reconstructed model is identical to the original, and the system is Poisson stable.

2. Models and Assumptions

Choose a system of assumptions to the formal description of nonlinear models, reconstructed from the experimental data.

Assume that the data obtained as a result of the operation of control systems are output processes of nonlinear affine dynamical system defined by the transformation $D \times U \to T_x X$:

$$\dot{x} = f\big(x(t),t\big) + h\big(x(t),t\big)u(t), \tag{2.1}$$

where the x — vector of system states from $\mathbb{R}^n$; $u(t)$ — p-dimensional control vector from the set of admissible controls $U \subseteq \mathbb{R}^p$; manifold X, usually identified with $\mathbb{R}^n$ or $\mathbb{R}^{n+1}$, in the latter case, the manifold includes an independent variable t (time); T_xX — the tangent space to X at x, defined by the admissible controls; $f(\cdot) = (f_1,\ \ldots.,\ f_n)$, $h(\cdot) = (h_1,\ \ldots.,\ h_n)$ — $\mathbb{C}^r$-smooth ($r \geq 1$) vector function defined in some domain $\mathrm{D} \subseteq \mathbb{R}^n$, which is regarded as the phase space (limited, unlimited or identical with Euclidean space $\mathbb{R}^n$.)

On the right side of the control system restrictions:

1. $x \mapsto f(x)$ — a smooth vector field on X for any fixed $u \in U$;
2. $(x,u) \mapsto f(x) + h(u)$ — continuous mapping $x \in X$, $u \in U$;
3. $(x,u) \mapsto \partial\big[f(x) + h(u)\big] / \partial x$ — continuous mapping in any local coordinates on X at $x \in X$, $u \in U$.
4. Admissible controls — measurable locally bounded mapping $u : t \mapsto u(t) \in U$.

Let depending on the choice of control signals, the system (2.1) can be converted to an equivalent autonomous, i.e., in the presence of control parameterizes the autonomous system. Thus, the helper object is an autonomous system of ordinary differential equations, written in the form

$$\dot{x} = f(x). \tag{2.2}$$

Differentiable mapping $\varphi : \tau \to D$, where τ — the interval on the axis of t, called a solution $x = \varphi(t)$ (2.1) if

$$\dot{\varphi}(t) = f(\varphi(t)) \text{ for all } t \in \tau. \tag{2.3}$$

By assumption, the conditions of Cauchy's theorem are satisfied, therefore, for any values $x_0 \in D$ and $t_0 \in \mathbb{R}^1$ there exists a unique solution φ, satisfying the initial condition

$$x_0 = \varphi(t_0). \tag{2.4}$$

Solution is defined on some interval (t_+, t_-),contains the value $t = t_0$. In general, the boundary points t_+ and t_- can take both finite and infinite values. Solutions of (2.2) have the following properties:

1. If $x = \varphi\ (t)$ — solution of system (2.2), it is obvious that $x. = \varphi\ (t + C)$ is also a solution defined on the interval $(t_+ - C, t_- - C)$.
2. Solutions $x = \varphi\ (t)$ and $x = \varphi\ (t + C)$ can be regarded as solutions corresponding to one initial point x_0, but different initial time t_0.
3. Solution satisfying (2.4) can be written as $x = \varphi\ (t - t_0, x_0)$, where $\varphi(0, x_0) = x_0$.
4. If $x_1 = \varphi\ (t_1 - t_0, x_0)$, then $\varphi\ (t - t_0, x_0) = \varphi\ (t - t_1, x_1)$. Denoting $t_1 - t_0$ a new t_1, at $t - t_2$ — by t_2, we obtain the so-called group property decisions:

$$\varphi(t_2,\ \varphi(t_1, x_0)) = \varphi(t_1 + t_2, x_0).$$

It is known that the solution $x_1 = \varphi\ (t_1 - t_0,\ x_0)$ of problem (2.3) for $\mathbb{C}^r$-smooth system (2.1) is smooth with respect to time and initial data x_0. The first derivative $\eta(t - t_0,\ x_0) \equiv \dfrac{\partial\varphi}{\partial x_0}$ — satisfies the so-called variational equation $\dot{\eta} = f\,'(\varphi(t - t_0, x_0))\eta$ with initial condition: $\eta(0, x_0) = I$ (I - identity matrix). Variational equation — a linear non-autonomous system, obtained by formal differentiation of (2.2).

For a discrete dynamical system

$$x(k+1) = f(x(k), k) + h\big(x(k), k\big)u(k),$$

a sequence $\{x(k)\}_{k=-\infty}^{+\infty}$ is called a trajectory of x_0.

A set A is called invariant under the dynamical system if $A = \varphi(t,\ A)$, for any t. In this expression denotes $\varphi(t,\ A)$ the set $\bigcup\limits_{\mathbf{x}\in A}\varphi(t,\ A)$. This definition implies that if $x \in A$ the trajectory $\varphi(t,\ A)$ lies in the set A.

By definition, the attractor Atr — this is a closed invariant set with a neighborhood (the absorbing region) $U\,(Atr)$, that the trajectory of an arbitrary point x, belonging to $U\,(Atr)$, satisfies

$$\rho(\varphi(t, x), Atr) \to 0 \text{ in } t \to +\infty,$$

where $\rho(x, Atr) = \inf\limits_{x_0 \in Atr} \|\, x - x_0 \,\|$.

Examples of attractors are equilibrium states, stable periodic orbits and stable invariant torus containing quasiperiodic trajectory. For strange attractors of chaotic systems that are invariant closed set composed of only unstable trajectories, the condition quasiminimal.

3. Group Properties for the Construction of Equivalent Mapping

Consider the research questions of the global structure of orbits of dynamical systems, which do not depend on the choice of coordinate system, i.e., admit a symmetry transformation. From a global point of view, the change of coordinates is a diffeomorphism (in the case of a smooth structure) or a homeomorphism (in the topological situation) phase spaces. Thus, we can introduce a natural equivalence relation between the dynamical systems associated with different classes of coordinate changes, and interpret the problem of describing the structure of orbits as a classification problem of dynamical systems up to these equivalence relations.

Given the equivalence of the trajectories on the torus [15], we can show that every periodic point of the map, whose spectrum does not contain a unit, defines certain $\mathbb{C}^1$-modules. Since these periodic orbits are separated from each other, their spectra can be outraged, regardless, at least, for any finite set of points. The modules from a variety of periodic orbits a independent.

On the other hand, at least in some cases, the local spectrum is a complete invariant with respect to a smooth conjugacy. Different approaches to the problem of local smooth conjugacy shown in [16].

For the global structure of orbits there are ways to build structures of independent modules associated with periodic orbits. In the case of infinitely many periodic orbits, like expanding mapping and hyperbolic automorphism of the torus, there are infinitely many invariants of the local $\mathbb{C}^1$-equivalence. For these two cases, which represent the simplest examples of hyperbolic systems, the spectra of periodic orbits form a complete system of invariants for $\mathbb{C}^1$- and $\mathbb{C}^\infty$-equivalences in the vicinity of these maps, respectively. Satisfactory description of the set of possible values for the eigen values of periodic points remains an open problem.

A variety of modules offer a significant, although not complete information about the smooth equivalence in the vicinity of the rotation. Every real analytic compression stores a unique way an affine structure. For the discrete map ψ, two structures identified near the ends of the segment are found in the middle. The transition functions between the two structures in any of the fundamental domain $[a, \psi(a)]$ generate infinite moduli space ψ [16]. In practice, this can be interpreted to mean that there are changes of coordinates that will ψ affine mapping from $[0, \psi^{-1}(0)]$ in the $[0, a]$ and $[\psi(a), 1]$ in the $[\psi^2(a), 1]$. Coordinates are uniquely determined up to two factors, one at each end. Then the map $[\psi(a), 1] \to [\psi^2(a), 1]$ can be normalized so that $\psi_a : [0,1] \to [0,1]$:

$$\psi_a = \frac{\psi(a + t(\psi(a) - a)) - \psi(a)}{\psi^2(a) - \psi(a)},$$

which can be extended to the whole real line by the formula

$$\psi_a(T+k) = \psi_a(T) + k, \quad k \in \mathbb{Z}.$$

Thus, the two diffeomorphism ψ_1 and ψ_2, for which $\psi_1(0) = \psi_2(0) = 0$ are equivalent if $\psi_2(T) = \psi_1(T+s) - \psi_1(s)$ for some s in [0, 1]. Constructed above the map transition depends on the factors that determine the linearization ψ in the vicinity of the endpoints, as well as the choice of the reference point a as well. It is obvious that the change does not alter the linearization map transition if a amended accordingly. Changing the reference point leads to the substitution mapping the transition to the equivalent.

The notion of smooth equivalence in accordance with [16] can be easily moved in case of continuous time. Flow equivalence is a contingency flows as a differentiable actions of a group $\mathbb{R}$ of real numbers. The structure of the orbits of the flow, unlike the case of discrete-time systems, has two distinct aspects: (1) The relative behavior of points on different orbits and (2) evolution of the initial condition along the orbit over time. There is a natural way to change the thread, keeping the first aspect of the structure of its orbits, and it does not change its orbit.

For the two streams φ^t and ε^t will define the time change by using the vector fields of infinitesimal generator of the symmetry group:

$$\xi = \left.\frac{d\varphi^t}{dt}\right|_{t=0} \quad \text{and} \quad \eta = \left.\frac{d\varepsilon^t}{dt}\right|_{t=0}.$$

From the uniqueness of solutions of differential equations, it follows that the zeros of the vector fields are fixed points of the corresponding flow. Thus, we conclude that $\xi(x) = 0$ if and only if $\eta(x) = 0$. In addition, if x is not a fixed point, then the tangent vector to the curve $\{\varphi^t(x)\}$ and $\{\varepsilon^t(x)\}$, does not vanish and have the same direction.

It is natural to try to describe all the changes of time given flow modulo trivial substitutions. This problem is essentially equivalent to the problem of describing the space of all sufficiently smooth positive functions up to adding features that are derivatives of other smooth functions in the direction of flow.

It makes sense when determining equivalence to consider the conservation of structural stability for flows. One way of determining is based on the equivalence of all the perturbations [16]. Not being fully degenerate, this requirement is rarely performed, for example, in the presence of periodic orbits of periods are modules in this sense. We assume that the local topological equivalence should preserve the structural stability, thus transforming the homeomorphism can be quite close to the identity for small perturbations.

For all the above concepts in this section of the compact phase spaces is immaterial. In addition, a natural way to modify these definitions for the cases where for some points of the dynamical system is defined only for a finite time interval. Such a generalization leads to the concepts of local and semilocal structural stability.

For any contraction phase space can not have a smooth structure, so that these concepts are not directly applicable. However, the contraction mapping of a small disc in the Euclidean space, structural stability, as well as a hyperbolic linear map in the vicinity of the fixed point.

Symmetry rotation is not structurally stable. Since the topological conjugacy preserves the periodic orbits, and the transformation of rotation by an irrational angle can not be associated with the transformation of rotation, for which the corresponding number is rational. But since the rational numbers, and the set of irrational numbers are dense, then some arbitrarily small perturbations of the turn to rational angle you can find the rotation by an irrational angle, and vice versa.

4. Geometric Method for the Reduction of Systems

4.1. Theorems on Central Invariant Manifold

According to [15] on centrally-stable manifold, the system around an equilibrium point O takes the form

$$\begin{aligned} \dot{y} &= By + f_1(x, y, z), \\ \dot{z} &= Cz + f_2(x, y, z), \\ \dot{x} &= Ax + \psi_0(x, y, z), \end{aligned} \tag{4.1}$$

where $x \in \mathbb{R}^m$, $y \in \mathbb{R}^k$, $z \in \mathbb{R}^{n-m-k}$ $(j = \overline{1,m})$; spectr $A = \{\lambda_{m+1}, ..., \lambda_k\}$, $\mathrm{Re}\,\lambda_j < 0$, $(j = \overline{m+1,k})$; spectr $C = \{\lambda_{k+1}, ..., \lambda_n\}$, $\mathrm{Re}\,\lambda_j > 0$, $(j = \overline{k+1,n})$; $\mathbb{C}^r$-functions f_1, f_2 and ψ_0 together with its first derivatives vanish at the origin.

In this case, the right side of the system may depend on the control u either continuously (in this case, smooth manifolds, discussed below, depend continuously on u), or smooth. In the latter case, u is included in the number of "central" variables x and, thus, further investigated the variety and the foliation will have the smoothness of u, equal to the smoothness of x.

Theorem 1. In a small neighborhood of the equilibrium state there exists a $(m + \mathrm{k})$-dimensional center-stable invariant manifold. Central-stable manifold is not uniquely defined, but for any two manifolds W_1^{sC} and W_2^{sC} functions ϕ_1^{sC} and ϕ_2^{sC} the class $\mathbb{C}^r$ that contains the point O and concerns at this point subspace $\{z = 0\}$, define the same the same symmetries group of the point O.

Proof. Without loss of generality, we consider the second-order system. Let the system defined on a manifold gives rise to a one-parameter transformation group generated by the operator

$$A = \xi_1(x)\frac{\partial}{\partial x_1} + \xi_2(x)\frac{\partial}{\partial x_2},$$

with infinitesimal operator

$$X = \eta_1(x)\frac{\partial}{\partial x_1} + \xi_2(x)\frac{\partial}{\partial x_2}.$$

We need to show that the operator

$$\tilde{A} = \tilde{\xi}_1(x',\tau)\frac{\partial}{\partial x_1} + \tilde{\xi}_2(x',\tau)\frac{\partial}{\partial x_2}.$$

system defined by W_2^{sC} an equivalent (τ - group option).

We write the transformation of the group as a Li series (the operator exponential):

$$x'_1 = e^{\tau X} x_1,$$
$$x'_2 = e^{\tau X} x_2,$$
$$x_1 = e^{\tau X} x'_1,$$
$$x_2 = e^{\tau X} x'_2.$$

We represent $\tilde{A}$ in the previous coordinates. For this we compute:

$$A = \left(\tilde{A}x_1\right)\Big|_{\dot{x}=x'(x)}\frac{\partial}{\partial x_1} + \left(\tilde{A}x_2\right)\Big|_{\dot{x}=x'(x)}\frac{\partial}{\partial x_2} = \left(\tilde{A}e^{-\tau X}x'_1\right)\Big|_{\dot{x}=x'(x)}\frac{\partial}{\partial x_1} + \left(\tilde{A}e^{-\tau X}x'_2\right)\Big|_{\dot{x}=x'(x)}\frac{\partial}{\partial x_2},$$

it follows that

$$\left(\tilde{A}e^{-\tau X}x'_1\right)\Big|_{\dot{x}=x'(x)} = \xi_1(x),$$
$$\left(\tilde{A}e^{-\tau X}x'_2\right)\Big|_{\dot{x}=x'(x)} = \xi_2(x).$$

Since $\xi_{1,2}$ do not depend on τ, as determined by the operator X, we have

$$\frac{d}{d\tau}\left(\tilde{A}e^{-\tau X}x'_1\right) = 0,$$

so

$$\frac{\partial \tilde{A}}{\partial \tau} e^{-\tau X} x'_1 - \tilde{A} X e^{-\tau X} x'_1 + X \tilde{A} e^{-\tau X} x'_1 = 0 .$$

A similar formula holds for the second coordinate, which together form a differential equation:

$$\frac{\partial \tilde{A}}{\partial \tau} = \tilde{A} X - X \tilde{A} = [\tilde{A}, X]$$

with initial condition

$$\tilde{A}(x', \tau)\Big|_{\tau=0} = A(x') .$$

Solution of the resulting Cauchy problem can be obtained by expanding the operator $\tilde{A}(x', \tau)$ in a Taylor series in powers of τ:

$$\tilde{A}(x', \tau) = \tilde{A}(x') + \tau \frac{\partial \tilde{A}}{\partial \tau}\Big|_{\tau=0} + \frac{\tau^2}{2!} \frac{\partial^2 \tilde{A}}{\partial \tau^2}\Big|_{\tau=0} + \dots ,$$

thus, we have:

$$\frac{\partial \tilde{A}}{\partial \tau}\Big|_{\tau=0} = [A, X].$$

Similarly,

$$\frac{\partial^2 \tilde{A}}{\partial \tau^2}\Big|_{\tau=0} = \big[[A, X], X\big].$$

Finally, a number will look like in the Hausdorff form:

$$\tilde{A} = A + \tau[A, X] + \frac{\tau^2}{2!}\big[[A, X], X\big], \dots .$$

By Theorem 5.6. from [15] function ϕ_1^{sC} and ϕ_2^{sC} determine the same expansion in Taylor series, which suggests fields commutation:

$$[A,X]=0,$$

i.e.

$$\tilde{A}=A.$$

Thus, the theorem is proved.

At the reference time $t\to -t$ of A, B and C go into respectively A, B and C. Thus, part of the spectrum of characteristic exponents corresponding to the variable z, is now left of the imaginary axis, and part of the spectrum, corresponding to the variable y — the right of it. To the system obtained from system (4.1) by time reversal, we can again apply the theorem about the Central stable manifold and get the following theorem on center-unstable manifold.

Theorem 2. In a small neighborhood of the equilibrium state O there exists an $(n-\mathrm{k})$-dimensional invariant manifold is a $\mathbb{C}^r$-smooth. Center unstable manifold includes all trajectories that remain in a small neighborhood of O for all negative values of time. For any two W_1^{usC} and W_2^{usC} functions ϕ_1^{usC} and ϕ_2^{usC}, the class $\mathbb{C}^r$ containing the point O and tangent at this point subspace $\{y=0\}$ define the same symmetries group.

Intersection of center-stable and unstable manifolds of a central $\mathbb{C}^r$-smooth m-dimensional invariant center manifold, defined by the equation of the form $(y, z)=\phi^C(x)$. Function ϕ^C with all derivatives, in particular, the Taylor expansion of functions ϕ^C at O is uniquely determined by the system.

Straightening of the central stable and center-unstable manifolds, as well as rectification of the strong stable and strong unstable invariant foliations on these manifolds leads to the good fact.

Theorem 3 [15]. With a $\mathbb{C}^{r-1}$-smooth transformation of the system (4.1) can locally be reduced to

$$\dot{y}=\left(A+F_1(x,y,z)\right)y,$$
$$\dot{z}=\left(C+F_2(x,y,z)\right)z,$$
$$\dot{x}=Bx+\Psi_0(x)+\Psi_1(x,y,z)y+\Psi_2(x,y,z)z,$$

where Ψ_0— is a $\mathbb{C}^r$-smooth function, which together with its first derivative vanishes at $x=0$; $F_{1,2}$ are functions that vanish at the origin: $\Psi_{1,2}\in\mathbb{C}^{r-1}$; $\Psi_1(x,y,0)=\Psi_2(x,0,z)=0$.

Here, the local center-unstable manifold given by the equation $\{y=0\}$, the local center-stable manifold — $\{z=0\}$, and a local center manifold — $\{y=0,\ z=0\}$. Strong stable foliation consists of surfaces $\{x=\text{const},\ z=0\}$, and the layers are strongly unstable foliation of the form $\{x=\text{const},\ y=0\}$.

A similar theory is constructed for discrete systems.

For systems that admit the group of symmetries and construct models of systems, reduced to a central invariant manifold results can be summarized as follows [17].

In the local region of the qualitative dynamic behavior of systems typologically equivalent system, reduced to the central manifold

$$\dot{x} = Ax(t) + \Psi_0(x), \tag{4.2}$$

where $\Psi_0(x)$— is a $\mathbb{C}^r$-smooth function ($\Psi_0(x_0) = \Psi_0'(x_0) = 0,\ x_0 = 0$), whose structure is determined by the symmetry transformation.

For a discrete system, the local center manifold is determined by the system:

$$x(t+1) = Ax(t) + \vartheta_0(x), \tag{4.3}$$

where $\vartheta_0(x)$— is a $\mathbb{C}^r$-smooth function ($\vartheta_0(x_0) = \vartheta_0'(x_0) = 0,\ x_0 = 0$) defined on the basis of symmetry groups constructed on the reconstructed attractor.

For the reconstruction of a nonlinear system in the form (4.2) (4.3) proposed the allocation of local regions of phase trajectories $\varphi_1(x)$ and $\varphi_2(x)$ that are close to periodic, and the construction of finite-transformations transform one area to another. That is, the construction of the symmetry group of phase trajectories, which is characterized by the transformation of graphs:

$$\text{graph}\ \{\varphi_1(x)\} \to \text{graph}\ \{\varphi_2(x)\}.$$

The resulting transformations determine the structure of the desired evolution equations.

5. Algorithms and Applications

The result of reconstruction algorithms [1] is a set of points belonging to the attractor of the system. The task is an automatic search for symmetries of the local sections of the phase trajectories. We seek the symmetry of translation, rotation, stretching and compression.

You must select the initial sequence collection sites so that when reducing them to a single scale, position and angle of rotation, they would be most similar to each other, as well as numerical indicators of transformations taking one piece to another without breaking of symmetry, to give a numerical estimate degree of symmetry breaking. The Fig. 5.1 shows a schematic illustration of the problem.

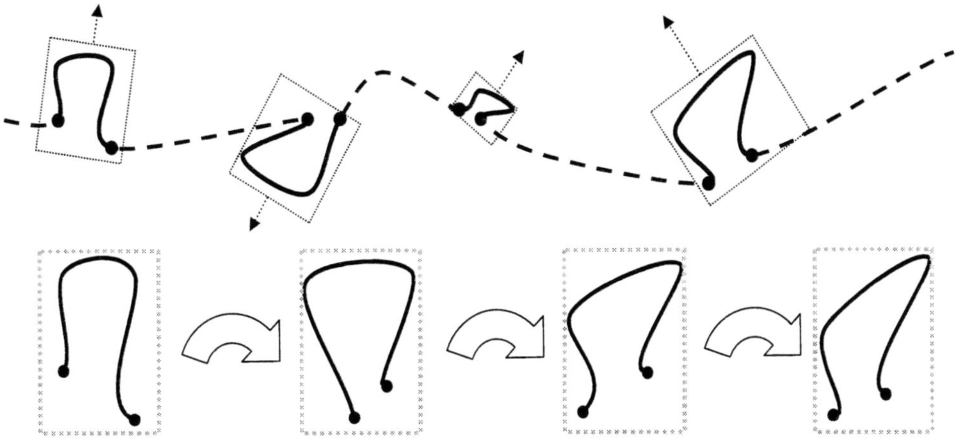

Figure 5.1. Schematic illustration of the problem..

The task of allocating a set of similar fragments in the circuit is defined as follows:

Input: sequence of *n* points $\langle v_1, v_2, \text{L } v_n \rangle$, the function $Dist(V_1, V_2)$, which gives an assessment of similarity between any two fragments of the original sequence $V_i = \langle v_j, v_{j+1}, \text{L } v_{j+p} \rangle$.

Output: a set of disjoint fragments of the original sequence $S = (V_1, V_2, \text{L } V_k)$, such that for any other set of fragments $S' = (V_1, V_2, \text{L } V_h)$ of the condition:

$$\forall S' \neq S: \max_{V_i', V_j' \in S'} \left(\frac{1}{1 + Dist(V_i', V_j')} \right) \cdot \sum_{i=1}^{k} |V_i'| \cdot \frac{1}{m} \geq \max_{V_i, V_j \in S} \left(\frac{1}{1 + Dist(V_i, V_j)} \right) \cdot \sum_{i=1}^{h} |V_i| \cdot \frac{1}{m}$$

Number of different fragments, which can be identified in the original sequence:

$$m = \frac{1}{2} n(n-1)$$

Number of arbitrary sets of fragments, determined by the size of the set of all subsets of the set $\langle V_1, V_2, \Lambda\ V_m \rangle$:

$$u = \text{P}(m) = 2^m$$

Based on this challenge $T(n) = \Theta(2^{n^2})$.

Marking. In the simplest case for the selection of fragments of trajectories satisfying the decision to consider the fragments located between all possible pairs of contour points. In most problems, this approach is not possible because of limited computer resources. So it makes sense to consider as a start / end fragment only some points of the contour, for example, extremes of functions of each of the coordinates and evenly distributed along the length of the markers on the slowly varying parts (Fig. 5.2). This approach provides a sufficient rational partition of the circuit, although it suffers from some redundancy.

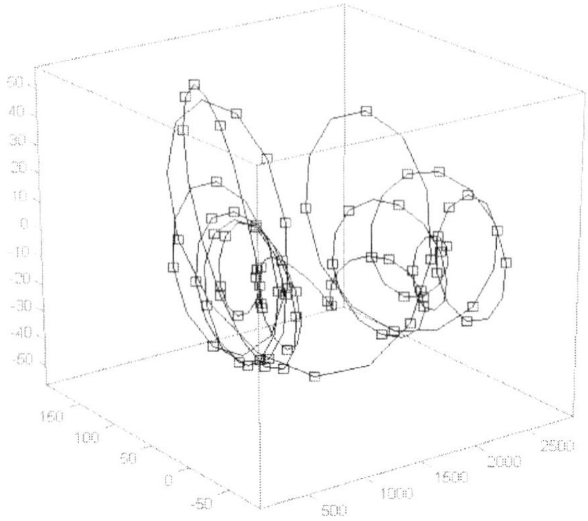

Figure 5.2. Example of marking.

Partitioning algorithm can be anything important to exclude monotonous plots and identify the markers characteristic sections of the circuit. Extracting fragments. To isolate fragments of contour to study the simplest case, you can use brute force pairs of markers. In this case, the number of fragments needed to be addressed:

$$F_{\text{полн}} = \frac{n^2 - n}{2},$$

where n — number of markers on the contour.

It makes sense to limit the length of a fragment in the original contour points as the top and bottom as well as fragments, the length of the approaching to the original contour, are not of interest to identify patterns of distribution symmetry, and too small fragments do not reflect the specific behavior of the system. In addition, the introduction of restrictions on the length of the above - increase the number of fragments for consideration is linear, as opposed to quadratic, with the full brute force.

Selected fragments for the possibility of a comparison between a need to interpolate in a way that they all described the same fixed number of points (Fig. 5.3).

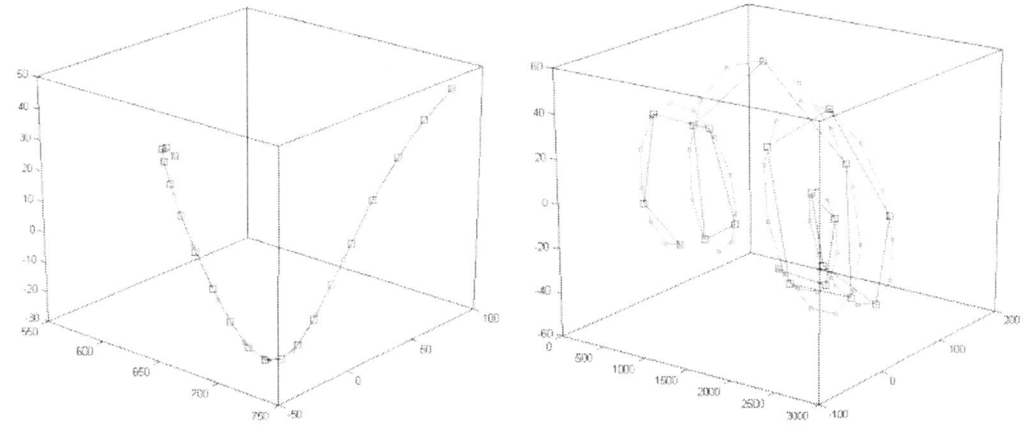

Figure 5.3. Interpolation of selected fragments.

Normalization. After reducing to one the number of points each of the fragments should be subjected to the normalization procedure. The aim of the procedure is to convert the fragment into a descriptor — the image that is invariant with respect to the transport, rotation, and scaling the original fragment, and obtaining the numerical parameters of this transformation (Fig. 5.4).

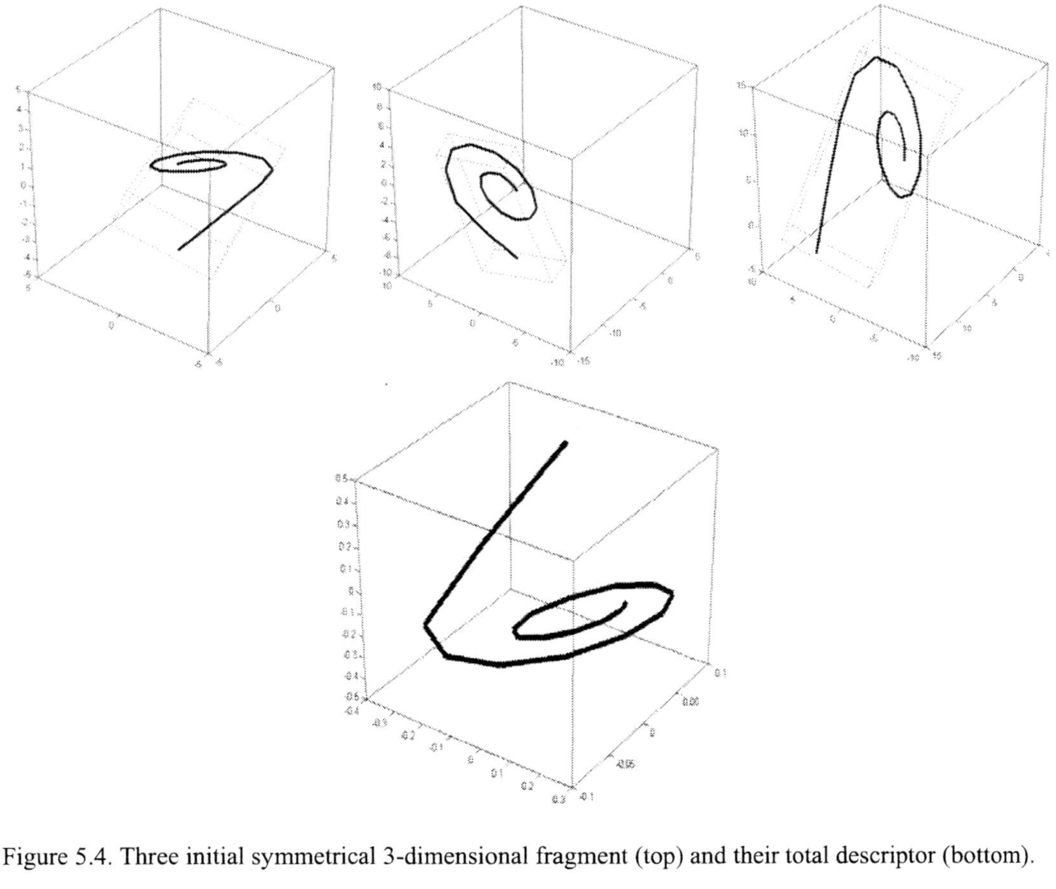

Figure 5.4. Three initial symmetrical 3-dimensional fragment (top) and their total descriptor (bottom).

The proposed normalization procedure is a sequence of the following:

Isolation in a fragment of the so-called "axis" - pairs of points, the distance between the maximum in the current space projection.

Rotation of the fragment so that its axis were parallel to the coordinate axes.

The construction restricting fragment of the n-dimensional parallelepiped and the shift of the fragment so that the center of the parallelepiped coincides with the center coordinates.

Scaling the fragment so that the length of its main axis was equal to 1.

Fragment of an n-dimensional the contour F consisting of m real points, represented as a matrix $m \times n$:

$$F_b = [X_1 \; X_2 \cdots X_n],$$

where $X_i = \left[x_{i,1}, x_{i,2}, \ldots, x_{i,m}\right]^T$.

For a compact description of the transformation fragment as a single matrix, we give it to homogeneous coordinates using matrix transformations, obtain:

$$F_c = [X_1 \; X_2 \cdots X_n \; 1]$$

In the normalization process uses three types of transformations: translation, scaling and rotation.

Translation transformation in a matrix form is:

$$M_{\text{sh}} = \begin{bmatrix} 1 & 0 & \cdots & 0 & 0 \\ 0 & 1 & & 0 & 0 \\ \vdots & & \ddots & 0 & \vdots \\ 0 & 0 & & 1 & 0 \\ -g_{\text{sh},1} & -g_{\text{sh},2} & \cdots & -g_{\text{sh},n} & 1 \end{bmatrix}, \tag{5.1}$$

where $g_{\text{sh},i} (i = 1, .., n)$ — the shift along the axis X_i.

Transformation of the scaling down and scaling up

$$M_{\text{sc}} = \begin{bmatrix} g_{\text{sc}} & 0 & \cdots & 0 & 0 \\ 0 & g_{\text{sc}} & & 0 & 0 \\ \vdots & & \ddots & 0 & \vdots \\ 0 & 0 & & g_{\text{sc}} & 0 \\ 0 & 0 & \cdots & 0 & 1 \end{bmatrix}, \tag{5.2}$$

where g_{sc} — a scaling factor.

Rotation transformation is:

$$M_{\mathrm{rt}} = \begin{bmatrix} 1 & 0 & \mathrm{L} & \mathrm{L} & 0 & 0 \\ 0 & \mathrm{O} & & & 0 & 0 \\ \mathrm{M} & & \cos(g_{\mathrm{rt}}) & \sin(g_{\mathrm{rt}}) & & \mathrm{M} \\ \mathrm{M} & & -\sin(g_{\mathrm{rt}}) & \cos(g_{\mathrm{rt}}) & & \mathrm{M} \\ 0 & 0 & & & \mathrm{O} & 0 \\ 0 & 0 & \mathrm{L} & \mathrm{L} & 0 & 1 \end{bmatrix}, \tag{5.3}$$

where g_{rt} — angle, a significant elements of the matrix are located in cells (i, i), (i, j), (j, i) и (j, j).

Normalization Algorithm is as Follows

[1] Initialize the total transformation matrix of normalization:

$$M_{\mathrm{norm}} = \begin{bmatrix} 1 & 0 & \mathrm{L} & 0 \\ 0 & 1 & & \mathrm{M} \\ \mathrm{M} & & \mathrm{O} & 0 \\ 0 & \mathrm{L} & 0 & 1 \end{bmatrix}.$$

[2] Setting the order of normalization $k = 1$.

[3] 3. $F^k = F_{\mathrm{r}}$.

[4] Finding the axis of the fragment A^k — the largest segment connecting two points of the loop F^k in space. With a simple enumeration pairs of points fragment of F^k sought a pair with the greatest distance:

$$A^k = \begin{bmatrix} x_{i,1} & x_{i,2} & \mathrm{L} & x_{i,n} & 1 \\ x_{i,1} & x_{j,2} & \mathrm{L} & x_{j,n} & 1 \end{bmatrix} : \max \sqrt{\sum_{d=k}^{n} \left(x_{d,j} - x_{d,i} \right)^2}$$

[5] Transfer A^k so that it coincided with the first point of origin. Conversion is performed by multiplying with the matrix translation $M_{\mathrm{sh},k}$ form (5.1) with coefficients $g_{\mathrm{sh},i} = A^k(1,i)$:

$$A^k_{\mathrm{sh}} = A^k \times M_{\mathrm{sh},k}.$$

[6] Complement the overall normalization of the current transformation matrix:

$$M_{\mathrm{norm}} = M_{\mathrm{norm}} \times M_{\mathrm{sh},k}.$$

[7] Rotation A^k_{sh} so that it coincided with the direction of the axis X_k.

a. Initialize rotation matrix:

$$M_{\mathrm{rt},k} = \begin{bmatrix} 1 & 0 & \cdots & 0 \\ 0 & 1 & & \vdots \\ \vdots & & \ddots & 0 \\ 0 & \cdots & 0 & 1 \end{bmatrix}$$

b. Setting the plane of rotation $p = n$.

c. $A^k_{\mathrm{rt},p} = A^k_{\mathrm{sh}}$

d. Calculating the angle between the projection $A^k_{\mathrm{rt},p}$ on the plane OX_1X_p and the axis X_1 of the formula:

$$\varphi_{k,p} = \begin{cases} \arctan\left(\dfrac{A^k(2,k)}{A^k(2,p)}\right), & A^k(2,k) > 0, A^k(2,p) > 0; \\ -\arctan\left(\dfrac{A^k(2,k)}{A^k(2,p)}\right) + \dfrac{\pi}{2}, & A^k(2,k) \le 0, A^k(2,p) > 0; \\ \arctan\left(\dfrac{A^k(2,k)}{A^k(2,p)}\right) + \pi, & A^k(2,k) \ge 0, A^k(2,p) \le 0; \\ -\arctan\left(\dfrac{A^k(2,k)}{A^k(2,p)}\right) + \dfrac{3\pi}{2}, & A^k(2,k) > 0, A^k(2,p) \le 0. \end{cases}$$

e. Rotation $A^k_{\mathrm{rt},p}$ with a matrix form (5.3) by an angle $g_{\mathrm{rt}} = -\varphi_{k,p}$:

$$A^k_{\mathrm{rt},p-1} = A^k_{\mathrm{rt},p} \times M_{\mathrm{rt},p}.$$

f. Complement the current rotation matrix transformation:

$$M_{\mathrm{rt},k} = M_{\mathrm{rt},k} \times M_{\mathrm{rt},p}.$$

g. Setting the next plane of rotation $p = p - 1$.
h. If $p \geq 1$, then go to step {d}.

8 Complement the overall normalization of the current transformation matrix:

$$M_{\mathrm{norm}} = M_{\mathrm{norm}} \times M_{\mathrm{rt},k}.$$

9 The transformation of the fragment by the current matrix of normalization:

$$F^{k+1} = F^k \times M_{\mathrm{norm}}$$

10 Installation of the next order of normalization $k = k + 1$.
11 If $k < n$, then go to step {4}.
12 Formation of the bounding parallelepiped fragment B is described by the matrix $2 \times n$:

$$B = \begin{bmatrix} b_{1,1} & b_{1,2} & \mathrm{L} & b_{1,n} \\ b_{2,1} & b_{2,2} & \mathrm{L} & b_{2,n} \end{bmatrix},$$

$$b_{1,i} = \min(X_i),\ b_{2,i} = \max(X_i).$$

13 Transfer fragment so that the center of the bounding parallelepiped it coincided with the origin. Conversion is performed by multiplying with the translation matrix M_{sh} form (5.3) with coefficients $g_{sh,i} = -(B(1,i) + B(2,i))/2$:

$$F_{\mathrm{cn}} = F^k \times M_{\mathrm{sc}}.$$

14 Scaling of the fragment in a way that restricts its parallelepiped fit into the unit cube. This conversion scaling matrix M_{sc} form (5.2) with a coefficient $g_{\mathrm{sc}} = B(2,i) - B(1,i)$:

$$F_{\text{norm}} = F_{\text{cn}} \times M_{\text{sc}}$$

15 Complement the overall normalization of the current transformation matrix:
$M_{\text{norm}} = M_{\text{norm}} \times M_{\text{cn}} \times M_{\text{sc}}$.

The result of the normalization procedure is the descriptor the fragment F_{norm}, the matrix M_{norm} is converted into a fragment of the descriptor and a set of indicators and all the intermediate transformation matrices: $n-1$ matrices $M_{\text{sh},k}$, $(n^2-n)/2$ matrices $M_{\text{rt},p}$ and one matrix M_{sc}.

In what is easily obtain the transformation matrix of one fragment to another:

$$M_{\text{trans}AB} = M_{\text{norm}A} \times M_{\text{norm}B}^{-1}.$$

Evaluation of symmetry breaking. After receiving the handles of the two supposedly symmetrical fragments F_A and F_B can give a numerical estimate of divergence between them D_{AB}, for it is proposed to use the properties of Fourier series. Also, this indicator can be used as an estimate of the degree of symmetry breakdown.

Successively translate the descriptor in both frequency representation, using the discrete Fourier transform for each point:

$$s_{d,k} = \sum_{p=1}^{m} x_{d,p} e^{\frac{-2\pi i}{m}(k-1)(p-1)}, 1 \le d \le n.$$

For an n-dimensional the descriptor in this case applies n-dimensional Fourier transforms independently for each coordinate. Result of the transformation will range following structure:

$$S_d = \left[s_{d,0} \quad s_{d,1} \quad \text{L} \quad s_{d,m} \right]^{\text{T}}, 1 \le d \le n.$$

Couples $(S_1 \quad S_{n-1}), (S_2 \quad S_{n-2}), \text{K}, (S_{(n-1)/2} \quad S_{n/2})$ are complex conjugate numbers. The transition from the spectrum to the original contour by using the inverse discrete Fourier transform of the form:

$$x_{d,k} = \frac{1}{m} \sum_{p=1}^{m} s_{d,p} e^{\frac{2\pi i}{m}(k-1)(p-1)}, 1 \le d \le n.$$

For the purpose of comparing the contours of the Fourier transformation is convenient because it provides a selection of principal components. The closer a pair of conjugate elements in the middle of the spectrum, the smaller, high frequency, especially circuit it describes. Each of the pairs of conjugate elements of the spectrum determines when restoring an ellipse, and all recovered shape is a superposition of these ellipses. Each of the next pair will be added to restore the image all the more finer details of the original contour. The first, odd element of the spectrum indicates the position of the center of the figure for comparison; the contours are not being used.

To calculate the distance between the fragments is proposed to summarize the relevant elements of the divergence of the spectra of these contours, after Fourier transformation. Higher-frequency pair (close to the center of the spectrum), the elements of the spectrum have less impact on the indicator closeness, than the low frequency.

Corresponding criterion is:

$$D_{A,B}=\sum_{i=1}^{q}\left(\left(\sqrt{\sum_{j=1}^{n}\left(\mathrm{Im}(S_{B,i,j})-\mathrm{Im}(S_{A,i,j})\right)^{2}}+\sqrt{\sum_{j=1}^{n}\left(\mathrm{Re}(S_{B,i,j})-\mathrm{Re}(S_{A,i,j})\right)^{2}}\right)\cdot\beta_i\right),$$

where q — the number of conjugate pairs of elements of the spectrum, n — the dimension of phase space; β_i — discount factors ($i = 1, .., q$), which determine the degree of importance of components of the spectrum of frequencies - the relevant elements of the spectrum contours A and B. Coefficients β_i are chosen empirically, based on the conditions of the problem.

Once the fragments have been normalized circuit, must choose from a variety of sequences of one the most satisfying the requirements of the problem - containing the most balanced between fragments. Number sequences that can be composed of non-overlapping fragments, even for small problems with the number of markers about 100 very large, so it is proposed to solve by using a genetic algorithm.

Example 1. Detection of Symmetries for a Model Example

Source data set consists of 300 points. These are the result of Ressler system simulation

$$\dot{x}_1=-(x_2+x_3),$$
$$\dot{x}_2=x_1+0.2x_2,$$
$$\dot{x}_3=0.2+x_3(x_1-2.6).$$

The observed parameter x3 reconstructed attractor (Fig. 5.5).

Attractor has been subjected to alignment markers. The total number of markers: 144. After labeling of the 10238 possible paths have been selected 1375, the others did not satisfy the restrictions in length: from 5 to 50 points of the original contour. Each segment was interpolated 60-th evenly spaced points along its length and subjected to the normalization procedure. After that was constructed adjacency matrix fragments (Fig. 5.6).

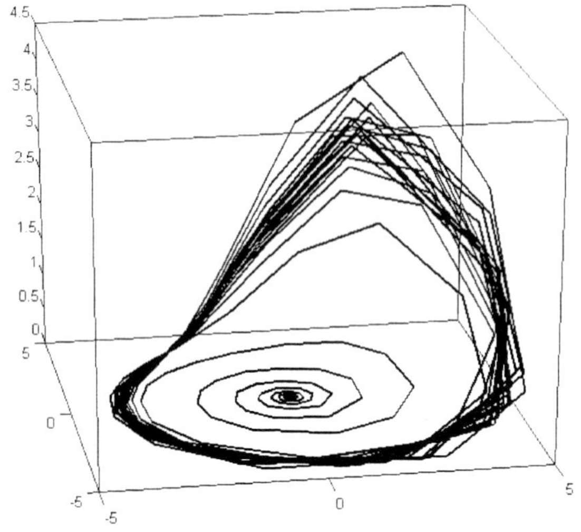

Figure 5.5. Attractor.

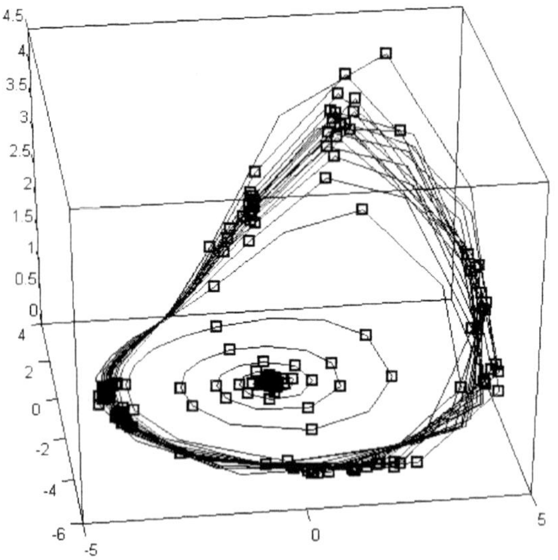

Figure 5.6. Attractor with markers.

Been carried out selection of solutions by using genetic algorithm (population size 200, the probability of mutation rates $\alpha = 0.3$ and $\beta = 0.1$, limiting the number of steps lengths: 5, an elite selection of a new generation, the type of parental choice - outbreeding, unique individual in the population) and selected the winning solution.

The time of the preprocessing step: 3 min. 27 sec. Time step of genetic selection decisions: 4 min. 36 sec., the population converged 5 times for 533 iterations, the decision of the winner: the similarity of the fragments 0.3917, length 197.

The results are shown in Fig. 5.7. Part of the trajectory for which were not symmetrical parts shown dotted.

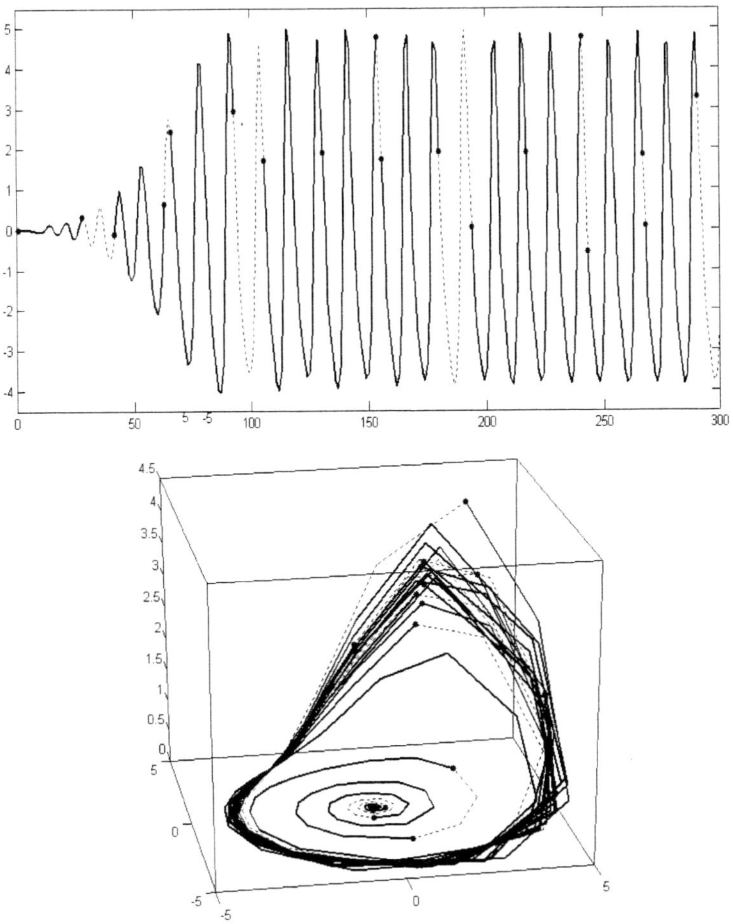

Figure 5.7. Decision-winner in the time series (left) and in the attractor (right).

Example 2. Reconstruction of the Model to Experimental Data

Original data set consists of 200 points. The data represent the traffic network. Smoothing was not applied. The reconstructed attractor has been subjected to alignment markers. The total number of markers: 192. After labeling of the 18153 possible paths have been selected 2633, the others did not satisfy the restrictions in length: from 15 to 30 points of the original contour. Each segment was interpolated 60-th evenly spaced points along its length and subjected to the normalization procedure in Fig. 5.8.

Was carried out the evolutionary selection of solutions (population size 200, the probability of mutation rates and limiting the number of steps lengths: 8, an elite selection of a new generation, the type of parental choice — outbreeding, unique individual in the population) and selected the winning solution. The time of the preprocessing stage: 12 min. 43 sec. Time step of genetic selection decisions: 6 min. 11 sec., The population converged 8 times for 652 iterations, the decision of the winner: the similarity of fragments of 0.576, length 125. The results are shown in Fig. 5.9.

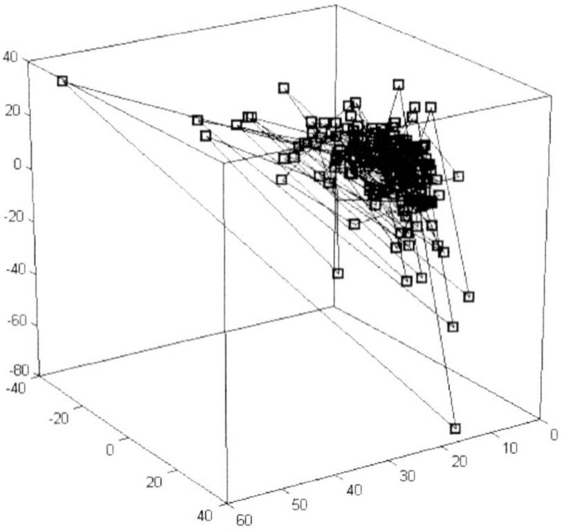

Figure 5.8. Labeled attractor.

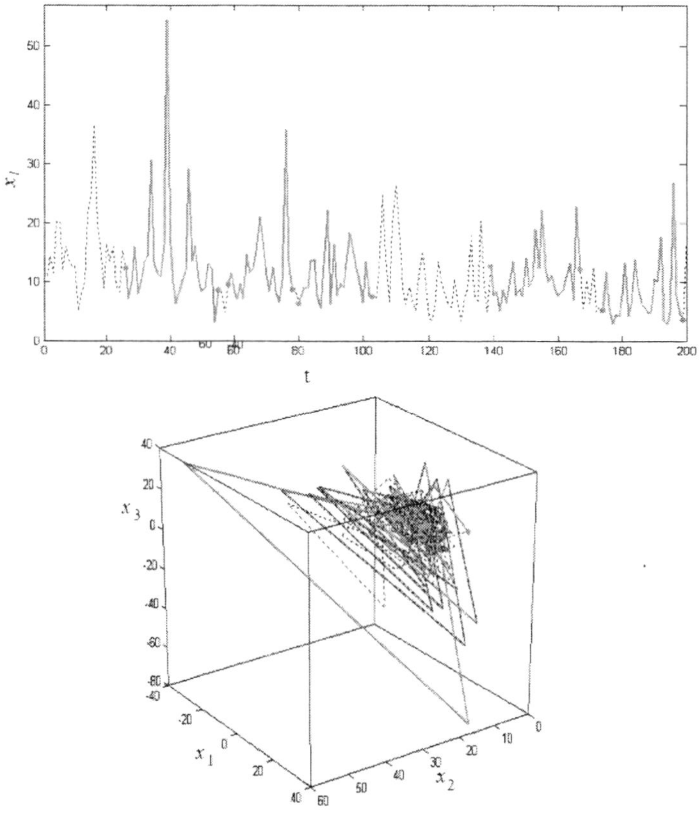

Figure 5.9. Decision-winner at baseline (left) and the attractor (right).

The presence of the symmetry of the system determined the structure of the form (4.3). Parameter identification system using the method of least squares, gives the following result:

$$A = \begin{bmatrix} 0.9413 & -0.1805 & 0.1164 & -0.0295 \\ -0.0545 & 0.8226 & 0.1622 & 0.1056 \\ 0.0014 & -0.0105 & -0.4455 & 0.8471 \\ -0.0062 & 0.0341 & -0.8860 & -0.5404 \end{bmatrix},$$

$$\Psi_0 = \begin{bmatrix} 0.0399 \\ 0.0463 \\ -0.4848 \\ -0.1851 \end{bmatrix} \left(\exp(t^{0.0001})\sin(t^{0.4})\right),$$

$$C = 10^4 \left[\, 2.1037 \quad \text{-}0.0124 \quad 0.1202 \quad \text{-}0.0302 \right].$$

Fig. 5.10 shows a comparison of the dynamics of the original system and the reconstructed model.

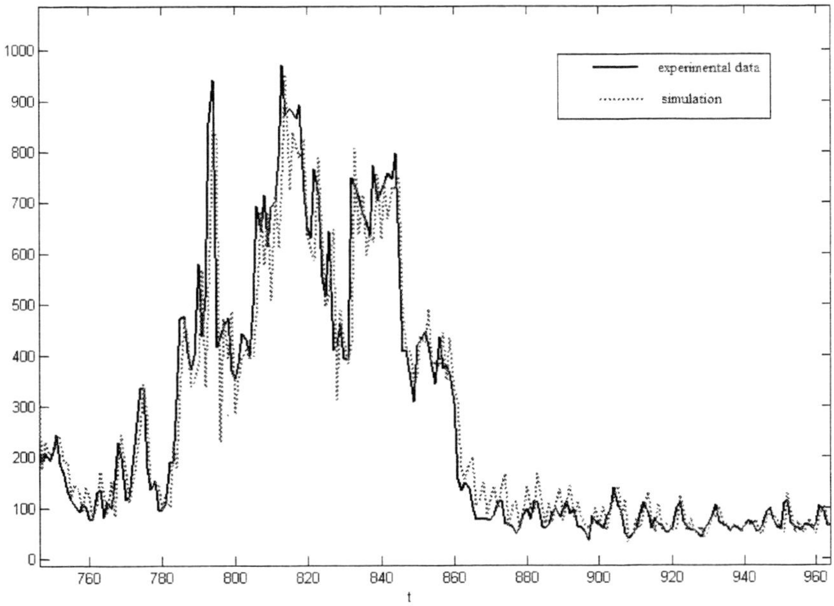

Figure 5.10. Comparison of dynamics.

Note that in determining the conversion makes sense to consider the conservation of structural stability for flows. Based on the equivalence of all the disturbances in model assumed that the local topological equivalence should preserve the structural stability, and the conversion can be close enough to the identity for small perturbations.

Resumes

The use of symmetric properties of evolution equations can be successfully used for the reconstruction of the system. Developed a constructive method for constructing evolutionary models based on geometric analysis of the experimentally obtained phase portrait of the system.

References

[1] Packard, N. H., Crutchfield, J. P., Farmer, J. D., Shaw, R. S. (1980) *Geometry from a time series. Phys. Rev. Lett.* 45, P. 712–716.

[2] Takens, F. (1981) *Detecting strange attractors in turbulence* In: Dynamical Syst. and Turbulence, Warwick 1980, *Lecture Notes in Mathematics*, 898 (eds. D. Rand and L.-S. Young). Springer. P. 366–381.

[3] Cremers, X., Hubler, A. (1987) *Construction of differential equation from experimental data.* Z. Naturforschung A., V. 42. P. 797–802.

[4] Crutchfield, J. P, McNamara, B. S. (1987) *Equations of motion from a data series. Complex Systems.* V. 1. P. 417–452.

[5] Janson, N. B., Pavlov A. N., Anisliclienko V. S. (2003) *Global reconstruction: application to biological data and secure communication* In: Chaos and its reconstruction (ed. by Gouesbet G., Meunier-Guttin-Cluzel S.). Nova Science publishers. P. 287–317.

[6] Aguirre, L. A., Mendes, E. M. (1996) *Global nonlinear polynomial models: structure, term clustering and fixed points Int. J. Bifurc. Chaos.* V. 6(2). P. 279–294.

[7] Aguirre, L. A., Billings, S. A. (1995) *Identification of models for chaotic systems from noisy data: implications for performance and nonlinear filtering. Physica D.* V. 85. P. 239–258.

[8] Cao, L. (1997) *Practical method for determining the minimum embedding dimension of a scalar time series. Physcai D.* V. 110. P. 43–50.

[9] Gouesbet G, Maquet X. (1992) *Construction of phenomenological models from numerical scalar time series. Physica D.* V. 58. P. 202–215.

[10] Rosenstein M. T., Collins J. J., De Luca C. J. (1994) *Reconstruction expansion as a geometry-based framework for choosing proper delay times Physica D.* V. 73. P. 82–98.

[11] Sauer, T. (1994) *Reconstruction of dynamical systems from interspike intervals. Phys. Rev. Lett.* V. 72. P. 3811–3814.

[12] Brawn R., Rulkov N. F., Tracy E. R. (1994) *Modelling and synchronizing chaotic systems from time-series data. Pthys. Rev. E.* V. 49. P. 3784–3800.

[13] Breeden J. L., Hubler A. (1990) *Reconstructing equations of motion from experimental data with unobserved variables Phys. Rev. A.* V. 42. N. 10. P. 5817–5826.

[14] Gouesbet, G., Letellier, C. (1994) *Global vector-field reconstruction by using a multivariate polynomial* L_2 *approximation on nets. Phys. Rev. E.* V.49. P. 4955–4972.

[15] Shilnikov L. P., Shilnikov, A. L., Turaev, D. V. And Chua, L, O. (1998, 2001) *Methods of qualitative theory in nonlinear dynamics.* Part 1, 2. World Scientific Publishing.

[16] Hasselblatt, B. and Katok A. (2003) *A first course in dynamics — With a panorama of recent developments* Cambridge Univ. Press.

[17] Nikulchev E.V. (2007) *Geometric method of reconstructing systems from experimental data. Technical Physics Letters*. V. 33(3). P. 267–269.

In: Evolution Equations
Editor: Arthur L. Claes
ISBN: 978-1-61209-652-0

Chapter 12

EVOLUTION OF COEFFICIENTS AND EIGENVALUES OF POLYNOMIALS AND RELATED GENERALIZED DYNAMICS

Robert M. Yamaleev*
Joint Institute for Nuclear Research, LIT, Dubna, Russia
Facultad de Estudios Superiores,
Universidad Nacional Autonoma de Mexico
Cuautitlán Izcalli, Campo 1, C.P.54740, México

Abstract

Invariant properties of the polynomials are studied. Class of polynomials with congruent set of eigenvalues is introduced. Evolution equations for eigenvalues and coefficients remaining the polynomial within proper class of polynomials are formulated. Evolution equations are given by equations for Weierstrass and Jacobian hyper-elliptic functions. The link between evolution equations for the coefficients of the polynomials and dynamical evolution equations of physical systems is established. Within the framework of generalized mechanics the inner and outer coordinates of momenta are interrelated as roots and coefficients of the characteristic polynomials. In particularly, in the case of $n = 2$ degree we deal with the relativistic dynamics. The case of $n = 3$ degree we arrive to new dynamics beyond the relativistic one. As an application the algorithm of finding of roots of the polynomials based on the evolution process is built.

Keywords: polynomials, elliptic functions, algorithm, evolution equation, dynamics, relativistic mechanics.

1. Introduction

The evolution equations of classical dynamics in $2p$ dimensional phase space are generated by single Hamiltonian closely related with the energy of the physical system. Y.Nambu [1] had suggested the system of evolution equations generated by the pair of Hamiltonian

*E-mail address: iamaleev@servidor.unam.mx

functions for some physical systems defined in $3p$ dimensional phase space. Many papers devoted to this topic had been published, investigating rather the mathematics than a physical sense of the formalism (see, for instance, Refs.[11] and references therein) .

The first physical model to which can be applied the Nambu's formalism has been indicated own by Y.Nambu in [1], this is the Euler model of free rigid body. In Refs.[14], [15] has been done the first physical example where the Nambu's formalism can be applied, the relativistic mechanics in a new interpretaton. Within the framework of this formulation of relativistic mechanics the Nambu's formalism of $3p$ dimensional space has received its interpretation and explanation. It has been shown [10] that the top model of classical mechanics formally coincides with the relativistic motion of a particle confined into oscillator potential. Furthermore, it has been established that the mechanics, where the Nambu's formalism of n-order can be applied, deals with a new kind of physical system possessing with n sub-energies and describing with the set of inner and outer coordinates of momenta (see, for instance, Refs.[13]-[16] and references therein). These kind of systems are characterized with n degree polynomials the roots and coefficients of which are related with inner and outer coordinates of the momenta, correspondingly.

In the present paper we demonstrate that the characteristic polynomials play a crucial role in construction of evolution of the dynamical systems. The process of construction of evolution equations is based on a new technique of solution of polynomial equations.

It is shown, the evolution equations for the coefficients can be identified with equations for Weierstrass hyper-elliptic functions, whereas the evolution equations for the squared roots are given by equations for Jacobian hyper-elliptic functions. Furthermore, it is demonstrated a link between evolution equations for the coefficients of the polynomials and the dynamic equations of generalized mechanics. Within the framework of n-order generalized mechanics the inner- and outer-momenta are interrelated as roots and coefficients of the characteristic n-degree polynomial. In particular $n = 2$ case, we recover the celebrated equations of motion of the relativistic particle inside stationary potential field where the role of characteristic quadratic polynomial plays the mass-shell equation.

Besides the Introduction the paper contains the following sections.

In Section 2, the equations of evolution for the coefficients of n-degree polynomial are formulated. In Section 3, some peculiarities of the cubic equation is explored. In Section 4, the elements of the relativistic dynamics based on quadratic characteristic polynomial is presented. In Section 5 elements of the generalized dynamics related with n-degree characteristic polynomial is constructed. In Section 6, the Algorithm of finding of eigenvalues of the n-degree polynomial is built.

2. Evolution equations for eigenvalues and coefficients of n-degree polynomial

If F is a field and $q_1, ..., q_n$ are algebraically independent over F, the polynomial

$$p(X) = \prod_{i=1}^{n}(X - q_i)$$

is referred to as *generic polynomial* over F of degree n. The polynomial equation of n-degree over field F is written in the form

$$p(X) := X^n + \sum_{k=1}^{n-1} (-)^k (n-k+1) P_k X^{n-k} + (-)^n P^2 = 0, \tag{2.1}$$

where the coefficients $P_k \in F(q_1, ..., q_n)$. In this paper we shall restrict our attention only to polynomials with real coefficients and with simple roots. The signs at the coefficients in Eq.(2.1) are changed from term to term which allows in Vièta formulae to keep only the positive signs. The expressions at the coefficients are included for a convenience and have no real bearing on the theory. The mapping from the set of eigenvalues onto the set of coefficients is given by Vièta formulae

$$(a)\ nP_1 = \sum_{i=1}^{n} q_i,\ \ (b)\ P^2 = \prod_{i=1}^{n} q_i,\ \ (c)\ P_k = \sum_{1\le r_1<...<r_k\le n} \prod_{j=1}^{k} q_{r_j},\ k = 2, ..., n-1, \tag{2.2}$$

here P_1, P_k, P^2 are defined via *elementary symmetric k-degree polynomials* of the eigenvalues. The number of monomials inside k-th elementary polynomial is equal to binomial coefficient:

$$C_n^k = \begin{pmatrix} k \\ n \end{pmatrix} = \frac{n!}{k!(n-k)!}. \tag{2.3}$$

Since the roots of the generic polynomial $p(X)$ are algebraically independent, this polynomial is, in some sense, the most general polynomial possible.

In $p(X)$ replace X with $X = Y + P_1$. This transformation will eliminate the $(n-1)$-degree term, resulting in a polynomial of the form

$$r(Y) := Y^n + \sum_{k=2}^{n-1} (-)^k R_k Y^{n-k} + (-)^n R_0 = 0. \tag{2.4}$$

The polynomials $p(X)$ and $r(Y)$ have the same splitting field and hence the same Galois group. Let E be the splitting field of $r(Y)$, let $y_k, k = 1, .., n$ be its roots in E and $G = G_F(E)$ be its Galois group.

Lemma 2.1

The coefficients $R_k, k = 2, ..., n-1,\ R_0$ of Eq.(2.4) remain unaltered with respect to simultaneous translations of the eigenvalues of Eq.(2.1)

The **Proof** of the statement it follows from formula $Y = X - P_1$ which in terms of the eigenvalues is expressed as follows

$$y_k = q_k - \frac{1}{n}\sum_{i=1}^{n} q_i = \frac{1}{n}\sum_{i\neq k}^{n} (q_k - q_i), k = 1, ..., n. \tag{2.5}$$

It is seen that the eigenvalues of Eq.(2.4) are represented by differences between the eigenvalues of Eq.(2.1), hence, they are invariants with respect to the simultaneous translations. Since the coefficients $R_0, R_k, k = 2, ..., n-1$, are sum of uniform monomials of $y_k, k = 1, ..., n$ they have same feature, namely, they are invariants with respect to simultaneous translations of the eigenvalues of Eq.(2.1).

End of Proof.

The polynomial $r(Y)$ we denominate as *invariant polynomial* (with respect to translations of the roots of (2.1)).

The main task of this section is to introduce evolution equations for the coefficients of Eq.(2.1) for which the coefficients of Eq.(2.4) $R_k, k = 2, ..., n-1, R_0$ serve as the first integrals. This result is given by the following

Theorem 2.2

Let $q_k, k = 1, ..., n$ be set of eigenvalues of polynomial equation of n- degree (2.1). Let differentials of all eigenvalues are equal to each other

$$dq_1 = dq_2 = \ldots = dq_k = \ldots = dq_n. \tag{2.6}$$

Then differentials of the coefficients satisfy the following system of equations:

$$2P_{n-1}dP_1 = dP^2, \tag{2.7}$$

$$dP_{n-k} = (k+2)P_{n-k-1}\, dP_1, \quad k = 1, ..., n-2; \tag{2.8}$$

Proof.

From (2.2a) it follows

$$dq_k = dP_1, \ k = 1, ..., n.$$

Coefficients of the polynomial (2.1) are symmetric forms of the eigenvalues where k-th coefficient $(n-k+1)P_k$ consists of C_n^k monomials of k-degree. Thus, the derivative of this coefficient $(n-k+1)dP_k$ contains kC_n^k monomials and, since the derivatives of all eigenvalues are equal to dP_1, the derivative of $(n-k+1)P_k$ is proportional to λdP_1 where the coefficient of proportionality consists of kC_n^k $(k-1)$-degree symmetric monomials. On the other hand, the symmetric polynomial of $(k-1)$-th degree can expressed only by C_n^{k-1} symmetric monomials of $(k-1)$-degree. This means the expression for λ consists of $kC_n^k/C_n^{k-1} = n-k+1$ same symmetric polynomials of $(k-1)$-degree which are equal to the $(k-1)$-th coefficient $(n-k+2)P_{k-1}$. The result is expressed as follows

$$(n-k+1)dP_k = (n-k+1)(n-k+2)P_{k-1}\, dP_1, \quad k = 2, 3, ..., n-1.$$

Differentiation of the last coefficient is obtained by direct calculation

$$dP^2 = 2P_{n-1}dP_1, \tag{2.9}$$

which completes the system of differential equations for the coefficients of Eq.(2.1).

End of Proof.

The following Lemma demonstrates an important role of the invariant polynomial in the evolution process.

Lemma 2.3

The first integrals of evolution equations (2.7)-(2.8) are given by coefficients of invariant polynomial (2.4).

Proof

In fact, the roots of Eq.(2.4), $y_k, \ k = 1, ..., n$, according to formulae (2.5), are the first integrals of evolution equations for eigenvalues (2.6). Equations (2.7)-(2.8) are consequences of Eqs.(2.6), hence coefficients R_k as algebraic functions of the solutions of evolution equations are first integrals of Eqs.(2.7)-(2.8).

End of Proof

Inversely by substituting Y with $X - P_1$ in Eq.(2.4) we come back from invariant equation (2.4) to equation (2.1). Substitute $X - P_1$ instead of Y and gather together powers of X, and the coefficients of the obtained polynomial compare with coefficients of Eq.(2.1). We shall see that the coefficients $P_k,\ k = 1, ..., n-1$ now are expressed as k-degree polynomials of P_1 with coefficients consisting of $R_k,\ k = 2, ..., n-1$. Especially notice, the invariant R_0 is defined by n-degree polynomial of P_1 the coefficients of which are certain polynomials of $R_k.k = 2, ..., n-1$. The task is to define a explicit form of the polynomial of P_1. The general form of the desired polynomial can be presented as follows

$$P_1^n + \sum_{k=2}^{n-1} P_1^{n-k} f_k(R_1, ..., R_k) + R_0 = P^2.$$

Now, the next task is to find an explicit form of the function f_k. For that purpose explore firstly the case $P^2 = 0$. In this case one of the solutions of Eq.(2.1) is equal to zero. Then the corresponding solution of the invariant polynomial is

$$y_1 = -P_1(P^2 = 0) = -P_1(0),$$

which satisfy Eq.(2.4). By replacing Y with $(-P_1(0))$ in Eq.(2.4) we come to the following equation for $P_1(0)$:

$$P_1^n(0) + \sum_{k=2}^{n-1} R_k P_1^{n-k}(0) + R_0 = 0. \tag{2.10}$$

Notice, here the signs at all coefficients are positive. Since we use evolution equations (2.7)-(2.8) in order to pass from Eq.(2.1) with coefficients $\{\ P_k,\ P^2 \neq 0\ \}$ to the polynomial with coefficients $\{\ \widetilde{P}_k,\ P^2 = 0\ \}$, these two polynomials possess with the same invariants as the original one. Hence, when $P^2 \neq 0$ equation (2.10) is transformed as follows

$$P_1^n + \sum_{k=2}^{n-1} R_k P_1^{n-k} + R_0 = P^2. \tag{2.11}$$

Notice, the situation is somewhat similar the conventional *Classical Invariant Theory of Polynomials* [9]. If classical invariant theory is a study of properties of a polynomial $p(x)$ that are unchanged under fractional linear transformations, within the framework of the present approach we study properties of polynomials under translational transformations.

Theorem 2.4

If evolution of the coefficients of Eq.(2.1) obey equations (2.7)-(2.8), then

$$(k+1)P_{n-k} = \frac{1}{k!}\frac{d^k}{dP_1{}^k}P^2. \tag{2.12}$$

Proof.

Differentiate equation (2.11) by taking into account that $R_0, R_k, k = 1, ..., n-1$ are constants. We get

$$dP_1\ (\ nP_1^{n-1} + \sum_{k=2}^{n-2}(n-k)R_k P_1^{n-k-1} + R_{n-1}\ ) = dP^2. \tag{2.13}$$

Compare this equation with Eq.(2.9). It is seen, the expression inside brackets at dP_1 is nothing else than the coefficient $2P_{n-1}$ expressed as a polynomial of P_1 with invariant coefficients. Hence,

$$2P_{n-1} = nP_1^{n-1} + \sum_{k=2}^{n-2}(n-k)R_k P_1^{n-k-1} + R_{n-1} = \frac{d}{dP_1}P^2. \tag{2.14}$$

Next, differentiate (2.14) with respect to P_1, this leads to the equation

$$dP_{n-1} = 3P_{n-2}\, dP_1, \tag{2.15}$$

where $3P_{n-2}$ is the following coefficient of Eq.(2.1):

$$3P_{n-2} = \frac{n(n-1)}{2}P_1^{n-2} + \sum_{k=2}^{n-3}\frac{(n-k)(n-k-1)}{2}R_k P_1^{n-k-2} + R_{n-2}. \tag{2.16}$$

Consequently,

$$3P_{n-2} = \frac{d^2}{dP_1{}^2}P^2.$$

The next step of this procedure leads to the equation

$$dP_{n-2} = 4P_{n-3}\, dP_1, \tag{2.17}$$

where

$$4P_{n-3} = \frac{n(n-1)(n-2)}{1\cdot 2\cdot 3}P_1^{n-3} + \sum_{k=2}^{n-4}\frac{(n-k)(n-k-1)(n-k-2)}{1\cdot 2\cdot 3}R_k P_1^{n-k-3} + R_{n-3}.$$

Hence,

$$4P_{n-3} = \frac{d^3}{dP_1{}^3}P^2.$$

By induction we come to the general formula for l-th coefficient P_{n-l}:

$$(l+1)P_{n-l} = \binom{l}{n}P_1^{n-l} + \sum_{k=2}^{n-l-1}\binom{l}{n-k}R_k P_1^{n-k-l} + R_{n-l} = \frac{1}{l!}\frac{d^l}{dP_1{}^l}P^2. \tag{2.18}$$

End of Proof.

This theorem has some interesting consequences.

Corollary 2.5

The following representation for polynomial $p(X)$ holds true

$$p(X) = \exp(-X\frac{d}{dP_1})P^2 = 0. \tag{2.19}$$

Proof

The generator of translation defined by Euler operator is represented by the following expansion

$$\exp(-X\frac{d}{dP_1}) = 1 - X\frac{d}{dP_1} + X^2\frac{1}{2!}\frac{d^2}{dP_1^2} + ... + X^n\frac{1}{n!}\frac{d}{dP_1^n} + ... \tag{2.20}$$

By differentiating Eq.(2.11) n-times with respect to P_1, we get

$$\frac{d}{dP_1^n}P^2 = n!. \tag{2.21}$$

Hence the sum in (2.20) is completed by this term. Thus,

$$\exp(-X\frac{d}{dP_1})P^2 = p(X). \tag{2.22}$$

Here the variable X means one of the roots of Eq.(2.1) in quality of which let us take q_n.

The derivative with respect to P_1 due to (2.6) can be expressed by the following sum

$$\frac{d}{dP_1} = \sum_{k=1}^{n} \frac{\partial q_k}{\partial P_1}\frac{\partial}{\partial q_k} = \sum_{k=1}^{n} \frac{\partial}{\partial q_k}. \tag{2.23}$$

Then,

$$\exp(-q_n\frac{d}{dP_1})P^2 = \prod_{k=1}^{n-1} \exp(-q_n\frac{\partial}{\partial q_k})\exp(-q_n\frac{\partial}{\partial q_n}). \tag{2.24}$$

Now, take into account Vièta formula for P^2 given by (2.2). The Euler operator (2.24) will translate each root by q_n. The last operator in (2.24) acts only upon n-th root resulting $q_n - q_n = 0$. Hence,

$$\exp(-X\frac{d}{dP_1})P^2 = p(X) = 0.$$

End of Proof.

3. Relativistic Lorentz-force equations and related quadratic polynomial

There exist an certain inter-relation between the evolution of the polynomials and the dynamic systems. The aim of this section is to demonstrate as evolution of the quadratic polynomial is related with the relativistic Lorentz-force equations. In the sequel, we shall use this example as a starting platform to pass to case of higher degree polynomials and related generalized dynamics. Let us start with generic quadratic polynomial

$$X^2 - 2P_0\, X + P^2 = 0, \tag{3.1}$$

with real coefficients $P_0^2 \geq P^2$, and real eigenvalues p_1^2, p_2^2. This polynomial closely related with the relativistic dynamics [14], [6]. In order to demonstrate this fact let us start from the Lorentz-force equations describing a motion of the charged particle inside external electromagnetic fields $\mathbf{E},\ \mathbf{B}$ [1]

$$\frac{d\mathbf{P}}{d\tau} = (\mathbf{E}\,P_0 + [\mathbf{P}\times\mathbf{B}]),\ \frac{dP_0}{d\tau} = (\mathbf{E}\cdot\mathbf{P}), \tag{3.2}$$

[1]Hereafter for the sake of simplicity we omit all parameters like charge, mass, light-velocity and other parameters regularizing physical dimensions taking them equal to unit.

$$\frac{d\mathbf{r}}{d\tau} = \mathbf{P}, \quad \frac{dt}{d\tau} = P_0. \tag{3.3}$$

Consider projection of equations (3.2) on direction of the motion defined by unit vector $\vec{n} = \mathbf{P}/P$:

$$\frac{dP}{d\tau} = (\mathbf{E} \cdot \mathrm{n})\, P_0, \quad \frac{dP_0}{d\tau} = (\mathbf{E} \cdot \mathrm{n})\, P, \tag{3.4}$$

Then one deals only with the lengths of the momenta P_0, P. Here parameter of evolution τ means a proper time, simplify Eqs.(3.4) by introducing a new evolution parameter by

$$\frac{dP}{ds} = P_0, \quad \frac{dP_0}{ds} = P, \tag{3.5}$$

where

$$\frac{ds}{d\tau} = (\mathbf{E} \cdot \mathrm{n}). \tag{3.6}$$

The first constant of motion is easily found

$$P_0^2 - P^2 = M^2. \tag{3.7}$$

Here M^2 is a constant of motion. Conventionally, this constant is interpreted as a *square of inertial mass* [10]. Inside stationary potential field, when $\mathbf{E} = -\nabla V(r)$, dynamic equations imply the other integral of motion, the energy,

$$\mathcal{E}_0 = P_0 + V(r). \tag{3.8}$$

At the rest state where $P = 0$ one obtains $P_0(P = 0) = M$. Consider the following linear combinations of two kinds of energies:

$$p_1^2 = P_0 - M, \quad p_2^2 = P_0 + M. \tag{3.9}$$

From formulae (3.7) and (3.9) it follows

$$P = p_1 p_2, \quad P_0 = \frac{1}{2}(p_2^2 + p_1^2), \quad M = \frac{1}{2}(p_2^2 - p_1^2). \tag{3.10}$$

Notice, the first two formulae of (3.10) mean Vièta formulae for quadratic polynomial equation (3.1). Substitution $X = Y + P_0$ in (3.1) leads to *invariant equation*

$$Y^2 = P_0^2 - P^2 = M^2, \tag{3.11}$$

the invariant coefficient of which is equal to the invariant of physical motion. Quadratic equation for P_0 is given by

$$P_0^2 - M^2 = P^2. \tag{3.12}$$

Differentiating this equation we derive evolution equations for the coefficients, which in the case of quadratic polynomial, of course, is a simple task:

$$2P_0\, P = \frac{d}{ds}P^2, \quad \frac{d}{ds}P_0 = P. \tag{3.13}$$

Solutions of these evolution equations are given by hyperbolic cosine-sine functions

$$P = M\ sinh(s), \ \ P_0 = M\ cosh(s). \tag{3.14}$$

Then the eigenvalues of Eq.(3.1) are expressed by hyperbolic cosine-sine functions of one-half argument

$$p_1^2 = \sqrt{2M}\ sinh(\frac{s}{2}), \ \ p_2^2 = \sqrt{2M}\ cosh(\frac{s}{2}). \tag{3.15}$$

Now, consider so-called the *effective potential representation*. With this purpose let us come back to the equations written with respect to proper-time τ. Then from the second equation of (3.13) it follows

$$\mathcal{E}_0 = P_0 + V(r).$$

Further, replace P_0 by $\mathcal{E}_0 - V(r)$ in the first equation of (3.13), this gives

$$\frac{d}{ds}\mathbf{P} = -\nabla V(r)\ (\mathcal{E}_0 - V(r) = -\nabla W(r, \mathcal{E}_0), \tag{3.16}$$

where the effective potential is defined by

$$W(r\mathcal{E}_0) = \mathcal{E}_0 V(r) - \frac{1}{2}V^2(r). \tag{3.17}$$

Eq.(3.16) is the Newtonian equation of motion with *effective potential* W and with the energy

$$\mathcal{E} = \frac{1}{2}P^2 + W(r, \mathcal{E}_0) = \frac{1}{2}(\mathcal{E}_0^2 - M^2). \tag{3.18}$$

4. Cubic polynomial equation and related dynamics

In the previous section we have demonstrated as the relativistic dynamics is related with the evolution of quadratic polynomial. This example provides us with appropriate tool in order to construct generalized scheme based on the evolution of polynomials of higher order. In this section we explore the case of cubic polynomial

$$p(X) = X^3 - 3P_1\, X^2 + 2P_2\, X - P^2 = 0, \tag{4.1}$$

where coefficients and eigenvalues are connected by the following formulae

$$3P_1 = q_1 + q_2 + q_3, \ 2P_2 = q_1q_2 + q_2q_3 + q_3q_1, \ P^2 = q_1q_2q_3. \tag{4.2}$$

Let us assume that $p(X)$ is irreducible. By replacing X with $X = Y + P_1$ we come to invariant polynomial

$$Y^3 + R_2\, Y - R_0 = 0, \tag{4.3}$$

where

$$(a)\ R_2 = 2P_2 - 3P_1^2, \ \ (b)\ P_1^3 + R_2\, P_1 + R_0 = P^2. \tag{4.4}$$

The eigenvalues and the coefficients of this polynomial are invariants with respect to simultaneous translations of the eigenvalues of Eq. (4.1). Obviously this statement is a consequence of the formula $Y = X - P_1$ from which it follows

$$3y_1 = e_2 - e_3, \ 3y_2 = e_3 - e_1, \ 3y_3 = e_1 - e_2, \tag{4.5}$$

where

$$e_1 = q_3 - q_2, e_2 = q_1 - q_3, e_3 = q_2 - q_1.$$

Evolution equations remaining unaltered coefficients R_0, R_2 are obtained directly by differentiating Eqs.(4.4). We get

$$dP_1\,(3P_1^2 + R_2) = 2P_2\,dP_1 = dP^2, \tag{4.6}$$

$$dP_2 = 3dP_1.$$

The following equations for the eigenvalues hold true

$$\frac{d}{ds}q_k = \frac{d}{ds}P_1 = P,\ k = 1, 2, 3. \tag{4.7}$$

The generalized dynamics of third order is an object of special interest because this dynamics is closely related with the classical elliptic functions. The solutions of the evolution equations for the eigenvalues and the coefficients of the cubic polynomial can be represented via elliptic Jacobi and Weierstrass functions, correspondingly [2]. On making use (4.7) from (4.4b) we come to the following differential equation

$$P_1^3 + R_1\,P_1 + R_0 = (\frac{dP_1}{ds})^2. \tag{4.8}$$

Write this equation in the following designations

$$(\frac{d\wp}{dz})^2 = 4\wp(z)^3 - g_2\wp(z) - g_3, \tag{4.9}$$

where $z = 4s$, $g_2 = -4R_1,\ g_3 = -4R_0$ and $\wp(2s) = P_1(s)$. The integral formula for $\wp(z)$ is given by

$$z = \int_{\wp}^{\infty} (4x^3 - g_2 x - g_3)^{-1/2} dx. \tag{4.10}$$

The functions $\wp(z)$ and $\wp'(z)$ are Weierstrass elliptic functions with periods $2\omega_1,\ 2\omega_2$. Define $\omega_3 = -\omega_1 - \omega_2$. Then the values

$$\wp(\omega_1),\ \wp(\omega_2),\ \wp(\omega_3)$$

are the roots of cubic equation

$$4x^3 - g_2 x - g_3 = 0. \tag{4.11}$$

Introduce new variables $p_k,\ k = 1, 2, 3$ by $p_k^2 = q_k$. Then $P = p_1 p_2 p_3$. For p_k evolution equations are derived from (4.7):

$$\frac{dp_1}{ds} = p_2 p_3,\ \frac{dp_2}{ds} = p_1 p_3,\ \frac{dp_3}{ds} = p_2 p_1. \tag{4.12}$$

Solutions of these equations are given by quotients of Jacobi elliptic functions. Let $sn(u), cn(u), dn(u)$ be a set of Jacobi elliptic functions. Define quotients of these functions (in Glaisher notations [2]) by

$$ns = -\frac{1}{sn},\ cs = -\frac{cn}{sn},\ ds = -\frac{dn}{sn},$$

which obey the following differential equations

$$\frac{d}{du}ns(u) = cs(u)ds(u), \frac{d}{du}cs(u) = ns(u)ds(u), \frac{d}{du}ds(u) = cs(u)ns(u), \quad (4.13)$$

with

$$ns^2 - cs^2 = 1, \; ds^2 - cs^2 = 1 - \mathrm{k}^2.$$

From Eqs.(4.12) it follows that the values

$$e_1 = p_3^2 - p_2^2, \; e_2 = p_1^2 - p_3^2,$$

are constants of motion. Hence, the solutions of Eqs.(4.12) are proportional to the Jacobi elliptic functions

$$p_3 = \sqrt{e_1} ns(u, \mathrm{k}), \; p_2 = \sqrt{e_1} cs(u, \mathrm{k}), \; p_1 = \sqrt{e_1} ds(u, \mathrm{k}), \; \mathrm{k}^2 = 1 - \frac{e_2}{e_1}. \quad (4.14)$$

The final result is formulated by the following

Statement:

If the evolution of squared roots of the eigenvalues of the cubic equation obey equations for Jacobi elliptic functions, then the evolution of the coefficients are governed by equations for Weierstrass elliptic functions.

For the cubic equation there exist also another possibility to express its eigenvalues, namely, by usally trigonometric functions. The eigenvalues of the invariant polynomial (4.3) are given by formulae:

$$y_1 = -\frac{2}{3}\sqrt{3R_1}\cos\theta, \; y_2 = \frac{1}{3}\sqrt{3R_1}(\cos\theta + \sqrt{3}\sin\theta), \; y_3 = \frac{1}{3}\sqrt{3R_1}(\cos\theta - \sqrt{3}\sin\theta). \quad (4.15)$$

It is easy to verify that

$$y_1 + y_2 + y_3 = 0, \; y_1 y_2 + y_2 y_3 + y_3 y_1 = -R_1.$$

Equation for the last coefficient, R_0, leads to trigonometric equation for $\cos(\theta)$:

$$-R_0 = y_1 y_2 y_3$$

$$= \cos\theta(\cos^2\theta - 3\sin^2\theta) = 4\cos^3\theta - 3\cos\theta = \frac{27}{2}\frac{R_0}{(\sqrt{3R_1})^3}.$$

By using the trigonometric formula

$$\cos(3\theta) = 4\cos^3\theta - 3\cos\theta$$

we come to simple trigonometric equation:

$$\cos(3\theta) = \frac{3\sqrt{3}}{2}\frac{R_0}{\sqrt{R_1^3}}. \quad (4.16)$$

This equation is valid if only if the cubic equation possesses with real solutions which provided by the following inequality

$$(\frac{R_0}{2})^2 < (\frac{R_1}{3})^3. \tag{4.17}$$

The algorithm elaborated in the previous section in the case of cubic equation (4.1) is simplified as follows:

$$2P_2 dP_1 = 2P_2 dx, \;\; 2P_2 dP_2 = 3P_1 dx, \tag{4.18}$$

with the initial data $P_1(x = P^2) = P_1$, $P_2(x = P^2) = P_2$. At the final stage when $x = P^2 = 0$ quantities $P_1(P^2 = 0) = \overset{(1)}{P}_1$, $P_2(P^2 = 0) = \overset{(1)}{P}_2$ satisfy the following relationships

$$R_2 = 2\overset{(1)}{P}_2 - 3\overset{(1)}{P}{}_1^2, \;\; \overset{(1)}{P}{}_1^3 + R_1 \overset{(1)}{P}_1 + R_0 = 0.$$

Thus the new equation possesses with same invariants as the original one,

$$X^3(0) - 3\overset{(1)}{P}_1 X^2(0) + 2\overset{(1)}{P}_2 X(0) = 0. \tag{4.19}$$

Equations (4.1) and (4.19) possess with congruent eigenvalues where, notice, Eq.(4.19) has one trivial solution. Therefore the problem is reduced to the problem of solution of quadratic equation

$$X^2(0) - 3\overset{(1)}{P}_1 X^2(0) + 2\overset{(1)}{P}_2 X(0) = 0. \tag{4.20}$$

From solutions of Eq.(4.20) to the solutions of Eq.(4.1) one comes simply by the set of translations:

$$q_1 = P_1 - \overset{(1)}{P}_1, \;\; q_2 = q_2(0) + P_1 - \overset{(1)}{P}_1, \;\; q_3 = q_3(0) + P_1 - \overset{(1)}{P}_1.$$

The dynamics related with the cubic polynomial, evidently, is characterized with two first constants of motion R_2, R_0 which do not depend of potential field. In quality of evolution equations we have to use Eqs.(4.6)-(4.7) written with respect to parameter of evolution defined in (3.6):

$$P_2 = \frac{d}{ds}P, \;\; \frac{d}{ds}P_2 = 3dP_1, \;\; \frac{d}{ds}P_1 = P. \tag{4.21}$$

Evolution equations for the eigenvalues, correspondingly, are given by

$$\frac{d}{ds}p_k^2 = P, \;\; k = 1, 2, 3. \tag{4.22}$$

From these equations it follows that the constants of motion are given by formulae

$$M_1 = p_2^2 - p_3^2, \; M_2 = p_3^2 - p_1^2, \; M_3 = p_1^2 - p_2^2. \tag{4.23}$$

In order to include into this scheme the potential field $V(r)$ we have to use equation (3.6):

$$\frac{ds}{d\tau} = (\mathrm{U} \cdot \mathrm{n}), \;\; \mathrm{U} = \nabla V(r), \;\; P\mathrm{n} = \mathbf{P}. \tag{4.24}$$

Furthermore, the set of evolution equations has to be completed with the interrelation between momentum and velocity with respect to time-like parameter:

$$\mathbf{P} = \frac{d\mathbf{r}}{d\tau}. \tag{4.25}$$

In these designations evolution equations (5.21) are transformed into the following dynamic equations

$$\frac{d}{d\tau}\mathbf{P} = -\nabla V(r)\, P_2, \; \frac{d}{d\tau}P_2 = -3(\mathbf{P}\cdot\nabla)V(r)P_1, \; \frac{d}{d\tau}P_1 = -(\mathbf{P}\cdot\nabla)V(r). \tag{4.26}$$

In the case of stationary potential, beside the first integrals R_2, R_0, the equations imply another integral of motion, the energy

$$\mathcal{E}_1 = P_1 + V. \tag{4.27}$$

By substituting the expression for P_1 from (4.27) into the second equation of Eqs.(4.26), we get

$$\mathcal{E}_2 = P_2 + 3\mathcal{E}_1 V - \frac{3}{2}V^2 = \frac{1}{2}(R_2 + 3\mathcal{E}_1^2). \tag{4.28}$$

This is second expression for the energy. Next, express from this formula P_2 and substitute into first equation of (4.26). This leads to Newtonian equation

$$\frac{d}{d\tau}\mathbf{P} = -\nabla W(r, \mathcal{E}_1), \tag{4.29}$$

with effective potential

$$2W(r, \mathcal{E}_1) = \mathcal{E}_2 V(r) - 3\mathcal{E}_1 V^2(r) + V^3(r). \tag{4.30}$$

From Eq.(4.29) one may find formula for the total energy

$$\mathcal{E} = \frac{1}{2}P^2 + W = \frac{1}{2}(R_0 + R_2\mathcal{E}_1 + \mathcal{E}_1^3). \tag{4.31}$$

5. Generalized dynamics related with evolution of n-degree polynomial

Now we have accumulated enough experience in order to be able to build a general form of the dynamics related with n-degree polynomials. Let us start with evolution equations (2.8) written for n-degree polynomial (2.1). Dynamic equations with respect to some time-like evolution parameter s describing a motion inside stationary potential field $V(r)$ are formulated in the following form [16]:

$$\frac{d\mathbf{P}}{d\tau} = \mathrm{U}\, P_{n-1}, \tag{5.1a}$$

$$\frac{dP_k}{d\tau} = (\mathrm{U}\cdot\mathbf{P})P_{k-1}(n-k+2), \;\; k = 2, \ldots, n-1, \tag{5.1b}$$

$$\frac{dP_1}{d\tau} = (\mathrm{U} \cdot \mathbf{P}). \tag{5.1c}$$

The set of evolution equations has to be completed with the interrelation between momentum and velocity given by Eq.(4.25). This system of dynamic equations remain invariant the coefficients $R_0, R_k, k = 1, ..., n-1$. The motion of a physical system obeying to these equations possesses with the set of *inner* and *outer* momenta. Evolution of the set of outer-momenta $\{P^2, P_k, k = 1, ..., n-1\}$ are given by Eqs.(5.1a,b,c), whereas the evolution of the inner-momenta $\{p_k, k = 1, ..., n\}$ obey

$$\frac{dp_k^2}{d\tau} = (\mathrm{U} \cdot \mathbf{P}),\ k = 1, ..., n. \tag{5.2}$$

The first integrals of this system are given by

$$M_{ik} = p_i^2 - p_k^2,\quad i \neq k. \tag{5.3}$$

From Eq.(5.1c) we find the first constant of integration, the energy $\mathcal{E}_1 = P_1 + V$, or $P_1 = \mathcal{E}_1 - V$. By substituting P_1 into $k = 2$-th equation of (5.1b) we find the other constant of integration (the second expression for the energy):

$$\mathcal{E}_2 = P_2 + nE_1\, V - \frac{n}{2}V^2.$$

Next, by substituting P_2 from the last expression into the following equation with $k = 3$ we shall find the third constant of integration:

$$\mathcal{E}_3 = P_3 - (n-1)E_2\, V + \frac{n(n-1)}{2}E_1\, V^2 - \frac{n(n-1)}{2 \cdot 3}V^3.$$

Continue this process up till $(n-1)$-th stage. Then, at $(n-1)$-th stage we shall find the expression for P_{n-1}. By using this formula in Eq.(5.1a) we come to Newtonian equation

$$\frac{d}{d\tau}\mathbf{P} = -\nabla W(r, \mathcal{E}_1), \tag{5.4}$$

where the effective potential is given by the following series

$$W(r, \mathcal{E}_1) = \frac{1}{2}\sum_{k=1}^{n}(k+1)\mathcal{E}_{n-k}V^k(r)(-1)^{k+1},\ \text{with}\ \mathcal{E}_0 = \frac{1}{n+1}. \tag{5.5}$$

From Eq.(5.4) it follows formula for the total energy

$$\mathcal{E} = \frac{1}{2}P^2 + W = \frac{1}{2}(R_0 + \sum_{k=2}^{n-1} R_k \mathcal{E}_1^{k-1} + \mathcal{E}_1^n). \tag{5.6}$$

6 Algorithm of finding the roots of n-degree polynomials

The problem of construction of solutions of the polynomial equations as a certain functions of the coefficients is one of the oldest mathematical problems. E.Galois and H.Abel had proved that the polynomial higher than fourth order, in general, does not admit a presentation of solutions by radicals [4]. This rigorous mathematical theory directed mathematicians to look for the other ways of solutions. In particularly, the eigenvalues of the polynomials can be expressed in analytical way as certain functions of the coefficients [3]. Ch.Hermite was first who found an elegant expression of eigenvalues of the quintic equation by modular functions [7]. The theory of elliptic functions originally was related with the problem of finding of eigenvalues of the cubic polynomial. In fact, the eigenvalues of the cubic equation admit presentation by Weierstrass elliptic functions at the periods [12]. It is clear, however, that for a search of analytical solutions of the $n > 5$-degree polynomials one needs of mathematical tools beyond the elliptic functions. In this context as hopeful tools one may consider the theories of hyper-elliptic functions [8] and multi-complex algebras [13].

In the previous sections we have found that the evolution governed by equations (2.7)-(2.8) remains the polynomials within certain class of polynomials with congruent set of eigenvalues. This fact prompts an idea to use the evolution equations in order to come to the simplified polynomial. The main purpose of this section is to elaborate an algorithm of solution of n-degree polynomial equation by reducing the degree of the polynomial equation. The evolution from original polynomial to the simplified polynomial is fulfilled in such a way which remains unaltered coefficients of the invariant polynomial. Hence, initial and final polynomials of this evolution process will possess with congruent set of eigenvalues, so that the solutions of the former can be obtained from the solutions of the latter by simultaneous transformations of translation.

The algorithm is directed in a such way that at the final point of evolution we shall come to the polynomial with $P^2 = 0$. The last coefficient P^2 is used as a parameter of the evolution. This evolution will remain unaltered the coefficients of Eq.(2.4), hence from solutions of the polynomial with $P^2 = 0$ we may come to the solutions of the original equation simply by translation of the set of solutions of the simplified polynomial with new coefficients.

Denote $x = P^2$. Re-write Eq.(2.8) with respect to x. We get

$$2P_{n-1}dP_1 = dx,$$

$$\frac{dP_{n-k}}{dx} = (k+2)P_{n-k-1}\frac{dP_1}{dx}, \quad k = 1, ..., n-3; \tag{6.1}$$

$$\frac{dP_2}{dx} = nP_1\frac{dP_1}{dx}.$$

This is a *Cauchy problem* with initial data $P_k(x = P^2)$, $k = 1, 2, 3, ..., n-1$. The variable x runs from $x = P^2$ up till $x = 0$. These equations usually are resolved by using the *Cauchy-Lipschitz* method of calculation [5]. The procedure is fulfilled by dividing the interval $(x_0, 0)$ into N parts:

$$\Delta x_0 = x_1 - x_0, \; \Delta x_i = x_{i+1} - x_i, \; \Delta x_{n-1} = x - x_{n-1}, \tag{6.2}$$

where $x_i < x_{i+1}$, $x_N = 0$.

In this way the continuous evolution process is transformed into discrete process consisting of N steps. At the last step of discrete evolution process we come to n-degree polynomial free of the last coefficient:

$$\overset{(1)}{X}{}^{n} + \sum_{k=1}^{n-1} (-)^k (n-k+1) \overset{(1)}{P_k} \overset{(1)}{X}{}^{k} = 0. \tag{6.4}$$

One of the solutions is trivial, excluding this solution we come to the polynomial of $(n-1)$-degree:

$$\overset{(1)}{X}{}^{n-1} + \sum_{k=1}^{n-1} (-)^k (n-k+1) \overset{(1)}{P_k} \overset{(1)}{X}{}^{k} + 2\overset{(1)}{P}_{n-1} = 0. \tag{6.5}$$

Let us mention that Eq.(6.4) possesses with same invariants as original one, i.e. Eq.(2.1). Suppose that the solutions of Eq.(6.5) are $\overset{(1)}{q}_k$, $k = 1, ..., n-1$. Complete this set of solutions by $\overset{(1)}{q}_n = 0$. Then the solutions of original equation one may find simply by using the following translations

$$q_k = \overset{(1)}{q_k} + P_1 - \overset{(1)}{P_1}, \ k = 1, ..., n. \tag{6.6}$$

If solution of the $(n-1)$-degree equation still is difficult problem, then one may apply the algorithm again in order to reduce the problem of solution of $(n-1)$-degree equation to the problem of solution $(n-2)$-degree equation. This process can be continued up till quadratic or linear equation. At each step of the iteration one will find an information on coefficient $\overset{(r)}{P}_1$. At the r-th iteration one deals with $(n-r)$-degree equation in the form

$$\overset{(r)}{X}{}^{n-r} + \sum_{k=1}^{n-r-1} (-)^k (n-k+1) \overset{(r)}{P_k} \overset{(r)}{X}{}^{k} + (r+1)\overset{(r)}{P}_{n-r} = 0. \tag{6.7}$$

At this stage the norm of the coefficient $\overset{(r)}{P}_{n-r}$ will fulfil the role of evolution parameter of the next evolution process. As soon as the solutions of the lowest polynomial is found, the inverse process of iterations will consist of translations of the known set of eigenvalues with known parameter of translation. Let the last step of iteration be a linear equation with unique solution:

$$\overset{(n-1)}{q_1} = n \overset{(n-1)}{P_1}.$$

Then the solutions of the original equation (2.1) are found as a result of the following set of translations

$$q_1 = n \overset{(n-1)}{P_1} + n \sum_{r=2}^{n} \frac{1}{r} (\overset{(n-r)}{P_1} - \overset{(n-r+1)}{P_1}), \tag{6.8a}$$

$$q_s = \sum_{r=s}^{n} \frac{1}{r} (\overset{(n-r)}{P_1} - \overset{(n-r+1)}{P_1} , \ s = 2, 3, ..., n; \tag{6.8b}$$

where $\overset{(0)}{P_1} = P_1$.

7. Concluding remarks

This is a prodigious fact that the evolution equations elaborated for polynomials serve as dynamic equations for the generalized dynamics. We have restricted our attention only with the dynamic equations over one-dimensional coordinate space. On the theory of the generalized dynamics in $4D$ space with physical units one may consult in Refs.[15] and [16] and the references therein.

Notice, the Algorithm of calculation of eigenvalues of the polynomials elaborated in this paper is distinct of the numerical methods of calculations which principally are based on iteration process with initial sampling, or tentative data of solutions, so that effectiveness of these algorithms depends of the initial data. The iteration process based on the present Algorithm uses in the quality of initial data the coefficients of the original polynomial. The method elaborated here can be considered also as a theory of functional connection between coefficients and eigenvalues of the polynomial expressed as one-valued Weierstrass-type and Jacobi-type hyper-elliptic functions. Evidently, the present method without any principal difficulties can be continued to the case of polynomials defined over the field of complex numbers.

References

[1] Y.Nambu, Generalized Hamiltonian Dynamics. *Phys.Rev.* D **7** (1973), 2405.

[2] Akhiezer N.I., Elements of theory of elliptic functions, Moscow, "Nauka", 1970.

[3] G.Belardinelli, Fonctions hypergeometriques de plusieurs variables et resolution analitique des equations algebriques generales. Paris: Gauthier-Villars, 1960.

[4] F.Cajoris, History of Mathematics, New York, 1919, pp.349-350.
I.Richards, An application of Galois theory to elementary arithmetic, *Advances in Mathematics* **13** (1974) 268-273.

[5] Harold T.Davis, Introduction to nonlinear differential and integral euations. Dover Publications, INC.,N.Y., 1968. (ISBN 0-486-60971-5)

[6] P.Fjelstad, S.Gal, Two dimensional geometries, topologies, trigonometries and physics generated by complex-type numbers, *Advances in Applied Clifford algebras*, **11(1)**, (2001), 81.

[7] Ch.Hermite, Sur la resolution de l'eqation du cinquieme degree C.R.Acad.Sci.Paris. - 1858. V.46.
Ch.Hermite, Sur l'eqation du cinquieme degree C.R.Acad.Sci.Paris. - 1865. V.46; 1866.-V.62.

[8] L.J.Mordell, On the rational solutions of the indeterminate equations of the 3-rd and 4-th degrees. *Proc.Camb.Phill.Soc.* **21** (1922) 179-192.

[9] P.J.Olver, Classical Invariant Theory, LondonMathematical Society Student Texts 44, Cambridge Univeristy Press, 1999.

[10] Al.Molgado, A.Rodriguez-Dominguez, Mapping betwween the dynamic and mechanical properties of relativistic oscillator and Euler free rigid body. *J.Nonlinear Math. Phys.* **14** No.4 (2007) 534-547.
A.R.Rodriguez-Dominguez, Lorentz-force equations as Heisenberg equations for 4D-quantum system. *Revista Mexicana de Fisica* **53** (4) (2007) 270-280.

[11] Foundations of Nambu mechanics. *Comm.Math.Phys.* **160** (1994) 295.

[12] H.Weber, Lehrbuch der Algebra.Bd.2. Gruppen. Lineare Gruppen. Anwendungen der Gruppen Theorie. Algebraische Zahlen.- Braunschweig, 1899.
H.Weber, Lehrbuch der Algebra.Bd.3. Elliptische Functionen und algebraische Zahlen.- Braunschweig, 1908.

[13] R.M.Yamaleev, Multicomplex algebras on polynomials and generalized Hamilton dynamics, *J.Math.Anal.Appl.*, **322**, (2006), 815-824.
R.M.Yamaleev, Complex algebras on N-order polynomials and generalizations of trigonometry, oscillator model and Hamilton dynamics. *Advances in Applied Clifford algebras*, **15(1)**, (2005), 123.

[14] R.M.Yamaleev, "Relativistic Equations of Motion within Nambu's Formalism of Dynamics", *Ann.Phys.* **285 (2)** (2000) 141-160.

[15] R.M.Yamaleev, "Generalized Newtonian Equations of Motion", *Ann.Phys.* **277 (1)** (1999) 1-18;
R.M.Yamaleev, "Elliptic and Hyperelliptic Deformed Mechanics in n-Dimesional Phase Space", *JINR Communications*,P2-94-109,Dubna, (1995);
R.M.Yamaleev, "Generalized Lorentz-Force Equations", *Ann.Phys.* **292 (1)** (2001) 157-178.

[16] R.M.Yamaleev, Extended Relativistic Dynamics of Charged Spinning Particle in Quaternionic Formulation, *Advances in Applied Clifford Algebras*, **13(2)**, (2003), 183-218.
R.M.Yamaleev, Ternary Electrodynamics, *Far East J.Dynamic Systems*, **9(2)** (2007) 307-324.

In: Evolution Equations
Editor: Arthur L. Claes
ISBN: 978-1-61209-652-0

Chapter 13

Master Equations in the Theory of Stochastic Processes

***Nanasaheb S. Patil*[a] *and Shambhu N. Sharma.*[*,b]**
[a]Department of Electrical Engineering, National Institute of Technology, Surat, India
[b]Instrumentation and Control Engineering Division, NSIT Delhi

Abstract

The stochastic processes, which are exploited to analyze the dynamical system in noisy environments, are white noise process, the Brownian motion process, the Ornstein-Uhlenbeck (OU) process. More notably, the OU process is a coloured noise process. In 1991, Karatzas and Shreve (I. Karatzas and S. E. Shreve, *Brownian Motion and Stochastic Calculus*, Springer-Verlag, New York, 1991) published an authoritative book discussing the Brownian motion process as well as the Brownian motion process-driven stochastic differential system in the rigorous mathematical framework, especially in the Itô setting. The estimation algorithm for the Itô stochastic differential equation (SDE) is developed using the Fokker-Planck equation, the master equation for the Itô SDE setting. The power spectral density of the stochastic process associated with the stochastic problems in physics, digital wireless communications is the frequency-dependent that justifies the usefulness of coloured noise processes in dynamical systems. The problem of analyzing the coloured noise-driven stochastic differential system can be accomplished by generalizing the Fokker-Planck equation. Master equations for coloured noise-driven stochastic differential systems involve the more general structures in contrast to the Fokker-Planck equation. For this reason, it seems quite worthwhile to summarize available results in literature, which are widespread, about the master equations. This paper encompasses six different master equations, which can be exploited to accomplish the estimation procedures for stochastic differential systems, especially where observations are not available. The structure of a master equation is dictated on the basis of the noise process associated with the dynamical system. The proofs of master equations of this paper can be accomplished using the functional calculus and perturbation-theoretic approaches to the system non-linearity-process noise coupling

---

* Instrumentation and Control Engineering Division, NSIT Delhi (on lien), (Presently with National Institute of Technology, Surat, India), snsvolterra@gmail.com

terms. Furthermore, this paper discusses briefly the general theory of two mathematical problems related to the estimation theory for coloured noise processes.

Keywords: coloured noise process, characteristic functional, cumulant generating functional, master equations, the OU process

1. Introduction

The Fokker-Planck equation, the evolution of conditional probability density for given initial states for a Markov process, has found its applications in developing prediction algorithms for stochastic problems arising from finance and technology. A nice exposition about the theory of stochastic processes and the stochastic process-driven dynamical systems can be found in Birkoff [3], Jazwinski [7], Revuz and Yors [13]. The Fokker-Planck equation for one variable, several variables, method of solutions and its applications, especially dealing with laser statistics, can be found in Risken [14]. The Fokker-Planck equation involves first-order time derivative, first-order as well as second-order spatial derivatives of the conditional probability density. Recently, one of us with another coauthor tried to summarize elementary proofs as well as proofs constructed from the standard theories of stochastic processes to arrive at the Fokker-Planck equation (Sharma and Patel [16]). An approximate solution method to the Fokker-Planck equation as well as a Fokker-Planck analysis were explained as well. An alternative proof of the FPE was given by AN Kolmogorov using the different method with different notations, which is also known as the Kolmogorov forward equation (Feller [4]). The standard structure of the stochastic differential system associated with the Fokker-Planck equation is the Brownian motion-perturbed stochastic differential system. More notably, the LTI system driven by the Brownian motion process generates the OU process. The OU process, a coloured noise process, is a Gauss-Markov process. The noise process associated with the dynamical system has finite, non-zero correlation time (Hänggi [6]). Secondly, the power spectral density of flicker noise, which is associated with the CMOS digital circuits, varies with $\frac{1}{f}$-fashion. The autocorrelation of the stationary coloured noise depends on the time-interval and the correlation time. The power spectral density of the coloured noise depends on frequency and correlation terms. The noise perturbation acting on the dynamical system chiefly decides the structure of the SDE. As a result of this, the corresponding master equation is formalized that can be exploited to develop the estimation algorithms for non-linear stochastic differential systems. Publications on the problem of analyzing non-linear dynamical systems in coloured noise environment arising from technology, especially from the Practitioners' view point, are relatively less known (Hänggi [6], Stratonovich [17]).

This paper attempts to explain six different structures of master equations for non-linear stochastic differential systems. This paper reviews the proofs of master equations using the functional calculus approach. The perturbation-theoretic expansion of the system non-linearity-noise perturbation term, which was pioneered by RL Stratonovich, to construct the proofs of master equations is explained in some detail as well. Subsequently, this paper discusses briefly the estimation theory for the coloured noise-driven 'vector' stochastic differential system by utilizing the corresponding master equation. As a result of this, two mathematical problems associated with estimation theory are stated. Some of the results of this paper, which are spread in literature, can be found in authoritative books, physics and probability Journals. On the other hand, some of the results of this paper are not sufficiently well known in literature.

2. Master Equations

The master equation can be defined as the evolution of conditional probability density for given initial states for a process satisfying a stochastic differential equation. The master equation for the Brownian motion-perturbed dynamical system is termed as the Fokker-Planck equation. On the other hand, the master equation for the coloured noise-driven stochastic differential system is termed as stochastic equation. The stochastic equation assumes the structure of an infinite series. In this section, we begin with the proof of the master equation involving the concept of functional derivative (Kowalski and Steeb [11]). The autocorrelation of the coloured noise process with zero mean and stationary property is stated as $\mathrm{cov}(x_t, x_s) = R_{xx}(\tau, \tau_{cor})$, where $\tau = t - s$ and τ_{cor} is the correlation time of the coloured noise process. Furthermore, after introducing the notion of Gaussianity and Markovian character, the coloured noise process becomes an exponentially correlated process, the OU process. Here, the Authors derive the master equation for the coloured noise-driven stochastic differential system,

$$\dot{x}_t = f(x_t) + g(x_t)\xi_t, \tag{1}$$

where the process ξ_t is coloured noise, the terms f, g denote the system non-linearity and the process noise coefficient respectively. Equation (1) can be described as the non-Markovian SDE, since the coloured noise process ξ_t generates the non-Markov process x_t. Define the conditional probability density $p_t(x) = \langle \delta(x_t - x) \rangle$, where $p_t(x) = p(t, x | x_{t_0}, t_0)$. The time derivative of the conditional probability density is

$$\dot{p}_t(x) = -\frac{\partial}{\partial x}\langle \delta(x_t - x)\dot{x}_t \rangle. \tag{2}$$

Equation (2) in combination with equation (1) leads to

$$\dot{p}_t(x) = -\frac{\partial}{\partial x}\langle \delta(x_t - x)(f(x_t) + g(x_t)\xi_t) \rangle$$

$$= -\frac{\partial f(x)p}{\partial x} - \frac{\partial g(x)}{\partial x}\langle \delta(x_t - x)\xi_t \rangle. \tag{3}$$

The cross correlation term $\langle \delta(x_t - x)\xi_t \rangle$ can be derived using the functional Taylor series expansion. The term $a[x+\eta]$ can be stated as

$$a[x+\eta] = \sum_{0\le n} \frac{1}{n!} \int x_{t_1} x_{t_2} .....x_{t_n} \frac{\delta^n a[\eta]}{\delta\eta_{t_1}\delta\eta_{t_2}....\delta\eta_{t_n}} dt_1 dt_2 ...dt_n,$$

multiplying both sides by the process x_t and taking expectation both sides, we have

$$\langle x_t a[x+\eta] \rangle = \sum_{0\le n} \frac{1}{n!} \int \langle x_t x_{t_1} x_{t_2} ....x_{t_n} \rangle \frac{\delta^n a[\eta]}{\delta\eta_{t_1}\delta\eta_{t_2}....\delta\eta_{t_n}} dt_1 dt_2 ...dt_n. \tag{4}$$

The term $\langle x_t x_{t_1} ...x_{t_n} \rangle$ has the interpretation as the $(n+1)$th moment. In the theory of stochastic processes, the $(n+1)$th order derivative of the characteristic functional $\chi[\alpha]$ evaluated at $\alpha = 0$ gives the $(n+1)$th order moment. As a result of this,

$$\langle x_t x_{t_1} ...x_{t_n} \rangle = \frac{\delta^n}{\delta\alpha_{t_1}\delta\alpha_{t_t}..\delta\alpha_{t_n}} \frac{\delta\chi[\alpha]}{\delta\alpha_t}\Big|_{\alpha=0}$$

$$= \frac{\delta^n}{\delta\alpha_{t_1}\delta\alpha_{t_t}..\delta\alpha_{t_n}} (\chi[\alpha] \frac{\delta \log \chi[\alpha]}{\delta\alpha_t})\Big|_{\alpha=0}$$

$$= \sum_{0\le m\le n} \binom{n}{m} (\frac{\delta^m}{\delta\alpha_{t_1}\delta\alpha_{t_2}..\delta\alpha_{t_m}} \frac{\delta \log \chi[\alpha]}{\delta\alpha_t}\Big|_{\alpha=0})$$

$$\times \frac{\delta^{n-m}\chi[\alpha]}{\delta\alpha_{t_{m+1}}\delta\alpha_{t_{m=2}}....\delta\alpha_{t_n}\delta\alpha_{t_t}..\delta\alpha_{t_n}}\Big|_{\alpha=0}$$

$$= \sum_{0\le m\le n} \frac{n!}{m!n-m!} C_{m+1}(t, t_1,......, t_m) m_{n-m}(t_{m+1}, t_{m+2},..., t_n), \tag{5}$$

where the terms $C_{m+1}(t, t_1,......, t_m)$, and $m_{n-m}(t_{m+1}, t_{m+2}, ..., t_n)$ are $(m+1)$th order cumulant and $(n-m)$th order moment respectively. After plugging equation (5) into equation (4), we have

$$\langle x_t a[x+\eta]\rangle = \sum_{0\le n}\int(\sum_{0\le m\le n}\frac{1}{m!n-m!}C_{m+1}(t,t_t,t_2,...,t_m)m_{n-m}(t_{m+1},t_{m+2},..,t_n)$$

$$\times\frac{\delta^n a[\eta]}{\delta\eta_{t_1}\delta\eta_{t_2}....\delta\eta_{t_n}})dt_1dt_2...dt_n,$$

$$=\sum_{0\le m}\frac{1}{m!}\int C_{m+1}(t,t_1,t_2,...,t_m)dt_1dt_2...dt_m(\frac{\delta^m}{\delta\eta_{t_1}\delta\eta_{t_2}...\delta\eta_{t_m}}$$

$$(\sum_{0\le r}\frac{1}{r!}\int m_r(t_1,t_2,....,t_r)\frac{\delta^r a[\eta]}{\delta\eta_{t_1}\delta\eta_{t_2}...\delta\eta_{t_r}}dt_1dt_2..dt_r)).$$

The term within the second bracket of the above expression can be regarded as $\frac{\delta^m\langle a[x+\eta]\rangle}{\delta n_{t_1}\delta\eta_{t_2}...\delta\eta_{t_m}}=\left\langle\frac{\delta^m a[x+\eta]}{\delta x_{t_1}\delta x_{t_2}...\delta x_{t_m}}\right\rangle$. As a result of this, we have

$$\langle x_t a[x+\eta]\rangle=\sum_{0\le m}\frac{1}{m!}\int C_{m+1}(t,t_1,t_2,...,t_m)\left\langle\frac{\delta^m a[x+\eta]}{\delta x_{t_1}\delta x_{t_2}...\delta x_{t_m}}\right\rangle dt_1dt_2..dt_m.$$

For $x_t=\xi_t$, and $a[x+\eta]=\delta(x_t-x)$, we arrive at

$$\langle\delta(x_t-x)\xi_t\rangle=\sum_{n=0}^{\infty}\frac{1}{n!}\int_{t_0}^{t}....\int_{t_0}^{t}C_{n+1}(t,t_1,t_2,.......,t_n)\left\langle\frac{\delta^n\delta(x_t-x)}{\delta\xi_{t_1}\delta\xi_{t_2}.....\delta\xi_{t_n}}\right\rangle dt_1dt_2....dt_n. \quad (6)$$

Equation (2) in conjunction with equation (6) leads to

$$\dot{p}(x)=-\frac{\partial f(x)p}{\partial x}-\frac{\partial g(x)}{\partial x}(\sum_{m=0}^{\infty}\frac{1}{m!}\int C_{m+1}(t,t_1,....,t_m)\left\langle\frac{\delta^m\delta(x_t-x)}{\delta\xi_{t_1}\delta\xi_{t_2}.....\delta\xi_{t_m}}\right\rangle dt_1dt_2....dt_m)$$

$$=-\frac{\partial f(x)p}{\partial x}-\langle\xi_t\rangle\frac{\partial g(x)p}{\partial x}-\frac{\partial g(x)}{\partial x}$$

$$\times(\sum_{m=1}^{\infty}\frac{1}{m!}\int C_{m+1}(t,t_1,.....,t_m)\left\langle\frac{\delta^m\delta(x_t-x)}{\delta\xi_{t_1}\delta\xi_{t_2}.....\delta\xi_{t_m}}\right\rangle dt_1dt_2....dt_m). \quad (7)$$

Note that equation (7) can be regarded as the *first* master equation of the paper, the master equation for the coloured noise-driven stochastic differential system. The proof of the master equation, equation (7), involving the concept of functional derivative is also given in Hänggi [6]. Equation (7) can be regarded as a parabolic linear non-homogeneous equation with constant coefficients of order one in partial differentiation. The non-homogeneity is attributed to the last term of the right-hand side of equation (7) involving the integral sign.

Here, the Authors state the *second* master equation of the paper, the master equation for the coloured noise process-driven stochastic differential system, where the input noise process is zero mean and stationary. As a result of this, we have the master equation for the coloured noise-driven stochastic differential system that can be recast as

$$\dot{p}_t(x) = -\frac{\partial f(x)p}{\partial x} - \frac{\partial g(x)}{\partial x}$$

$$(\sum_{m=1}^{\infty}\frac{1}{m!}\int C_{m+1}(t_k - t_{k-1})\left\langle \frac{\delta^m \delta(x_t - x)}{\delta\xi_{t_1}\delta\xi_{t_2}.....\delta\xi_{t_m}}\right\rangle dt_1 dt_2....dt_m), \tag{8}$$

where the interval $\tau = t_k - t_{k-1}, 1 \leq k \leq m$ and $t_0 = t$. Note that equation (8) is a consequence of equation (7).

Consider the input noise process is the weakly coloured noise process with zero mean and stationary property. The master equation for the weakly coloured noise-driven stochastic differential system can be derived by taking the Taylor series expansion of the term $\frac{\delta^m (x_t - x)}{\delta\xi_{t_1}\delta\xi_{t_2}....\delta\xi_{t_m}}$ in powers of $t_{k-1} - t_k$ upto the first-order expansion term. However, the simplified structure of the master equation seems to be difficult because of the m th order functional derivative associated with the equation. An alternative and a popular structure of the master equation for the weakly coloured noise-driven stochastic differential system can be obtained by considering a special case of the stochastic equation that can be found in the succeeding discussion of the paper, see equation (20) of the paper.

The OU process has received considerable attention as well as its long standing history. The OU process is a stationary, Gauss-Markov process. Because of the Gaussianity, the second-order cumulant becomes covariance and higher-order cumulants vanish. After introducing the zero mean and stationary conditions, i.e. cov $(t,s) = R_{\xi\xi}(t-s), t > s$, ignoring higher-order cumulants, equation (8) reduces to the following:

$$\dot{p}_t(x) = -\frac{\partial f(x)p}{\partial x} - \frac{\partial g(x)}{\partial x}\int R_{\xi\xi}(t-s)\left\langle \frac{\delta\, \delta(x_t - x)}{\delta\xi_s}\right\rangle ds$$

$$= -\frac{\partial f(x)p}{\partial x} + \frac{\partial g(x)}{\partial x}\frac{\partial}{\partial x}\int R_{\xi\xi}(t-s)\left\langle \delta(x_t - x)\frac{\delta(x_t - x)}{\delta\xi_s}\right\rangle ds. \tag{9}$$

Equation (9) can be regarded as the *third* master equation of this paper, the master equation for the OU process-driven stochastic differential system. Equation (9) can be further rewritten by replacing the term $R_{\xi\xi}(t-s)$ with the autocorrelation of the OU process.

Here, we derive the master equation for the OU process-driven stochastic differential system, where the OU process is a weakly coloured noise process that can be regarded as the *fourth* master equation of this paper. The weakly coloured noise can be defined as a noise process with finite, non-zero and smaller correlation time. We exploit the definition, zero mean, stationary, Gauss-Markov properties of the weakly coloured noise process as well as the first-order Taylor series expansion of the functional derivative $\frac{\delta(x_t - x)}{\delta\xi_s}$. The functional derivative $\frac{\delta(x_t - x)}{\delta\xi_s}$ of the state x_t with respect to the noise process ξ_s can be stated as (Hänggi [6], Sharma and Patel [16])

$$\frac{\delta(x_t - x)}{\delta\xi_s} = g(x_t)\exp\left(\int_s^t (f'(x_\tau) - g'(x_\tau)\frac{f(x_\tau)}{g(x_\tau)})d\tau\right). \tag{10}$$

The first-order Taylor series expansion of the functional derivative $\frac{\delta(x_t - x)}{\delta\xi_s}$ in powers of $(s-t)$, where $s \le t$, in combination with equation (10) can be stated as

$$\frac{\delta(x_t - x)}{\delta\xi_s} = \frac{\delta x_t}{\delta\xi_t} + (s-t)\frac{\partial}{\partial s}\left(\frac{\delta x_t}{\delta\xi_s}\right)\Big|_{s=t}$$

$$= g(x_t) + (t-s)\, g(x_t)(f'(x_t) - g'(x_t)\frac{f(x_t)}{g(x_t)})$$

$$= g(x_t)(1 + (t-s)\, (\frac{f'(x_t)g(x_t) - g'(x_t)f(x_t)}{g(x_t)})). \tag{11}$$

The second-order cumulant of the OU process can be stated as

$$C_2(t, s) = R_{\xi\xi}(t-s) = \frac{D}{\tau_{cor}} e^{-\frac{|t-s|}{\tau_{cor}}}. \tag{12}$$

After combining equations (11)-(12) with equation (9), we have

$$\dot{p}(x) = -\frac{\partial}{\partial x} f\, p + D\frac{\partial}{\partial x}(g\frac{\partial}{\partial x} g(1+\tau_{cor} g(\frac{f}{g})')p).$$

Here, we explain briefly an alternative approach to arrive at the master equation for the coloured noise-driven stochastic differential equation. The alternative approach to arrive at the master equation is different in two senses: (i) exploiting the notion of the transition probability density, conditional characteristic function, the perturbation-theoretic expansion of the system non-linearity-noise process coupling term in contrast to the functional derivative concept (ii) adopting the different notations. Here, the Authors will not delve into a greater detail of arriving at the alternative proof of the master equation for the coloured noise-driven stochastic differential system. A greater detail can be found in Stratonovich ([17], p. 91). The evolution of conditional probability density $p(x)$ for the scalar stochastic differential equation, the *fifth* master equation of the paper, can be stated as

$$\dot{p}(x) = \sum_{1\le n} \frac{(-1)^n}{n!} \frac{\partial^n k_n(x) p(x)}{\partial x^n}, \tag{13}$$

where $k_n(x) = \underset{\tau\to 0}{Lim} \frac{E((dx_t)^n | x_t = x)}{\tau}$. Making the use of the perturbation-theoretic expansion of the term $F(x_t, \xi_t)$, considering the input noise process with zero mean, stationary property and subsequently, taking the special case of the term $F(x_t, \xi_t)$, i.e. $f(x_t) + g(x_t)\xi_t$, we arrive at the relatively simpler expression of the term $k_n(x)$ of the master equation. As a result of these, we have

$$k_1(x) = f + \frac{\mu_2}{2} gg' + \frac{\mu_3}{6} g(gg')' + \mu_1 g^2 g'(\frac{f}{g})',$$

$$k_2(x) = \mu_2 g^2 + \mu_3 g^2 g' + 2\mu_1 g^3 (\frac{f}{g})',\ k_3(x) = \mu_3 g^3, \tag{14}$$

where the prime sign denotes the spatial derivative. The contributions to the coefficients μ_1, μ_2 and μ_3 of the above set of equations come from the input process ξ_t as well as its autocorrelations. The explicit expressions for the terms μ_1, μ_2 and μ_3 can be found in equation (4.180) of Stratonovich [17]. Despite involving different structures of the master equations, equations (8) and (13), the both describe the master equation for the stochastic differential system of equation (1). The difference in the structure is attributed to the difference in the method adopted to arrive at the expressions. It is interesting to note that the special cases of the both equations w'd be same under some assumptions and approximations, e.g. the OU process by treating it as a weakly coloured noise input process.

Consider the input process ξ_t is a weakly coloured noise process with zero mean and stationary process, see equation (1). The weakly coloured noise process has smaller correlation time. As a result of this, the master equation for the weakly coloured noise process-driven stochastic differential system reduces to the Fokker-Planck equation, i.e.

$$\dot{p}(x) = \sum_{1 \le n \le 2} \frac{(-1)^n}{n!} \frac{\partial^n k_n(x) p(x)}{\partial x^n}, \tag{15}$$

where

$$k_1(x) = f + \frac{\mu_2}{2} gg' + \mu_1 g^2 g' (\frac{f}{g})', \; k_2(x) = \mu_2 g^2 + 2\mu_1 g^3 (\frac{f}{g})'. \tag{16}$$

Equation (16) is a special case of equation (14). The term $k_n(x)$ of equation (15), $n \geq 3$, vanishes for the weakly coloured noise input process, a greater detail is given in Stratonovich ([17], p. 98). Consider

$$k_1(x) = M(x) + \frac{1}{4}\frac{dk(x)}{dx}, \; k_2(x) = k(x) \tag{17}$$

as well as exploiting equation (16), we have

$$\frac{dk(x)}{dx} = 2\mu_2 gg' + 2\mu_1 (3g^2 g'(\frac{f}{g})' + g^3 (\frac{f}{g})''),$$

$$\frac{1}{4}\frac{dk(x)}{dx} = \frac{\mu_2 gg'}{2} + \frac{1}{2}\mu_1 (3g^2 g'(\frac{f}{g})' + g^3 (\frac{f}{g})'')$$

$$= \frac{1}{2}\mu_2 gg' + \frac{3}{2}\mu_1 g^2 g'(\frac{f}{g})' + \frac{\mu_1}{2} g^3 (\frac{f}{g})'', \tag{18}$$

$$M(x) = f - \frac{\mu_1}{2} g^2 g'(\frac{f}{g})' - \frac{\mu_1}{2} g^3 (\frac{f}{g})''.$$

A simple calculation shows that the Fokker-Planck equation, equation (15), can be rewritten as

$$\dot{p}(x) = -\frac{\partial}{\partial x} M(x)p(x) + \frac{1}{4}\frac{\partial}{\partial x}(k(x)\frac{\partial p}{\partial x} + \frac{\partial k(x)p(x)}{\partial x}). \tag{19}$$

The above equation can be regarded as a consequence of equations (15) and (17). Equation (19) in combination with equation (18) leads to

$$\dot{p}(x) = -\frac{\partial}{\partial x} M(x)p(x) + \frac{1}{4}\frac{\partial}{\partial x}(k(x)\frac{\partial p}{\partial x} + \frac{\partial k(x)p(x)}{\partial x})$$

$$= -\frac{\partial}{\partial x} f\,p + \frac{1}{2}\mu_1 \frac{\partial}{\partial x} g^2 g'(\frac{f}{g})' p + \frac{\mu_1}{2}\frac{\partial}{\partial x} g^3 (\frac{f}{g})'' p$$

$$+\frac{1}{4}\mu_2 \frac{\partial}{\partial x}(g^2 \frac{\partial p}{\partial x}) + \frac{1}{2}\mu_1 \frac{\partial}{\partial x}(g^3 (\frac{f}{g})' \frac{\partial p}{\partial x})$$

$$+\frac{1}{4}\mu_2 \frac{\partial^2 g^2 p}{\partial x^2} + \frac{1}{2}\mu_1 \frac{\partial^2}{\partial x^2} g^3 (\frac{f}{g})' p, \tag{20}$$

where the term $(\frac{f}{g})$ of equation (20) is twice differentiable. The above equation can be considered as the *sixth* master equation of this paper. Fox [5] can be consulted for the Fokker-Planck analysis of the weakly coloured noise-driven stochastic differential system.

Here, the Authors explain the master equation for the Brownian motion-driven stochastic differential system, which can be regarded as a special case of equation (13). Consider the stochastic differential system of the form $dx_t = f(x_t)dt + g(x_t)dB_t$. The terms

$$k_1(x) = \underset{\tau\to 0}{Lim} \frac{E((dx_t)|x_t = x)}{\tau} = f(x),$$

$$k_2(x) = \underset{\tau\to 0}{Lim} \frac{E((dx_t)^2|x_t = x)}{\tau} = g^2(x),$$

where $dt = \tau$. The higher-order terms vanish, which are the consequence of the Itô theory. Thus, the infinite series structure of the master equation stated in equation (13) reduces to a finite series involving the first-order and second-order spatial derivatives. The resulting evolution equation is

$$\dot{p}(x) = -\frac{\partial f(x)p(x)}{\partial x} + \frac{1}{2}\frac{\partial^2 g^2(x)p(x)}{\partial x^2}.$$

The above equation is termed as the Kolmogorov-Fokker-Planck equation, the master equation for the Brownian motion-driven stochastic differential system.

3. Two Mathematical Problems in Coloured Noise Process

Here, the Authors explain two mathematical problems, which merit considerable attention on the part of investigators. The both problems are about the estimation theory for the coloured noise-driven *vector* stochastic differential equation from the prediction as well as filtering viewpoints.

(i) The first problem is to develop the exact as well as approximate prediction equations for the coloured noise-driven stochastic differential equation. The prediction equations involve the conditional mean and variance evolution equations without accounting observations. On the other hand, filtering equations involve the conditional mean and variance evolution equations accounting observations. Equation (13) describes the master equation for the coloured noise-driven stochastic differential system, the stochastic equation. The master equation for the coloured noise-driven *vector* stochastic differential system can be stated by introducing the component-wise contribution, i.e.

$$\dot{p}(x) = \sum_{1\le p} \sum_{(i_1,i_2,i_3,\ldots,i_p)\in s_p} \frac{(-1)^p}{p!} \frac{\partial^p k_{i_1,i_2,\ldots,i_p}(x)p(x)}{\partial x_{i_1}\partial x_{i_2}\ldots\partial x_{i_p}}, \tag{21}$$

where $1 \le i_p \le n$, $1 \le p$ and the size of the state vector is n. Suppose the operator associated with the above has the adjoint properties, then the backward operator can be stated as the term

$$\sum_{1\le p} \sum_{(i_1,i_2,i_3,\ldots,i_p)\in s_p} \frac{k_{i_1,i_2,\ldots,i_p}(x)}{p!} \frac{\partial^p(.)}{\partial x_{i_1}\partial x_{i_2}\ldots\partial x_{i_p}}.$$

Note that the backward operator acts on the scalar function. The vector stochastic equation (21) assumes an infinite series structure as well as the contribution to term $k_{i_1,i_2,\ldots,i_p}$ comes from the system non-linearity-noise perturbation coupling terms. As a results of these, the problem of deriving the exact evolution of conditional moment exploiting equation (21) as well as approximate evolution of conditional moment, which can be regarded as the *prediction* equations, seems intractable that merits further investigations.

(ii) Here the Authors explain the second potential problem, which is about the filtering theory for the coloured noise-driven *vector* stochastic differential system. Most notably, the first step to develop filtering theory is to derive the expression of the filtering density $p(x|Y_t)$ and its stochastic evolution. The filtering density evolution for the Itô stochastic differential system, where the Brownian noise is the observation noise, can be formalized as a stochastic partial-integro differential equation (Liptser and Shiryaev ([12], p. 318), Sharma [15]). On the other hand, the filtering theory for the dynamical system embedded in the coloured noise environment becomes quite difficult problem, especially the case considering the coloured noise as the observation noise. More precisely, consider the filtering model assuming the structure, i.e.

$$\dot{x}_t = f(x_t) + g(x_t)\xi_t, \ \dot{z}_t = h(x_t) + \varsigma_t. \tag{22}$$

Equation (22) describes the dynamical system driven by the coloured noise process ξ_t and the observation noise is the coloured noise ς_t. The filtering density for the filtering model, i.e. equation (22), can be stated as

$$p(x|Y_t) = \underset{\substack{(t_{k+1}-t_k)\to 0 \\ n\to\infty}}{LIM} p(x|z_{\tau_n}, z_{\tau_{n-1}}, \ldots\ldots, z_{\tau_2}, z_{\tau_1}, z_{\tau_0}).$$

Furthermore, the filtering density

$$p(x|z_{\tau_n}, z_{\tau_{n-1}}, \ldots\ldots, z_{\tau_2}, z_{\tau_1}, z_{\tau_0}) = \frac{p(z_{\tau_n}, z_{\tau_{n-1}}, \ldots\ldots, z_{\tau_2}, z_{\tau_1}, z_{\tau_0}|x_{\tau_n} = x)p(x)}{p(z_{\tau_n}, z_{\tau_{n-1}}, \ldots\ldots, z_{\tau_2}, z_{\tau_1}, z_{\tau_0})}$$

and the term

$$p(z_{\tau_n}, z_{\tau_{n-1}}, \ldots\ldots, z_{\tau_2}, z_{\tau_1}, z_{\tau_0}|x_{\tau_n} = x) =$$

$$E(p(z_{\tau_n}, z_{\tau_{n-1}}, \ldots\ldots, z_{\tau_2}, z_{\tau_1}, z_{\tau_0}|x_{\tau_n}, x_{\tau_{n-2}}, \ldots, x_{\tau_2}, x_{\tau_1}, x_{\tau_0})|x_{\tau_n} = x).$$

The closed-form expression under the limiting case for the conditional expectation of the conditional probability density $p(z_{\tau_n}, z_{\tau_{n-1}}, \ldots\ldots, z_{\tau_2}, z_{\tau_1}, z_{\tau_0}|x_{\tau_n}, x_{\tau_{n-2}}, \ldots, x_{\tau_2}, x_{\tau_1}, x_{\tau_0})$ seems to be intractable, since the observation noise is not the Gaussian process. This introduces the formidable complexity into deriving the closed-form expression for the filtering density as well as its evolution. For this reason, filtering theory for the above filtering model merits considerable attention. Contributions to the coloured noise filtering theory from the *dynamical systems*' viewpoint will advance the topics in the theory of stochastic processes, find their applications for stochastic problems arising from diverse field as well as reveal many-sided relationships.

The motivation for stating two mathematical problems associated with filtering theory and stochastic processes of this paper hinges on the tradition set by David Hilbert for advancing mathematics, announcing open problems in multi-varied branches of mathematics. The Hilbert's viewpoint confirms unity of mathematics, impossibility of dividing it among different branches. A trend for posing general questions in sciences can be also found in Kakutani's famous talk [9]delivered in the 1950 International Congress of Mathematics, see [8] as well. Furthermore, AN Kolmogorov suggested 'analyzing specific problems, especially in dynamical systems and mechanics, generates joining points between the different branches of mathematics' (Abraham and Marsden [1], pp. 741-757), see the celebrated book, *Mathematical Methods of Classical Mechanics*, authored by Arnold [2] as well.

4. Conclusion

In this paper, the Authors have compiled six different structures of master equations and their proofs using the functional calculus as well as the perturbation-theoretic approaches. More notably, the structure of a master equation is adjudged on the basis of the noise perturbation driving the dynamical system. The master equations of this paper can be explored in a greater detail for developing estimation algorithms for dynamical systems embedded in noisy environments. Some of the results of the paper, especially equations (8), (20)-(21), are not widely known in the literature as well.

Another contribution of this paper is to highlight two mathematical problems associated with 'estimation theory for coloured noise processes' by discussing briefly their general theory. Furthermore, these two open problems deserve considerable attention and resolving them will lead to exciting connections as well as applications in different branches of science. Contributions to the two mathematical problems of this paper will advance the theory of stochastic processes.

References

[1] R. Abraham, J. E. Marsden, *Foundations of Mechanics* (Second Edition), Addison-Wesley Publishing Company, Inc., Redwood City, CA, 1987.

[2] V. I. Arnold, *Mathematical Methods of Classical Mechanics* (Graduate Text in Mathematics), Springer, USA, 1989.

[3] G. D. Birkoff, Dynamial systems, *Colloq. Publ.IX, Amer. Math. Soc.*, Providence, R. I., 1966.

[4] W. Feller, *An Introduction to Probability Theory and its Applications*, John Wiley and Sons, New York and Chichester, vol.2, 2000.

[5] R. F. Fox, Uniform convergence to an effective Fokker-Planck equation for weakly colored noise, *Physical Review A* 34 (1986), pp. 4525-4527.

[6] P. Hänggi, The functional derivative and its use in the description of noisy dynamical systems, *Stochastic Processes Applied to Physics* (L. Pesquera and M. Rodriguez, Eds.), World Scientific, Heyden, Philadelphia, 1995, pp. 69–95.

[7] A. H. Jazwinski, *Stochastic Processes and Filtering Theory*, Academic Press, New York and London, 1970.

[8] S. Kakutani, *Proc. Natl. Acad. Sci. U.S.A, 28*(1), 1942, pp. 16-21.

[9] S. Kakutani, Ergodic theory, *Proc. Intern. Congr. Math.* 2(1950), pp.128-142.

[10] I. Karatzas, S. E. Shreve, *Brownian Motion and Stochastic Calculus* (graduate text in mathematics), Springer, New York, 1991.

[11] K. Kowalski, W-H. Steeb *Non-Linear Dynamical Systems and Carleman Linearization*, World Scientific, Singapore, New Jersy, 1991.

[12] R. S. Liptser, A. N. Shiryaev, *Statistics of Random Processes* (vol.1), Springer, Berlin 1977.

[13] D. Revuz, M. Yor, *Continuous Martingales and Brownian Motion*, Springer-Verlag, Berlin, Heidelberg, 1991.

[14] H. Risken, *The Fokker-Planck Equation: Method of Solutions and Applications*, Springer-Verlag, Berlin, 1984.
[15] Shambhu N. Sharma, A Kushner approach for small random perturbations of a stochastic Duffing-van der Pol system, *Automatica* (a Journal of IFAC, International Federation of Automatic Control) 45 (2009), pp. 1097-1099.
[16] Shambhu N. Sharma, Hiren G Patel, The Fokker-Planck equation, *Stochastic Control* (Chris Myers, Ed.), Sycio Science Publisher, Vienna, Rizeka, August 2010, pp. 1-20.
[17] R. L. Stratonovich, *Topics in the Theory of Random Noise* (vol 1), Gordan and Breach, New York 1963.

In: Evolution Equations
Editor: Arthur L. Claes
ISBN: 978-1-61209-652-0

Chapter 14

NONLINEAR EVOLUTION INCLUSIONS BASED ON YOSIDA APPROXIMATIONS

Ram U. Verma*
Texas A&M University, Department of Mathematics
Kingsville, Texas, U.S.

Abstract

Based on a general model for the Yosida approximation, the solvability of a general class of first-order nonlinear evolution inclusions is investigated. The notion of the $H-$maximal monotonicity plays a greater role in developing a general framework for the existing Yosida approximations and its characterizations in the literature.

2000 Mathematics Subject Classifications: 49J40, 65B05

Keywords and phrases: Generalized first-order evolution inclusions; variational problems; maximal monotone mapping; $H-$maximal monotone mapping; generalized resolvent operator; general Yosida approximations.

1. Preliminaries

Consider a real separable Hilbert space with the norm $\|\cdot\|$ and the inner product $\langle\cdot,\cdot\rangle$. We study a general class of first-order nonlinear evolution inhomogeneous inclusions of the form:

$$u'(t) + Mu(t) \ni f(t)\, for\, almost\, all\, t \in (0,T), \qquad u(0) = u_0, \tag{1.1}$$

where $M : X \to 2^X$ is a multivalued mapping on X, $f \in W_2^1(0,T;X)$, T is fixed, $0 < T < \infty$, and $u : [0,\infty) \to X$ is a continuous function such that the above inclusion problem holds. Note that the inclusion problem for the Banach space version for the above inclusion problem differs only in the sense that the notions of the monotonicity and

*E-mail address: verma99@msn.com

maximal monotonicity properties are replaced, respectively, by the accretivity and maximal accretivity for the set-valued mapping M. In addition, the solution framework will be different in a generalized way.

Definition 1.1. Let $M : X \to 2^X$ be a set-valued mapping on a real Hilbert space X, and let $H : X \to X$ be $(r)-$strongly monotone. Then M is said to be accretive if $R^M_{\rho,H}$ is single-valued and $(\frac{1}{r})-$Lipschitz continuous for $r > 0$. Furthermore, M is $m-$accretive if M is accretive and $R^M_{\rho,H}$ exists for every $\rho > 0$ on X, where $R^M_{\rho,H}$ is the resolvent of M.

Lemma 1.1. Let $M : X \to 2^X$ be a set-valued mapping on a real Hilbert space X. Then following properties are equivalent:

(i) M is monotone.

(ii) M is accretive.

Lemma 1.2. Let $M : X \to 2^X$ be a set-valued mapping on a real Hilbert space X. Let $H : X \to X$ be $(r)-$strongly monotone. Then we have the following implications that are equivalent:

(i) M is $H-$maximal monotone.

(ii) M is monotone and $R(H + \rho M) = X$.

(iii) M is $m-$accretive.

We intend in this chapter to examine the solvability of the inclusion problem (1.1) based on the notion of the $H-$ maximal monotonicity [2] and the generalized Yosida approximations. The generalized Yosida approximation turns out to be Lipschitz continuous, while we explore the inclusion problem (1.1). The obtained results seem to be application-enhanced to problems arising from other fields, including optimization and control theory, variational inequality and variational inclusion problems, and unify a large class of results relating to nonlinear first-order evolution equations/inclusions. There are also some detailed results that are investigated on the generalized Yosida approximations empowered by the $H-$ maximal monotonicity frameworks. Furthermore, the results are general in nature. For more details, we refer the reader [1-15].

The content of this submission is organized as follows: Section 1 deals, as usual, with introductory and preliminary materials on first-order nonlinear evolution equations/inclusions based on the Yosida approximation. In Section 2, the $H-$ maximal monotonicity and related auxiliary results are discussed, while in Section 3, the Yosida approximation is generalized to case of the resolvent operator based on $H-$ maximal monotonicity models with several results presented, especially for the solvability of the generalized first-order nonlinear evolution inclusions.

2. Auxiliary Results

Definition 2.1. Let $H : X \to X$ be an $(r)-$strongly monotone single-valued mapping and $M : X \to 2^X$ be a set-valued mappings. The map $M : X \to 2^X$ is said to be $H-$maximal monotone if

(i) M is monotone,

(ii) $R(H + \rho M) = X$ for $\rho > 0$.

Definition 2.2. Let $H : X \to X$ be a single-valued mapping and $M : X \to 2^X$ be a set-valued mapping. Let H be $(r)-$ strongly monotone. The map M is said to be accretive iff $(H+\rho M)^{-1}$ is single-valued and $(H+\rho M)^{-1}$ is Lipschitz continuous for all $\rho > 0$.

Proposition 2.1. [2] Let $H : X \to X$ be a single-valued mapping, and $M : X \to 2^X$ be a set-valued mapping such that $D(H) \cap D(M) \neq \emptyset$. Let H be $(r)-$ strongly monotone, and let M be an $H-$maximal monotone mapping. Then the generalized resolvent operator associated with M and defined by

$$R^M_{\rho,H}(u) = (H + \rho M)^{-1}(u)\,\forall\, u \in X,$$

is $(\frac{1}{r})-$ Lipschitz continuous.

Next, we generalize the Yosida approximation M_ρ by $M_\rho = \rho^{-1}(I - HR^M_{\rho,H})$, where $H : X \to X$ is an $(r)-$strongly monotone mapping on X for $\rho > 0$, and for $R^M_{\rho,H} = (H + \rho M)^{-1}$, which reduces to the Yosida approximation of M for $H = I$:

$$M_\rho = \rho^{-1}(I - R^M_\rho),$$

where I is the identity and $R^M_\rho = (I + \rho M)^{-1}$.

Lemma 2.1. Let $H : X \to X$ and $M : X \to 2^X$ be mappings such that $D(H) \cap D(M) \neq \emptyset$. Let H be $(r)-$ strongly monotone and $HoR^M_{\rho,H}$ be cocoercive, and let M be an $H-$maximal monotone mapping. Then the generalized Yosida approximation M_ρ of M defined by

$$M_\rho = \rho^{-1}(I - HR^M_{\rho,H}),$$

where

$$R^M_{\rho,H}(u) = (H + \rho M)^{-1}(u)\,\forall\, u \in D(H) \cap D(M),$$

is $(\frac{1}{\rho})-$ Lipschitz continuous.

Proof. For any $u, v \in X$, we have

$$\begin{aligned} & \langle M_\rho(u) - M_\rho(v), u - v\rangle \\ = \; & \langle M_\rho(u) - M_\rho(v), \rho[M_\rho(u) - M_\rho(v) - (M_\rho(u) - M_\rho(v))] + u - v\rangle \\ = \; & \rho\|M_\rho(u) - M_\rho(v)\|^2 \end{aligned}$$

$$\begin{aligned}
&- \langle \rho^{-1}[u - v - (HR^M_{\rho,H}(u) - HR^M_{\rho,H}(v))], -(HR^M_{\rho,H}(u) - HR^M_{\rho,H}(v))\rangle \\
&= \rho\|M_\rho(u) - M_\rho(v)\|^2 + \rho^{-1}\langle u - v, HR^M_{\rho,H}(u) - HR^M_{\rho,H}(v)\rangle \\
&- \rho^{-1}\langle HR^M_{\rho,H}(u) - HR^M_{\rho,H}(v), HR^M_{\rho,H}(u) - HR^M_{\rho,H}(v)\rangle \\
&\geq \rho\|M_\rho(u) - M_\rho(v)\|^2 + \rho^{-1}\|HR^M_{\rho,H}(u) - HR^M_{\rho,H}(v)\|^2 \\
&- \rho^{-1}\|HR^M_{\rho,H}(u) - HR^M_{\rho,H}(v)\|^2 \\
&\geq \rho\|M_\rho(u) - M_\rho(v)\|^2.
\end{aligned}$$

For $H = I$, Lemma 2.1, it reduces to

Lemma 2.2. Let $M : X \to 2^X$ be a set-valued mapping. Let M be a maximal monotone mapping. Then the Yosida approximation M_ρ of M defined by

$$M_\rho = \rho^{-1}(I - R^M_\rho),$$

where

$$R^M_\rho(u) = (I + \rho M)^{-1}(u)\,\forall\, u \in D(M),$$

is $(\frac{1}{\rho})-$ Lipschitz continuous.

Proposition 2.2. Let $H : X \to X$ and $M : X \to 2^X$ be mappings such that $D(H) \cap D(M) \neq \emptyset$. Let H be $(r)-$ strongly monotone and $HoR^M_{\rho,H}$ be cocoercive, and let M be an $H-$maximal monotone mapping. Then the generalized Yosida approximation M_ρ of M defined by $M_\rho = \rho^{-1}(I - HR^M_{\rho,H})$, satisfies:

(i) $M_\rho(u) \in MR^M_{\rho,H}(u)$.

(ii) M_ρ is $H-$maximal monotone.

(iii) $(M_\lambda)_\mu = M_{\lambda+\mu}$.

Proof. To prove (i), consider

$$\begin{aligned}
w = R^M_{\rho,H}(u) &\Rightarrow u \in (H + \rho M)(w) \\
&\Rightarrow \rho M_\rho(u) = u - H(w) \in \rho M(w).
\end{aligned}$$

The proofs of (ii) and (iii) follow, respectively, from the Lipschitz continuity of the generalized resolvent $R^M_{\rho,H}(u)$ and the definition of M_ρ.

Proposition 2.3. Let $H : X \to X$ and $M : X \to 2^X$ be mappings such that $D(H) \cap D(M) \neq \emptyset$. Let H be $(r)-$ strongly monotone and $HoR^M_{\rho,H}$ be cocoercive, and let M be an $H-$maximal relaxed monotone mapping. Then the generalized Yosida approximation M_ρ of M defined by $M_\rho = \rho^{-1}(I - HR^M_{\rho,H})$ satisfies: for all $u \in D(M)$,

$$M_\rho(u) \to M_0(u), \|M_\rho(u)\| \uparrow \|M_0(u)\|\ for\ \rho \downarrow 0$$

$$\|M_\rho(u) - M_0(u)\|^2 \leq \|M_0(u)\|^2 - \|M_\rho(u)\|^2\ for\ all\ \rho > 0.$$

3. Generalized First-Order Evolution Inclusions

Let $H : X \to X$ be a single-valued mapping, and $M : X \to 2^X$ be a multivalued mapping. In this section, we consider the solvability of first-order nonlinear evolution inclusions of the form

$$u'(t) + Mu(t) \ni f(t) \; for\, almost\, all\, t \in (0, T)$$
$$u(0) = u_0, \tag{3.1}$$

where $M : X \to 2^X$ is a multivalued mapping on X, $f \in W_2^1(0, T; X)$, T is fixed, $0 < T < \infty$, and $u : [0, \infty) \to X$ is a continuous function such that (3.1) holds. Here M is $H-$maximal monotone and the Yosida approximation of M is defined by

$$M_\rho = \rho^{-1}(I - H(R_{\rho,H}^M)).$$

We generalize the results on the first-order evolution equations to the case of the first-order evolution inclusions based on the $H-$maximal monotonicity frameworks and the generalized Yosida approximations.

Theorem 3.1. Let $H : X \to X$ be $(r)-$strongly monotone, and let $M : X \to 2^X$ be $H-$maximal monotone on a separable Hilbert space X. Let $HoR_{\rho,H}^M$ be cocoercive, where $R_{\rho,H}^M) = (H + \rho M)^{-1}$ for $\rho > 0$. Suppose that the given

$$u_0 \in D(M), f \in W_2^1(0, T; X)$$

are fixed. Then (3.1) has exactly one solution $u \in W_2^1(0, T; X)$ such that $M : X \to 2^X$ is $H-$maximal monotone.

Proof. The proof is based on the results from Section 2, especially Lemma 2.1 and Proposition 2.1. First, we consider the regularized problems

$$u_\rho^{'}(t) + M_\rho u_\rho(t) = f(t), u_\rho(0) = u_0, \rho > 0. \tag{3.2}$$

As the function f is continuous on [0,T] by the hypotheses, and M_ρ is $(\frac{1}{\rho})-$Lipschitz continuous by Lemma 2.1, problems (3.2) can be solved as for first-order evolution equations. To achieve that goal, we need to arrive at *a priori* estimate

$$\|u_\rho^{'}(t)\| \leq C \, \forall \rho > 0, t \in [0, T]. \tag{3.3}$$

Now we differentiate (3.2) by setting $g_\rho(t) = M_\rho o u_\rho(t)$ as follows:

$$u_\rho^{''}(t) + g_\rho^{'}(t) = f'(t) \; for\, almost\, all\, t. \tag{3.4}$$

Under the hypotheses all the derivatives exist. Since $HoR_{\rho,H}^M$ is cocoercive (and hence $HoR_{\rho,H}^M$ is nonexpansive), it implies

$$\langle H(R_{\rho,H}^M(u)) - H(R_{\rho,H}^M(u)), u - v\rangle \leq \|u - v\|^2.$$

It follows that M_ρ is monotone, and thus we have

$$\langle M_\rho u_\rho(t+h) - M_\rho u_\rho(t), u_\rho(t+h) - u_\rho(t)\rangle \geq 0.$$

It follows that

$$\langle g_\rho'(t), u_\rho'(t)\rangle \geq 0.$$

Therefore, we have

$$\begin{aligned} &\langle u_\rho''(t), u_\rho'(t)\rangle \leq -\langle g_\rho'(t), u_\rho'(t)\rangle + \langle f'(t), u_\rho'(t)\rangle \\ \leq\ & \langle f'(t), u_\rho'(t)\rangle \\ \leq\ & \|f'(t)\|\,\|u_\rho'(t)\| \\ \leq\ & \frac{1}{2}\|f'(t)\|^2 + \frac{1}{2}\|u_\rho'(t)\|^2. \end{aligned}$$

Applying the integration by parts to $\langle u_\rho''(t), u_\rho'(t)\rangle$, we have

$$\begin{aligned} & 2\int_0^t \langle u_\rho''(s), u_\rho'(s)\rangle ds \\ =\ & \|u_\rho'(t)\|^2 - \|u_\rho'(0)\|^2 \\ \leq\ & \|f\|_Y^2 + \int_0^t \|u_\rho'(s)\|^2 ds, \end{aligned}$$

where $Y = W_2^1(0, T; X)$.
This is equivalent to

$$\|u_\rho'(t)\|^2 - \|u_\rho'(0)\|^2 \leq \|f\|_Y^2 + \int_0^t \|u_\rho'(s)\|^2 ds,$$

where $Y = W_2^1(0, T; X)$.

Now by Gronwall lemma,

$$\|u_\rho'(t)\|^2 \leq c(\|u_\rho'(0)\|^2 + \|f\|_Y^2).$$

Finally, using (3.2), we have $u_\rho'(0) = -M_\rho(u_0) + f(0)$ and $\|M_\rho u_0\| \leq \|M_0 u_0\|$, and thus, it follows that (3.3) holds.

Corollary 3.1. Let $M : X \to 2^X$ be maximal monotone on a separable Hilbert space X. Suppose that the given

$$u_0 \in D(M), f \in W_2^1(0, T; X)$$

are fixed. Then (3.1) has exactly one solution $u \in W_2^1(0, T; X)$ such that $M : X \to 2^X$ is maximal monotone.

4. Concluding Remark

The obtained results on the first-order evolution inclusions seem to be applicable to the case of a real Banach space setting in terms of accretivity and $m-$accretivity (or maximal accretivity), while the solution concept also changes to an integral solution based on the difference method belonging to (3.1) as backward differences. The proof for the uniqueness assures that each classical solution of (3.1) is also an integral solution.

References

[1] Ahmed, N.U., Optimization and Identification of Systems Governed by Evolution Equations on Banach Spaces, *Pitman Research Notes in Mathematics*, Volume 184, Longman Scientific and Technical, 1988.

[2] Fang, Y.P., Huang, N.J., $H-$ monotone operators and system of variational inclusions, *Communications on Applied Nonlinear Analysis* **11** (1) (2004), 93–101.

[3] Komura, Y., Nonlinear semigroups in Hilbert space, *Journal of Mathematical Society of Japan* **19** (1967), 493–507.

[4] K. Nakajo, Takahashi, W., Strong convergence theorems for nonxpansive mappings and nonexpansive semigroups, *Journal of Mathematical Analysis and Applications* **279** (2003), 372–379.

[5] Verma, R.U., $A-$ monotonicity and applications to nonlinear inclusion problems, *Journal of Applied Mathematics and Stochastic Analysis* **17** (2004), 193–195.

[6] Verma, R.U., $A-$ monotonicity and its role in nonlinear variational inclusions, *Journal of Optimization Theory and Applications* **129** (3)(2006), 457–467.

[7] Verma, R.U., First-order evolution equations and Yosida approximations based on the relative maximal monotonicity, *International Journal of Evolution Equations* **5** (2) (2010), 1–19.

[8] Verma, R.U., General system of $A-$ monotone nonlinear variational inclusions with applications, *Journal of Optimization Theory and Applications* **131** (1) (2006), 151–157.

[9] Verma, R.U., Approximation solvability of a class of nonlinear set-valued inclusions involving $(A, \eta)-$ monotone mappings, *Journal of Mathematical Analysis and Applications* **337** (2008), 969–975.

[10] Yosida, K., On differentiability and the representation of one-parameter semigroups of linear operators, *Journal of Mathematical Society of Japan* **1** (1948), 15–21.

[11] Yosida, K., *Functional Analysis,* Springer-Verlag, Berlin, 1965.

[12] Zeidler, E., *Nonlinear Functional Analysis and its Applications* **I**, Springer-Verlag, New York, New York, 1984.

[13] Zeidler, E., *Nonlinear Functional Analysis and its Applications* **II/A**, Springer-Verlag, New York, New York, 1990.

[14] Zeidler, E., *Nonlinear Functional Analysis and its Applications* **II/B**, Springer-Verlag, New York, New York, 1990.

[15] Zeidler, E., *Nonlinear Functional Analysis and its Applications* **III**, Springer-Verlag, New York, New York, 1985.

In: Evolution Equations
Editor: Arthur L. Claes
ISBN: 978-1-61209-652-0

Chapter 15

WEIGHTED PSEUDO ALMOST AUTOMORPHIC SOLUTIONS OF NEUTRAL FUNCTIONAL DIFFERENTIAL EQUATIONS

Syed Abbas*
School of Basic Sciences
Indian Institute of Technology Mandi,
Mandi, H.P. - 175001, India

Abstract

In this chapter we study a non-autonomous neutral functional differential equation in a Banach space. Using the theory of semigroups of operators to evolution equations and contraction mapping theorem we establish the existence and uniqueness of a weighted pseudo almost automorphic solution of the problem under consideration.

2000 Math Subject Classification: 34K40, 34K14, 20M20, 47H20, 47D60
Keywords:Neutral differential equation, almost automorphic solutions, exponentially stable semigroup, semigroup of linear operators, C_0-semigroup.

1 Intoduction

The concept of almost automorphic functions was first introduced by Bochner [6, 7]. The almost automorphic functions are more general than almost periodic functions. For more details on this topic we refer to the recent book by N'Guérékata [19] where the author gave an elegant overview of the almost automorphic functions and their applications to the differential equations. The pseudo almost automorphic functions are the natural generalization of almost automorphic functions. Recently these functions are further generalized

*E-mail address::sabbas.iitk@gmail.com

by N'Guérékata et.al. [5] and named weighted pseudo almost automorphic. The authors [5] have proved some very important properties of these functions like the composition theorem and completeness property. The study of weighted pseudo almost automorphic solutions are very new topic. For more details on theory and applications of these functions we refer to [2, 5, 18].

Functional differential equations arise as models in several physical phenomena, for example, reaction diffusion equations, climate models, coupled oscillators, population ecology, neural networks, the propagation of waves etc. A class of functional differential equations called neutral differential equations arises in many phenomena as in the study of oscillatory systems and also in modelling of several physical problems [1, 22]. A good guide to the literature for neutral functional differential equations is the Hale and Lunel book [15]. Partial neutral differential equations arise, for instance, in transmission line theory. Neutral differential equations also arise in the theory of heat conduction. Under the condition that the internal energy and the heat flux depends linearly on the temperature u and on its gradient Δu, the evolution of the temperature in different types of materials can be very well described by classical heat equation. But this description is not very much satisfactory in materials with fading memory. The theory developed by Gurtin [14] the internal energy of the system and the heat flux are described as functionals of u and u_x. To describe this phenomena the following system has been frequently used:

$$\begin{aligned} \frac{d}{dt}\Big(u(t,x)+\int_{-\infty}^{t} k_1(t-s)u(s,x)ds\Big) &= c\Delta u(t,x)+\int_{-\infty}^{t} k_2(t-s)\Delta u(s,x)ds, \\ & u(t,x)=0, x\in\partial\Omega, \qquad (1) \end{aligned}$$

where Ω is an open and bounded subset $\mathbb{R}^n$ with smooth boundary. The temperature in x at the time t is given by $u(t,x)$ and c represents a physical constant. The functions $k_i : \mathbb{R} \to \mathbb{R}$, $i = 1, 2$, are the internal energy and the heat flux relaxation respectively. Assuming the history of the solution u is known that is solution is known on $(-\infty, 0]$, the above system can be transformed into the following abstract form

$$\frac{d}{dt}(u(t)+f(t,u_t)) = Au(t)+g(t,u_t).$$

There exists an extensive literature on ordinary neutral differential equations in the theory of population dynamics. It is a natural tendency of biological populations to migrate from high population density regions to regions with minor density. So the above abstract form can be used to describe spatial diffusion phenomena very well.

In a recent paper, Liu and Song [18] studied existence of weighted pseudo almost automorphic solution of the following differential equation

$$\begin{aligned} \frac{du(t)}{dt} &= A(t)u(t)+f(t,u(t)), \\ & t\in\mathbb{R}, \quad u\in X. \qquad (2) \end{aligned}$$

Motivated by the work of Liu and Song, in this chapter we consider a nonautonomous neutral functional differential equations and study the existence of weighted pseudo almost automorphic solution of neutral functional differential equations. In this work we consider the following equation in a complex Banach space X,

$$\frac{du(t)}{dt} = A(t)u(t) + \frac{d}{dt}F_1(t, u_t) + F_2(t, u_t), \quad t \in \mathbb{R}, \quad u \in X, \tag{3}$$

where the family $\{A(t) : t \in \mathbb{R}\}$ of operators in X generates an exponentially stable evolution system $\{U(t, s), t \geq s\}$. We also discuss the differential equation

$$\frac{du(t)}{dt} = Au(t) + \frac{d}{dt}F_1(t, u_t) + F_2(t, u_t), \quad t \in \mathbb{R}, \quad u \in X, \tag{4}$$

where $u_t(\theta) = u(t + \theta)$, $\theta \leq 0$ and $F_1 : \mathbb{R} \times X \to X$, $F_2 : \mathbb{R} \times X \to X$. We assume that A is the infinitesimal generator of an C_0−semigroup $\{T(t), t \geq 0\}$.

The organization of this chapter is as follows: In section 2 we describe the basic definitions of weighted pseudo almost automorphic functions. In section 3 we study the existence and uniqueness of the weighted pseudo almost automorphic solutions of equations (3) and (4).

2 Preliminaries

Let X be a complex Banach space endowed the norm $\|\cdot\|_X$. We denote by $B(X)$ the Banach space of all bounded linear operators from X to itself endowed with the operator norm given by

$$\|L\|_{B(X)} := \sup\{\|Lx\|_X : x \in X, \ \|x\|_X \leq 1\}, \quad L \in B(X).$$

Denote $\mathbb{U}$ the collections of all positive locally integrable functions $\rho : \mathbb{R} \to X$. For each $\rho \in \mathbb{U}$ define

$$m(r, \rho) = \int_{-r}^{r} \rho(t)dt.$$

Denote
$\mathbb{U}_\infty$: The set of all $\rho \in \mathbb{U}$ such that $\lim_{r\to\infty} m(r, \rho) = \infty$
$\mathbb{U}_b$: The set of all bounded $\rho \in \mathbb{U}_\infty$ such that $\inf_{t\in\mathbb{R}} \rho(t) > 0$.

Definition 2.1 *A continuous function $f : \mathbb{R} \to X$ is said to be almost automorphic if for every sequence of real number $\{t_n\}_{n\in\mathbb{N}}$ there exists a subsequence $\{t_{nk}\}_{k\in\mathbb{N}}$ such that*

$$f(t + t_{nk}) \to g(t)$$

and

$$g(t - t_{nk}) \to f(t)$$

uniformly. This is also equivalent to

$$\lim_{l\to\infty} \lim_{k\to\infty} f(t + t_{nk} - t_{nl}) = f(t)$$

for each $t \in \mathbb{R}$.

Denote $AA(X)$ be the set of all almost automorphic functions from $\mathbb{R}$ to X. Then $(AA(X), \|\cdot\|_{AP(X)})$ is a Banach space with the supremum norm given by

$$\|u\|_{AA(X)} = \sup_{t\in\mathbb{R}} \|u(t)\|_X.$$

Every almost automorphic functions $f : \mathbb{R} \to X$ are bounded [19].

Definition 2.2 *A continuous mapping $f : \mathbb{R} \times X \to X$ is said to be almost automorphic if $f(t, x)$ is almost automorphic for each $t \in \mathbb{R}$ uniformly for x in a compact subset of X. This is equivalent to that for every sequence of real number $\{t_n\}_{n\in\mathbb{N}}$ there exists a subsequence $\{t_{nk}\}_{k\in\mathbb{N}}$ such that*

$$f(t + t_{nk}, x) \to g(t, x)$$

and

$$g(t - t_{nk}, x) \to f(t, x)$$

uniformly for each $t \in \mathbb{R}$ and $x \in K$ (compact subset of X).

The set of all such functions are denoted by $AA(\mathbb{R} \times X, X)$.

A bounded continuous function $g : \mathbb{R} \times X \to X$ is said to be in $PAA_0(\mathbb{R} \times X, \rho)$ for $\rho \in \mathbb{U}_\infty$ if

$$\lim_{r\to\infty} \frac{1}{m(r,\rho)} \int_{-r}^{r} \|g(t, x)\|_X \rho(t)dt = 0$$

uniformly for $x \in X$.

Definition 2.3 *A continuous function $f : \mathbb{R}\times X \to X$ is said to be weighted pseudo almost automorphic if it can be decomposed as a sum of two functions*

$$f = f_1 + f_2,$$

where f_1 is almost automorphic and $f_2 \in PAA_0(\mathbb{R} \times X, \rho)$.

Example: Consider the function

$$f(t) = \sin \frac{1}{2 + \cos t + \cos \sqrt{2}t} + e^{\alpha t}$$

It is well known that the function $\sin \frac{1}{2+\cos t+\cos \sqrt{2}t}$ is almost automorphic. Now consider the weight function ρ defined by $\rho(t) = 1 \quad t < 0$ and $\rho(t) = e^{-\beta t} \quad t \geq 0$ for some $\beta > 0$. It is easy to verify that

$$m(r, \rho) = \int_{-r}^{r} \rho(t)dt = \int_{-r}^{0} \rho(t)dt + \int_{0}^{r} \rho(t)dt = r + \frac{1 - e^{-\beta r}}{\beta}.$$

Thus $\lim_{r\to\infty} m(r, \rho) = \infty$ which implies that $\rho \in \mathbb{U}_\infty$. Further

$$\begin{aligned}\int_{-r}^{r} e^{\alpha t}\rho(t)dt &= \int_{-r}^{0} e^{\alpha t}dt + \int_{0}^{r} e^{\alpha t}e^{-\beta t}dt \\ &= \frac{1 - e^{-\alpha r}}{\alpha} + \int_{0}^{r} e^{(\alpha-\beta)t}dt \\ &= \frac{1 - e^{-\alpha r}}{\alpha} + \frac{e^{(\alpha-\beta)r} - 1}{\alpha - \beta}. \qquad (5)\end{aligned}$$

Thus for $\alpha \leq \beta$, we have

$$\lim_{r\to\infty} \frac{1}{m(r,\rho)} \int_{-r}^{r} e^{\alpha t}\rho(t)dt = 0.$$

Hence $e^{\alpha t} \in PAA_0(\mathbb{R}, \rho)$ and so $f(t) \in WPAA(\mathbb{R})$. It is also interesting to note that

$$\lim_{r\to\infty} \frac{1}{2r} \int_{-r}^{r} e^{\alpha t}dt = \lim_{r\to\infty} \frac{e^{\alpha r} - e^{-\alpha r}}{2\alpha r} \to \infty.$$

This implies that $f(t)$ does not belongs to $PAA(\mathbb{R})$, the space of all pseudo almost automorphic functions.

Definition 2.4 *[20] A family of bounded operators $\{(U(t,s))_{t,s\in\mathbb{R}}, t \geq s\}$ on a Banach space X is called a (strongly continuous) evolution family if*

(i) $U(t,s) = U(t,r)U(r,s)$ and $U(s,s) = I$ for $t \geq r \geq s$ and $t, r, s \in \mathbb{R}$,

(ii) the mapping $\{(\tau,\sigma) \in \mathbb{R}^2 : \tau \geq \sigma\} \ni (t,s) \rightarrowtail U(t,s)$ is strongly continuous.

Evolution families are also called evolution systems, evolution operators, evolution processes, propagators or fundamental solution.

Definition 2.5 *By a weighted pseudo almost automorphic solution $u : \mathbb{R} \to X$ of equation (3) we mean that $u \in WPAA(X)$, $u(t)$ is continuously differentiable and satisfies (3) on $\mathbb{R}$.*

Definition 2.6 *By an weighted pseudo almost automorphic mild solution $u : \mathbb{R} \to X$ of the differential equation (3) we mean that $u \in WPAA(X)$, and $u(t)$ satisfies*

$$\begin{aligned} u(t) - F_1(t,u_t) &= U(t,a)(u(a) - F_1(a,u_a)) \\ &+ \int_a^t U(t,\xi)F_2(\xi,u_\xi)d\xi, \end{aligned} \tag{6}$$

for $t \geq a$.

An evolution family $U(t,s)$ is called exponentially stable if there exists $M \geq 1$ and $\delta > 0$ such that

$$\|U(t,s)\|_{B(X)} \leq Me^{-\delta(t-s)}, \quad t \geq s.$$

Throughout this chapter we assume that $U(t,s)$ is an exponentially stable evolution family.

It is easy to see that if $U(t,s)$ is exponentially stable then the relation (6) can be replaced by

$$u(t) = F_1(t,u_t) + \int_{-\infty}^{t} U(t,\xi)F_2(\xi,u_\xi)d\xi.$$

In [3] Acquistapace and Terreni gave conditions on $A(t), t \in \mathbb{R}$, which ensure the existence of an evolution family $\{U(t,s), t \geq s > -\infty\}$ on X such that,

$$u(t) = U(t,0)u(0) + \int_0^t U(t,\xi)f(\xi)d\xi, \tag{7}$$

where $u(t)$ satisfies,

$$\frac{du(t)}{dt} = A(t)u(t) + f(t), \quad t \in \mathbb{R}. \tag{8}$$

These conditions, now known as the Acquistapace-Terreni conditions (ATCs), are as follows.

(ATCs): There exist a constant $K_0 > 0$ and a set of real numbers $\alpha_1, \alpha_2, \ldots, \alpha_k, \beta_1, \ldots, \beta_k$ with $0 \leq \beta_i < \alpha_i \leq 2$, $i = 1, 2, \ldots, k$, such that

$$\|A(t)(\lambda - A(t))^{-1}(A(t)^{-1} - A(s)^{-1})\|_{B(X)} \leq K_0 \sum_{i=1}^{k} (t-s)^{\alpha_i} |\lambda|^{\beta_i - 1},$$

for $t, s \in \mathbb{R}$, $\lambda \in S_{\theta_0} \backslash \{0\}$, where

$$\rho(A(t)) \supset S_{\theta_0} = \{\lambda \in \mathbb{C} : |arg\lambda| \leq \theta_0\} \cup \{0\}, \quad \theta_0 \in (\frac{\pi}{2}, \pi)$$

and there exists a constant $M \geq 0$ such that

$$\|(\lambda - A(t))^{-1}\|_{B(X)} \leq \frac{M}{1 + |\lambda|}, \quad \lambda \in S_{\theta_0}.$$

If (ATCs) are satisfied, then from Theorem 2.3 of [4], there exists a unique evolution family $\{U(t,s), t \geq s > -\infty\}$ on X, which governs the linear version of (8).

In order to prove our results, we assume the following:

Assumptions:

A1 : The functions F_1, F_2 are Lipschitz continuous, that is, there exist positive numbers L_{F_1}, L_{F_2} such that

$$\|F_i(t, \phi) - F_i(s, \psi)\|_X \leq L_{F_i}(|t - s| + \|\phi - \psi\|_{WPAA(X)})$$

for all $t, s \in \mathbb{R}$ and for each $\phi, \psi \in WPAA(X)$ and $i = 1, 2$;

A2 : $A(t)$, $t \in \mathbb{R}$, satisfy Acquistapace and Terreni conditions (ATCs);

A3 : $U(t,s), t \geq s$, satisfy the condition that, for each of real number $\{t_n\}_{n \in \mathbb{N}}$ there exists a subsequence $\{t_{nk}\}_{k \in \mathbb{N}}$ and for any fixed $s \in \mathbb{R}$, $\epsilon > 0$ there exists an $n \in \mathbb{N}$ such that

$$\|U(t + t_{nk}, s + t_{nk}) - U(t,s)\|_{B(X)} < \epsilon e^{-\frac{\delta}{2}(t-s)}$$

and

$$\|U(t - t_{nk}, s - t_{nk}) - U(t,s)\|_{B(X)} < \epsilon e^{-\frac{\delta}{2}(t-s)}$$

for each $t, s \in \mathbb{R}$ and $t \geq s$.

A4 : The functions $F_i(t, u)$ are weighted pseudo almost automorphic for u weighted pseudo almost automorphic.

For $u \in WPAA(X)$, $F_i(t, u)$ is weighted pseudo almost automorphic, hence it is uniformly bounded. We assume that

$$\|F_i\|_{WPAA(X)} \leq M_i, \quad i = 1, 2.$$

3 Existence and uniqueness results

We define a mapping F by

$$(Fu)(t) = F_1(t, u_t) + \int_{-\infty}^{t} U(t, s)F_2(s, u_s))ds.$$

We show the following

Lemma 3.1 *For u weighted pseudo almost automorphic, the operator Fu is weighted pseudo almost automorphic.*

Proof: First we observe that F is bounded. Consider

$$\begin{aligned} \|(Fu)(t)\|_X &\leq \|F_1(t, u_t)\|_X \\ &\quad + \int_{-\infty}^{t} \|U(t, s)\| \|F_2(s, u_s)\|_X ds \\ &\leq M_1 + \int_{-\infty}^{t} Me^{-\delta(t-s)} M_2 ds \\ &\leq M_1 + M_2 \frac{M}{\delta} < \infty. \end{aligned}$$

Hence Fu is bounded. Now we show that $(Fu)(t)$ is weighted pseudo almost automorphic with respect to $t \in \mathbb{R}$. Note that the functions F_1 and F_2 are weighted pseudo almost automorphic. We can decompose F_i into two part

$$F_i = G_i + H_i \quad i = 1, 2$$

such that for every sequence of real number $\{t_n\}_{n \in \mathbb{N}}$ there exists a subsequence $\{t_{nk}\}_{k \in \mathbb{N}}$ such that

$$G_i(t + t_{nk}, x) \to g_i(t, x)$$

and

$$g_i(t - t_{nk}, x) \to G_i(t, x)$$

uniformly for each $t \in \mathbb{R}$ and $x \in K$ and $H_i \in PAA_0(\mathbb{R} \times X)$. Now we define

$$Fu(t) = G_1(t, u_t) + \int_{-\infty}^{t} U(t, \xi)G_2(\xi, u_\xi)d\xi$$

and

$$\bar{F}u(t) = H_1(t, u_t) + \int_{-\infty}^{t} U(t, \xi)H_2(\xi, u_\xi)d\xi$$

Now we show that Fu is almost automorphic and $\bar{F}u \in PAA_0(X)$. For the sequence of real number $\{t_n\}_{n \in \mathbb{N}}$ there exists a subsequence $\{t_{nk}\}_{k \in \mathbb{N}}$ such that

$$\begin{aligned} (Fu)(t + t_{nk}) &= G_1(t + t_{nk}, u_{t+t_{nk}}) + \int_{-\infty}^{t+t_{nk}} U(t + t_{nk}, \xi)G_2(\xi, u_\xi)d\xi \\ &= G_1(t + t_{nk}, u_{t+t_{nk}}) \\ &\quad + \int_{-\infty}^{t} U(t + t_{nk}, \xi + t_{nk})G_2(\xi + t_{nk}, u_{\xi+t_{nk}})d\xi \qquad (9) \end{aligned}$$

Define

$$Iu(t) = g_1(t, u_t) + \int_{-\infty}^{t} U(t,\xi) g_2(\xi, u_\xi) d\xi$$

Consider

$$\begin{aligned}
&\|(Fu)(t+t_{nk}) - (Iu)(t)\|_X \\
&\leq \|G_1(t+t_{nk}, u_{t+t_{nk}}) - g_1(t, u_t)\|_X \\
&+ \int_{-\infty}^{t} \Big(\|U(t+t_{nk}, \xi + t_{nk}) - U(t,\xi)\|_{B(X)} \|G_2(\xi + t_{nk}, u_{\xi + t_{nk}})\|_X \\
&+ \|U(t,s)\|_{B(X)} \|G_2(\xi + t_{nk}, u_{\xi+t_{nk}}) - g_2(\xi, u_\xi)\|_X \Big) d\xi \\
&\leq \|G_1(t+t_{nk}, u_{t+t_{nk}}) - g_1(t, u_t)\|_X \\
&+ M_2 \epsilon \int_{-\infty}^{t} e^{-\frac{\delta}{2}(t-\xi)} d\xi \qquad \text{(since } \|G_2\| \leq \|F_2\| \leq M_2) \\
&+ M \int_{-\infty}^{t} e^{-\delta(t-\xi)} \|G_2(\xi + t_{nk}, u_{\xi+t_{nk}}) - g_2(\xi, u_\xi)\|_X d\xi \\
&\leq \epsilon + M_2 \frac{2}{\delta} \epsilon + \frac{M}{\delta} \epsilon \\
&\leq \epsilon.
\end{aligned} \tag{10}$$

Hence we conclude that $Fu(t + t_{nk}) \to Iu(t)$. Now consider

$$\begin{aligned}
&\|(Iu)(t - t_{nk}) - (Fu)(t)\|_X \\
&\leq \|g_1(t - t_{nk}, u_{t-t_{nk}}) - G_1(t, u_t)\|_X \\
&+ \int_{-\infty}^{t} \|U(t - t_{nk}, \xi - t_{nk}) - U(t,\xi)\|_{B(X)} \|g_2(\xi - t_{nk}, u_{\xi - t_{nk}})\|_X \\
&+ \|U(t,s)\|_{B(X)} \|g_2(\xi - t_{nk}, u_{\xi - t_{nk}}) - G_2(\xi, u_\xi)\|_X d\xi
\end{aligned}$$

$$\begin{aligned}
&\leq \|g_1(t - t_{nk}, u_{t-t_{nk}}) - G_1(t, u_t)\|_X \\
&+ \|g_2\|_{AA(X)} \epsilon \int_{-\infty}^{t} e^{-\frac{\delta}{2}(t-\xi)} d\xi \\
&+ M \int_{-\infty}^{t} e^{-\delta(t-\xi)} \|g_2(\xi - t_{nk}, u_{\xi - t_{nk}}) - G_2(\xi, u_\xi)\|_X d\xi \\
&\leq \epsilon + \|g_2\|_{AA(X)} \frac{2}{\delta} \epsilon + \frac{M}{\delta} \epsilon \\
&\leq \epsilon.
\end{aligned} \tag{11}$$

Thus one can conclude that $Iu(t - t_{nk}) \to Fu(t)$. The above analysis shows that Fu is almost automorphic.

Next we show that $\bar{F}u \in PAA_0(X)$. Consider

$$\lim_{r\to\infty} \frac{1}{m(r,\rho)} \int_{-r}^{r} \|\bar{F}u(t)\|_X \rho(t) dt$$

$$\leq \lim_{r\to\infty} \frac{1}{m(r,\rho)} \int_{-r}^{r} \|H_1(t,u_t)\|_X \rho(t)dt$$
$$+ \lim_{r\to\infty} \frac{1}{m(r,\rho)} \int_{-r}^{r} \int_{-\infty}^{t} \|U(t,\xi)\|_X \|H_2(\xi,u_\xi)\|_X \rho(t)d\xi dt$$
$$\leq \lim_{r\to\infty} \frac{1}{m(r,\rho)} \int_{-r}^{r} \|H_1(t,u_t)\|_X \rho(t)dt$$
$$+ \lim_{r\to\infty} \frac{M}{m(r,\rho)} \int_{-r}^{r} \int_{-\infty}^{-r} e^{-\delta(t-\xi)} \|H_2(\xi,u_\xi)\|_X \rho(t)d\xi dt$$
$$+ \lim_{r\to\infty} \frac{M}{m(r,\rho)} \int_{-r}^{r} \int_{-r}^{t} e^{-\delta(t-\xi)} \|H_2(\xi,u_\xi)\|_X \rho(t)d\xi dt \quad (12)$$

Using the Fubini theorem the first integral in the above inequality become

$$\lim_{r\to\infty} \frac{M}{m(r,\rho)} \int_{-r}^{r} \int_{-\infty}^{-r} e^{-\delta(t-\xi)} \|H_2(\xi,u_\xi)\|_X \rho(t)d\xi dt$$
$$\leq \lim_{r\to\infty} \frac{M}{m(r,\rho)} \int_{-\infty}^{-r} e^{\delta\xi} \|H_2(\xi,u_\xi)\|_X d\xi \int_{-r}^{r} e^{-\delta t}\rho(t)dt$$
$$\leq \lim_{r\to\infty} \frac{M}{m(r,\rho)} \|H_2\|_{PAA_0(X)} \|\rho\|_{L^1(\mathbb{R})} \frac{e^{-\delta r} - e^{\delta r}}{\delta} \int_{-\infty}^{-r} e^{\delta\xi} d\xi$$
$$\leq \lim_{r\to\infty} \frac{M}{m(r,\rho)} \|H_2\|_{PAA_0(X)} \|\rho\|_{L^1(\mathbb{R})} \frac{e^{-2\delta r} - 1}{\delta^2} \to 0, \quad (13)$$

because $m(r,\rho) \to \infty$ as $r \to \infty$. Now from the second integral we have

$$\lim_{r\to\infty} \frac{M}{m(r,\rho)} \int_{-r}^{r} \int_{t}^{-r} e^{-\delta(t-\xi)} \|H_2(\xi,u_\xi)\|_X \rho(t)d\xi dt$$
$$\leq \lim_{r\to\infty} \frac{M}{m(r,\rho)} \int_{-r}^{r} \|H_2(\xi,u_\xi)\|_X \rho(t)dt \int_{-r}^{t} e^{-\delta(t-\xi)} d\xi$$
$$\leq \lim_{r\to\infty} \frac{M}{m(r,\rho)} \frac{1-e^{-\delta(t+r)}}{\delta} \int_{-r}^{r} \|H_2(\xi,u_\xi)\|_X \rho(t)dt \to 0. \quad (14)$$

It is easy to observe that

$$\lim_{r\to\infty} \frac{1}{m(r,\rho)} \int_{-r}^{r} \|H_1(t,u_t)\|_X \rho(t)dt \to 0,$$

since $H_1 \in PAA_0(\mathbb{R} \times X)$.

Thus we conclude that $\bar{F}u \in PAA_0(X)$. Which implies that F acts on $WPAA(X)$.

Theorem 3.2 *Suppose F_1, F_2 satisfies assumptions (A1)-(A4). Let*

$$L_{F_1} + \frac{M}{\delta} L_{F_2} < 1,$$

then equation (3) has a weighted pseudo almost automorphic solution.

Proof: For $u, v \in WPAA(X)$ we have

$$\begin{aligned} \|Fu(t) - Fv(t)\| &\leq \|F_1(t, u_t) - F_1(t, v_t)\| \\ &+ \int_{-\infty}^{t} \|U(t, s)\| \times \|F_2(s, u_s) - F_2(s, v_s)\| ds \\ &\leq L_{F_1}\|u_t - v_t\| + L_{F_2}\|u - v\|\frac{M}{\delta} \\ &\leq (L_{F_1} + L_{F_2}\frac{M}{\delta})\|u - v\|. \end{aligned} \tag{15}$$

Thus we have

$$\|Fu - Fv\| \leq (L_{F_1} + L_{F_2}\frac{M}{\delta})\|u - v\|.$$

Hence from contraction principle we get the existence of weighted almost automorphic solution.

Remark: Using similar arguments one can deal with the existence and uniqueness of a weighted pseudo almost automorphic solution for (4). Here our evolution operator becomes

$$U(t, s) = T(t - s)$$

In this case we have the mild solution given by

$$u(t) = F_1(t, u_t) + \int_{-\infty}^{t} T(t - s)F_2(s, u_s)ds.$$

One can easily see that the case $\rho = 1$ corresponds to the existence and uniqueness of pseudo almost automorphic solutions of the problems (3) and (4).

References

[1] Abbas, S., Bahuguna, D., Almost periodic solutions of neutral functional differential equations, *Comp. Math. Appl.*, 55-11 (2008), 2593-2601.

[2] Abbas, S., Weighted pseudo almost automorphic sequences and their applications *Elect. J. Diff. Eqns.,* Vol. 2010(2010), No. 121, 1-14.

[3] Acquistapace, P., Terreni, B., Aunified approach to abstract linear parabolic equations, *Rend. Sem. Math. Uni. Padova*, 78 (1987), 47-107.

[4] Acquistapace, P., Evolution operators and strong solution of abstract linear parabolic equations, *Differential Integral Equations* , 1 (1988), 433-457.

[5] Blot, J., Mophu, G.M., N'Guérékata, G. M., Pennequin, D., Weightedd pseudo almost automorphic functions and applications to abstarct differential equations , *Nonlinear Analysis*, Vol.71, (2009), 903-909.

[6] Bochner,S., Neumann, Von, J., On compact solutions of operational-differential equations. I, *Annals of Mathematics,* vol. 36, no. 1 (1935), 255-291.

[7] Bochner,S., Continuous mappings of almost automorphic and almost periodic functions, *PNAS, USA*, vol. 52 (1964), 907-910.

[8] Boukli-Hacene, N., Ezzinbi, K., Weighted pseudo almost periodic solutions for some partial functional differential equations, *Nonlinear Anal.* 71 (2009), 3612-3621.

[9] Diagana, T., Weighted pseudo almost periodic solutions to some differential equations, *Nonlinear Anal.* 68 (2008), 2250-2260.

[10] Diagana,T., Existence of weighted pseudo almost periodic solutions to some classes of hyperbolic evolution equations, *J. Math. Anal. Appl.* 350 (2009), 18-28.

[11] Ezzinbi, K., N'Guérékata, G. M., Almost automorphic solutions for some partial functional differential equations, *J. Math. Anal. Appl.* 328 (1) (2007), 344-358.

[12] Ezzinbi, K., Fatajou, S., N'Guérékata, G. M., Pseudo almost automorphic solutions to some neutral partial functional differential equations in Banach spaces, *Nonlinear Anal.* 70 (2009), 1641-1647.

[13] Goldstein, J., Semigroups of Linear Operators and Applications. *Oxford Science Publications*, 1985.

[14] Gurtin, M. E.; Pipkin, A. C. A general theory of heat conduction with finite wave speeds. *Arch. Rat. Mech. Anal.* 31 (1968), 113-126.

[15] Hale, J. K., Lunel, V., Sjoerd M., Introduction to functional differential equations. Applied mathematical sciences, 99. Springer-Verlag, New York, (1993).

[16] Islam, Mohammad,N., Raffoul, Youssef, N., Periodic solutions of neutral nonlinear system of differential equations with functional delay. *J. Math. Anal. Appl.*, 331 (2007), 1175-1186.

[17] Krasnosel'skii, M.A., Lusnikov, A.B., Regular fixed points and stable invariant subsets of monotone operators *Funct. Anal. Appl.*, 30 (1996), 174-183.

[18] Liu, Jing-huai, Song,Xiao-Qiu, Almost automorphic and weighted pseudo almost automorphic solutions of semilinear evolution equations, *Journal of Functional Analysis* Article in Press, 2009.

[19] N'Guérékata, G. M., Topic in almost automorphy, *Springer-Verlag* New York, 2005.

[20] Pazy, A., *Semigroups of Linear Operators and Applications to Partial Differential Equations*, Springer-Verlag, 1983.

[21] Raffoul, Y., N., Periodic solutions for neutral nonlinear differential equations with functional delay. *Elec. J. Diffirential Equation.*, 102 (2003), 1-7.

[22] Rubanik, V.P., Oscillations of quasilinear systems with retardation. *Nauka, Moscow*, 1969.

[23] Zhang, L., Xu, Y., Weighted pseudo-almost periodic solutions of a class of abstract differential equations, *Nonlinear Anal.* 71 (2009), 37053714.

In: Evolution Equations
Editor: Arthur L. Claes

ISBN: 978-1-61209-652-0

Chapter 16

Elliptic Perturbations of Some Parabolic and Hyperbolic Problems

Narcisa Apreutesei[1,*] ***and Behzad Djafari Rouhani***[2,†]
[1]Department of Mathematics, Technical University "Gh. Asachi",
Iasi, Romania
[2]Department of Mathematical Sciences, University of Texas at El Paso,
El Paso, TX, USA

Abstract

The goal of this chapter is to put together some recent results concerning applications of monotone second order differential equations to singularly perturbed problems of elliptic - parabolic and elliptic - hyperbolic type. More exactly, the solution v of the heat equation or of the telegraph system is compared with the solution v_ε of an elliptic regularization. This elliptic regularization is a perturbed problem written with the aid of a small parameter $\varepsilon > 0$. It is a particular case of some second order differential equations governed by a maximal monotone operator in the Hilbert space $L^2(\Omega)$. Under some specific hypotheses, we construct a zero order asymptotic approximation for v_ε making use of the boundary layer function method of Vishik and Lyusternik. The higher order regularity of the solutions to both perturbed and unperturbed problems is investigated. The order of accuracy of the difference $v_\varepsilon - v$ is also established in some appropriate function spaces. Thus, the solution v of the heat equation (or telegraph system) is approximated by the solution v_ε of its elliptic regularization, which is a more regular function. This is a motivation for the study of the above mentioned second order evolution equations associated with monotone operators. This study can involve different unperturbed problems: semilinear heat equation, linear heat equation with nonlinear boundary conditions, semilinear telegraph system, nonlinear telegraph system with nonlinear boundary conditions, etc.

Introduction

Elliptic regularizations with a small parameter can be employed to approximate the solutions of parabolic and hyperbolic problems on a finite time interval $[0, T]$. In this chapter,

*E-mail address: napreut@gmail.com
†E-mail address: behzad@utep.edu

we are concerned with the asymptotic analysis of some partial differential equations representing the semi-linear heat equation with Dirichlet boundary condition, the linear heat equation with nonlinear boundary conditions, and the semi-linear telegraph system with homogeneous boundary conditions. Denote the three problems by (A_0), (P_0) and (T_0), respectively and by (A_ε), (P_ε), (T_ε) their elliptic regularizations, where $\varepsilon > 0$ is a small parameter. One constructs zero order asymptotic approximations for the solutions of these problems, employing the boundary layer function method of Vishik and Lyusternik ([16], [17], [18], [19], [20]). According to this method, one searches the asymptotic approximation like a sum of the solution of the unperturbed problem (heat equation or telegraph system) with a boundary layer function, which can be found explicitly. The role of the boundary layer function is to compensate the difference between the solutions of the two problems (perturbed and unperturbed) in the neighbourhood of the final point T.

We will prove that the solutions of (A_0), (P_0), (T_0) are approximated by the solutions of the perturbed problems (A_ε), (P_ε), (T_ε) respectively, which are smoother functions. This justifies the advantage of the elliptic regularizations. The perturbed problems can be written like two-point boundary value problems. Therefore, the existence of their solutions is assured by the results from the preliminary section. In each case, we also find the order of accuracy of the asymptotic approximations.

Problems (A_ε) and (P_ε) are singularly perturbed problems of the elliptic-parabolic type, while (T_ε) is an elliptic-hyperbolic singularly perturbed problem.

The structure of the chapter is the following. First section is devoted to some basic notions and results on the existence of the solution to some first order and second order evolution equations associated to maximal monotone operators in Hilbert spaces.

In the second section, we compare the solution v of the semilinear heat equation with the solution v_ε of an elliptic regularization. Some bilocal boundary conditions are attached to this equation. Under some specific hypotheses, we construct a zero order asymptotic approximation for the solution v_ε and find the order of accuracy of the difference $v_\varepsilon - v$. The problem is singularly perturbed of elliptic-parabolic type. The case of the linear heat equation with nonlinear boundary conditions is the subject of Section 3. The fourth section deals with the semilinear telegraph system and its corresponding elliptic regularization. Thus the singular perturbation problem is of elliptic-hyperbolic type. Last section presents briefly the conclusion.

1. First and Second Order Differential Equations of Monotone Type

In this section we give some basic concepts concerning maximal monotone operators in Hilbert spaces and recall some existence results for nonlinear evolution equations of first and of second order.

1.1. Maximal Monotone Operators

Along this presentation, H will design a real Hilbert space with the scalar product $(\cdot, \cdot)$ and the associated norm $||\cdot||$. For a given operator A, we denote by $D(A)$ its domain and by

$\overline{D(A)}$ the closure of the domain $D(A)$. We consider here only single-valued (univoque) operators.

Definition 1.1. An operator $A : D(A) \subseteq H \to H$ is called *monotone operator in* H if

$$(x_1 - x_2, Ax_1 - Ax_2) \geq 0, \ (\forall)\, x_1, x_2 \in D(A). \tag{1.1}$$

Definition 1.2. The operator $A : D(A) \subseteq H \to H$ is called *maximal monotone in* H if A is monotone and, regarded as a subset of $H \times H$, it is not properly included in any other monotone subset of $H \times H$.

We introduce now two single-valued operators which are important in the theory of maximal monotone operators.

Definition 1.3. We define *the resolvent of the maximal monotone operator* A as the operator $J_\lambda = (I + \lambda A)^{-1}$. *The Yosida approximation of* A is the operator $A_\lambda = (I - J_\lambda)/\lambda$, where $\lambda > 0$ is a given scalar.

Remark that $D(J_\lambda) = D(A_\lambda) = H$, $(\forall)\, \lambda > 0$. Some properties of the resolvent J_λ and of Yosida approximation A_λ are collected below.

Theorem 1.4. *For a maximal monotone operator* $A : D(A) \subseteq H \to H$ *and a scalar* $\lambda > 0$, *we have:*

(i) J_λ *is nonexpansive (Lipschitz continuous with the Lipschitz constant* 1*);*

(ii) $A_\lambda x \in A(J_\lambda x)$, $(\forall)\, x \in H$;

(iii) A_λ *is a monotone and Lipschitz continuous operator with the Lipschitz constant* $1/\lambda$;

(iv) $\lim_{\lambda \to 0} A_\lambda x = A^0 x$, $(\forall)\, x \in D(A)$.

Definition 1.5. We say that an operator $A : D(A) \subseteq H \to H$ is *coercive* if there exists $\omega > 0$ such that $(x, Ax) \geq \omega \|x\|^2$, for all $x \in D(A)$.

Theorem 1.6. *Every maximal monotone and coercive operator is surjective, that is* $R(A) = H$.

Definition 1.7. The function $\varphi : H \to (-\infty, +\infty]$ is *proper* if it is not identically $+\infty$, i.e. there is an $x \in H$ such that $\varphi(x) < +\infty$.

Definition 1.8. The function $\varphi : H \to (-\infty, +\infty]$ is *convex* if for all $x, y \in H$ and $0 \leq \lambda \leq 1$,

$$\varphi((1-\lambda)x + \lambda y) \leq (1-\lambda)\varphi(x) + \lambda\varphi(y).$$

Definition 1.9. A function $\varphi : H \to (-\infty, +\infty]$ is said to be *lower semicontinuous at* $x \in H$ if $\liminf_{y \to x} \varphi(y) = \varphi(x)$. Function φ is *lower semicontinuous on* H if it is lower semicontinuous at every point $x \in H$. The set $D(\varphi) = \{x \in H;\ \varphi(x) < +\infty\}$ is called *the effective domain of* φ.

We now introduce the notion of subdifferential mapping.

Definition 1.10. Given a proper and convex function $\varphi : H \to (-\infty, +\infty]$ and an element $x \in H$, we define *the subdifferential of* φ *at* x the set

$$\partial\varphi(x) = \{v \in H;\ \varphi(x) - \varphi(y) \leq (v, x - y),\ (\forall)\, y \in H\}. \tag{1.2}$$

The (possibly multivalued) application $x \mapsto \partial\varphi(x)$ is called *the subdifferential of* φ.

Theorem 1.11. *If $\varphi : H \to (-\infty, +\infty]$ is proper, convex and lower semicontinuous on H, then $\partial\varphi$ is a maximal monotone operator and $\overline{D(\partial\varphi)} = \overline{D(\varphi)}$.*

For more details about the theory of maximal monotone operators and their applications to evolution equations, the reader may refer to [7].

1.2. First Order Evolution Equations in Hilbert Spaces

Now we recall briefly some existence results for the Cauchy problem

$$\begin{cases} u'(t) + Au(t) \ni f(t), \ 0 < t < T, \\ u(0) = a, \end{cases} \tag{1.3}$$

where A is a given operator on the Hilbert space H, $a \in H$ and $f \in L^1(0, T; H)$. Let's begin with the definition of the strong solution.

Definition 1.12. A function $u \in C([0, T]; H)$ is called *a strong solution* (or simply a *solution*) of problem (1.3) if u is absolutely continuous on every compact from $(0, T)$, $u(t) \in D(A)$ for a.a. $t \in (0, T)$, u satisfies (1.3) for a.a. $t \in (0, T)$.

We now give an existence and regularity result for the solution of problem (1.3).

Theorem 1.13. ([7], p. 131, 133) *If $A : D(A) \subseteq H \to H$ is a maximal monotone operator, $a \in D(A)$ and $f \in W^{1,1}(0, T; H)$, then problem (1.3) has a unique strong solution $u \in W^{1,\infty}(0, T; H)$.*

In the case of the subdifferential mappings one can improve this result ([7], p.189).

Theorem 1.14. *If $f \in L^2(0, T; H)$, $A = \partial\varphi$ (where $\varphi : H \to (-\infty, +\infty]$ is convex, proper and lower semicontinuous on H) and $a \in \overline{D(A)}$, then problem (1.3) has a (unique) strong solution $u \in C([0, T]; H)$ which satisfies in addition $u \in W^{1,2}(\delta, T; H)$, $(\forall)\, \delta \in (0, T)$, $u(t) \in D(A)$ for a.a. $t \in (0, T)$, $\varphi(u) \in L^1(0, T)$. Moreover, if $a \in D(\varphi)$, then $u' \in L^2(0, T; H)$ and $\varphi(u) \in L^\infty(0, T)$.*

We go on with an existence result ([11]) for the Cauchy problem

$$\begin{cases} u'(t) + \partial\varphi(u(t)) + B(t, u(t)) = 0, \\ u(0) = a, \end{cases} \tag{1.4}$$

Theorem 1.15. *Let $\varphi : H \to (-\infty, +\infty]$ be a convex, proper and lower semicontinuous function and $B : [0, T] \times \overline{D(\varphi)} \to H$ be a function satisfying the following hypotheses:*

(i) $(\exists)\, \omega \geq 0$ such that

$$||B(t, x_1) - B(t, x_2)|| \leq \omega ||x_1 - x_2||, \ (\forall) t \in [0, T], \ (\forall) x_1, x_2 \in \overline{D(\varphi)};$$

(ii) $(\forall) x \in \overline{D(\varphi)}$, the application $t \mapsto B(t, x)$ belongs to $L^2(0, T; H)$.

Then, for every $a \in \overline{D(\varphi)}$, there is a unique strong solution of the above Cauchy problem such that $\sqrt{t}u'(t) \in L^2(0, T; H)$.

Another useful aspect we will use in the third section is the following. Let V, H be two Hilbert spaces with the norms $||\cdot||_V$ and $||\cdot||$, respectively. One identifies H with its own dual. Let V' be the dual of V, $(\cdot, \cdot)$ be the scalar product in H and $< \cdot, \cdot >$ be the dual pair between V and V'. One supposes that $V \subset H$ with a continuous and dense injection. Hence we can write $V \subset H \subset V'$ algebraically and topologically.

We are given a family of bilinear forms $a(t,u,\psi): V \times V \to \mathbb{R}, t \in [0,T]$ such that
(i) for every $u, \psi \in V$, the function $t \longmapsto a(t,u,\psi)$ is measurable;
(ii) $|a(t,u,\psi)| \leq M||u||_V.||\psi||_V$, a.e. $t \in [0,T]$, $(\forall)\, u, \psi \in V$;
(iii) $a(t,u,u) + c||u||^2 \geq \omega ||u||_V^2$, a.e. $t \in [0,T]$, $(\forall)\, u \in V$, where $\omega > 0$ and $c \in \mathbb{R}$ are constants.

Theorem 1.16. [15] *If $(i)-(iii)$ hold, $l \in L^2(0,T;V')$ and $u_0 \in H$, then the problem*

$$\begin{cases} < u'(t), \psi > + a(t,u(t),\psi) = < l(t), \psi >, \\ \text{a.e. } t \in [0,T], \ (\forall)\psi \in V \\ u(0) = u_0 \end{cases}$$

has a unique solution $u \in L^2(0,T;V) \cap C([0,T];H)$ with $u' \in L^2(0,T;V')$.

Replacing (i) by

$$(iv) \quad \text{for every } u, \psi \in V, \text{ the function } t \longmapsto a(t,u,\psi) \text{ is of class } C^1[0,T],$$

we can improve the above result (see [15]) :

Theorem 1.17. *Under the assumptions $(ii)-(iv)$, if $l \in L^2(0,T;H)$, there exists one and only one function u such that*

$$u \in L^2(0,T;V), \quad u' \in L^2(0,T;H),$$

$$\begin{cases} < u'(t), \psi > + a(t,u(t),\psi) = < l(t), \psi >, \\ \text{a.e. } t \in [0,T], \ (\forall)\psi \in V \\ u(0) = 0. \end{cases}$$

For first order evolution equations governed by time-dependent operators $A(t)$,

$$u'(t) + A(t)u(t) = 0 \text{ a.e. on } (0,T),\ u(0) = a, \tag{1.5}$$

Kato has stated the following theorem.

Theorem 1.18. [14] *Let $A(t): D \subseteq H \to H, t \in [0,T]$, be a family of single-valued maximal monotone operators with the domain $D(A(t)) = D$ independent of t. Assume there is $L > 0$ such that for all $x \in D$ and $s,t \in [0,T]$,*

$$||A(t)x - A(s)x|| \leq L|t-s|(1 + ||x|| + ||A(s)x||).$$

Then, for every initial value $a \in D$, problem (1.5) admits a unique strong solution $u \in W^{1,1}(0,T;H)$, with $u(t) \in D$, $(\forall)t \in [0,T]$.

Recall now the notion of semigroup of contractions.

Definition 1.19. A *continuous semigroup of contractions defined on the nonempty closed set $C \subseteq H$* is a family of operators $\{S(t) : C \to C,\ t \geq 0\}$ verifying the conditions:
$a)$ $S(t+s)x = S(t)S(s)x$, $(\forall)\, x \in C$, $(\forall)\, t, s \geq 0$;
$b)$ $S(0)x = x$, $(\forall)\, x \in C$;
$b)$ $(\forall)\, x \in C$, the application $t \mapsto S(t)x$ is continuous on $[0,\infty)$;
$d)$ $||S(t)x - S(t)y|| \leq ||x-y||$, $(\forall)\, x, y \in C$, $(\forall)\, t \geq 0$.

Definition 1.20. The *infinitesimal generator of the semigroup of contractions* $\{S(t) : C \to C,\ t \geq 0\}$ is defined as the operator A given by $Ax = \lim_{h\to 0_+} (S(h)x - x)/h$ strongly in H, with the domain $D(A) = \{x \in C;\ (\exists)$ the above limit$\}$. We also say that *A generates the semigroup $S(t)$*.

Now we are interested in linear evolution equations of the form

$$\begin{cases} u'(t) + Au(t) = f(t) \text{ a.e. } t \in (0,T) \\ u(0) = a, \end{cases} \tag{1.6}$$

where $A : D(A) \subseteq H \to H$ is a linear operator such that $-A$ is the infinitesimal generator of a C_0-semigroup $\{S(t) : H \to H,\ t \geq 0\}$, $f \in L^1(0,T;H)$ and $a \in H$.

Definition 1.21. The function $u \in C([0,T];H)$ given by

$$u(t) = S(t)a + \int_0^t S(t-s) f(s)\,ds \tag{1.7}$$

is said to be the *mild solution* of problem (1.6).

Theorem 1.22. [13] *Assume that $f \in W^{1,1}(0,T;H)$, $a \in D(A)$ and $A : D(A) \subseteq H \to H$ is a linear operator such that $-A$ is the infinitesimal generator of a C_0-semigroup $\{S(t) : H \to H,\ t \geq 0\}$. Then the mild solution u from (1.7) is the unique strong solution of (1.6) and for a.a. $t \in (0,T)$,*

$$u'(t) = -S(t)Aa + S(t)f(0) + \int_0^t S(t-s) f'(s)\,ds.$$

1.3. Second Order Evolution Equations of Monotone Type

Now we turn our attention to the second order evolution equation below, to which some bilocal boundary conditions have been associated:

$$\begin{cases} u''(t) = Au(t) + f(t),\ t \in [0,T] \\ u(0) = a,\ u(T) = b. \end{cases} \tag{1.8}$$

Theorem 1.23. [6] *If $A : D(A) \subseteq H \to H$ is a maximal monotone operator in the Hilbert space H, $a, b \in D(A)$, and $f \in L^2(0,T;H)$, then problem (1.8) has a unique solution $u \in W^{2,2}(0,T;H)$.*

Remark 1.24. We notice that the solution u of problem (1.8) can be approximated by the unique solution u_λ of the problem

$$\begin{cases} u_\lambda'' = A_\lambda u_\lambda + f, \text{ a.e. on } (0,T) \\ u_\lambda(0) = a,\ u_\lambda(T) = b, \end{cases}$$

namely $u_\lambda \to u$ in $C([0,T];H)$. Moreover, $u_\lambda' \to u'$ in $C([0,T];H)$ and $u_\lambda'' \rightharpoonup u''$ in $L^2(0,T;H)$.

In the sequel, we find an estimate of u'' in a weighted function space. Let $\gamma : [0,T] \to \mathbb{R}$ be the function defined by $\gamma(t) = \min\{t, T-t\}$, $t \in [0,T]$ and let $L^2_\gamma(0,T;H)$ be the space $L^2(0,T;H)$ with the weight function γ. This means that the scalar product in $L^2_\gamma(0,T;H)$ is

$$(u,v)_\gamma = \int_0^T \gamma(t)(u(t),v(t))\,dt$$

and the norm $|.|_\gamma$ is given by

$$|u|_\gamma^2 = \int_0^T \gamma(t)\,||u(t)||^2 dt.$$

Similarly is defined the weighted space $L^2_{\gamma^3}(0,T;H)$. Denote by $(.,.)_{\gamma^3}$ and $|\cdot|_{\gamma^3}$ the scalar product and the associated norm in this space. Observe that

$$L^2(0,T;H) \subseteq L^2_\gamma(0,T;H) \subseteq L^2_{\gamma^3}(0,T;H).$$

Proposition 1.25. *Under the hypotheses of Theorem 1.23, if* $u \in W^{2,2}(0,T;H)$ *is the solution of* (1.8), *then* $|u''|_{\gamma^3} \leq |f|_{\gamma^3} + 3|u'|_\gamma$.

A generalization of the above theorem for the case when $a,b \in \overline{D(A)}$, is given by the Bruck's theorem.

Theorem 1.26. [12] *If* $A : D(A) \subseteq H \to H$ *is a maximal monotone operator on* H, $a,b \in \overline{D(A)}$, *and* $f \in L^2(0,T;H)$, *then problem* (1.8) *has a unique solution* $u \in C([0,T];H) \cap W^{2,2}_{loc}(0,T;H)$. *Furthermore,* $u' \in L^2_\gamma(0,T;H)$ *and* $u'' \in L^2_{\gamma^3}(0,T;H)$.

A more general problem is studied in the sequel (see [1]).

Proposition 1.27. *Let* $A : D(A) \subseteq H \to H$ *be a maximal monotone operator in* H, $a,b \in D(A)$, $f \in L^2(0,T;H)$ *and* $p,r : [0,T] \to \mathbb{R}$ *be continuous functions,* $p(t) \geq c > 0$, $(\forall)\, t \in [0,T]$. *Then problem*

$$\begin{cases} p(t)\,u''(t) + r(t)\,u'(t) \in Au(t) + f(t), \text{ a.e. on } [0,T] \\ u(0) = a, \quad u(T) = b \end{cases} \tag{1.9}$$

has a unique solution $u \in W^{2,2}(0,T;H)$.

Some examples which we need during next sections will illustrate the theoretical results. Let Ω be an open bounded subset of $\mathbb{R}^N$ with the boundary $\partial\Omega$ smooth enough and let H be the Hilbert space $L^2(\Omega)$. Denote by $H_0^1(\Omega)$ and $H^2(\Omega)$ the usual Sobolev spaces and by Δ the Laplace operator.

Example 1. Consider the boundary value problem

$$\begin{cases} p\dfrac{\partial^2 v}{\partial t^2} + r\dfrac{\partial v}{\partial t} = -\Delta v + \beta(v) - f, \text{ a.e. on } Q = (0,T)\times\Omega \\ v(t,x) = 0 \text{ a.e. on } \Sigma = (0,T)\times\partial\Omega \\ v(0,x) = a(x),\ v(T,x) = b(x) \text{ a.e. on } \Omega. \end{cases} \tag{1.10}$$

Suppose that $\beta : \mathbb{R} \to \mathbb{R}$ is a nonlinear nondecreasing function, $\beta \in C^1(\mathbb{R})$, $f \in L^2(0,T;L^2(\Omega))$, $p,r : [0,T] \to \mathbb{R}$, $p(t) \geq c > 0$ on $[0,T]$ and $a,b \in H_0^1(\Omega) \cap H^2(\Omega)$. Observe that

$$Av = -\Delta v + \beta(v),\ D(A) = H_0^1(\Omega) \cap H^2(\Omega)$$

is a maximal monotone operator in $L^2(\Omega)$ (see [7], p. 89). Moreover, A is the subdifferential $\partial\varphi$ of the proper, convex and lower semicontinuous function $\varphi : L^2(\Omega) \to (-\infty, +\infty]$,

$$\varphi(v) = \begin{cases} \frac{1}{2}\int_\Omega \|\nabla v\|^2 dx + \int_\Omega j(v)\, dx, \\ v \in H_0^1(\Omega),\ j(v) \in L^1(\Omega) \\ +\infty, \text{ otherwise,} \end{cases}$$

where $j : \mathbb{R} \to (-\infty, +\infty]$ is the function with the property $\beta = \partial j$.

Problem (1.10) can be written in the form (1.9) with A defined above. Therefore, we can state the following existence result.

Proposition 1.28. *Under the above hypotheses, problem* (1.10) *admits a unique solution* $v \in W^{2,2}(0, T; L^2(\Omega))$.

Example 2. We now study the boundary value problem

$$\begin{cases} p\dfrac{\partial^2 v}{\partial t^2} + r\dfrac{\partial v}{\partial t} = -\Delta v - f, \text{ a.e. on } Q = (0,T) \times \Omega \\ -\dfrac{\partial v}{\partial \eta} = \beta(v) \text{ a.e. on } \Sigma = (0,T) \times \partial\Omega \\ v(0,x) = a(x),\ v(T,x) = b(x) \text{ a.e. on } \Omega, \end{cases} \tag{1.11}$$

where $\partial/\partial\eta$ denotes the outward normal derivative. Let A be the operator

$$Av = -\Delta v,\ D(A) = \left\{ v \in H^2(\Omega),\ -\frac{\partial v}{\partial \eta} = \beta(v) \text{ a.e. on } \partial\Omega \right\}.$$

Then the above problem can be written as (1.9). Mention must be made that A is maximal monotone in $L^2(\Omega)$ ([10]) and, if $j : \mathbb{R} \to (-\infty, +\infty]$ is the lower semicontinuous convex proper function with the property $\beta = \partial j$, then A coincides with the subdifferential $\partial\varphi$ of function $\varphi : L^2(\Omega) \to (-\infty, +\infty]$,

$$\varphi(v) = \begin{cases} \frac{1}{2}\int_\Omega \|\nabla v\|^2 dx + \int_{\partial\Omega} j(v)\, dx, \\ v \in H^1(\Omega),\ j(v) \in L^1(\partial\Omega) \\ +\infty, \text{ otherwise.} \end{cases}$$

Applying Proposition 1.27 for problem (1.11) we obtain:

Proposition 1.29. *If* $\beta : \mathbb{R} \to \mathbb{R}$ *is a nonlinear maximal monotone univoque and densely defined operator,* $p, r : [0,T] \to \mathbb{R}$, $p(t) \geq c > 0$ *on* $[0,T]$, $f \in L^2(0,T; L^2(\Omega))$ *and* $a, b \in H^2(\Omega)$, *with* $-\partial a/\partial\eta = \beta(a)$, $-\partial b/\partial\eta = \beta(b)$, *then problem* (1.11) *has a unique solution* $v \in W^{2,2}(0,T; L^2(\Omega))$.

We now establish some existence theorems for some second order differential equations with a small parameter. They will be the key for proving in the next sections the higher order regularity results for the perturbed problems. Consider the evolution equation with two-points boundary conditions

$$\begin{cases} \varepsilon u'' - u' \in Au + B(t,u), \text{ a.e. } t \in (0,T) \\ u(0) = a,\ u(T) = b. \end{cases} \tag{1.12}$$

One states an existence result for this problem.

Theorem 1.30. [2] *Let A be a maximal monotone densely defined operator on H $(\overline{D(A)} = H)$, $a, b \in D(A)$, where $D(A)$ is the domain of A, $\varepsilon > 0$ a small parameter and $B : [0,T] \times \overline{D(A)} \longrightarrow H$ be an application with the following properties:*

(i) there is $\omega \geq 0$ such that, for every $t \in [0,T]$ and $x_1, x_2 \in \overline{D(A)}$,

$$||B(t, x_1) - B(t, x_2)|| \leq \omega ||x_1 - x_2||;$$

(ii) for every $x \in \overline{D(A)}$, the function $t \longmapsto B(t, x)$ is in $L^2(0,T;H)$.

Then problem (1.12) has a unique solution $u \in W^{2,2}(0,T;H)$.

Proof. Suppose in the beginning that A is strongly monotone, i.e. there exists $k > 0$ such that

$$(y_1 - y_2, x_1 - x_2) \geq k ||x_1 - x_2||^2,$$

$(\forall)\, y_i \in Ax_i,\, x_i \in D(A)$, $i = 1, 2$. First we show the existence. Let $\alpha \in L^2(0,T;H)$ be fixed. Taking into account (i), for each given $x \in H = \overline{D(A)}$, we have $||B(t, \alpha(t))|| \leq \omega ||\alpha(t) - x|| + ||B(t, x)||$, so with the aid of (ii), we get $B(t, \alpha(t)) \in L^2(0,T;H)$. Then the auxiliary problem

$$\begin{cases} \varepsilon u'' - u' \in Au + B(t, \alpha), \text{ a.e. } t \in (0,T) \\ u(0) = a, \ u(T) = b \end{cases} \tag{1.13}$$

has a unique solution $u_\alpha \in W^{2,2}(0,T;H)$ (see Proposition 1.27).

Let $\mathcal{L}^2$ be the Hilbert space $L^2(0,T;H)$ with the weight function e^{Mt}, where M is a real constant which will be suitable chosen. Therefore, the scalar product and the norm in $\mathcal{L}^2$ are

$$<< u, v >>= \int_0^T e^{Mt} (u(t), v(t))\, dt; \ |u|^2_{\mathcal{L}^2} = \int_0^T e^{Mt} ||u(t)||^2 dt.$$

The spaces $\mathcal{L}^2$ and $L^2(0,T;H)$ are equivalent algebraically and topologically.

Consider the operator $C : L^2(0,T;H) \to L^2(0,T;H)$, $C\alpha = u_\alpha$. We verify that C is a contraction on $\mathcal{L}^2$. To do this, we take $\alpha, \beta \in L^2(0,T;H)$ arbitrarily fixed and denote by u_α and u_β the corresponding solutions of (1.13). Subtracting the equations for α and β, multiplying by $e^{Mt}(u_\alpha - u_\beta)$ and integrating over $[0,T]$, we obtain via the strong monotonicity of A,

$$\varepsilon e^{Mt} \left(u'_\alpha - u'_\beta, u_\alpha - u_\beta\right) |_0^T - \varepsilon M \int_0^T e^{Mt} \left(u'_\alpha - u'_\beta, u_\alpha - u_\beta\right) dt -$$

$$-\varepsilon \int_0^T e^{Mt} ||u'_\alpha - u'_\beta||^2 dt - \frac{e^{Mt}}{2} ||u_\alpha - u_\beta||^2 |_0^T +$$

$$+\frac{M}{2} \int_0^T e^{Mt} ||u_\alpha - u_\beta||^2 dt - \int_0^T e^{Mt} (B(t, \alpha) - B(t, \beta), u_\alpha - u_\beta)\, dt$$

$$\geq k \int_0^T e^{Mt} ||u_\alpha - u_\beta||^2 dt,$$

which is equivalent to

$$\varepsilon \int_0^T e^{Mt} ||u_\alpha' - u_\beta'||^2 dt + \left(k - \frac{M}{2} - \frac{\varepsilon M^2}{2}\right) \int_0^T e^{Mt} ||u_\alpha - u_\beta||^2 dt \leq$$

$$\leq \omega \left(\int_0^T e^{Mt} ||\alpha - \beta||^2 dt\right)^{1/2} \left(\int_0^T e^{Mt} ||u_\alpha - u_\beta||^2 dt\right)^{1/2}. \quad (1.14)$$

A simple calculation shows that for small $\varepsilon > 0$, there exists $M \in \mathbb{R}$ such that $k - \frac{M}{2} - \frac{\varepsilon M^2}{2} > 0$ and $L = \omega / \left(k - \frac{M}{2} - \frac{\varepsilon M^2}{2}\right) < 1$. Hence (1.14) implies

$$|u_\alpha - u_\beta|_{\mathcal{L}^2} \leq L \, |\alpha - \beta|_{\mathcal{L}^2}, \text{ with } L < 1.$$

Thus we have proved that C is a contraction of $L^2(0,T;H)$ (which is equivalent to $\mathcal{L}^2$). By Banach fixed point theorem, we deduce that (1.12) has a solution $u \in W^{2,2}(0,T;H)$.

In order to prove the uniqueness, let u, v be two solutions of (1.12). Subtracting the corresponding equations, multiplying by $e^{Mt}(u-v)$ and following the same method as in the proof of the existence, we obtain as in (1.14),

$$\varepsilon \int_0^T e^{Mt} ||u' - v'||^2 dt + \left(k - \frac{M}{2} - \frac{\varepsilon M^2}{2} - \omega\right) \int_0^T e^{Mt} ||u - v||^2 dt \leq 0.$$

Since M is chosen such that $k - \frac{M}{2} - \frac{\varepsilon M^2}{2} > \omega$, the above inequality implies the uniqueness.

If A is maximal monotone, but not strongly monotone, we write (1.12) in the form

$$\begin{cases} \varepsilon u'' - u' \in \widetilde{A}u + \widetilde{B}(t,u), \text{ a.e. } t \in (0,T) \\ u(0) = a, \; u(T) = b, \end{cases}$$

where $\widetilde{A} = A + I$ and $\widetilde{B}(t,u) = B(t,u) - u$. Then, applying the first step of the proof for $k = 1$ and $\omega + 1$ instead of ω, one obtains the existence and uniqueness of the solution in the general case. The proof is complete.

It is known by Proposition 1.27 that problem

$$\begin{cases} \varepsilon u'' - u' \in Au + f, \text{ a. e. } t \in (0,T) \\ u(0) = a, \; u(T) = b, \end{cases} \quad (1.15)$$

with the initial values $a, b \in D(A)$ has a unique solution in $W^{2,2}(0,T;H)$. We prove an extension of this result for $a, b \in \overline{D(A)}$.

Theorem 1.31 [4]. *If $A : D(A) \subseteq H \to H$ is a maximal monotone operator in the Hilbert space H, $f \in L^2(0,T;H)$ and $a, b \in \overline{D(A)}$, then problem (1.15) has a unique solution $u \in C([0,T];H) \cap W^{2,2}_{loc}(0,T;H)$. Furthermore, $u' \in L^2_\gamma(0,T;H)$ and $u'' \in L^2_{\gamma^3}(0,T;H)$, where $\gamma(t) = \min\{t, T-t\}$, $t \in [0,T]$. Here $L^2_\gamma(0,T;H)$ is the space $L^2(0,T;H)$ with the weight γ.*

To prove this theorem, we take sequences $a_n, b_n \in D(A)$, such that $a_n \to a, b_n \to b$ as $n \to \infty$, and consider the problem

$$\begin{cases} \varepsilon u_n'' - u_n' \in Au_n + f, \text{ a. e. } t \in (0,T) \\ u_n(0) = a_n,\ u_n(T) = b_n. \end{cases} \tag{1.16}$$

The existence and uniqueness of the solution $u_n \in W^{2,2}(0,T;H)$ to (1.16) is known from Proposition 1.27. However, since we want to show that $(u_n'')_{n\geq 1}$ is bounded in $L^2_{\gamma^3}(0,T;H)$, we need some estimates on $\int_0^T \gamma^3 |u_\lambda''|^2\, dt$, where $u_\lambda \in W^{2,2}(0,T;H)$ is the unique solution of the approximate equation

$$\begin{cases} \varepsilon u_\lambda'' - u_\lambda' = A_\lambda u_\lambda + f, \text{ a. e. } t \in (0,T) \\ u_\lambda(0) = a \in D(A),\ u_\lambda(T) = b \in D(A). \end{cases} \tag{1.17}$$

Here $A_\lambda = (I - J_\lambda)/\lambda$ denotes as usual the Yosida approximation of A and $J_\lambda = (I + \lambda A)^{-1}$ is the resolvent of A.

Some estimates on u_n and u_λ are stated in the following lemmas.

Lemma 1.32. *We have:*

$$|u_n(t) - u_m(t)|^2 \leq \max\left(|a_n - a_m|^2, |b_n - b_m|^2\right),\ 0 \leq t \leq T. \tag{1.18}$$

Proof. Since A is monotone, we can easily obtain

$$\varepsilon\left(|u_n - u_m|^2\right)'' - \left(|u_n - u_m|^2\right)' \geq 2\varepsilon\left|u_n' - u_m'\right|^2 \geq 0, \tag{1.19}$$

so

$$\frac{d}{dt}\left(e^{-t/\varepsilon}\left(|u_n(t) - u_m(t)|^2\right)'\right) \geq 0.$$

Now two cases can happen: either $\left(|u_n(t) - u_m(t)|^2\right)' \leq 0$, for all $0 \leq t \leq T$, and in this case $|u_n(t) - u_m(t)|^2 \leq |a_n - a_m|^2$ for all t; or otherwise, there exists $t_0 \in [0,T]$ such that $\left(|u_n - u_m|^2\right)'(t_0) > 0$. In this situation,

$$0 \leq \int_{t_0}^t \frac{d}{dt}\left(e^{-t/\varepsilon}\left(|u_n(t) - u_m(t)|^2\right)'\right) dt,\ t \geq t_0$$

implies that $e^{-t/\varepsilon}\left(|u_n(t) - u_m(t)|^2\right)' \geq e^{-t_0/\varepsilon}\left(|u_n - u_m|^2\right)'(t_0),\ (\forall)\, t \geq t_0$, hence $\left(|u_n - u_m|^2\right)'(t) > 0,\ (\forall)\, t \geq t_0$. Then $|u_n(t) - u_m(t)|^2 \leq |b_n - b_m|^2$ for all $t \in [t_0, T]$. If $t_0 = 0$, this estimate holds on the whole $[0,T]$. If $t_0 > 0$ and the derivative $\left(|u_n - u_m|^2\right)'(t)$ takes negative values for $t < t_0$, then the first case applies and we have the estimate by $|a_n - a_m|^2$ on $[0, t_0)$. Therefore, in both cases the inequality (1.18) holds, as claimed.

Remark. Form this inequality, we can get that u_n converges uniformly on $[0,T]$ to a function $u \in C([0,T];H)$ satisfying $u(0) = a$ and $u(T) = b$.

Lemma 1.33. *Let* $\gamma(t) = \min\{t, T-t\}$, $t \in [0,T]$. *Then, the following estimate holds:*

$$2\varepsilon \int_0^T \gamma(t) \left|u_n' - u_m'\right|^2 dt \leq (2\varepsilon + T/2) \max\left(\left|a_n - a_m\right|^2, \left|b_n - b_m\right|^2\right).$$

In particular this shows that $u_n' \to u'$ *in* $L_\gamma^2(0,T;H)$, *as* $n \to \infty$.

Proof. Let $\gamma_\eta(t) = \min\{t-\eta, T-t-\eta\}$, for $t \in (0,T)$ and fixed $\eta \in (0, T/2)$. Multiplying (1.19) by γ_η and integrating on $[\eta, T-\eta]$, we are led to

$$2\varepsilon \int_\eta^{T-\eta} \gamma_\eta \left|u_n' - u_m'\right|^2 dt \leq -2\varepsilon \left|u_n(T/2) - u_m(T/2)\right|^2 + \varepsilon \left|u_n(\eta) - u_m(\eta)\right|^2 +$$

$$+\varepsilon \left|u_n(T-\eta) - u_m(T-\eta)\right|^2 + \int_\eta^{T/2} \left|u_n - u_m\right|^2 dt - \int_{T/2}^{T-\eta} \left|u_n - u_m\right|^2 dt.$$

By Lemma 1.32, we derive that

$$2\varepsilon \int_\eta^{T-\eta} \gamma_\eta \left|u_n' - u_m'\right|^2 dt \leq (2\varepsilon + T/2 - \eta) \max\left(\left|a_n - a_m\right|^2, \left|b_n - b_m\right|^2\right).$$

The claim follows by letting $\eta \to 0$ and using Fatou's lemma.

We now prove some estimates for the solution u_λ of the approximate equation (1.17).

Lemma 1.34. *Denote* $||u_\lambda''||_2 = ||u_\lambda''||_{L^2(0,T;H)}$. *Then* $\{||u_\lambda''||_2\}_{\lambda>0}$ *is bounded.*

Proof. Multiplying (1.17) by u_λ'' and integrating on $[0,T]$, we get

$$\varepsilon \int_0^T \left|u_\lambda''\right|^2 dt \leq \frac{1}{2}\left|u_\lambda'(T)\right|^2 + \left(A_\lambda b, u_\lambda'(T)\right) - \left(A_\lambda a, u_\lambda'(0)\right) + \int_0^T \left(f, u_\lambda''\right) dt. \quad (1.20)$$

Since $(\forall)\, x, y \in H$,

$$\left(x, u_\lambda'(T)\right) - \left(y, u_\lambda'(0)\right) = \int_0^T \left(\frac{t}{T}x + \frac{T-t}{T}y, u_\lambda''\right) dt + \frac{1}{T}(x-y, b-a),$$

one can write

$$\left|\left(A_\lambda b, u_\lambda'(T)\right) - \left(A_\lambda a, u_\lambda'(0)\right)\right| = \left|\int_0^T \left(\frac{t}{T}A_\lambda b + \frac{T-t}{T}A_\lambda a, u_\lambda''\right) dt + \right.$$

$$\left. + \frac{1}{T}\left(A_\lambda b - A_\lambda b, b-a\right)\right| \leq \sqrt{T}\left(\left|A^0 a\right| + \left|A^0 b\right|\right) ||u_\lambda''||_2 + \frac{b-a}{T}\left(\left|A^0 a\right| + \left|A^0 b\right|\right).$$

On the other hand, the inequality for $y = 0$,

$$\left(x, u_\lambda'(T)\right) \leq \frac{|x|}{T}\left(\int_0^T t^2 dt\right)^{1/2} ||u_\lambda''||_2 + \frac{b-a}{T}|x|$$

implies that

$$\left|u_\lambda'(T)\right| = \sup_{|x|\leq 1}\left(x, u_\lambda'(T)\right) \leq \sqrt{T/3}\, ||u_\lambda''||_2 + \frac{b-a}{T}.$$

Hence by (1.20) we obtain

$$\varepsilon \int_0^T \left|u_\lambda''\right|^2 dt = \varepsilon \left\|u_\lambda''\right\|_2^2 \leq \frac{T}{6} \left\|u_\lambda''\right\|_2^2 + K_1 \left\|u_\lambda''\right\|_2 + K_2,$$

where $K_1 = |b-a| / \sqrt{3T} + \sqrt{T} \left(\left|A^0 a\right| + \left|A^0 b\right| + \|f\|_2\right)$ and

$$K_2 = |b-a| / T \left(|b-a| / 2T + \left|A^0 a\right| + \left|A^0 b\right|\right).$$

This implies that $(\varepsilon - T/6) \left\|u_\lambda''\right\|_2^2 \leq K_1 \left\|u_\lambda''\right\|_2 + K_2$, i. e. for $\varepsilon > T/6$, $\left\{\left\|u_\lambda''\right\|_2\right\}_{\lambda>0}$ is bounded.

Now we can divide the interval $(0, T)$ into finitely many subintervals of equal length $T' < 6\varepsilon$. In this way, we get the boundedness of $\left\{\left\|u_\lambda''\right\|_2\right\}_{\lambda>0}$ on each of these subintervals, and therefore we conclude that $\left\{\left\|u_\lambda''\right\|_2\right\}_{\lambda>0}$ is bounded. The lemma is proved.

Lemma 1.35. *The solution u_λ to (1.17) converges uniformly on $[0, T]$ to the unique solution u of the problem*

$$\begin{cases} \varepsilon u'' - u' \in Au + f, \text{ a. e. } t \in (0, T) \\ u(0) = a \in D(A),\ u(T) = b \in D(A). \end{cases} \tag{1.21}$$

Moreover, $u_{\lambda_n}' \to u'$ in $C([0, T]; H)$ for some sequence $\lambda_n \to 0^+$ and $u_\lambda'' \rightharpoonup u''$ in $L^2(0, T; H)$ as $\lambda \to 0^+$.

Proof. The proof of the previous lemma shows that $\{\sup_{0\leq t\leq T} |u_\lambda'(t)|\}_{\lambda>0}$ and hence $\left\{\|u_\lambda'\|_2\right\}_{\lambda>0}$ are bounded. Then, in view of (1.17), the sequence $\{\|A_\lambda u_\lambda\|_2\}_{\lambda>0}$ is also bounded, say by k. Now we have $\left(u_\lambda' - u_\mu', u_\lambda - u_\mu\right)' = \left(u_\lambda'' - u_\mu'', u_\lambda - u_\mu\right) + \left|u_\lambda' - u_\mu'\right|^2$. This implies that

$$\begin{aligned} \varepsilon \int_0^T \left|u_\lambda' - u_\mu'\right|^2 dt &= -\varepsilon \int_0^T \left(u_\lambda'' - u_\mu'', u_\lambda - u_\mu\right) dt = \\ &= -\int_0^T \left(u_\lambda' + A_\lambda u_\lambda - u_\mu' - A_\mu u_\mu, u_\lambda - u_\mu\right) dt. \end{aligned}$$

Since $u_\lambda = \lambda A_\lambda u_\lambda + J_\lambda u_\lambda$, $A_\lambda x \in A(J_\lambda x)$ and A is monotone, we deduce that

$$\varepsilon \int_0^T \left|u_\lambda' - u_\mu'\right|^2 dt = -\int_0^T \left(A_\lambda u_\lambda - A_\mu u_\mu, \lambda A_\lambda u_\lambda - \mu A_\mu u_\mu\right) dt -$$

$$-\int_0^T \left(A_\lambda u_\lambda - A_\mu u_\mu, J_\lambda u_\lambda - J_\mu u_\mu\right) dt \leq (\lambda + \mu) \left(\|A_\lambda u_\lambda\|_2 + \|A_\mu u_\mu\|_2\right)^2 \leq 4(\lambda + \mu) k^2.$$

Hence we proved that $\varepsilon \left\|u_\lambda' - u_\mu'\right\|_2^2 \leq 4(\lambda + \mu) k^2 \to 0$ as $\lambda, \mu \to 0^+$. Therefore,

$$|u_\lambda(t) - u_\mu(t)|^2 = \left|\int_0^t \left(u_\lambda'(s) - u_\mu'(s)\right) ds\right|^2 \leq T \int_0^T \left|u_\lambda'(s) - u_\mu'(s)\right|^2 ds \leq 4\frac{k^2 T}{\varepsilon} (\lambda + \mu).$$

This shows that u_λ converges to some function, say $u_\lambda \to u$ uniformly on $[0, T]$ and $u_\lambda' \to u'$ in $L^2(0, T; H)$, as $\lambda \to 0^+$. Moreover, since u_λ'' is bounded in $L^2(0, T; H)$ and

$u''_\lambda \to u''$ in $\mathcal{D}'(0,T;H)$, then $u''_\lambda \rightharpoonup u''$ (weakly) in $L^2(0,T;H)$ as $\lambda \to 0^+$. Let us show now the existence of a sequence $\lambda_n \to 0^+$ such that $u'_{\lambda_n} \to u'$ in $C([0,T];H)$. Indeed, since $u'_\lambda \to u'$ in $L^2(0,T;H)$, there exists $\lambda_n \to 0^+$ such that $u'_{\lambda_n}(t) \to u'(t)$ as $n \to \infty$, a. e. $t \in (0,T)$. Let $t_0 \in (0,T)$ be such that $u'_{\lambda_n}(t_0) \to u'(t_0)$. Then we have

$$\left\|u'_{\lambda_n}(t) - u'(t)\right\|^2 = \left\|u'_{\lambda_n}(t_0) - u'(t_0)\right\|^2 + 2\int_{t_0}^{t} \left(u'_{\lambda_n}(s) - u'(s), u''_{\lambda_n}(s) - u''(s)\right) ds.$$

This implies that $u'_{\lambda_n} \to u'$ in $C([0,T];H)$ as $n \to \infty$.

Denoting by $\mathcal{A}$ the realization of A in $L^2(0,T;H)$, since $\mathcal{A}$ is demiclosed, $J_\lambda u_\lambda \to u$ in $L^2(0,T;H)$ and $\varepsilon u''_\lambda - u'_\lambda - f = \mathcal{A}_\lambda u_\lambda \in \mathcal{A}(J_\lambda u_\lambda)$, letting $\lambda \to 0^+$, we conclude that u is a solution to (1.21). This completes the proof of the lemma.

Proposition 1.36. *If $u \in W^{2,2}(0,T;H)$ is the solution of (1.21), then*

$$|u''|_{\gamma^3} \le \frac{1}{\varepsilon}|f|_{\gamma^3} + \left(\frac{K}{\varepsilon} + 3\right)|u'|_\gamma,$$

where K is the constant given by the inclusion $L^2_\gamma \hookrightarrow L^2_{\gamma^3}$, and $\gamma(t) = \min\{t, T-t\}$, $t \in [0,T]$.

Proof. Consider again the approximate problem (1.17). A simple computation leads to

$$(A_\lambda u_\lambda + u'_\lambda + f, A_\lambda u_\lambda) \le \varepsilon (u'_\lambda, A_\lambda u_\lambda)',$$

so

$$\left|A_\lambda u_\lambda + \frac{u'_\lambda}{2} + \frac{f}{2}\right|^2 - \left|\frac{u'_\lambda}{2} + \frac{f}{2}\right|^2 \le \varepsilon (u'_\lambda, A_\lambda u_\lambda)'.$$

Multiplying by γ^3, integrating over $[0,T]$ and using the equalities $\gamma(0) = \gamma(T) = 0$, we get

$$\int_0^T \gamma^3 \left|A_\lambda u_\lambda + \frac{u'_\lambda}{2} + \frac{f}{2}\right|^2 dt - \frac{1}{4}\int_0^T \gamma^3 |u'_\lambda + f|^2 dt \le -3\varepsilon[\int_0^{T/2} \gamma^2 (u'_\lambda, A_\lambda u_\lambda)\, dt -$$

$$-\int_{T/2}^T \gamma^2 (u'_\lambda, A_\lambda u_\lambda)\, dt] \le 3\varepsilon \int_0^T \gamma^2 |(u'_\lambda, A_\lambda u_\lambda)|\, dt$$

$$= 3\varepsilon \int_0^T \gamma^2 \left|\left(u'_\lambda, A_\lambda u_\lambda + \frac{u'_\lambda + f}{2}\right) - \left(u'_\lambda, \frac{u'_\lambda + f}{2}\right)\right| dt$$

$$\le 3\varepsilon \left(\int_0^T \gamma |u'_\lambda|^2 dt\right)^{1/2} \left[\left(\int_0^T \gamma^3 \left|A_\lambda u_\lambda + \frac{u'_\lambda + f}{2}\right|^2 dt\right)^{1/2} + \frac{1}{2}\left(\int_0^T \gamma^3 |u'_\lambda + f|^2 dt\right)^{1/2}\right].$$

Hence we proved that

$$\left(\int_0^T \gamma^3 \left|A_\lambda u_\lambda + \frac{u'_\lambda + f}{2}\right|^2 dt\right)^{1/2} \le \frac{1}{2}\left(\int_0^T \gamma^3 |u'_\lambda + f|^2 dt\right)^{1/2} + 3\varepsilon \left(\int_0^T \gamma |u'_\lambda|^2 dt\right)^{1/2}.$$

Using the equality $A_\lambda u_\lambda = \varepsilon u''_\lambda - u'_\lambda - f$, the left hand side becomes $\left(\int_0^T \gamma^3 \left|\varepsilon u''_\lambda - \frac{u'_\lambda + f}{2}\right|^2 dt\right)^{1/2}$. Then,

$$\varepsilon \left|u''_\lambda\right|_{\gamma^3} - \frac{1}{2}\left|u'_\lambda + f\right|_{\gamma^3} \leq \left|\varepsilon u''_\lambda - \frac{u'_\lambda + f}{2}\right|_{\gamma^3} \leq \frac{1}{2}\left|u'_\lambda + f\right|_{\gamma^3} + 3\varepsilon\left|u'_\lambda\right|_\gamma,$$

so $\varepsilon\left|u''_\lambda\right|_{\gamma^3} \leq |f|_{\gamma^3} + (K + 3\varepsilon)\left|u'_\lambda\right|_\gamma$, where $K > 0$ is the constant given in the inclusion $L^2_\gamma \hookrightarrow L^2_{\gamma^3}$. By Lemma 1.35 it follows that $u'_\lambda \to u'$ in $L^2_\gamma(0, T; H)$ and $u''_\lambda \rightharpoonup u''$ in $L^2_{\gamma^3}(0, T; H)$, as $\lambda \to 0^+$. The proof is now complete by letting $\lambda \to 0^+$ in the above inequality.

Proof of Theorem 1.31. As in Lemma 1.32, we find that, if u_1 and u_2 are two solutions of problem (1.15), corresponding to the boundary conditions (a_1, b_1) and (a_2, b_2), respectively, then

$$|u_1(t) - u_2(t)|^2 \leq \max\left(|a_1 - a_2|^2, |b_1 - b_2|^2\right), \ 0 \leq t \leq T.$$

This proves the uniqueness of the solution to (1.15). To show the existence, consider the sequences $a_n, b_n \in D(A)$ such that $a_n \to a$, $b_n \to b$ as $n \to \infty$ and the system

$$\begin{cases} \varepsilon u''_n - u'_n \in A u_n + f, \text{ a. e. } t \in (0, T) \\ u_n(0) = a_n, \ u_n(T) = b_n. \end{cases}$$

This problem has a unique solution $u_n \in W^{2,2}(0, T; H)$ and u_n converges in $C([0, T]; H)$ to a function $u \in C([0, T]; H)$ satisfying $u(0) = a$ and $u(T) = b$. From Lemma 1.33 and Proposition 1.36, we know that $u'_n \to u'$ in $L^2_\gamma(0, T; H)$ and $u''_n \rightharpoonup u''$ in $L^2_{\gamma^3}(0, T; H)$. Therefore, $u \in W^{2,2}_{loc}(0, T; H)$, $u' \in L^2_\gamma(0, T; H)$, and $u'' \in L^2_{\gamma^3}(0, T; H)$.

Now let $0 < \delta < T/2$ be given and let $\mathcal{A}$ be the operator defined by

$$\mathcal{A}u = \left\{v \in L^2(\delta, T - \delta; H); \ v(t) \in Au(t) \text{ a. e. on } (0, T)\right\}.$$

The operator $\mathcal{A}$ is maximal monotone, hence demiclosed. Then

$$\begin{cases} u_n \to u \text{ in } L^2(\delta, T - \delta; H) \\ u'_n \to u' \text{ in } L^2(\delta, T - \delta; H) \\ u''_n \rightharpoonup u'' \text{ in } L^2(\delta, T - \delta; H) \end{cases}, \text{ as } n \to \infty.$$

Therefore, letting $n \to \infty$ and using the demiclosedness of $\mathcal{A}$, it follows that u is a solution of the equation $\varepsilon u'' - u' \in Au + f$ a. e. on $(\delta, T - \delta)$. Since $0 < \delta < T/2$ is arbitrary, we conclude that u satisfies the equation from (1.15). The theorem is proved.

2. Asymptotic Approximations for the Semi-linear Heat Equation

Along the next three sections, we give some applications of the second order evolution equations presented before to some singularly perturbed problems.

In this section we are concerned with problems (A_0) and (A_ε) below. The results exposed here are mainly obtained in [3]. Consider the nonlinear boundary value problem

$$\begin{cases} \dfrac{\partial v}{\partial t} - \Delta v + \beta(v) = f, \text{ a.e. on } Q = (0,T) \times \Omega \\ v(t,x) = 0 \text{ a.e. on } \Sigma = (0,T) \times \partial\Omega \\ v(0,x) = a(x) \text{ a.e. on } \Omega, \end{cases} \tag{A$_0$}$$

where Ω is an open bounded subset of $\mathbb{R}^N$, with sufficiently smooth boundary $\partial\Omega$, β is a nonlinear nondecreasing function from $\mathbb{R}$ to $\mathbb{R}$, $f \in L^2(0,T;L^2(\Omega))$ and $a \in H_0^1(\Omega) \cap H^2(\Omega)$. This problem has a unique strong solution $v \in W^{1,\infty}(0,T;L^2(\Omega)) \cap L^\infty(0,T;H_0^1(\Omega) \cap H^2(\Omega))$ ([9]).

For the small parameter $\varepsilon > 0$, we take the elliptic regularization of (A_0):

$$\begin{cases} \varepsilon\dfrac{\partial^2 v_\varepsilon}{\partial t^2} - \dfrac{\partial v_\varepsilon}{\partial t} = -\Delta v_\varepsilon + \beta(v_\varepsilon) - f, \text{ a.e. on } Q \\ v_\varepsilon(t,x) = 0 \text{ a.e. on } \Sigma = (0,T) \times \partial\Omega \\ v_\varepsilon(0,x) = a(x),\ v_\varepsilon(T,x) = b(x) \text{ a.e. on } \Omega. \end{cases} \tag{A$_\varepsilon$}$$

If $b \in H_0^1(\Omega) \cap H^2(\Omega)$, under the above hypotheses, problem (A_ε) has a unique solution $v_\varepsilon \in W^{2,2}(0,T;L^2(\Omega))$ (Proposition 1.28).

We compare the solution v of (A_0) with the solution v_ε of (A_ε).

Since v does not generally satisfy the boundary condition $v(T,.) = b$, problem (A_ε) is a singularly perturbed one and, at least in a neighbourhood $(\delta,T] \times \Omega$ of the final point $\{T\} \times \Omega$ (where $\delta > 0$ is an arbitrarily small, but fixed number), the difference $v_\varepsilon - v$ in the norm of $C([\delta,T];L^2(\Omega))$ does not tend to zero as $\varepsilon \to 0$. The set $(\delta,T] \times \Omega$ is a boundary layer. We prove that $v_\varepsilon - v$ tends to zero in $C([0,\delta];L^2(\Omega))$ and we construct an asymptotic approximation V_ε of v_ε, which is valid in the whole domain $Q = (0,T) \times \Omega$.

Moreover, we show that $|v_\varepsilon - v|_{L^2(Q)} = o(\varepsilon^{1/2})$ and $|v_\varepsilon - V_\varepsilon|_{C([0,T];L^2(\Omega))} = o(\varepsilon^{1/4})$.

2.1. An Elliptic-Parabolic Singularly Perturbed Problem

Suppose that f, β, a, b are smooth enough. We will specify later the conditions verified by these functions. We begin with an extension of Definition 1.8.2.

Definition 2.1. The function $V_\varepsilon(t,x)$ is called *an asymptotic approximation* for the solution v_ε of (A_ε) with respect to the parameter ε in the subdomain $Q_1 = (0,\delta] \times \Omega \subseteq Q$ $(0 < \delta < T)$, if

$$|v_\varepsilon - V_\varepsilon|_{C([0,\delta];L^2(\Omega))} \to 0 \text{ as } \varepsilon \to 0.$$

Furthermore, if

$$|v_\varepsilon - V_\varepsilon|_{C([0,\delta];L^2(\Omega))} = o\left(\varepsilon^k\right),$$

then V_ε is called an *asymptotic approximation* of v_ε in Q_1 with the order of accuracy ε^k. We recall that the symbol $\alpha(\varepsilon) = o(\varepsilon^k)$ means that $(\exists)\, c, \varepsilon_0 > 0$ such that $|\alpha(\varepsilon)| \le c\varepsilon^k$, for all $0 \le \varepsilon \le \varepsilon_0$.

Since v does not satisfy the boundary condition $v\,(T,.) = b$ (verified by v_ε), at least in a neighbourhood $(\delta, T] \times \Omega$ of $\{T\} \times \Omega$ the solution v of (A_0) is not an asymptotic approximation for the solution v_ε of the singularly perturbed problem (A_ε). This set $(\delta, T] \times \Omega$ is a boundary layer.

We introduce a boundary layer function $\Pi\,(\tau, x)$, whose role is to compensate for the discrepancy introduced by v in the boundary condition $v_\varepsilon\,(T, x) = b\,(x)$ a.e. on Ω. So we are looking for an asymptotic approximation of v_ε of the form

$$V_\varepsilon\,(t, x) = v\,(t, x) + \Pi\,(\tau, x)\,, \; \tau = \frac{T - t}{\varepsilon}, \tag{2.1}$$

where Π approaches zero as $\tau \longrightarrow \infty$. This means that

$$\Pi(\infty, x) = 0 \text{ a.e. on } \Omega. \tag{2.2}$$

Here τ is called the boundary layer variable. Let r_ε be the remainder term, i.e.

$$r_\varepsilon(t, x) = v_\varepsilon(t, x) - V_\varepsilon(t, x) = v_\varepsilon(t, x) - v(t, x) - \Pi(\tau, x). \tag{2.3}$$

Substituting formally v_ε from (2.3) into (A_ε), we have

$$\begin{aligned} &\varepsilon\left(\frac{\partial^2 v}{\partial t^2} + \frac{1}{\varepsilon^2}\frac{\partial^2 \Pi}{\partial \tau^2} + \frac{\partial^2 r_\varepsilon}{\partial t^2}\right) - \left(\frac{\partial v}{\partial t} - \frac{1}{\varepsilon}\frac{\partial \Pi}{\partial \tau} + \frac{\partial r_\varepsilon}{\partial t}\right) = \\ &= -\Delta v - \Delta \Pi - \Delta r_\varepsilon + \beta(v + \Pi + r_\varepsilon) - f \text{ a.e. on } Q, \end{aligned}$$

$$v + \Pi + r_\varepsilon = 0 \text{ a.e. on } \Sigma,$$

$$\begin{aligned} &v(0, x) + \Pi(\infty, x) + r_\varepsilon(0, x) = a(x), \\ &v(T, x) + \Pi(0, x) + r_\varepsilon(T, x) = b(x) \text{ a.e. on } \Omega. \end{aligned}$$

Identifying the coefficients of the same powers of ε, separately for the terms depending on t and on τ, respectively and using relation (2.2), we derive that Π is the solution of the boundary value problem

$$\frac{\partial^2 \Pi}{\partial \tau^2} + \frac{\partial \Pi}{\partial \tau} = 0 \text{ a.e. on } \tilde{Q} = (0, \infty) \times \Omega$$

$$\Pi = 0 \text{ a.e. on } \tilde{\Sigma} = (0, \infty) \times \partial\Omega$$

$$\Pi(0, x) = b(x) - v(T, x), \;\; \Pi(\infty, x) = 0 \text{ a.e. on } \Omega$$

and r_ε is the solution of

$$\begin{aligned} \varepsilon\frac{\partial^2 r_\varepsilon}{\partial t^2} - \frac{\partial r_\varepsilon}{\partial t} &= -\Delta r_\varepsilon + \beta(v + \Pi + r_\varepsilon) - \\ &\quad - \beta(v) - \left(\Delta \Pi + \varepsilon\frac{\partial^2 v}{\partial t^2}\right) \text{ a.e. on } Q \end{aligned} \tag{2.4}$$

$$r_\varepsilon = 0 \text{ a.e. on } \Sigma \tag{2.5}$$

$$r_\varepsilon(0, x) = r_\varepsilon(T, x) = 0 \text{ a.e. on } \Omega. \tag{2.6}$$

First, observe that the problem in Π has a unique solution,

$$\Pi(\tau, x) = e^{-\tau}[b(x) - v(T, x)].$$

This implies that

$$|\,\Pi\,|^2_{L^2(Q)} = \frac{\varepsilon}{2}(1 - e^{-2T/\varepsilon})\,|\,b - v(T,\cdot)\,|^2_{L^2(\Omega)}; \tag{2.7}$$

$$|\,\Delta\Pi\,|^2_{L^2(Q)} = \frac{\varepsilon}{2}(1 - e^{-2T/\varepsilon})\,|\,\Delta(b - v(T,\cdot))\,|^2_{L^2(\Omega)}; \tag{2.8}$$

$$|\,\Pi\left(\frac{T-t}{\varepsilon}, x\right)|_{C([0,T];L^2(\Omega))} = |\,b - v(T,\cdot)\,|_{L^2(\Omega)}\,. \tag{2.9}$$

In the sequel, we prove that under additional assumptions the second derivative $\partial^2 v/\partial t^2$ from (2.4) makes sense and belongs to $L^2\,(Q)$. Hence $v \in W^{2,2}\,(0, T; L^2\,(\Omega))$. Next, we show that v_ε is also more regularly. Thus v is approximated by the smoother function v_ε. Finally, we study the existence of the solution to the time dependent boundary value problem $(2.4) - (2.6)$ and find the order of $v_\varepsilon - V_\varepsilon$ in different function spaces.

2.2. Regularity Results

We show that, under stronger conditions on data, the solutions of problems (A_0) and (A_ε) are more regularly.

We are firstly concerned with the boundary value problem (A_0) under the assumptions:
(A1) $f \in W^{1,2}(0, T; L^2(\Omega))$;

(A2) $a \in H^1_0(\Omega) \cap H^2(\Omega)$;

(A3) $\beta \in C^1(\mathbb{R}),\;\; \beta' \geq 0,\;\; \beta' \in L^\infty(\mathbb{R})$;

(A4) $d = f(0,\cdot) + \Delta a - \beta(a) \in H^1_0(\Omega)$.

Condition $(A4)$ is a high order compatibility condition. If $a \in H^2_0(\Omega) \cap H^3(\Omega)$, $f(0,\cdot) \in H^1_0(\Omega)$, $\beta(0) = 0$ and β is as in $(A3)$, then $(A4)$ is satisfied.

We show that, under hypotheses $(A1) - (A4)$, the problem (A_0) has a solution in $W^{2,2}(0, T; L^2(\Omega))$. Recall that v is in $C([0, T]; L^2(\Omega)) \cap L^\infty(0, T; H^1_0(\Omega) \cap H^2(\Omega))$ and $\partial v/\partial t \in L^\infty(0, T; L^2(\Omega))$.

Theorem 2.2. *Suppose that $(A1) - (A4)$ hold. Then, the solution v of (A_0) is in $W^{2,2}(0, T; L^2(\Omega))$.*

Proof. Consider the auxiliary problem

$$\begin{cases} \dfrac{\partial w}{\partial t} - \Delta w + \beta'(v)w = \dfrac{\partial f}{\partial t} & \text{a.e. on } Q \\ w = 0 & \text{a.e. on } \Sigma \\ w(0, x) = d(x) & \text{a.e. on } \Omega, \end{cases} \tag{2.10}$$

which can be written in the form (1.4), with d instead of a, $B(t,w) = (\partial f/\partial t)(t,\cdot) - \beta'(v(t,\cdot))w$, $D(B) = [0,T] \times L^2(\Omega)$ and

$$\varphi(w) = \begin{cases} \frac{1}{2}\int\limits_{\Omega} \|\nabla w\|^2 dx, & w \in H_0^1(\Omega) \\ +\infty, & \text{otherwise.} \end{cases}$$

Since the function $t \to B(t,w)$ is in $L^2(0,T;L^2(\Omega))$ and $\beta' \in L^\infty(\mathbb{R})$, we can apply Theorem 1.15 to deduce that the problem (2.10) has a unique solution $w \in C([0,T];L^2(\Omega))$. Moreover, since $d \in D(\varphi)$, from Theorem 1.14, we deduce that $\partial w/\partial t \in L^2(0,T;L^2(\Omega))$.

Let us show now that $w = \partial v/\partial t$. Indeed, let $\{S(t),\ t \geq 0\}$ be the semigroup of linear contractions, generated by Δ. Then, Theorem 1.22 assures that the solution w of (2.10) is given by

$$w(t) = S(t)d + \int_0^t S(t-s)\left[\frac{\partial f}{\partial s}(s) - \beta'(v(s))w(s)\right] ds, \tag{2.11}$$

while the solution v of (A_0) can be written as

$$v(t) = S(t)a + \int_0^t S(s)[f(t-s) - \beta(v(t-s))]ds.$$

Since $a \in D(-\Delta)$ and $f - \beta \circ v \in W^{1.2}(0,T;L^2(\Omega))$, we can differentiate this equality to obtain that

$$\begin{aligned} \frac{\partial v}{\partial t}(t) &= S(t)\Delta a + S(t)(f(0,\cdot) - \beta(a)) + \\ &+ \textstyle\int_0^t S(s)\left[\frac{\partial f}{\partial s}(t-s) - \beta'(v(t-s)) \cdot \frac{\partial v}{\partial s}(t-s)\right] ds = \\ &= S(t)d + \textstyle\int_0^t S(t-s)\left[\frac{\partial f}{\partial s}(s) - \beta'(v(s)) \cdot \frac{\partial v}{\partial s}(s)\right] ds. \end{aligned} \tag{2.12}$$

We have applied again Theorem 1.22. Taking into account the boundedness of β' and the fact that $S(t)$ is a contraction, from (2.11) and (2.12) it follows that there exists a positive constant M, such that

$$\left| \frac{\partial v}{\partial t}(t) - w(t) \right| \leq M \int_0^t \left| \frac{\partial v}{\partial s}(s) - w(s) \right| ds.$$

Therefore, by the Gronwall-Bellman lemma we can see that $w = \partial v/\partial t$, hence $\partial^2 v/\partial t^2 \in L^2(0,T;L^2(\Omega))$. This completes the proof.

Now we introduce the assumption

(A5) $b \in H_0^1(\Omega) \cap H^2(\Omega)$.

We study the high order regularity of the solution of problem (A_ε). To this end, we first establish a regularity result for the bilocal problem

$$\begin{cases} u'' = Au + g \text{ a.e. } t \in (0,T) \\ u(0) = a,\ u(T) = b. \end{cases} \tag{2.13}$$

Suppose that:

(i) A is a univoque, linear, self-adjoint, densely-defined and coercive operator (and, consequently maximal monotone) in a real Hilbert space H;

(ii) $a, b \in D(A)$;

(iii) $g \in W^{1,2}(0, T; H)$.

Lemma 2.3. *Under hypotheses (i), (ii), (iii), the solution u of (2.13) is in $W^{3,2}_{loc}(0, T; H)$. Moreover, $u''' \in L^2_{\gamma^3}(0, T; H)$, where $\gamma(t) = \min\{t, T-t\}, t \in [0, T]$.*

Proof. We formally derive (2.13) with respect to t and, denoting $w = u'$, one obtains the boundary value problem

$$\begin{cases} w'' = Aw + g', \text{ a.e. } t \in (0, T) \\ w(0) = u'(0), \quad w(T) = u'(T). \end{cases} \tag{2.14}$$

Since $u'(0), u'(T) \in \overline{D(A)} = H$ and $g' \in L^2(0, T; H)$, we can apply Theorem 1.26 to obtain that (2.14) has a unique solution $w \in W^{2,2}_{loc}(0, T; H)$ and $w' \in L^2_\gamma(0, T; H)$, $w'' \in L^2_{\gamma^3}(0, T; H)$.

Our purpose is to prove that $u' = w$. In this way, we will arrive at the conclusion of the lemma.

To this end, let us approach the solutions u and w of (2.13) and (2.14) by the solutions u_λ, w_λ of the problems

$$\begin{cases} u''_\lambda = A_\lambda u_\lambda + g \text{ a.e. } t \in (0, T) \\ u_\lambda(0) = a, \quad u_\lambda(T) = b \end{cases}$$

and

$$\begin{cases} w''_\lambda = A_\lambda w_\lambda + g' \text{ a.e. } t \in (0, T) \\ w_\lambda(0) = u'(0), \quad w_\lambda(T) = u'(T), \end{cases}$$

respectively. It is known that $u_\lambda \to u$, $u'_\lambda \to u'$ in $C([0, T]; H)$, $u''_\lambda \rightharpoonup u''$ in $L^2(0, T; H)$ (Remark 1.24 of Theorem 1.23) and $w_\lambda \to w$ in $C([0, T]; H)$, $w'_\lambda \to w'$ in $C([\alpha, T - \alpha]; H)$, with $\alpha > 0$ arbitrarily chosen (Proposition 1.25 and Theorem 1.26). We wish to prove that

$$\lim_{\lambda \to 0_+} \left[u'_\lambda(t) - w_\lambda(t)\right] = 0, \text{ uniformly on } (0, T). \tag{2.15}$$

To do this, denote $\theta_\lambda = u'_\lambda - w_\lambda$ and observe that, since A_λ is linear and bounded, $g \in W^{1,2}(0, T; H)$ and $u_\lambda \in W^{2,2}(0, T; H)$, θ_λ verifies the problem

$$\begin{cases} \theta''_\lambda = A_\lambda \theta_\lambda \text{ a.e. } t \in (0, T) \\ \theta_\lambda(0) = u'_\lambda(0) - u'(0), \quad \theta_\lambda(T) = u'_\lambda(T) - u'(T) \end{cases}$$

and it belongs to $W^{2,2}(0, T; H)$.

But $(\|\theta_\lambda\|^2)'' = 2(\theta_\lambda, \theta''_\lambda) + 2\|\theta'_\lambda\|^2 \geq 0$ on $(0, T)$.

This implies that the function $t \to \|\theta_\lambda(t)\|^2$ is convex, so

$$\|\theta_\lambda(t)\| \leq \max(\|u'_\lambda(0) - u'(0)\|, \ \|u'_\lambda(T) - u'(T)\|).$$

Then $\theta_\lambda \to 0$ in $C([0, T]; H)$ as $\lambda \to 0_+$, i.e. (2.15).

Therefore, $u' = w$ and this leads to the conclusions of the lemma.

With the aid of this lemma, we can prove the high regularity for the solution of problem (A_ε).

Theorem 2.4. *Under the assumptions (A1)-(A3), (A5), for every given* $\varepsilon > 0$, *the solution* v_ε *of the problem* (A_ε) *is in* $W^{3,2}_{loc}(0,T;L^2(\Omega))$. *Moreover,* v_ε *satisfies*

$$\frac{\partial^3 v_\varepsilon}{\partial t^3} \in L^2_{\gamma^3}(0,T;L^2(\Omega)),$$

where $\gamma(t) = \min\{t, T-t\}$ *on* $[0,T]$.

Proof. Since the unique solution v_ε of (A_ε) belongs to $W^{2,2}\left(0,T;L^2\left(\Omega\right)\right)$, observe that the function $g = \partial v_\varepsilon/\partial t + \beta(v_\varepsilon) - f \in W^{1,2}(0,T;L^2(\Omega))$. Thus, problem (A_ε) is a particular case of the two-point boundary value problem (2.13) with $Au = -\Delta u$, $D\left(A\right) = H_0^1\left(\Omega\right) \cap H^2\left(\Omega\right)$ and $H = L^2\left(\Omega\right)$. By Lemma 2.3, we reach the conclusion of the theorem.

2.3. The Asymptotic Algorithm

Suppose that hypotheses (A1)-(A5) hold. We turn our attention to problem $(2.4)-(2.6)$. Since $\Delta\Pi + \varepsilon\left(\partial^2 v/\partial t^2\right) \in L^2\left(Q\right)$ and β satisfies (A3), it follows that the operator B : $[0,T] \times L^2\left(\Omega\right) \to L^2\left(\Omega\right)$,

$$B\left(t, r_\varepsilon\right) = \beta\left(v + \Pi + r_\varepsilon\right) - \beta\left(v\right) - \left(\Delta\Pi + \varepsilon\frac{\partial^2 v}{\partial t^2}\right)$$

satisfies the assumptions of Theorem 1.30 and therefore problem $(2.4)-(2.6)$ has a unique solution $r_\varepsilon \in W^{2,2}\left(0,T;L^2\left(\Omega\right)\right)$.

We are ready to state the main result.

Theorem 2.5. [3] *Suppose that conditions (A1)-(A5) hold. For sufficiently small* ε, *the function* V_ε *in* (2.1) *is an asymptotic approximation for the solution* v_ε *of* (A_ε) *in the entire domain Q and the following estimates yield:*

$$|v_\varepsilon - v|_{L^2(Q)} = o(\varepsilon^{1/2}), \quad |v_\varepsilon - v - \Pi|_{C([0,T];L^2(\Omega))} = o(\varepsilon^{1/4}). \tag{2.16}$$

Moreover,

$$|\nabla(v_\varepsilon - v - \Pi)|^2_{(L^2(Q))^N} = o(\varepsilon), \quad |\frac{\partial}{\partial t}(v_\varepsilon - v - \Pi)|_{L^2(Q)} = o(\varepsilon^0). \tag{2.17}$$

Proof. Denote

$$\int_Q \left(\frac{\partial^2 v}{\partial t^2}\right)^2 dtdx = (M_1)^2 > 0. \tag{2.18}$$

First, observe that (2.8) and (2.18) lead to the existence of a positive constant M_2 such that

$$\int_Q \left(\Delta\Pi + \varepsilon\frac{\partial^2 v}{\partial t^2}\right)^2 dtdx \le \varepsilon(M_2)^2.$$

One multiplies (2.4) by r_ε and one integrates over Q. Applying Green's formula, we infer

$$\varepsilon\int_Q [\frac{\partial}{\partial t}\left(r_\varepsilon\frac{\partial r_\varepsilon}{\partial t}\right)-\left(\frac{\partial r_\varepsilon}{\partial t}\right)^2]dtdx-$$

$$-\frac{1}{2}\int_Q\frac{\partial}{\partial t}(r_\varepsilon^2)dtdx=\int_Q\|\nabla r_\varepsilon\|^2dtdx+$$

$$+\int_Q r_\varepsilon[\beta(v+\Pi+r_\varepsilon)-\beta(v)]dtdx-\int_Q r_\varepsilon(\Delta\Pi+\varepsilon\frac{\partial^2 v}{\partial t^2})dtdx$$

or, via (2.6) and (2.18),

$$\varepsilon\int_Q\left(\frac{\partial r_\varepsilon}{\partial t}\right)^2dtdx+\int_Q\|\nabla r_\varepsilon\|^2dtdx$$

$$\leq-\int_Q(\Pi+r_\varepsilon)[\beta(v+\Pi+r_\varepsilon)-\beta(v)]dtdx+$$

$$+\int_Q\Pi[\beta(v+\Pi+r_\varepsilon)-\beta(v)]dtdx+\sqrt{\varepsilon}M_2\left(\int_Q r_\varepsilon^2dtdx\right)^{1/2}. \tag{2.19}$$

But β is a monotone function, so $(\Pi+r_\varepsilon)[\beta(v+\Pi+r_\varepsilon)-\beta(v)]\geq 0$ a.e. on Q. Also, it follows that there is an element $\tilde{v}\in H^2(\Omega)\cap H_0^1(\Omega)$ such that

$$\beta(v+\Pi+r_\varepsilon)-\beta(v)=(\Pi+r_\varepsilon)\beta'(\tilde{v}),$$

so, from (2.19) and the boundedness of β', we derive

$$\varepsilon\int_Q\left(\frac{\partial r_\varepsilon}{\partial t}\right)^2dt\,dx+\int_Q\|\nabla r_\varepsilon\|^2dt\,dx\leq$$

$$\leq M_3\int_Q\Pi(\Pi+r_\varepsilon)dtdx+\sqrt{\varepsilon}M_3|r_\varepsilon|_{L^2(Q)}, \tag{2.20}$$

where M_3 is a positive constant. But (2.7) and (2.20) imply that

$$\varepsilon\int_Q\left(\frac{\partial r_\varepsilon}{\partial t}\right)^2dtdx+\int_Q\|\nabla r_\varepsilon\|^2dt\,dx\leq\varepsilon M_4+\sqrt{\varepsilon}M_5|r_\varepsilon|_{L^2(Q)}. \tag{2.21}$$

The Poincaré inequality implies

$$|r_\varepsilon|_{L^2(Q)}^2\leq C|\nabla r_\varepsilon|_{(L^2(Q))^N}^2\leq\sqrt{\varepsilon}M_6|r_\varepsilon|_{L^2(Q)}+\varepsilon M_7,$$

so

$$|r_\varepsilon|_{L^2(Q)} \leq \sqrt{\varepsilon} M_8. \tag{2.22}$$

Then, in view of (2.22) and (2.7), we get $|v_\varepsilon - v|_{L^2(Q)} \leq M_9\sqrt{\varepsilon}$, which is the first relation in (2.16). Now, from (2.22) and (2.21), we find some positive constants, M_{10}, M_{11} such that

$$|\nabla r_\varepsilon|^2_{(L^2(Q))^N} \leq \varepsilon M_{10}$$

and

$$\left|\frac{\partial r_\varepsilon}{\partial t}\right|^2_{L^2(Q)} \leq M_{11}. \tag{2.23}$$

These inequalities prove (2.17).

Finally, remark that (2.22), (2.23) and

$$\int_\Omega r_\varepsilon^2(t,x)dx = 2\int_0^t \int_\Omega r_\varepsilon(s,x)\frac{\partial r_\varepsilon}{\partial s}(s,x)dsdx \text{ a.e. } t \in (0,T),$$

lead to

$$|r_\varepsilon|_{C([0,T];L^2(\Omega))} = o(\varepsilon^{1/4}).$$

This is the second equality (2.16). Thus, the theorem is completely proved.

Corollary 2.6. *Under the hypotheses of Theorem 2.5, we have two cases:*
1) If $b(x) = v(T,x)$ a.e. on Ω, then $|v_\varepsilon - v|_{C([0,T];L^2(\Omega))} = o(\varepsilon^{1/4})$. The problem (P_ε) is regularly perturbed in Q and v is an asymptotic approximation of v_ε in the entire domain Q.
2) Otherwise, for each δ, $0 < \delta < T$, $|v_\varepsilon - v|_{C([0,\delta];L^2(\Omega))} = o(\varepsilon^{1/4})$. The problem (P_ε) is singularly perturbed and v is an asymptotic approximation of v_ε in $Q_1 = (0,\delta] \times \Omega$.

Proof. Indeed, with the aid of (2.9) and (2.16), we find

$$|v_\varepsilon - v|_{C([0,T];L^2(\Omega))} \leq M_{12}\varepsilon^{1/4} + |b - v(T,\cdot)|_{L^2(\Omega)},$$

where M_{12} is a positive constant. So, in the first case, $|v_\varepsilon - v|_{C([0,T];L^2(\Omega))} = o(\varepsilon^{1/4})$.

In the second case, v is not an asymptotic approximation for v_ε on Q, but for every $\delta \in (0,T)$, we deduce that

$$|v_\varepsilon - v|_{C([0,\delta];L^2(\Omega))} \leq M_{12}\varepsilon^{1/4} + |b - v(T,\cdot)|_{L^2(\Omega)} e^{-\frac{T-\delta}{\varepsilon}}.$$

3. Asymptotic Approximation for the Heat Equation with Nonlinear Boundary Conditions

In this section we are interested in the linear heat equation with nonlinear boundary conditions ([5]). We study the nonlinear boundary value problem

$$\begin{cases} \varepsilon\dfrac{\partial^2 v_\varepsilon}{\partial t^2} - \dfrac{\partial v_\varepsilon}{\partial t} = -\Delta v_\varepsilon - f, \ (t,x) \in (0,T) \times \Omega = Q \\ -\partial v_\varepsilon / \partial\eta = \beta(v_\varepsilon), \ (t,x) \in (0,T) \times \partial\Omega = \Sigma \\ v_\varepsilon(0,x) = a(x), \ v_\varepsilon(T,x) = b(x), \ x \in \Omega, \end{cases} \tag{P_ε}$$

where $\varepsilon > 0$ is a small parameter, and the corresponding reduced problem

$$\left\{\begin{array}{c} \dfrac{\partial v}{\partial t} - \Delta v = f,\ (t,x) \in (0,T) \times \Omega = Q \\ -\partial v/\partial \eta = \beta(v),\ (t,x) \in (0,T) \times \partial\Omega = \Sigma \\ v(0,x) = a(x),\ x \in \Omega. \end{array}\right. \tag{P$_0$}$$

Here $\partial/\partial\eta$ denotes the outward normal derivative. Observe that problem (P_0) consists of the heat equation, a nonlinear boundary condition and an initial condition and (P_ε) is an elliptic regularization of problem (P_0).

Suppose that Ω is a bounded open subset of $\mathbb{R}^N$, with the boundary $\partial\Omega$ smooth enough, β is a nonlinear maximal monotone univoque graph in $\mathbb{R} \times \mathbb{R}$ such that its domain $D(\beta)$ is dense in $\mathbb{R}$, f is a function in $W^{1,2}(0,T;L^2(\Omega))$, and $a, b \in H^2(\Omega)$, with $-\partial a/\partial\eta = \beta(a)$, $-\partial b/\partial\eta = \beta(b)$.

Assume that

(H1) $f \in W^{1,2}(0,T;L^2(\Omega))$;

(H2) $\beta : \mathbb{R} \to \mathbb{R}$, $\beta \in C^1(\mathbb{R})$, $\beta' \geq 0$, $\beta' \in L^\infty(\mathbb{R})$;

(H3) $a \in H^2(\Omega)$, $-\partial a/\partial\eta = \beta(a)$;

(H4) $b \in H^2(\Omega)$, $-\partial b/\partial\eta = \beta(b)$.

By Proposition 1.29, problem (P_ε) has a unique solution

$$v_\varepsilon \in W^{2,2}(0,T;L^2(\Omega)).$$

It is known that (see [10]) the operator

$$Av = -\Delta v,\ D(A) = \{v \in H^2(\Omega), -\partial v/\partial\eta = \beta(v) \text{ a.e. on } \partial\Omega\}$$

is maximal monotone in the Hilbert space $H = L^2(\Omega)$. Moreover, if function $j : \mathbb{R} \to (-\infty, +\infty]$ is the lower semicontinuous convex proper function with the property $\partial j = \beta$, then A coincides with the subdifferential $\partial\varphi$ of the lower semicontinuous convex proper function $\varphi : L^2(\Omega) \to (-\infty, +\infty]$,

$$\varphi(v) = \left\{\begin{array}{ll} \dfrac{1}{2}\int\limits_\Omega |\nabla v|^2\,dx + \int\limits_{\partial\Omega} j(v)\,dx, & \\ if\ v \in H^1(\Omega),\ j(v) \in L^1(\partial\Omega) & \\ +\infty, & \text{otherwise.} \end{array}\right.$$

Using the above notation, we can write (P_0) as

$$\left\{\begin{array}{c} v' + Av = f,\ t \in (0,T) \\ v(0) = a. \end{array}\right.$$

We know that (see [7], Chapter 4), if $f \in L^2(0,T;L^2(\Omega))$ and $\partial f/\partial t \in L^1(0,T;L^2(\Omega))$, then (P_0) has a unique strong solution

$$v \in W^{1,\infty}(0,T;L^2(\Omega)) \cap L^\infty(0,T;H^2(\Omega)).$$

We investigate the behavior of the solution v_ε of (P_ε) as $\varepsilon \to 0$. Since the condition $v(T,x) = b(x)$ a.e. on Ω is necessary for the convergence

$$v_\varepsilon \to v \text{ as } \varepsilon \to 0, \text{ in } C([0,T];L^2(\Omega)),$$

it follows that v_ε has a singular behavior with respect to ε, in a neighbourhood $(\delta, T] \times \Omega$ of the set $\{(T, x), \ x \in \Omega\}$. This neighbourhood is a *boundary layer*. Problem (P_ε) is a singularly perturbed problem of the elliptic-parabolic type. A similar problem can be find in [8].

3.1. The Asymptotic Approximation

Along this subsection, suppose that f, β, a, b are smooth enough. Our aim is to construct an asymptotic expansion for the solution v_ε of (P_ε), which is valid both in the boundary layer $(\delta, T] \times \Omega$ and outside it. Using the boundary function method of Vishik and Lyusternik, we are looking for an asymptotic expansion of v_ε of the form

$$v_\varepsilon(t, x) = v(t, x) + \Pi(\tau, x) + r_\varepsilon(t, x), \tag{3.1}$$

where $\tau = (T - t)/\varepsilon$ is the boundary layer variable, Π is the boundary layer function and r_ε is the remainder. Function Π is introduced to compensate the singular behavior of v_ε with respect to ε, in the boundary layer. So Π must be small enough outside the boundary layer:

$$\Pi(\infty, x) = 0, \ x \in \Omega. \tag{3.2}$$

One formally substitutes v_ε from (3.1) into (P_ε). We will prove in the next section that $v \in W^{2,2}(0, T; L^2(\Omega))$. We have

$$\begin{cases} \varepsilon(\dfrac{\partial^2 v}{\partial t^2} + \dfrac{1}{\varepsilon^2}\dfrac{\partial^2 \Pi}{\partial \tau^2} + \dfrac{\partial^2 r_\varepsilon}{\partial t^2}) - (\dfrac{\partial v}{\partial t} - \dfrac{1}{\varepsilon}\dfrac{\partial \Pi}{\partial \tau} + \dfrac{\partial r_\varepsilon}{\partial t}) = \\ \quad = -\Delta(v + \Pi + r_\varepsilon) - f, \ (t, x) \in Q. \\ -\dfrac{\partial v}{\partial \eta} - \dfrac{\partial \Pi}{\partial \eta} - \dfrac{\partial r_\varepsilon}{\partial \eta} = \beta(v + \Pi + r_\varepsilon), \ (t, x) \in \Sigma \\ v(0, x) + \Pi(\infty, x) + r_\varepsilon(0, x) = a(x), \ x \in \Omega \\ v(T, x) + \Pi(0, x) + r_\varepsilon(T, x) = b(x), \ x \in \Omega. \end{cases}$$

Identifying the coefficients of the same powers of ε, separately for the terms depending on t and on τ and using (3.2), we find that Π and r_ε verify the boundary value problems

$$\begin{cases} \dfrac{\partial^2 \Pi}{\partial \tau^2} + \dfrac{\partial \Pi}{\partial \tau} = 0, \ (\tau, x) \in (0, \infty) \times \Omega = Q' \\ \Pi(0, x) = b(x) - v(T, x), \ \Pi(\infty, x) = 0, \ x \in \Omega \end{cases} \tag{3.3}$$

and, respectively,

$$\begin{cases} \varepsilon\dfrac{\partial^2 r_\varepsilon}{\partial t^2} - \dfrac{\partial r_\varepsilon}{\partial t} = -\Delta r_\varepsilon - (\Delta\Pi + \varepsilon\dfrac{\partial^2 v}{\partial t^2}), \ (t, x) \in Q \\ -\dfrac{\partial r_\varepsilon}{\partial \eta} = \beta\,(v + \Pi + r_\varepsilon) - \beta\,(v) + \\ +e^{-\tau}\left[\beta\,(v\,(T, .)) - \beta\,(b)\right], \ (t, x) \in \Sigma \\ r_\varepsilon(0, x) = r_\varepsilon(T, x) = 0, \ x \in \Omega. \end{cases} \tag{3.4}$$

Problem (3.3) has a unique solution

$$\Pi(\tau, x) = e^{-\tau}[b(x) - v(T, x)], \ x \in \Omega. \tag{3.5}$$

It is easy to check that

$$\int_Q \Pi^2(\tau, x)\, dtdx = \frac{\varepsilon}{2}|b - v(T, .)|^2_{L^2(\Omega)}\left(1 - e^{-2T/\varepsilon}\right).$$

Therefore,

$$|\Pi|^2_{L^2(Q)} = o(\varepsilon). \tag{3.6}$$

Analogously,

$$\begin{cases} |\Delta\Pi|^2_{L^2(Q)} = o(\varepsilon), \\ |\Pi|^2_{L^2(\Sigma)} = o(\varepsilon), \\ |e^{-\tau}[\beta(v(T, .)) - \beta(b)]|^2_{L^2(\Sigma)} = o(\varepsilon). \end{cases} \tag{3.7}$$

In order to study problem (3.4), we first give a regularity result for (P_0).

3.2. Auxiliary Results

We study the existence and uniqueness of the solution to the auxiliary boundary value problem

$$\begin{cases} \dfrac{\partial u}{\partial t} - \Delta u = h, \ (t, x) \in Q = (0, T) \times \Omega \\ \dfrac{\partial u}{\partial \eta} + \alpha u = g, \ (t, x) \in \Sigma = (0, T) \times \partial\Omega \\ u(0, x) = u_0(x), \quad x \in \Omega, \end{cases} \tag{3.8}$$

where α and g are given functions defined on Σ, and h is given on Q.

To this end, we use Theorem 1.16. One takes $H = L^2(\Omega)$, $V = H^1(\Omega)$ and

$$a(t, u(t), \psi) = \int_\Omega \nabla u \cdot \nabla\psi dx + \int_{\partial\Omega} \alpha u\psi dx,$$

$$< l(t), \psi >= \int_\Omega h\psi dx + \int_{\partial\Omega} g\psi dx,$$

for every $\psi \in V = H^1(\Omega)$. Here $< .,. >$ denotes the dual pair between an element of V and an element of V'. In order to have conditions (ii) and (iii), it is enough to impose $\alpha \geq 0$, $\alpha \in L^\infty(\Sigma)$ and $g \in L^2\left(0, T; H^{-1/2}(\partial\Omega)\right)$. For (ii) one uses the Trace Theorem. From Theorem 1.16, it follows

Lemma 3.1. *If* $h \in L^2\left(0, T; H^{-1}(\Omega)\right)$, $g \in L^2\left(0, T; H^{-1/2}(\partial\Omega)\right)$, $\alpha \geq 0$, $\alpha \in L^\infty(\Sigma)$ *and* $u_0 \in L^2(\Omega)$, *then problem* (3.8) *has a unique solution* $u \in L^2\left(0, T; H^1(\Omega)\right) \cap C\left([0, T]; L^2(\Omega)\right)$ *with* $\partial u/\partial t \in L^2\left(0, T; H^{-1}(\Omega)\right)$.

Now we use Theorem 1.17 to improve this result provided that $h \in L^2(0,T;L^2(\Omega))$, $g \in L^2(0,T;H^{1/2}(\partial\Omega))$ and $u_0 \in H^2(\Omega)$. To this end, one transforms problem (3.8) into a problem with the zero initial condition, with the aid of the change of function

$$\widetilde{u}(t,x) = u(t,x) - u_0(x),\ \ (t,x) \in Q.$$

Thus problem (3.8) becomes

$$\begin{cases} \dfrac{\partial \widetilde{u}}{\partial t} - \Delta \widetilde{u} = h + \Delta u_0, & (t,x) \in Q \\ \dfrac{\partial \widetilde{u}}{\partial \eta} = g - \alpha u - \dfrac{\partial u_0}{\partial \eta}, & (t,x) \in \Sigma \\ \widetilde{u}(0,x) = 0, & x \in \Omega. \end{cases} \tag{3.9}$$

It can be written under the form

$$a(t,\widetilde{u}(t),\psi) + < \widetilde{u}'(t),\psi > = < l(t),\psi >, \text{ a.e. } t \in [0,T],$$

for each $\psi \in H^1(\Omega)$, where

$$a(t,\widetilde{u}(t),\psi) = a(\widetilde{u}(t),\psi) = \int_\Omega \nabla\widetilde{u}\cdot\nabla\psi dx$$

and

$$< l(t),\psi > = \int_\Omega (h + \Delta u_0)\psi dx + \int_{\partial\Omega} (g - \alpha u - \frac{\partial u_0}{\partial \eta})\psi dx,$$

for every $\psi \in H^1(\Omega)$. Since the solution u of (3.8) belongs to $L^2(0,T;H^1(\Omega))$, we have that $g - \alpha u - \dfrac{\partial u_0}{\partial \eta} \in L^2(0,T;H^{1/2}(\partial\Omega)) \subset L^2(0,T;L^2(\partial\Omega))$, so l is well defined.

The hypotheses of Theorem 1.17 are verified for $H = L^2(\Omega)$, $V = H^1(\Omega)$ (note that $a(t,\widetilde{u}(t),\psi)$ is independent of t) and $h \in L^2(0,T;L^2(\Omega))$. Therefore (3.9) admits a unique solution $\widetilde{u} \in L^2(0,T;H^1(\Omega))$, with the derivative $\dfrac{\partial \widetilde{u}}{\partial t} \in L^2(0,T;L^2(\Omega))$. Moreover, we can deduce that $\widetilde{u} \in L^2(0,T;H^2(\Omega))$. Coming back to (3.8), we can give now the following regularity result.

Lemma 3.2. *Suppose that* $\alpha \in L^\infty(\Sigma)$, $\alpha \geq 0$, $u_0 \in H^2(\Omega)$ *and*

$$h \in L^2(0,T;L^2(\Omega)),\quad g \in L^2\left(0,T;H^{1/2}(\partial\Omega)\right).$$

Then the solution u *of problem* (3.8) *has the properties*

$$u \in L^2(0,T;H^2(\Omega)) \cap C([0,T];L^2(\Omega))$$

$$\text{and } \frac{\partial u}{\partial t} \in L^2(0,T;L^2(\Omega)).$$

3.3. Regularity Results

One proves that, under additional conditions on $f,\ a,\ b,\ \beta$, the solutions v of (P_0) and v_ε of (P_ε) are more regular. We begin with the regularity of v.

Theorem 3.3. *Under the hypotheses* $(H1)-(H3)$ *and*

$$(H5) \qquad a \in H^4(\Omega),\ f(0,.) \in H^2(\Omega),$$

the solution v *of* (P_0) *satisfies in addition the conditions*

$$\partial v/\partial t \in L^2(0,T;H^2(\Omega)),\ \ \partial^2 v/\partial t^2 \in L^2(Q). \tag{3.10}$$

Proof. We differentiate problem (P_0) with respect to t in the sense of distributions and denote $w = \partial v/\partial t$. Hence w satisfies the problem

$$\begin{cases} \dfrac{\partial w}{\partial t} - \Delta w = \dfrac{\partial f}{\partial t}, \ \ (t,x) \in Q \\ -\dfrac{\partial w}{\partial \eta} = \beta'(v)w, \ \ (t,x) \in \Sigma \\ w(0,x) = w_0(x), \ \ x \in \Omega, \end{cases} \tag{3.11}$$

where $w_0 = \Delta a + f(0,\cdot) \in H^2(\Omega)$, via (H5). This problem has the form (3.8), with $g = 0,\ h = \partial f/\partial t$ and $\alpha = \beta'(v)$. By Lemma 3.2 it follows that (3.11) has a unique solution $w \in C([0,T];L^2(\Omega)) \cap L^2(0,T;H^2(\Omega))$, with $\partial w/\partial t \in L^2(Q)$.

Since this solution of (3.11) coincides with the solution in distributions, we have $w = \partial v/\partial t$ and consequently v has the properties (3.10), as claimed.

In order to show the regularity of v_ε, we first give a general existence result concerning the boundary value problem

$$\begin{cases} \varepsilon\dfrac{\partial^2 u}{\partial t^2} - \dfrac{\partial u}{\partial t} = -\Delta u + h, \ (t,x) \in (0,T) \times \Omega \\ -\dfrac{\partial u}{\partial \eta} = \partial j\left(\gamma + u\right) + g, \ (t,x) \in (0,T) \times \partial\Omega \\ u\left(0,x\right) = u_0\left(x\right),\ u\left(T,x\right) = u_1\left(x\right),\ x \in \Omega. \end{cases} \tag{3.12}$$

Here $\varepsilon > 0$ is a small parameter and Ω is an open bounded set in $\mathbb{R}^N$, with its boundary $\partial\Omega$ of class C^2. Suppose that the following conditions hold:

(1) $h \in L^2(0,T;L^2\left(\Omega\right))$;
(2) $g \in L^\infty(0,T;H^{1/2}\left(\partial\Omega\right))$;
(3) $j : \mathbb{R} \to \mathbb{R}$ is a proper, convex and lower semicontinuous function;
(4) $u_0, u_1 \in H^2\left(\Omega\right),\ -(\partial u_0/\partial\eta) = \partial j\left(\gamma\left(0,.\right) + u_0\right) + g\left(0,.\right)$,
$-\left(\partial u_1/\partial\eta\right) = \partial j\left(\gamma\left(T,.\right) + u_1\right) + g\left(T,.\right)$ a.e. on $\partial\Omega$;
(5) $\gamma \in L^\infty\left(0,T;H^{3/2}\left(\partial\Omega\right)\right)$.

Theorem 3.4. *If the above assumptions hold, then problem* (3.12) *has a unique solution* $u \in W^{2,2}\left(0,T;L^2\left(\Omega\right)\right) \cap L^2\left(0,T;H^2\left(\Omega\right)\right)$.

Proof. We show that (3.12) is equivalent to a variational problem and thus it has a solution, namely a point of minimum for a certain convex, proper and lower semicontinuous function (F below).

Problem (3.12) can be written equivalently under the form

$$\begin{cases} \varepsilon e^{t/\varepsilon}\dfrac{\partial}{\partial t}\left(e^{-t/\varepsilon}\dfrac{\partial u}{\partial t}\right) = -\Delta u + h, \ (t,x) \in (0,T) \times \Omega \\ -\dfrac{\partial u}{\partial \eta} = \partial j(\gamma + u) + g, \ (t,x) \in (0,T) \times \partial\Omega \\ u(0,x) = u_0(x), \ u(T,x) = u_1(x), \ x \in \Omega. \end{cases} \tag{3.13}$$

Observe that the time-dependent operator

$$\begin{cases} A(t)\,u = -\Delta u, D(A(t)) = \\ = \left\{u \in H^2(\Omega), \ -\dfrac{\partial u}{\partial \eta} = \partial j(\gamma + u) + g, \text{ a.e. on } \partial\Omega\right\} \end{cases}$$

is maximal monotone in $L^2(\Omega)$. Moreover it coincides with the subdifferential $\partial\phi(t,u)$ of the proper, convex and lower semicontinuous function

$$\phi(t,u) = \begin{cases} \dfrac{1}{2}\int\limits_{\Omega} |\nabla u|^2 dx + \int\limits_{\partial\Omega} j(\gamma + u)\,dx + \int\limits_{\partial\Omega} gudx, \\ \text{if } u \in H^1(\Omega), \ j(\gamma + u) \in L^1(\partial\Omega), \\ +\infty, \qquad \text{otherwise.} \end{cases}$$

Denote by $\mathcal{L}^2$ the space $L^2(0,T;L^2(\Omega))$ with the weight $e^{-t/\varepsilon}$ and by $\partial\widetilde{\phi}$ the realization of $\partial\phi$ in $\mathcal{L}^2$, i.e. $\left(\partial\widetilde{\phi}(u)\right)(t) = \partial\phi(t, u(t))$ a.e. on $[0,T]$. Let B be the maximal monotone operator (in $\mathcal{L}^2$) (see section 2.3) given by

$$\begin{cases} Bu = -\varepsilon e^{t/\varepsilon}\dfrac{\partial}{\partial t}\left(e^{-t/\varepsilon}\dfrac{\partial u}{\partial t}\right), D(B) = \\ \{u \in \mathcal{L}^2, \ u', u'' \in \mathcal{L}^2, \ u(0,.) = u_0, \ u(T,.) = u_1 \text{ on } \Omega\}. \end{cases}$$

Thus problem (3.13) can be written in $\mathcal{L}^2$ as

$$Bu + \partial\widetilde{\phi}(u) + h = 0. \tag{3.14}$$

We want to show that

$$Bu + \partial\widetilde{\phi}(u) + h = \partial F, \tag{3.15}$$

where $F : \mathcal{L}^2 \to (-\infty, \infty]$,

$$F(u) = \begin{cases} \frac{1}{2}\int_0^T \int\limits_\Omega e^{-t/\varepsilon}\left[\varepsilon\left(\frac{\partial u}{\partial t}\right)^2 + |\nabla u|^2\right] dtdx + \\ + \int_0^T \int\limits_{\partial\Omega} e^{-t/\varepsilon} j(\gamma + u)\, dtdx + \\ + \int_0^T \int\limits_{\partial\Omega} e^{-t/\varepsilon} gu\, dtdx + \int_0^T \int\limits_\Omega e^{-t/\varepsilon} hu\, dtdx + \\ + \varepsilon\psi(u(0,.) - u_0) + \varepsilon e^{-T/\varepsilon}\psi(u(T,.) - u_1), \\ \text{if } u \in W^{1,2}(0,T;L^2(\Omega)) \cap L^2(0,T;H^1(\Omega)) \\ \text{and } j(\gamma+u) \in L^1(0,T;L^1(\partial\Omega)), \\ +\infty, \quad \text{otherwise} \end{cases}$$

and

$$\psi(x) = \begin{cases} 0, & x = 0 \\ +\infty, & \text{otherwise.} \end{cases}$$

It is enough to prove the inclusion

$$Bu + \partial\widetilde{\phi}(u) + h \subseteq \partial F, \ (\forall)\, u \in D(B) \cap D\left(\widetilde{\partial\phi}\right). \tag{3.16}$$

Since both $B + \partial\widetilde{\phi} + h$ and ∂F are maximal monotone (as in the proof of Theorem 4.1.1), inclusion (3.16) implies equality (3.15).

Denote by $< .,. >_{\mathcal{L}^2}$ the scalar product in $\mathcal{L}^2$. Let $u \in D(B) \cap D\left(\widetilde{\partial\phi}\right)$ be given and $w \in L^2(0,T;L^2(\Omega))$ be arbitrary. If $w \in D(F)$, we have

$$\begin{aligned} &< Bu + \partial\widetilde{\phi}(u) + h,\ u - w >_{\mathcal{L}^2} = \\ &= -\varepsilon e^{-T/\varepsilon}\int\limits_\Omega \frac{\partial u}{\partial t}(T,x)\,[u_1(x) - w(T,x)]dx + \\ &+ \varepsilon\int\limits_\Omega \frac{\partial u}{\partial t}(0,x)\,[u_0(x) - w(0,x)]\,dx + \\ &+ \varepsilon\int_0^T\int\limits_\Omega e^{-t/\varepsilon}\frac{\partial u}{\partial t}\left(\frac{\partial u}{\partial t} - \frac{\partial w}{\partial t}\right) dtdx + \\ &+ \int_0^T\int\limits_\Omega e^{-t/\varepsilon}\nabla u\,(\nabla u - \nabla w)\, dtdx + \\ &+ \int_0^T\int\limits_{\partial\Omega} e^{-t/\varepsilon}\left[\partial j(\gamma + u) + g\right]\cdot(u - w)\, dtdx + \\ &+ \int_0^T\int\limits_\Omega e^{-t/\varepsilon} h\cdot(u - w)\, dtdx. \end{aligned} \tag{3.17}$$

On the other hand,

$$F(u) - F(w) \leq \varepsilon \int_0^T \int_\Omega e^{-t/\varepsilon} \frac{\partial u}{\partial t}\left(\frac{\partial u}{\partial t} - \frac{\partial w}{\partial t}\right) dtdx +$$

$$+ \int_0^T \int_\Omega e^{-t/\varepsilon} \nabla u (\nabla u - \nabla w) \, dtdx +$$

$$+ \int_0^T \int_{\partial\Omega} e^{-t/\varepsilon} \partial j (\gamma + u)(u - w) \, dtdx$$

$$+ \int_0^T \int_{\partial\Omega} e^{-t/\varepsilon} g \cdot (u - w) \, dtdx +$$

$$+ \int_0^T \int_\Omega e^{-t/\varepsilon} h \cdot (u - w) \, dtdx$$

$$+ \varepsilon [\psi (u(0,.) - u_0) - \psi (w(0,.) - u_0)] +$$

$$+ \varepsilon e^{-T/\varepsilon} [\psi (u(T,.) - u_1) - \psi (w(T,.) - u_1)]. \tag{3.18}$$

We have used the inequality $\frac{1}{2}(x^2 - y^2) \leq x(x - y)$. From (3.17) and (3.18) we get

$$F(u) - F(w) \leq < Bu + \partial\widetilde{\phi}(u) + h, \; u - w >_{\mathcal{L}^2}, \tag{3.19}$$

for every $w \in D(F)$. If $w \notin D(F)$, then (3.19) is obvious. Hence (3.16) is proved.

Now, u is a solution for problem (3.14) if and only if u is a minimum point of F. Therefore (3.14) has a solution $u \in D(F)$, i.e. $u \in W^{1,2}(0,T;L^2(\Omega))$, with $u \in L^2(0,T;H^1(\Omega))$.

Since $D(F) = D(B) \cap D\left(\partial\widetilde{\phi}\right)$, it follows that u is in $W^{2,2}(0,T;L^2(\Omega))$. By (3.12), we have also that $u \in L^2(0,T;H^2(\Omega))$. The uniqueness can be easily proved.

As a consequence of this result, we deduce the existence of the solution of problem (3.4). To do this, it is sufficient to observe that, if we take in (3.12), $h = -\Delta\Pi - \varepsilon\frac{\partial^2 v}{\partial t^2}$, $g = -\beta(v) + e^{-\tau}[\beta(v(T,\cdot)) - \beta(b)]$, $\gamma = v + \Pi$ and $u_0 = u_1 = 0$, we obtain problem (3.4). These functions satisfy conditions $(1) - (5)$ above. Therefore we have:

Corollary 3.5. *If hypotheses $(H1) - (H4)$ hold, then problem (3.4) has a unique solution $r_\varepsilon \in W^{2,2}(0,T;L^2(\Omega)) \cap L^2(0,T;H^2(\Omega))$.*

3.4. Estimates of the Remainder Term

Let us return to the asymptotic expansion of the solution v_ε of (P_ε). As we have seen in section 3.1, it has the form (3.1), where the boundary layer function Π is given by (3.5) and the remainder term r_ε is the solution of (3.4). Under the assumptions $(H1) - (H5)$, the solution v of (P_0) is in $W^{2,2}(0,T;L^2(\Omega))$ (Theorem 3.3) and problem (3.4) has a unique solution $r_\varepsilon \in W^{2,2}(0,T;L^2(\Omega)) \cap L^2(0,T;H^2(\Omega))$ (Corollary 3.5).

We show that $r_\varepsilon \to 0$ in $C([0,T];L^2(\Omega))$, $\nabla r_\varepsilon \to 0$ in $(L^2(Q))^N$, as $\varepsilon \to 0$ and we find the order of accuracy of $v_\varepsilon - v$. More exactly, we have

Theorem 3.6. [5] *Suppose that hypotheses $(H1)-(H5)$ hold. Then we have the estimates:*

$$\begin{cases} |v_\varepsilon - v|_{L^2(Q)} = o\left(\varepsilon^{1/2}\right), \\ |v_\varepsilon - v - \Pi|_{C([0,T];L^2(\Omega))} = o\left(\varepsilon^{1/4}\right), \\ |\nabla(v_\varepsilon - v)|_{(L^2(Q))^N} = o\left(\varepsilon^{1/2}\right). \end{cases} \tag{3.20}$$

Proof. Multiplying the equation in (3.4) by r_ε and integrating over Q, one obtains

$$\varepsilon \int_Q \left[\frac{\partial}{\partial t}\left(\frac{\partial r_\varepsilon}{\partial t} r_\varepsilon\right) - \left(\frac{\partial r_\varepsilon}{\partial t}\right)^2\right] dtdx - \frac{1}{2}\int_Q \frac{\partial\left(r_\varepsilon^2\right)}{\partial t} dtdx =$$

$$= \int_Q |\nabla r_\varepsilon|^2 dtdx - \int_\Sigma \frac{\partial r_\varepsilon}{\partial \eta} r_\varepsilon dtdx - \int_Q r_\varepsilon\left(\Delta\Pi + \varepsilon\frac{\partial^2 v}{\partial t^2}\right) dtdx.$$

Via the third relation (3.4), this implies

$$\varepsilon \int_Q \left(\frac{\partial r_\varepsilon}{\partial t}\right)^2 dtdx + \int_Q |\nabla r_\varepsilon|^2 dtdx +$$

$$+ \int_\Sigma (r_\varepsilon + \Pi)\left[\beta(v + \Pi + r_\varepsilon) - \beta(v)\right] dtdx =$$

$$= \int_\Sigma \Pi\left[\beta(v + \Pi + r_\varepsilon) - \beta(v)\right] dtdx -$$

$$- \int_\Sigma e^{-T} r_\varepsilon \left[\beta(v(T,x)) - \beta(b)\right] dtdx + + \int_Q r_\varepsilon\left(\Delta\Pi + \varepsilon\frac{\partial^2 v}{\partial t^2}\right) dtdx.$$

Using the monotonicity of β and writing

$$\beta(v + \Pi + r_\varepsilon) - \beta(v) = \beta'(\widetilde{v})(\Pi + r_\varepsilon), \tag{3.21}$$

where $\widetilde{v} \in L^2(0,T;L^2(\partial\Omega))$, we get

$$\varepsilon \int_Q \left(\frac{\partial r_\varepsilon}{\partial t}\right)^2 dtdx + \int_Q |\nabla r_\varepsilon|^2 dtdx \le \int_\Sigma \beta'(\widetilde{v})\,\Pi^2 dtdx +$$

$$+\left(\int_{\Sigma}\beta'(\widetilde{v})\,\Pi^2 dtdx\right)^{1/2}\left(\int_{\Sigma}\beta'(\widetilde{v})r_\varepsilon^2 dtdx\right)^{1/2}+$$

$$+\left(\int_{\Sigma} r_\varepsilon^2 dtdx\right)^{1/2}\left(\int_{\Sigma} e^{-2\tau}\left[\beta\left(v\left(T,x\right)\right)-\beta\left(b\right)\right]^2 dtdx\right)^{1/2}+$$

$$+\left(\int_{Q} r_\varepsilon^2 dtdx\right)^{1/2}\left(\int_{Q}\left(\Delta\Pi+\varepsilon\frac{\partial^2 v}{\partial t^2}\right)^2 dtdx\right)^{1/2}.$$

By (3.7) and $(H2)$, we deduce

$$\varepsilon\int_{Q}\left(\frac{\partial r_\varepsilon}{\partial t}\right)^2 dtdx+\int_{Q}\left|\nabla r_\varepsilon\right|^2 dtdx\le c_1\varepsilon+$$

$$+c_2\varepsilon^{1/2}\left(\int_{\Sigma} r_\varepsilon^2 dtdx\right)^{1/2}+c_3\varepsilon^{1/2}\left(\int_{Q} r_\varepsilon^2 dtdx\right)^{1/2}.$$

According to the Trace Theorem,

$$\varepsilon\int_{Q}\left(\frac{\partial r_\varepsilon}{\partial t}\right)^2 dtdx+\int_{Q}\left|\nabla r_\varepsilon\right|^2 dtdx\le c_1\varepsilon+$$

$$+c_2\varepsilon^{1/2}\left(\int_{Q}(r_\varepsilon^2+\left|\nabla r_\varepsilon\right|^2)dtdx\right)^{1/2}+c_3\varepsilon^{1/2}\left(\int_{Q} r_\varepsilon^2 dtdx\right)^{1/2}. \tag{3.22}$$

Now we multiply the equation of (3.4) by $e^{-t}r_\varepsilon$ and integrate over Q, to find

$$\varepsilon\int_{Q} e^{-t}\left[\frac{\partial}{\partial t}\left(\frac{\partial r_\varepsilon}{\partial t}r_\varepsilon\right)-\left(\frac{\partial r_\varepsilon}{\partial t}\right)^2\right]dtdx-\frac{1}{2}\int_{Q} e^{-t}\frac{\partial\left(r_\varepsilon^2\right)}{\partial t}dtdx=$$

$$=\int_{Q} e^{-t}\left|\nabla r_\varepsilon\right|^2 dtdx-\int_{\Sigma} e^{-t}\frac{\partial r_\varepsilon}{\partial\eta}r_\varepsilon dtdx-\int_{Q} e^{-t}r_\varepsilon\left(\Delta\Pi+\varepsilon\frac{\partial^2 v}{\partial t^2}\right)dtdx.$$

Using again (3.21) and the monotonicity of β, this leads to

$$\varepsilon\int_{Q} e^{-t}\left(\frac{\partial r_\varepsilon}{\partial t}\right)^2 dtdx+\int_{Q} e^{-t}\left|\nabla r_\varepsilon\right|^2 dtdx+\frac{1}{2}\int_{Q} e^{-t}r_\varepsilon^2 dtdx\le$$

$$\leq \varepsilon \left(\int_Q e^{-t} \left(\frac{\partial r_\varepsilon}{\partial t} \right)^2 dtdx \right)^{1/2} \left(\int_Q e^{-t} r_\varepsilon^2 dtdx \right)^{1/2} +$$

$$+ \int_\Sigma e^{-t} \beta' \left(\widetilde{v} \right) \Pi \left(\Pi + r_\varepsilon \right) dtdx +$$

$$+ \left(\int_\Sigma e^{-t} r_\varepsilon^2 dtdx \right)^{1/2} \cdot (\int_\Sigma e^{-2T-t} \left[\beta \left(v \left(T, x \right) \right) - \beta \left(b \right) \right]^2 dtdx)^{1/2} +$$

$$+ \left(\int_Q e^{-t} r_\varepsilon^2 dtdx \right)^{1/2} \cdot \left(\int_Q e^{-t} \left(\Delta \Pi + \varepsilon \frac{\partial^2 v}{\partial t^2} \right)^2 dtdx \right)^{1/2} .$$

Since $\beta' \in L^\infty \left(\mathbb{R} \right)$ (say $\left| \beta' \left(r \right) \right| \leq M, \left(\forall \right) r \in \mathbb{R}$), we obtain $\varepsilon \int_Q e^{-t} \left(\frac{\partial r_\varepsilon}{\partial t} \right)^2 dtdx +$

$$+ \int_Q e^{-t} \left| \nabla r_\varepsilon \right|^2 dtdx + \frac{1}{2} \int_Q e^{-t} r_\varepsilon^2 dtdx \leq$$

$$\leq \varepsilon \left(\int_Q e^{-t} \left(\frac{\partial r_\varepsilon}{\partial t} \right)^2 dtdx \right)^{1/2} \left(\int_Q e^{-t} r_\varepsilon^2 dtdx \right)^{1/2}$$

$$+ M \int_\Sigma e^{-t} \Pi^2 dtdx + M \left(\int_\Sigma e^{-t} \Pi^2 dtdx \right)^{1/2} \left(\int_\Sigma e^{-t} r_\varepsilon^2 dtdx \right)^{1/2} +$$

$$+ \left(\int_\Sigma e^{-t} r_\varepsilon^2 dtdx \right)^{1/2} \cdot (\int_\Sigma e^{-2T-t} \left[\beta \left(v \left(T, x \right) \right) - \beta \left(b \right) \right]^2 dtdx)^{1/2} +$$

$$+ \left(\int_Q e^{-t} r_\varepsilon^2 dtdx \right)^{1/2} \left(\int_Q e^{-t} \left(\Delta \Pi + \varepsilon \frac{\partial^2 v}{\partial t^2} \right)^2 dtdx \right)^{1/2} . \qquad (3.23)$$

Denoting $A^2 = \int_Q \left(\partial r_\varepsilon / \partial t \right)^2 dtdx,\ B^2 = \int_Q r_\varepsilon^2 dtdx,\ C^2 = \int_Q \left| \nabla r_\varepsilon \right|^2 dtdx$, we deduce from (3.22) and (3.23) (with the aid of (3.7))

$$\varepsilon A^2 + C^2 \leq c_1 \varepsilon + c_2 \varepsilon^{1/2} \left(B + C \right) + c_3 \varepsilon^{1/2} B \qquad (3.24)$$

and

$$\varepsilon e^{-T} A^2 + \frac{1}{2} e^{-T} B^2 + e^{-T} C^2 \leq$$

$$\leq \varepsilon AB + c_4\varepsilon + c_5\varepsilon^{1/2}(B+C) + c_6\varepsilon^{1/2}B, \tag{3.25}$$

respectively. From (3.24) we infer

$$C \leq c_7\varepsilon^{1/2} + c_8\varepsilon^{1/4}B^{1/2}. \tag{3.26}$$

Using (3.26) into (3.24), one finds

$$A \leq c_9 + c_{10}\varepsilon^{-1/4}B^{1/2} + c_{11}\varepsilon^{-1/8}B^{1/4}. \tag{3.27}$$

Taking into account (3.26), (3.27) and (3.25), we get

$$B^2 \leq c_{12}\varepsilon^{3/4}B^{3/2} + c_{13}\varepsilon^{7/8}B^{5/4} + \\ + c_{14}\varepsilon^{1/2}B + c_{15}\varepsilon^{3/4}B^{1/2} + c_{16}\varepsilon,$$

hence $B = o\left(\varepsilon^{1/2}\right)$. Using this in (3.26) and (3.27), we infer $C = o\left(\varepsilon^{1/2}\right)$ and $A = o\left(\varepsilon^{0}\right)$, so

$$|r_\varepsilon|_{L^2(Q)} = o\left(\varepsilon^{1/2}\right), \ |\nabla r_\varepsilon|_{(L^2(Q))^N} = o\left(\varepsilon^{1/2}\right). \tag{3.28}$$

Moreover, since

$$\int_\Omega r_\varepsilon^2(t,x)\,dx = \int_0^t \int_\Omega \frac{\partial}{\partial t}\left(r_\varepsilon^2\right) dtdx \leq 2\,|r_\varepsilon|_{L^2(Q)} \left|\frac{\partial r_\varepsilon}{\partial t}\right|_{L^2(Q)} \leq c_{17}\varepsilon^{1/2},$$

one deduces that

$$|r_\varepsilon|_{C([0,T];L^2(\Omega))} = o\left(\varepsilon^{1/4}\right). \tag{3.29}$$

Now, from (3.28) and (3.29), we obtain (3.20) with the aid of (3.6), (3.5).

Corollary 3.7. *Under the hypotheses of Theorem 3.6, we have two cases:*

1) *If $b = v(T,.)$, then (P_ε) is a regularly perturbed problem and*

$$|v_\varepsilon - v|_{C([0,T];L^2(\Omega))} = o\left(\varepsilon^{1/4}\right).$$

2) *If $b \neq v(T,.)$, then (P_ε) is a singularly perturbed problem of elliptic-hyperbolic type and, for every $\delta \in (0,T)$, we have*

$$|v_\varepsilon - v|_{C([0,\delta];L^2(\Omega))} = o\left(\varepsilon^{1/4}\right).$$

4. Asymptotic Approximation for the Semi-linear Telegraph System

The results from this section refer to the telegraph system and have been proved in [4]. Consider the problem (T_0) given by the nonlinear hyperbolic system (S), the boundary conditions (BC) and the initial conditions (IC) below:

$$\begin{cases} \dfrac{\partial u}{\partial t} + \dfrac{\partial v}{\partial x} + F(x, u(t,x)) = f(t,x), \ (t,x) \in Q = (0,T)\times(0,1) \\ \dfrac{\partial v}{\partial t} + \dfrac{\partial u}{\partial x} + G(x, v(t,x)) = g(t,x), \ (t,x) \in Q = (0,T)\times(0,1), \end{cases} \tag{S}$$

$$u(t,0) = u(t,1) = 0,\ 0 < t < T \tag{BC}$$

$$u(0,x) = a_1(x),\ v(0,x) = a_2(x),\ 0 < x < 1. \tag{IC}$$

Assume that:

(C1) $F, G : [0,1] \times \mathbb{R} \to \mathbb{R}$, $F(\cdot,\xi), G(\cdot,\xi) \in L^2(0,1)$, $(\forall)\,\xi \in \mathbb{R}$;

(C2) functions $\xi \mapsto F(x,\xi)$, $\xi \mapsto G(x,\xi)$ are continuous and nondecreasing for a.a. $x \in (0,1)$;

(C3) $a_1 \in H_0^1(0,1)$, $a_2 \in H^1(0,1)$;

(C4) $(f,g) \in W^{1,1}\left(0,T; L^2(0,1)^2\right)$.

As a consequence of Theorem 5.1.1 from [13], we have the following existence result:

Theorem 4.1. *If conditions $(C1)-(C4)$ hold, then problem (T_0) has a unique strong solution*

$$(u,v)^T \in W^{1,\infty}\left(0,T; L^2(0,1)^2\right) \text{ with } u_x, v_x \in L^\infty\left(0,T; L^2(0,1)\right).$$

Here the upper symbol T from α^T means the transposed of α.

Denote by H the real Hilbert space $L^2(0,1)^2$ with the scalar product

$$\left(\begin{pmatrix} p_1 \\ p_2 \end{pmatrix}, \begin{pmatrix} q_1 \\ q_2 \end{pmatrix}\right)_H = \int_0^1 \left[p_1(x)\, q_1(x) + p_2(x)\, q_2(x)\right] dx.$$

The corresponding norm is

$$\left\|\begin{pmatrix} p_1 \\ p_2 \end{pmatrix}\right\|_H = \left[|p_1|^2_{L^2(0,1)} + |p_2|^2_{L^2(0,1)}\right]^{1/2}.$$

We mention that (T_0) (even in a more general case) is a mathematical model of many problems arising from physics and technical sciences. For example, electrical transmission line phenomena and integrated circuits can be described by such a mathematical model. Problem (T_0) can also represent a model for some problems from mechanics and hydraulics: the unsteady fluid flow through pipes and the fluid flow through a tree-structured system of transmission pipelines.

We compare the solution $(u,v)^T$ of (T_0) with the solution $(u^\varepsilon, v^\varepsilon)^T$ of its elliptic regularization (T_ε), consisting of the system (S_ε), the boundary conditions (BC_ε) and the initial conditions (IC_ε) below:

$$\begin{cases} \varepsilon \dfrac{\partial^2 u^\varepsilon}{\partial t^2} - \dfrac{\partial u^\varepsilon}{\partial t} = \dfrac{\partial v^\varepsilon}{\partial x} + F(x, u^\varepsilon(t,x)) - f(t,x),\ (t,x) \in Q \\ \varepsilon \dfrac{\partial^2 v^\varepsilon}{\partial t^2} - \dfrac{\partial v^\varepsilon}{\partial t} = \dfrac{\partial u^\varepsilon}{\partial x} + G(x, v^\varepsilon(t,x)) - g(t,x),\ (t,x) \in Q, \end{cases} \tag{S$_\varepsilon$}$$

$$u^\varepsilon(t,0) = u^\varepsilon(t,1) = 0,\ 0 < t < T \tag{BC$_\varepsilon$}$$

$$\begin{cases} u^\varepsilon(0,x) = a_1(x),\ v^\varepsilon(0,x) = a_2(x),\ 0 < x < 1 \\ u^\varepsilon(T,x) = b_1(x),\ v^\varepsilon(T,x) = b_2(x),\ 0 < x < 1. \end{cases} \tag{IC$_e$}$$

As a particular case, Proposition 5.1.1 from [13] becomes

Proposition 4.2. *If conditions* $(C1)$, $(C2)$ *are satisfied, then the operator* $A : D(A) \subset H \to H$ *defined by*

$$A\begin{pmatrix} u \\ v \end{pmatrix} = \begin{pmatrix} v' + F(\cdot, u) \\ u' + G(\cdot, v) \end{pmatrix}, \; D(A) = H_0^1(0,1) \times H^1(0,1)$$

is maximal monotone in $H = L^2(0,1)^2$.

Denoting $w^\varepsilon = \begin{pmatrix} u^\varepsilon \\ v^\varepsilon \end{pmatrix}$, $h = \begin{pmatrix} f \\ g \end{pmatrix}$, $w_0 = \begin{pmatrix} a_1 \\ a_2 \end{pmatrix}$, $w_T = \begin{pmatrix} b_1 \\ b_2 \end{pmatrix}$, problem (T_ε) can be written as

$$\begin{cases} \varepsilon \dfrac{\partial^2 w^\varepsilon}{\partial t^2} - \dfrac{\partial w^\varepsilon}{\partial t} = Aw^\varepsilon - h, \; 0 < t < T \\ w^\varepsilon(0) = w_0, \; w^\varepsilon(T) = w_T. \end{cases}$$

We add the following condition for the boundary values b_1, b_2 :

(C5) $b_1 \in H_0^1(0,1)$, $b_2 \in H^1(0,1)$.

Applying Proposition 1.27 for problem in w^ε under hypotheses $(C1) - (C5)$, we get the existence of a unique solution of problem (T_ε), namely $(u^\varepsilon, v^\varepsilon)^T \in W^{2,2}(0,T;H)$.

We show that (T_ε) is a singularly perturbed problem, we find a boundary layer and construct an asymptotic approximation for $(u^\varepsilon, v^\varepsilon)^T$, which is valid in the whole domain Q. One finds also the order of the accuracy of this asymptotic approximation.

4.1. An Elliptic-Hyperbolic Singularly Perturbed Problem

Since the solution $w = (u,v)^T$ of the unperturbed problem (T_0) does not satisfy the boundary conditions $u(T,\cdot) = b_1$, $v(T,\cdot) = b_2$, it follows that (T_ε) is a singularly perturbed problem. It is of elliptic-hyperbolic type. At least in some neighbourhood $(\delta, T] \times (0,1)$ of $\{T\} \times (0,1)$ (with $\delta > 0$ arbitrarily small number), the difference $w^\varepsilon - w$ does not tend to 0 in $C([\delta, T]; H)$ as $\varepsilon \to 0$. Thus, the set $(\delta, T] \times (0,1)$ is a boundary layer.

Using the Vishik and Lyusternik method again, we search an asymptotic approximation of $(u^\varepsilon, v^\varepsilon)^T$ of the form

$$\begin{cases} U^\varepsilon(t,x) = u(t,x) + \Pi(\tau, x) \\ V^\varepsilon(t,x) = v(t,x) + \Phi(\tau, x), \end{cases} \tag{4.1}$$

where $\tau = (T-t)/\varepsilon$ is the boundary layer variable and $(\Pi, \Phi)^T$ is the boundary layer function. Denoting by $(r^\varepsilon, s^\varepsilon)^T$ the remainder term, we can write

$$\begin{cases} u^\varepsilon(t,x) = u(t,x) + \Pi(\tau, x) + r^\varepsilon(t,x) \\ v^\varepsilon(t,x) = v(t,x) + \Phi(\tau, x) + s^\varepsilon(t,x), \end{cases} \tag{4.2}$$

for $(t,x) \in (0,T) \times (0,1)$. The role of Π and Φ is to remove the discrepancy which exists between $(u^\varepsilon, v^\varepsilon)^T$ and $(u,v)^T$ in the neighbourhood of T. Since the solutions of (T_ε) and (T_0) should be closed outside the boundary layer, we impose the conditions

$$\Pi(\infty, x) = \Phi(\infty, x) = 0, \; 0 < x < 1.$$

Replacing u^ε and v^ε from (4.2) into (T_ε), one obtains

$$\begin{cases} \varepsilon\left(\dfrac{\partial^2 u}{\partial t^2}+\dfrac{1}{\varepsilon^2}\dfrac{\partial^2\Pi}{\partial\tau^2}+\dfrac{\partial^2 r^\varepsilon}{\partial t^2}\right)-\left(\dfrac{\partial u}{\partial t}-\dfrac{1}{\varepsilon}\dfrac{\partial\Pi}{\partial\tau}+\dfrac{\partial r^\varepsilon}{\partial t}\right)= \\ \quad =\dfrac{\partial v}{\partial x}+\dfrac{\partial\Phi}{\partial x}+\dfrac{\partial s^\varepsilon}{\partial x}+F\left(x,u+\Pi+r^\varepsilon\right)-f, \\ \varepsilon\left(\dfrac{\partial^2 v}{\partial t^2}+\dfrac{1}{\varepsilon^2}\dfrac{\partial^2\Phi}{\partial\tau^2}+\dfrac{\partial^2 s^\varepsilon}{\partial t^2}\right)-\left(\dfrac{\partial v}{\partial t}-\dfrac{1}{\varepsilon}\dfrac{\partial\Phi}{\partial\tau}+\dfrac{\partial s^\varepsilon}{\partial t}\right)= \\ \quad =\dfrac{\partial u}{\partial x}+\dfrac{\partial\Pi}{\partial x}+\dfrac{\partial r^\varepsilon}{\partial x}+G\left(x,v+\Phi+s^\varepsilon\right)-g, \end{cases}$$

$$\begin{cases} u(t,0)+\Pi(\tau,0)+r^\varepsilon(t,0)=0 \\ u(t,1)+\Pi(\tau,1)+r^\varepsilon(t,1)=0, \end{cases}$$

$$\begin{cases} u(0,x)+\Pi(\infty,x)+r^\varepsilon(0,x)=a_1(x),\ v(0,x)+\Phi(\infty,x)+s^\varepsilon(0,x)=a_2(x) \\ u(T,x)+\Pi(0,x)+r^\varepsilon(T,x)=b_1(x),\ v(T,x)+\Phi(0,x)+s^\varepsilon(T,x)=b_2(x), \end{cases}$$

for $0<t<T, 0<\tau<\infty, 0<x<1$. One equates the coefficients of the same powers of ε, separately for the terms depending on t and on τ. Therefore Π and Φ verify the boundary value problems:

$$\begin{cases} \dfrac{\partial^2\Pi}{\partial\tau^2}+\dfrac{\partial\Pi}{\partial\tau}=0,\ \tau>0,\ 0<x<1 \\ \Pi(0,x)=b_1(x)-u(T,x),\ \Pi(\infty,x)=0,\ 0<x<1 \end{cases} \tag{4.3}$$

and

$$\begin{cases} \dfrac{\partial^2\Phi}{\partial\tau^2}+\dfrac{\partial\Phi}{\partial\tau}=0,\ \tau>0,\ 0<x<1 \\ \Phi(0,x)=b_2(x)-v(T,x),\ \Phi(\infty,x)=0,\ 0<x<1. \end{cases} \tag{4.4}$$

These problems have unique solutions, namely

$$\begin{cases} \Pi(\tau,x)=e^{-\tau}\left[b_1(x)-u(T,x)\right] \\ \Phi(\tau,x)=e^{-\tau}\left[b_2(x)-v(T,x)\right]. \end{cases} \tag{4.5}$$

Observe that

$$\Pi(\tau,0)=\Pi(\tau,1)=0,\ \tau>0. \tag{4.6}$$

It is easy to check that

$$\begin{cases} |\Pi|^2_{L^2(Q)}=(\varepsilon/2)\left(1-e^{-2T/\varepsilon}\right)|b_1-u(T,\cdot)|^2_{L^2(0,1)} \\ |\partial\Pi/\partial x|^2_{L^2(Q)}=(\varepsilon/2)\left(1-e^{-2T/\varepsilon}\right)|b_1'-u'(T,\cdot)|^2_{L^2(0,1)} \\ |\Pi|_{C([0,T];L^2(0,1))}=|b_1-u(T,\cdot)|_{L^2(0,1)}, \end{cases}$$

$$\begin{cases} |\Phi|^2_{L^2(Q)}=(\varepsilon/2)\left(1-e^{-2T/\varepsilon}\right)|b_2-v(T,\cdot)|^2_{L^2(0,1)} \\ |\partial\Phi/\partial x|^2_{L^2(Q)}=(\varepsilon/2)\left(1-e^{-2T/\varepsilon}\right)|b_2'-v'(T,\cdot)|^2_{L^2(0,1)} \\ |\Phi|_{C([0,T];L^2(0,1))}=|b_2-v(T,\cdot)|_{L^2(0,1)}. \end{cases}$$

Taking into account that $(u,v)^T$ is the solution of problem (T_0), Π, Φ satisfy (4.3), (4.4), and (4.6), it follows that r^ε, s^ε verify the problem

$$\begin{cases} \varepsilon \dfrac{\partial^2 r^\varepsilon}{\partial t^2} - \dfrac{\partial r^\varepsilon}{\partial t} = \dfrac{\partial s^\varepsilon}{\partial x} + F\left(x, u + \Pi + r^\varepsilon\right) - F\left(x,u\right) + \dfrac{\partial \Phi}{\partial x} - \varepsilon \dfrac{\partial^2 u}{\partial t^2} \\ \varepsilon \dfrac{\partial^2 s^\varepsilon}{\partial t^2} - \dfrac{\partial s^\varepsilon}{\partial t} = \dfrac{\partial r^\varepsilon}{\partial x} + G\left(x, v + \Phi + s^\varepsilon\right) - G\left(x,v\right) + \dfrac{\partial \Pi}{\partial x} - \varepsilon \dfrac{\partial^2 v}{\partial t^2} \\ r^\varepsilon\left(t,0\right) = r^\varepsilon\left(t,1\right) = 0 \\ r^\varepsilon\left(0,x\right) = r^\varepsilon\left(T,x\right) = s^\varepsilon\left(0,x\right) = s^\varepsilon\left(T,x\right) = 0, \end{cases} \tag{4.7}$$

for $0 < t < T$, $0 < x < 1$. In this system, the second derivatives $\partial^2 u/\partial t^2$ and $\partial^2 v/\partial t^2$ appear, but we do not know yet anything about them. In the next subsection we show that, under additional conditions, they belong to $L^1\left(0,T;L^2\left(0,1\right)\right)$.

4.2. Regularity of the Solutions

We first obtain a higher regularity result for problem (T_0). Since its solution will be compared with the solution of (T_e), it is appropriate to prove a higher regularity result for (T_ε) too. It will be done in the second theorem of this subsection.

Suppose that $(C1)$, $(C3)$ and $(C5)$ hold and in addition we have the following hypotheses which are stronger than $(C2)$ and $(C4)$:

(C6) F, G are differentiable in ξ and $0 \le F_\xi \le M$, $0 \le G_\xi \le M$ $(M > 0)$ $x \in (0,1)$, $\xi \in \mathbb{R}$. Here F_ξ, G_ξ are the derivatives of F, G with respect to ξ.

(C7) F_ξ, G_ξ are Lipschitz continuous in ξ : there exists $L > 0$ such that $(\forall)\,(x,\xi)$, $(x,\eta) \in (0,1) \times \mathbb{R}$,

$$\left|F_\xi\left(x,\xi\right) - F_\xi\left(x,\eta\right)\right| \le L\left|\xi - \eta\right|, \ \left|G_\xi\left(x,\xi\right) - G_\xi\left(x,\eta\right)\right| \le L\left|\xi - \eta\right|.$$

(C8) $f, g \in W^{2,\infty}\left(0,T;L^2\left(0,1\right)\right)$.

(C9) Denote $U_0 = f\left(0,\cdot\right) - a_2' - F\left(\cdot,a_1\right)$, $V_0 = g\left(0,\cdot\right) - a_1' - G\left(\cdot,a_2\right)$ and assume that $U_0 \in H_0^1\left(0,1\right)$, $V_0 \in H^1\left(0,1\right)$.

Theorem 4.3. *Under hypotheses* $(C1)$, $(C3)$, $(C6) - (C9)$, *the solution* $(u,v)^T$ *of problem* (T_0) *belongs to* $W^{2,1}\left(0,T;L^2\left(0,1\right)^2\right)$ *and* $\partial u/\partial t$, $\partial v/\partial t \in L^\infty\left(0,T;H^1\left(0,1\right)\right)$.

Proof. We use the classical idea of formally differentiating (T_0) with respect to t, establishing some regularity for the new problem, and then returning to the original problem to derive higher regularity for its solution.

Deriving formally (T_0) with respect to t and denoting $U = \partial u/\partial t$, $V = \partial v/\partial t$, we arrive at the problem:

$$\begin{cases} \dfrac{\partial U}{\partial t} + \dfrac{\partial V}{\partial x} + F_\xi\left(x, u\left(t,x\right)\right) U = \dfrac{\partial f}{\partial t}, \ (t,x) \in Q \\ \dfrac{\partial V}{\partial t} + \dfrac{\partial U}{\partial x} + G_\xi\left(x, v\left(t,x\right)\right) V = \dfrac{\partial g}{\partial t}, \ (t,x) \in Q, \end{cases} \tag{4.8}$$

$$U\left(t,0\right) = U\left(t,1\right) = 0, \ 0 < t < T \tag{4.9}$$

$$U\left(0,x\right) = U_0\left(x\right), \ V\left(0,x\right) = V_0\left(x\right), \ 0 < x < 1. \tag{4.10}$$

We first prove that problem $(4.8)-(4.10)$ admits a unique solution $(U,V)^T \in W^{1,1}\left(0,T;L^2(0,1)^2\right)$ and then we show that it coincides with $(\partial u/\partial t, \partial v/\partial t)^T$. In order to do this, denote $W=\begin{pmatrix}U\\V\end{pmatrix}$, $W_0=\begin{pmatrix}U_0\\V_0\end{pmatrix}$ and

$$B(t,W)=\begin{pmatrix}\partial V/\partial x+F_\xi(x,u)U-\partial f/\partial t\\ \partial U/\partial x+G_\xi(x,v)V-\partial g/\partial t\end{pmatrix},$$

$$D(B(t,\cdot))=H_0^1(0,1)\times H^1(0,1).$$

Then, the above problem becomes

$$\begin{cases} W'+B(t,W)=0,\ 0<t<T\\ W(0)=W_0.\end{cases}$$

The time-dependent operator $B(t,\cdot)$ is maximal monotone in H, for every fixed $t\in[0,T]$ and has the domain independent of t. In addition, we have the estimate

$$\|B(t,W)-B(s,W)\|_H^2\le 2\int_0^1\{|F_\xi(x,u(t,x))-F_\xi(x,u(s,x))|^2U^2+$$

$$+|G_\xi(x,v(t,x))-G_\xi(x,v(s,x))|^2V^2+$$

$$+\left|\frac{\partial f}{\partial t}(t,x)-\frac{\partial f}{\partial t}(s,x)\right|^2+\left|\frac{\partial g}{\partial t}(t,x)-\frac{\partial g}{\partial t}(s,x)\right|^2\}dx.$$

Since $(u,v)^T\in W^{1,\infty}(0,T;H)$, we obtain with the aid of $(C7)$ and $(C8)$:

$$|F_\xi(x,u(t,x))-F_\xi(x,u(s,x))|\le L|u(t,x)-u(s,x)|\le c_1|t-s|,$$

$$\left|\frac{\partial f}{\partial t}(t,x)-\frac{\partial f}{\partial t}(s,x)\right|\le c_2|t-s|$$

and similarly for G and g. Hence, $(\forall)\ s,t\in[0,T]$,

$$\|B(t,W)-B(s,W)\|_H\le c|t-s|\le c|t-s|\left[1+\|W\|_H+\|B(s,W)\|_H\right].$$

By Kato's Theorem for time-dependent problems (Theorem 1.18), we derive that problem $(4.8)-(4.10)$ has a unique strong solution $(U,V)^T\in W^{1,1}(0,T;H)$ with $U,V\in L^\infty(0,T;H^1(0,1))$.

We deduce now that $U=\partial u/\partial t$ and $V=\partial v/\partial t$. Indeed, denote by $\{S(t):H\to H\}$ the C_0-semigroup of bounded linear operators on $H=L^2(0,1)^2$ generated by $-A$, where A is the linear and maximal monotone operator

$$A\begin{pmatrix}U\\V\end{pmatrix}=\begin{pmatrix}V'\\U'\end{pmatrix},\ D(A)=H_0^1(0,1)\times H^1(0,1).$$

Then the solution $\begin{pmatrix}U\\V\end{pmatrix}$ of problem $(4.8)-(4.10)$ can be written as (see Theorem 1.22 and Definition 1.21):

$$\begin{pmatrix}U(t)\\V(t)\end{pmatrix}=S(t)\begin{pmatrix}U_0\\V_0\end{pmatrix}+\int_0^t S(t-s)\begin{pmatrix}(\partial f/\partial t)(s,x)-F_\xi(x,u(s,x))U\\(\partial g/\partial t)(s,x)-G_\xi(x,v(s,x))V\end{pmatrix}ds,\quad (4.11)$$

while the solution $(u,v)^T$ of problem (T_0) is given by

$$\begin{pmatrix} u(t) \\ v(t) \end{pmatrix} = S(t)\begin{pmatrix} a_1 \\ a_2 \end{pmatrix} + \int_0^t S(s)\begin{pmatrix} f(t-s,x) - F(x,u(t-s,x)) \\ g(t-s,x) - G(x,v(t-s,x)) \end{pmatrix} ds.$$

Since $\begin{pmatrix} a_1 \\ a_2 \end{pmatrix} \in D(A)$ and $\partial f/\partial t - F_\xi(\cdot,u)U$ and $\partial g/\partial t - G_\xi(\cdot,v)V$ belong to $W^{1,1}(0,T;L^2(0,1))$, we can differentiate the last equality with respect to t, obtaining thus (via Theorem 1.22)

$$\begin{pmatrix} \partial u/\partial t \\ \partial v/\partial t \end{pmatrix} = S(t)\begin{pmatrix} f(0,x) - a_2' - F(x,a_1) \\ g(0,x) - a_1' - G(x,a_2) \end{pmatrix} +$$

$$+\int_0^t S(t-s)\begin{pmatrix} (\partial f/\partial t)(s,x) - F_\xi(x,u(s,x))(\partial u/\partial t) \\ (\partial g/\partial t)(s,x) - G_\xi(x,v(s,x))(\partial v/\partial t) \end{pmatrix} ds. \tag{4.12}$$

Subtracting (4.11) and (4.12), one finds the estimate

$$\left\|\begin{pmatrix} U(t) \\ V(t) \end{pmatrix} - \begin{pmatrix} \partial u/\partial t \\ \partial v/\partial t \end{pmatrix}\right\|_H \le \left\|\int_0^t S(t-s)\begin{pmatrix} -F_\xi(x,u(s,x))(U - \partial u/\partial t)(s) \\ -G_\xi(x,v(s,x))(V - \partial v/\partial t)(s) \end{pmatrix} ds\right\|_H.$$

Since $S(t)$ is a contraction and F_ξ, G_ξ are bounded (hypothesis $(C6)$), we infer

$$|U(t) - (\partial u/\partial t)(t)|^2_{L^2(0,1)} + |V(t) - (\partial v/\partial t)(t)|^2_{L^2(0,1)} \le$$

$$\le M\int_0^t \left(|U(s) - (\partial u/\partial t)(s)|^2_{L^2(0,1)} + |V(s) - (\partial v/\partial t)(s)|^2_{L^2(0,1)}\right) ds,$$

so by Gronwall-Bellman Lemma it follows that $U = \partial u/\partial t, V = \partial v/\partial t$.

Combining with the first part of the proof, we arrive at the conclusion that $(u,v)^T \in W^{2,1}(0,T;H)$ and $u,v \in W^{1,\infty}(0,T;H^1(0,1))$, as claimed.

Making use of Theorem 1.31, we can also derive a higher regularity result for the solution of problem (T_ε).

Theorem 4.4. *If $(C1)$, $(C3)$, $(C5)-(C8)$ hold, then for every fixed $\varepsilon > 0$, the solution $(u^\varepsilon, v^\varepsilon)^T$ of problem (T_ε) belongs to $W^{3,2}_{loc}(0,T;H)$. More exactly,*

$$\frac{\partial^3 u^\varepsilon}{\partial t^3}, \frac{\partial^3 v^\varepsilon}{\partial t^3} \in L^2_{\gamma^3}(0,T;L^2(0,1)),$$

where $\gamma(t) = \min\{t, T-t\}, t \in [0,T]$.

Proof. Consider the auxiliary boundary value problem

$$\begin{cases} \varepsilon\dfrac{\partial^2 U^\varepsilon}{\partial t^2} - \dfrac{\partial U^\varepsilon}{\partial t} = \dfrac{\partial V^\varepsilon}{\partial x} + F_\xi(x,u^\varepsilon(t,x))\dfrac{\partial u^\varepsilon}{\partial t} - \dfrac{\partial f}{\partial t}, & (t,x) \in Q \\ \varepsilon\dfrac{\partial^2 V^\varepsilon}{\partial t^2} - \dfrac{\partial V^\varepsilon}{\partial t} = \dfrac{\partial U^\varepsilon}{\partial x} + G_\xi(x,v^\varepsilon(t,x))\dfrac{\partial v^\varepsilon}{\partial t} - \dfrac{\partial g}{\partial t}, & (t,x) \in Q, \end{cases} \tag{4.13}$$

$$U^\varepsilon(t,0) = U^\varepsilon(t,1) = 0,\ 0 < t < T \tag{4.14}$$

$$\begin{cases} U^{\varepsilon}(0,x) = \dfrac{\partial u^{\varepsilon}}{\partial t}(0,x),\ V^{\varepsilon}(0,x) = \dfrac{\partial v^{\varepsilon}}{\partial t}(0,x),\ 0 < x < 1 \\ U^{\varepsilon}(T,x) = \dfrac{\partial u^{\varepsilon}}{\partial t}(T,x),\ V^{\varepsilon}(T,x) = \dfrac{\partial v^{\varepsilon}}{\partial t}(T,x),\ 0 < x < 1. \end{cases} \tag{4.15}$$

Denoting $Z^{\varepsilon} = \begin{pmatrix} U^{\varepsilon} \\ V^{\varepsilon} \end{pmatrix}$, $Z_0 = \begin{pmatrix} \frac{\partial u^{\varepsilon}}{\partial t}(0,x) \\ \frac{\partial v^{\varepsilon}}{\partial t}(0,x) \end{pmatrix}$, $Z_T = \begin{pmatrix} \frac{\partial u^{\varepsilon}}{\partial t}(T,x) \\ \frac{\partial v^{\varepsilon}}{\partial t}(T,x) \end{pmatrix}$, $B\begin{pmatrix} U \\ V \end{pmatrix} = \begin{pmatrix} V' \\ U' \end{pmatrix}$ with $D(B) = H_0^1(0,1) \times H^1(0,1)$, and

$$\widetilde{h}(t) = \begin{pmatrix} F_{\xi}(x, u^{\varepsilon}(t,x)) \dfrac{\partial u^{\varepsilon}}{\partial t} - \dfrac{\partial f}{\partial t} \\ G_{\xi}(x, v^{\varepsilon}(t,x)) \dfrac{\partial v^{\varepsilon}}{\partial t} - \dfrac{\partial g}{\partial t} \end{pmatrix},$$

the above problem can be written in the form:

$$\begin{cases} \varepsilon \dfrac{\partial^2 Z^{\varepsilon}}{\partial t^2} - \dfrac{\partial Z^{\varepsilon}}{\partial t} = BZ^{\varepsilon} + \widetilde{h},\ 0 < t < T \\ Z^{\varepsilon}(0) = Z_0,\ Z^{\varepsilon}(T) = Z_T. \end{cases} \tag{4.16}$$

Proposition 4.2 with $F = G = 0$ assures that B is maximal monotone in $H = L^2(0,1)^2$. Since $Z_0, Z_T \in \overline{D(B)} = H$ and $\widetilde{h} \in L^2(0,T;H)$, by Theorem 1.31 we infer that problem $(4.13) - (4.15)$ admits a unique solution $Z^{\varepsilon} = (U^{\varepsilon}, V^{\varepsilon})^T$ in $W_{loc}^{2,2}(0,T;H)$, with $\dfrac{\partial^2 U^{\varepsilon}}{\partial t^2}, \dfrac{\partial^2 V^{\varepsilon}}{\partial t^2} \in L_{\gamma^3}^2(0,T;L^2(0,1))$.

To finish the proof of the theorem, it is sufficient to show that $\dfrac{\partial u^{\varepsilon}}{\partial t} = U^{\varepsilon}$ and $\dfrac{\partial v^{\varepsilon}}{\partial t} = V^{\varepsilon}$.

To this end, we approach the solutions $Z^{\varepsilon} = (U^{\varepsilon}, V^{\varepsilon})^T$ and $w^{\varepsilon} = (u^{\varepsilon}, v^{\varepsilon})^T$ of $(4.13) - (4.15)$ (or equivalently (4.16)) and (T_{ε}), by the solutions $Z_{\lambda}^{\varepsilon} = (U_{\lambda}^{\varepsilon}, V_{\lambda}^{\varepsilon})^T$ and $w_{\lambda}^{\varepsilon} = (u_{\lambda}^{\varepsilon}, v_{\lambda}^{\varepsilon})^T$ of the problem

$$\begin{cases} \varepsilon \dfrac{\partial^2 Z_{\lambda}^{\varepsilon}}{\partial t^2} - \dfrac{\partial Z_{\lambda}^{\varepsilon}}{\partial t} = B_{\lambda} Z_{\lambda}^{\varepsilon} + \widetilde{h},\ 0 < t < T \\ Z_{\lambda}^{\varepsilon}(0) = Z_0,\ Z_{\lambda}^{\varepsilon}(T) = Z_T, \end{cases}$$

and

$$\begin{cases} \varepsilon \dfrac{\partial^2 w_{\lambda}^{\varepsilon}}{\partial t^2} - \dfrac{\partial w_{\lambda}^{\varepsilon}}{\partial t} = B_{\lambda} w_{\lambda}^{\varepsilon} + R(w_{\lambda}^{\varepsilon}) - h,\ 0 < t < T \\ w_{\lambda}^{\varepsilon}(0) = w_0,\ w_{\lambda}^{\varepsilon}(T) = w_T \end{cases},$$

respectively. Here B_{λ} is the Yosida approximation of the operator B, and R is given by

$$R(w) = R\begin{pmatrix} u \\ v \end{pmatrix} = \begin{pmatrix} F(\cdot, u) \\ G(\cdot, v) \end{pmatrix}.$$

These problems have unique solutions $Z_{\lambda}^{\varepsilon} \in W_{loc}^{2,2}(0,T;H)$, $w_{\lambda}^{\varepsilon} \in W^{2,2}(0,T;H)$. With the aid of Lemma 1.35 ($Z_0, Z_T, w_0, w_T \in \overline{D(B_{\lambda})} = H$), we can find a subsequence $\lambda_n \to 0^+$ such that

$$\begin{cases} w_{\lambda_n}^{\varepsilon} \to w^{\varepsilon} \text{ in } C([0,T];H) \\ (w_{\lambda_n}^{\varepsilon})' \to (w^{\varepsilon})' \text{ in } C([0,T];H) \\ (w_{\lambda_n}^{\varepsilon})'' \rightharpoonup (w^{\varepsilon})'' \text{ in } L^2(0,T;H) \end{cases}$$

and

$$\begin{array}{c} Z^{\varepsilon}_{\lambda_n} \to Z^{\varepsilon} \text{ in } C([0,T];H) \\ (Z^{\varepsilon}_{\lambda_n})' \to (Z^{\varepsilon})' \text{ in } C([0,T];H) \end{array}.$$

We want to prove that

$$\lim_{n\to\infty} \left[(w^{\varepsilon}_{\lambda_n})'(t) - Z^{\varepsilon}_{\lambda_n}(t) \right] = 0 \text{ uniformly on } (0,T). \tag{4.17}$$

To do this, denote $\theta_\lambda = \left[(w^{\varepsilon}_{\lambda})'(t) - Z^{\varepsilon}_{\lambda}(t) \right]$. Since B_λ is linear and bounded and $w^{\varepsilon}_{\lambda} \in W^{2,2}(0,T;H)$, observe that θ_λ verifies the problem

$$\begin{cases} \varepsilon\theta''_\lambda - \theta'_\lambda = B_\lambda\theta_\lambda,\ 0<t<T \\ \theta_\lambda(0) = (w^{\varepsilon}_{\lambda})'(0) - Z_0,\ \theta_\lambda(T) = (w^{\varepsilon}_{\lambda})'(T) - Z_T \end{cases}$$

and $\theta_\lambda \in W^{2,2}_{loc}(0,T;H)$. Like in Lemma 1.32 we can show that the map $t \mapsto \|\theta_\lambda(t)\|^2$ is convex on $[0,T]$, so that

$$\|\theta_\lambda(t)\| \le \max\left(\|\theta_\lambda(0)\|, \|\theta_\lambda(T)\|\right) = \max\left(\left\|(w^{\varepsilon}_{\lambda})'(0) - Z_0\right\|, \left\|(w^{\varepsilon}_{\lambda})'(T) - Z_T\right\|\right).$$

Hence $\theta_{\lambda_n}(t) \to 0$ in $C([0,T];H)$, i. e. (4.17) holds. This implies that $(w^{\varepsilon})' = Z^{\varepsilon}$ or $\dfrac{\partial u^{\varepsilon}}{\partial t} = U^{\varepsilon}$ and $\dfrac{\partial v^{\varepsilon}}{\partial t} = V^{\varepsilon}$. By $\dfrac{\partial^2 U^{\varepsilon}}{\partial t^2}, \dfrac{\partial^2 V^{\varepsilon}}{\partial t^2} \in L^2_{\gamma^3}(0,T;L^2(0,1))$, it follows the regularity for u^{ε} and v^{ε}. This completes the proof.

4.3. The Order of the Approximation

We study the existence of the solution to problem (4.7) and establish some convergence results for this solution.

Suppose conditions $(C1)$, $(C3)$, $(C5)-(C9)$ are satisfied. Observe first that (4.7) can be written as (1.12), where A and B are given by

$$A\begin{pmatrix} p \\ q \end{pmatrix} = \begin{pmatrix} q' \\ p' \end{pmatrix},\ D(A) = H^1_0(0,1) \times H^1(0,1),$$

$$B\left(t, \begin{pmatrix} p \\ q \end{pmatrix}\right) = \begin{pmatrix} F(x,u+\Pi+p) - F(x,u) + \partial\Phi/\partial x - \varepsilon\left(\partial^2 u/\partial t^2\right) \\ G(x,v+\Phi+q) - G(x,v) + \partial\Pi/\partial x - \varepsilon\left(\partial^2 v/\partial t^2\right) \end{pmatrix}.$$

It is easy to check that A and $B(t,\cdot)$ satisfy the hypotheses of Theorem 1.30. Therefore problem (4.7) has a unique solution $(r^{\varepsilon}, s^{\varepsilon}) \in W^{2,2}\left(0,T;L^2(0,1)^2\right)$.

Making use of Theorem 1.31, we can establish the following approximation result for (T_0) and (T_ε).

Theorem 4.5. *Assume $(C1)$, $(C3)$, $(C5)-(C9)$ hold. Then, for sufficiently small $\varepsilon > 0$, function $(U^{\varepsilon}, V^{\varepsilon})$ in (4.1) with Π, Φ from (4.5) is an asymptotic approximation for the solution $(u^{\varepsilon}, v^{\varepsilon})^T$ of (T_ε) in the entire domain Q. Moreover, we get the following estimates:*

$$|u^{\varepsilon} - u|_{L^2(Q)} = o\left(\varepsilon^{1/2}\right),\ |v^{\varepsilon} - v|_{L^2(Q)} = o\left(\varepsilon^{1/2}\right),$$

$$|u^{\varepsilon} - u - \Pi|_{C([0,T];L^2(0,1))} = o\left(\varepsilon^{1/4}\right),\ |v^{\varepsilon} - v - \Phi|_{C([0,T];L^2(0,1))} = o\left(\varepsilon^{1/4}\right).$$

Proof. Multiplying the first equation in (4.7) by $e^{-t}r^{\varepsilon}$ and the second one by $e^{-t}s^{\varepsilon}$, summing them up and integrating on Q, we get

$$\varepsilon\int_Q e^{-t}\left[\frac{\partial}{\partial t}\left(r^{\varepsilon}\frac{\partial r^{\varepsilon}}{\partial t}\right)-\left(\frac{\partial r^{\varepsilon}}{\partial t}\right)^2\right]dtdx-\frac{1}{2}\int_Q e^{-t}\frac{\partial\left(r^{\varepsilon}\right)^2}{\partial t}dtdx+$$

$$+\varepsilon\int_Q e^{-t}\left[\frac{\partial}{\partial t}\left(s^{\varepsilon}\frac{\partial s^{\varepsilon}}{\partial t}\right)-\left(\frac{\partial s^{\varepsilon}}{\partial t}\right)^2\right]dtdx-\frac{1}{2}\int_Q e^{-t}\frac{\partial\left(s^{\varepsilon}\right)^2}{\partial t}dtdx=\int_Q e^{-t}\left[r^{\varepsilon}\frac{\partial s^{\varepsilon}}{\partial x}+s^{\varepsilon}\frac{\partial r^{\varepsilon}}{\partial x}\right]dtdx+$$

$$+\int_Q e^{-t}r^{\varepsilon}\left[F\left(x,u+\Pi+r^{\varepsilon}\right)-F\left(x,u\right)\right]dtdx+\int_Q e^{-t}s^{\varepsilon}\left[G\left(x,v+\Phi+s^{\varepsilon}\right)-G\left(x,v\right)\right]dtdx+$$

$$+\int_Q e^{-t}\left[r^{\varepsilon}\left(\frac{\partial\Phi}{\partial x}-\varepsilon\frac{\partial^2 u}{\partial t^2}\right)+s^{\varepsilon}\left(\frac{\partial\Pi}{\partial x}-\varepsilon\frac{\partial^2 v}{\partial t^2}\right)\right]dtdx.$$

Integrating by parts and using the initial conditions from (4.7), one derives that

$$\frac{\varepsilon}{2}\int_Q e^{-t}\frac{\partial}{\partial t}\left[\left(r^{\varepsilon}\right)^2+\left(s^{\varepsilon}\right)^2\right]dtdx=\varepsilon\int_Q e^{-t}\left[\left(\frac{\partial r^{\varepsilon}}{\partial t}\right)^2+\left(\frac{\partial s^{\varepsilon}}{\partial t}\right)^2\right]dtdx+\frac{1}{2}\int_Q e^{-t}\left[\left(r^{\varepsilon}\right)^2+\left(s^{\varepsilon}\right)^2\right]dtdx+$$

$$+\int_Q e^{-t}\left[\left(\Pi+r^{\varepsilon}\right)^2\frac{\partial F}{\partial\xi}\left(x,\widetilde{u}\right)+\left(\Phi+s^{\varepsilon}\right)^2\frac{\partial G}{\partial\xi}\left(x,\widetilde{v}\right)\right]dtdx-\int_Q e^{-t}\Pi\left(\Pi+r^{\varepsilon}\right)\frac{\partial F}{\partial\xi}\left(x,\widetilde{u}\right)dtdx-$$

$$-\int_Q e^{-t}\Phi\left(\Phi+s^{\varepsilon}\right)\frac{\partial G}{\partial\xi}\left(x,\widetilde{v}\right)dtdx+\int_Q e^{-t}\left[r^{\varepsilon}\left(\frac{\partial\Phi}{\partial x}-\varepsilon\frac{\partial^2 u}{\partial t^2}\right)+s^{\varepsilon}\left(\frac{\partial\Pi}{\partial x}-\varepsilon\frac{\partial^2 v}{\partial t^2}\right)\right]dtdx. \tag{4.18}$$

With the aid of the initial conditions, integrating by parts, this implies

$$\varepsilon\int_Q e^{-t}\left[\left(\frac{\partial r^{\varepsilon}}{\partial t}\right)^2+\left(\frac{\partial s^{\varepsilon}}{\partial t}\right)^2\right]dtdx+\frac{1-\varepsilon}{2}\int_Q e^{-t}\left[\left(r^{\varepsilon}\right)^2+\left(s^{\varepsilon}\right)^2\right]dtdx\le M\int_Q e^{-t}\left(\Pi^2+\Phi^2\right)dtdx+$$

$$+\left[M\left(\int_Q e^{-t}\Pi^2 dtdx\right)^{1/2}+\left(\int_Q e^{-t}\left(\frac{\partial\Phi}{\partial x}-\varepsilon\frac{\partial^2 u}{\partial t^2}\right)^2 dtdx\right)^{1/2}\right]\left(\int_Q e^{-t}\left(r^{\varepsilon}\right)^2 dtdx\right)^{1/2}+$$

$$+\left[M\left(\int_Q e^{-t}\Phi^2 dtdx\right)^{1/2}+\left(\int_Q e^{-t}\left(\frac{\partial\Pi}{\partial x}-\varepsilon\frac{\partial^2 v}{\partial t^2}\right)^2 dtdx\right)^{1/2}\right]\left(\int_Q e^{-t}\left(s^{\varepsilon}\right)^2 dtdx\right)^{1/2},$$

so by the estimates for Π and Φ, it follows that

$$\varepsilon\int_Q e^{-t}\left[\left(\frac{\partial r^{\varepsilon}}{\partial t}\right)^2+\left(\frac{\partial s^{\varepsilon}}{\partial t}\right)^2\right]dtdx+\frac{1-\varepsilon}{2}\int_Q e^{-t}\left[\left(r^{\varepsilon}\right)^2+\left(s^{\varepsilon}\right)^2\right]dtdx\le$$

$$\leq M_1\varepsilon + M_2\varepsilon^{1/2}\left[\left(\int_Q e^{-t}\left(r^\varepsilon\right)^2 dtdx\right)^{1/2} + \left(\int_Q e^{-t}\left(s^\varepsilon\right)^2 dtdx\right)^{1/2}\right].$$

Here M, M_1, M_2 are positive constants. This implies that $\left(\int_Q \left(r^\varepsilon\right)^2 dtdx\right)^{1/2} = o\left(\varepsilon^{1/2}\right)$ and $\left(\int_Q \left(s^\varepsilon\right)^2 dtdx\right)^{1/2} = o\left(\varepsilon^{1/2}\right)$, that is

$$|u^\varepsilon - u - \Pi|_{L^2(Q)} = o\left(\varepsilon^{1/2}\right),\ |v^\varepsilon - v - \Phi|_{L^2(Q)} = o\left(\varepsilon^{1/2}\right).$$

The claim follows via the same estimates for Π and Φ, under (4.6).

Corollary 4.6. *Suppose the above hypotheses are fulfilled.*

1) If $u(T,\cdot) = b_1$ and $v(T,\cdot) = b_2$, then $(u,v)^T$ is an asymptotic approximation of $(u^\varepsilon, v^\varepsilon)^T$ in the whole domain Q and

$$|u^\varepsilon - u|_{C([0,T];L^2(0,1))} = o\left(\varepsilon^{1/4}\right),\ |v^\varepsilon - v|_{C([0,T];L^2(0,1))} = o\left(\varepsilon^{1/4}\right).$$

In this case, problem (T_ε) is regularly perturbed.

2) Otherwise, for every $\delta \in (0,T)$, the pair $(u,v)^T$ is an asymptotic approximation of $(u^\varepsilon, v^\varepsilon)^T$ in the subdomain $Q_1 = (0,\delta] \times (0,1)$ and

$$|u^\varepsilon - u|_{C([0,\delta];L^2(0,1))} = o\left(\varepsilon^{1/4}\right),\ |v^\varepsilon - v|_{C([0,\delta];L^2(0,1))} = o\left(\varepsilon^{1/4}\right).$$

Problem (T_ε) is a singularly perturbed problem of elliptic-hyperbolic type.

5. Conclusion

In this chapter we have approximated the solutions of the heat equation and of the telegraph system by the solutions of some elliptic regularizations with a small parameter $\varepsilon > 0$. Bilocal boundary conditions have been associated to these equations. Employing the Vishik and Lyusternik method, we found the boundary layers and determined the order with respect to ε of the approximations in the spaces $L^2(0,T;H)$ or $C([0,T];H)$, where H is the Hilbert space $L^2(\Omega)$ or $L^2(0,1) \times L^2(0,1)$. In the general case, the problems studied in Sections 2 and 3 are elliptic- parabolic, while the problem analyzed within Section 4 is elliptic-hyperbolic. In this way, the solutions of the heat equation and of the telegraph system are approximated by the solution of their elliptic regaularizations, which are more regular functions.

References

Aftabizadeh, A. R.; Pavel, N.H. Nonlinear boundary value problems for some ordinary and partial differential equations associated with monotone operators. *J. Math. Anal. Appl.* 1991, 156, 535–557.

Apreutesei, N.. First order asymptotic approximation for a class of elliptic equations. *Comm. Appl. Nonlinear Anal.* 2001, 8, 89-101.

Apreutesei, N. An asymptotic approximation for a semi–linear elliptic equation. *Comm. Appl. Anal.* 2003, 7, no. 1, 67-77.

Apreutesei, N.; Djafari Rouhani, B. Elliptic regularization for the semi-linear telegraph system. *Nonlinear Anal., Theory Methods Appl.* 2010, Ser. A, 72, no. 6, 3049-3061.

Apreutesei, N. Volpert, V. Elliptic regularization for the heat equation with nonlinear boundary conditions. *Asympt. Anal.* 2003, 35, no. 2, 151-164.

Barbu, V. A class of boundary problems for second–order abstract differential equations. *J. Fac. Sci. Univ. Tokyo Sect. I* 1972, 19, 295–319.

Barbu, V. *Nonlinear Semigroups and Differential Equations in Banach Spaces* , Noordhoff, Leyden, 1976.

Barbu, L., Cosma, E. Elliptic regularizations for the nonlinear heat equation. *J. Math. Anal. Appl.* 2009, 351, no. 1, 392-399.

Brézis, H. Propriétés régularisantes de certains semi–groupes nonlinéaires. *Israel J. Math.* 1971, 9, 513–534.

Brézis, H. Monotonicity methods in Hilbert spaces and some applications to nonlinear partial differential equations, Contributions to Nonlinear Functional Analysis, Ed. E. Zarantonello; Academic Press 1971.

Brézis, H. *Opérateurs Maximaux Monotones et Semi–groupes de Contractions dans les Espaces de Hilbert*, North–Holland, Amsterdam, 1973.

Bruck, R. E. Periodic forcing of solutions of a boundary value problem for a second order differential equation in Hilbert space. *J. Math. Anal. Applic.* 1980, 76, 159–173.

Hokkanen, V. M.; Moroşanu, G. *Functional Methods in Differential Equations* , Chapman & Hall /CRC, Boca Raton, London, New York, Washington D.C., 2002.

Kato, T. Nonlinear semigroups and evolution equations. *J. Math. Soc. Japan* 1967, 19, no. 4, 508-520.

Lions, J. L. *Equations Différentielles Opérationnelles* , Springer-Verlag, 1961.

Lions, J. L. *Perturbations Singulières dans les Problèmes aux Limites et en Contrôle Optimal*, Springer–Verlag, Berlin, Heidelberg, New York, 1973.

Vasilieva, A.; Butuzov, V.; Kalachev, L. *The Boundary Function Method for Singular Perturbation Problems*, SIAM Philadelphia, 1995.

Vishik, M.; Lyusternik, L. Regular degeneration and boundary layer for linear differential equations with small parameter multiplying the highest derivatives. *Usp. Mat. Nauk* 1957, 12, 3-122 (in Russian), *Amer. Math. Soc. Transl., Ser. 2* 1962, vol. 20, 239-364.

Volevich, L. R. The Vishik-Lyusternik method in general elliptic boundary value problems with small parameter. Preprint No. 26. 2002, Institute of Applied Mathematics, Moscow.

Volevich, L. R. The Vishik-Lyusternik method in elliptic problems with a small parameter. *Trudy Mosk. Mat. Obshch.* 2006, 67, (in Russian), *Trans. Moscow Math. Soc. 2006*, 87-125.

INDEX

D

E

F

G

H

I

J

K

L

T

U

V

W

Y